# LABORATORY TECHNIQUES IN ELECTROANALYTICAL CHEMISTRY

## MONOGRAPHS IN ELECTROANALYTICAL CHEMISTRY AND ELECTROCHEMISTRY

**consulting editor**
**Allen J. Bard**
*Department of Chemistry*
*University of Texas*
*Austin, Texas*

Electrochemistry at Solid Electrodes, *Ralph N. Adams*

Electrochemical Reactions in Nonaqueous Systems, *Charles K. Mann and Karen Barnes*

Electrochemistry of Metals and Semiconductors: The Application of Solid State Science to Electrochemical Phenomena, *Ashok K. Vijh*

Modern Polarographic Methods in Analytical Chemistry, *A. M. Bond*

Laboratory Techniques in Electroanalytical Chemistry, *edited by Peter T. Kissinger and William R. Heineman*

*Other Volumes in Preparation*

# LABORATORY TECHNIQUES IN ELECTROANALYTICAL CHEMISTRY

*Editors*

PETER T. KISSINGER
*Purdue University*
*and*
*Bioanalytical Systems, Inc.*
*West Lafayette, Indiana*

WILLIAM R. HEINEMAN
*University of Cincinnati*
*Cincinnati, Ohio*



MARCEL DEKKER, INC. New York and Basel

Library of Congress Cataloging in Publication Data
Main entry under title:

Laboratory techniques in electroanalytical chemistry.

Includes index.
1. Electrochemical analysis—Laboratory manuals.
I. Kissinger, Peter T., [date]. II. Heineman, William R.
QD115.L23 1984 543'.0871 84-4274
ISBN 0-8274-1864-X



MARCEL DEKKER, INC.
270 Madison Avenue, New York, New York 10016

Current printing (last digit):
10 9 8 7 6 5 4 3 2 1

PRINTED IN THE UNITED STATES OF AMERICA

# Preface

The text that had the major influence on the education of most American electrochemists between the late 1950s and the late 1960s was Delahay's *New Instrumental Methods in Electrochemistry* (1954). For well over a decade this book played a dominant role in placing finite current electroanalytical chemistry on its modern course—beyond the titrations, coulometry, polarography, and electrogravimetry of an earlier age. The membrane-covered oxygen electrode developed by Clark in 1955 rapidly became one of the more important analytical tools. Modern engineering mathematics began to conquer mass transfer problems in the 1950s and toward the end of that decade operational amplifiers revolutionized experimental electrochemistry.

In the 1960s modern organic, inorganic, and biological electroanalytical chemistry advanced rapidly behind the advent of cyclic voltammetry, pulse chronoamperometry, rotating ring-disk electrodes, and the use of ESR and optical measurements coupled to electrochemical experiments. After a burst of tremendous activity, it became clear during this period that chronopotentiometry was a thing of the past. In the second half of the 1960s thin-layer electrochemistry had its heyday and electrodes other than mercury began to attract serious attention. It was discovered that platinum was not bad for studies in nonaqueous solvents and even the likes of glassy carbon and carbon paste could actually solve chemical problems, to say nothing of thin films of metals and semiconductors deposited on glass! Ralph Adams' book, *Electrochemistry at Solid Electrodes* (1969), surprised more than a few who did not believe it was possible. Anodic stripping voltammetry became a commercial reality and even was able to compete, for some elements, with atomic spectroscopy. Polarography gradually diminished in importance for practical analysis through the early 1960s. A little company in Princeton, New Jersey, brought it out of hibernation by making differential pulse polarography commercially available via the first significant op-amp-based electrochemical instruments.

The 1970s saw electroanalytical chemistry refine its fundamental basis and develop more powerful and more reliable instrumentation. Chemistry became more important than technique. For the first time electrochemists had a significant clientele of users whose primary interest was not electrochemistry per se. Two important fundamental new directions in the 1970s were toward chemically modified electrodes and photoelectrochemistry. A significant new development of practical analytical importance was the use of electrochemical flow cells for liquid chromatography, a development that introduced thousands of pharmacologists, clinical chemists, and toxicologists to electrochemistry for the first time. A little company in West Lafayette, Indiana, made this methodology easily accessible to non-electrochemists.

Thus, this book was conceived in 1970 by Peter Kissinger to provide a means of enabling a neophyte in electroanalytical chemistry to get started in the lab. During the decade over which this effort has spanned, the emphasis has been expanded to include a pedagogical component.

The information in research papers is just too hard for beginners to dig out and in most cases is devoid of any meaningful experimental insights. The review chapters are often turgid—replete with unexplained and generally confusing jargon and mathematics. The textbooks on instrumental methods have their good practical moments; but these moments are usually short, incomplete, and often out of date. None of these really do justice to more than a very limited selection of material for the "bench chemist" wanting to get things moving in a hurry.

The emphasis of this book is entirely on *analytical*, mechanistic (homogeneous), kinetic (homogeneous), and synthetic (laboratory scale) applications. Physical electrochemistry per se is not a direct concern. Equilibrium methods (potentiometry) are intentionally omitted. There is no attempt to include specific chemical examples except where these are particularly illustrative and have pedagogical value. No extensive review of the original literature is included. References to key reviews and papers of historical interest are emphasized. Authors have selected experimental approaches that work best and have commented freely on outmoded or underdeveloped methods. The authors and editors have made value judgments that undoubtedly will disappoint some readers. Electrochemistry is a very broad subject. Those interested in batteries, fuel cells, corrosion, membrane potentials, and so forth will not satisfy their needs here.

The organization of the book flows from principles through methodology to applications. Chapters 2 through 5 are devoted to what the editors perceive as the most commonly used techniques in electroanalytical chemistry today. These techniques have passed the development stage and can now be considered routine. The approach in these chapters is designed to give the reader an intuitive understanding of each technique without mathematical rigor. This is achieved by considering the excitation signal for each technique and the resulting concentration-distance profiles which determine the consequent response signal. Value judgments are given to permit an educated selection of techniques to use in a given situation. Chapter 2 provides the fundamental concepts that are important throughout the book. It is the editors' opinion that Chaps. 2 through 5 are suitable for use in graduate-level introductory courses in electroanalytical chemistry. Serious students who are interested in more rigor and depth are encouraged to progress to the excellent book by Allen Bard and Larry Faulkner, *Electrochemical Methods* (Wiley, New York, 1980).

Instrumentation for selected aspects of electroanalytical chemistry is covered in Chaps. 6 through 8. Although computerization is making a tremendous impact on electroanalytical instrumentation, many aspects of these chapters are timeless. The basic configurations of a potentiostat have not changed since the early 1960s, although the electronic components themselves are dramatically different! Learn to build your own potentiostat in Chap. 6, then see how to fine tune it in Chap. 7.

Chapters 9 through 15 deal with some practical aspects of electroanalytical chemistry. There are probably some helpful topics for everyone scattered through these chapters, which are aimed at giving the neophyte some insight into the nuts and bolts of electrochemical cells and solutions. The reader is also referred to Don Sawyer and Julian Roberts' book on *Experimental Electrochemistry for Chemists* (Wiley, New York, 1974).

Many students have learned about digital simulation from Joseph Maloy's "underground chapter", which is now seeing print as Chap. 16.

Chapter 17 again is for the neophyte trying to determine which technique can be useful for unraveling mechanism. Many of the techniques described in Chap. 3 are illustrated here.

Chapters 18 through 24 are a potpourri of electroanalytical techniques and applications. Some aspects of hybrid techniques in which electroanalytical chemistry is combined with other instrumental methods, such as spectroscopy or chromatography, are discussed.

While it is risky to predict the future, the 1980s will continue the trends of the seventies—more chemistry; more practical organic, inorganic, and biochemical problem solving; more detailed knowledge of the chemical nature of the interface and how it can be manipulated to human advantage. We can almost certainly count on those working on chemically modified electrodes to come up with something spectacular: While the big payoff seems far away, forebodings of success appear in the literature day by day. Publication of this volume coincides with the 150th anniversary of Faraday's law. There seems no end of uses for this important principle describing the relationship between chemistry and electricity.

We hope that this book is of some use to you in the course of your research, for which we wish you good luck—it certainly helps.

Peter T. Kissinger
William R. Heineman

# Contents

# Contributors

WAYNE E. BRITTON Programs in Chemistry, The University of Texas at Dallas, Richardson, Texas

MARVIN A. BROOKS Hoffman-La Roche, Inc., Nutley, New Jersey

DAVID J. CURRAN Department of Chemistry, University of Massachusetts, Amherst, Massachusetts

GLENN DRYHURST Department of Chemistry, University of Oklahoma, Norman, Oklahoma

CHRISTIE G. ENKE Department of Chemistry, Michigan State University, East Lansing, Michigan

LARRY R. FAULKNER School of Chemical Sciences, University of Illinois at Champaign-Urbana, Urbana, Illinois

STEVEN N. FRANK* Department of Chemistry, University of Texas at Austin, Austin, Texas

ALBERT J. FRY Department of Chemistry, Wesleyan University, Middletown, Connecticut

ZBIGNIEW GALUS Department of Chemistry, University of Warsaw, Warsaw, Poland

WILLIAM E. GEIGER Department of Chemistry, University of Vermont, Burlington, Vermont

*Present address: Texas Instruments Inc., Dallas, Texas

IRA B. GOLDBERG Rockwell International Science Center, Thousand Oaks, California

FRED M. HAWKRIDGE Department of Chemistry, Virginia Commonwealth University, Richmond, Virginia

M. DALE HAWLEY Department of Chemistry, Kansas State University, Manhattan, Kansas

WILLIAM R. HEINEMAN Department of Chemistry, University of Cincinnati, Cincinnati, Ohio

F. JAMES HOLLER* Department of Chemistry, Michigan State University, East Lansing, Michigan

CSABA P. KESZTHELYI Department of Chemistry, Louisiana State University, Baton Rouge, Louisiana

PETER T. KISSINGER Department of Chemistry, Purdue University, and Bioanalytical Systems, Inc., West Lafayette, Indiana

J. T. MALOY† Department of Chemistry, West Virginia University, Morgantown, West Virginia

HARRY B. MARK, JR., Department of Chemistry, University of Cincinnati, Cincinnati, Ohio

DAVID L. McALLISTER‡ Department of Chemistry, University of Oklahoma, Norman, Oklahoma

TED M. McKINNEY Consultant, Riverside, California

SU-MOON PARK Department of Chemistry, University of New Mexico, Albuquerque, New Mexico

CARL R. PREDDY§ Department of Chemistry, Purdue University, West Lafayette, Indiana

DAVID K. ROE Department of Chemistry, Portland State University, Portland, Oregon

DARYL A. ROSTON Department of Chemistry, Northern Illinois University, DeKalb, Illinois

---

*Present address: Department of Chemistry, University of Kentucky, Lexington, Kentucky
†Present address: Department of Chemistry, Seton Hall University, South Orange, New Jersey
‡Present address: Great Lakes Chemical Corporation, West Lafayette, Indiana
§Present address: Kodak Research Labs, Eastman Kodak Company, Rochester, New York

RONALD E. SHOUP Bioanalytical Systems, Inc., West Lafayette, Indiana

HIROYASU TACHIKAWA* Department of Chemistry, University of Illinois at Champaign-Urbana, Urbana, Illinois

NICHOLAS WINOGRAD Department of Chemistry, The Pennsylvania State University, University Park, Pennsylvania

JOHN A. WISE† Department of Chemistry, University of Cincinnati, Cincinnati, Ohio

*Present address: Department of Chemistry, Jackson State University, Jackson, Mississippi
†Present address: The Goodyear Tire & Rubber Company, Akron, Ohio

# LABORATORY TECHNIQUES IN ELECTROANALYTICAL CHEMISTRY

# 1
# An Overview

PETER T. KISSINGER / Department of Chemistry, Purdue University, and Bioanalytical Systems, Inc., West Lafayette, Indiana



## I. SOME PHILOSOPHY

Much of the initial progress in science derives from passive observation of a natural system. This approach is severely limited in that one remains at the mercy of the system for what qualitative and quantitative properties it chooses to reveal. Nevertheless, passive observation is the first step in all areas of research and remains of ultimate importance in some (e.g., astronomy). A more pragmatic approach is often one in which the experimenter plays an active role by forcing information out of the object of his or her affection. The experimenter first tickles the system with a forcing function thought to be appropriate by passive observation and intuition. Every so often the experimenter will even be inclined to try anything. This is not as bad as it sounds because the "anything" *selected* will be inseparable from previous experience and will invariably be better than doing nothing. The observed response function to the chosen excitation signal is analyzed and this information is then used to design new active probes. Such a closed-loop modus operandi is shown in Figure 1.1.

Experiments of this kind are most easily interpreted when the system behaves linearly. Linear performance is recognized by the validity of two simple rules:

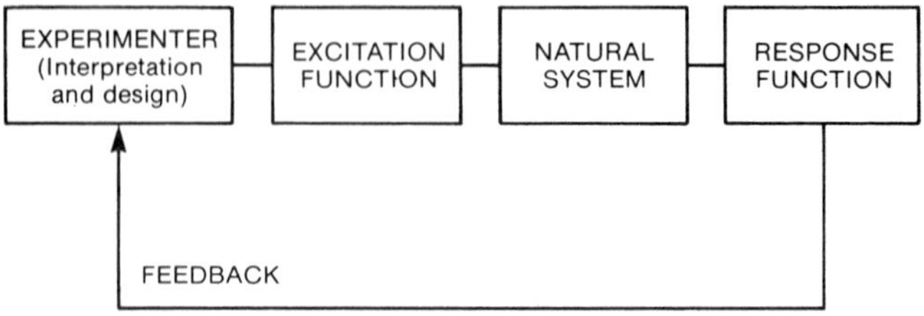


**Figure 1.1** Block diagram illustrating modification of experimental design using feedback from previous experiments.

1. The amplitudes of the response and excitation functions are directly proportional (i.e., an increase in excitation amplitude results in the same *relative* increase in response amplitude).
2. The response to a collection of excitations is given by the sum of responses to the individual excitations (i.e., superposition theory holds).

Problems (nonlinearity) will certainly develop if the system under study significantly influences the excitation signal applied to it. It is often therefore essential to monitor (via negative feedback) the excitation *at its site of application* to ensure that this is not the case. In some systems, an error in the excitation function may be related directly to the response function. This circumstance permits compensation of the error by a positive feedback link between the two critical functions. These modifications are depicted in Figure 1.2.

The simple concepts we have just discussed can be given a sound mathematical basis which is useful in the design of new experiments and instrumentation systems. The practice of interfacial electrochemistry involves analysis of the response of a phase boundary to various stimuli. In this book we examine the most frequently chosen perturba-

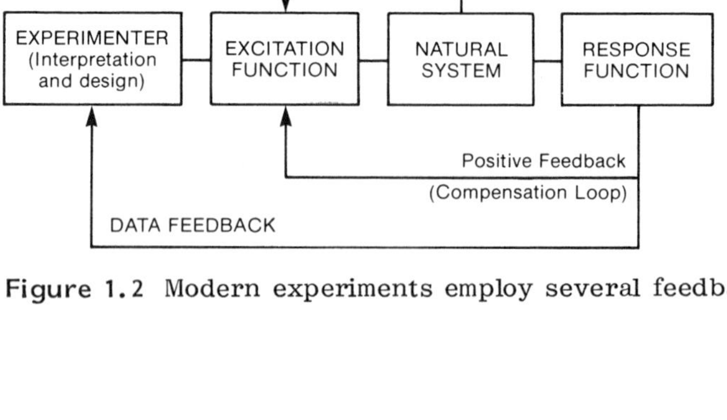


**Figure 1.2** Modern experiments employ several feedback loops.

tions and their experimental implementation. Some understanding of the properties of the phase boundary between an electrode and a solution can help tie together the seemingly endless variety of electrochemical techniques.

## II. PROGRESS AND PROGNOSIS

Electrochemical progress has benefited tremendously from the application of feedback principles. It is essential for the modern electrochemist to be aware of them and use the opportunities they provide to design improved experiments. Unfortunately, the technological imbalance in our world has afforded these opportunities to relatively few of us. Even in the more fortunate economies, investigators have been tied down to methods (e.g., dc polarography) which by themselves provide insufficient data for many purposes. Although economic constraints are still extant, the availability of inexpensive integrated circuits has largely eliminated any excuse for not applying state-of-the-art instrumentation for most electroanalytical techniques.

A deficient knowledge of electronics remains a significant problem. For the electrochemist this becomes a trivial excuse because integrated subsystems (linear and nonlinear, analog and digital) for building highly sophisticated electrochemical apparatus can be effectively utilized with *working* knowledge of only a few basic laws of electricity. In fact, practitioners of our art need know far less physical electronics than did their predecessors only a decade ago. Many students find that there is a considerable psychological barrier which must be exceeded before the "fear of electronics" is overcome. Once this is accomplished, the rest is easy.

In the text that follows we examine the methodology of electrochemistry as it may be usefully applied to a wide variety of chemical problems. In short, when a molecule can be coupled to an electrode reaction, that reaction may be used for analytical, mechanistic, or kinetic studies involving this substance. We make no pretense about the fact that this is not a book for physical electrochemists. Those who wish to direct their efforts toward unraveling the microsteps in heterogeneous electrode processes per se are best served by other monographs. Those focused on generation of electrical power from chemical reactions will also find little here of direct interest.

The great facility we have for dealing with electricity leads us to convert or transduce information in virtually every other data domain (e.g., chemical, optical, magnetic) into an easily processed electrical signal. Similarly, there is much virtue in controlling some event in a nonelectrical domain with an electrical signal. Faraday's discovery of the one-to-one equivalence that exists between chemistry and electricity provides the basis for all electroanalytical techniques. The electrode in solution is in many respects an ideal transducer between the chemical and electrical domains. The role of the electrode may either be to monitor species in solution or to generate new species that will interact with the medium in an interesting or perhaps useful way.

Molecular specificity is afforded by control of the potential energy difference across the electrode solution interface. The selectivity or resolution of electrochemistry is not nearly as great as we would like, and the inability to distinguish between two or more species can be a serious weakness. Solution electrochemistry suffers by comparison with mass spectrometry in the same sense that optical spectra in solution suffer by comparison with vapor-phase spectra. Nevertheless, it is an immutable fact of nature that every molecule participating in the electron transfer reaction will be related to an integral number of electrons in the external circuitry. Often this fact can be used to great advantage.

Before considering instrumentation in some detail (Chaps. 6 to 15) it will be helpful to outline the kinds of experiments that we wish to implement electronically. It is useful to characterize electroanalytical techniques as either static or dynamic. Static methods are philosophically akin to the passive observation mentioned earlier. They entail measurements of potential difference at zero current such that the system defined by the solid-solution interphase is not disturbed and nernstian equilibrium is maintained. Although such potentiometric measurements (e.g., pH, pM) are of great practical importance, our focus here will be on the dynamic techniques. Here the system is intentionally disturbed from equilibrium by excitation signals consisting of a wide variety of potential and current programs.

While dc polarography remains the only dynamic electroanalytical technique familiar to most chemists, others are becoming increasingly important. The recent teaming of many of these methods with spectroscopic techniques (both optical and magnetic) and the developing commercial and academic role of electrosynthesis lends support to this trend. Of particular importance is the realization that great benefits result from studies at stationary (both solid and liquid) electrodes. Consequently, polarography has been relegated to second place in most modern electroanalytical laboratories and in many ways is becoming a thing of the past.

This text places electrochemical techniques in perspective by introducing the most promising methods and evaluating their relative merits for specific applications. Although an almost endless variety of excitation functions have been applied to the electrode-solution interface, most of these have had little following even among electrochemists, and some have been superseded by more modern ideas. Emphasis is given to practical considerations. Although the important phenomenological aspects are covered, a detailed mathematical development has been avoided in deference to the extant literature.

The 1950s and 1960s were characterized by an enormous exploratory effort in the theory and methodology of electrochemical techniques and their possible application to chemical problems. Chemistry seemed ancillary to much of this work in that the model systems examined were more often than not uninteresting to chemists. This is not to say that no exciting things happened electro*chemically* (e.g., electrochemiluminescence). It has just taken a while for the technologists and chemists to get together. This is beginning to happen and it appears that dynamic electrochemistry will, finally, have a secure place in studies of solution chemistry. It is still necessary to recognize the limitations. The competitive position vis-a-vis optical, magnetic, and other analytical probes must be kept in mind for a specific chemical problem.

## III. SO MUCH NOMENCLATURE, SO MUCH JARGON

### A. Techniques

The nonequilibrium behavior attendant on dynamic electrochemical techniques is largely controlled by diffusion which may, or may not, be coupled with chemical reactions in solution or on the surface. The voltage, current, and time variables may be adjusted (although not independently) by the experimenter to obtain maximum information from the system at hand. The system response to a given voltage or current excitation program can sometimes be predicted by a linear systems approach utilizing mathematical transforms. Many individual circumstances are not amenable to this approach and, in fact, few of the differential equations describing real chemical situations are analytically soluble. Nevertheless, individual techniques are best recognized by their excitation-response characteristics. As we shall see later, the various dynamic electrochemical methods both mathematically and experimentally fall into two convenient categories—large amplitude and small amplitude—related to the magnitude of the excitation signal.

Within these broad categories two principal classes may be distinguished depending on whether current or voltage is the controlled parameter. The individuals in each class are often described by an operational nomenclature consisting of an independent-variable part followed by a dependent-variable part (i.e., volt-ammetry, chronopotentiometry) with some system-specific modifers (i.e., rotating disk voltammetry). Unfortunately, the whims of history have left the electrochemical nomenclature in a rather confused state. The operational approach has been only partially adopted but seems to be gaining in popularity.

In general, electrochemists apply an electrical *excitation* signal (often the independent variable) to some *system* (particular electrode-solution composition and geometry) and then monitor a *response* signal (dependent variable) which together with the excitation allows some description of the properties of the system at hand. The nomenclature ideally should incorporate enough information about these three items to describe the experiment (technique) to some reasonable degree without being cumbersome. The nomenclature cannot be expected to provide a phenomenological explanation of the response beyond the individual's ability to understand the experiment thereby described. The operational approach implicitly ignores the subtleties of individual techniques, but does describe the experiment.

The term "polarography" provides a particular bone of contention. The continued use of "polarography" as a general appellation for all faradaic electrochemistry is very undesirable. Polarography is actually a relatively old technique (i.e., dc voltammetry at the dropping mercury electrode). The term "polarography" is generally accepted in connection with techniques employing a dropping mercury electrode; however, more general usage is strongly discouraged. Fig-

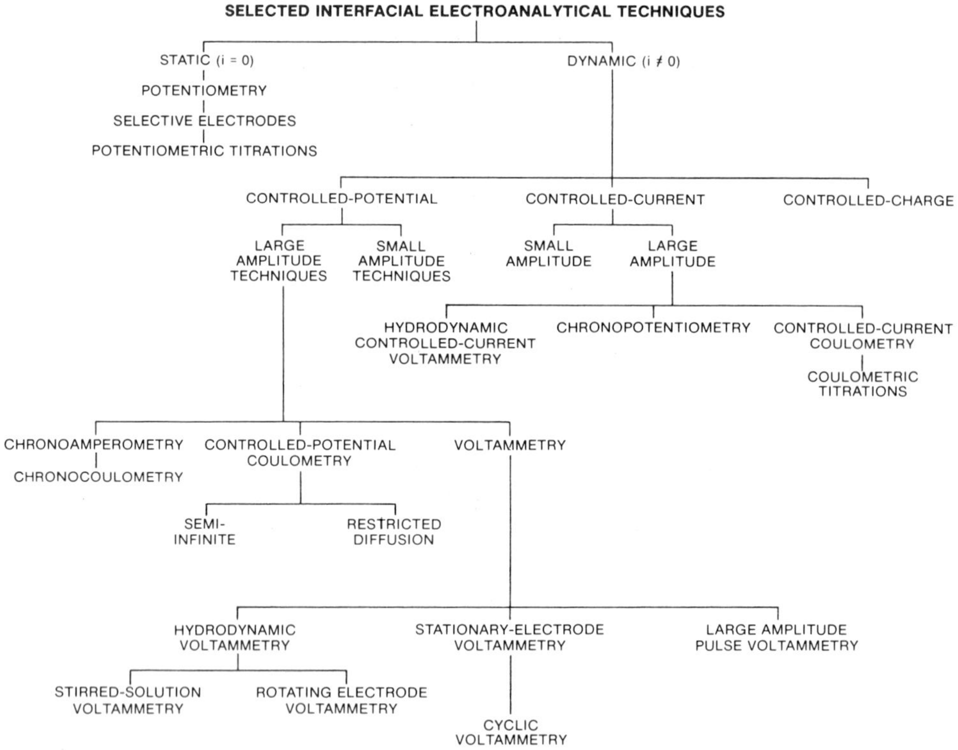


Figure 1.3 Family tree of electroanalytical techniques.

ure 1.3 illustrates a family tree relating the more important methods. The reader unfamiliar with electroanalytical chemistry should be able to see the logic in these associations after completing Chapters 3 to 5. It would be a good self-test to be able to identify the essential characteristics of each technique in Figure 1.3.

### B. Chemical Complications

For introducing techniques and how to instrument them, we will assume a generalized system

$$O + ne \rightleftarrows R$$

where O and R are the oxidized and reduced forms of the *electroactive species*, in a polar solvent containing a relatively high concentration of inert *supporting electrolyte*, and the heterogeneous electron-transfer (ET) rate constant is large (fast, reversible). For dynamic experiments the sample is often either completely in the O or R form, whereas for static experiments both forms are usually present.

In the real world, the simple redox couple may be perturbed by finite ET rates, by adsorption of O and/or R on the electrode surface, and by homogeneous (i.e., in solution) chemical kinetics involving O and/or R. Various combinations of heterogeneous ET steps (E) with homogeneous chemical steps (C) are encountered. It should be clear that if one or more species in equilibrium in solution are electroactive, electrochemistry can be used to perturb the equilibrium and study the solution chemistry.

Many other complications occur, including heterogeneous chemical steps and homogeneous ET steps. Our understanding of these complications when built into real chemical systems is at a very primitive stage. There is much work to be done before evaluating such problems becomes a routine matter. The literature often displays more optimism than is warranted by the facts. This has caused many disappointments among practical problem solvers. Some investigators have thus come to view analytical electrochemistry with considerable skepticism. One should recognize that electrochemistry will always be more useful as a quantitative tool than as a diagnostic tool because of its relatively low resolution. There is a tendency in some circles to overinterpret electrochemical results without stating the appropriate reservations. Whenever possible, supporting evidence from other sources (usually spectroscopic or chromatographic techniques) should be brought to bear on any mechanistic conclusions suggested by electrochemical data. A large percentage of the extant literature has ignored this principle and should therefore be examined with appropriate caution.

# 2
# Fundamental Concepts of Analytical Electrochemistry

PETER T. KISSINGER / Department of Chemistry, Purdue University, and Bioanalytical Systems, Inc., West Lafayette, Indiana

CARL R. PREDDY* / Department of Chemistry, Purdue University, West Lafayette, Indiana

RONALD E. SHOUP / Bioanalytical Systems, Inc., West Lafayette, Indiana

WILLIAM R. HEINEMAN / Department of Chemistry, University of Cincinnati, Cincinnati, Ohio



## I. INTRODUCTION

The common thread binding all electroanalytical methods is heterogeneity. The very act of placing an electrode in contact with a solution creates a phase boundary that differentiates otherwise identical

---

*Present address: Kodak Research Labs, Eastman Kodak Company, Rochester, New York

solute molecules into two types; those at a distance from the electrode, and those close enough to participate in the fascinating mutual interactions known collectively as electrochemistry. This is not a trivial distinction, for often it is the bulk-phase properties alone which are of analytical concern. Unlike most spectroscopic methods, electrochemical measurements are actually made on only a minute fraction of the sample confined to a highly inhomogeneous environment, the electrode-solution interphase. The coupling and interplay within this region of such phenomena as interfacial charge transfer, diffusional mass transfer, adsorption, chemisorption, homogeneous-phase chemical reaction, convection, and dissolution can cloud the interpretation of electrochemical data and discourage the practically minded analyst. On the other hand, electrochemistry offers an invaluable tool for fundamental investigation of these processes, each of importance in its own right. In either case, the ultimate success of the experimenter will depend on a firm grasp of the underlying physical principles.

We have chosen to dwell on certain theoretical aspects only to the extent that they will serve as a foundation for the methodology presented in the ensuing chapters. It is practice rather than theory that is of primary concern here. We acknowledge extensive, intentional bias and abridgement in derivations and discussion. Our intent is to convey the feeling that much may be gained from electrochemical investigation without recourse to an entanglement of mathematical expressions, provided that one develops an insight into the physical realities of electrode processes.

In many respects, the simplest and best understood process influencing electrochemistry is diffusion. Diffusion is a factor in virtually every type of electroanalytical measurement, yet it is most often introduced as a set of elemental laws, devoid of physical significance. The governing mathematical relationships are apt to seem abstruse, but are in fact quite elegant and readily grasped. Their derivation follows from simple finite mathematics, using as a model a collection of inert particles in random motion. This microscopic-level approach relies more on statistical rather than thermodynamic concepts, and although less rigorous, Fick's well-known laws arise directly with few assumptions and without losing sight of physical reality.

## II. LINEAR DIFFUSION

### A. Impulse Relaxation

As with the familiar theoretical treatment of ideal gases, the logical starting point for understanding the process of diffusion is consideration of the motion of a single solute molecule or ion. In fluid media, a molecule undergoes frequent collisions with neighboring molecules. This continuous rebounding imparts randomness to each molecule's trajectory. In a closed, isotropic system (no phase boundaries or thermal, electromagnetic, or concentration gradients), there is no favored direction of motion. An imaginary plane dividing the system will be continuously traversed by many molecules moving from both sides, but on any significant time interval, the number that have passed from one direction will exactly equal the number from the other. Thus there will be no *net* movement of material from one side to the other. The concentration, or more specifically the chemical potential, of the solute is uniform throughout, and the system is therefore at equilibrium.

When regions of dissimilar chemical potential are created, solute molecules will move between them until a homogeneous condition is restored. This relaxation process involves a temporary net *mass transfer* across some imaginary plane. The transfer of matter from a region of higher chemical potential to one of lower chemical potential is the process of diffusion. The motive force behind this movement is maximization of entropy. The fully relaxed system is in its most random configuration.

The transition between regions of high and low chemical potential in real systems cannot be infinitely sharp. There must be a zone over which the concentration varies with distance. A place in which $dC/dx$ is nonzero is said to exhibit a *concentration gradient.* A plot of concentration as a function of distance is called a *concentration profile,* so the gradient is the first derivative of the profile function. Since diffusional mass transfer is always from high to low concentration, the direction of movement will be determined by the sign of the gradient. One imagines that molecules diffuse *down* their respective concentration gradient, meaning that direction in which the sign of $dC/dx$ is negative from the molecules' frame of reference. This is a point worth pondering. It is fundamental to the understanding of unrestricted motion under the influence of any type of potential field: thermal, electrical, or gravitational. The mathematical description of the transfer of heat along temperature gradients is identical to mass diffusion in every respect. Just as objects at high temperature cool more rapidly than when they are near room temperature, so it is reasonable to expect the rate of mass transfer to be related to the magnitude of the concentration gradient. The form of this relationship will be examined presently.

Another useful concept is that of *flux.* Flux is defined as the number of molecules penetrating a unit area of an imaginary plane in a unit of time. The usual units are $mol/(cm^2\ s)$, and the sign is identified with the direction of motion, positive toward and negative away from the plane. The prior assertion that equilibrium demands no net mass transfer is equivalent to a requirement that the sum of the fluxes of all components is exactly zero at any test plane within the system. Flux is a measure of the rate of mass transfer at a fixed point. Its electrochemical relevance stems from the direct relationship it holds to electrode current.

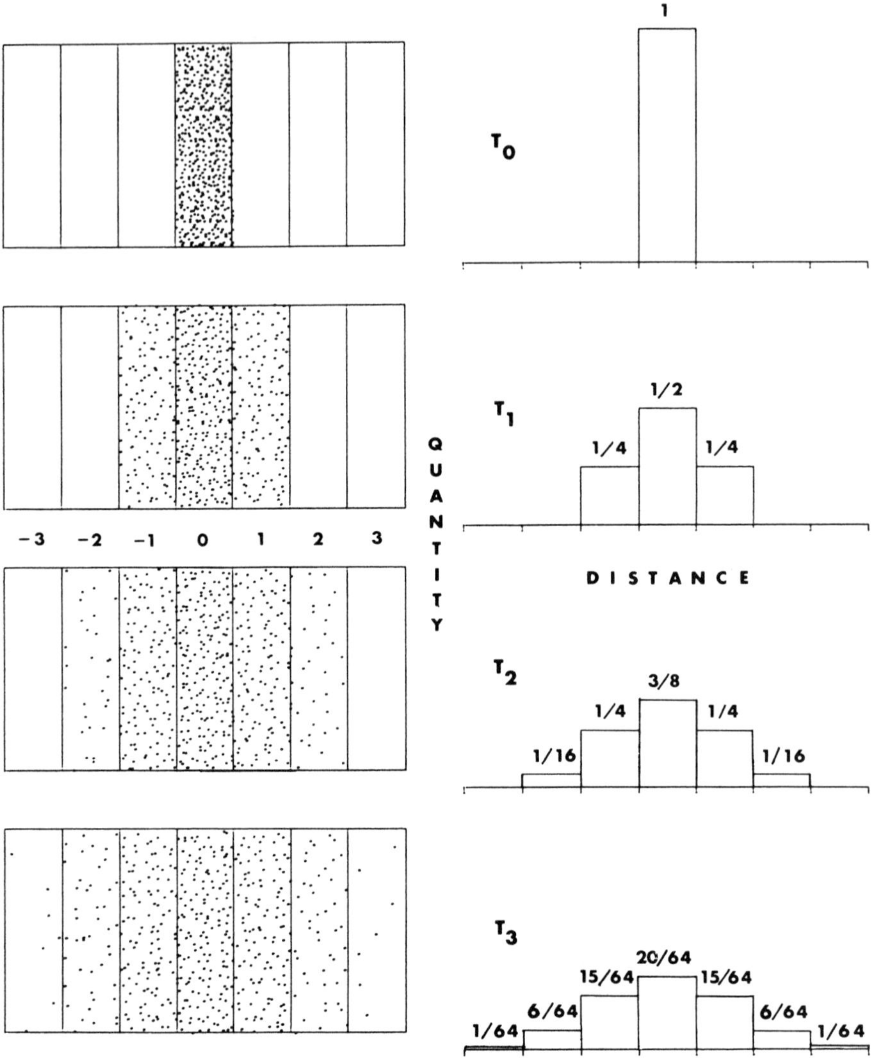


Figure 2.1 Schematic representation of the diffusion process.

Consider an infinite volume of pure solvent at equilibrium. If we imagine a portion of this system divided into a number of thin, adjacent slabs of uniform thickness $\Delta x$, the picture resembles the left side of Figure 2.1. Into the central box, arbitrarily labeled 0, we introduce a number $N_0$ of solute molecules at time $t = 0$. At the starting time the concentration profile across this region exhibits a sharp spike at the center. The right side of Figure 2.1 depicts the profile as a histogram of a quantity of molecules versus distance from the center.

Since the concentration profile is a step function at either boundary of box 0, the gradient at these points is some huge negative value. According to our notion of molecular behavior, this impulse of concentration should be short-lived when left to itself. Bombardment by solvent molecules and collision with each other will cause the molecules to wander out randomly from the central box. Isotropy assures that motion to the left is no more probable than motion to the right. The fluxes across either boundary should be equal at all times. After some time $T_1$, half of the original $N_0$ molecules will have left the 0th box. One-fourth will be found in box 1, and an equal number will have made their way into box −1. The situation is now as depicted in the second set of Figure 2.1. The model system now consists of three nonempty boxes with $N_0/2$ molecules in the center and $N_0/4$ in each of the neighboring boxes. Each box is entirely independent of the other in the sense that each will behave exactly as box 0 did in the first step. At some later time $T_2$, half the original molecules in each box will have meandered into the adjacent boxes. The distribution ratio across the boxes has progressed from 1:2:1 after the first time increment, to 1:4:6:4:1 after the second, then to 1:6:15:20:15:6:1 at some later time $T_3$. The fraction of molecules in the ith box after the kth step is given by the familiar binomial distribution function

$$\frac{N_{i,k}}{N_0} = \frac{2k!}{(k+i)!\,(k-i)!} \times 2^{-2k}$$

$$k = 0, 1, \ldots ; i = 0, \ldots, \pm k \qquad (2.1)$$

It should be emphasized that the time steps are *not* evenly spaced.

If the total number of molecules is very large and the distance increments, $\Delta x$, become infinitesimally small ($\Delta x \to dx$), the distribution will be described by a smooth function (a Gaussian distribution) following the familiar normal curve of error,

$$\frac{dN}{N_0} = \frac{1}{(4\pi Dt)^{1/2}} \exp\left(\frac{-x^2}{4Dt}\right)dx \qquad (2.2)$$

where $dN/N_0$ is the fraction of the total population of molecules located between x and (x + dx), and D is a constant known as the diffusion coefficient and has units of $cm^2/s$. The preexponential factor in the normal distribution is usually described in terms of $[1/\sigma(2\pi)^{1/2}]$, where $\sigma$, the population standard deviation, is a measure of distribution width. In the present case

$$\sigma = \sqrt{2Dt} \qquad (2.3)$$

If one considers a unit cross-sectional area (i.e., 1 $cm^2$), dN/dx is a measure of concentration in the region of space between x and (x + dx), and

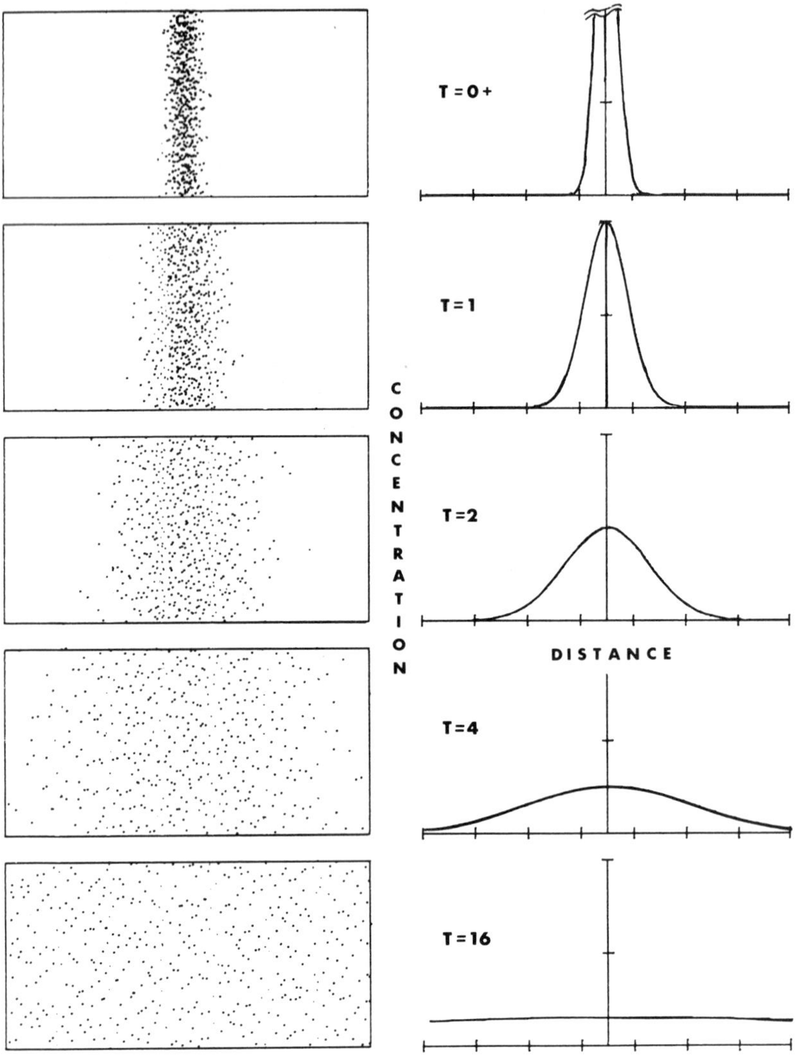


Figure 2.2 Spreading of an impulse of molecules at various times after creation.

$$C_{x,t} = \frac{N_0}{(4\pi Dt)^{1/2}} \exp\left(\frac{-x^2}{4Dt}\right) \tag{2.4}$$

Figure 2.2 illustrates a plot of concentration vs. distance profiles at various times following the creation of an impulse of solute molecules at

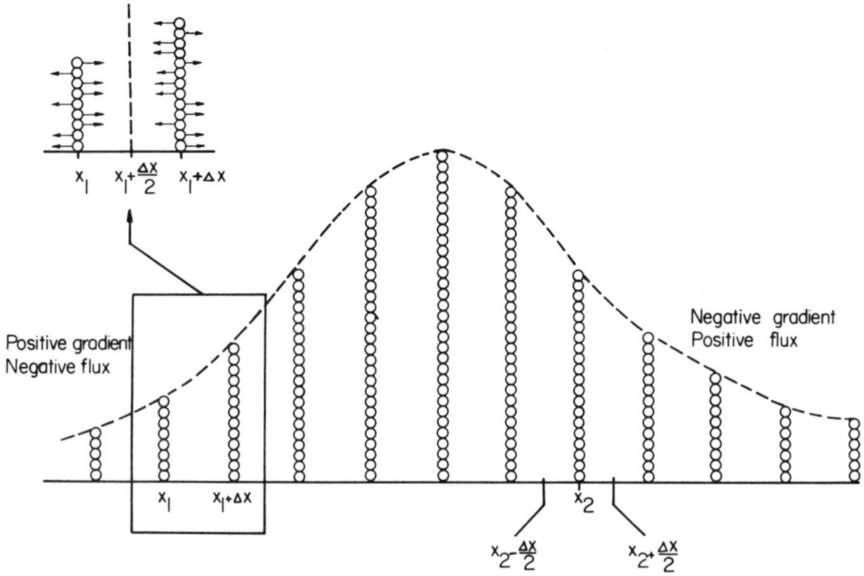


Figure 2.3 Schematic representation of a Gaussian concentration profile by closely spaced spikes (or impulses) of molecules.

$t = 0$. Note that the concentration gradient $dC_{x,t}/dx$ is zero at $x = 0$ but nonzero in the neighboring solution. The driving force for the diffusion process is nonuniformity of chemical potential, so in the limit as $t \to \infty$, $dC/dx$ will be zero throughout.

Examining a given concentration profile in greater detail, we may construct the model illustrated in Figure 2.3, which approximates the bell shape by many hypothetical impulse functions of different heights located $\Delta x$ distance increments apart. Simple diffusion is a linear process which fits the two criteria discussed earlier (Chap. 1). According to the superposition principle, the impulse functions may be treated completely independently if there is no solute-solute interaction (e.g., dimerization). The future of the total profile will be described accurately by a summation of the behavior of each independent impulse.

The net movement of molecules per unit area in a diffusion process at a point x as a function of time t is described as the flux per unit area, $J_{x,t}$, which has units of mol/(cm$^2$ s). Referring to the inset in Figure 2.3, which shows a magnified view of the volume region from $x_1$ to $(x_1 + \Delta x)$, one can see that, in time $\Delta t$, one-half of the molecules in column $x_1$ on the average, will have migrated to position $(x_1 + \Delta x)$, and one-half will have gone in the opposite direction, to $(x_1 - \Delta x)$. Similarly, one-half of the molecules at $x_1 + \Delta x$ will migrate to $x_1$, and so on. Therefore, one can define the flux at $(x_1 + \Delta x/2)$ at a time t (denoted $J_{x_1+\Delta x/2,t}$) as the net flow from $x_1$ to $x_1 + \Delta x$. Thus

$$J_{x_1+\Delta x/2,t} = \frac{1}{\Delta t}\left(\frac{1}{2} N_{x_1,t} - \frac{1}{2} N_{x_1+\Delta x,t}\right) \qquad (2.5)$$

describes the flow of molecules from $x_1$ to $(x_1 + \Delta x)$ minus that from $(x_1 + \Delta x)$ to $x_1$ across the hypothetical plane located at $(x_1 + \Delta x/2)$. This then gives the flux at point $(x_1 + \Delta x/2)$, where $\Delta t$ is the time required to traverse the distance $\Delta x$. In terms of concentration $(N/\Delta x)$,

$$J_{x_1+\Delta x/2,t} = \frac{\Delta x}{2\,\Delta t}(C_{x_1,t} - C_{x_1+\Delta x,t}) \qquad (2.6)$$

or

$$J_{x_1+\Delta x/2,t} = \frac{(\Delta x)^2}{2\,\Delta t}\;\frac{C_{x_1,t} - C_{x_1+\Delta x,t}}{\Delta x} \qquad (2.7)$$

Such a partial differential equation, expressed in terms of discrete units of x and t, is called a *finite difference equation*. It is subject to some "discretization error" depending on the magnitude of the finite differences chosen. In the limit as x and t become infinitesimal, the equation converges to reality (i.e., the discretization error becomes zero) and thus

$$\lim_{\substack{\Delta x \to 0 \\ \Delta t \to 0}} J_{x_1+\Delta x/2,t} = \lim_{\substack{\Delta x \to 0 \\ \Delta t \to 0}} \frac{(\Delta x)^2}{2\,\Delta t}\;\frac{C_{x_1,t} - C_{x_1+\Delta x,t}}{\Delta x} \qquad (2.8a)$$

$$J_{x_1,t} = D\left(-\frac{\partial C_{x,t}}{\partial x}\right)_{x=x_1} \qquad (2.8b)$$

$$= -D\left(\frac{\partial C_{x,t}}{\partial x}\right)_{x=x_1} \qquad (2.8c)$$

Note that D is defined by the Einstein equation using finite differences as $D = (\Delta x)^2/2\,\Delta t$. Equation 2.8c is known as Fick's first law and $\partial C/\partial x$ is, of course, the concentration gradient. It should be clear that the net transfer of solute mass per unit time across a plane intersecting a concentration profile will be proportional to the steepness of the profile and in the downhill direction. This fact must be intuitively obvious to anyone who wishes to use electrochemical techniques effectively.

It is often of interest to examine how the concentration at a given point changes as a function of time. Generally, it is not possible to investigate diffusion characteristics of a system under conditions of constant concentration gradient, since this implies the existence of a steady state. It is usually more convenient to evaluate such parameters from a determination of the change of concentration with time, as it is caused by diffusion in the system in question. Although qualitatively useful, Fick's first law of diffusion is of less value to the practical experimenter in that it does not directly relate to the easily measurable quantities of concentration, time, and distance. Fick's second law of diffusion, to be derived presently, partially alleviates these difficulties.

Referring to Figure 2.4, one can see that as long as $J_{x_2-\Delta x/2,t} = J_{x_2+\Delta x/2,t}$, the concentration of material bounded by the walls of unit cross-sectional area at $(x_2 \pm \Delta x/2)$ will be time invariant. On the other hand, if $J_{x_2-\Delta x/2,t} > J_{x_2+\Delta x/2,t}$, a net increase of material into the box would be expected, and concentration would increase with time. Since J is defined as the flux per unit area, dividing J by the distance between the walls yields the time rate of change of concentration within the box. More formally,

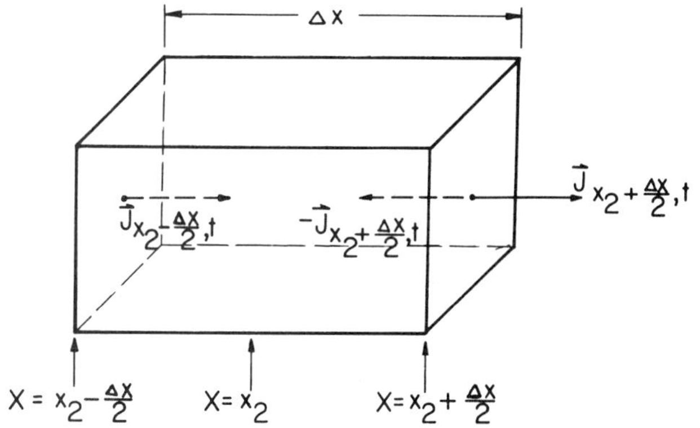


Figure 2.4 Diagram of portion of solution of length x and unit cross-sectional area, located about $x = x_2$.

$$\left(\frac{\Delta C}{\Delta t}\right)_{x=x_2} = \frac{J_{x_2-\Delta x/2,t} - J_{x_2+\Delta x/2,t}}{\Delta x} \tag{2.9}$$

Substituting for the flux terms using Equation 2.7 gives us

$$\left(\frac{\Delta C}{\Delta t}\right)_{x=x_2} = \frac{(\Delta x)^2}{2\Delta t}\left(\frac{C_{x_2-\Delta x,t} - C_{x_2,t}}{\Delta x} - \frac{C_{x_2,t} - C_{x_2+\Delta x,t}}{\Delta x}\right)\Bigg/\Delta x \tag{2.10a}$$

or

$$\left(\frac{\Delta C}{\Delta t}\right)_{x=x_2} = D\left[\frac{C_{x_2+\Delta x,t} - 2C_{x_2,t} + C_{x_2-\Delta x,t}}{(\Delta x)^2}\right] \tag{2.10b}$$

In the limit as $\Delta x \to 0$ and $\Delta t \to 0$, the finite difference equation becomes

$$\left(\frac{\partial C}{\partial t}\right)_{x=x_2} = D\left(\frac{\partial^2 C}{\partial x^2}\right)_{x=x_2} \tag{2.11}$$

which is known as Fick's second law of diffusion. As can easily be verified, Equation 2.4 describing $C_{x,t}$ is an analytical solution of Fick's second law. As predicted earlier, at the point of inflection on the instantaneous concentration profile curve, where $J_{x_2-\Delta x/2,t} = J_{x_2+\Delta x/2,t}$, the second partial derivative is zero, so the concentration is invariant. Above the point of inflection, the second derivative is negative; the concentration of material at points in this region are decreasing as time goes on. Similarly, the concentration at points below the inflection point are increasing with time.

If the diffusion process is *coupled* with other influences (chemical reactions, adsorption at an interface, convection in solution, etc.), additional concentration dependences will be added to the right side of Equation 2.11, often making it analytically insoluble. In such cases it is profitable to retreat to the finite difference representation and model the experiment on a digital computer. Modeling of this type, when done properly, is not unlike carrying out the experiment itself (provided that the discretization error is equal to or smaller than the accessible experimental error). The method is known as *digital simulation* and the result obtained is the finite difference solution. A more detailed description of this approach is given in Chapter 16.

Analytical solutions of Fick's laws are most facilely derived by an approach utilizing Laplace transforms, a subject described in every undergraduate book on differential equations. The solution of diffusion equations has fascinated academic electroanalytical chemists for years and they naturally have a tendency to expound on them at the slightest provocation. Fortunately, the chemist using electrode reactions can accomplish a great deal without more than a cursory appreciation of the mathematics. Our intention here is to provide this qualitative appreciation on a level sufficient to understand laboratory techniques.

## B. Step Response

A frequently encountered diffusion experiment is initiated with a step functional change in concentration. In this case we envision a barrier separating a region of finite solute concentration from a region of zero concentration as shown in Figure 2.5a. When the hypothetical barrier is carefully removed so as not to cause convection, the sharp boundary is gradually made more diffuse as solute moves to the right. The flux decreases with time as the concentration gradient becomes more gradual, as indicated in Figure 2.5b.

Simple diffusion is a linear process. Therefore, the original step function may be broken into a series of spatially separated impulses (Fig. 2.5c). Each impulse can be treated independently according to the results described above (Fig. 2.5d). The total concentration profile will be described by the summation of response functions to the original excitation pulses. Thus, making use of Equation 2.4 for each impulse, we have

$$C_{x,t} = \frac{C^\circ}{(4\pi Dt)^{1/2}} \sum_{n=0}^{m} \exp\left[\frac{-(x - n\,\Delta x)^2}{4Dt}\right]\Delta x \qquad (2.12)$$

for the total, which in the limit as $\Delta x \to dx$ and $m \to \infty$ becomes*

---

*Equation 2.13 is commonly expressed in terms of the complementary error function such that

$$C_{x,t} = \frac{C^\circ}{2}\,\mathrm{erfc}\left(\frac{-x}{\sqrt{4Dt}}\right)$$

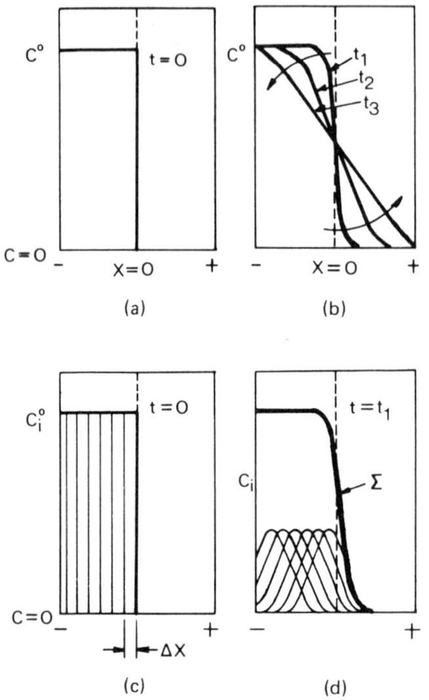


**Figure 2.5** Diffusion from region of finite concentration into a region of zero concentration. Diffusional response to a step-function excitation under linear conditions.

$$C_{x,t} = \frac{C^\circ}{2\pi^{1/2}}\left[\pi^{1/2} - 2\int_0^{(4Dt)^{1/2}} \exp\left(\frac{-x^2}{4Dt}\right) d\left(\frac{x}{4Dt}\right)^{1/2}\right] \tag{2.13}$$

and then differentiation with respect to x yields the concentration gradient

$$\left(\frac{\partial C}{\partial x}\right)_{x,t} = \frac{C^\circ}{(4\pi Dt)^{1/2}} \exp\left(\frac{-x^2}{4Dt}\right) \tag{2.14}$$

This expression has the form of the normal curve of error and therefore we anticipate the *gradients* in this case to have a Gaussian shape.

It is important to recognize the unique relationship that exists between the responses to an impulse and step change in concentration.

The derivative of the step response (Eq. 2.14) is identical to the impulse response (Eq. 2.4), and the integral of the impulse response is identical to the step response. This reciprocity is an important property of linear systems in general. The reader should now appreciate that under linear conditions the time dependence of *any* concentration profile can be treated by adding the response functions for its component impulses.

## C. Barriers to Free Diffusion

Thus far we have examined diffusion under infinite conditions, where no phase boundaries exist. Some practical situations may be adequately described by the treatment above. More frequently, the diffusion process will be initiated in the neighborhood of one or more phase boundaries as, for example, in chromatography and electrochemistry. The phase boundaries may be either permeable or impermeable to the diffusing solute. In electrochemical techniques the boundary (e.g., the working electrode) is usually impermeable; however, this is not always so (e.g., some ion-selective electrodes, membranes, liquid-liquid interfaces). In the most interesting cases the diffusion barrier is not passive, but somehow interacts with the solute (e.g., an electrochemical reaction occurs).

Let us first consider what would have resulted had the impulse described above been created in the neighborhood of an impenetrable inert barrier. Such a barrier will behave as a reflector. In Figure 2.6a an impulse of molecules is suddenly created in solution a short distance away from the interface. The dashed line is the profile that would result in the absence of the barrier. The dotted line results from "folding" this concentration profile about the interfacial plane. Because diffusion is a linear process, the actual concentration profile in the presence of the barrier will be the sum of the dotted and dashed lines, as shown in Figure 2.6b.

If the impulse is created *at* the interface as in Figure 2.6c, then the hypothetical (dashed) concentration profile results in an actual profile (Fig. 2.6d) with twice the amplitude of what would have been extant had the barrier not been present. The results for later times can be arrived at by the same logic. Note that at infinite time the profile will have completely "relaxed" away. Diffusion under these conditions is most commonly encountered in electrochemistry and is described as being semi-infinite. We can, in fact, create a close approximation to an impulse of molecules by generating product molecules in response to a pulse excitation (potential, current, or charge). A good example is when the surface itself is the reactant, such as when metal ions are suddenly released from a metal surface by oxidative dissolution.

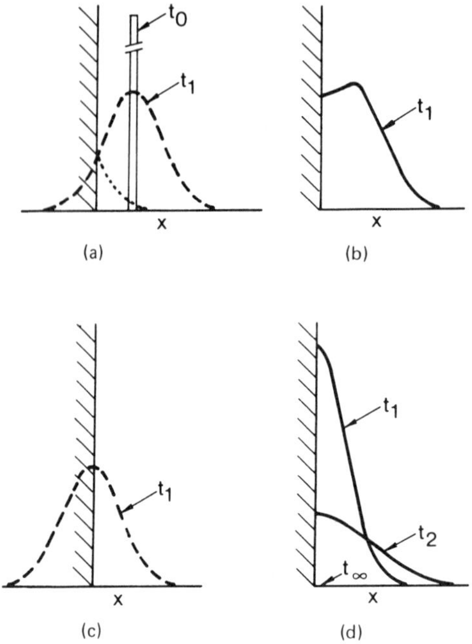


**Figure 2.6** Diffusional spreading of an impulse of molecules created near (a)-(b) and at (c)-(d) an impenetrable barrier.

Occasionally (e.g., thin-layer electrochemistry, porous bed electrodes, metal atoms dissolved in a mercury film), diffusion may be further confined by a second barrier. Figure 2.7 illustrates the case of *restricted diffusion* when the solution is confined between two

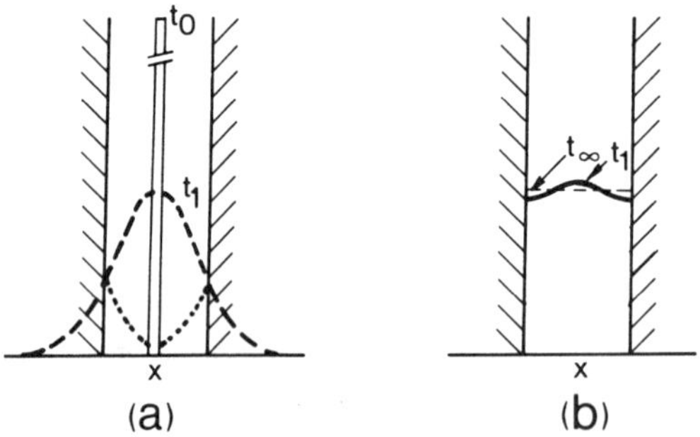


**Figure 2.7** Diffusional spreading of an impulse of molecules created between two barrier plates.

parallel barrier plates. Once again the folding technique quickly enables a prediction of the actual result. In this case, complete relaxation of the profile results in a uniform finite concentration across the slab of solution, in distinct contrast to the semi-infinite case. When the slab thickness $\ell$ is given, the time for the average molecule to diffuse across the slab is calculable from the Einstein equation such that

$$t = \frac{\ell^2}{2D} \tag{2.15}$$

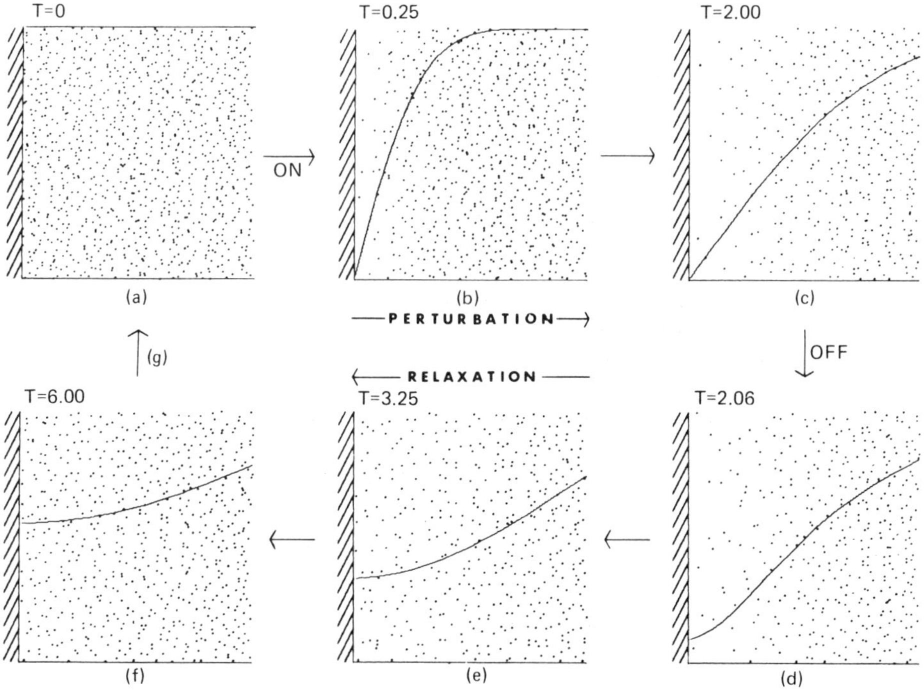


Figure 2.8 Diffusional relaxation following momentary creation of a vacancy at an inert barrier. Semi-infinite conditions prevail. A density map and concentration-distance profiles are shown. (a) Initial condition; (b) vacancy creation; (c) vacancy extends farther into bulk of medium; (d) relaxation begins; (e) relaxation continues; (f) relaxation continues; (g) initial condition restored.

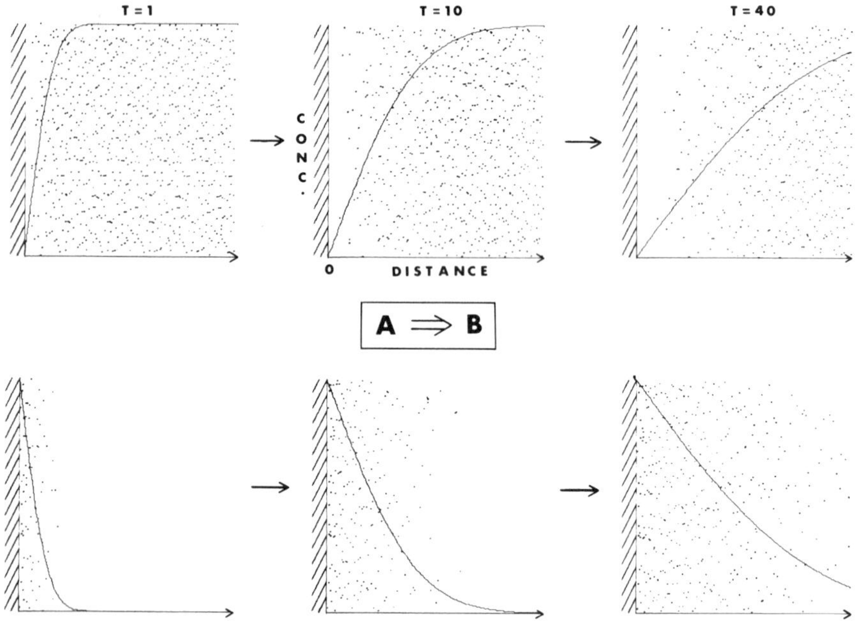


Figure 2.9 Diffusion of a reactant toward a surface (top) occurs simultaneously with diffusion of product away from the surface (bottom).

Thus the relaxation time for a concentration profile in the thin layer will vary with the thickness squared.

Although we will not belabor this point here, the folding technique can easily be adapted to the mathematical description. The reader should recognize that the presence of the barrier does not invalidate either of Fick's laws but does influence the boundary conditions chosen to obtain a specific solution, whether analytically or by the finite difference approach.

It is also important to recognize that the considerations above apply if we create a solute *vacancy* (i.e., a region of zero concentration) rather than an impulse. For example, it is possible to create a vacancy in a layer of solution adjacent to an electrode by instantly converting the solute electrochemically. This is illustrated in Figure 2.8 for a semi-infinite situation. The initial condition (Fig. 2.8a) is perturbed (2.8b and c) and the resulting vacancy then relaxes as indicated in Figure 2.8d-f until the initial condition is restored at infinite time (2.8a).

There is an important electrochemical technique called chronoamperometry (i.e., current measured as a function of time), where a vacancy is maintained at the surface (Chap. 3). The electrode potential is controlled at a value sufficient to react immediately any sample molecule that diffuses to the surface. This results in the sequence shown in Figure 2.9. The top drawings indicate consumption of the reactant at three times following the application of the potential. The corresponding concentration profiles are also shown. The bottom sequence indicates creation of product at the same three times. The sample molecules try their best to fill in the vacancy, but the electrode reaction prevents this. If the electrode is disconnected so no further reaction is possible, the initial condition will be restored.

The diffusion problems encountered in chronoamperometry are described in more detail in Chapter 3. The reader should be able to make mental pictures of the transport of molecules and ions to and from an electrode surface. In our experience, this is one of the most difficult concepts to grasp for beginning students of electrochemistry. Many hours of practice at sketching concentration-distance profiles is highly recommended. A review of this section after reading Chapters 3 to 5 would probably prove worthwhile.

## III. THE CHARGED INTERPHASE

Every chemist at one time or another has encountered the Debye-Hückel theory of electrolyte solutions. It is essential to understand the nature of ions in homogeneous solution in order to appreciate the interaction of these same ions with a charged surface. If one considers the assumptions of the Debye-Hückel theory for a central ion (i.e., the solvated ion is a hard nonpolarizable sphere, a point charge, in a continuous medium of uniform dielectric constant), it is reasonable to apply both the assumptions and the resulting conclusions to a charged surface. After all, the concept of a planar charged surface is no different from that of an ion of very large radius in contact with much smaller ions and solvent molecules. Theories that deal with either situation on a microscopic level have been grossly approximate. Nevertheless, simple electrostatic models are available in which the attraction between an ion and a surface is countered by thermal agitation.

Let us begin by going back to some very fundamental ideas from physics. First consider two parallel inert metallic conductors separated by an evacuated space, d. If an electrical potential difference is applied across the plates (Fig. 2.10), the difference in potential between the bulk conductor phases implies that an electric field exists between them of magnitude $\Delta\phi_m/d$ (V/cm). Any molecule or ion in free space between the plates may sense the field ("potential gradient") but not the absolute magnitude of the inner potentials, $\phi_m$, for each plate.

If we place a neutral molecule in the field, very little will happen unless the field strength is extremely great (e.g., $10^6$-$10^9$ V/cm). On such occasions there will be significant orientation of dipoles, polarization of orbitals, and possibly even ionization. In the latter

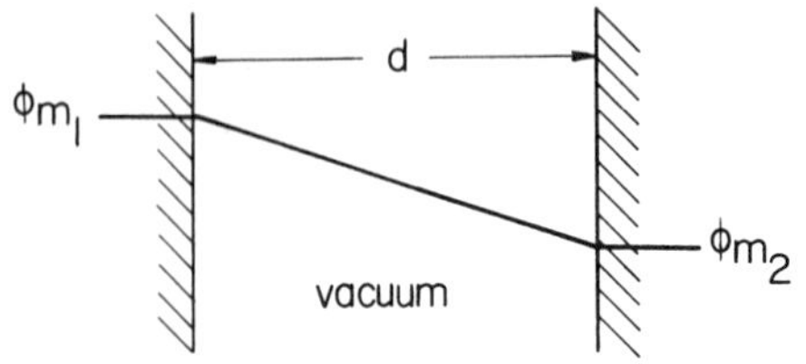


**Figure 2.10** Linear field between two conducting plates with a vacuum between them.

case, an organic molecule loses an electron from its highest filled molecular orbital forming a cation radical. Field ionization in this manner is extremely difficult to achieve in free space. On the other hand, the ionization process can often be stimulated by radiation (e.g., photoionization) or the nearness of a metal electrode (e.g., field ionization mass spectrometry). In either case an electron leaves the molecule and is accelerated toward the positive electrode (anode) while the cation radical accelerates toward the negative electrode (cathode). Both particles pick up kinetic energy along the way, depending only on their charge and the potential difference through which they fall. Gas-phase redox reactions are quite familiar in field ionization mass spectrometry. Molecules passing through a strong field near a fine wire or sharp-edged anode are oxidized to cations with little internal energy and therefore little tendency to fragment. The cations are focused and accelerated into the analyzer region of the mass spectrometer and then counted at a detector.

Suppose now that we fill the space between our planar electrodes with a solution. First let us choose a pure solvent of low dielectric constant (e.g., hexane) with no charge carriers present. How does this compare with the previous situation? First, we are limited in the field we can achieve before breakdown in the dielectric occurs. It is virtually impossible to field ionize a molecule in such a medium. On the other hand, photoionization can be accomplished with the field providing an impetus to charge separation. As in a vacuum the photoionized molecule and the electron are accelerated in opposite directions, but now a terminal velocity is readily achieved depending on the viscous drag of each charged particle. The solvated photoelectron will, of course, move far more rapidly than the ion.

The complexity of the situation can be increased significantly by replacing the hexane with a medium of high dielectric constant (e.g., water) containing many charge carriers. Now the potential difference we can apply at zero current is limited to a few electron volts (at most). A greater difference would result in electrochemical reaction of either the solvent and/or the ions present. Why is this so?

If we consider the potential gradient across the present cell we find that it exists only in the narrow *interphase* regions near the two

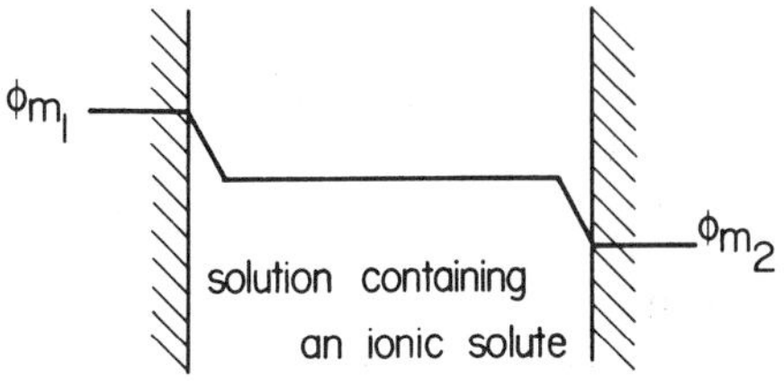


**Figure 2.11** Potential gradients are near the plates when a solution containing ions is placed between them.

surfaces (Fig. 2.11). Furthermore, the thickness of the potential gradient is related to the ionic strength. At high ionic strengths the potential will decay away more quickly at each interphase. This is what should be expected from the Debye-Hückel theory, since the potential falls away from a central ion faster at higher ionic strength. This should make sense if we consider the competition between ordering by electrostatic attraction and disordering by thermal motion in the medium. At high ionic strength the large number of ions permits a *time-averaged distribution* closer to the surface, whereas at low ionic strength the thermal agitation competes more effectively and the time-averaged distribution of an ion excess (or ion imbalance) extends farther from the surface. In the bulk of the solution, electroneutrality exists and the potential, $\phi_S$, is independent of distance.

The interphase between an electrolyte solution and an electrode has come to be known as the electrical double layer. It was recognized early that the interphase behaves like a capacitor in its ability to store charge. Helmholtz therefore proposed a simple electrostatic model of the interphase based on charge separation across a constant distance as illustrated in Figure 2.12. This parallel-plate capacitor model survives principally in the use of the term "double layer" to describe a situation which is quite obviously far more complex. Helm-

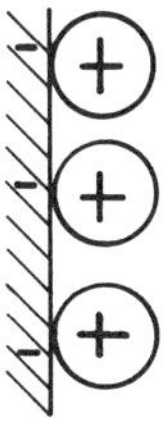

**Figure 2.12** Simple capacitor model of electrode solution interface as a charged double layer (original Helmholtz model). Negatively charged surface. Positively charged ions attracted to the surface forming an electrically neutral interphase.

holtz was unable to account for the experimentally observed potential dependence and ionic strength dependence of the capacitance. For an *ideal* capacitor Q = CV and C, the capacitance, is not a function of V.

Figure 2.13 illustrates what is presently a widely accepted model of the electrode-solution interphase. This model has evolved from simpler models which first considered the interphase as a simple capacitor (Helmholtz), then as a Boltzmann distribution of ions (Gouy-Chapman). The electrode is covered by a sheath of oriented solvent molecules (water molecules are illustrated). Adsorbed anions or molecules, A, contact the electrode directly and are not fully solvated. The plane that passes through the center of these molecules is called the inner Helmholtz plane (IHP). Such molecules or ions are said to be *specifically adsorbed* or *contact adsorbed*. The next layer of molecules carry their primary (hydration) shell and are separated from the electrode by the monolayer of oriented solvent (water) molecules adsorbed on the electrode. The plane passing through the center of these solvated molecules or ions is referred to as the outer Helmholtz plane (OHP). Beyond the compact layer defined by the OHP is a Boltzmann distribution of ions determined by electrostatic interaction between the ions and the potential at the OHP and the random jostling of ions and solvent molecules. This region is referred to as the *diffuse layer*. In the diffuse layer the random thermal motion tends to distribute the ions evenly throughout the solution, whereas the electrostatic forces tend to attract or dispel ions from the surface depending on their charge. These two tendencies counterbalance each other resulting in a nonuniform distribution of ions near the surface. As one

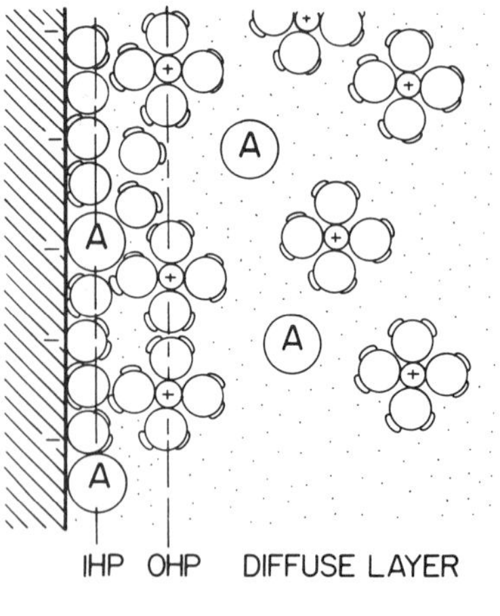


**Figure 2.13** Model of the electrode-solution interphase as described by Bockris, Devanathan, and Muller (see J. O'M. Bockris and A. K. N. Reddy, *Modern Electrochemistry*, Vol. 2, Chap. 7, Plenum, New York, 1970).

looks beyond the diffuse layer, *ions in the homogeneous bulk solution phase cannot feel the presence of the electrode.* In fact, the potential diminishes exponentially with distance from the surface and the field strength becomes zero very close (usually considerably less than $10^{-6}$ m) to the electrode.

As a result of the availability of charge carriers, all of the potential difference between two electrodes is dropped across the two interphases for an electrolyte solution and not across the bulk solution phase. When a current passes across the solution, there is a possibility that a potential difference will develop due to the finite conductivity of the solution. In most electroanalytical experiments this is very small compared to the interfacial potential difference and always results in a comparatively weak electric field (small potential dropped across a large distance). This matter will be dealt with beginning in Chapter 6.

Knowledge of the structure of the interphase is necessary before a detailed understanding of the electron transfer process is possible. A great deal of progress has been made in the last 50 years. Nevertheless, most of the details remain beyond the reach of current experimental techniques.

## IV. THE NERNST EQUATION AND ELECTROCHEMICAL REVERSIBILITY

Let us consider the surface electron-transfer reaction occurring at a single electrode maintained at a fixed potential, where both O and R are soluble and the reaction shown is rate controlling (i.e., no other processes limit this heterogeneous reaction):

$$\mathrm{O} + n\mathrm{e} \underset{k_{b,h}}{\overset{k_{f,h}}{\rightleftharpoons}} \mathrm{R} \tag{2.16}$$

The forward and backward heterogeneous rate constants, $k_{f,h}$ and $k_{b,h}$, are *formal* constants, which implies that activity coefficients are assumed to be unity. In normal homogeneous solution kinetics, for example,

$$\mathrm{O} \underset{k_b}{\overset{k_f}{\rightleftharpoons}} \mathrm{R} \tag{2.17}$$

the formal rate equation is given by

$$\frac{d[R]_t}{dt} = -\frac{d[O]_t}{dt} = k_f[O]_t - k_b[R]_t \tag{2.18}$$

where the rate constants are expressed in units of $s^{-1}$ and the concentrations in mol/L. On the other hand, in electrode kinetics the reaction occurs only near the interface between the electrode and the solution, where the electric field strength is great. It is therefore convenient to deal in terms of mol/cm$^2$ ($\underline{N}$) transformed per unit time [i.e., mol/(cm$^2$ s)] such that

$$\frac{d\underline{N}_R}{dt} = -\frac{d\underline{N}_O}{dt} = k_{f,h}C_O - k_{b,h}C_R \tag{2.19}$$

where $C_O$ and $C_R$ are the surface concentrations in mol/cm$^3$ (note that we use brackets for mol/L and C for mol/cm$^3$ and an underlined symbol to indicate per unit area). $C_O$ and $C_R$ are first assumed to be the same as the bulk concentrations, and independent of cell current and time. It should be clear that the heterogeneous rate constants will have units of cm/s. The *net* transformation of O to R ($-d\underline{N}_O/dt$) can be expressed as

$$\frac{d\underline{N}_{O,f}}{dt} - \frac{d\underline{N}_{O,b}}{dt} \tag{2.20}$$

or [mol/(cm$^2$ s) of O lost in the forward reaction] − [mol/(cm$^2$ s) of O gained via the reverse reaction].

From Faraday's law, Q = nFN, the number of coulombs passed per square centimeter in the forward reaction is $\underline{Q}_f = nF\underline{N}_{O,f}$ and in the reverse reaction $\underline{Q}_b = nF\underline{N}_{O,b}$ where $\underline{N}_{O,f}$ and $\underline{N}_{O,b}$ are the mol/cm$^2$ of O lost and gained. The current density $\underline{i}$ (A/cm$^2$) is given by $d\underline{Q}/dt$, so that the partial current densities for the forward and reverse reactions are

$$\underline{i}_f = \frac{d\underline{Q}_f}{dt} = nF\,\frac{d\underline{N}_{O,f}}{dt} \tag{2.21}$$

and

$$\underline{i}_b = \frac{d\underline{Q}_b}{dt} = nF\,\frac{d\underline{N}_{O,b}}{dt} \tag{2.22}$$

By analogy, the net current density (A/cm$^2$) is given by

$$\underline{i}_{net} = -nF \frac{d\underline{N}_O}{dt} = \underline{i}_f - \underline{i}_b = nF\left(\frac{d\underline{N}_{O,f}}{dt} - \frac{d\underline{N}_{O,b}}{dt}\right)$$
$$= nF(k_{f,h}C_O - k_{b,h}C_R) \qquad (2.23)$$

The actual current (amperes) measured is

$$i_{net} = \underline{i}_{net}A = nFA(k_{f,h}C_O - k_{b,h}C_R) \qquad (2.24)$$

The *formal rate constants* $k_{f,h}$ and $k_{b,h}$ are potential dependent. At any given potential, in order for the transition from the oxidized form to occur, it will be necessary to pass over an *activation free-energy barrier,* $\Delta G_f^\ddagger$, as illustrated by the Morse curves in Figure 2.14. *The rate of reaction will be proportional to* $\exp(-\Delta G_f^\ddagger/RT)$.

At equilibrium, $\Delta G_f^\ddagger = \Delta G_b^\ddagger$ as shown in Figure 2.15, where the probability of electron transfer is the same in each direction, $-\underline{i}_b = \underline{i}_f = \underline{i}_0$, where $\underline{i}_0$ is the *exchange current density.* Under nonequilibrium conditions favoring the cathodic reaction (O → R), $\Delta G_f^{\ddagger\prime} < \Delta G_b^{\ddagger\prime}$ by an amount of free energy given by $-nF(E - E^{\circ\prime})$, where $E^{\circ\prime}$ is the formal potential of the couple O/R.

Now let us idealize the intersection region of the overlapping Morse potential energy curves as shown below. In the nonequilibrium situation, we arbitrarily define a and b so that the change in barrier height for the reverse reaction is

$$b = \Delta G_b^{\ddagger\prime} - \Delta G_0^\ddagger \qquad (2.25)$$

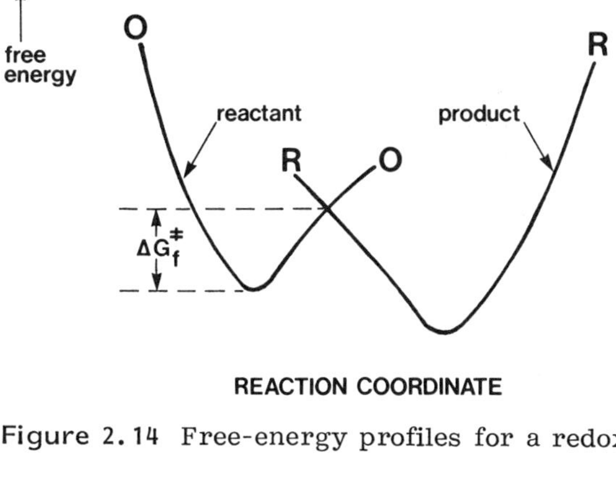


Figure 2.14 Free-energy profiles for a redox reaction.

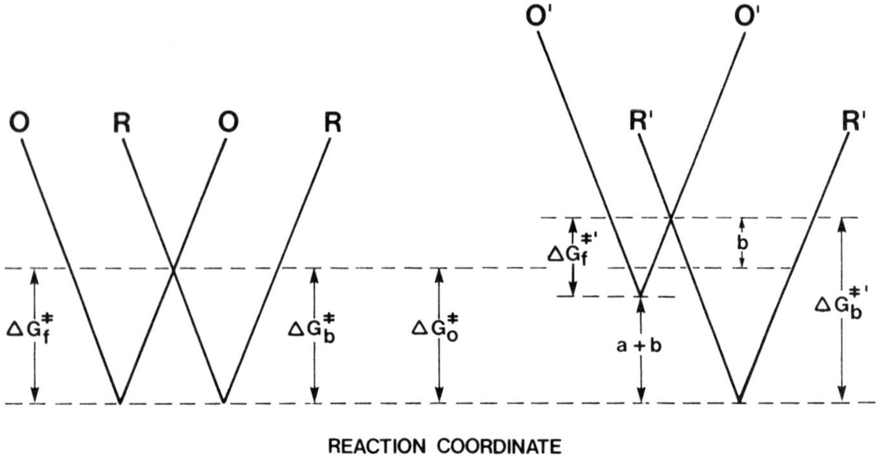


$\Delta G_f^{\ddagger} = \Delta G_b^{\ddagger} = \Delta G_0^{\ddagger}$

$E = E^{o\prime}$

$\Delta(\Delta G^{\ddagger}) = 0$

$i_{net} = 0$

Equilibrium situation (if $C_O = C_R$)

$\Delta G_f^{\ddagger} < \Delta G_b^{\ddagger\prime}$

$E \neq E^{o\prime}$

$\Delta(\Delta G^{\ddagger}) = a + b = -nF(E - E^{o\prime})$

$i_{net} \neq 0$

Nonequilibrium situation (if $C_O = C_R$)

Figure 2.15 Linearized Morse curves.

and

$$a + b = \Delta G_b^{\ddagger\prime} - \Delta G_f^{\ddagger\prime} \tag{2.26}$$

In order to move from the state of equilibrium on the left to the nonequilibrium situation depicted in Figure 2.15 on the right, we apply a new potential E which causes more O to be transformed into R per unit time than R is transformed into O (i.e., a net reduction). This shifts the potential curves to a new relative position such that

$$\Delta G_b^{\ddagger\prime} - \Delta G_f^{\ddagger\prime} = -nF(E - E^{o\prime}) \tag{2.27}$$

$$\Delta G_f^{\ddagger\prime} - nF(E - E^{o\prime}) = \Delta G_b^{\ddagger\prime} = \Delta G_0^{\ddagger\prime} + b \tag{2.28}$$

and

$$-nF(E - E^{\circ\prime}) = a + b \tag{2.29}$$

If we define the *transfer coefficient* $\alpha$ as $a/(a + b)$, then

$$a = -\alpha nF(E - E^{\circ\prime}) \tag{2.30}$$

and

$$b = -(1 - \alpha)nF(E - E^{\circ\prime}) \tag{2.31}$$

The transfer coefficient depends on the symmetry of the two potential curves and for the idealized curves shown, $\alpha = 0.5$. Now since

$$\Delta G_f^{\ddagger\prime} = \Delta G_0^{\ddagger} + b + nF(E - E^{\circ\prime}) = \Delta G_0^{\ddagger} + b - (a + b)$$

$$= \Delta G_0^{\ddagger} - a \tag{2.32}$$

then by substitution

$$\Delta G_f^{\ddagger\prime} = \Delta G_0^{\ddagger} + \alpha nF(E - E^{\circ\prime}) \tag{2.33}$$

and

$$\Delta G_b^{\ddagger\prime} = \Delta G_0^{\ddagger} + b = \Delta G_0^{\ddagger} - (1 - \alpha)nF(E - E^{\circ\prime}) \tag{2.34}$$

Using the Arrhenius equation,

$$k = Ae^{-\Delta G^{\ddagger}/RT} \tag{2.35}$$

where $A$ is a constant preexponential factor, we have

$$k_{f,h} = Ae^{-\Delta G_f^{\ddagger\prime}/RT} = Ae^{-\Delta G_0^{\ddagger}/RT}e^{-\alpha nF(E-E^{\circ\prime})/RT} \tag{2.36}$$

and

$$k_{b,h} = Ae^{-\Delta G_b^{\ddagger\prime}/RT} = Ae^{-\Delta G_0^{\ddagger}/RT}e^{(1-\alpha)nF(E-E^{\circ\prime})/RT} \tag{2.37}$$

At the standard potential, $k_{f,h} = k_{b,h} = k_{s,h}$, for all reactants in the standard state, where $k_{s,h}$ is the standard heterogeneous rate constant, defined according to the Arrhenius equation as

$$k_{s,h} = Ae^{-\Delta G_0^{\ddagger}/RT} \tag{2.38}$$

Substituting Equation 2.38 into Equations 2.36 and 2.37, we have

$$k_{f,h} = k_{s,h}e^{-\alpha nF(E-E^{\circ\prime})/RT} \tag{2.39}$$

and

$$k_{b,h} = k_{s,h}e^{(1-\alpha)nF(E-E^{\circ\prime})/RT} \tag{2.40}$$

Thus we have succeeded in expressing the individual heterogeneous rate constants in terms of a standard rate constant and a potential dependent exponential term. Since $\underline{i}_f = nFC_Ok_{f,h}$ and $\underline{i}_b = nFC_Rk_{b,h}$, the individual current densities can now be written as follows:

$$\underline{i}_f = nFC_Ok_{s,h}e^{-\alpha nF(E-E^{\circ\prime})/RT} \tag{2.41}$$

and

$$\underline{i}_b = nFC_Rk_{s,h}e^{(1-\alpha)nF(E-E^{\circ\prime})/RT} \tag{2.42}$$

The net current was shown earlier to be given by $A(\underline{i}_f - \underline{i}_b)$. Now we can explicitly include the potential dependence such that

$$i_{net} = nFAk_{s,h}[C_Oe^{-\alpha nF(E-E^{\circ\prime})/RT} - C_Re^{(1-\alpha)nF(E-E^{\circ\prime})/RT}] \tag{2.43}$$

This important expression is widely known as the Eyring equation. Note that $i_{net}$ depends on both the potential and the surface concentration of each form of the couple. For example, a high concentration of R and a very positive potential combine to result in a large anodic current.

At equilibrium, $i_{net} = 0$, so

$$C_O e^{-\alpha nF(E_{eq}-E^{\circ\prime})/RT} = C_R e^{(1-\alpha)nF(E_{eq}-E^{\circ\prime})/RT} \tag{2.44}$$

or

$$\frac{C_O}{C_R} = e^{nF(E_{eq}-E^{\circ\prime})/RT} \tag{2.45}$$

and

$$2.303 \log \frac{C_O}{C_R} = \frac{nF(E_{eq} - E^{\circ\prime})}{RT} \tag{2.46}$$

Happily then, in agreement with Nernst,

$$E_{eq} = E^{\circ\prime} + 2.303\left(\frac{RT}{nF}\right)\log \frac{C_O}{C_R} \tag{2.47}$$

It is important to recognize that the Nernst equation is valid only at the equilibrium condition, determined by specifying $i_{net} = 0$. Most electrochemical techniques (chronoamperometry, chronocoulometry, voltammetry, etc.) involve nonequilibrium conditions and therefore cannot be expected to exhibit a nernstian response unless the rates are very fast and equilibrium is quickly reestablished at the surface.

To derive the equation for $\underline{i}_0$ (and obtain the Butler-Volmer equation), substitute $E^{\circ\prime} = E_{eq} - (RT/nF) \ln (C_O/C_R)$ from the Nernst equation into equations 2.41 and 2.42.

$$\underline{i}_f = nFk_{s,h}C_O \exp\left(\frac{-\alpha nF}{RT}E + \frac{\alpha nF}{RT}E_{eq} - \frac{\alpha nF}{RT}\frac{RT}{nF}\ln\frac{C_O}{C_R}\right) \tag{2.48}$$

$$= nFk_{s,h}C_O \exp\left[\frac{-\alpha nF}{RT}(E - E_{eq}) + \ln\left(\frac{C_O}{C_R}\right)^{-\alpha}\right] \tag{2.49}$$

$$= nFk_{s,h}C_O^{(1-\alpha)}C_R^{\alpha}e^{-\alpha nF\eta/RT} \tag{2.50}$$

where the overpotential $\eta = E - E_{eq}$. Similarly,

$$i_b = nFk_{s,h}C_O^{(1-\alpha)}C_R^{\alpha}e^{(1-\alpha)nF\eta/RT} \tag{2.51}$$

At $E = E_{eq}$, $\eta = 0$,

$$i_{f(eq)} = i_{b(eq)} = i_0 = nFk_{s,h}C_O^{(1-\alpha)}C_R^{\alpha} \tag{2.52}$$

At $E \neq E_{eq}$, $i_{net} = A(i_f - i_b)$,

$$i_{net} = nFAk_{s,h}C_O^{(1-\alpha)}C_R^{\alpha}[e^{-\alpha nF\eta/RT} - e^{(1-\alpha)nF\eta/RT}] \tag{2.53}$$

$$= i_0A[e^{-\alpha nF\eta/RT} - e^{(1-\alpha)nF\eta/RT}] \tag{2.54}$$

Two limiting forms of the Butler-Volmer equation of experimental interest are concerned with the current response of the system at both small and large overpotentials. For small overpotentials, $\eta < 8$ mV/n, the exponential terms may be linearized (remember that $e^{-x} = 1 - x$ for small x), so that

$$i_{net} = i_0[(1 - \alpha nf\eta) - (1 - nf\eta - \alpha nf\eta)] \tag{2.55}$$

or

$$i_{net} = i_0 nf\eta \tag{2.56}$$

Under this small-amplitude condition, current is directly proportional to overpotential, as shown in Figure 2.16. By analogy with Ohms' law ($i = E/R = EY$), $i_0nf$ defines a transfer admittance and $1/i_0nf$ an impedance. The latter is often called the *faradaic resistance*, meaning that near the equilibrium potential an electrochemical reaction gives a linear current vs. potential curve as shown below, with zero current intercept and slope equal to the transfer admittance. Systems with higher admittances demonstrate faster heterogeneous electron transfer than those with lower values for $i_0nf$. The other limiting form of the Butler-Volmer equation deals with large overpotentials ($\eta > 120$ mV/n), where it may be assumed that the rate of one of the reactions becomes negligible. For example,

$$i_{net} = i_0e^{-\alpha nf\eta} \tag{2.57}$$

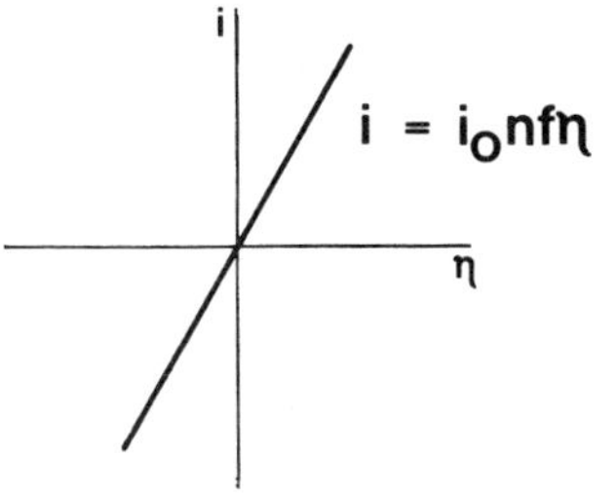


Figure 2.16 Ohmic behavior of heterogeneous electron transfer at low overpotentials.

or

$$\ln i_{net} = \ln i_0 - \alpha nf\eta \qquad (2.58)$$

which is known as the Tafel equation. Differentiation gives

$$\frac{d \ln i_{net}}{d\eta} = -\alpha fn \qquad (2.59)$$

Thus for large amplitudes the current is logarithmically related to overpotential as shown below. Tafel plots (Fig. 2.17) are frequently employed by physical electrochemists to determine exchange currents and transfer coefficients. There are many other ways to obtain these

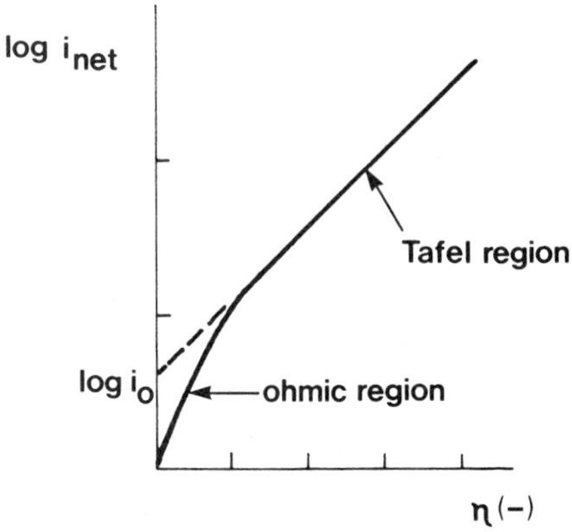


Figure 2.17 Tafel plot.

parameters experimentally, but such numbers are rarely of interest to the analytical chemist. As we will see later, the rate of the heterogeneous electron transfer *relative* to other controlling factors (e.g., diffusion and coupled chemical reactions) is of critical importance to most experiments.

The complete Eyring equation (good for all values of $\eta$) is plotted in Figure 2.18, illustrating that the net current is the sum of two opposing exponential branches, one for the cathodic (reduction) reaction and one for the anodic (oxidation) reaction. Notice that the $i_{net}$ curve passes through the origin where $\eta = 0$ since the partial currents are equal and opposite at this point. This is where the electric field at the interface is such that $\Delta G^{\ddagger}$ for the forward and reverse reactions is the same. This does not imply that there is no electric field!

Once again, it is important to recognize that these current-voltage curves obtain only for the situation in which the activation parameters of the electron transfer itself are controlling. This circumstance is unusual for most analytical electrochemical techniques where it is not desirable to be rate limited by the actual exchange of electrons at the surface.

At the same time, it is significant that the rate constants for electrochemical reactions are an exponential function of the potential. Thus the electrochemist has a great deal of control over kinetics. The exponential potential dependence also means that there is not a sharp threshold for electron transfer, so that the resolution between different chemical species is not nearly as good as one would hope. The sharpness of transitions in atomic (gas phase) spectroscopy or m/e discrimination in mass spectrometry is so much greater than in electrochemistry that the latter must be considered a low-resolution technique by comparison. A great many compounds will not exchange electrons with

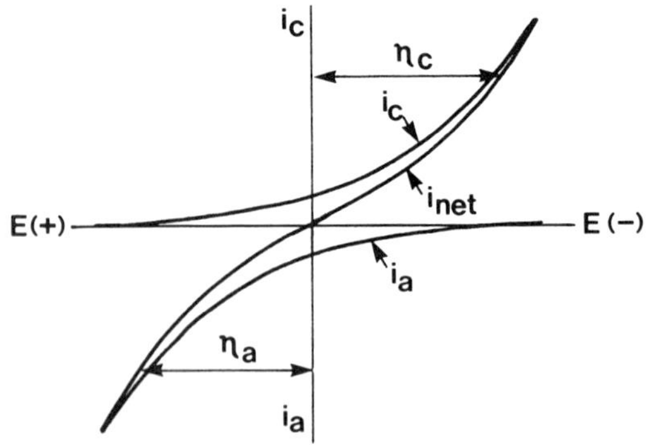


Figure 2.18 Example of current-potential curve assuming that electron transfer is rate limiting and concentrations of O and R are equal.

a metal in ionic solution due to the limited accessible range of electron free energy. Reaction of the medium and/or electrode material itself limits the available range to ca. 2-5 eV, depending on the specific medium and electrode chosen. Many practical examples of this limitation will be described in later chapters.

## V. COUPLED CHEMICAL REACTIONS AND CHEMICAL REVERSIBILITY

One of the most intriguing aspects of electrochemistry involves the homogeneous chemical reactions that often accompany heterogeneous electron-transfer processes occurring at the electrode-solution interface. The addition or removal of an electron from a molecule generates a new redox state, which can be chemically reactive. A variety of mechanisms, some of which involve complicated sequences of electrode and chemical reactions, have been characterized. Several of the more common mechanisms with examples of applicable chemical systems are described below.

One of the simplest electrode reactions is the *EC mechanism* (also called a *following chemical reaction*) in which the electrogenerated species (R) rearranges or reacts with some other solution component (Z) at a rate characterized by the rate constant k. The EC mechanism is summarized by the following reaction sequence, in which the labels E and C identify the heterogeneous electron transfer reaction (electrode reaction) and the subsequent homogeneous solution reaction (chemical reaction), respectively:

$$\text{Electrode reaction, E:} \quad \mathrm{O} + e \rightleftharpoons \mathrm{R} \tag{2.60a}$$

$$\text{Chemical reaction, C:} \quad \mathrm{R} + \mathrm{Z} \xrightarrow{k} \text{product(s)} \tag{2.60b}$$

An example of the EC mechanism is the oxidation of ascorbic acid [1]:

E: 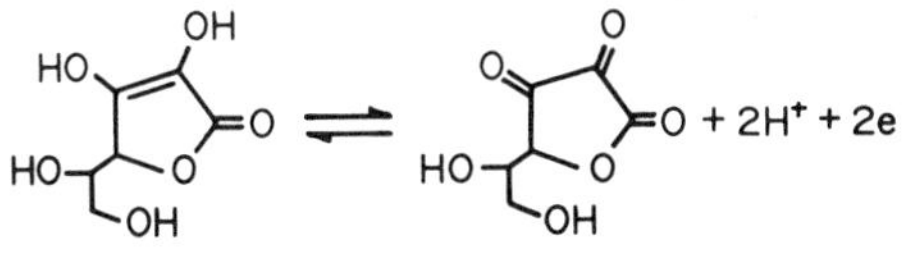 (2.61a)



C: 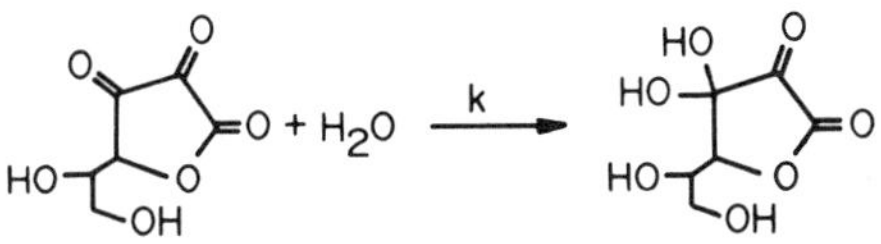 (2.61b)

Other examples of the EC mechanism include nucleophilic addition of Z (such as $H_2O$ or $CN^-$) to electrogenerated organic radical cations, ligand exchange reactions in the case of coordination compounds, and redox reactions between R and Z (or O and Z in the case of a reduction).

The *catalytic regeneration* mechanism is a variation of the EC mechanism in which the initial electroactive species is regenerated by the homogeneous chemical reaction, as follows:

$$\text{E:}\quad R \rightleftharpoons O + e \qquad (2.62a)$$

$$\text{C:}\quad O + Z \xrightarrow{k} R + Z^{+} \qquad (2.62b)$$

An interesting aspect of this mechanism is the implicit unreactivity of Z at the electrode, although it reacts with electrogenerated O. This situation has been frequently encountered in the reaction of certain types of biological materials such as heme proteins. An example is the reduction of cytochrome c by electrogenerated methyl viologen cation radical, $MV^{+\cdot}$ [2]:

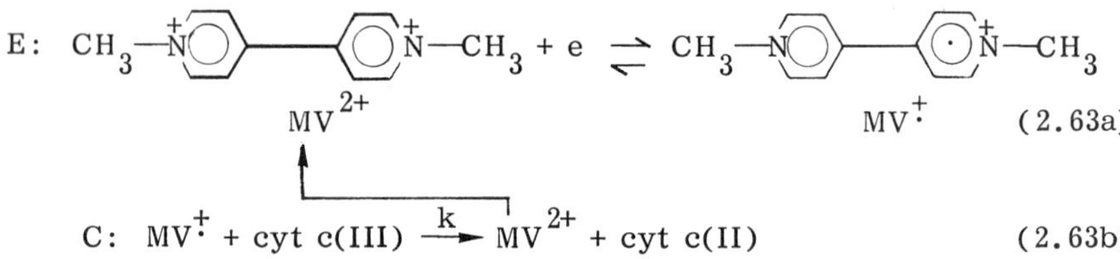

(2.63a)
(2.63b)

Cytochrome c contains a heme iron, which is reduced by $MV^{+\cdot}$ from Fe(III) to Fe(II). The unreactivity of cytochrome c at the electrode is possible due to physical isolation of the iron redox center from the electrode by the surrounding protein structure.

A more complicated mechanism involving regeneration of starting material is the *ECC or half-regeneration* mechanism:

$$\text{E:}\quad R \rightleftharpoons O + e \qquad (2.64a)$$

$$\text{C:}\quad O + Z \rightarrow OZ \qquad (2.64b)$$

$$\text{C:}\quad OZ + O \rightarrow OZ^{+} + R \qquad (2.64c)$$

The oxidation of 9,10-diphenylanthracene (DPA) in acetonitrile in the presence of nucleophiles such as water and pyridine (Py) has been shown to follow this mechanism [3]:

$$\text{E:}\quad \text{DPA} \rightleftharpoons \text{DPA}^{\overset{+}{\cdot}} + e \tag{2.65a}$$

$$\text{C:}\quad \text{DPA}^{\overset{+}{\cdot}} + \text{Py} \xrightarrow{k_1} \text{DPA(Py)}^{\overset{+}{\cdot}} \tag{2.65b}$$

$$\text{C:}\quad \text{DPA(Py)}^{\overset{+}{\cdot}} + \text{DPA}^{\overset{+}{\cdot}} \xrightarrow{k_2} \text{DPA(Py)}^{2+} + \text{DPA} \tag{2.65c}$$

$\text{DPA(Py)}^{2+} \rightarrow$ further reaction

The *ECE mechanism* involves electrochemical generation of a species that then reacts with some other component in solution. The product of this reaction is more easily oxidized/reduced than the starting material and so is immediately electrolyzed. The sequence of reactions can be summarized as follows:

$$\text{E:}\quad \text{R} \rightleftharpoons \text{O} + e \tag{2.66a}$$

$$\text{C:}\quad \text{O} + \text{Z} \xrightarrow{k} \text{OZ} \tag{2.66b}$$

$$\text{E:}\quad \text{OZ} \rightarrow \text{OZ}^{+} + e \tag{2.66c}$$

This mechanism is encountered in the oxidation of organic compounds in the presence of a nucleophile. A typical example is the oxidation of a hydroquinone in the presence of thiol [4]:

E: 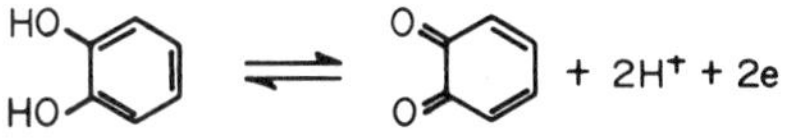

(2.67a)

C: 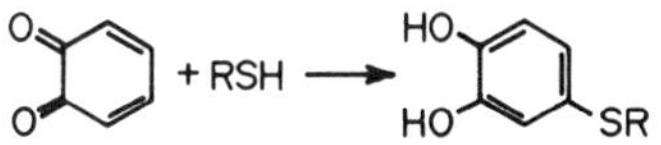

(2.67b)

E: 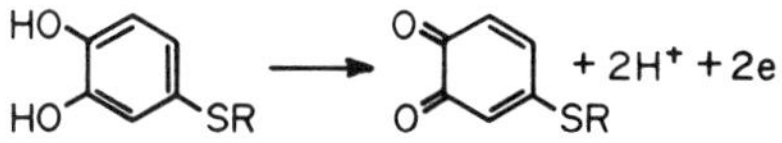

(2.67c)

Another variation of an ECE mechanism involves a reproportionation redox reaction:

$$\text{E:}\quad \text{R} \rightleftharpoons \text{O}^{2+} + 2e \tag{2.68a}$$

$$C:\ R + O^{2+} \rightleftharpoons 2O^{+} \quad (2.68b)$$

$$E:\ O^{+} \rightleftharpoons O^{2+} + e \quad (2.68c)$$

This mechanism can be found in the electrochemistry of compounds with two or more oxidation states. Oxidation of the tantalum cluster compound $Ta_6Br_{12}^{2+}$ is a typical case [5]:

$$E:\ Ta_6Br_{12}^{2+} \rightleftharpoons Ta_6Br_{12}^{4+} + 2e \quad (2.69a)$$

$$C:\ Ta_6Br_{12}^{2+} + Ta_6Br_{12}^{4+} \rightleftharpoons 2Ta_6Br_{12}^{3+} \quad (2.69b)$$

$$E:\ Ta_6Br_{12}^{3+} \rightleftharpoons Ta_6Br_{12}^{4+} + e \quad (2.69c)$$

In the *CE mechanism* (or preceding chemical reaction) the electroactive species is generated from an electroinactive species by a chemical reaction:

$$C:\ A \xrightarrow{k} O \quad (2.70a)$$

$$E:\ O + e \rightleftharpoons R \quad (2.70b)$$

Thus the rate of electrolysis of O is determined by the magnitude of k in concert with the diffusion process. Metal complexes such as cadmium nitrilotriacetic acid $Cd(NTA)^-$, in which the metal complex first dissociates to form hydrated metal ion that then undergoes reduction, fall into this category [6]:

$$C:\ Cd(NTA)^{-} \rightleftharpoons Cd^{2+} + NTA^{3-} \quad (2.71a)$$

$$E:\ Cd^{2+} + 2e \rightleftharpoons Cd \quad (2.71b)$$

The interest in characterization of electrode mechanisms has motivated the development of a multitude of electrochemical techniques. These techniques enable the sequence of reactions to be determined and the rate constant(s) of the homogeneous chemical reactions to be measured. Such information is deduced from the effect of the coupled chemical reactions on the response signal to a particular excitation signal that is impressed on the electrode. (This is a subject of considerable discussion in subsequent chapters.) Spectroscopic techniques (UV-visible-IR, NMR, ESR) have been effectively coupled with electrochemistry to enable monitoring of homogeneous chemical reactions. Intermediates and products can sometimes be identified from their spectra.

Electrochemistry has proven to be a valuable technique for generating reactive oxidation states and studying the attendant solution chemistry of such electrogenerated species. The ease of oxygen removal from electrochemical cells greatly facilitates the study of oxygen-sensitive species which are electrogenerated. Small quantities of valuable material can be studied with ease. Very rapid reactions can be monitored since some electrochemical techniques are capable of measuring reactions up to the limit of diffusion-controlled rates. Electrochemistry often has an additional advantage over traditional chemical approaches in that the solution is not complicated by a reagent added to generate the redox state of interest.

## VI. LIQUID-SOLID ADSORPTION

### A. Types of Adsorption

Many species dissolved in solution exhibit a tendency to adsorb on the electrode surface, a phenomenon that can markedly affect the results of electrochemical experiments. For example, the course of an electrode reaction can be altered or the rate of electron exchange enhanced or virtually stopped. Adsorption is responsible for much "unusual" electrochemical behavior and is frequently blamed for unexplained results. Thus it is important for the chemist using electrochemical techniques to recognize phenomena that are attributable to adsorption and to realize which techniques are useful for studying adsorbed species.

Adsorption can be considered in terms of the equilibrium (after Anson [7]) in Equation 2.72

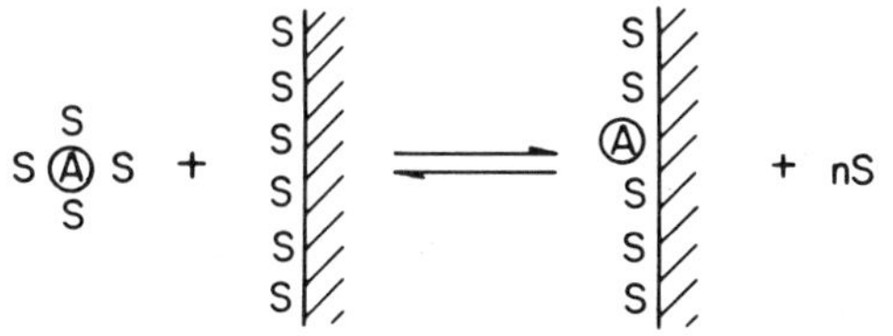


(2.72)

where the adsorbing molecule or ion, A, loses some or all of its solvate molecules, S, and displaces oriented solvent molecules from the electrode surface so that contact between electrode and adsorbate is made.

The driving forces for Reaction 2.72 are varied: hydrophobicity of adsorbate, electrostatic attraction between an ionic adsorbate and a charged electrode, and adsorbate-electrode bonding. Anson has classified adsorbing species into five categories [7].

*Class 1A* consists of simple inorganic anions and cations that are so weakly solvated that electrostatic attraction between ionic charge and an oppositely charged electrode surface pulls Reaction 2.72 to the right. Examples of class 1A ions are $ClO_4^-$, $NO_3^-$, $H_2PO_4^-$, $PF_6^-$, $Cs^+$, and $R_4N^+$. The hydrophobic character of some class 1A anions is attributed to a tendency to disrupt the local solvent structure when they are dissolved in water.

*Class 1B* ions adsorb to a much greater extent than do class 1A ions because the attraction between adsorbate and electrode extends beyond simple coulombic forces to covalent bonding with the electrode via donation of adsorbate electrons to orbitals on the electrode surface. Examples of class 1B ions in order of increasing strength of interaction with the electrode are the anions $Cl^- < Br^- < I^-$; $SO_4^{2-} \ll S_2O_3^{2-}$; $NCO^- \ll NCS^-$.

Most cations are too strongly solvated in water to follow Reaction 2.72 and contact the electrode directly and therefore are not specifically adsorbed.

*Class II* consists of neutral organic molecules that adsorb on the electrode surface primarily because of hydrophobicity in aqueous solutions. Generally, the less soluble a molecule is, the more strongly it adsorbs. Bonding between the electrode surface and $\pi$ electrons and nitrogen lone-pair electrons of the molecule can also enhance adsorption. Class II molecules can be found in virtually every category of organic compound. Typical examples are camphor, n-butanol, and quinoline.

Figure 2.19 shows the dependence of adsorption of class 1A, 1B, and II species on the electronic charge density of the electrode. Adsorption of anions and cations is enhanced by electrostatic attraction with an oppositely charged electrode. The greater adsorption of 1B anions compared to 1A anions is evident. Neutral organic molecules adsorb to the greatest extent on an uncharged electrode. As the electrode becomes charged (positively or negatively) the neutral molecules are displaced by oriented water molecules whose dipole charges become attracted to the electrode or by specifically adsorbed anions (for a positively charged electrode). Some organic molecules have been shown to orient in a specific geometry on the electrode surface. Aromatic molecules tend to lie with the plane of the ring(s) parallel to the electrode surface, enabling the $\pi$ electrons to interact most strongly with a positively charged electrode. Some molecules apparently exhibit more than one orientation depending on the electrode charge. For example, $\beta$-napthoquinoline is thought to adsorb perpendicularly to the electrode at negative potentials and parallel (flat) at positive potentials [8].

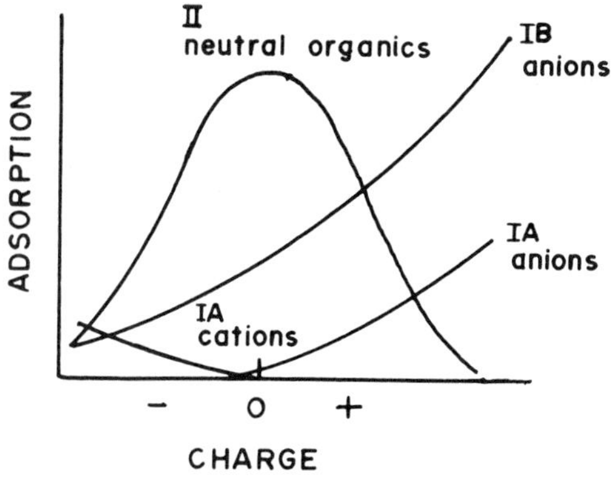


Figure 2.19 Dependence of adsorption of class IA, IB, and II adsorbates on electrode charge. [From Ref. 8, adapted with permission from F. C. Anson, *Acc. Chem. Res.* *8*:400 (1975). Copyright 1975 American Chemical Society.]

Class 1 and II species constitute the classical examples of adsorption that are typically encountered by the electrochemist. The next classes have been recognized more recently and are quite different from the classical cases.

*Class III* adsorbates are cations such as Zn(II), Cd(II), Tl(I), and Pb(II), which do not adsorb when in solution with nonadsorbing anions but which exhibit an "induced adsorption" by strongly adsorbing class 1B anions such as $I^-$. Such induced adsorption involves the formation of surface complexes or crystalline bilayers with the adsorbed anion.

*Class IV* adsorbates are transition metal cations whose adsorption is induced by complexation with $NCS^-$ and $N_3^-$ anions, both of which adsorb strongly on mercury.

*Class V* adsorbates exhibit metal-metal bonding between an adsorbed complex such as $Ru(en)_2^+$ and a mercury electrode.

Electrodes can become coated with interfering films via two processes that are quite different from the classes of adsorption described above.

1. Rapid polymerization of electrochemically generated species such as organic radicals can cover an electrode with a polymeric film. Such films are sometimes impenetrable and difficult to remove, which results in passivation of the electrode surface. A typical case is the oxidation of 1,2-diaminobenzene [9]. The modification of electrode properties by coating with thin polymer films is presently an area of very active investigation [10-12].

2. The oxidation of an electrode such as mercury in the presence of an anion such as bromide forms a film of insoluble $Hg_2Br_2$ on the electrode surface. The formation of analogous films with sulfur-containing compounds causes substantial difficulties when mercury electrodes are used in biological media, since the films tend to passivate the electrode surface.

Recently, adsorption has been used as a means of purposely modifying the electrode surface to form a "chemically modified electrode" [10,11].

## B. Effects of Adsorption

The effects of adsorption on an electrochemical response signal are varied, ranging from complete passivation of the electrode to catalytic enhancement of a redox process to the appearance of new electrochemical response signals. Commonly encountered situations are dealt with briefly below. The reader is referred to the works by Bauer [13] and Reilley and Stumm [14] for a more detailed discussion.

### Inhibition of Electrode Processes

A common phenomenon of adsorption is physical blocking of the electrode surface by the adsorbate, forcing a redox reactant to penetrate an adsorbed film before exchanging electrons with the electrode. This is particularly characteristic of strongly adsorbed organic molecules. The effect on a redox process can vary from mild distortion of the electrochemical signal to complete inhibition of the electrode process. A typical example is the effect of adsorbed camphor on the reduction of the Cu(II)-EDTA complex. Figure 2.20A shows an electrocapillary curve [a plot of the drop time of a dropping mercury electrode (Chap. 9) as a function of potential] of supporting electrolyte with and without camphor. The depression of drop time indicates adsorption of camphor in the potential range $-0.2$ to $-1.1$ V vs. SCE. Figure 2.20B shows strong inhibition of the polarographic reduction (see Chap. 3, Sec. III.D, for a discussion of polarography) of copper within the potential range of camphor adsorption. The chapter by Reilley and Stumm provides more detailed descriptions of penetration of adsorbed films [14].

### Adsorption Waves

Electrochemical response signals attributable to adsorbed species are observable with most electroanalytical techniques. An example of an adsorption postwave is shown in Figure 2.21 for the voltammogram (see Chap. 3, Secs. III.A and III.B, for a discussion of this type of voltammetry) of $Pb^{2+}$ in the presence of a large excess of $I^-$ [15]. The first, rounded peak is due to the reduction of solution-soluble lead diffusing to the electrode; the second, sharp peak is attributed to reduction of

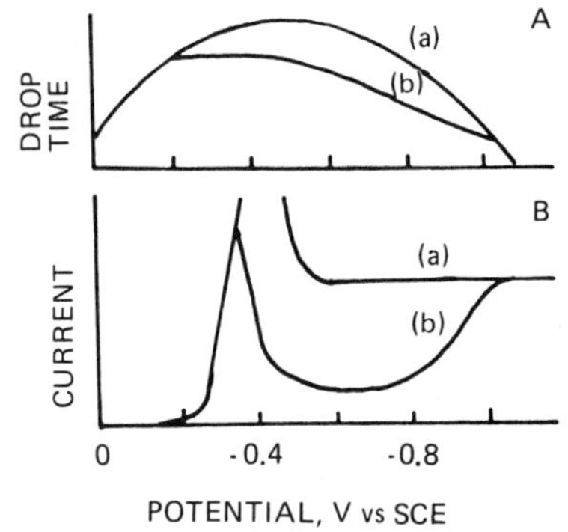


Figure 2.20 Effect of adsorbed camphor on the electrocapillary curve of mercury and on the polarographic reduction of Cu(II)-EDTA. (A) Electrocapillary curve of camphor in acetate buffer: a, without; b, with $3 \times 10^{-4}$ M camphor. (B) Polarograms of a solution containing 0.009 M EDTA, 0.001 M Cu(II), 0.1 M acetate buffer, pH 4.5: a, without; b, with $3 \times 10^{-4}$ M camphor. (Adapted from Ref. 14, p. 91.)

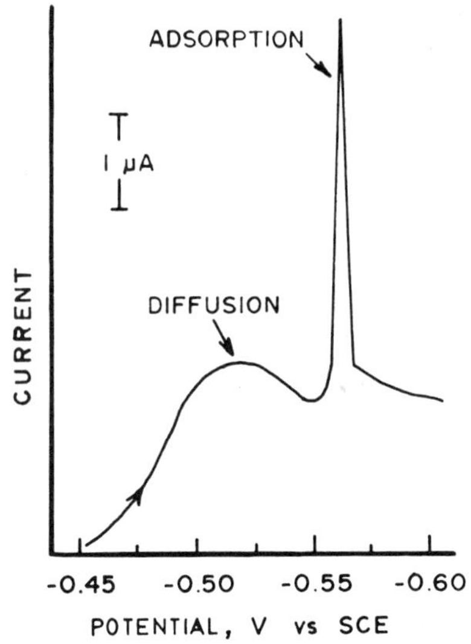


Figure 2.21 Stationary electrode voltammogram at a hanging mercury drop electrode for 0.495 mM $Pb^{2+}$ in 1.0 M NaI, 0.01 M $HClO_4$, scan rate = 0.005 V/s. [From Ref. 15, reprinted with permission from R. W. Murray and D. J. Gross, *Anal Chem.* *38*:392 (1966). Copyright 1966 American Chemical Society.]

lead-iodide complex that is adsorbed on the electrode surface. Such adsorption waves are frequently very sharp "spikes" compared to the diffusion-controlled waves. This is due both to the immediate availability of material for reaction at the surface (i.e., mass transfer is not needed) and the limited amount of reactant present. Since the adsorbed species is stabilized by its interaction with the electrode, it is more difficult to reduce than the solution species. Hence the adsorption wave is a postwave, occurring at a more negative potential than the diffusion waves. Adsorption of the *product* of the electrode reaction typically results in a prewave, since electron transfer is then energetically easier. Such pre- and postwaves are often observable under the condition of large surface coverage by a strongly adsorbed electroactive species.

Current spikes that are attributable to rapid adsorption or desorption of an adsorbate are sometimes observable for strongly adsorbing, but electroinactive, species such as camphor at a mercury electrode. The spike is a nonfaradaic current caused by the change in capacitance resulting from the sudden alteration in double-layer structure when the molecule adsorbs or desorbs.

#### Potential of the OHP

The presence of adsorbed ions can alter the potential of the OHP and, thereby, influence the rate of charge transfer at the electrode. For example, a specifically adsorbed anion causes the potential at the OHP to be more negative than in the absence of adsorption.

### C. Techniques

Adsorption phenomena have been studied by means of virtually every electrochemical technique including the recently developed spectroelectrochemical methods. Electrocapillary methods and measurements of double-layer capacitance have played a central role in the understanding of adsorption. Ac studies have also been very useful and are very sensitive to adsorption effects. More recently, chronocoulometry (Chap. 3, Sec. II.C) has been applied effectively to the measurement of quantities of adsorbed electroactive species. The interested reader is referred to the sections that deal with these techniques for more detailed information.

## VII. CONCLUSION

In the present chapter we have given an overview of several of the more important fundamental concepts that will time and again reappear in the chapters that follow. The reader wishing to have a better feeling for how these concepts influence electroanalytical experiments would do well to return to Chapter 2 after gaining an appreciation of how the experiments are actually carried out in the laboratory.

## REFERENCES

1. S. P. Perone and W. J. Kretlow, *Anal. Chem. 38*:1760 (1966).
2. E. Steckhan and T. Kuwana, *Ber. Bunsenges. Phys. Chem. 78*:253 (1974).
3. J. F. Evans and H. N. Blount, *J. Am. Chem. Soc. 100*:4191 (1978).
4. K. T. Findley, in *The Chemistry of the Quinoid Compounds*, Part 2 (S. Patai, ed.), Wiley, New York, 1974, pp. 877-1144.
5. N. Winograd and T. Kuwana, *J. Am. Chem. Soc. 92*:224 (1970); *93*:4343 (1971).
6. J. Koryta, *Coll Czech. Chem. Commun. 24*:3057 (1959).
7. F. C. Anson, *Acc. Chem. Res. 8*:400 (1975).
8. H. H. Bauer, *Electrodics*, Georg Thieme, Stuttgart, West Germany, 1972, p. 77.
9. A. M. Yacynych and H. B. Mark, Jr., *J. Electrochem. Soc. 123*:1346 (1976).
10. R. W. Murray, *Acc. Chem. Res. 13*:135 (1980).
11. R. W. Murray, in *Electroanalytical Chemistry*, Vol. 13 (A. J. Bard, ed.), Marcel Dekker, New York, 1984, Chap. 3.
12. A. J. Bard, *J. Chem. Educ. 60*:302 (1983).
13. Ref. 8, Chap. 5.
14. C. N. Reilley and W. Stumm, in *Progress in Polarography*, Vol. 1 (P. Zuman and I. M. Kolthoff, eds.), Wiley-Interscience, New York, 1962.
15. R. W. Murray and D. J. Gross, *Anal. Chem. 38*:392 (1966).

# 3
# Large-Amplitude Controlled-Potential Techniques

WILLIAM R. HEINEMAN / Department of Chemistry, University of Cincinnati, Cincinnati, Ohio

PETER T. KISSINGER / Department of Chemistry, Purdue University, and Bioanalytical Systems, Inc., West Lafayette, Indiana

# I. INTRODUCTION

## A. Classification of Dynamic Techniques via Large- and Small-Amplitude Excitations

In Chapter 2 the concept of faradaic impedance was introduced whereby a heterogeneous electron transfer reaction can be modeled as an equivalent electrical circuit. In the simplest case, *at small overpotentials,* we saw how it was possible to represent the electrochemical reaction kinetics by a single resistor since the small net current is linearly related to $\eta$ and is therefore said to be *ohmic.* Electrochemical techniques in which the current and potential fall within this linear region are referred to as *small-amplitude techniques* and will be discussed briefly in Chapter 5. Mathematical and experimental advantages are sometimes associated with such techniques, particularly in the study of the heterogeneous electron transfer process per se. In the present and following chapter we will consider *large-amplitude* controlled-potential and controlled-current techniques. These techniques operate in the nonlinear region, where current is exponentially related to overpotential. The large-amplitude techniques prove to be useful for certain analytical problems and for studies of coupled chemical reactions.

# II. POTENTIAL-STEP TECHNIQUES IN STATIONARY SOLUTION

## A. Potential-Step Excitation

The electroanalytical techniques *chronoamperometry, chronocoulometry,* and *chronoabsorptometry* are all based on the same excitation function of one or more potential steps that are applied to an electrode immersed in a nonstirred solution. The system response is thus identical for all three techniques. They differ only in the data domain of the monitored response. Consequently, the common excitation aspect is dealt with here, whereas each monitored response is considered individually in subsequent sections.

The important concept in these dynamic electrochemical methods is *diffusion-controlled* oxidation or reduction. Consider a planar electrode that is immersed in a quiescent solution containing O as the only electroactive species. This situation is illustrated in Figure 3.1A, where the vertical axis represents concentration and the horizontal axis represents distance from the electrode-solution interface. This interface or boundary between electrode and solution is indicated by the vertical line. The dashed line is the initial concentration of O, which is homogeneous in the solution; the initial concentration of R is zero. The excitation function that is impressed across the electrode-solution interface consists of a potential step from an initial value $E_i$, at which there is no current due to a redox process, to a second

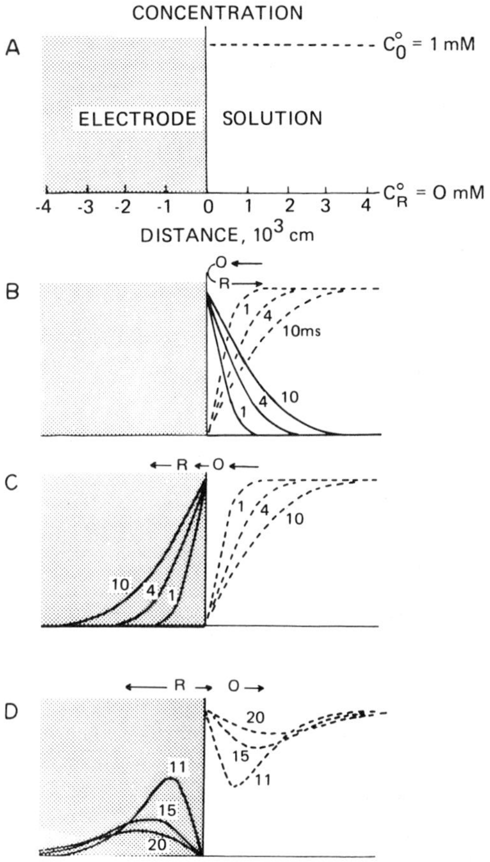


**Figure 3.1** Concentration-distance profiles during diffusion-controlled reduction of O to R at a planar electrode. $D_O = D_R$. (A) Initial conditions prior to potential step. $C_O^\circ = 1$, $C_R^\circ = 0$. (B) Profiles for O (dashed line) and R (solid line) at 1, 4, and 10 ms after potential step to $E_s$. R is soluble in solution. (C) Profiles for O and R at 1, 4, and 10 ms after potential step to $E_s$. R is soluble in the electrode. (D) Potential stepped from $E_s$ to $E_f$ at $\tau = 10$ ms so that oxidation of R to O is now diffusion controlled. Profiles are for 11, 15, and 20 ms after step to $E_s$.

potential $E_s$, as shown in Figure 3.2. The value of this second potential is such that essentially all of O at the electrode surface is instantly reduced to R as in the generalized system of Reaction 3.1:

$$O + ne \longrightarrow R \tag{3.1}$$

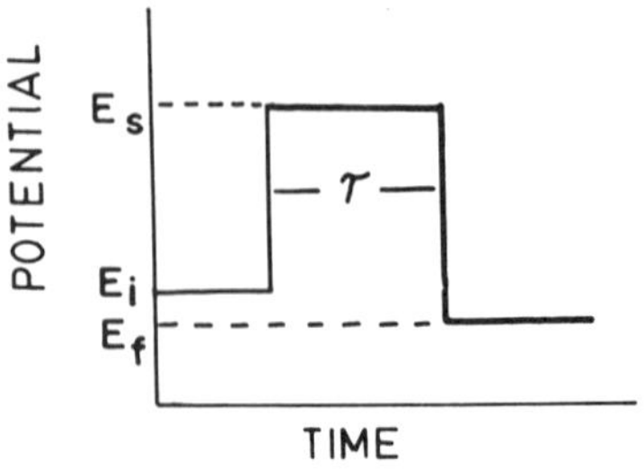


Figure 3.2 Generalized excitation signal for potential-step techniques. Step from initial potential ($E_i$) to step potential ($E_s$) to final potential $E_f$. $\tau$ is the duration of potential step at $E_s$.

The electrochemical depletion of O in the solution immediately adjacent to the electrode prompts a net diffusion of O from the bulk solution into the depleted region. However, as rapidly as O diffuses to the electrode, it is converted to R. Thus the rate of loss of O and the consequent rate of formation of R are both controlled only by the rate of diffusion of O to the electrode surface under the influence of a concentration gradient. This depletion of O and accumulation of R as a function of time is graphically represented in Figure 3.1B by the concentration-distance profiles for the two species at different times following application of the potential step. It is important that the reader feel comfortable with this figure. Note that a net mass transfer always occurs from a region of high concentration to a region of low concentration. In other words, the species diffuse down their respective concentration profiles. The profiles shown in the figure are for the case in which the diffusion coefficients of O and R are equal. It is also assumed that sufficient supporting electrolyte is present in solution to eliminate migration.

Figure 3.1C illustrates what happens when the species R is soluble in the electrode material rather than in the solution. Such is the situation during the reduction of certain metal ions (e.g., $Pb^{2+}$, $Cd^{2+}$) to neutral atoms at a mercury electrode; the metal atoms dissolve readily in the mercury. Consequently, the concentration-distance profiles for R extend *into the electrode* rather than into the solution. The area under a particular R profile is proportional to the amount of R generated up to that specific time; the area above a particular O profile is proportional to the amount of O consumed by electrolysis.

The duration of the potential step (the time interval $\tau$ of $E_s$) is usually determined by the type of information that the experimenter desires of the particular system. The time may vary from as little as 10 $\mu$s to as long as several seconds. The minimum time is limited by the ability of the potentiostat to charge the electrode, whereas the maximum time is determined by the onset of convection from vibrations or density gradients. Special shielded electrodes enable times on the order of 100s to be reached with no apparent disturbance of the diffusional process.

The potential step $E_s$ is generally terminated by switching the potential to some final value $E_f$ at which R is now oxidized back to O. If this final potential is sufficiently positive, the concentration of R at the electrode surface is made to be essentially zero. Consequently, accumulated R now diffuses to the electrode, where it is consumed by oxidation back to O, the original species in solution. This is illustrated in Figure 3.1D. Since the concentration of R is zero at both the surface and in the bulk of the electrode, R diffuses both toward and away from the interface under the influence of both downhill sides of its concentration-distance profile. For this reason, all of the R originally generated is not oxidized back to O unless the potential is maintained for a considerable length of time.

The exact ratio of O and R concentrations at the *electrode surface* is related to the imposed potential by the Nernst equation for a reversible couple:

$$E = E^{o\prime} - \frac{0.059}{n} \log \frac{C_R^s}{C_O^s} \qquad (3.2)$$

To impose the diffusion-controlled conversion of O to R as described above, the potential E impressed across the electrode-solution interface must be a value such that the ratio $C_R^s/C_O^s$ is large. Table 3.1

**Table 3.1** Relationship of $C_R^s/C_O^s$ to E for $E^{o\prime}_{O,R} = 0$ for a Reversible System

| $C_R^s/C_O^s$ | E (mV) n = 1 | E (mV) n = 2 |
|---|---|---|
| 1/10,000 | 236 | 118 |
| 1/1000 | 177 | 88.5 |
| 1/100 | 118 | 59 |
| 1/10 | 59 | 29.5 |
| 1 | 0 | 0 |
| 10/1 | −59 | −29.5 |
| 100/1 | −118 | −59 |
| 1000/1 | −177 | −88.5 |
| 10,000/1 | −236 | −118 |

shows the potentials that must be applied to the electrode to achieve various ratios of $C_R^S/C_O^S$ for the case in which $E^{o\prime}_{O,R} = 0$. For *practical* purposes, $C_R^S/C_O^S = 1000$ is equivalent to reducing the concentration of O to zero at the electrode surface. According to Table 3.1, an applied potential of −177 mV (vs. $E^{o\prime}$) for n = 1 or −88.5 mV for n = 2 will achieve this ratio. Similar arguments apply to the selection of the final potential. On the reverse step, a small $C_R^S/C_O^S$ is desired to cause diffusion-controlled oxidation of R. Impressed potentials of +177 mV beyond the $E^{o\prime}$ for n = 1 and +88.5 mV for n = 2 correspond to $C_R^S/C_O^S = 10^{-3}$. These calculations are valid only for reversible systems. Larger potential excursions from $E^{o\prime}$ are necessary for irreversible systems. Also, the effects of iR drop in both the electrode and solution must be considered and compensated for as described in Chapter 6.

## B. Chronoamperometry

The excitation signal in chronoamperometry is a square-wave voltage signal, as shown in Figure 3.3A, which steps the potential of the working electrode from a value at which no faradaic current occurs, $E_i$, to a potential, $E_s$, at which the surface concentration of the electroactive species is effectively zero. The potential can either be maintained at $E_s$ until the end of the experiment or be stepped to a final potential $E_f$ after some interval of time $\tau$ has passed. The latter experiment is termed double-potential-step chronoamperometry. The reader is referred to Section A for a detailed description of the resulting physical phenomena that occur in the vicinity of the electrode.

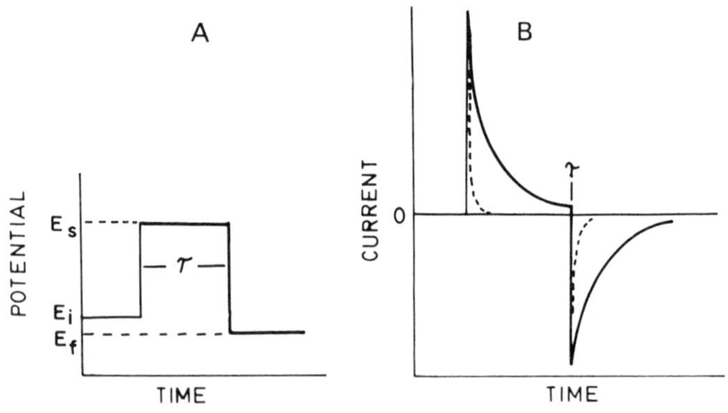


Figure 3.3 Chronoamperometry. (A) Potential excitation signal for double potential step. (B) Current-time response signal (chronoamperogram).

Current as a function of time is the system response as well as the monitored response in chronoamperometry. A typical double-potential-step chronoamperogram is shown by the solid line in Figure 3.3B. (The dashed line shows the background response to the excitation signal for a solution containing supporting electrolyte only. This current decays rapidly when the electrode has been charged to the applied potential.) The potential step initiates an instantaneous current as a result of the reduction of O to R. The current then drops as the electrolysis proceeds.

Understanding the shape of the chronoamperogram requires consideration of concentration-distance profiles for a potential-step excitation in conjunction with Faraday's law. Faraday's law is so fundamental to dynamic electrochemical experiments that it cannot be emphasized too much. It is important to keep in mind that the charge Q passed across the interface is related to the *amount* of material that has been converted, and the current i is related to the instantaneous *rate* at which this conversion occurs. Current is physically defined as the rate of charge flow; therefore,

$$Q = nFN \tag{3.3}$$

where N is the number of moles converted, and the instantaneous current at time t is

$$i_t = \left(\frac{dQ}{dt}\right)_t = nF\left(\frac{dN}{dt}\right)_t \tag{3.4}$$

The rate of conversion, dN/dt, is directly proportional to the electrode area and to the flux of material to the electrode as described by the following equation (which is derived from Fick's first law):

$$i_t = nFAD_O\left(\frac{\partial C_O}{\partial x}\right)_{x=0,t} \tag{3.5}$$

where

$i_t$ = current at time t, A
n = number of electrons, eq/mol
F = Faraday's constant, 96, 485 C/eq
A = electrode area, $cm^2$
C = concentration of O, $mol/cm^3$ (not mol/L!)
D = diffusion coefficient of O, $cm^2/s$

t = time, s
x = distance from the electrode, cm

The product $D_O(\partial C_O/\partial x)_{x=0,t}$ is the flux or the number of moles of O diffusing per unit time to unit area of the electrode in units of $mol/(cm^2 \cdot s)$. (The reader should perform a dimensional analysis on the equations to justify the units used.) Since $(\partial C_O/\partial x)_{x=0,t}$ is the slope of the concentration-distance profile for species O at the electrode surface at time t, the expected behavior of the current during the chronoamperometry experiment can be determined from the behavior of the slope of the profiles as shown in Figure 3.1B. Examination of the profiles for O at x = 0 reveals a decrease in the slope with time, which means a decrease in current. In fact, the current decays smoothly from an expected value of ∞ at t = 0 and approaches zero with increasing time as described by the Cottrell equation for a planar electrode.

$$i_t = \frac{nFAC_O^\circ D_O^{1/2}}{\pi^{1/2} t^{1/2}} = Kt^{-1/2} \tag{3.6}$$

where the units are identical with those for Equation 3.5. The current is inversely proportional to the square root of time.

The Cottrell equation states that the product $it^{1/2}$ should be a constant K for a diffusion-controlled reaction at a planar electrode. Deviation from this constancy can be caused by a number of situations, including nonplanar diffusion, convection in the cell, slow charging of the electrode during the potential step, and coupled chemical reactions. For each of these cases the variation of $it^{1/2}$ when plotted against t is somewhat characteristic.

Nonplanar electrodes such as spheres and cylinders exhibit an increase in $it^{1/2}$ with increasing t. However, planar diffusion can be closely approximated at sufficiently short times. A common example is the use of the hanging mercury drop electrode. At times less than about 1 s, this electrode follows the Cottrell equation for planar diffusion with negligible variation of $it^{1/2}$ with t. At such short times, the curvature of the electrode surface is negligible relative to the depth of the diffusion layer. When time becomes sufficiently long that the effects of spherical diffusion become significant, $it^{1/2}$ begins to increase. The time restriction for the approximation of planar diffusion is determined for electrodes of different geometries by the time at which departure from $it^{1/2}$ constancy occurs. (An interesting application of this concept involves the dual-area nature of a minigrid electrode [1].) In order to retain the relative simplicity of planar diffusion mathematics, it is often desirable to operate within such a time restriction when nonplanar electrodes are used. The Cottrell equation has been appropriately modified for electrodes that are spherical, cylindrical, and so on [2].

Positive deviations of $it^{1/2}$ with increasing time can also be evidence for convection within an electrochemical cell. Convection can be caused by external vibrations or by density gradients created by the local concentration differences resulting from the electrochemical perturbation. While the influence of external vibrations can be largely eliminated by isolation of the cell with a damped table, the natural convection due to unequal densities of O and R is an unavoidable consequence of the experiment, the importance of which depends on the particular species involved. The effect of natural convection at planar electrodes is most serious when the surface is mounted vertically. It is therefore desirable to carry out electrochemical experiments at surfaces facing up or down whenever possible.

Variations in $it^{1/2}$ can signify slow attainment of the imposed potential by the electrode. Although the input voltage to a potentiostat is a square wave for a potential-step experiment, the actual charging of the electrode to the new potential may lag significantly because of iR drop within the cell or limitations in the potentiostat itself. (Details of these important problems are considered in Chapter 7.) The result of slow charging is diminished faradaic current until the correct potential (i.e., the state of diffusion control) is achieved. Consequently, $it^{1/2}$ is correspondingly diminished during this time. Thus the onset of negative $it^{1/2}$ deviation at shorter values of t is roughly characteristic of the time required to achieve diffusion control at the electrode. As a point of reference, charging times on the order of microseconds have been reported for optimized systems [3].

In all of the evaluations above it is prudent to first employ a redox system that is known to be reversible and free from the complications of coupled chemical reactions and adsorption. Deficiencies in the electronics and/or cell design will thus be revealed before the critical experiments are begun. Typical examples of reversible systems are the reduction of $Tl^+$ at a mercury electrode [4] and the ferri-ferrocyanide system in 1-3 M KCl at solid electrodes [5].

Chronoamperometry has proven useful for the measurement of diffusion coefficients of electroactive species. An average value of $it^{1/2}$ over a range of time is determined at an electrode the area of which is accurately known and with a solution of known concentration. The diffusion coefficient can then be calculated from $it^{1/2}$ via the Cottrell equation. Although the electrode area can be physically measured, a common practice is to measure it electrochemically by performing the chronoamperometric experiment on a redox species whose diffusion coefficient is known [6]. The value of A is then calculated from $it^{1/2}$. Such an electrochemically measured surface area takes into account any unusual surface geometry that may be difficult to measure geometrically.

If the *heterogeneous* electron transfer of the redox species with the electrode itself is slow, the current after the potential step is necessarily less than in a system in which the electron transfer is rapid. This aspect of chronoamperometry has been used for the measurement of heterogeneous rate constants [7].

The behavior of $it^{1/2}$ as a function of time can be influenced substantially by the presence of chemical reactions that are coupled to the electrode process (see Chap. 2). Consequently, characteristic variations of $it^{1/2}$ vs. t have been effectively utilized for the quantitative study of such homogeneous chemical reactions. The ECE reaction in which a chemical step exists between two electron transfer steps is one mechanism that has been investigated by means of chronoamperometry:

$$\text{E:}\quad O + e \rightarrow R \tag{3.7a}$$

$$\text{C:}\quad R \xrightarrow{k} X \tag{3.7b}$$

$$\text{E:}\quad X + e \rightarrow P \tag{3.7c}$$

As shown in the reaction sequence above, a rate-determining chemical step is interposed between the two electrode reactions. (See Chap. 2 for an explanation and an example of this mechanism.) The two dashed lines in Figure 3.4A show hypothetical chronoamperograms for

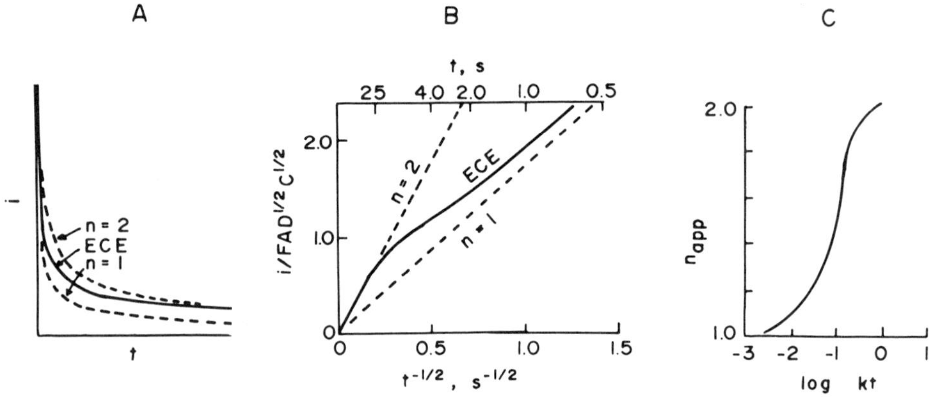


Figure 3.4 Chronoamperometry for ECE mechanism. (A) Current-time response. (B) Data plotted as i vs. $t^{-1/2}$. (C) Calculated working curve for obtaining rate constant k from experimental value for $n_{app}$. [From Ref. 8, adapted with permission from G. S. Alberts and I. Shain, *Anal. Chem.* *35*:1859 (1963). Copyright 1963 American Chemical Society.]

the 1e reduction of O to R and for the direct 2e reduction of O to P with no kinetic complications. The solid line shows a typical chronoamperogram for an ECE mechanism. The current is intermediate between the 1e and 2e reductions, since the reduction of X to P is controlled by the rate of the chemical reaction of R to generate X. The exact position of the solid line is determined by the value of the rate constant k.

Figure 3.4B displays the data in terms of normalized i vs. $t^{-1/2}$. For the 1e and 2e cases, the two straight lines are simple plots of the Cottrell equation (3.6) for which the slope is determined by the respective n values. The shape of the ECE curve can be understood in terms of the amount of time allowed for the chemical reaction of R. At short times (large $t^{-1/2}$) the ECE curve approaches the 1e Cottrell plot since insufficient time is allowed for R to generate a measurable amount of X. Hence on this time scale the mechanism is essentially the 1e reduction of O to R. At long times (small $t^{-1/2}$), ample time is allowed for R to generate X, which is then reduced to P. Consequently, on this time scale the mechanism is indistinguishable from a simple 2e reduction of O directly to P. The transition between these two extremes occurs during the chronoamperometry experiment when the time is comparable to the half-life of the chemical reaction. It is in this time frame that the effect of the ECE mechanism is evident.

Figure 3.4C is a calculated working curve that relates the apparent value for n ($n_{app}$) calculated from $it^{1/2}$ using the Cottrell equation to the rate constant k and the time t during the chronoamperogram. Such a curve is used to calculate the value of k. For example, the value of $n_{app}$ for a particular time t is determined and the corresponding value of kt is obtained from the working curve. Knowing t, one can then calculate k. Chronoamperometry has been applied to the study of several electrode mechanisms [8,9]. Such studies should be undertaken with an awareness that different mechanisms may exhibit similar responses [9,10]. It is always best to obtain mechanistic information from several independent sources. The reader is referred to Chapter 17 for further examples of the use of chronoamperometry in studying mechanisms of electrode processes.

Double-potential-step chronoamperometry is particularly suited for studying systems that follow EC [11] or dimerization [12] mechanisms.

## C. Chronocoulometry

Chronocoulometry is simply chronoamperometry in which the current is integrated so that the monitored response is now charge, Q [13-15]. As such, the excitation is a potential step, as shown in Figure 3.5A, the details of which are dealt with in Section II.A. It should be emphasized that the system response to this excitation is still current,

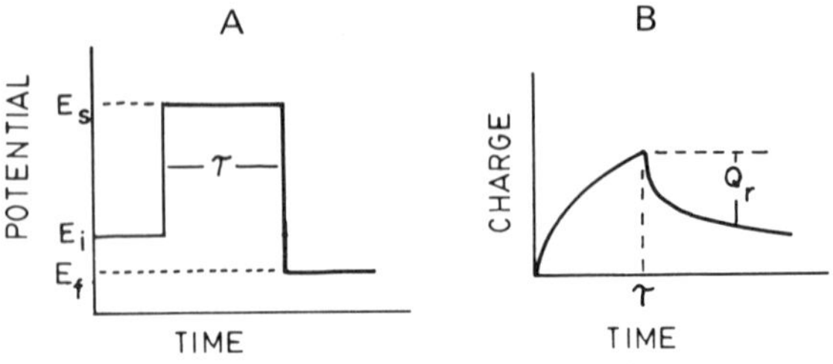


Figure 3.5 (A) Potential excitation for double potential step. (B) Charge-time monitored response (chronocoulogram).

as explained in Section II.B. However, by integrating the current and presenting the charge as a function of time (Fig. 3.5B), some of the information obtained in the current response becomes more easily extractable. Although the integration of the current could be achieved by manually integrating a chronoamperometric i-t curve, the task is usually accomplished by analog integration with an operational amplifier or by digital integration following analog-to-digital conversion of the data.

The measured charge represents contributions from three possible *sinks*:

1. Electrolysis of electroactive species in solution at a rate that is controlled by diffusion to the electrode,
2. Electrolysis of electroactive species that is adsorbed on the electrode surface
3. Charging of the electrode-electrolyte double-layer capacitance to the new potential

In a simplified mathematical treatment, the three sinks of charge are considered separately as additive functions,

$$\text{Total charge} = \text{diffusing component} + \text{adsorbed component} + \text{double-layer charging} \tag{3.8a}$$

$$Q_{total} = Q_{diff} + Q_{ads} + Q_{d\ell} \tag{3.8b}$$

$$Q_{total} = \int_0^t i\,dt = \frac{2nFAC_O^{\circ}D_O^{1/2}t^{1/2}}{\pi^{1/2}} + nFA\Gamma_O + Q_{d\ell} \tag{3.8c}$$

where the total charge $Q_{total}$ and the double-layer charge $Q_{d\ell}$ are in coulombs, $\Gamma_O$ is the amount of adsorbed reactant in mol/cm$^2$, and the other terms are identified with Equation 3.5. (The reader is referred

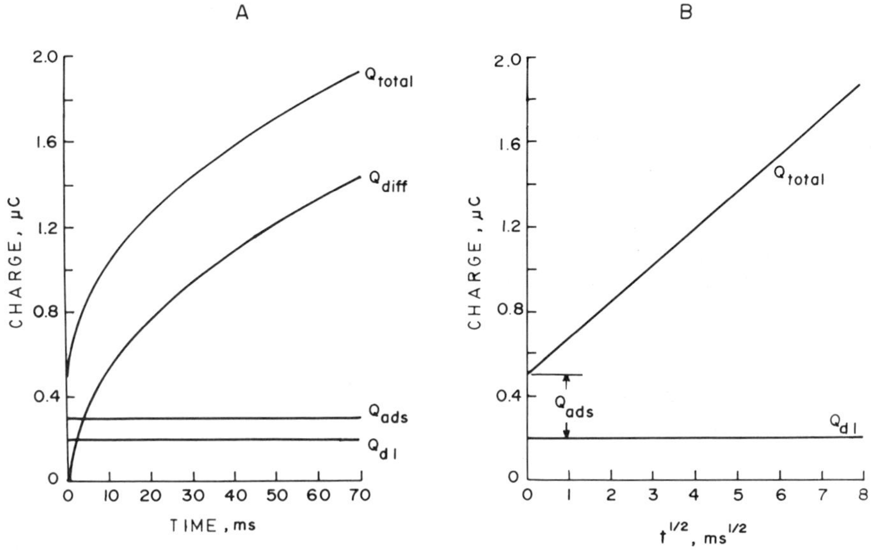


Figure 3.6 Effect of adsorption of electroactive species on chronocoulometry. (A) Charge-time curves for individual components contributing to total charge ($Q_{total}$) as in Equation 3.8b. (B) Plots of charge vs. $t^{1/2}$ for $Q_{total}$ and $Q_{d\ell}$.

to Chapter 2 for a pertinent discussion of adsorption and double-layer charging.) The expression for the diffusing component of charge is simply the Cottrell equation integrated with respect to time.

A typical chronocoulometric Q-t curve is illustrated in Figure 3.6A for a system in which the electroactive species O adsorbs on the electrode surface. The contribution of the three individual components of Equation 3.8b is also shown. It should be apparent that the total charge required for double-layer charging and for reducing the adsorbed O is "instantly" consumed whereas the charge required for reducing the O in solution exhibits a diffusion-controlled time dependency. This is also expressed mathematically in Equation 3.8c, which shows that a plot of Q vs. $t^{1/2}$ should be a straight line with an intercept equal to ($nFA\Gamma_O + Q_{d\ell}$). Such a plot of Q vs. $t^{1/2}$ is shown in Figure 3.6B.

That the adsorption and double-layer components can be so easily separated from the diffusional component by plotting Q vs. $t^{1/2}$ is the forte of chronocoulometry. Since $Q_{d\ell}$ can usually be measured in a separate experiment on the supporting electrolyte alone, the contribution of the adsorbed species can be determined and $\Gamma_O$ calculated. This is illustrated in Figure 3.6B, in which the difference between the intercepts of Q and $Q_{d\ell}$ is the amount of charge required to reduce the adsorbed species, $Q_{ads}$. An implied assumption is that $Q_{d\ell}$ is the same in the presence and absence of the adsorbed species. Thus chronocoulometry is a most useful technique for determining the surface excess $\Gamma$ of an *electroactive* species. The information is also contained in a chronoamperometry experiment, but in much less accessible form. The adsorption of many electroactive inorganic ions, metal complexes, and organic compounds at electrode surfaces has been investigated by means of chronocoulometry [15-17].

Chronocoulometry as described above involves measurement of charge after what is frequently referred to as the forward potential step. A logical extension of the method is to step the potential to its original value or some other value $E_f$ at which the redox process is reversed and monitor the resulting charge behavior, Figure 3.5 [13].

Chronocoulometry can also be used for the study of homogeneous chemical reactions that are coupled to the electrode reaction. The technique has not yet seen extensive application to this end, although the theory is quite highly developed [18].

## D. Chronoabsorptometry

The development of electrodes that exhibit optical transparency has enabled spectral observations to be made directly through the electrode simultaneously with electrochemical perturbations [19-23]. These electrodes typically consist of a very thin film of conductive material such as Pt, Au, carbon, or a semiconductor such as doped tin oxide that is deposited on a glass or quartz substrate. Miniature metal screens, minigrid electrodes in which the presence of very small holes (6-40 μm) lends transparency, have also been used. Optically transparent electrodes (OTE) and the cells that incorporate them are discussed in Chapters 11 and 12.

A beam of light can be directed perpendicularly through the transparent electrode and the solution as shown in Figure 3.7A. The electrochemical cell is now also a spectral cell (one optical face being

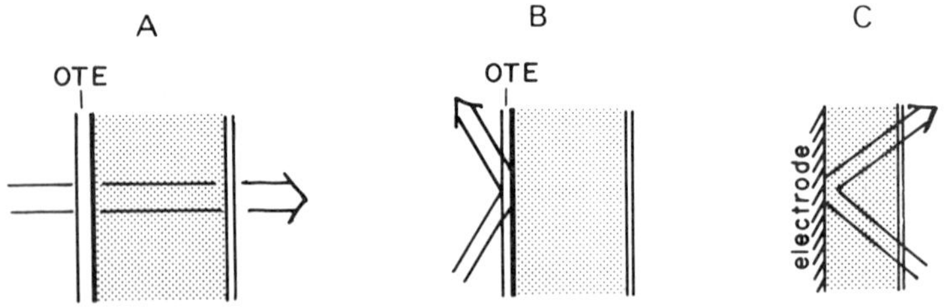


Figure 3.7 Spectroelectrochemical techniques: (A) transmission, (B) internal reflectance, (C) specular reflectance.

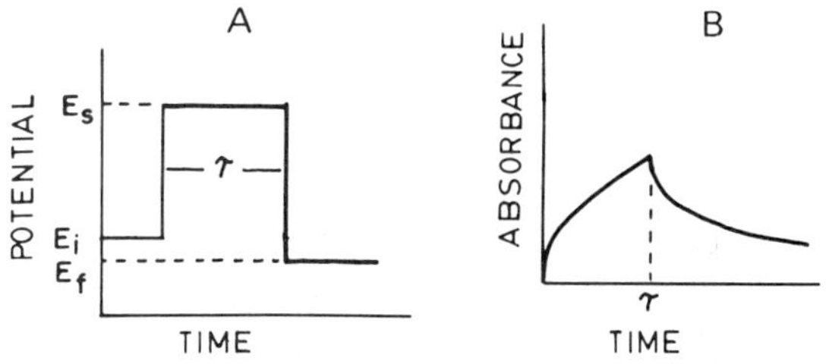


**Figure 3.8** Chronoabsorptometry. (A) Potential excitation signal for double potential step. (B) Absorbance-time monitored response (chronoabsorptogram).

the OTE) that is positioned in the light beam of a spectrometer. This electrode transparency enables changes in absorbance that occur in the solution adjacent to the electrode to be monitored during an electrochemical perturbation. If such optical monitoring is performed in conjunction with a potential step as the electrochemical excitation signal (Fig. 3.8A), the technique is termed *potential-step chronoabsorptometry*. The resulting absorbance-time response is illustrated by the following example. (The reader is referred to Section II.A for a discussion of the potential-step perturbation, which is most useful at this point.) Let us again assume that species O is present in solution and is reducible to R at the electrode surface according to Equation 3.1. In this case we shall also assume that R absorbs light at the visible wavelength of 600 nm. Consequently, any R that is formed at the electrode will be detected by a change in absorbance of a 600-nm light beam passing through the electrode and the solution as in Figure 3.7A. When the potential of the transparent electrode is stepped so that the formation of R is diffusion controlled, the resulting increase in absorbance is proportional to the accumulation of R. The amount of R that has accumulated at any time t is proportional to the area beneath the concentration-distance profile for R at that particular time, as shown in Figure 3.1B.

The actual change in absorbance is related to concentration and optical path length by Beer's law,

$$A = \varepsilon bc \tag{3.9}$$

For chronabsorptometry this law must be couched in a form that takes into account the continuous increase in the thickness of the diffusion layer, which is the optical cell, and the fact that the concentration of the absorbing species is not homogeneous within this layer. The optical integration of the concentration can be expressed mathematically by Equation 3.10.

$$A_t = \varepsilon_R \int_0^\infty C_{R,x,t}\, dx \qquad (3.10)$$

where $\varepsilon_R$ is the molar absorptivity ($M^{-1}cm^{-1}$) of R at the wavelength being monitored and $C_{R,x,t}$ is the concentration-distance profile of R, which is changing during the electrogeneration. This equation is equivalent to Beer's law. Substitution of the diffusion equations that describe the behavior of $C_{R,x,t}$ gives the following absorbance-time relationship:

$$A_t = \frac{2}{\pi^{1/2}} \varepsilon_R C_O^\circ D_O^{1/2} t^{1/2} \qquad (3.11)$$

where $C_O^\circ$ and $D_O$ are the molar concentration and diffusion coefficient of O and t is time in seconds. The reader should note that the concentration and the diffusion coefficient are for the oxidized species O, even though it is the reduced species R that is being monitored optically; the formation of R is controlled by the diffusion of O to the electrode. A typical absorbance-time curve is illustrated in Figure 3.8B. As described by Equation 3.11, A increases at a $t^{1/2}$ rate.

It is apparent from Equation 3.11 that a plot of A vs. $t^{1/2}$ is a straight line. The value for $\varepsilon_R$ can be obtained from the slope of this line since $C_O^\circ$ is usually known and $D_O$ can be measured by means of chronoamperometry. Thus this method is useful for obtaining molar absorptivities of electrogenerated species.

The analogy between chronoabsorptometry and chronocoulometry is evident from the $t^{1/2}$ response of both A and Q and the similarity of Equations 3.11 and 3.8c. Whereas chronoabsorptometry optically integrates the *amount of R present* at any time in accordance with Beer's law, chronocoulometry integrates the amount of R that has been *generated* at any time by integrating the current required to form R.

Chronoabsorptometry is useful for the study of homogeneous chemical reactions that involve an electrogenerated species. For example, consider the situation in which an additional species, Z, is also present in solution and is capable of a rapid homogeneous chemical reaction with R at a rate characterized by the second-order rate constant k to form products P (see Chap. 2 for a discussion of the EC mechanism).

$$\text{E:}\quad O + e \rightleftharpoons R \qquad (3.12)$$

$$\text{C:}\quad R + Z \xrightarrow{k} P \qquad (3.13)$$

In this case, the electrogenerated R will encounter Z as it diffuses away from the electrode and will react to form P. This *diffusional mixing* of two reactants (R and Z in this example) is the basis for chronoabsorptometry (and other electrochemical techniques) as a kinetic method. The homogeneous chemical reaction of R will perturb its accumulation by a magnitude that is proportional to both k and the concentration of Z. The influence of this perturbation is apparent from the concentration-distance profiles for R, which are illustrated for a rate constant of $10^7$ L/(mol·s) in Figure 3.9.

Comparison of the areas under the profiles for R in Figures 3.1B and 3.9 shows a significant decrease in the amount of R accumulating in solution in Figure 3.9 because of its reaction with Z. Note that in both cases R is being generated at the same rate. Since the absorbance response of the monitored species R is proportional to the area beneath the profile $C_{R,x,t}$ for R, this kinetic perturbation is also manifested in the absorbance-time response when R is monitored optically. The A-t curves for different rate constants are shown in Figure 3.10. It is apparent that the absorbance response is quite sensitive to the magnitude of the rate constant, k. Such a rate constant can be determined by fitting an experimental curve to calculated curves such as those in Figure 3.10 or by fitting to a dimensionless working curve [21,22]. This has been demonstrated for several types of homogeneous reactions [19,21,22].

The transparency of the electrode also enables spectra to be recorded of electrogenerated species as well as of any species produced as a result of a homogeneous chemical reaction. Such spectra have been recorded with rapid scanning spectrometers that are capable of recording as many as 100 or more spectra per second in the UV-visible range [24]. Spectra can be useful for structural identification of

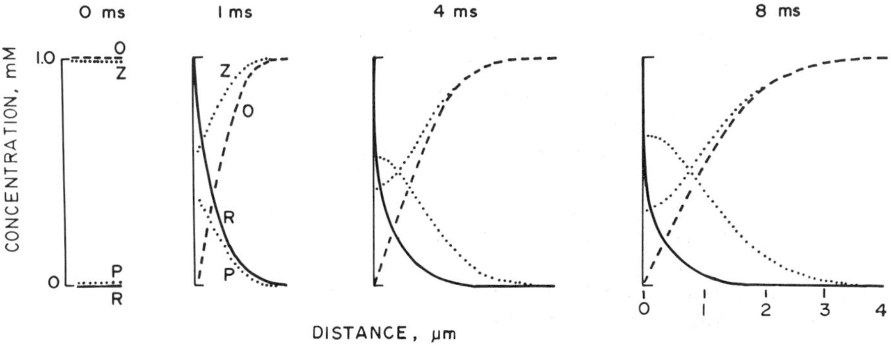


Figure 3.9 Concentration-distance profiles for EC mechanism with k = $10^7$ L/(mol s). See Figure 3.1 legend for conditions. [Adapted from R. F. Broman, W. R. Heineman, and T. Kuwana, *Faraday Discuss. Chem. Soc.* *56*:16 (1974).]

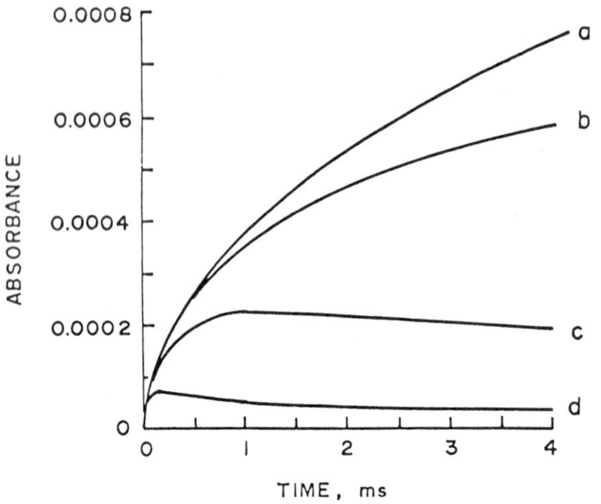


Figure 3.10 Absorbance-time curves for transmission spectroelectrochemistry. EC mechanism for rate constants k of (a) 0, (b) $10^5$, (c) $10^6$, and (d) $10^7$ L/(mol s). R monitored optically, $D_0 = 10^{-6}$ $cm^2$/s, $\varepsilon_R = 10^4$ L/(mol cm), concentrations of A and Z = 1.0 mM. [Adapted from R. F. Broman, W. R. Heineman, and T. Kuwana, *Faraday Discuss. Chem. Soc. 56*:16 (1974).]

intermediate components in the reaction sequence and for the determination of the wavelengths of maximum absorbance for such components so that set wavelength experiments can be performed at maximum sensitivity to measure kinetic parameters as described above.

Optical measurements through the electrode can also be made by *internal reflectance spectroscopy* (IRS). In this technique the light beam is introduced to the electrode at an angle of incidence $\theta_i$ as illustrated in Figure 3.7B. When this angle is greater than the critical angle $\theta_c$ the beam will be totally reflected at the electrode-solution interface as shown. However, a small standing wave component will exist within the solution itself. The penetration depth of this component into the solution is typically 2000-4000 Å for visible light. Light-absorbing species in solution interact with this evanescent wave so that spectra can be recorded by varying the wavelength of the incident beam and analyzing the reflected beam. Consequently, spectroelectrochemical experiments analogous to the types described above for transmission chronoabsorptometry can also be performed with IRS chronoabsorptometry. Because of the limited penetration depth of the IRS wave, the *optical* cell is a very thin layer of solution adjacent to the electrode. Thus the optical advantages of an optically transparent thin-layer cell (see Sec. II.E) are realized, but the disadvantages (high solution resistance, poor current distribution) are not. The IRS mode offers an enhanced sensitivity in absorbance change compared to transmission and is therefore useful for very fast reactions in which the absorbance change is quite small. Signal-averaging techniques are used for these fast reactions in order to detect the small absorbance changes. Rate constants for chemical reactions that approach diffusion control have been measured by this method [19,21,22].

Spectroelectrochemical measurements can be made at conventional, nontransparent electrodes by specular reflectance [25,26]. The optical beam is passed through the electrolyte and reflected from the electrode surface as shown in Figure 3.7C. This technique has been used effectively to study electrode mechanisms and to observe changes in the electrode surface itself.

Raman spectroelectrochemistry has been developed recently [27, 28]. In this technique a laser excitation beam is directed at the electrode and the resulting Raman scattering is observed. A particularly important aspect of Raman spectroelectrochemistry is the structural information about electrogenerated species that is contained in a Raman spectrum.

## E. Potential-Step Excitation in Thin Layers of Solution

An operational definition of thin-layer electrochemistry is that area of electrochemical endeavor in which special advantage is taken of restricting the diffusional field of electroactive species and products. Typically, the solution under study is confined to a well-defined layer, less than 0.2 mm thick, trapped between an electrode and an inert barrier, between two electrodes, or between two inert barriers with an electrode in between. Diffusion under this restricted condition has been described in Chapter 2 (Sec. II.C). Solution trapped in a porous-bed electrode will have qualitatively similar electrochemical properties; however, geometric complexities make this configuration less useful for analytical purposes. The variety of electrical excitation signals applicable to thin-layer electrochemical work is large. Three reviews of the subject have appeared [29-31].

All of the principles of semi-infinite potential-step experiments discussed above apply to thin-layer work. Some modification in the quantitative response is necessary to account for the presence of a diffusion barrier. Figure 3.11 illustrates the diffusion phenomena occurring during a chronoamperometric experiment in a thin-layer cell of typical dimensions. It will be useful to compare this figure with the semi-infinite situation depicted in Figure 3.1. Notice that the supply of reactant in the bulk solution phase is effectively infinite in Figure 3.1, whereas in Figure 3.11 there is no bulk solution and *all* of the available reactant is readily consumed. The two ex-

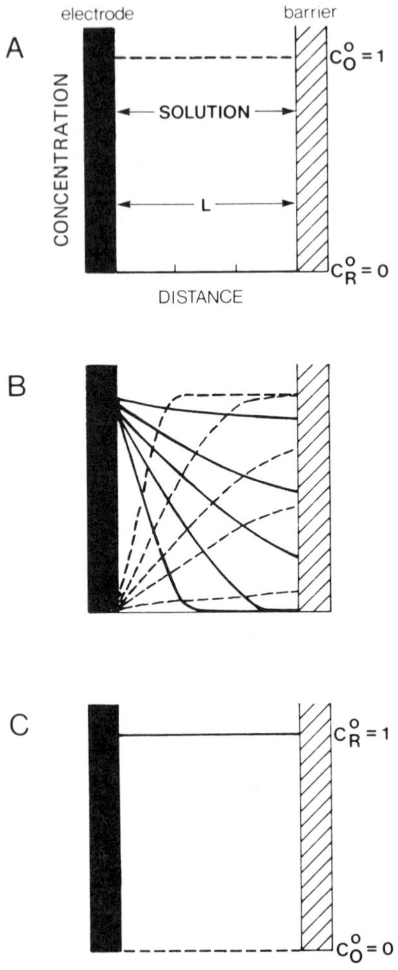


Figure 3.11 Thin-layer system response to potential-step, diffusion-controlled reduction of O to R with $D_O = D_R$. (A) Initial condition. (B) Profiles for O (dashed lines) and R (solid line) after potential step. (C) Final condition.

periments are, of course, identical until the concentration profiles strike the barrier. If at any time the potential is stepped back to a value where $R \rightarrow O$ occurs at a diffusion-controlled rate, the condition shown in Figure 3.11A would be restored. For stable species the complete interconversion of O to R and R to O could be repeated indefinitely. The reader should attempt to sketch the concentration profiles extant when the barrier plate in Figure 3.11 is replaced by another electrode identical to the first, a situation sometimes used in practice.

The measured i-t response and Q-t response to a potential step are shown in Figure 3.12 for an electrode both with and without a diffusion barrier. The mathematical description is somewhat more complex for the restricted diffusion case. The simple boundary value problem describing this experiment is commonly solved in elementary texts on differential equations under the guise of "heat conduction in

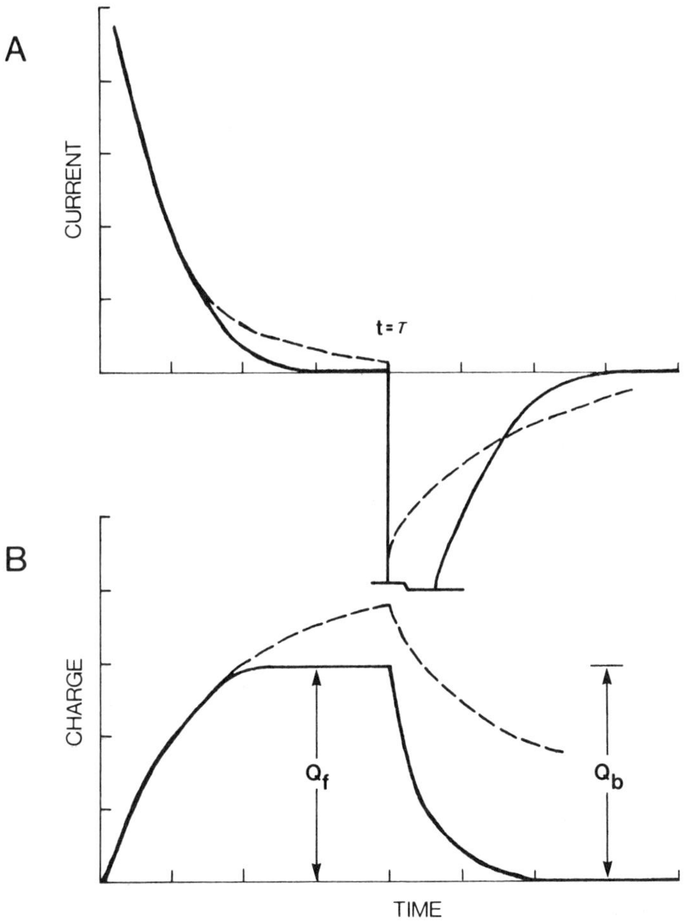


Figure 3.12 Double potential-step thin-layer electrochemistry. (A) Chronoamperometry. (B) Chronocoulometry. Dashed line, same experiment for semi-infinite situation.

a slab of finite width" [31,32]. Differentiation of the concentration profile, $C_{O,x,t}$, at the electrode surface immediately affords the chronoamperometric response.

$$i_t = nFAD_O \left( \frac{\partial C_{O,x,t}}{\partial x} \right)_{x=0,t} = \frac{2nFAC_O^\circ D_O}{L} \sum_{k=0}^{\infty} \exp (P) \qquad (3.14)$$

where $P = -D_O(2k + 1)^2\pi^2 t/4L^2$. Integration of this series readily provides an expression for the coulometric response

$$Q_t = \int_0^t i\, dt = nFALC_O^\circ \left[ 1 - \frac{8}{\pi^2} \sum_{k=0}^{\infty} \frac{1}{(2k + 1)^2} \exp (P) \right] \qquad (3.15)$$

In the limit as $t \to \infty$, Equation 3.15 converges to Faraday's law,

$$Q_{inf} = nFALC_O^\circ = nFVC_O^\circ = nFN \qquad (3.16)$$

where V is the volume of the thin-layer cavity.

While Equations 3.14 and 3.15 provide an understanding of the time response, to date they have had no practical significance. There is no advantage to be gained from data on the rising part of a thin-layer chronocoulogram, since semi-infinite chronocoulometry can provide the same information more efficaciously. The important datum from the present technique is $Q_{inf}$, which can be used to calculate n, V, or $C_O^\circ$ when the other parameters are known. The number of electrons involved in a process is essential for elucidating mechanisms. Thin-layer coulometry is eminently suitable for this purpose. Considering that V is typically less than 10 μL, such experiments require very little material and very little time (typically a few seconds). A characteristic time constant for the theoretical response is $L^2/D$ seconds, which is the time at which the electrolysis will be 93% complete. This is one-half the mean time for a molecule to diffuse across L as given by the Einstein equation (Chap. 2, Sec. II.C). The challenging details of constructing useful thin-layer cells are described in Chapter 12.

A typical example of the use of thin-layer coulometry to determine an n value is for the drug chlorpromazine (CPZ). A thin-layer coulogram for the oxidation of CPZ by a potential step to 620 mV vs. SCE is shown in Figure 3.13. A coulogram is also shown for the same potential step repeated in supporting electrolyte.

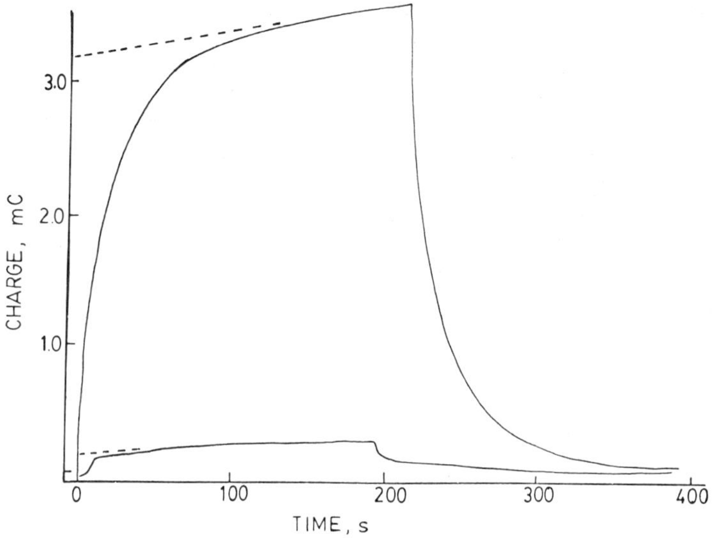


Figure 3.13 Thin-layer controlled-potential coulometry of chlorpromazine (CPZ) oxidation in an optically transparent thin-layer cell. $E_i$ = 250 mV; $E_s$ = 575 mV; $E_f$ = 250 mV vs. SCE. $4.8 \times 10^{-4}$ M CPZ, 3 M $H_2SO_4$; and 3 M $H_2SO_4$ alone. (From T. B. Jarbawi, Ph.D. dissertation, University of Cincinnati, 1981.)

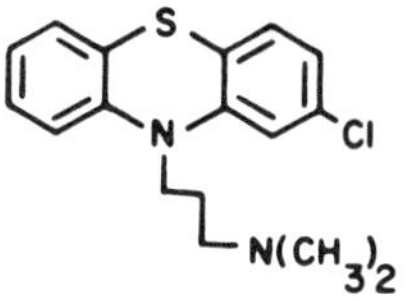


Chlorpromazine (CPZ)

The total charge $Q_T$ and the background charge $Q_B$ are measured by extrapolating the linear portion of the Q-t curve back to t = 0 s. The difference in $Q_T$ and $Q_B$ is the faradaic charge from which a value for n is calculated from Equation 3.16. (The thin-layer cell is precalibrated with respect to its volume V by performing a similar experiment on a species of known n and C° and calculating V.) In the case of CPZ oxidation, a calculated value of 1e suggests oxidation to $CPZ^{\dot{+}}$ [33].

Under ideal conditions, charge consumed by the double-layer capacitance and adsorbed reactants will follow the same time course as discussed above for ordinary chronocoulometry. Since the ratio of electrode area to solution volume is larger for thin-layer experiments, charge thus accounted for may represent a much greater proportion of the total. This fact points to an advantage of restricted diffusion experiments for studying some surface phenomena.

One area in which the thin-layer approach has been moderately popular is spectroelectrochemistry [20,22,23]. Transparent or reflecting electrodes permit spectral observation of reaction product(s) without interference from starting material [34]. The stability of product species may be assessed either spectrally or by comparing the ratio of $Q_b$ to $Q_f$ (Fig. 3.12B) at various switching times $\tau$. In semi-infinite spectroelectrochemical experiments the light beam must pass

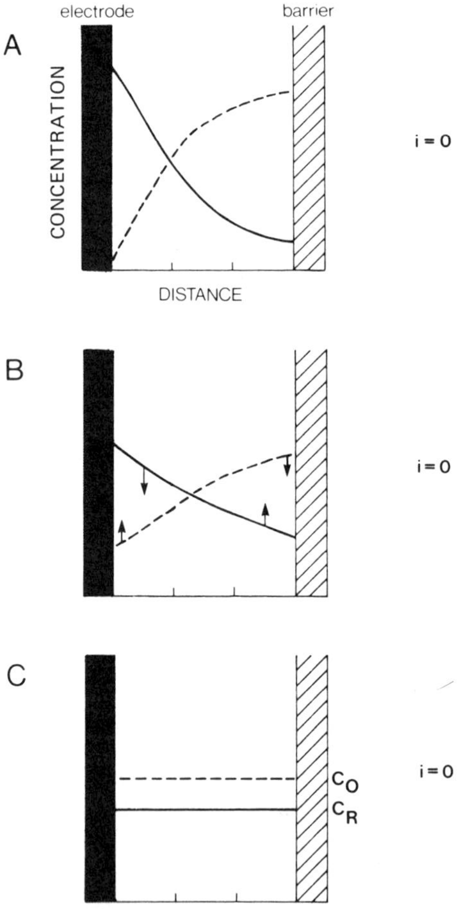


Figure 3.14 Concentration profile relaxation in thin-layer cell. (A) Initial condition (solid line, product; dashed line, reactant). (B) Relaxation in progress. (C) Final condition.

through a large volume of solution before reaching the diffusion layer unless, of course, IRS techniques are used (see Sec. II.D). This means that it will be very difficult to monitor *reactant* since the concentration (and absorbance) change will be very small relative to the bulk solution phase. This is not the case in thin-layer spectroelectrochemistry, and either reactant or product can be monitored with equal facility. Because of the difficulty in designing thin-layer electrodes that behave ideally at very short times, the semi-infinite experiment is preferred for following rapid kinetics. Thin-layer cells are, however, most effective for measuring the spectra of products and of reasonably stable intermediates using conventional or rapid scanning spectrophotometers.

An important characteristic of thin-layer cells is rapid diffusional mixing. For example, consider Figure 3.11B. If the experiment is interrupted at this point by disconnecting the working electrode, the concentrations of O and R will rapidly become homogeneous throughout the cell as illustrated in Figure 3.14. If either species is subject to a chemical reaction that is slow in relation to this concentration profile relaxation, the kinetics may be treated as for an initially homogeneous concentration of reactant. This greatly simplifies the mathematics when contrasted to the usual electrochemical experiment, where the reacting species are not uniformly distributed. Naturally, rate processes examined in this way will have to be much slower than those accessible by semi-infinite experiments.

The oxidation of 5,6-diaminouracil provides a good example of the use of thin-layer spectroelectrochemistry to determine the rate constant of an electrogenerated intermediate [35]. The reaction sequence is as follows:

E: 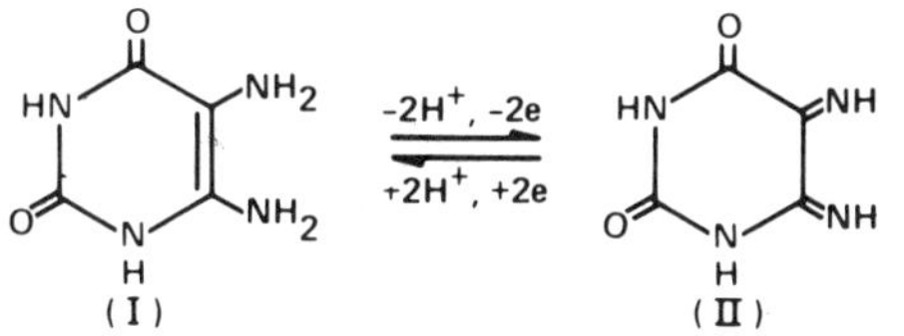 (3.17a)

C: 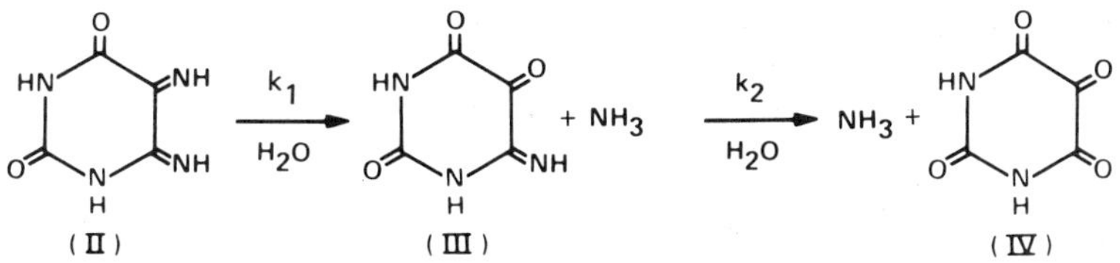

(3.17b)

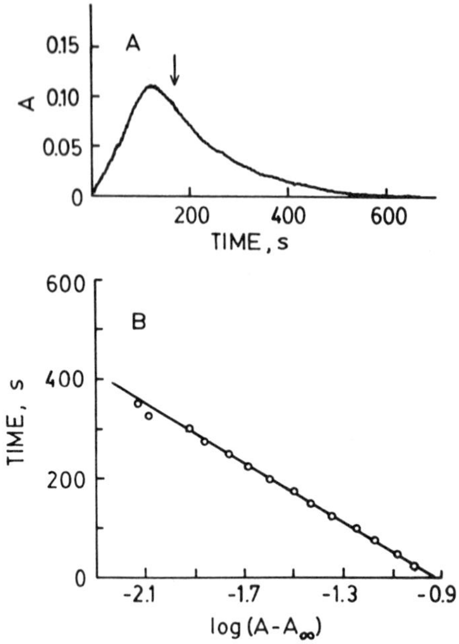


Figure 3.15 Thin-layer chronoabsorptometry of electrogenerated reactant in an optically transparent thin-layer electrode. (A) Variation of the absorbance of the intermediate species ($A_{320nm}$) during the electrooxidation of 10 mM 5,6-diaminouracil in McIlvaine buffer pH 5 at 0.35 V. Arrow indicates time at which all 5,6-diaminouracil has been electrolyzed. (B) Kinetic plot of time vs. the absorbance at 330 nm of intermediate species. Absorbance data were taken from curve (A) setting time = 0 at the point where the concentration of 5,6-diaminouracil was zero. (From Ref. 35.)

A thin-layer chronoabsorptometry experiment in which the diimine intermediate (species II) is monitored at 330 nm is shown in Figure 3.15A. The absorbance increases when the potential is stepped to 0.35 V to electrogenerate the imine and then decreases due to the hydrolysis reaction. A first-order kinetic plot of the absorbance decrease (Fig. 3.15B) gives a pseudo-first-order value for k of $7.6 \times 10^{-3}$ $s^{-1}$.

Since the redox active species in a thin-layer cell rapidly equilibrate with the potential applied to the electrode, precise values of $E^{o\prime}$ and n can be obtained from a spectropotentiostatic technique [23, 36]. In this method a series of potentials is applied to the cell and a spectrum is recorded at each potential after equilibration has been achieved. A typical example is the study of $[Tc(diars)_2Cl_2]ClO_4$ in dimethylformamide [37].

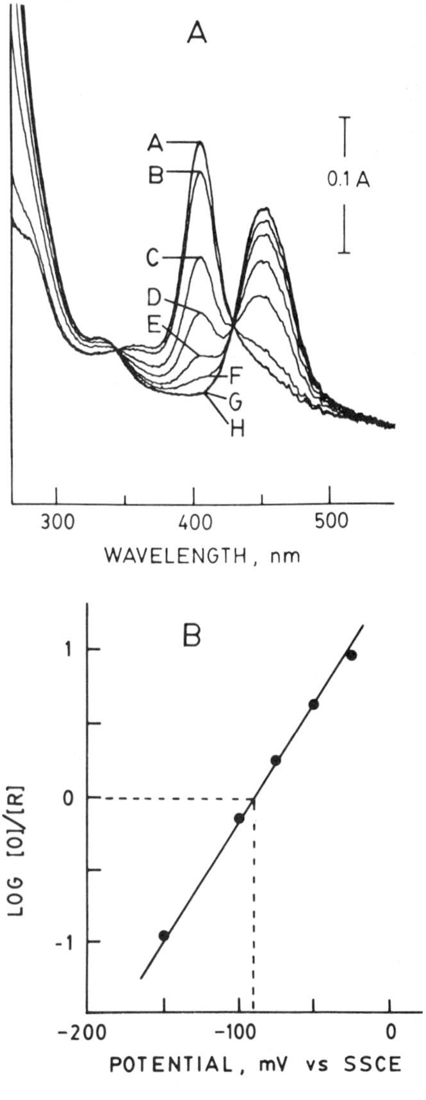


Figure 3.16 (A) Spectra recorded during spectropotentiostatic experiment in optically transparent thin-layer electrode on 0.87 mM $[Tc(diars)_2Cl_2]^+$, 0.5 M TEAP in dimethylformamide. Applied potentials: A, −250; B, −150; C, −100; D, −75; E, −50; F, 25; G, 100; H, 250 mV vs. SSCE. (B) Nernst plot at 403 nm. [From Ref. 37, reprinted with permission from R. W. Hurst, W. R. Heineman, and E. Deutsch, *Inorg. Chem.* *20*:3298 (1981). Copyright 1981 American Chemical Society.]

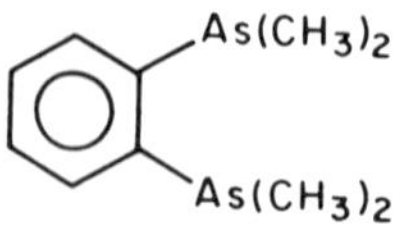


**diars**

o-Phenylenebis(dimethylarsine)
(diars)

Spectra at several potentials are shown in Figure 3.16A. Spectrum A (at −0.250 V vs. SSCE) corresponds to the Tc(III) complex, whereas spectrum H (at 0.250 V vs. SSCE) corresponds to the Tc(IV) complex. (Note: SSCE denotes an SCE prepared with NaCl instead of KCl to avoid precipitation of $KClO_4$ at the reference electrode-dimethylformamide junction.) The ratio of [Tc(III)]/[Tc(IV)] for the intermediate potentials can be calculated from the spectra and a Nernst plot obtained (Fig. 3.16B). The intercept gives $E^{o\prime} = -91$ mV vs. SSCE and the slope gives n = 0.99.

## III. POTENTIAL-SCAN TECHNIQUES IN STATIONARY SOLUTION

Some of the most useful electroanalytical techniques are based on the concept of continuously varying the potential that is applied across the electrode-solution interface and measuring the resulting current. The present section describes large-amplitude techniques in this category for a working electrode immersed in a stationary solution.

Two voltammetric techniques, stationary electrode voltammetry (SEV) and cyclic voltammetry (CV), are among the most effective electroanalytical methods available for the *mechanistic* probing of redox systems. In part, the basis for their effectiveness is the capability for rapidly observing redox behavior over the entire potential range available. Since CV is an extension of SEV, many points pertinent to CV are discussed in the SEV section.

The polarographic techniques, which are also discussed in this section, are important in that they are among the more sensitive, reproducible, and easily used electroanalytical methods for *analysis*.

### A. Stationary-Electrode Voltammetry

In stationary-electrode voltammetry (SEV) the potential applied across the electrode-solution interface is scanned linearly from an initial value $E_i$ to a final value $E_f$ at a constant rate $\nu$ (V/s) [6,7]. This potential excitation is illustrated by the E-time profiles in Figure 3.17 for several rates of scan and several values of $E_i$ and $E_f$. An excursion

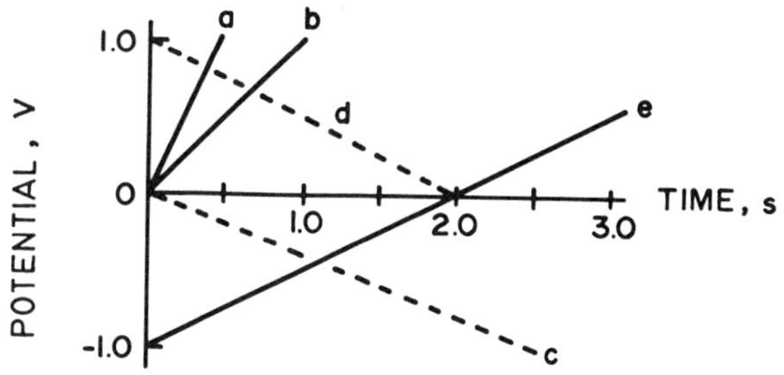


Figure 3.17 Potential-time profiles for stationary electrode voltammetry.

| Scan | $E_i$ (V) | $E_f$ (V) | $\nu$ (V/s) | Scan direction |
|---|---|---|---|---|
| a | 0 | 1.0 | 2.0 | Positive |
| b | 0 | 1.0 | 1.0 | Positive |
| c | 0 | −1.0 | −0.4 | Negative |
| d | 1.0 | 0 | −0.5 | Negative |
| e | −1.0 | 0.5 | 0.5 | Positive |

in which the potential becomes increasingly positive (i.e., is heading in the positive direction) is termed a positive scan (*not* an *anodic* scan). Since the time required for scan a is 0.5 s, the scan rate is +1.0 V/0.5 s, or +2.0 V/s. Scan d is termed a negative scan since the potential is becoming less positive (i.e., is heading in the negative direction, even though in this example the actual potential at any time is positive).

To obtain a stationary-electrode voltammogram, the current is measured during the potential scan in an unstirred solution. A voltammogram that was obtained with a Pt working electrode in a 1.00 mM $Fe^{2+}$ solution with 1.0 M HCl supporting electrolyte is shown in Figure 3.18. The potential scan that was applied to the working electrode to obtain this voltammogram is identical with excitation signal b in Figure 3.17. The potential axis in Figure 3.17 is the horizontal axis of Figure 3.18. Note that the experiment is completed in only 1 s.

In SEV one generally wants to initiate the potential scan at an $E_i$ at which no electrolysis occurs and then scan through the standard electrode potential(s) of the species in solution. Thus the composition of the reaction layer will not be altered in an unknown manner at the beginning of the experiment by application of $E_i$. The formal electrode potential of the $Fe^{3+}/Fe^{2+}$ couple in 1 M HCl is ca. 0.46 V vs. SCE. Since the solution to be examined contained iron in the +2 oxidation

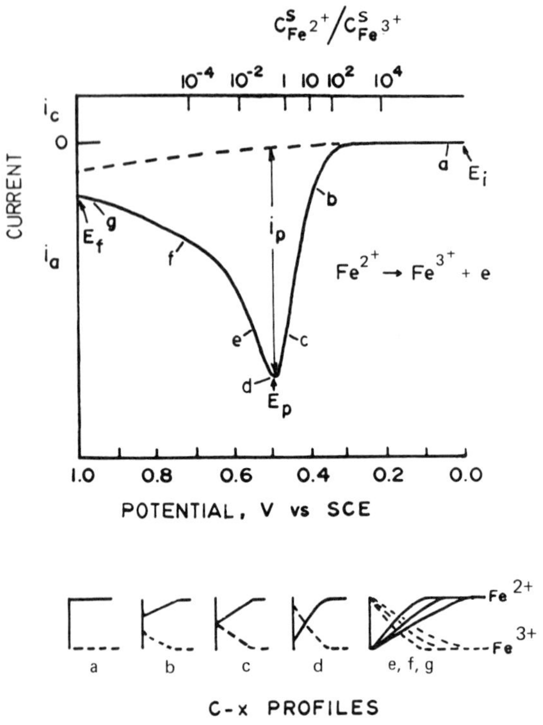


Figure 3.18 Stationary-electrode voltammogram for 1 mM $Fe^{2+}$ in 1 M HCl at a Pt electrode. Electrode reaction is $Fe^{2+} \rightarrow Fe^{3+} + e$. Dashed line, SEV for 1 M HCl without $Fe^{2+}$. Scan rate = 1.0 V/s. Concentration-distance profiles during potential scan shown below voltammogram.

state, an $E_i$ that was *negative* of this potential was selected. This selection enabled the $E_i$ to be applied across the electrode-solution interface without substantially altering the redox state of iron at the electrode surface. In order to do this, the $E_i$ should be negative of the standard electrode potential by at least 177 mV for an n = 1 reversible system. (The reader is referred to Table 3.1 and the accompanying text for correlation of ratios of redox states at the electrode surface with applied potential via the Nernst equation.) Note that selection of an $E_i$ positive of the standard electrode potential (e.g., 1.0 V) would cause immediate conversion of $Fe^{2+}$ to $Fe^{3+}$ at the electrode surface (chronoamperometry) before the potential scan was initiated. Also note that in the case in which the iron is initially present in solution as $Fe^{3+}$, 1.0 V would then be a good selection for $E_i$.

In a practical vein, one does not always know what either the standard redox potential or the redox state is for the species in solution. In this case a potential can usually be found (by trial and error or by measuring the open-circuit potential) which when applied does not cause faradaic current to occur. Such a potential is generally suitable for $E_i$ since the absence of current (other than charging current) can be taken as evidence that no significant electrolytic perturbation of the reaction layer is occurring. One can then "find" the redox couple by scanning in either direction from this $E_i$.

In the Figure 3.18 example, after imposing $E_i$ across the electrode-solution interface, the potential is scanned positively toward the standard redox potential of the $Fe^{3+}/Fe^{2+}$ couple. The ratio of iron redox states that must exist at the electrode surface at several potentials during the scan is shown on the potential axis. These values are dictated by the Nernst equation for a reversible system. Since the solution initially contained "only" $Fe^{2+}$, the $Fe^{3+}$ required to satisfy the Nernst equation is obtained from $Fe^{2+}$ by oxidation, causing anodic current.

The physical situation in the solution adjacent to the electrode during the potential scan is illustrated by the concentration-distance profiles included in Figure 3.18 for selected potentials [38]. The C-x profiles labeled "a" are for $Fe^{2+}$ and $Fe^{3+}$ when $E_i$ is imposed. Note that the application of $E_i$ does not measurably alter the concentration of $Fe^{2+}$ at the electrode surface as compared to the solution bulk. As the potential is scanned positively, the concentration of $Fe^{2+}$ at the electrode surface decreases in order to establish an $Fe^{2+}/Fe^{3+}$ ratio that satisfies the Nernst equation for the applied potential at any particular instant. This is illustrated by profiles b-e. Note that profile c (for which the concentration of $Fe^{2+}$ at the electrode surface equals the concentration of $Fe^{3+}$) correponds to an $E_{applied}$ that is the formal electrode potential (vs. SCE) of the couple.

Profiles e-g correspond to potentials sufficiently positive of the formal electrode potential that the concentration of $Fe^{2+}$ is *effectively* zero at the electrode surface. The conditions for these profiles are analogous to those for chronoamperometry (Sec. II.A, Fig. 3.1). Once the potential has reached a value sufficient for a zero reactant-surface concentration, the potential and its rate of change become immaterial to the diffusion controlled current. In other words, should the scan be stopped at f, the current will follow the same time course as if the scan had been continued.

The behavior of the current during the potential scan can be understood by carefully examining the C-x profiles in Figure 3.18. Recalling that Equation 3.5 shows the current to be determined by the slope of the C-x profile for the electroactive species at the electrode surface, one can correlate the magnitude of the slopes of the profiles in Figure 3.18 with current on the voltammogram at those potentials. The slope of profile a is zero and negligible current exists at that potential. As the potential is then scanned positively, $(\partial C/\partial x)_{x=0}$ increases for profiles b-d and the current in Figure 3.18 increases correspondingly. When profile d is reached, $(\partial C/\partial x)_{x=0}$ then decreases, as shown by profiles e through g because of depletion of $Fe^{2+}$ near the electrode. At profile e the chronoamperometric condition of diffusion-controlled conversion of $Fe^{2+}$ is achieved (i.e., $C^s_{Fe^{2+}} \approx 0$). Correspondingly, the current now drops with a $t^{-1/2}$ dependence as in chronoamperometry. Thus the observed current behavior for the voltammogram is an increase to a peak current with a subsequent decay due to depletion of electroactive species near the electrode.

Two important parameters in SEV are the peak potential $E_p$ and the peak current $i_p$, which are the potential and current at the characteristic voltammogram peak as shown in Figure 3.18. For a reversible system the peak current is defined by the Randles-Sevcik equation,

$$i_p = (2.69 \times 10^5)n^{3/2}AD^{1/2}C^\circ\nu^{1/2} \qquad (3.18)$$

at 25°C where $i_p$ is the peak current (A), n is the number of electrons transferred, A is the electrode area ($cm^2$), D is the diffusion coefficient of the species being oxidized/reduced ($cm^2/s$), C° is the concentration of this same species in the bulk solution ($mol/cm^3$), and $\nu$ is the scan rate (V/s). The peak potential is defined (for a reduction) by

$$E_p = E^{\circ\prime}_{O,R} - \frac{0.029}{n} \qquad (3.19)$$

at 25°C where $E_p$ and $E^{\circ\prime}_{O,R}$ are expressed in volts, $D_O = D_R$, and $E^{\circ\prime}_{O,R}$ is the formal electrode potential corrected to the reference electrode being used.

According to Equation 3.18, $i_p$ is proportional to $\nu^{1/2}$ and a plot of $i_p$ vs. $\nu^{1/2}$ should be a straight line for a reversible system. Although the peak current increases with scan rate, the potential at which the peak occurs is invariant with scan rate as indicated by Equation 3.19. The reason for the increase in current with faster scan rate is illustrated in Figure 3.19. Figure 3.19 shows voltammograms and the C-x profiles for SEV at two scan rates. Comparison of the profiles for the faster scan rate b with the profiles for scan a reveals a difference in the slope of the profiles at x = 0. For any given potential $(\partial C/\partial x)_{x=0}$ is larger for the faster scan. Consequently, the current is larger. In both cases, the value of the concentra-

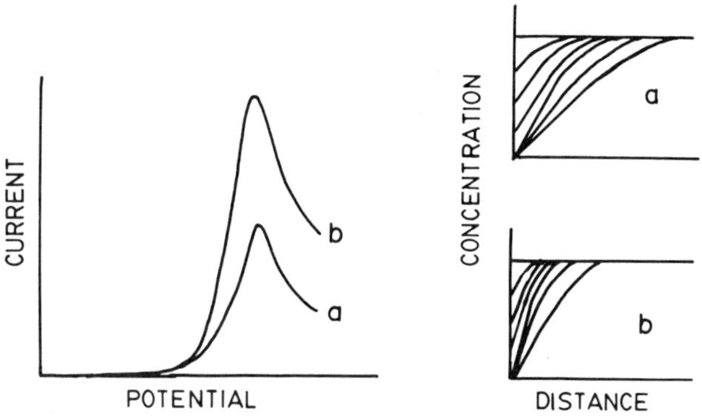


**Figure 3.19** Voltammograms and concentration-distance profiles for fast scan rate (b) and slow scan rate (a).

tion at the electrode surface (x = 0) is determined by the Nernst equation and is therefore independent of the scan rate. However, in the faster scan less time is available for consequent depletion of the adjacent solution, resulting in steeper profiles. [If the current for both scans is integrated where $Q = \int_0^t i\,dt$, for which scan rate will Q (the total charge passed) be larger?]

In the case of an irreversible system, the equation for $i_p$ is

$$i_p = (2.99 \times 10^5)n(\alpha n_a)^{1/2} A D^{1/2} C^{\circ} \nu^{1/2} \tag{3.20}$$

where $\alpha$ is the transfer coefficient and $n_a$ is the number of electrons in the rate-determining step of the electrode process. The peak potential $E_p$ is no longer independent of scan rate, as shown by Equation 3.21.

$$E_p = E^{\circ\prime} + \frac{RT}{\alpha n_a F}\left[-0.78 + \ln \frac{k_1}{D_O^{1/2}} - \frac{1}{2} \ln \left(\frac{\alpha n_a F}{RT} \nu\right)\right] \tag{3.21}$$

where $k_1$ is related to the heterogeneous rate constant of the electron transfer reaction. In general, irreversible behavior gives voltammograms that are more spread out and flatter.

An important aspect of voltammetry is the background current, which is the current obtained on a solution containing all ingredients except the electroactive species of interest. This current is designated by the dashed line in Figure 3.18. It is composed of (1) residual current, oxidation or reduction of adventitious electroactive material (e.g., dissolved oxygen, the solvent, or the electrode itself), and (2) charging current, nonfaradaic current required to charge the electrode to a given potential (behavior analogous to a capacitor). The peak current should be measured in a manner that does not include the background current. A common technique is to extrapolate the baseline current that precedes the faradaic current and measure $i_p$ by difference. Alternatively, a separate scan (at the *same* scan rate) can be made on the supporting electrolyte solution. Here $i_p$ is taken as the difference between the peak current (measured from zero current) and the background current at the same potential.

The current due to charging the electrode to a given potential is directly proportional to the scan rate as given by Equation 3.22:

$$i_c = C \frac{dE}{dt} = AC_{d\ell}\nu \tag{3.22}$$

where $i_c$ is the charging current, A is the electrode area ($cm^2$), and $C_{d\ell}$ is the capacitance of the double layer per square centimeter. Consequently, the correction for this component of the background current becomes increasingly significant as the scan rate increases. It is important to recognize that $C_{d\ell}$ is a function of potential and therefore $i_c$ is a function of E as well as of $\nu$. Also, $C_{d\ell}$ is not necessarily a linear function of E, so that the linear extrapolation of the baseline over a large distance (potential) can lead to significant error in the measurement of $i_p$.

SEV has a reasonable detection limit, being useful at $10^{-5}$ M in some cases. Sensitivity can be improved by increasing the scan rate, which increases the magnitude of $i_p$ as in Equation 3.18. However, charging current becomes a problem at higher scan rates since $i_c$ increases more rapidly with $\nu$ than does $i_p$ ($i_c \propto \nu$, $i_p \propto \nu^{1/2}$). This significant charging current can be compensated for by a differential technique in which simultaneous scans are performed on two matched electrodes, one immersed in sample and one in supporting electrolyte. The background current from the supporting electrolyte scan is electronically subtracted from the voltammogram of the sample. The detection limit of the rapid scan and differential SEV techniques is better than that of conventional polarography but not as good as that of the current sampled pulse techniques that are discussed later in this chapter and in Chapter 5.

The detection limit of SEV for analysis is greatly extended by its use in conjunction with an electrochemical preconcentration step. This technique, termed dc stripping voltammetry, is discussed in Chapter 19.

SEV is an effective means of probing homogeneous chemical reactions that are coupled to electrode reactions, especially when it is

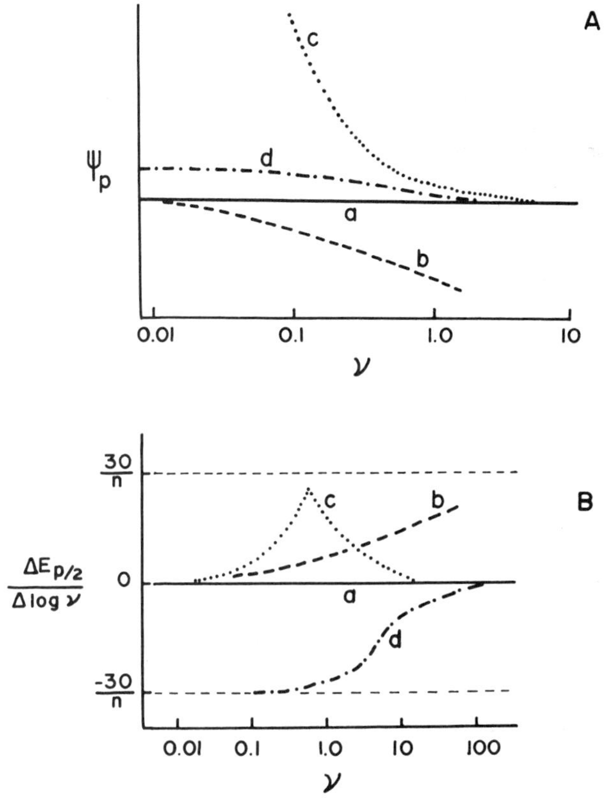


Figure 3.20 The influence of scan rate on (A) peak current function and (B) half-peak potential for several electrode processes with reversible electron transfer.

| | |
|---|---|
| a. Reversible electron transfer | E: O + e $\rightleftharpoons$ R |
| b. Preceding chemical reaction | C: Z $\rightleftharpoons$ O |
| | E: O + e $\rightleftharpoons$ R |
| c. Catalytic regeneration | E: O + e $\rightleftharpoons$ R |
| | C: R + Z $\rightarrow$ O |
| d. EC mechanism | E: O + e $\rightleftharpoons$ R |
| | C: R $\rightarrow$ Z |

[From Ref. 39, reprinted with permission from R. S. Nicholson and I. Shain, *Anal. Chem.* *36*:706 (1964). Copyright 1964 American Chemical Society.]

extended to cyclic voltammetry as described in the next section. Considerable information can be obtained from the dependence of $i_p$ and $E_p$ on the rate of potential scan. Figure 3.20 illustrates the behavior of $i_p$ and $E_p$ with variation in scan rate for a reversible heterogeneous electron transfer reaction that is coupled to various types of homogeneous chemical reactions. The current function $\Psi_p$ is proportional to $i_p$ according to the equation

$$\Psi_p = \frac{i_p}{nFAD^{1/2}C^\circ(nF\nu/RT)^{1/2}} \tag{3.23}$$

Different types of coupled chemical reactions (or mechanisms) can be identified by their characteristic current and potential behavior with varying scan rate as shown in Figure 3.20. Theory and diagnostic criteria have been developed for a variety of kinetic cases in which the homogeneous chemical reactions are coupled to either reversible or irreversible heterogeneous electron exchange [39]. Adsorption of the electroactive species has also been considered [40]. In general, conclusions based on current measurements are more reliable diagnostics of electrode mechanism than those based on potential measurements because experimental factors (iR drop, junction potentials, state of electrode surface) can shift the potential axis significantly.

In view of the fact that SEV is a transient technique and that the potential axis may also be thought of as a time axis, it has often been called chronoamperometry with potential sweep. The entire sweep may be applied during the lifetime of a single mercury drop in a polarographic experiment. Historically, this hybrid technique was termed single-sweep oscillographic polarography or cathode-ray polarography. The IUPAC suggests single-sweep polarography.

## B. Cyclic Stationary-Electrode Voltammetry

Cyclic stationary-electrode voltammetry, usually called cyclic voltammetry (CV), is perhaps the most effective and versatile electroanalytical technique available for the mechanistic study of redox systems [39,41-44]. It enables the electrode potential to be rapidly scanned in search of redox couples. Once located, a couple can then be characterized from the potentials of peaks on the cyclic voltammogram and from changes caused by variation of the scan rate. CV is often the first experiment performed in an electrochemical study. Since cyclic voltammetry is a logical extension of stationary-electrode voltammetry (SEV), some important aspects of CV were treated in the preceding section.

The repetitive triangular potential excitation signal for CV causes the potential of the working electrode to sweep back and forth between two designated values (the switching potentials). The potential excitation that is applied across the electrode-solution interface in order to obtain a cyclic voltammogram is illustrated by the potential-time profiles in Figure 3.21. The labeled segments in Figure 3.21 can be interpreted in the following way:

a. Positive potential scan from +0.25 to +0.75 V
b. Direction of scan reversed at switching potential +0.75 V
c. Negative potential scan from +0.75 to +0.25 V
d. Termination of the first cycle

Note that up to point b the excitation signal is that of SEV, as described in the preceding section (see Fig. 3.17).

Although the potential scan is frequently terminated at the end of the first cycle (point d), it can be continued for any number of cycles, hence the terminology *cyclic* voltammetry. The dotted line in Figure 3.21 denotes a second cycle. A scan in which the potential is becoming increasingly positive (i.e., heading in a positive direction) is termed a positive scan. A scan in which the potential is becoming increasingly negative is a negative scan (even though the potential may actually be positive as is the case in this example). Since the time required for the first cycle is 10 s, the scan rate is 1 V/10 s or 0.1 V/s. The scan profile used in a particular experiment is generally determined by the location of the redox couple of interest.

To obtain a cyclic voltammogram, the current at the working electrode in an unstirred solution is measured during the potential

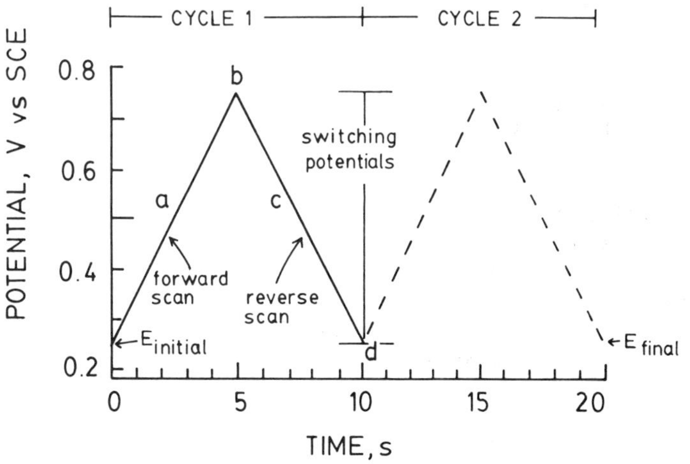


**Figure 3.21** Typical potential-time excitation signal for cyclic voltammetry.

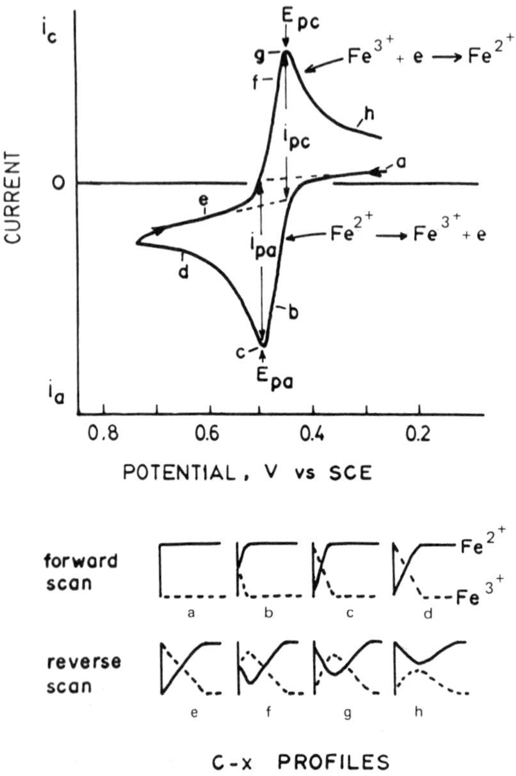


Figure 3.22 Cyclic voltammogram of $Fe^{2+}$ in 1 M $H_2SO_4$. Concentration-distance profiles during potential scan shown below voltammogram.

scan. A cyclic voltammogram that was obtained with a carbon paste working electrode immersed in a 1.00 mM $Fe^{2+}$ solution with 1.0 M $H_2SO_4$ supporting electrolyte is shown in Figure 3.22.

The potential scan that was applied across the electrode-solution interface to obtain this cyclic voltammogram is identical with the profile depicted in Figure 3.21. Correlate the potential of Figure 3.21 with the potential (horizontal) axis of Figure 3.22. Note that the positive scan from +0.25 to +0.75 V in Figure 3.22 corresponds to segment a in Figure 3.21 and that the reverse (negative) scan corresponds to segment c. The arrows designate the direction of the potential scan.

During the scan from +0.25 to +0.75 V, the applied potential becomes sufficiently positive at +0.4 V to cause oxidation of $Fe^{2+}$ to occur at the electrode surface. This oxidation is accompanied by anodic current, which increases rapidly until the surface concentration of $Fe^{2+}$ approaches zero as signaled by peaking of the current at point c in Figure 3.22. The current then decays (after c) as the solution surrounding the electrode is depleted of $Fe^{2+}$ due to its conversion to $Fe^{3+}$. This depletion of $Fe^{2+}$ and accumulation of $Fe^{3+}$ near the electrode is depicted by concentration-distance profiles a-e in Figure 3.22. The magnitude of the current is related to the slope of the C-x profile for $Fe^{2+}$ as described by Equation 3.5. The reader is referred to Section III.A for further explanation of the characteristic current behavior during this positive scan as well as a discussion concerning the selection of the initial potential, $E_i$.

During the positive scan in which $Fe^{2+}$ is oxidized to $Fe^{3+}$, the depletion of $Fe^{2+}$ in the vicinity of the electrode is accompanied by an accumulation of $Fe^{3+}$. This can be seen by the concentration-distance profiles at various potentials in Figure 3.22. After the direction of potential scan is switched at +0.75 V to a negative scan, oxidation continues (as is evident by the anodic current and the C-x profile e) until the applied potential becomes sufficiently negative to cause reduction of the accumulated $Fe^{3+}$. Reduction of $Fe^{3+}$ is signaled by the appearance of cathodic current. Once again, the current increases as the potential moves increasingly negative until the reduction of $Fe^{3+}$ sufficiently depletes the region near the electrode of $Fe^{3+}$ to cause the current to peak and then to decrease. Thus the physical phenomena that caused a current peak during the oxidation cycle also cause a current peak during the reduction cycle. This can be seen by comparing the concentration-distance profiles for the two scans.

Simply stated, in the forward scan $Fe^{3+}$ is electrochemically generated as indicated by the anodic current. In the reverse scan this $Fe^{3+}$ is reduced back to $Fe^{2+}$ as indicated by the cathodic current. Thus CV is capable of rapidly generating a new species during the forward scan and then probing its fate on the reverse scan. This is a very important aspect of the technique. (Would a cathodic current be expected if the same experiment were performed in a stirred solution?)

The important parameters of a cyclic voltammogram are the magnitudes of the peak currents, $i_{pa}$ and $i_{pc}$, and the potentials at which the peaks occur, $E_{pa}$ and $E_{pc}$. These parameters are labeled in Figure 3.22. One method for measuring $i_p$ involves extrapolation of a baseline current. The establishment of a correct baseline is essential for the accurate measurement of peak currents. This is not always easy, particularly for more complicated systems. The second sweep generally poses the main problem since the baseline is not the same as the residual current obtained by an identical experiment in supporting electrolyte. Difficulty in obtaining accurate peak currents is perhaps the biggest liability of CV. The reader is referred to Adam's monograph for a more detailed discussion of this sticky situation [6, pp. 152-158].

A redox couple in which both species rapidly exchange electrons with the working electrode is termed an *electrochemically* reversible couple. Such a couple can be identified from a cyclic voltammogram by measurement of the potential difference between the two peak potentials. Equation 3.24 applies to a system that is both electrochemically and chemically reversible:

$$\Delta E_p = E_{pa} - E_{pc} \approx \frac{0.059}{n} \tag{3.24}$$

where n is the number of electrons transferred and $E_{pa}$ and $E_{pc}$ are the anodic and cathodic peak potentials, respectively, in volts. This 0.059/n separation of peak potentials is independent of scan rate for a reversible couple, but is slightly dependent on switching potential and cycle number (see Ref. 7, pp. 229-230). The potential midway between the two peak potentials is the formal electrode potential (corrected for the reference electrode being used) of the couple:

$$E^{\circ\prime} = \frac{E_{pa} + E_{pc}}{2} \tag{3.25}$$

When the scan rate $\nu$ is increased, $i_{pa}$ and $i_{pc}$ both increase in proportion to $\nu^{1/2}$. The same reasoning for this current increase in SEV (see p. 82) applies here. Plots of $i_{pa}$ and $i_{pc}$ vs. $\nu^{1/2}$ should be linear with intercepts at the origin for a reversible couple.

The values of $i_{pa}$ and $i_{pc}$ are similar in magnitude for a reversible couple with no kinetic complications. That is,

$$\frac{i_{pa}}{i_{pc}} \approx 1 \tag{3.26}$$

(This equation is strictly applicable after several cycles. The first voltammogram is not quite the same as the reproducible curve obtained after several cycles.) However, the ratio of peak currents can be significantly influenced by chemical reactions coupled to the electrode process. Figure 3.23 illustrates variations in $i_{pa}/i_{pc}$ as a function of scan rate for several electrode mechanisms [39]. Such characteristic behavior in peak currents makes cyclic voltammetry a powerful technique for studying mechanisms of electrode reactions. Examples are given in Chapter 17.

Electrochemical irreversibility is caused by slow electron exchange of the redox species with the working electrode. (The reader is referred to Chapter 2 for a discussion of irreversibility.) In this case Equation 3.24 is not applicable. Electrochemical irreversibility

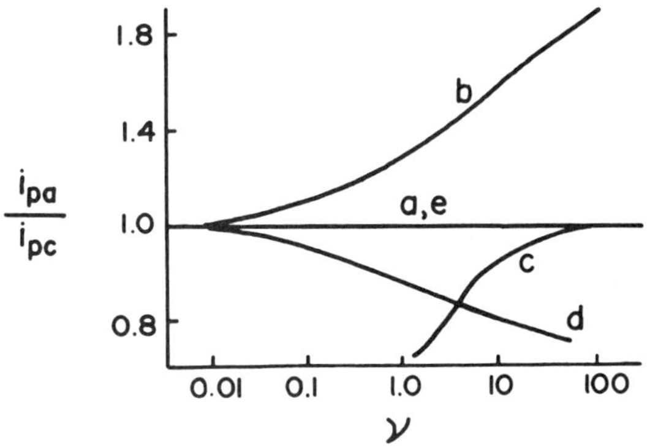


Figure 3.23 Variation in the ratio of anodic to cathodic peak currents as a function of scan rate for several electrode processes with reversible electron transfer. [From Ref. 39, reprinted with permission from R. S. Nicholson and I. Shain, *Anal. Chem. 36*:722 (1964). Copyright 1964 American Chemical Society.]

| | | |
|---|---|---|
| a. Reversible electron transfer | E: | $O + e \rightleftharpoons R$ |
| b. Preceding chemical reaction | C: | $Z \rightleftharpoons O$ |
| | E: | $O + e \rightleftharpoons R$ |
| c. EC mechanism | E: | $O + e \rightleftharpoons R$ |
| | C: | $R \rightarrow Z$ |
| d. EC mechanism | E: | $O + e \rightleftharpoons R$ |
| | C: | $R \rightleftharpoons Z$ |
| e. Catalytic regeneration | E: | $O + e \rightleftharpoons R$ |
| | C: | $R + Z \rightarrow O$ (O regenerated) |

[From Ref. 39, reprinted with permission from R. S. Nicholson and I. Shain, *Anal. Chem. 36*:706 (1964). Copyright 1964 American Chemical Society.]

is characterized by a separation of peak potentials that is greater than 0.059/n volts and that is *dependent* on the scan rate. Figure 3.24A illustrates voltammograms for reversible and irreversible systems.

The dependence of $\Delta E_p$ on the heterogeneous rate constant $k_s$ and the scan rate $\nu$ is shown in Figure 3.24B. It is apparent that at a given scan rate $\Delta E_p$ increases as the heterogeneous rate constant $k_s$ of the system decreases. Alternatively, for a system with a given

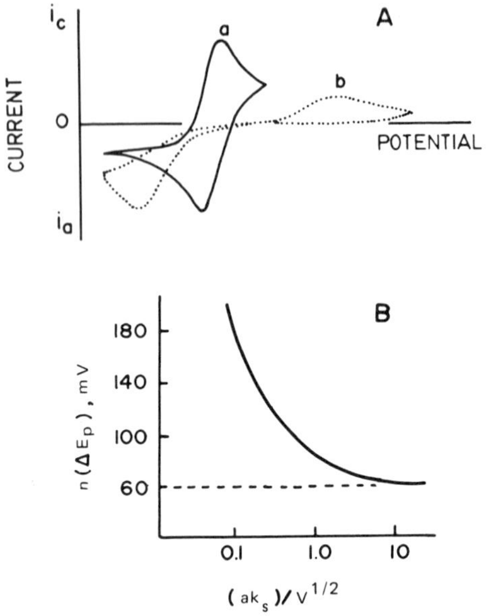


Figure 3.24 (A) Cyclic voltammograms exhibiting electrochemical: (a) reversibility—1.00 mM $Fe^{3+}$, 1.0 F HCl, 1.0 F $CaCl_2$, carbon paste electrode; (b) irreversibility—1.00 mM $Fe^{3+}$, 1.0 M $H_2SO_4$, same. (From Ref. 6.) (B) Dependence of $\Delta E_p$ on $k_s$ and $\nu^{1/2}$. [From Ref. 45, reprinted with permission from R. S. Nicholson, *Anal. Chem.* 37:667 (1965). Copyright 1965 American Chemical Society.]

rate constant, the peak separation increases with increasing scan rate. The magnitude of $k_s$ can be determined from a plot such as in Figure 3.24B if $\Delta E_p$ is known for a given scan rate. Figure 3.24B also points out that a system that appears reversible at one scan rate can "become irreversible" at faster scan rates. This illustrates the somewhat arbitrary nature of reversibility. Uncompensated iR drop also causes peak separation to increase with $\nu$. Elimination of iR drop by one of the methods described in Chapters 6 and 7 is crucial to the measurement of "correct" $\Delta E_p$ values. It is important to recognize that electrochemical irreversibility also influences the peak current ratio. The more irreversible a couple becomes, the smaller will be the $i_p$ on the reverse scan. This is often due both to the smaller $k_s$ and the fact that significant product has diffused away from the surface by the time the reverse $E_p$ is reached. Because of these complications, it can be dangerous to infer quantitative information from the $i_p$ ratio for extremely irreversible couples.

In most CV experiments there is little advantage to be gained by carrying on the potential scan for more than two to three cycles. Data are typically obtained via XY recorder at slow scans (i.e., less than 500 mV/s) and storage oscilloscope or computer at faster rates. Scan rates up to 20,000 V/s have been used; however, rates faster than 100 V/s are rarely practicable because of iR drop and charging current.

With the advent of digital implementation of electroanalytical experiments came the technique called "staircase cyclic voltammetry," wherein the triangular wave is approximated by a series of small potential steps. The reason for such an approximation is partly due to the impossibility of digitally generating a "pure" ramp and more importantly to the realization that substantial improvements accrue from sampling the current at the end of each step, where double-layer charging has decayed away. If the steps are small, the data will fit theory based on pure ramps quite well.

CV has become increasingly popular in all fields of chemistry as a means of studying redox states [42]. The method enables a wide potential range to be rapidly scanned for reducible or oxidizable species. This capability together with its variable time scale and good sensitivity make this the most versatile electroanalytical technique thus far developed. It must, however, be emphasized that its merits are largely in the realm of qualitative or "diagnostic" experiments. Quantitative measurements (of rates or concentrations) are *best* obtained via other means (e.g., step, pulse, or hydrodynamic techniques). Because of the kinetic control of many CV experiments, some caution is advisable when evaluating the results in terms of thermodynamic parameters (e.g., measurement of $E^{\circ\prime}$ for irreversible couples).

Perhaps the most useful aspect of CV is its application to the qualitative diagnosis of electrode reactions which are coupled to homogeneous chemical reactions [41-44]. The forte of CV is its ability to generate a species during one scan and then probe its fate with subsequent scans. Examples of CV usage in unraveling electrode mechanisms are presented in Chapter 17.

A chemical system should be examined using several carefully chosen electrochemical techniques in combination with spectroscopic information and good chemical intuition. It is dangerous to draw too many conclusions from voltammetric information alone. Chapter 17 is devoted to selected examples from the literature which illustrate this theme.

### C. Derivative Stationary-Electrode Voltammetry

The excitation signal for derivative stationary-electrode voltammetry (derivative SEV) is identical with that for SEV (i.e., a potential scan). The system response signal, current, is also the same. However, in derivative SEV this response signal is electronically differentiated,

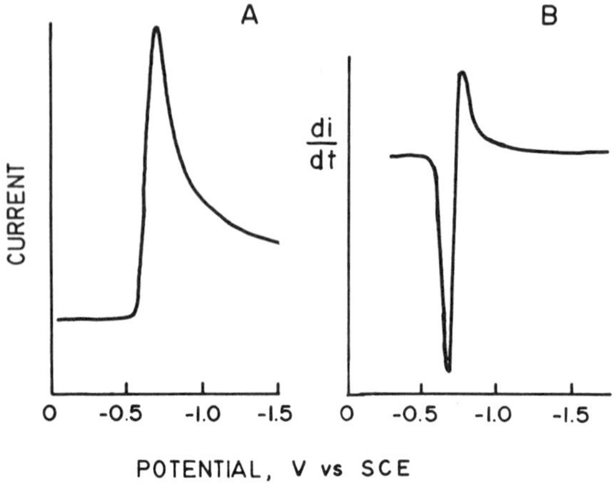


Figure 3.25 (A) Stationary electrode voltammogram of 2.00 mM $Cd^{2+}$ in 0.2 M KCl, $\nu$ = 0.0327 V/s, HMDE. (B) Derivative stationary electrode voltammogram, identical conditions. [From Ref. 46, reprinted with permission from S. P. Perone and T. R. Mueller, *Anal. Chem.* *37*:2 (1965). Copyright 1965 American Chemical Society.]

and the resulting derivative, di/dt, is recorded as a function of applied potential. Second and third derivatives can also be displayed.

Figure 3.25 illustrates voltammograms obtained for SEV and derivative SEV for reduction of $Cd^{2+}$ at a HMDE [46].

The first derivative peak height is directly proportional to the concentration of the electroactive species. Several advantages accrue from monitoring the derivative response, including increased sensitivity, resolution, and ease of obtaining rate parameters of coupled chemical reactions [47,48]. Enhanced signal/noise derives from the relatively small variation in double-layer charging current with potential, making its contribution to the derivative small. This advantage becomes important at high scan rates where charging current grows in relation to the faradaic response.

Although derivative SEV does exhibit these advantages over SEV, the technique has been used very little. The additional complexity of the differentiation apparently is not justified by the small improvements in performance.

## D. DC Polarography

Polarography is the classical electrochemical technique involving the dropping mercury electrode (DME). The excitation signal is a slow potential scan and the resulting current is displayed as a function of this applied potential. Although the term "polarography" has been used over the years in connection with a variety of electrochemical techniques, "polarography" is now reserved as a name for voltammetric experiments in which the DME is employed (see Chap. 9). Many texts have been written on the subject [49-52].

A typical polarogram is illustrated in Figure 3.26. In this case the solution contains the electroactive species $Pb^{2+}$ and $Cd^{2+}$ as well as 0.1 M KCl supporting electrolyte.

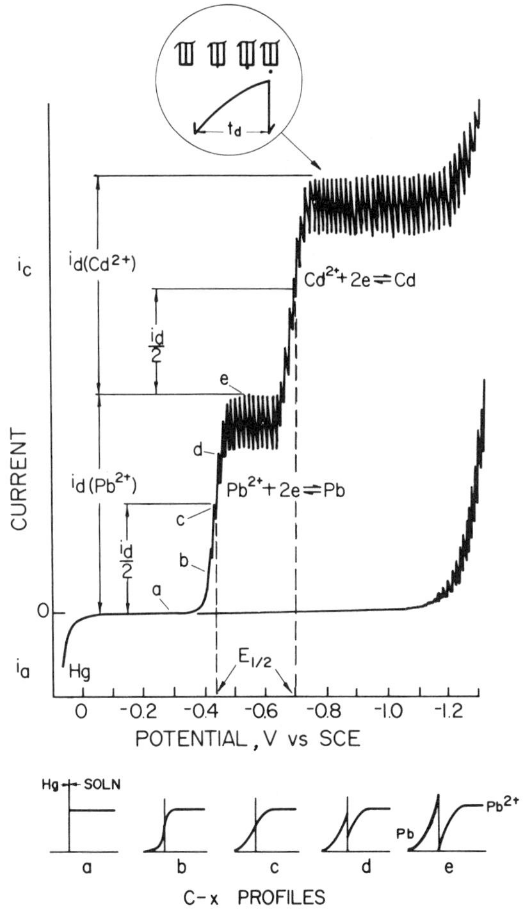


Figure 3.26 Polarogram of 1.0 mM $Pb^{2+}$, 1.0 mM $Cd^{2+}$, and 0.1 M KCl. $O_2$ removed. Concentration-distance profiles during potential scan shown below polarogram.

As the applied potential is scanned negatively from an initial value of 0.0 V vs. SCE, the current changes only slightly until a sufficiently negative potential is reached to cause reduction of $Pb^{2+}$ to $Pb^{\circ}$. As the potential becomes more negative, the current for this reduction increases. The increase is caused by the greater slope for the concentration-distance profile for $Pb^{2+}$ at the electrode surface [i.e., $(\partial C/\partial x)_{x=0}$ becomes greater]. This is shown by C-x profiles a to e for various potentials. The current reaches a plateau when the potential becomes sufficiently negative that the concentration of $Pb^{2+}$ is effectively zero at the electrode surface. Now the rate of $Pb^{2+}$ reduction (which determines the current) is governed (limited) by the rate of diffusion of $Pb^{2+}$ to the electrode surface. Consequently, the current plateau is referred to as the diffusion current ($i_d$) or the limiting current ($i_\ell$).

The insert in Figure 3.26 correlates the repetitive current fluctuation that is characteristic of the polarogram with the "life" of a mercury drop at the DME. The current increases as the drop grows, declining abruptly the moment the drop falls. This current-time behavior is determined by (1) the changing surface area of the drop and (2) the diffusion of $Pb^{2+}$ to the electrode surface. The electrode area increases at a $t^{2/3}$ rate, which would cause current (which is proportional to electrode area) to also increase at $t^{2/3}$. Counterbalancing this increase is the $t^{-1/2}$ decrease in current caused by depletion of $Pb^{2+}$ in the adjacent solution under diffusion-controlled conditions. (The diffusional current of polarography can be considered in terms of chronoamperometry at an expanding mercury sphere since the potential does not vary significantly during the life of a single drop. See Equation 3.6 for the $t^{-1/2}$ relationship.) Combining these two phenomena gives a current that increases at a $t^{1/6}$ rate during the lifetime of a drop. When the drop falls, the electrode area instantly decreases, causing the current to drop abruptly. Each falling drop stirs the solution sufficiently to erase almost all concentration depletion caused by electrolysis at the drop.

The diffusion current (over the lifetime of a drop) is expressed by the Ilkovic equation,

$$i_d(\max) = 708nm^{2/3}D^{1/2}t_d^{1/6}C^{\circ} \tag{3.27}$$

where

$i_d(\max)$ = diffusion current, μA
n = number of electrons transferred per molecule or ion
m = flow rate of mercury, mg/s
D = diffusion coefficient of electroactive species, $cm^2/s$

$t_d$ = drop time, s
$C^o$ = concentration of electroactive species, mM

The applied potential is related to the polarographic current for a reversible system by the Nernst equation as expressed in the form

$$E = E_{1/2} - \frac{0.059}{n} \log \frac{i}{i_d - i} \tag{3.28}$$

where i is the current at a particular applied potential E. The half-wave potential $E_{1/2}$ is defined as the potential at which the current is equal to $i_d/2$. It is also related to the formal electrode potential, $E^{o\prime}$, of the electroactive species according to

$$E_{1/2} = E^{o\prime} - \frac{0.059}{n} \log \frac{D_O^{1/2}}{D_R^{1/2}} \tag{3.29}$$

where $D_O$ and $D_R$ are the diffusion coefficients of the oxidized and reduced species, respectively. Because $E_{1/2}$ is directly related to the formal reduction potential of the redox couple, it is useful for qualitative analysis of electroactive species. Tables of $E_{1/2}$ values of redox couples in a variety of media have been compiled for the purpose of identification.

As Equation 3.27 shows, the diffusion current $i_d$ is directly proportional to the concentration of the electroactive species. Consequently, polarography has been used extensively for the analysis of solutions containing electroactive materials. The optimum concentration range for determinations by conventional polarography as described here is $10^{-2}$-$10^{-4}$ M. Reproducibility is generally ±3%.

The diffusion current is typically measured from a baseline which is obtained by extrapolating the residual current prior to the wave. Alternatively, the baseline can be recorded in a separate experiment on a solution of deoxygenated supporting electrolyte. The residual current arises from capacitive current required to maintain the expanding drop at the applied potential and from reduction of trace electroactive impurities.

The diffusion currents of electroactive species are additive as is shown for $Pb^{2+}$ and $Cd^{2+}$ in Figure 3.26. When the applied potential becomes sufficiently negative, reduction of $Pb^{2+}$ and $Cd^{2+}$ occurs simultaneously and the resulting current reflects the sum of these two processes. The baseline for measuring the $i_d$ of the second process is obtained by extrapolation of the $i_d$ of the first process. Thus the determination of mixtures of electroactive species is possible.

The "potential window" of the DME is limited by hydrogen evolution or reduction of supporting electrolyte cation on the negative extreme and by oxidation of mercury on the positive extreme. In certain cases the oxidation of mercury can be used effectively for the determination of species that form complexes (e.g., $CN^-$, EDTA) or insoluble salts (e.g., $Cl^-$ $Br^-$) with mercurous or mercuric ion. In this case the potential is scanned in the positive direction. The resulting anodic waves are a result of the oxidation of the mercury electrode (not of the ligand or anion) at a rate controlled by the diffusion of the *anion* or ligand (A) to the electrode surface. The general electrode reaction can be expressed as

$$Hg + A \rightleftharpoons HgA^{n+} + ne \qquad (3.30)$$

In this case $i_d$ is proportional to the concentration of A in solution. The variation in half-wave potentials corresponds to the variation in stabilities of the HgA compounds (i.e., $K_{sp}$ or $K_f$).

Many organic compounds undergo reduction or oxidation at a DME. Consequently, polarographic techniques have been used extensively for determinations of organic compounds and for studying the mechanisms of their electrode reactions. In aqueous solution, the reduction of organic compounds is frequently a 2e process accompanied by protonation as in Equation 3.31:

$$O + 2e + 2H^+ \longrightarrow RH_2 \qquad (3.31)$$

Polarograms for the quinone-hydroquinone couple are shown in Figure 3.27. The electrode mechanism for quinone reduction is

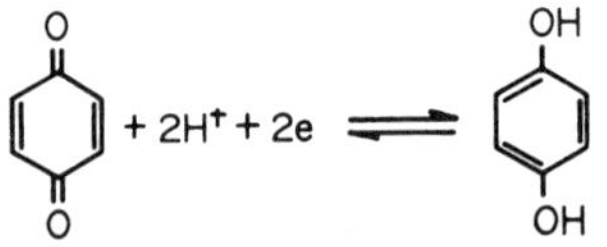


Q H$_2$Q (3.32)

The polarograms were obtained on solutions of Q, $H_2Q$ and an equimolar mixture of Q, $H_2Q$. In the latter case, both reduction and oxidation waves are obtained.

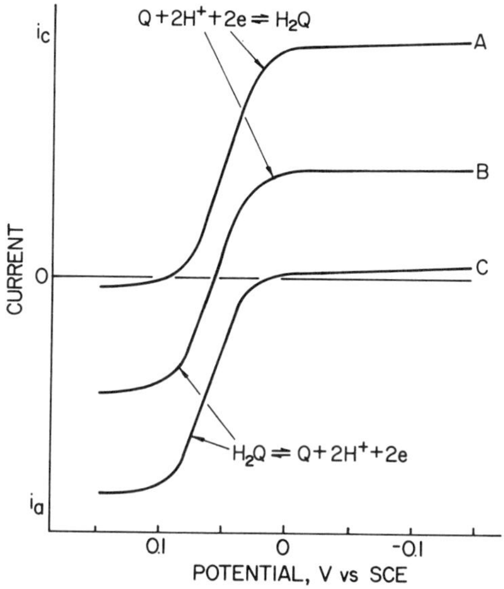


Figure 3.27 Polarograms for reduction of quinone (Q) and oxidation of hydroquinone, ($H_2Q$) in buffered solution: A, quinone; B, quinone-hydroquinone equimolar mixture; C, hydroquinone.

The $E_{1/2}$ for an electrode reaction of this type is pH dependent. As the pH increases, $E_{1/2}$ shifts negatively due to decreasing availability of protons. A typical $E_{1/2}$-pH plot is shown in Figure 3.28. The magnitude of negative shift per pH unit is related to the acid dissociation constants, $K_{a_1}$ and $K_{a_2}$, for the species $RH_2$ for a mechanism as shown by Equation 3.31.

In nonaqueous solvents, nonprotonated species can be generated. Consequently, many electrochemical studies of organic compounds employ nonaqueous solvents such as acetonitrile, dimethylformamide, and dimethyl sulfoxide [53]. Electrochemistry in nonaqueous solvents is addressed in Chapters 13 to 15.

In addition to its extensive use for analytical purposes, polarography has been widely employed for the measurement of formation constants of complexes by the effect of complexation on $E_{1/2}$, the kinetics of homogeneous chemical reactions that are coupled to the electrode process, the kinetics of heterogeneous electron transfer between a mercury electrode and an electroactive species, and the effects of adsorption of both electroactive and nonelectroactive species on the electrode surface. Polarography is also useful for detecting end points in titrations that involve electroactive species (amperometric titrations). Many of these measurements can now be made

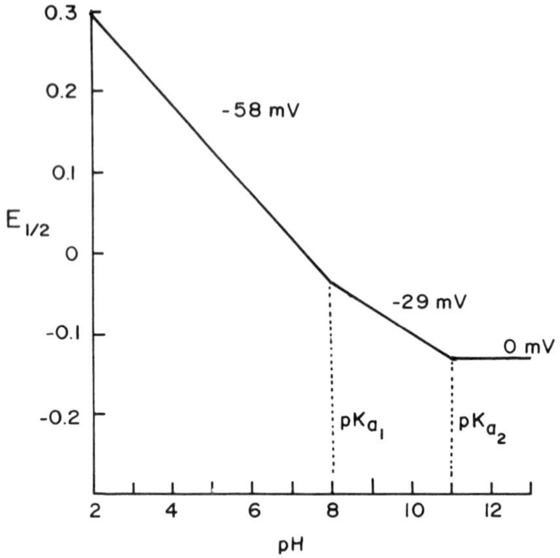


Figure 3.28 Dependence of $E_{1/2}$ on pH for typical 2e reduction of an organic compound. $O + 2e + 2H^+ \rightarrow RH_2$.

much more effectively by the more recently developed potential step and potential sweep methods. Polarography derives great advantage from the clean and reproducible electrode surface of pure mercury (and its excellent hydrogen overpotential) obtainable from the DME, thus eliminating many of the difficulties attendant on stationary electrodes. This is almost its only advantage; however, it remains of such significance that the chance of polarography completely disappearing is scant indeed. It has nevertheless lost its position as the paramount electroanalytical technique.

## E. Current-Sampled Polarography

Current-sampled polarography (sometimes called Tast polarography) is a variation of dc polarography in which the response current is only registered during a small interval of time (typically 16.7 ms) during the latter part of drop life for each drop. This enables the contribution of the charging current to the overall current to be minimized as shown in Figure 3.29A. As we have already noted, the drop surface area increases with $t^{2/3}$. Because capacitance is directly proportional to area and the charging current at a fixed voltage is related to the rate of change of capacitance (since $Q = CV$, $i_{d\ell} = V\,dC/dt$, and $i_{d\ell} \propto dA/dt$), then $i_{d\ell} \propto t^{-1/3}$. The double-

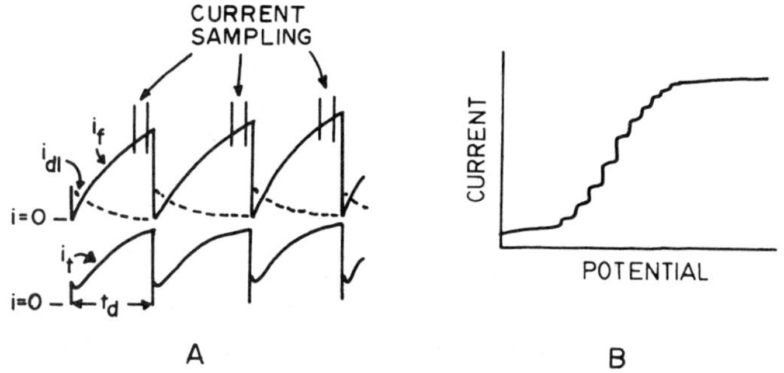


Figure 3.29 (A) Current-time curves illustrating the sampling sequence for current-sampled polarography. $i_f$, diffusion-controlled faradaic current; $i_{d\ell}$, double-layer charging current; $i_t = i_f + i_{d\ell}$. (B) Current-sampled polarogram.

layer charging current will therefore be large early in drop life and fall off as the drop grows.

A sample-and-hold circuit maintains the current value of the previous sampling interval and integrates out noise and interference. This gives the polarogram a staircase appearance as shown in Figure 3.29B. The slower the scan rate and the faster the drop rate, the more samples will be taken to represent a given polarographic wave and the smoother the wave will appear. The improved detection limit realized by minimization of charging current is slight; however, many workers prefer the sampled data presentation because it eliminates the annoying current fluctuations associated with drop growth. This saves ink as well as wear and tear on laboratory recorders.

It is important to recognize that interesting adsorption phenomena can have a profound influence on events early in drop life [54]. Data of this sort would be missed in a sampled experiment. Therefore, it is best to first look at the response without sampling. If no unusual behavior is observed, further experiments may be performed in the sampling mode.

In this technique the excitation and fundamental system response is identical to that for dc polarography. The sample-and-hold measuring scheme is like that used to great advantage in LAPP (see Sec. III.F).

## F. Large-Amplitude Pulse Voltammetry

In large amplitude pulse voltammetry (LAPV, often called "normal" pulse voltammetry for historical reasons) the excitation waveform consists of successive pulses of gradually changing amplitude between which a constant "initial" potential is applied. The initial potential is usually chosen in a region where none of the sample components are electrochemically reactive. The current response may be measured either during the forward step, after return to the initial condition, or both. Usually, the current is sampled only at the end of the forward step and the remainder of the response waveform is ignored. The sampled current is then plotted against the pulse height (thus "voltammetry"). Sometimes the difference between successive sampled currents is plotted as a function of potential. As we shall see, this results in a response having a derivative shape. It is important to recognize that all variations of LAPV are in essence a series of potential pulse chronoamperometry experiments.

Figure 3.30 illustrates the simplest example of LAPV at a stationary electrode. In this case we assume that the time delay $t_d$ between pulses is such that the initial condition is restored. This implies that all concentration profiles are completely relaxed before the next pulse is applied and every pulse initiates a new chronoamperometric experiment. The length of $t_d$ required to accomplish this will depend on the chemical reversibility of the system, being shorter for reversible reactions and longer for irreversible reactions.

For convenience, let us assume a fully (chemically and electrochemically) reversible system which gives the response to a ramp excitation illustrated in Figure 3.30. Although many pulses will be applied, we have selected three for closer examination.

*Response to pulse A*: Here the pulse amplitude is insufficient to initiate significant faradaic reaction and the sample concentration profile is not disturbed. The current response is then primarily due to charging the double-layer capacitance to the new potential ($E_i + \Delta E$). The current sampled just before $\tau$ is negligible, for by that time the double layer is fully charged.

*Response to pulse B*: Now the pulse amplitude extends to the half-wave potential for the reactive species, concentration gradients will form, and significant faradaic current will be sampled. Note that the double-layer charging current will have decayed away when the sample is taken. This time-domain discrimination against $i_{d\ell}$ is a principal advantage of pulse voltammetry. The faradaic current will follow a modified Cottrell equation:

$$i = nFAC^\circ \left(\frac{D}{\pi t}\right)^{1/2} \left[1 + \left(\frac{C_O}{C_R}\right)_{x=0}\right]^{-1} \tag{3.33}$$

where for a reversible system

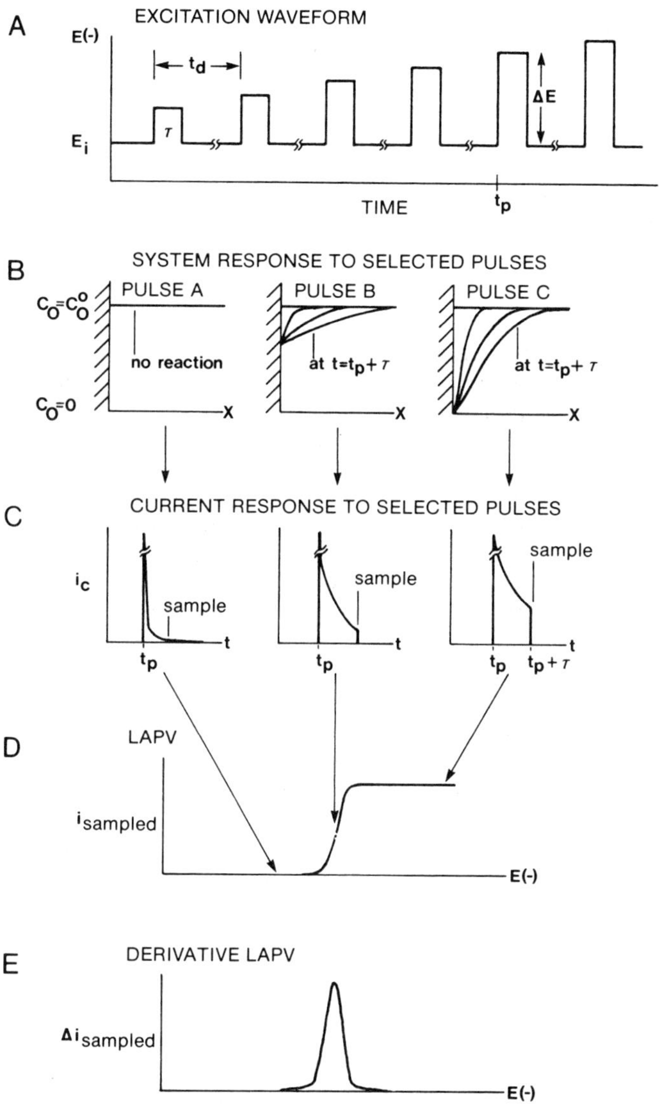


Figure 3.30 Introduction to stationary electrode LAPV.

$$\frac{C_O^s}{C_R^s} = \exp\left(\frac{nF}{RT}\right) [(E_i + \Delta E) - E_{1/2}] \tag{3.34}$$

In the case of pulse B, $(E_i + \Delta E) = E_{1/2}$, so $(C_O^s/C_R^s) = 1$ and $(1 + C_O^s/C_R^s)^{-1} = 0.5$.

*Response to pulse C*: Now we have the *limiting* case of chronoamperometry. The surface concentration has become effectively zero and the response will follow the Cottrell equation.

Plotting the sampled current versus the applied potential results in the complete pulse voltammogram shown in Figure 3.30. The reader should determine what effect taking current samples earlier or later would have on this response. Note that there is a trade-off between sensitivity and discrimination against the charging current! If the difference between successive samples is plotted (Fig. 3.30), a peaked response is obtained.

When LAPV is carried out with a dropping mercury electrode the technique is large-amplitude pulse polarography (usually referred to as "normal" pulse polarography). One pulse is applied per drop at a fixed time late in the life of the drop. The current is sampled just before the drop is dislodged by mechanical means, a time at which the rate of change of area is smallest. Because the pulse widths are usually in the 50-ms region, the electrode is effectively stationary and of fixed area during each pulse. Since each drop is a new electrode, the time between pulses $t_d$ is made to be the drop time. The timing circuitry that is used to generate the pulses also functions to trigger an electromechanical "drop knocker" or a static mercury drop electrode (see Chap. 9).

Sometimes it is useful to apply identical pulses to several drops before increasing the amplitude. This allows for time averaging and some improvement in the signal-to-noise ratio. LAPP is analytically useful, being sensitive to $10^{-6}$ M in some circumstances.

In an interesting variation of LAPP (originally ascribed to Kalousek) the potential applied to a DME is alternately switched from an initial potential (which is sufficiently negative of the $E_{1/2}$ for a reducible species to cause a limiting cathodic current) to a series of potential pulses changing in height from values more positive than $E_{1/2}$ toward more negative values. The current is sampled late in the life of each pulse, as indicated in Figure 3.31A. At first, the reduction product formed at the initial potential is oxidized during each pulse (unless the couple is totally irreversible) as shown by the anodic wave in Figure 3.31B. After the pulse heights pass through $E_{1/2}$ and reach potentials for reduction, a cathodic current is recorded. The relative values of $i_a$ and $i_c$ will depend on the reversibility of the reaction and the specific pulse widths and sampling times employed.

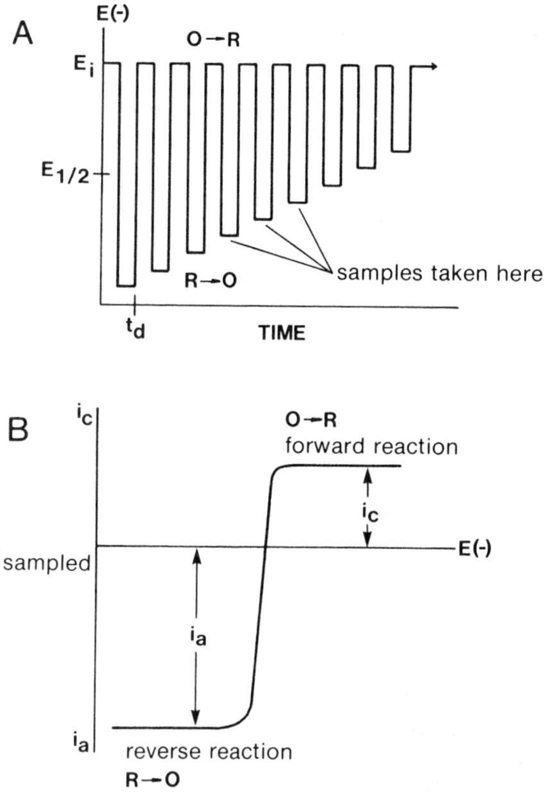


Figure 3.31 In one of many possible variations of LAPP the current is sampled on a reverse step just prior to drop fall. (A) excitation, (B) response.

In yet another variation the current is recorded at the initial potential following a series of pulses increasing in amplitude from the initial potential toward more negative values (i.e., the same excitation shown in Fig. 3.30A). When the pulses reach sufficiently negative potentials to cause reduction of the species, the product of the reduction is oxidized at $E_i$, and the sampled anodic current is recorded unless the couple is totally irreversible. This is equivalent to a series of potential pulse chronoamperometry experiments with samples taken on the reverse step. Techniques of this sort have been used occasionally in reversibility studies and for identification of unstable products of electrode processes. Although potentially useful, such methods have been largely ignored until very recently, when a number of such "reverse pulse" experiments have been reported [55].

We owe a great deal to the imaginative ideas conceived by Kalousek some 40 years ago. In many respects there is a close relationship between large- and small-amplitude pulse methods. The latter, generally the more popular of the two, are described in Chapter 5.

## G. Thin-Layer Voltammetry

Thin-layer voltammetry (TLV) is a very slow technique and this fact is used to great advantage in quantitative studies of electrode reactions with small heterogeneous rate constants. This area has been pioneered and reviewed by Hubbard [30,31]. TLV is also applicable to examination of metal deposition and adsorption. The principal work here was carried out by Schmidt and co-workers [56]. The technique has thus far not been particularly useful for analytical determinations; however, in combination with optical monitoring of products, it has proved helpful in mechanism studies of redox controlled solution chemistry.

The number of coulombs required to oxidize a substance R confined within a thin layer cell is given by Faraday's law,

$$Q = nFVC_R^\circ \tag{3.35}$$

where $C_R^\circ$ is the initial (t = 0) concentration of R and V is the volume of the thin-layer cavity. If the experiment is carried out slowly enough, $C_{R,t}$ will be uniform across the cell due to rapid diffusional mixing in thin solution films (see Sec. II.E). The current at any time will be given simply by

$$i = \frac{dQ}{dt} = nFV \frac{dC_{R,t}}{dt} \tag{3.36}$$

where $nV(dC_{R,t}/dt)$ is the number of equivalents consumed per second. We know that as R is depleted the product, O, is formed. Since the amount of material in the cavity is constant,

$$C_{O,t} = C_R^\circ - C_{R,t} \tag{3.37}$$

For a reversible system, the potential will be defined by the Nernst equation, such that

$$E_t = E^{o\prime} + \frac{RT}{nF} \ln \frac{C_{O,t}}{C_{R,t}} \qquad (3.38)$$

Substituting for $C_{O,t}$, we have

$$E_t = E^{o\prime} + \frac{RT}{nF} \ln \frac{C_R^o - C_{R,t}}{C_{R,t}} \qquad (3.39)$$

It is important to recognize that this equation will hold true regardless of whether R is converted to O in response to a small current, a slowly varying potential, or even via a photooxidation reaction.

For a slow potential ramp excitation

$$\frac{dE_t}{dt} = \nu \qquad (3.40)$$

where $\nu$ is the scan rate in V/s. If we differentiate Equation 3.39 with respect to time, substitute $\nu$ for dE/dt and i/nFV for $dC_R/dt$, and combine the result with the exponential form of 3.39, we have

$$i = -\frac{n^2F^2VC_R^o\nu \exp[(nF/RT)(E - E^{o\prime})]}{RT\{1 + \exp[(nF/RT)(E - E^{o\prime})]\}^2} \qquad (3.41)$$

The current is maximum at $E = E^{o\prime}$, so

$$i_p = -\frac{n^2F^2VC_R^o\nu}{4RT} \qquad (3.42)$$

A plot of i vs. E (Eq. 3.41) defines the ideal thin-layer voltammogram shown in Figure 3.32. The contrast with voltammetry under quasi-infinite conditions should be evident. Note that the peak is symmetrical about $E^{o\prime}$, and that the current drops to zero after the potential passes through $E_p$, and that $i_p$ is *directly* proportional to scan rate. The number of coulombs passed in fully traversing the peak is given by Equation 3.35 since the conversion of R to O is complete.

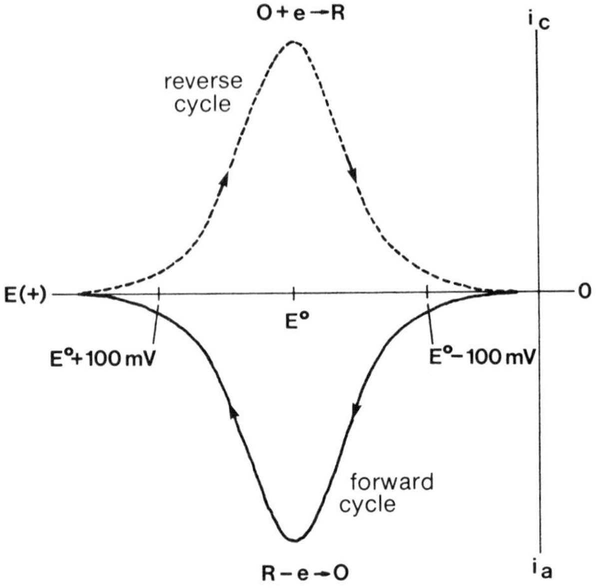


Figure 3.32 Thin-layer voltammetry for a reversible one-electron oxidation.

If we reverse the scan direction, a reduction wave is obtained (dotted line) for the reverse reaction which is an exact mirror image of the oxidation wave. This result only obtains for a fully reversible electrode reaction. When the heterogeneous rate constant is very slow, the forward and reverse waves will lose their symmetry and the peak potentials will separate.

Thin-layer CV is frequently used in conjunction with an optically transparent thin-layer electrode to obtain spectra, E°', and n for redox couples by the "spectropotentiostatic" technique and thin-layer coulometry, as described earlier in this chapter. CV is used initially to locate the redox couple and give an estimate of the E°'. Once the CV is obtained, appropriate potentials can be selected for the spectropotentiostatic experiment and potential-step coulometry. A typical thin-layer cyclic voltammogram for the Schiff base complex Co(sal$_2$en) in nonaqueous solvent is shown in Figure 3.33.

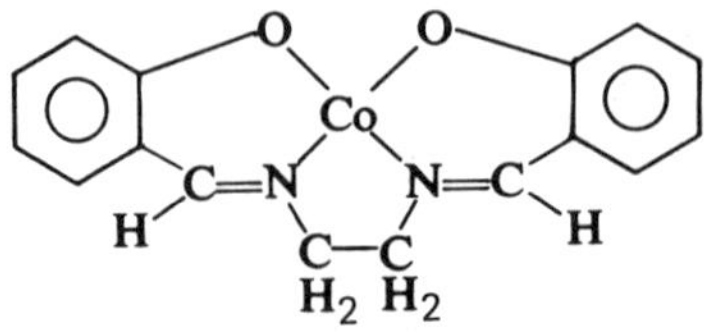

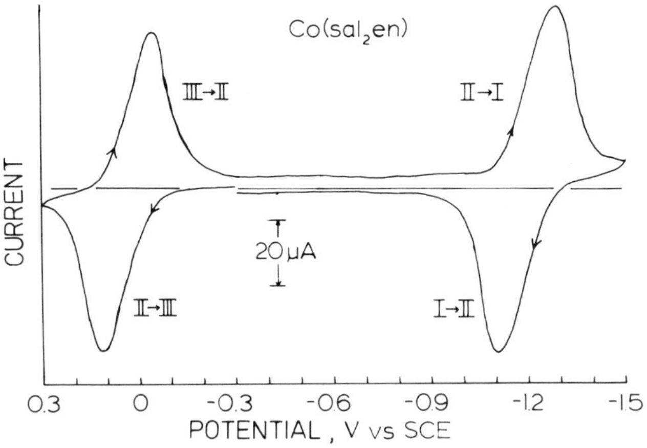


Figure 3.33 Thin-layer cyclic voltammogram of 1.3 mM Co($sal_2en$) in 0.5 M TEAP/DMF. Scan rate 2.0 mV/s, initial potential −0.3 V vs. SCE. (From Ref. 57.)

The waves for the Co(III)/Co(II) and Co(II)/Co(I) couples are well formed. The peak potentials, which should be identical for the oxidation and reduction halves of a couple, are split due to iR drop in the cell.

A unique form of thin-layer voltammetry was devised by Anderson and Reilley [58]. In this case two opposing working electrodes are used and the potential difference across each interface is controlled independently. The electronics necessary to accomplish this are described in Chapter 6. It is thus possible to achieve a cyclic steady-state condition whereby one electrode serves as cathode and the other as anode. Both forms of the redox couple are continuously interconverted. Figure 3.34 illustrates how this condition can be achieved. The cell initially contains only O and the potential of both electrodes is such that only O is stable. As $E_1$ is made more negative, conversion of O to R begins at $W_1$. $E_2$ is kept sufficiently positive to reconvert R back to O at $W_2$. Thus a steady-state current, $i_{ss} \neq f(t)$, is achieved. The flux of O toward $W_1$ will equal the flux of R toward $W_2$, with each molecule moving down its respective concentration profile. The profiles will have different slopes if the diffusion coefficients are different, since the flux is equal to D(dC/dx) for each species. Under limiting conditions both surface concentrations become zero and the current, $i_{ss}^{lim}$, is independent of potential.

Suppose that one species (e.g., R) of a redox couple is subjected to a chemical reaction leading to nonelectroactive products. If this reaction occurs during a limiting steady-state condition (Fig. 3.35),

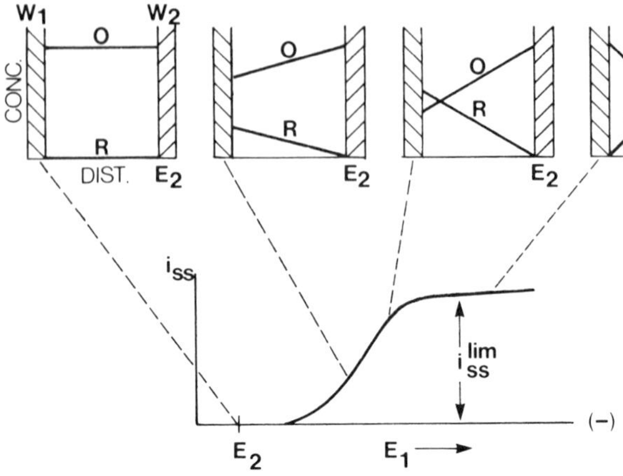


Figure 3.34 Thin-layer twin-electrode voltammetry. (From Ref. 58.)

$i_{ss}^{lim}$ will decay away until both R and O are completely exhausted. The rate constant for the coupled reaction can be calculated from the decay curve [59].

The twin-electrode technique has numerous potential applications both to analysis and mechanism studies [29]; however, because of the difficulty of manufacturing cells, experiments fall well behind the theoretical developments. One recent application utilizes the twin

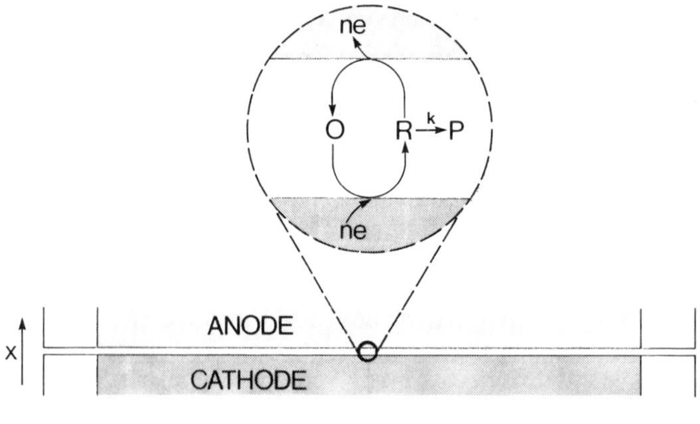


Figure 3.35 Destruction of a steady-state redox cycle by a coupled chemical reaction. [From Ref. 59, reprinted with permission from B. McDuffie, L. B. Anderson, and C. N. Reilley, *Anal. Chem. 38*:883 (1966). Copyright 1966 American Chemical Society.]

electrodes to selectively deposit Zn on one electrode and Cu on the other electrode to avoid the formation of an interfering intermetallic compound of Cu-Zn [60]. In general, though, there is little activity with twin electrode TLV at this time.

The detection limit for TLV has been improved substantially by using differential pulse voltammetry (Chap. 5). For example, detection limits in the $10^{-8}$ M range and below have been demonstrated for metal ions and electroactive drugs in thin-layer cells requiring less than 100 μL of sample [61,62].

## IV. CONTROLLED-POTENTIAL TECHNIQUES IN FLOWING SOLUTION

### A. Introduction to Hydrodynamic Voltammetry

Several electroanalytical techniques involve the hydrodynamic flow of electrolyte over the electrode surface. These methods take advantage of enhanced sensitivity resulting from the enhanced mass transfer of electroactive substance to the electrode that occurs under hydrodynamic conditions.

Some understanding of liquid flow along a solid-solution interface is useful in understanding these techniques. Consider a stationary platinum electrode immersed in a stirred solution. Three regions of solution flow can be identified. (1) *Turbulent flow* comprises the solution bulk. (2) As the electrode surface is approached a transition to *laminar flow* occurs. This is a nonturbulent flow in which adjacent layers slide by each other parallel to the electrode surface. (3) The rate of this laminar flow decreases near the electrode due to frictional forces until a thin layer of *stagnant solution* is present immediately adjacent to the electrode surface. It is convenient, although not entirely correct, to consider this thin layer of stagnant solution as having a discrete thickness $\delta$, called the Nernst diffusion layer.

Figure 3.36 illustrates the Nernst diffusion layer in terms of concentration-distance profiles for a solution containing species O. As pointed out previously, the concentration of redox species in equilibrium at the electrode-solution interface is determined by the Nernst equation. Figure 3.36A illustrates the concentration-distance profile for O under the condition that its surface concentration has not been perturbed. Either the cell is at open circuit or a potential that is sufficiently positive of $E^{o'}_{O,R}$ not to measurably alter the surface concentrations of the O,R couple has been applied.

The profiles in Figure 3.36B represent the situation when a potential is applied that requires equal concentrations of O and R at the electrode surface to satisfy the Nernst equation (i.e., $E = E^{o'}_{O,R}$). To fulfill this requirement, the electrode electrolyzes O to R at the rate required to maintain equal concentrations of O and R at the sur-

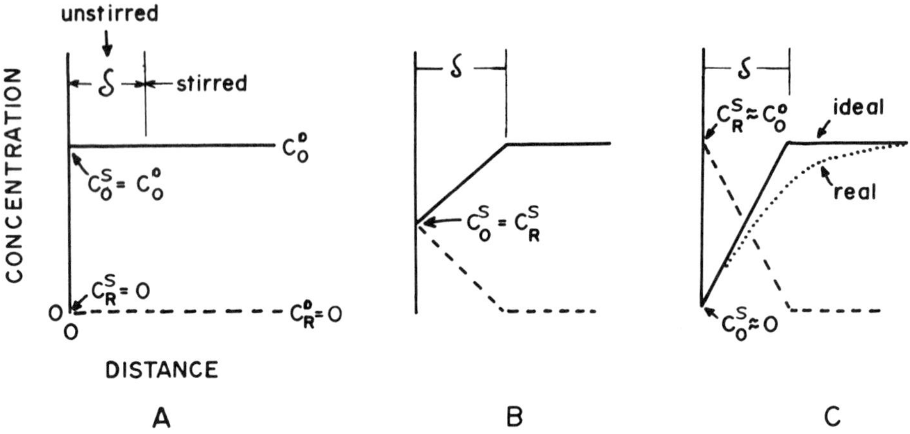


Figure 3.36 Concentration-distance profiles for voltammetry in stirred solution. $D_O = D_R$.

face. If this potential is maintained, a continuous electrolysis of O to R is necessary to maintain surface concentrations because R diffuses away from the interface across the stagnant layer and is then swept away by the laminar flow.

Two regions of this profile can be considered.

1. At distances greater than $\delta$, concentrations are maintained homogeneous by the stirring action. As long as the electrode area is small (microelectrode) relative to the solution volume and the experiment is not prolonged, the bulk concentrations will not be altered appreciably by the electrolytic conversion of O to R at the surface.
2. The removal of O at the electrode surface sets up a concentration gradient across the stagnant solution layer. Species O diffuses across this layer to the electrode surface where it it electrolyzed to R, which then diffuses back across the stagnant layer to the bulk solution.

Thus the electrolysis process is controlled by a combination of (1) mass transfer of O to the edge of the stagnant layer by the laminar flow and (2) subsequent diffusion of O across the stagnant layer under the influence of the concentration gradient caused by electrolysis of O to R at the electrode surface to satisfy the Nernst equation.

Figure 3.36C shows the profiles that result when the applied potential is sufficiently negative that the concentration of O at the electrode surface is effectively zero. In this case, essentially all of O at the electrode surface must be electrolyzed to R in order to satisfy the Nernst equation. Consequently, O is converted to R as rapidly as it can diffuse to the electrode surface. Since this is the limiting condition, application of even more negative potentials causes no measurable change in the profiles.

Although the transition between stagnant and flowing solution is considered to be abrupt in this example, the transition is in reality gradual. Consequently, the profiles will be rounded as shown by the dotted line in Figure 3.36C. However, the hypothetical situation of an abrupt transition is a useful approximation in mathematical treatments as shown below. (See Ref. 6, Chap. 4, for a critique of the Nernst diffusion layer.)

This discussion has focused on a stationary electrode immersed in a stirred or flowing solution. Similar arguments apply to the case of a rotating electrode.

## B. Stirred-Solution Voltammetry

Stirred-solution voltammetry utilizes current-voltage relationships which are obtained at a stationary electrode immersed in a stirred solution. In order to understand this aspect of electrochemistry it is extremely useful to consider a typical current-voltage curve (voltammogram) in terms of the concept of concentration-distance profiles presented in the preceding section. The discussion will consider the potential as the controlled variable rather than current.

The current resulting from an applied potential is determined by the slope of the concentration-distance profile of the reactant at the electrode surface as described by Equation 3.5. This equation can be expressed in terms of the Nernst diffusion layer concept by simply approximating $\partial x$ by $\delta$ and $\partial C$ by $C_O^\circ - C_O^s$ for the case in Figure 3.36.

$$i = nFAD_O \frac{C_O^\circ - C_O^s}{\delta} \tag{3.43}$$

Examination of the profiles in Figure 3.36 show that this is a valid substitution for the slope of the profile at the electrode surface for the ideal case of an abrupt transition at $x = \delta$.

Figure 3.37 illustrates a typical current-voltage curve that would be obtained for a solution containing equal concentrations of O and R. The shape of this curve can be understood by considering the slopes of the concentration-distance profiles that are depicted for several representative potentials. During oxidation the current is determined by the slope of the profile for R, whereas the profile for O is the determining factor during reduction. A limiting current ($i_\ell$) is reached when the surface concentration $C^s$ of the controlling species

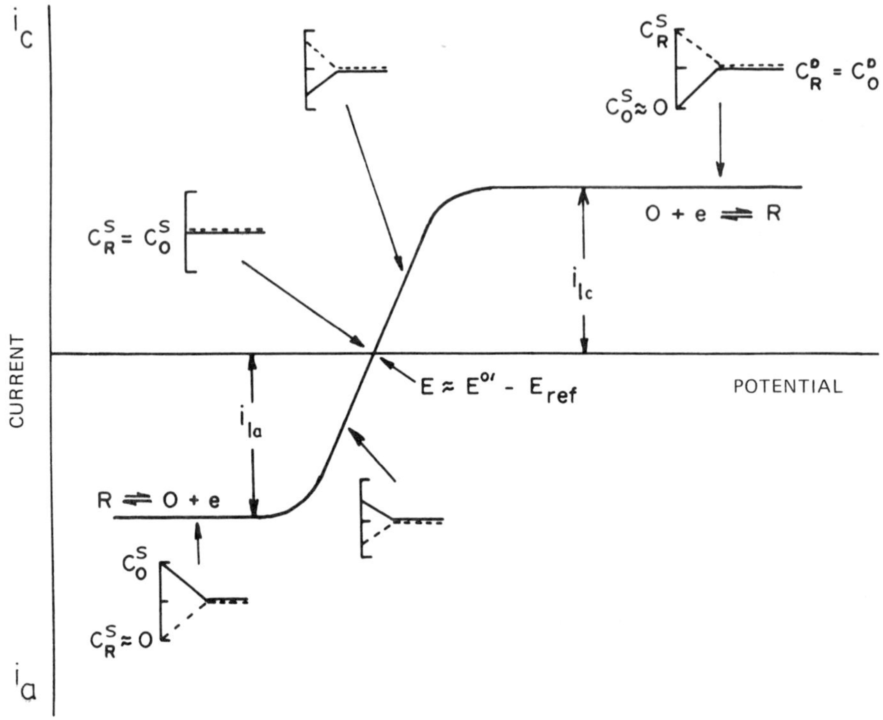


**Figure** 3.37 Voltammogram with representative concentration-distance profiles for a solution containing equal concentrations of O and R. $D_O = D_R$.

becomes effectively zero. Substitution of $C_O^S = 0$ into Equation 3.43 gives the equation for the limiting current.

$$i_\ell = \frac{nFADC_O^\circ}{\delta} \tag{3.44}$$

It is important to recognize that all forms of hydrodynamic voltammetry are *steady-state methods* in that the current at a given potential is independent of both the scan direction and time. In fact, identical voltammograms can be obtained by controlling the current and monitoring the potential response. This is, however, rarely done in practice.

## C. Rotating Disk Voltammetry

Stirred-solution voltammetry per se has been largely superseded by rotating disk voltammetry in which the hydrodynamics are more easily controlled [63-67]. A typical rotating disk electrode is shown in Figure 3.38. The hydrodynamic flow pattern resulting from rapid rotation of the disk moves liquid horizontally out and away from the center of the disk with a consequent upward axial flow to replenish liquid at the surface as shown by the arrows in Figure 3.38. A completely rigorous hydrodynamic treatment has been given by Levich [63,64]. The equation for the limiting current that is derived from this treatment is

$$i_\ell = 0.620 nFAC^\circ D^{2/3} \nu^{-1/6} \omega^{1/2} \tag{3.45}$$

where

$\omega$ = angular velocity of the disk ($\omega = 2\pi N$, N = rps)
$\nu$ = kinematic viscosity of the fluid, $cm^2/s$
$i_\ell$ = limiting current, A
$C^\circ$ = solution concentration, $mol/cm^3$

Good electrode design is necessary for strict adherence to this equation.

The shapes of rotating disk voltammograms are identical to the stirred-solution voltammograms. Thus a rotating disk voltammogram

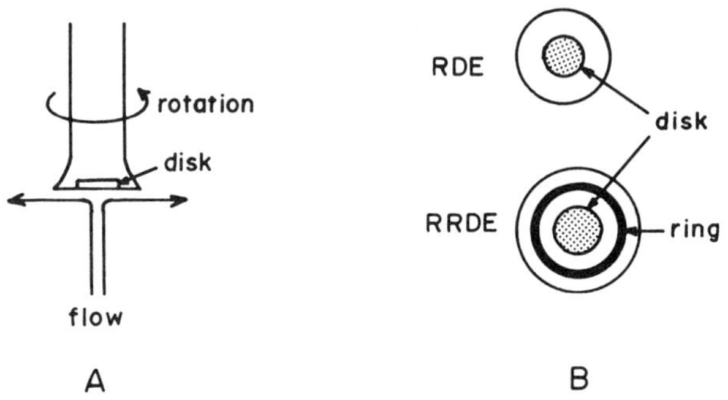


Figure 3.38 (A) Rotating disk electrode with hydrodynamic flow pattern. (B) Bottom view of rotating disk electrode (RDE) and rotating ring-disk electrode (RRDE).

for a solution containing equal concentrations of O and R would be identical in shape to the voltammogram in Figure 3.37.

Considerable information can be obtained about a redox system by rotating disk voltammetry. An initial voltammogram will locate redox couples in solution, giving information about redox potentials and reversibility. Electrode mechanisms which involve coupled homogeneous chemical reactions are sometimes amenable to study by the rotating disk. Generally, mechanisms and rate constants are deduced from variations in the limiting current as a function of the rotation speed of the disk. A classical example of such a study is the ECE mechanism. The "apparent n" values obtained from the limiting current via Equation 3.45 for a series of angular velocities of the disk are shown in Figure 3.39. At low angular velocities, the entire reaction sequence has time to occur as reactant is swept past the disk by laminar flow and the apparent n value is 2. However, as the angular velocity is increased, species B has less time to react before it is swept past the electrode surface. At sufficiently fast rotational rates, all of species B is swept past the surface before any significant reaction has occurred and the apparent n is 1. The magnitude of the rate constant k can be determined from the angular velocity required for the transitional region from n = 1 to n = 2. A variety of other mechanisms are also amenable to study by rotating disk electrodes [65,66].

Since the limiting current is proportional to concentration, an RDE can be used for analytical purposes (but almost never is in actual practice). In summary, the rotating disk electrode is one of the valuable techniques available for the fundamental study of electrode reactions as evidenced by its substantial popularity.

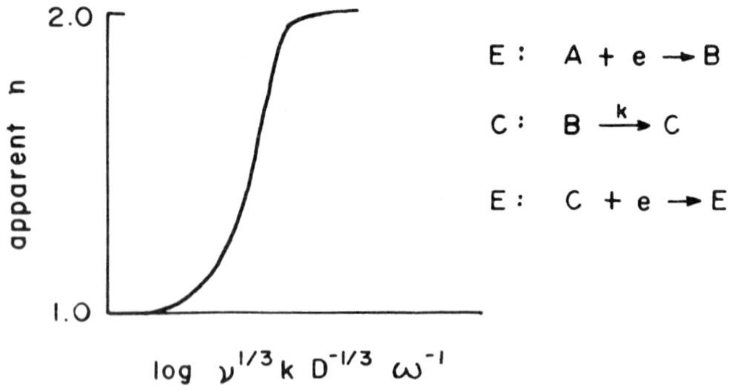


Figure 3.39 Variation in "apparent n" as a function of angular velocity of an RDE for the ECE mechanism.

## D. Rotating Ring-Disk Voltammetry

The rotating ring-disk electrode is an extension of the previously discussed disk electrode in that the disk is encircled by a ring which functions as a second working electrode. The configuration is shown in Figure 3.38B. The idea here is to generate electrochemically a reactive species at the disk and then to electrochemically monitor the species as it is swept past the ring by laminar flow. This technique has proven to be exceedingly useful in the study of reaction mechanisms since electroactive intermediates of coupled homogeneous chemical reactions can be monitored at the ring electrode.

The reduction of oxygen illustrates the use of rotating ring-disk voltammetry [6,68]. Figure 3.40A shows the voltammogram obtained at the disk for the reduction of oxygen. Figure 3.40B shows the resulting ring "voltammogram" when the potential of the ring is maintained at a value for oxidation of $H_2O_2$ and the resulting current plotted as a function of the potential applied to the *disk*. In the potential region in which the disk is generating $H_2O_2$, anodic current is recorded at the ring due to its oxidation of the $H_2O_2$ that is being swept past by laminar flow. When the potential applied to the disk becomes sufficiently negative that $O_2$ is being reduced to $OH^-$ at the disk, the ring current diminishes since $OH^-$ rather than $H_2O_2$ is being swept past. Such experiments can confirm the existence of a particular intermediate, such as $H_2O_2$ in this case, in an electrode reaction.

The rotating ring-disk electrode has been applied to a variety of electrode mechanisms [66,67]. It is one of the most useful techniques for studying electrode reactions.

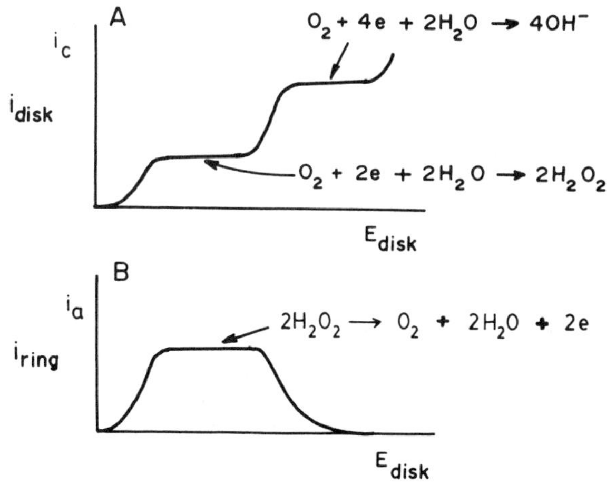


Figure 3.40 Disk (A) and ring (B) currents for reduction of oxygen at the rotating ring-disk electrode. [Adapted from Ref. 6.]

## E. Voltammetry in a Channel

Voltammetry experiments are occasionally undertaken in the form of a tubular or rectangular channel through which the electrolyte solution is pumped at a more or less constant velocity. The electrode may form the channel itself or be embedded in the wall of an inert material which defines the flow pattern. Sometimes the channel is packed with small particles of electrode material in contact with each other. The latter situation is designed to improve the conversion efficiency of the cell. When all of the electroactice molecules are converted during passage through such a porous bed, the efficiency is 100% and the cell is said to be operating coulometrically (see Sec. IV.F).

It should be clear from Faraday's law that the limiting current monitored for a coulometric flow cell will be given by

$$i_\ell = nFC_i u \tag{3.46}$$

where $C_i$ is the concentration of reactant and u is the volume flow rate in $cm^3/s$. Note that the current is independent of the diffusion coefficient for the reactant. In most flow cells the efficiency is far below 100%, and the operation is said to be amperometric. In such cases the current will be given by

$$i_\ell = nF(C_i - C_f)u \tag{3.47}$$

where $C_f$ is the reactant concentration at the outlet side. The problem is not so simple, however, because the conversion efficiency $[(C_f/C_i) \times 100\%]$ is a complicated function of u and the cell geometry. Nevertheless, understanding the essential characteristics of voltammetry under these conditions is perfectly analogous with our earlier discussion for stirred solutions and rotating electrodes. The mass transfer controlled steady-state current is the same function of potential and the voltammetric curves have the same characteristic shape.

Voltammetry experiments are not often performed in flow cells for analytical purposes. One reason for this is the special problem of ohmic potential losses (iR drops) at an electrode in a confined stream. Another reason is the problem of precisely pumping solution at a carefully controlled velocity. In general, rotating electrodes are more easily controlled and do not involve serious plumbing problems. On the other hand, flow cells operated at a fixed potential (i.e., at one point along the steady-state voltammetric curve) are eminently useful for electrosynthesis, chromatographic detection, and automated analysis systems. These features will be described in later chapters.

## F. Controlled-Potential Coulometry

Coulometric techniques involve the determination of the quantity of material electrolyzed from the amount of charge passed through an electrochemical cell during electrolysis. Faraday's law relates the measured charge to the amount of material electrolyzed.

$$N = \frac{Q}{nF} \tag{3.48}$$

where N is the number of moles of substance being electrolyzed, Q is the total charge passed (coulombs), n is the number of electrons transferred per molecule, and F is Faraday's constant (96,485 C/eq).

In controlled-potential coulometry the potential of a working electrode is applied at a value such that complete electrolysis of the desired species occurs. The total charge passed during the exhaustive electrolysis is obtained by integrating (usually electronically) the current. The completion of electrolysis is indicated when the current becomes indistinguishable from the residual current. The charge due to the residual current must be subtracted from the total measured charge in order to obtain the faradaic charge for the electrode reaction of interest. This residual charge is estimated by repeating the experiment on supporting electrolyte and integrating the current over the same time interval. There is no guarantee, however, that the processes responsible for the residual current are completely independent of the presence of sample and/or product.

Selection of the correct potential to apply can be made by examining a voltammogram of the species being electrolyzed. The applied potential should be well on the plateau of limiting current $i_\ell$ for the redox process in order to effect rapid and complete electrolysis.

Maximizing the magnitude of the current during electrolysis minimizes the time required for the complete electrolysis. Equation 3.44 relates the concentration of the electroactive species to the magnitude of the limiting current obtained during a controlled-potential electrolysis at an electrode immersed in a stirred solution. The current is usually increased by increasing the surface area of the electrode, A, and by minimizing the diffusion layer $\delta$ with efficient stirring. Of course, as the electrolysis proceeds, $C^\circ$ decreases toward zero, causing $i_\ell$ to drop. The current at any time t during electrolysis is given by

$$i = i_0 e^{-(AD/V\delta)t} \tag{3.49}$$

where $i_0$ is the limiting current at the onset of electrolysis, V is the solution volume, and the other terms were identified with Equation 3.44.

It is important to consider the parameters above to cause the electrolysis to proceed at a reasonable rate.

Controlled potential coulometry has been applied extensively to precision determinations [69] and to the establishment of n values for electrode reactions. The technique has also been used for the elucidation of electrode reactions [70,71]. It is a very useful technique when used in conjunction with a thin-layer cell where complete electrolysis is rapid (see Sec. II.E).

A variation of controlled potential coulometry is the indirect coulometric titration with simultaneous spectral monitoring (i.e., spectrocoulometry) [72]. This approach has been applied to the spectrocoulometric titration of biological redox components such as cytochrome c, cytochrome c oxidase, and ferredoxin [73-75]. Many of these biological species undergo direct electrolysis at an electrode at a negligible or very slow rate. Thus coulometric titration by means of an EC catalytic regeneration mechanism (see Chap. 2, Sec. V) is a particularly useful technique for these systems. An example is the titration of cytochrome c oxidase with electrogenerated methyl viologen radical cation ($MV^{\dot{+}}$).

E: 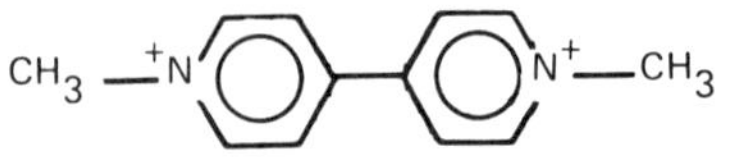 $+ e \rightleftharpoons MV^{\dot{+}}$ (3.50a)

$(MV^{2+})$

C: $4MV^{\dot{+}} + cyt_{ox} \rightleftharpoons 4MV^{2+} + cyt_{red}$ (3.50b)

The titrant $MV^{\dot{+}}$ is generated by controlled potential coulometry. Since $MV^{\dot{+}}$ reacts with the cytochrome, the amount of charge passing through the cell can be related to reduction of cytochrome c oxidase. Since cytochrome c oxidase contains two units of heme a and two atoms of copper, four electrons are required to reduce the oxidized form completely (two electrons to reduce the two atoms of $Fe^{3+}$ in the hemes and two electrons to reduce the two atoms of $Cu^{2+}$ to $Cu^{+}$). Figure 3.41A shows spectra recorded during the titration at an optically transparent electrode. The reduction of the two hemes causes the absorbance increase at 605 nm. The equivalence point (e.p.) of the titration can be determined by monitoring the absorbance changes at 605 nm during the titration as shown in Figure 3.41B. Titrations of this sort have been used mainly to measure n values of biological redox systems, rather than concentrations. The cell shown in Figure 12.13 has been particularly useful for such spectrocoulometric titrations.

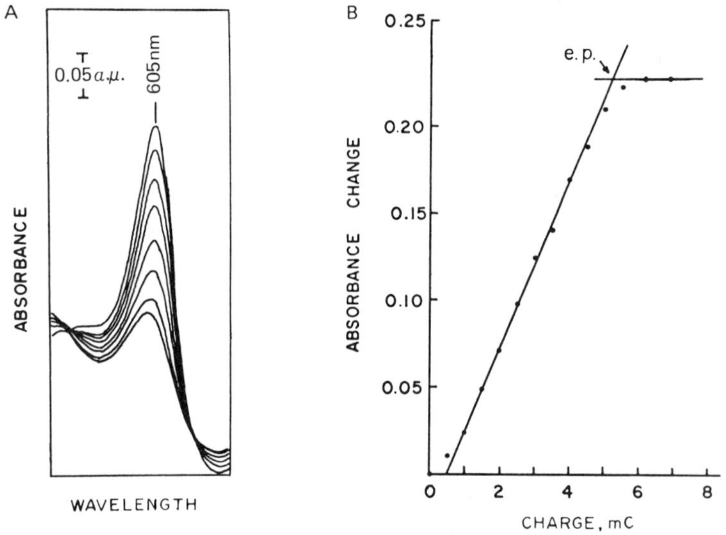


Figure 3.41 Indirect coulometric titration of cytochrome c oxidase by incremental generation of $MV^{+\cdot}$ at $SnO_2$ optically transparent electrode. (A) Spectra recorded during titration. (B) Titration curve; $MV^{+\cdot}$ added in increments of 0.25 mC. [Adapted from W. R. Heineman, T. Kuwana, and C. R. Hartzell, *Biochem. Biophys. Res. Commun.* *49*:1 (1972).]

## G. Preconcentration by Electrodeposition

Preconcentration by electrodeposition is one of the more valuable techniques for electroanalytical trace analysis. In this method the electroactive species to be determined is concentrated by electrodeposition into or onto an electrode placed in a stirred (usually) sample solution. For example, $Pb^{2+}$ can be concentrated by reducing it to $Pb^{\circ}$ at a hanging mercury drop electrode or at a thin mercury film electrode. This is an electrochemical extraction, the $Pb^{2+}$ being extracted from an aqueous solution into the liquid mercury phase as $Pb^{\circ}$. Since the volume of the mercury electrode is usually substantially less than that of the dilute aqueous solution being analyzed, the concentration of $Pb^{\circ}$ in the mercury becomes much greater than the concentration of $Pb^{2+}$ in the water as the electrodeposition proceeds. A concentration enhancement of as great as $10^6$ can take place. After a sufficiently high concentration level is reached in the drop, the stirring is stopped, and the electrodeposited $Pb^{\sigma}$ is electrochemically "stripped" out of the mercury and back into the aqueous solution as $Pb^{2+}$ by a positive potential scan from the deposition potential. The resulting current signal is proportional to the concentration of electrodeposited material, which in turn is proportional to the concentration of electroactive species in solution.

The combination of preconcentration by electrodeposition with stripping by voltammetry is probably the most sensitive electroanalytical method in common use today. Consequently, this popular technique is discussed in more detail in Chapter 19. Preconcentration of material by an electrode reaction has been used in sample preparation for atomic absorption, neutron activation, x-ray fluorescence, microprobe, and several other spectroscopic techniques. This subject is also reviewed in Chapter 19.

## H. Amperometric Titrations

A titration in which the endpoint is determined by the current resulting from a potential applied across two electrodes is an amperometric titration [76].

### Working Electrode-Reference Electrode

In the case in which the potential of a working electrode is controlled relative to a reference electrode, the potential is applied so that a limiting current which is proportional to the concentration of one or more of the reactants or products of the titration is measured. A titration curve is obtained by plotting the limiting current as a function of volume of titrant added. The shape of the titration curve can be predicted from hydrodynamic voltammograms of the solution obtained at various stages of the titration. Figure 3.42A shows voltammograms recorded during the titration of $Cl^-$ with $Ag^+$. The titration reaction is

$$Cl^-(\text{sample}) + Ag^+(\text{titrant}) \rightleftharpoons \underline{AgCl} \tag{3.51}$$

The working electrode is silver metal. At 0% titration, the anodic limiting current is controlled by the mass transport of $Cl^-$ to the electrode for the oxidation of Ag to form AgCl. As the titration proceeds, the decrease in this limiting current reflects the decrease in $Cl^-$ concentration due to addition of $Ag^+$ titrant until the current becomes zero at the equivalence point (100% titration). The appearance of a cathodic limiting current after the equivalence point reflects the reduction of excess $Ag^+$ titrant. Figure 3.42B shows resulting amperometric titration curves for two values of applied potential. Their shapes are determined by the behavior of the limiting current of the voltammogram at the particular potential during the titration.

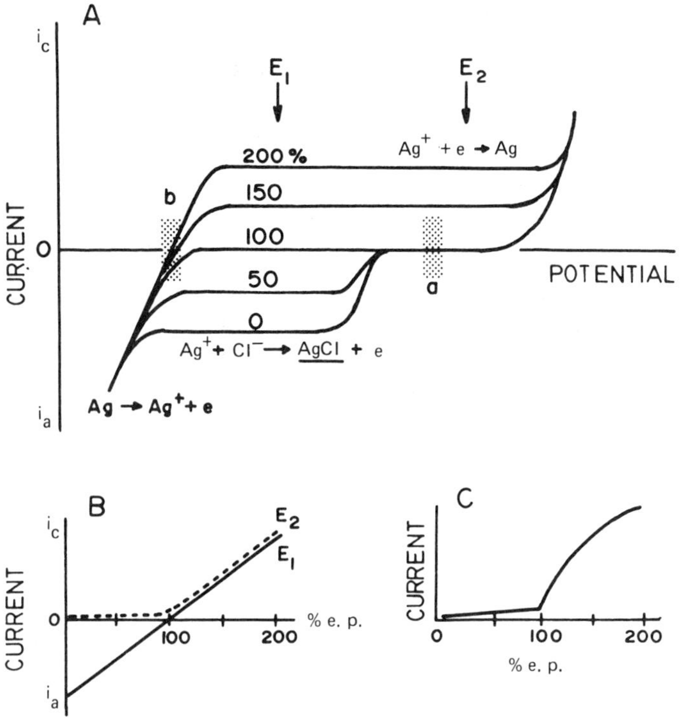


Figure 3.42 Titration of $Cl^-$ with $Ag^+$. ($Cl^- + Ag^+ \rightleftarrows \underline{AgCl}$). (A) Voltammograms at Ag electrode for 0, 50, 100, 150, and 200% of the equivalence point during the titration. (B) Amperometric titration curve for applied potential at $E_1$ and $E_2$. (C) Amperometric titration curve for two Ag electrodes. $\Delta E$ indicated by the shaded areas in (A).

## Two Working Electrodes

A useful variation of the amperometric titration involves measuring the current resulting from a small fixed potential applied across two working electrodes. One electrode functions as an anode and the other as a cathode. Once again, the expected current behavior during a titration can be explained by means of hydrodynamic voltammograms. Position a in Figure 3.42A shows the small potential, $\Delta E$, applied across two Ag electrodes for the example of $Cl^-$ titrated with $Ag^+$. Before the equivalence point is reached, the measured current is only the small residual current for reduction at the more negative electrode and oxidation at the more positive electrode. (Note that for a significantly larger current, the value for $\Delta E$ would have to be increased to a value where $H_2$ would evolve at the cathode and AgCl would deposit on the anode.) However, after the equivalence point is reached, the potential of the two electrodes shifts positively to position b, where the value of $\Delta E$ is sufficiently great to cause reduction of $Ag^+$ at the cathode and oxidation of Ag at the anode. The current is limited by the mass transport of $Ag^+$ to the cathode surface, which increases as additional excess $Ag^+$ is added. The resulting amperometric titration curve is shown in Figure 3.42C.

It is important to realize that the actual potential of the two electrodes is not fixed with respect to any reference electrode. Consequently, in Figure 3.42A the $\Delta E$ increment is free to move on the potential axis. It locates itself at whatever potential is necessary for the current at the anode to balance the current at the cathode. This means that the $\Delta E$ increment will always straddle the potential at which the current voltage curve shows zero current (intercepts the potential axis).

One advantage of the amperometric titration is its ease of automation. A titrator can be signaled to shut off when a specified current level is reached. The advantage of the two working electrode variation is the elimination of a reference electrode, which can be troublesome in nonaqueous solvents.

## REFERENCES

1. M. Petek, T. E. Neal, and R. W. Murray, *Anal. Chem. 43*:1069 (1971).
2. P. Delahay, *New Instrumental Methods in Electrochemistry*, Interscience, New York, 1954.
3. J. E. Davis and N. Winograd, *Anal. Chem. 44*:2152 (1972).
4. I. Shain and K. J. Martin, *J. Phys. Chem. 65*:254 (1961).
5. M. von Stackelberg, M. Pilgram, and V. Toome, *Z. Elektrochem. 57*:342 (1953).
6. R. N. Adams, *Electrochemistry at Solid Electrodes*, Marcel Dekker, New York, 1969.
7. A. J. Bard and L. R. Faulkner, *Electrochemical Methods: Fundamentals and Applications*, Wiley, New York, 1980.
8. G. S. Alberts and I. Shain, *Anal. Chem. 35*:1859 (1963).
9. M. D. Hawley and S. W. Feldberg, *J. Phys. Chem. 70*:3459 (1966).
10. S. Feldberg, *J. Phys. Chem. 73*:1238 (1969).
11. W. M. Schwarz and I. Shain, *J. Phys. Chem. 69*:30 (1965).
12. M. L. Olmstead and R. S. Nicholson, *Anal. Chem. 41*:851 (1969).
13. F. C. Anson, *Anal. Chem. 38*:54 (1966).
14. J. H. Christie, R. A. Osteryoung, and F. C. Anson, *J. Electroanal. Chem. 13*:236 (1967).

15. F. C. Anson and R. A. Osteryoung, *J. Chem. Educ. 60*:293 (1983).
16. B. Case and F. C. Anson, *J. Phys. Chem. 71*:402 (1967).
17. G. W. O'Dom and R. W. Murray, *Anal. Chem. 39*:51 (1967).
18. J. H. Christie, *J. Electroanal. Chem. 13*:79 (1967).
19. T. Kuwana and W. R. Heineman, *Acc. Chem. Res. 9*:241 (1976).
20. W. R. Heineman, *Anal. Chem. 50*:390A (1978).
21. T. Kuwana and N. Winograd, in *Electroanalytical Chemistry*, Vol. 7 (A. J. Bard, ed.), Marcel Dekker, New York, 1974, Chap. 1.
22. W. R. Heineman, F. M. Hawkridge, and H. N. Blount, in *Electroanalytical Chemistry*, Vol. 13 (A. J. Bard, ed.), Marcel Dekker, New York, 1984, Chap. 1.
23. W. R. Heineman, *J. Chem. Educ. 60*:305 (1983).
24. J. W. Strojek, G. A. Gruver, and T. Kuwana, *Anal. Chem. 41*:481 (1969).
25. A. W. B. Aylmer-Kelly, A. Bewick, P. R. Cantrill, and A. M. Tuxford, *Faraday Discuss. Chem. Soc. 56*:96 (1974).
26. J. D. E. McIntyre, in *Advances in Electrochemistry and Electrochemical Engineering*, Vol 9 (R. H. Muller, ed.), Wiley-Interscience, New York, 1973.
27. R. P. Van Duyne, M. R. Suchanski, J. M. Lakovits, A. R. Siedle, K. D. Parks, and T. M. Cotton, *J. Am. Chem. Soc. 101*:2832 (1979).
28. D. L. Jeanmaire and R. P. Van Duyne, *J. Electroanal. Chem. 84*:1 (1977).
29. C. N. Reilley, *Rev. Pure Appl. Chem. 18*:137 (1968).
30. A. T. Hubbard and F. C. Anson, in *Electroanalytical Chemistry*, Vol. 4 (A. J. Bard, ed.), Marcel Dekker, New York, 1970.
31. A. T. Hubbard, *CRC Crit. Rev. Anal. Chem. 3*:201 (1973).
32. D. M. Oglesby, S. H. Omang, and C. N. Reilley, *Anal. Chem. 37*:1312 (1965).
33. T. B. Jarbawi and W. R. Heineman, *J. Electroanal. Chem. 132*:323 (1982).
34. P. T. Kissinger and C. N. Reilley, *Anal. Chem. 42*:12 (1970).
35. J. L. Owens and G. Dryhurst, *J. Electroanal. Chem. 80*:171 (1977).
36. T. P. DeAngelis and W. R. Heineman, *J. Chem. Educ. 53*:594 (1976).
37. R. W. Hurst, W. R. Heineman, and E. Deutsch, *Inorg. Chem. 20*:3298 (1981).
38. J. T. Maloy, *J. Chem. Educ. 60*:285 (1983).
39. R. S. Nicholson and I. Shain, *Anal. Chem. 36*:706 (1964).
40. M. H. Hulbert and I. Shain, *Anal. Chem. 42*:162 (1970).
41. D. H. Evans, *Acc. Chem. Res. 10*:313 (1977).
42. W. R. Heineman and P. T. Kissinger, *Am. Labor. 14*(11):29 (1982).

43. D. H. Evans, K. M. O' onnell, R. A. Peterson, and M. J. Kelly, *J. Chem. Educ. 60*:290 (1983).
44. P. T. Kissinger and W. R. Heineman, *J. Chem. Educ. 60*:702 (1983).
45. R. S. Nicholson, *Anal. Chem. 37*:1351 (1965).
46. S. P. Perone and T. R. Mueller, *Anal. Chem. 37*:2 (1965).
47. S. P. Perone and C. V. Evins, *Anal. Chem. 37*:1061 (1965).
48. S. P. Perone and C. V. Evins, *Anal. Chem. 37*:1643 (1965).
49. I. M. Kolthoff and J. J. Lingane, *Polarography*, 2nd ed., Wiley-Interscience, New York, 1952.
50. L. Meites, *Polarographic Techniques*, 2nd ed., Wiley-Interscience, New York, 1965.
51. J. Heyrovsky and J. Kuta, *Principles of Polarography*, Academic Press, New York, 1966.
52. A. M. Bond, *Modern Polarographic Methods in Analytical Chemistry*, Marcel Dekker, New York, 1980.
53. C. K. Mann, in *Electroanalytical Chemistry*, Vol. 3 (A. J. Bard, ed.), Marcel Dekker, New York, 1969.
54. C. N. Reilley and W. Stumm, in *Progress in Polarography*, Vol. 1 (P. Zuman and I. M. Kolthoff, eds.), Wiley-Interscience, New York, 1962.
55. S. A. Borman, *Anal. Chem. 54*:698A (1982).
56. E. Schmidt and H. R. Gygax, *J. Electroanal. Chem. 12*:300 (1966).
57. D. F. Rohrbach, E. Deutsch, and W. R. Heineman, in *Characterization of Solutes in Nonaqueous Solvents* (G. Mamantov, ed.), Plenum Press, New York, 1978.
58. L. B. Anderson and C. N. Reilley, *J. Electroanal. Chem. 10*: 295, 538 (1965).
59. B. McDuffie, L. B. Anderson, and C. N. Reilley, *Anal. Chem. 38*:883 (1966).
60. D. A. Roston, E. E. Brooks, and W. R. Heineman, *Anal. Chem. 51*:1728 (1979).
61. T. P. DeAngelis, R. E. Bond, E. E. Brooks, and W. R. Heineman, *Anal. Chem. 49*:1792 (1977).
62. T. B. Jarbawi, W. R. Heineman, and G. J. Patriarche, *Anal. Chim. Acta 126*:57 (1981).
63. V. G. Levich, *Acta Physicochim. URSS 17*:257 (1942); *19*:133 (1944).
64. V. G. Levich, *Zh. Fiz. Khim. 18*:355 (1944); *22*:575 (1948); *22*:711 (1948).
65. F. Opekar and P. Beran, *J. Electroanal. Chem. 69*:1 (1976).
66. S. Bruckenstein and B. Miller, *Acc. Chem. Res. 10*:54 (1977).
67. W. J. Albery and M. L. Hitchman, *Ring-Disc Electrodes*, Oxford University Press, London, 1971.
68. A. Frumkin, L. Nekrasov, B. Levich, and Ju. Ivanov, *J. Electroanal. Chem. 1*:84 (1959).

69. J. E. Harrar, in *Electroanalytical Chemistry*, Vol. 8 (A. J. Bard, ed.), Marcel Dekker, New York, 1975.
70. A. J. Bard and K. S. V. Santhanam, in *Electroanalytical Chemistry*, Vol. 4 (A. J. Bard, ed.), Marcel Dekker, New York, 1970.
71. J. Janata and H. B. Mark, Jr., in *Electroanalytical Chemistry*, Vol. 3 (A. J. Bard, ed.), Marcel Dekker, New York, 1969.
72. R. Szentrimay, P. Yeh, and T. Kuwana, in *Electrochemical Studies of Biological Systems* (D. T. Sawyer, ed.), ACS Symposium Series No. 38, American Chemical Society, Washington, D.C., 1977, Chap. 9.
73. F. M. Hawkridge and T. Kuwana, *Anal. Chem. 45*:1021 (1973).
74. J. L. Anderson, T. Kuwana, and C. R. Hartzell, *Biochemistry 15*:3847 (1976).
75. L. H. Rickard, H. L. Landrum, and F. M. Hawkridge, *Bioelectrochem. Bioenerg.* 5:686 (1978).
76. R. W. Murray and C. N. Reilley, in *Treatise on Analytical Chemistry*, Part I, Vol. 4 (I. M. Kolthoff and P. J. Elving, eds.), Wiley-Interscience, New York, 1963, pp. 2187-2191.

# 4
# Large-Amplitude Controlled-Current Techniques

WILLIAM R. HEINEMAN / Department of Chemistry, University of Cincinnati, Cincinnati, Ohio

PETER T. KISSINGER / Department of Chemistry, Purdue University, and Bioanalytical Systems, Inc., West Lafayette, Indiana



## I. CONTROLLED-CURRENT TECHNIQUES IN STATIONARY SOLUTION

### A. Current-Step Excitation

In electrochemical techniques based on current excitation, a current signal is impressed on a quiescent electrochemical cell, and a response signal, usually the potential of the working electrode, is measured as a function of time. The most commonly used excitation signal is the application of a constant-current-step function to the cell as shown in Figure 4.1.

As was the case for techniques based on potential excitation, current-excitation methods are best understood by studying the time-dependent concentration changes in solution caused by the excitation signal applied to the electrode. Concentration-distance profiles for the case of species O being reduced to R by a current-step excitation

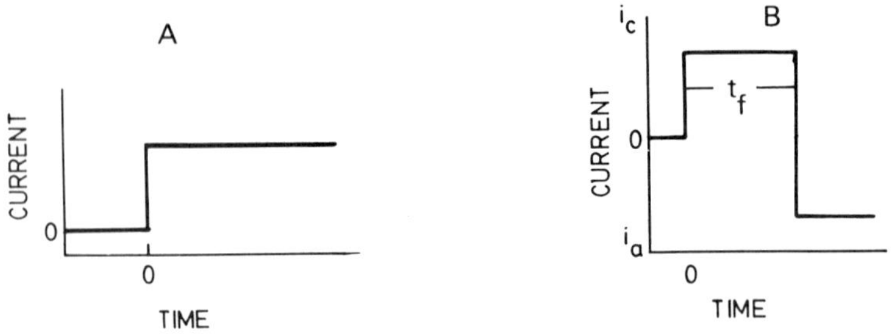


Figure 4.1 General excitation signals for controlled-current techniques. (A) Current step. (B) Double current step.

excitation signal (application of constant current to the cell) are shown in Figure 4.2. Consider first the profiles in Figure 4.2A for the reactant, O. An important concept is the relationship between the applied current and the slope of the profile at the electrode surface as expressed by

$$-\frac{i}{nFA} = -D_O\left(\frac{\partial C_O}{\partial x}\right)_{x=0} = J \tag{4.1}$$

which has been described previously (Eq. 3.5). As the equation shows, the slope of the profile at the electrode surface is controlled by the magnitude of the applied current i. In other words, the flux J of reactant to the electrode surface is the exact amount necessary to support the applied current. Note that while $(\partial C_O/\partial x)_{x=0}$ remains constant, the profiles in Figure 4.1A drop as solution near the electrode is depleted of O by the constant-current electrolysis. When the profile intersects the origin, the surface concentration of O is negligibly small. At this time, referred to as the transition time $\tau$, the rate of reduction of O becomes independent of the magnitude of i and is now controlled by the rate of diffusion of O to the electrode. The slope of the profile at the electrode surface now falls off in a manner analogous to the profiles in Figure 3.1. Since the flux of O to the electrode is no longer sufficiently great to support the applied current, some other reducible species in solution must begin to react so that all of the impressed current can pass through the cell. Now the total flux is defined as

$$-\frac{i}{nFA} = -\left[D_O\left(\frac{\partial C_O}{\partial x}\right)_{x=0} + D_{O'}\left(\frac{\partial C_{O'}}{\partial x}\right)_{x=0}\right] = J \tag{4.2}$$

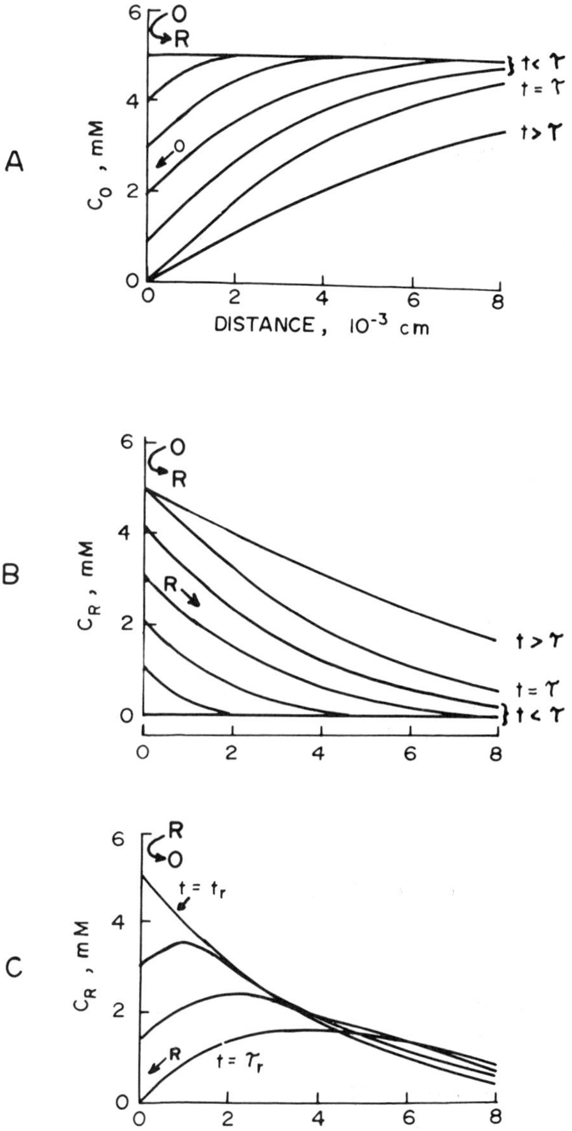


Figure 4.2 Concentration-distance profiles during reduction of O to R at a planar electrode by current-step excitation. Quiescent solution. $D_O = D_R = 10^{-5}$ cm$^2$/s, $C_O^o = 5$ mM, $C_R^o = 0$ mM, $i = 10^{-2}$ A/cm$^2$. (A) C-x profiles for O. Curves for 0, 0.07, 0.3, 0.7, 1.2, and 1.8 s. (B) C-x profiles for R. Curves for 0, 0.07, 0.3, 0.7, 1.2, and 1.8 s. (C) C-x profiles for R with current reversed at $t = \tau$. Curves for 0, 0.007, 0.3, and 0.6 s after current reversal. (Adapted from Ref. 1.)

where O' is a more difficult to reduce species, which only reacts after $\tau$. Since the flux of O is diffusion controlled after $\tau$ is reached, the value of $(\partial C_O/\partial x)_{x=0}$ will decrease with time as described by the Cottrell equation presented earlier (Eq. 3.6). This means that $(\partial C_{O'}/\partial x)_{x=0}$ must increase by a proportionate amount in order to supply reactant (O and O') to the electrode surface at the rate demanded by i.

The concentration profile of product R is determined by the flux of R away from the electrode surface. The flux of R is controlled by the flux of O to the electrode, so that

$$D_R \left(\frac{\partial C_R}{\partial x}\right)_{x=0} = -D_O \left(\frac{\partial C_O}{\partial x}\right)_{x=0} \tag{4.3}$$

If $D_O = D_R$, the concentration-distance profiles of O and R will have slopes of the same magnitude but opposite sign as shown in Figure 4.2B. At $\tau$ all of the reactant O at the electrode surface is converted to product R.

The effect of reversing the polarity of the current excitation signal on the concentration-distance profiles is shown in Figure 4.2C. The original reactant O is regenerated at a rate now determined by the flux of R back to the electrode. When the surface concentration of R becomes zero, a reverse transition time designated $\tau_r$ occurs. At this time another species must undergo reduction to enable the cell to accept the applied current.

## B. Chronopotentiometry

Chronopotentiometry is the electrochemical technique in which a current step is impressed across an electrochemical cell containing unstirred solution [1-5]. The resulting potential response of the working electrode is measured as a function of time (thus, chronopotentiometry), giving a chronopotentiogram as shown in Figure 4.3.

The shape of the potential-time response is determined by the concentration changes of O and R at the electrode surface during electrolysis. The potential is related to $C_O^S/C_R^S$ via the Nernst equation for a reversible system. The initial potential before current application is simply the rest potential or open-circuit potential ($E_{oc}$) of the solution, which reflects the initial $C_O^o/C_R^o$ in solution. At the instant of current application, this ratio becomes finite and the potential changes to a value consistent with the Nernst equation. Early in the experiment the potential changes dramatically from the rest potential to a value that is in the "range" of $E_{O,R}^{o'}$ vs. the reference electrode.

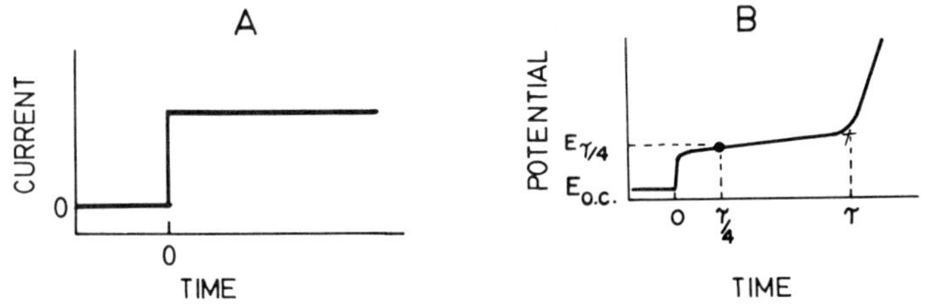


Figure 4.3 Chronopotentiometry. (A) Current excitation signal. (B) Potential response signal (chronopotentiogram).

The subsequent gradual change of potential during continued electrolysis reflects the conversion of O to R and the consequent decrease of $C_O^s/C_R^s$ at the electrode surface as shown in Figures 4.2 A and B. At the transition time $\tau$ the surface concentration of O approaches zero, causing the potential to shift rapidly to a more negative value so that electrolysis of another electroactive component can occur to support the constant current. If a second reducible species is present in solution, a second chronopotentiometric wave will occur. If not, the potential will shift to the reduction of solvent or supporting electrolyte. Oxidation of solvent, supporting electrolyte, or the electrode itself (e.g., Hg) occurs when an anodic current is applied.

For a reversible couple, the surface concentrations of the electroactive species for a simple reduction can be substituted into the Nernst equation to give an expression describing the potential-time response curve. At x = 0, $C_O^s/C_R^s$ can be related to time by $(\tau^{1/2} - t^{1/2})/t^{1/2}$, which gives

$$E = E_{\tau/4} - \frac{RT}{nF} \ln \frac{t^{1/2}}{\tau^{1/2} - t^{1/2}} \tag{4.4}$$

where RT/nF have their usual significance, t is time (s), and $\tau$ is the transition time (s). The value $E_{\tau/4}$ is analogous to the voltammetric $E_{1/2}$ and is defined by

$$E_{\tau/4} = E^{\circ\prime} - \frac{RT}{nF} \ln \frac{D_O^{1/2}}{D_R^{1/2}} \tag{4.5}$$

where D is the diffusion coefficient. $E_{\tau/4}$ corresponds to the potential at which the surface concentrations of the redox couple species are equal (i.e., $C_O^s = C_R^s$).

Under conditions of linear diffusion with no kinetic complications, the relationship between the applied current and the transition time is given by the *Sand equation*,

$$i\tau^{1/2} = \frac{nFA\pi^{1/2}D^{1/2}C^\circ}{2} \tag{4.6}$$

where i is the applied current (A), $\tau$ is the transition time (s), A is the electrode area ($cm^2$), D is the diffusion coefficient of electroactive species ($cm^2/s$), $C^\circ$ is the bulk concentration ($mol/cm^3$), n is the number of electrons transferred per ion/molecule, and F is Faraday's constant (96,485 C/eq).

The product $i\tau^{1/2}$ is a significant diagnostic parameter in chronopotentiometry. It is apparent from the Sand equation that the quantity $i\tau^{1/2}$ is a constant for a given concentration of electroactive species. The application of a larger i results in a smaller $\tau$ due to the more rapid electrolysis. The magnitude of $i\tau^{1/2}$ can be used analytically for concentration measurements down to about $10^{-4}$-$10^{-5}$ M. However, chronopotentiometry is rarely used for analysis, the voltammetric techniques being more popular because of their lower detection limits and the difficulty in measuring $\tau$, vide infra.

A very important feature of chronopotentiometry is the characteristic variation in $i\tau^{1/2}$ that occurs for certain electrode reactions when $\tau$ is varied over a wide range by varying i. The behavior of $i\tau^{1/2}$ can be effectively used to diagnose certain mechanisms of electrode reactions. Diagnostic curves for $i\tau^{1/2}$ vs. i are shown in Figure 4.4 for several mechanistic situations. A constant value of $i\tau^{1/2}$ over a wide range of $\tau$ (Fig. 4.4A) is characteristic of an uncomplicated, diffusion-controlled electrode reaction with no kinetic or adsorption phenomena at a planar electrode.

An electron-transfer reaction that is coupled to a *preceding chemical reaction* is characterized by a decrease in $i\tau^{1/2}$ as i increases (Fig. 4.4B). At very large i (short $\tau$), $i\tau^{1/2}$ is determined by the equilibrium concentration of O, since $\tau$ is reached before X can generate O by chemical reaction. As $\tau$ is increased by decreasing i, $i\tau^{1/2}$ begins to increase when the rate constant k is sufficiently large to generate additional O by the homogeneous chemical reaction. Further decrease in i causes further increase in $i\tau^{1/2}$ as more time is allowed for X to generate O. Extrapolation of the $i\tau^{1/2}$ increase to zero current gives an $i\tau^{1/2}$ intercept that is related to the sum of the concentrations of X and O, since sufficient time is allowed for all of X to convert to O as i approaches zero ($\tau \to \infty$). Note that the steepness of the $i\tau^{1/2}$ decrease with increasing i decreases as k increases. The magnitude of k can be calculated from this slope [1]. Curves of this type would be obtained for the reduction of $Cd(NTA)^-$, which is the example of a CE mechanism described in Chapter 2.

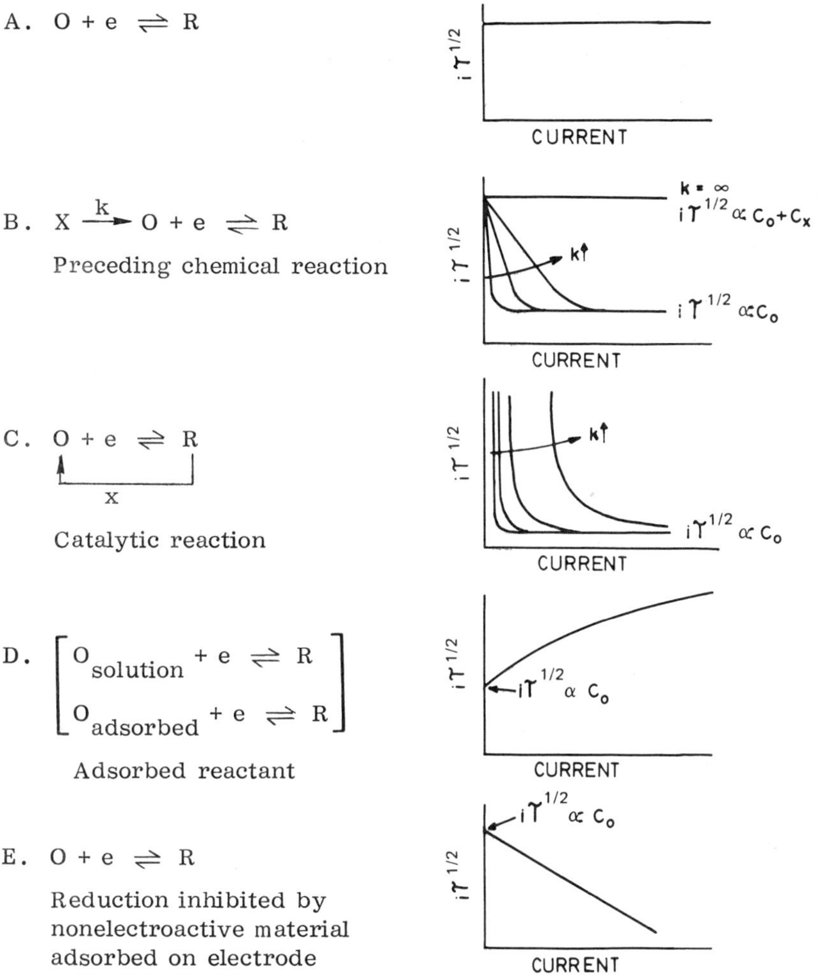


Figure 4.4 Chronopotentiometric $i\tau^{1/2}$ variations for a range of impressed currents as a diagnostic for electrode mechanisms.

In the case of the *catalytic chemical reaction*, $i\tau^{1/2}$ increases to infinity as i decreases, since no transition time is ever reached if R is given enough time to recycle O (Fig. 4.4C). The constant $i\tau^{1/2}$ that is reached with a sufficiently large current is proportional to the equilibrium concentration of O in solution and reflects the fact that the electrolysis time before $\tau$ is too short for significant reaction to occur. An example of this mechanism is the reduction of Ti(IV) in the presence of hydroxylamine [6].

In the case in which the electroactive species O is adsorbed on the electrode, $i\tau^{1/2}$ increases with increasing i (Fig. 4.4D). The Sand equation considers only those molecules of O that have reached the surface by diffusion. As $\tau$ decreases, the charge contribution to the reduction of $O_{ads}$ consumes an increasingly large fraction of the total current, causing $i\tau^{1/2}$ to increase. A typical example of this behavior is the reduction of Alizarin Red S at mercury [7].

Inhibition of electrode reactions caused by material adsorbed on the electrode surface causes $i\tau^{1/2}$ to decrease with increasing i (Fig. 4.4E). As i increases, less time is allowed for the electroactive species to penetrate the adsorbed layer. This behavior is analogous to the inhibition of the polarogram shown in Figure 2.20. The reduction of a copper tartrate complex at a mercury electrode in the presence of brucine, which adsorbs on the electrode, exhibits this typical behavior [8].

Chronopotentiometry suffers from the difficulty in experimentally determining the transition time and the fact that the charging current is not easily resolved from the faradaic response, since the electrode potential is varying throughout the experiment. Various extrapolation procedures have been devised, but usually without rigorous theoretical justification [1-5].

## C. Current-Reversal Chronopotentiometry

If the polarity of the applied current in an ordinary chronopotentiometry experiment is reversed during the recording of the chronopotentiogram, the product R of the initial electrochemical reaction may now undergo the reverse reaction to give a current-reversal chronopotentiogram as shown in Figure 4.5 [1–5]. A reverse transition time $\tau_r$ will result when the concentration of R becomes zero at the electrode

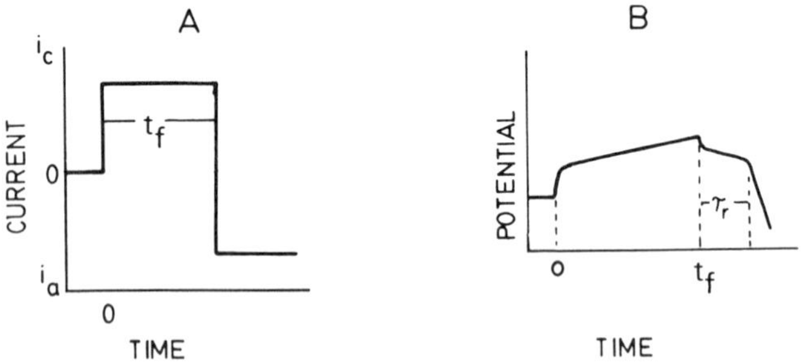


Figure 4.5 Current-reversal chronopotentiometry. (A) Current excitation signal. (B) Potential response.

surface (see Fig. 4.2.C). Such reverse potential-time curves can be treated quantitatively for reversible and irreversible couples.

For a system devoid of coupled chemical reactions and adsorption phenomena, the relationship between the forward electrolysis time $t_f$ and the reverse transition time $\tau_r$ is

$$\tau_r = \frac{1}{3} t_f \tag{4.7}$$

where $t_f$ is the time of current reversal. The current must be reversed before or at the forward transition time $\tau$ for this relationship to be valid. Two-thirds of the electrogenerated R is "lost" by diffusion away from the electrode.

Electrode mechanisms with following chemical reactions (EC mechanism; see Chap. 2) are particularly amenable to study by current-reversal chronopotentiometry. A good example is the oxidation of p-aminophenol (PAP), which undergoes the following reaction sequence [9]:

E: 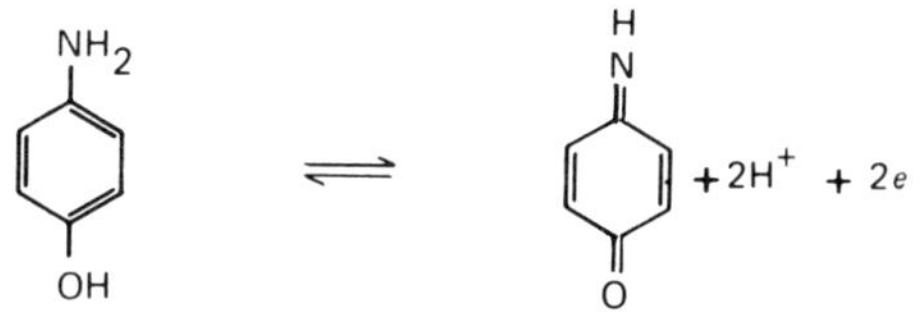


(4.8)

PAP (p-aminophenol) QI (quinoneimine)

C: 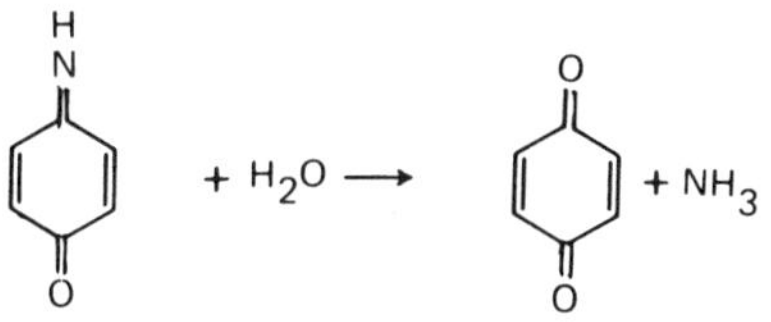


(4.9)

QI Q (quinone)

A current-reversal chronopotentiogram is shown in Figure 4.6. The application of an anodic current for the period $t_f$ generates QI at the electrode by oxidation of PAP. When the current is reversed, QI is reduced back to PAP during the reverse transition labeled $\tau_r$. The chemical reaction of QI to form Q causes $\tau_r/t_f$ to be less than 1/3 by

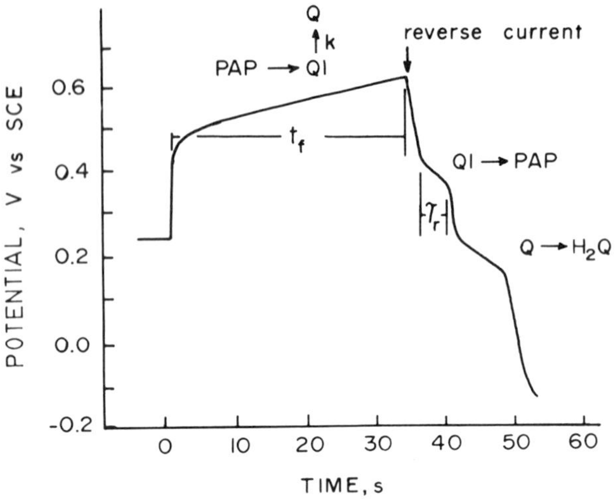


Figure 4.6 Current-reversal chronopotentiogram for the oxidation of 1 mM PAP in 0.1 M $H_2SO_4$. Pt electrode, i = 100 $\mu A/cm^2$. [From Ref. 9, adapted with permission from A. C. Testa and W. H. Reinmuth, *Anal. Chem.* *32*:1512 (1960). Copyright 1960 American Chemical Society.]

an amount that depends on the magnitude of the rate constant k. Figure 4.7 shows a calculated working curve which relates $\tau_r/t_f$ with $kt_f$. The rate constant can be calculated from this curve after the experimental value of $\tau_r$ is determined for a given $t_f$. The second reverse transition time in Figure 4.6 is due to reduction of the quinone, Q, to hydroquinone, $H_2Q$, at the electrode (see Eq. 3.32).

Current-reversal chronopotentiometry is also useful for detecting adsorption of the product generated during the forward electrolysis time. Complete adsorption of the product on the electrode surface causes $\tau_r$ to equal $t_f$ since no product is lost by diffusion from the electrode. This approach has been used to determine the amount of adsorbed material formed during the reduction of riboflavin and the oxidation of iodide [10].

## D. Cyclic Chronopotentiometry

In cyclic chronopotentiometry, the current is continually reversed at potentials corresponding to the forward and reverse transition times as shown in Figure 4.8 [11]. For a system with no kinetic or adsorption complications, the forward transition time $\tau$ decreases while $\tau_r$ increases until finally $\tau = \tau_r$ in the limit, at steady state. (Because the convergence rate is slow, equality of $\tau$ and $\tau_r$ is not commonly achieved

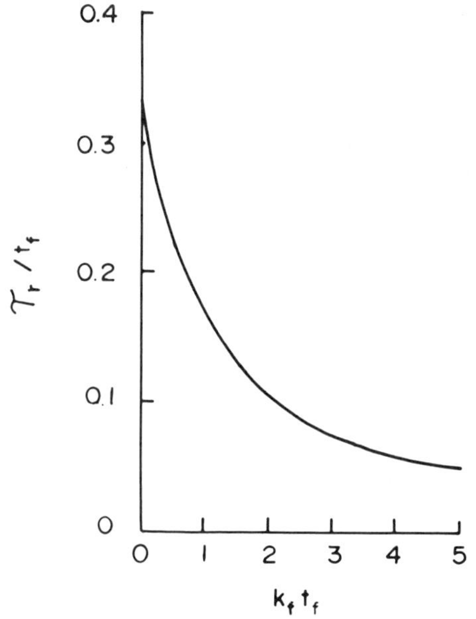


Figure 4.7 Working curve for the calculation of the rate constant $k_f$ for a following chemical reaction from current-reversal chronopotentiometry. [From Ref. 9, adapted with permission from A. C. Testa and W. H. Reinmuth, *Anal. Chem.* *32*:1512 (1960). Copyright 1960 American Chemical Society.]

experimentally before the onset of natural convection and nonplanar diffusion effects.) Quantitative treatments for single component systems, multicomponent systems, stepwise reactions, and systems involving chemical kinetics have been derived. The technique has not been used extensively.

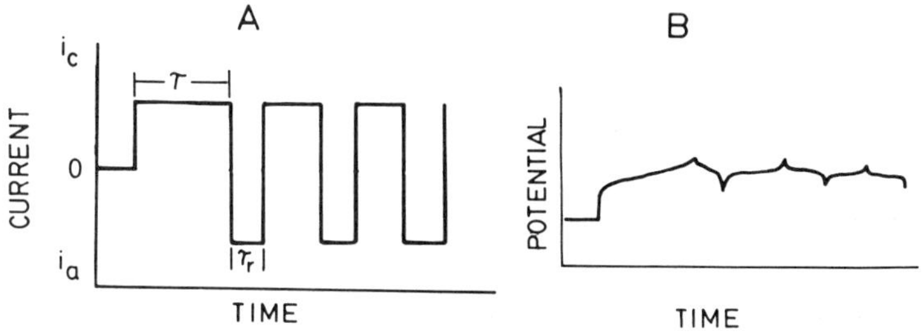


Figure 4.8 Cyclic chronopotentiometry. (A) Current excitation signal. (B) Potential response signal.

## E. Derivative Chronopotentiometry

In derivative chronopotentiometry the potential response signal of a normal chronopotentiometry experiment is electronically differentiated, and this rate of change of potential with time, dE/dt, is recorded as a function of time as shown in Figure 4.9 [12]. The minimum in a derivative chronopotentiogram is quantitatively related to the transition time. Thus for a reversible couple,

$$\left(\frac{dE}{dt}\right)_{min} = -\frac{27RT}{8nF\tau} \qquad (4.10)$$

Calculating $\tau$ this way is somewhat advantageous since the rate of charging of the double layer is minimal (not zero) at $[dE/dt]_{min}$. Derivative chronopotentiometry has not been widely utilized, probably because the improvement over the ordinary E vs. t response is insufficient to justify the more complex instrumentation required. A more serious criticism is that the minimum dE/dt occurs at a time that is dependent on the kinetics of the heterogeneous reaction. For fast (reversible) reactions $t_{min} = 4\tau/9$ and for very slow (irreversible) reactions $t_{min} = \tau/4$. Unfortunately, a great many reactions fall somewhere in between and the relation of $t_{min}$ to $\tau$ is not likely to be clearly defined.

## F. Programmed Current Chronopotentiometry

In general, chronopotentiometry consists of the application of a programmed current function to an electrode, the potential of which is measured as a function of time [13]. Thus the current program may be any function that is electronically feasible. Examples of potentially useful current functions in addition to those described in the preceding sections are ramp current chronopotentiometry for the elucidation of

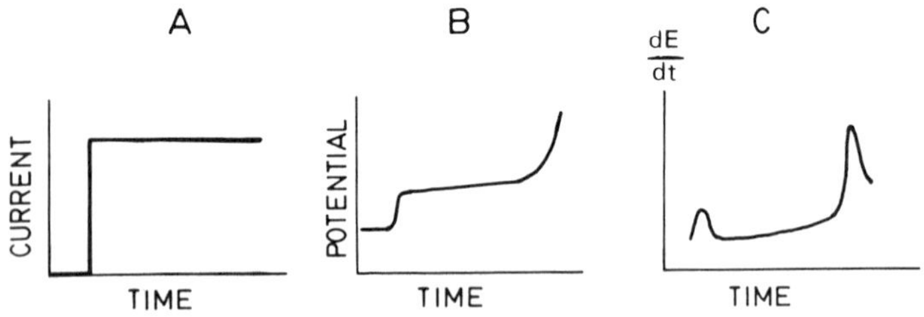


Figure 4.9 Derivative chronopotentiometry. (A) Current excitation signal. (B) Potential response signal. (C) Monitored response: dE/dt.

adsorption mechanisms and square-root-of-time current chronopotentiometry for which concentration is directly proportional to the transition time:

$$\tau = \frac{2nFD^{1/2}C^{\circ}}{\beta\pi^{1/2}} \tag{4.11}$$

where $\beta$ is a proportionality factor. In current-cessation chronopotentiometry, a constant current is applied, then abruptly discontinued, and the resulting potential decay is observed.

## G. Controlled-Current Chronoabsorptometry

Controlled-current chronoabsorptometry involves the simultaneous optical monitoring of the product or other redox component in the electrode mechanism during a chronopotentiometry experiment [14]. Although this technique has been demonstrated with $SnO_2$ optically transparent electrodes, it has generally received little use, since the resistance effects in thin-film electrodes can give unequal current densities across the electrode face. This results in distorted potential-time and absorbance-time responses. Consequently, the more prevalent spectroelectrochemical methods utilize potential as the excitation signal rather than current.

## H. Current-Step Excitation in Thin Layers of Solution

Although chronopotentiometric experiments were included among the earliest thin-layer studies, there has been very little work in recent years even though the theory is well developed [15]. Difficulties with iR losses in thin-layer cells compound the problem of defining the transition time and greatly limit the current and thus the time scale which can be studied. At this point in time there is little to recommend in thin-layer chronopotentiometry. Even the instrumental advantage of constant current for stationary solution coulometric experiments is no longer clear since excellent current integrators are available and controlled potential coulometry can be carried out with ease. The constant-current excitation is most valuable for the hydrodynamic experiments referred to in the following section. Thin-layer flow cells operated at constant current are useful for precise generation of micromolar titrants.

## II. CONTROLLED-CURRENT TECHNIQUES IN FLOWING SOLUTION

Controlled-current electrolysis in flowing solution has been extremely useful for analytical purposes. The prevalent techniques are constant-current coulometry and coulometric titrations, which are discussed in Chapter 20.

## REFERENCES


1. P. Delahay, *New Instrumental Methods in Electrochemistry*, Interscience, New York, 1954.
2. D. G. Davis, in *Electroanalytical Chemistry*, Vol. 1 (A. J. Bard, ed.), Marcel Dekker, New York, 1966.
3. M. Paunovic, *J. Electroanal. Chem. 14*:447 (1967).
4. P. Delahay, in *Treatise in Analytical Chemistry*, Part I, Vol. 4 (I. M. Kolthoff and P. J. Elving, eds.), Wiley-Interscience, New York, 1963, p. 2233.
5. R. N. Adams, *Electrochemistry at Solid Electrodes*, Marcel Dekker, New York, 1969, Chap. 6.
6. P. Delahay, C. C. Mattax, and T. Berzins, *J. Am. Chem. Soc. 76*:5319 (1954).
7. H. A. Laitinen and L. M. Chambers, *Anal. Chem. 36*:5 (1964).
8. M. Kodama and R. W. Murray, *Anal. Chem. 37*:1638 (1965).
9. A. C. Testa and W. H. Reinmuth, *Anal. Chem. 32*:1512 (1960).
10. H. B. Herman, S. V. Tatwawadi, and A. J. Bard, *Anal. Chem. 35*:2210 (1963).
11. H. B. Herman and A. J. Bard, *Anal. Chem. 35*:1121 (1963); *36*:510, 971 (1964).
12. D. G. Peters and S. L. Burden, *Anal. Chem. 38*:530 (1966).
13. R. W. Murray and C. N. Reilley, *J. Electroanal. Chem. 3*:64, 182 (1962).
14. W. N. Hansen, T. Kuwana, and R. A. Osteryoung, *Anal. Chem. 38*:1810 (1966).
15. A. T. Hubbard and F. C. Anson, in *Electroanalytical Chemistry*, Vol. 4 (A. J. Bard, ed.), Marcel Dekker, New York, 1970, pp. 129-214.

# 5

# Small-Amplitude and Related Controlled-Potential Techniques

PETER T. KISSINGER / Department of Chemistry, Purdue University, and Bioanalytical Systems, Inc., West Lafayette, Indiana



## I. INTRODUCTION

### A. What Is a Small-Amplitude Technique?

All of the techniques described in the preceding two chapters involve an initial condition in which the concentration of the reacting solute is uniform as a function of distance from the electrode surface. The experiments begin by application of a relatively large amplitude potential or current excitation signal. The initial flat reactant concentration profile is thereby dramatically perturbed. The perturbation of the electrode-solution interphase in these techniques is sufficient to require use of the entire exponential rate expression introduced in

Chapter 2 (see Sec. IV). As was described in Chapter 2, when the applied signal is of relatively small magnitude, the fundamental exponential rate expression can be linearized. For example, in a controlled-potential experiment if the excitation signal has an amplitude of less than about 10 mV/n, then the linearized version of the rate equation is adequately precise for essentially all experimental purposes.

The definition of a "small-amplitude" technique is obviously not precise because depending on the experimental objective, one might get away with a somewhat larger excitation signal. For example, when small-amplitude techniques are used for analytical purposes, the definition can be stretched somewhat in comparison to when they are used for careful studies of the physical chemistry of the electrode-solution interphase region.

While the small-amplitude perturbation can be made from an initial condition identical to that used for the large-amplitude techniques, it is usual to combine the two experiments together. One can, for example, add a small-amplitude perturbation on top of a large-amplitude signal to provide more complete information about the chemical system. Under normal circumstances the large-amplitude signal is applied for a longer period of time and the small amplitude is added momentarily. Looking at this another way, we use the large-amplitude signal to establish a new surface concentration for the oxidized and reduced forms of a redox couple and then move away from this condition slightly by rapid application of a small potential or current change. The large-amplitude technique therefore sets the initial condition for the small-amplitude technique.

Let us consider the very simple situation following a step change in applied potential. Refer back to Chapter 3 (see Sec. II.A and II.B) for an introduction to chronoamperometry. Consider Figure 5.1A in which a potential-step excitation is illustrated. A large-amplitude step is applied at $E_{1/2}$ for a period of time and then an additional small-amplitude perturbation is added. The total current response signal would look as is given in Figure 5.1B. A current-time transient would result as the surface concentration of reactant and product achieve a value consistent with the Nernst equation ($C_O{}^s = C_R{}^s$). When the additional small step is applied later in the experiment, the current will again respond as a transient piggybacked on the original current-time signal. This small-amplitude response can be considered to have been initiated at a condition defined by the concentration profile as it existed just prior to application of the small step. The concentration profile from the initial experiment (and therefore the reactant flux) is "wiggled" a small amount near the surface. Sometimes the surface concentration which exists prior to a small-amplitude perturbation is referred to as the "dc surface concentration." The time scale of the small-amplitude experiment is so much shorter than that for the large-amplitude signal that the latter for practical purposes is not changing during the course of a given small-amplitude perturbation.

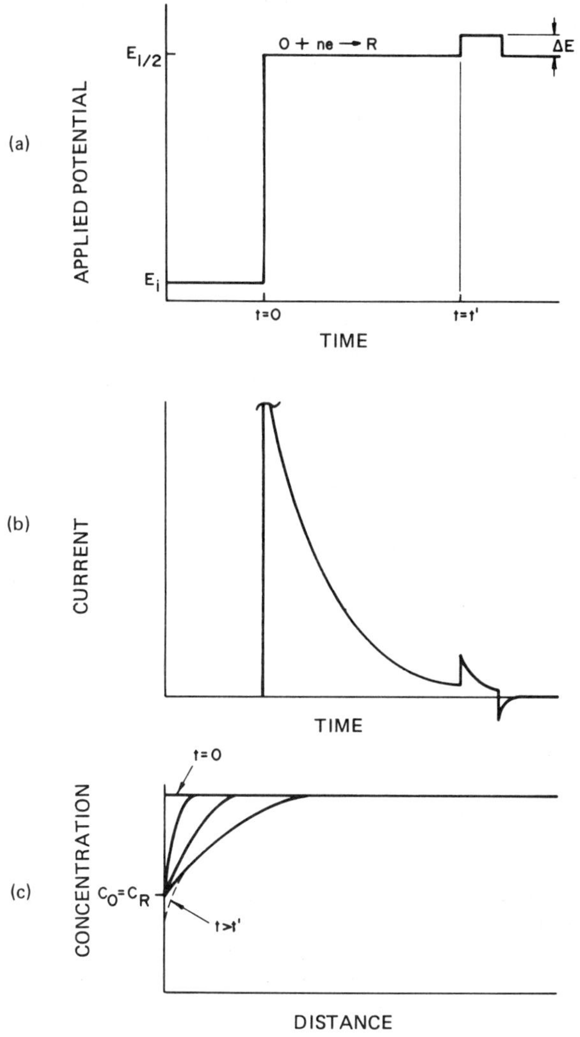


**Figure 5.1** A small amplitude pulse applied following a large amplitude potential step. (a) excitation signal. (b) current response. (c) concentration-distance response.

## B. Small-Amplitude Controlled-Potential Voltammetry

Over the years an extraordinarily large number of techniques have been explored based on the addition of a small-amplitude perturbation to a linear potential sweep. All sorts of different waveforms have been used for the small-amplitude part of the excitation signal. Simple sine waves, square waves, triangular waves, and sawtooth waves are among those which have been considered. In the early forms of these techniques the small-amplitude (so-called ac part) was simply summed with a slowly moving ramp (so-called dc part). The excitation signal would look something like that given in Figure 5.2a. While this type of excitation signal can be applied to all types of electrodes in both stationary and moving solutions, let us consider the situation in which the dc current response would be sigmoidal in shape as illustrated in Figure 5.2b.

The variation in small-amplitude current around the dc response will depend very much on the relation between the dc excitation and the half-wave potential. This is illustrated in Figure 5.2c. Prior to initiation of significant dc reduction current, the small-amplitude excitation signal does not result in any significant faradaic process. Well beyond the half-wave potential (on the limiting current plateau), the small-amplitude perturbation will again contribute very little due to the fact that we are already at a point where the surface concentration of the reactant is zero for all practical purposes. On the rising part of the "dc" voltammogram we anticipate a significant variation in the current due to the small-amplitude perturbation. The amplitude of *this* current response would be maximum at the point where the dc surface concentration is the greatest function of the applied potential (i.e., at the half-wave point). As a result, when we plot the small-amplitude current response as a function of the dc applied potential we achieve a peak-shaped curve with a maximum value at $E_{1/2}$ (at which point the small-amplitude potential is varying symmetrically around the $E_{1/2}$ value). Whenever a time-variant signal is impressed across the interface a capacitive current will result regardless of the presence of a redox active substance. As with large-amplitude techniques, separation of faradaic from nonfaradaic events is a major experimental goal.

Small-amplitude techniques can be conveniently divided into those which operate in the frequency domain and those which operate in the time domain. In principle, the information available from these different approaches is identical, but there are significant experimental factors which may make one or the other more desirable in a particular experiment. The frequency-domain techniques normally use analog instrumentation such as various impedance bridges and tuned amplifiers which can be made phase selective (such as the very popular "lock-in amplifier"). Phase selectivity provides discrimination against double-layer charging current.

The time-domain techniques in their present form are often instrumented by use of digital computers. They require the sampling of current signals at various times in response to the small-amplitude perturbation. Often the sampled currents are mathematically processed to minimize the inevitable charging current. By far the most popular technique of this type is so-called differential pulse voltammetry, but there are others on the horizon.

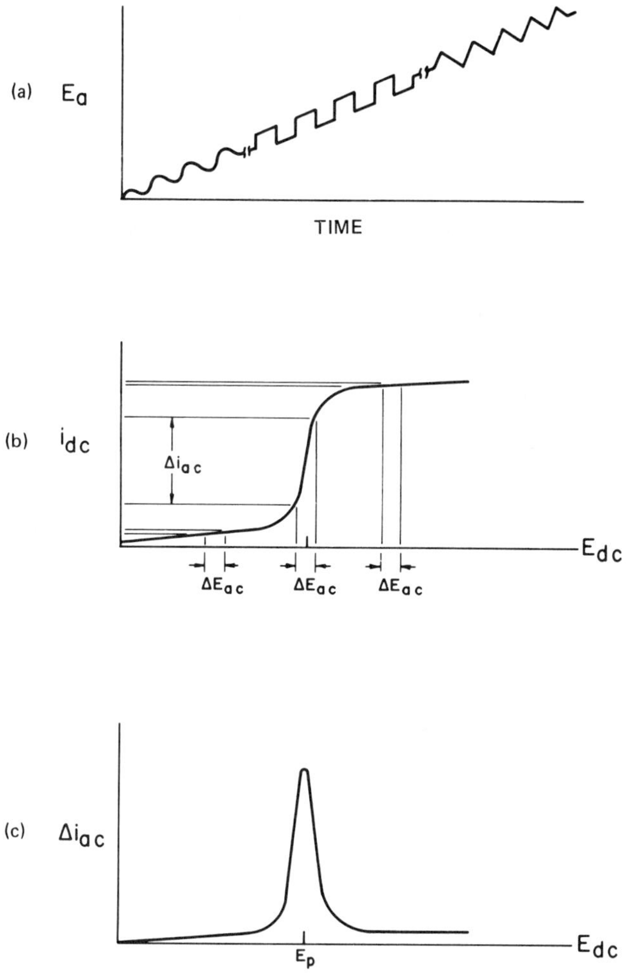


**Figure 5.2** Small amplitude voltammetric techniques. (a) Various small amplitude waveforms are imposed on a "dc" ramp (normally only one waveform is used in a given experiment). (b) The sigmoidal shaped dc response is typical of dc polarography and hydrodynamic voltammetry. The greatest amplitudes for the small amplitude current ($\Delta i_{ac}$) are achieved on the rising part of the dc current, where the small amplitude voltage signal causes the greatest change in the surface concentrations. (c) Small amplitude current response vs. applied dc potential.

Small-amplitude voltammetry techniques do provide some advantage for both fundamental and analytical studies. Physical electrochemists like the mathematical simplification afforded by linearization. In addition, the small-amplitude frequency-domain experiments can be carried out in a steady-state mode which improves the precision of the measurements. Analytical chemists frequently like these methods because the format of the response is convenient and frequently both the detection limits and selectivity are improved somewhat in comparison to the corresponding "dc" technique. We will now introduce in a very qualitative fashion the most popular frequency-domain and time-domain variants. The literature on this subject is extremely vast and only a few leading reviews and recent key papers will be cited.

## II. FREQUENCY-DOMAIN TECHNIQUES

### A. Impedance Measurements

Any system that responds to an electrical excitation signal can, in principle, be represented by an equivalent circuit constructed from various resistors, capacitors, and inductors. An ideal equivalent circuit would behave exactly in the same manner as the system it represents. The extremely complex nature of the electrode-solution interface makes it very difficult to arrive at an equivalent circuit which fits under all circumstances. Nevertheless, very useful models are available. A very simple representation of a three-electrode cell is given in Figure 5.3. The reference electrode probe is represented as a battery in series with a resistance (there is also, of course, a capacitance associated with the reference electrode). The working and auxiliary electrodes are considered as a faradaic impedance (a resistance in simplest terms) which parallels a capacitor. The representation of an electrochemical reaction as an impedance and the ionic structure of the interface as a capacitor are introduced in Chapter 2 (Secs. III and IV). There is also a resistance associated with the solution contacting the auxiliary, reference, and working electrodes. The solution resistance between the reference probe and the working electrode is referred to as the uncompensated resistance and this is emphasized in Figure 5.3c, the equivalent circuit for the working electrode alone. For our purposes it can be assumed that resistance and capacitance associated with the external leads can be ignored. The magnitude of the uncompensated resistance $R_u$ can be made very small in many circumstances, but can be an important factor in experiments which are either extremely fast or involve the use of nonaqueous solvents. The impact of $R_u$ on the electronics and cell design for electrochemical experiments will be considered further in Chapters 6, 7, and 12.

The three-element equivalent circuit for the working electrode is clearly rather simplistic. The solution does not behave in the same

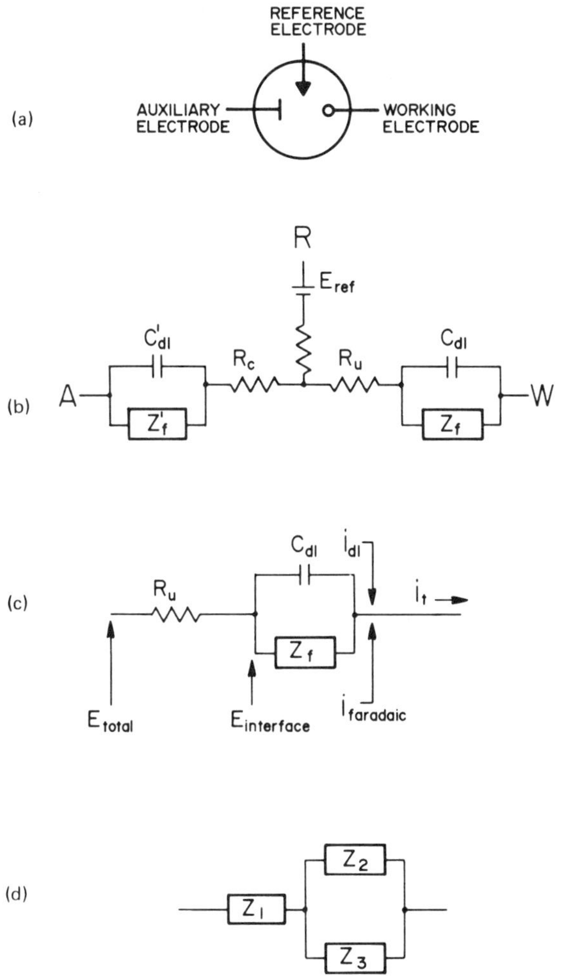


**Figure 5.3** Schematic representation of an electrochemical cell. (a) three electrodes. (b) equivalent circuit for three electrode cell. (c) equivalent circuit for the working-electrode interphase. (d) a "solution" impedance in series with two parallel "surface" impedances.

manner as a simple resistor. The interface does not fit the properties of a simple physical capacitor (the capacitance here does depend on applied potential, so this element is, in actual fact, nonlinear). In spite of these limitations, the model works adequately well under many circumstances. The variation in the double-layer capacitance as a function of potential is one of the more serious problems. Often, however, conditions are used here where the variation in the faradaic impedance with potential is much greater than that of the capacitance and a constant value for the latter can be assumed.

The three-element circuit can be represented as a generalized network as indicated in Figure 5.3d. The total impedance for the interface $Z_t$ consists of a simple parallel network in series with $Z_1$. The total impedance may then be represented as follows:

$$Z_t = Z_1 + \frac{Z_2 Z_3}{Z_2 + Z_3}$$

The experimentalist and theoretician alike have as a goal to relate the values of these impedances to phenomena taking place at the electrode in particular experimental circumstances. In classical electrochemical experiments, measurements of interfacial impedances were carried out using an impedance bridge such as that illustrated in Figure 5.4. A variable-frequency ac potential is applied across the bridge and components in the bridge circuit are varied until a balance is achieved. One advantage of this kind of experiment is the fact that it is carried out under steady-state circumstances. The dc potential of the two electrode cell is varied by changing the ratio of the oxidized and reduced forms of the couple under study. By the combination of a change in frequency with a change in composition, a great deal of very precise data can be obtained about a given redox process. This method of measuring impedances, while still among the very best available with regard to precision, is very awkward and time consuming in practice. Analytical electrochemists have very little interest in such

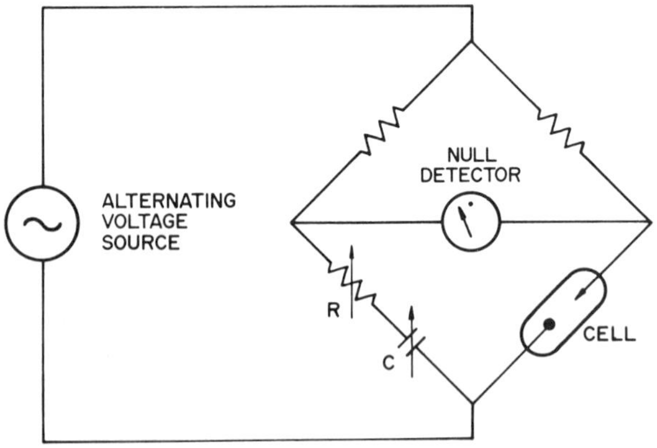


Figure 5.4 Classical impedance bridge for a two-electrode cell.

work and therefore we will not consider it in detail here (see Refs. 1 and 4).

## B. Sinusoidal Alternating-Current Voltammetry

Alternating-current voltammetry requires an excitation signal based on adding a small-amplitude sinusoidal potential modulation to a ramp of very low slope. In a typical experiment, the ramp has a slope of 5 mV/s or less and the sinusoid is a few millivolts in amplitude at several hundred cycles per second. For all practical purposes, the ramp is at 0 Hz and does not vary significantly during any given cycle. As described in an earlier section, the ramp sets the "dc surface concentration" and the sinusoid "wiggles" it more or less, depending on the relationship between the ramp potential and the $E^{o\prime}$ of the electrochemical reaction. Near $E^{o\prime}$ the sinusoid will have its maximum impact on surface concentrations (and therefore on the current). Because the frequency of the sinusoid is so much faster than the ramp, it is a simple matter to isolate the small-amplitude (ac) current response from the large-amplitude (dc) current response. The ac signal is converted to a voltage and then full-wave rectified and plotted on a conventional XY recorder versus the ramp potential.

In this frequency-domain technique, it is a simple matter to vary the frequency to test the kinetics of the electrochemical events. This is equivalent to varying the scan rate in cyclic voltammetry, the pulse width in square-wave voltammetry (described below), or the time at which data is acquired during a chronoamperometry experiment. If one of these time-domain techniques must be carried out at high speed to "see" a given phenomenon, relatively high frequencies will be necessary for the corresponding frequency-domain techniques.

In time-domain "pulse" techniques, the influence of double-layer charging current is minimized by recording data at longer times. In ac techniques, there is a different phase relationship between the excitation and response signals for capacitive and faradaic currents. Considering a fast (reversible) electron transfer reaction, the faradaic current is phase shifted 45° and the capacitive current 90° relative to the applied sinusoidal potential. It is therefore possible to separate these two contributions to the total response with a phase-sensitive detection system. In the real world, it is not so simple. Slow kinetics, adsorption, coupled chemical reactions, and solution resistance can all contribute to the relative phase angle between the faradaic and capacitive currents. These influences can usually be dealt with effectively but the required mathematics are beyond the scope of this text.

There are excellent theoretical [1-4] and practical [1-5] treatments of ac voltammetry and polarography. Advanced topics include the use of higher harmonics and even Fourier transform techniques to study the response to multiple frequencies applied simultaneously.

The term "tensammetry" applies to the use of ac voltammetry to study adsorption/desorption phenomenon which influence double-layer capacitance. Tensammetric waves can be used to study surfactants which are not undergoing electron-transfer reactions.

We intentionally give these techniques short shrift because they currently are not nearly as attractive as they once were in competition with time-domain methods. The latter are now much easier to carry out than in the past due to the availability of low-cost computers which make precisely timed excitation-response measurements routine.

### C. Cyclic Alternating-Current Voltammetry

Alternating-current voltammetry can be made more useful by cycling the direction of the potential ramp as in cyclic voltammetry (Chap. 3). As long as a large difference is maintained between the time scale of the CV scan (long) and the period of the sinusoid (short), it is possible to obtain the full information from both techniques in a single experiment. The CV part is most useful for diagnosing mechanisms and the ac part is ideal for quantitative work. For a rapid electron-transfer reaction where both forms of a redox couple are stable during the time scale of the dc scan, an identical forward and reverse ac response is obtained. This should make sense because the surface concentrations of O and R at any given potential are controlled only by the Nernst equation and not by the direction scanned to arrive at this potential. For slower electron-transfer kinetics, the relative positions of the peak maxima shift. For rapid electron transfers with a chemical reaction consuming the product, the peak height will be lower on the reverse scan. This relatively new technique is given more detailed coverage in Bard and Faulkner [1] and in a series of pioneering publications by Bond and co-workers [5-7].

## III. TIME-DOMAIN TECHNIQUES

### A. Differential Pulse Voltammetry

By the 1960s dc polarography was for all practical purposes a dying technique in American laboratories. The method suffered greatly from its slow speed, awkward apparatus, unusual data presentation, and poor detection limits. The latter two problems were largely corrected by the advent of differential pulse polarography (DPP), a technique that first became readily available in the late 1960s due to the efforts of Princeton Applied Research Corp. This technique overcame certain deficiencies of the dropping mercury electrode by (1) recording data only at the end of the drop life, (2) maximizing flux by using a potential pulse, (3) discriminating against charging current by using a timed sample-and-hold, and (4) by subtracting out the largest portion of the dc background current. The results were a substantial improvement in the minimum detectable concentration and a data presentation (Fig. 5.2c) more acceptable to practicing analysts. Later, the same approach was found to be useful for solid electrodes and mercury film electrodes and thus differential pulse voltammetry (DPV) became one of the most popular electroanalytical techniques in the 1970s. Stripping voltammetry based on DPV is currently a favorite (Chap. 19).

All of these things are accomplished by using the waveform shown in Figure 5.5, which may be conveniently thought of as the sum of a staircase (step height $\Delta E_s$) with a pulse train. The staircase potential establishes the dc surface concentrations (per Sec. I). Just prior to application of the small-amplitude pulse a current sample is taken at $s_1$, and a second sample, $s_2$, is taken at the end of the pulse. In the usual case, the sample is registered over a 16.7-ms window to nullify the impact of 60-Hz line interference (16.7-ms period). The pulse width $t_p$ has traditionally been 50 ms (a fixed parameter in some commercial instruments).

When a dropping mercury electrode is used, the pulse is applied just prior to the end of the drop life ($t_d = T$ in Fig. 5.5). When a new drop is created the experiment is identical to dc polarography until the pulse is applied. The current prior to $s_1$ is not recorded (eliminating most of the charging current associated with the increase in electrode surface area). The pulse is applied only during the last 1-10% of drop life (typically 0.5-5.0 s). The charging current associated with the chronoamperometry response to the pulse is likewise not recorded. The second sample $s_2$ ideally consists only of faradaic current. A plot of $(i_{s_2} - i_{s_1})$ vs. applied potential (usually the potential at $s_1$ is plotted rather than the potential at $s_2$).

For analytical purposes, the pulse heights used in DPV frequently exceed the limits imposed by the small-amplitude restriction defined in Chapter 2 and the introduction to this chapter (e.g., 100-mV pulse amplitudes are quite common). Large amplitudes provide an increase in response which must be balanced against the loss in resolution and the increase in charging current, which ultimately limits the minimum

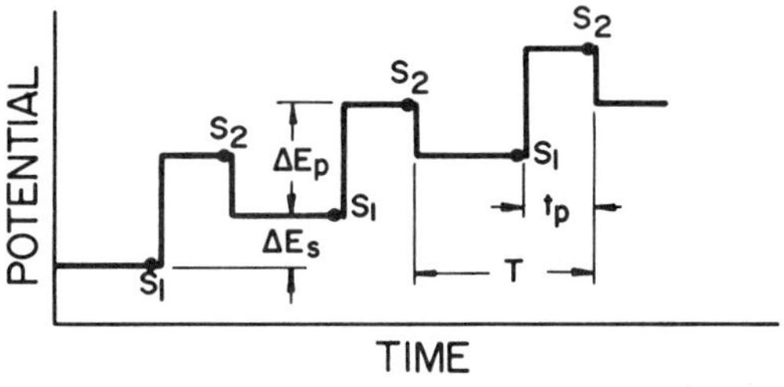


Figure 5.5 Excitation signal for differential pulse voltammetry.

detectable concentration. For most cases, a $\Delta$Ep of 50-100 mV is optimum. At greater amplitudes the relative increase in response falls off and performance is degraded.

A primary problem with DPV, which greatly limits its usefulness, is the very detrimental impact of slow electron-transfer kinetics (irreversibility). This lowers the response (peak current vs. sample concentration) and therefore raises the minimum detectable concentration. It also increases the peak width, making resolution a problem in some cases. Finally, slow electron-transfer kinetics (high activation energy) tends to make an electrochemical process very dependent on other substances present in the solution. When the samples under study are especially "dirty" (e.g., environmental samples), variation in the matrix can alter the DPV response for the analyte(s) of interest and make calibration of the procedure less reliable.

There are several good sources of more detailed information on differential pulse techniques [1,5,8]. As of this writing, instrumentation is available from Amel, Bioanalytical Systems, E. H. Sargent, IBM Instruments, Princeton Applied Research, and Tacussel. Application notes from these manufacturers can often provide invaluable practical information. Quality data can usually be obtained at $10^{-7}$ M for routine purposes and at lower concentrations under ideal circumstances.

## B. Alternate Drop Differential Pulse Polarography

When differential pulse voltammetry is carried out at a conventional dropping mercury electrode a small error is created by the fact that the electrode surface area (and therefore the capacitance) is slightly different between each of the two current samples taken during the life of a single drop. This problem has been cleverly addressed by Osteryoung and co-workers by using one drop (no pulse applied) to establish a true dc current sample at the same time the chronoamperom-

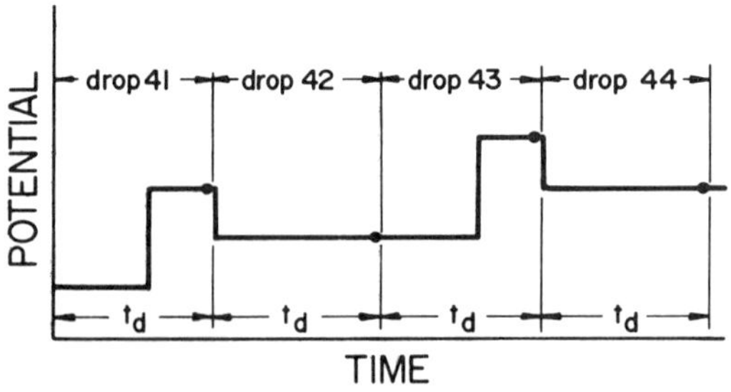


Figure 5.6 Excitation signal for alternate drop differential pulse polarography.

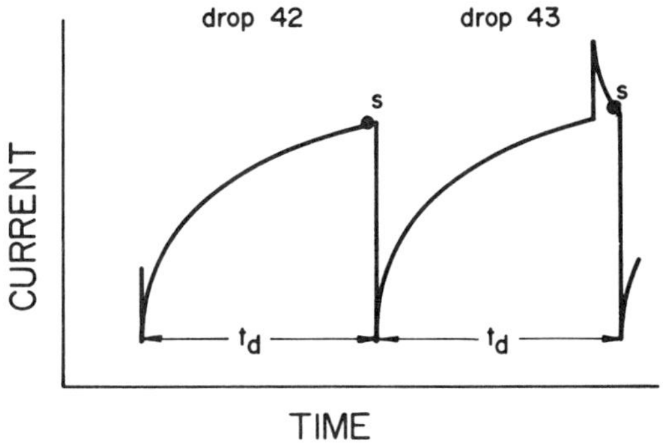


Figure 5.7 Total current response for alternate drop differential pulse polarography.

etry current is sampled when a pulse is applied (next drop) [5,9,10]. This is illustrated in Figures 5.6 and 5.7, where drop number 42 establishes the sampled background ("residual") current to be subtracted from the sampled chronoamperometry current recorded for drop number 43.

## C. Differential "Normal Pulse" Voltammetry

Large-amplitude ("normal") pulse voltammetry techniques were introduced in Chapter 3. The differential "normal pulse" (DNP) method combines several features of both the small- and large-amplitude pulse techniques. The rather odd waveform depicted in Figure 5.8 is

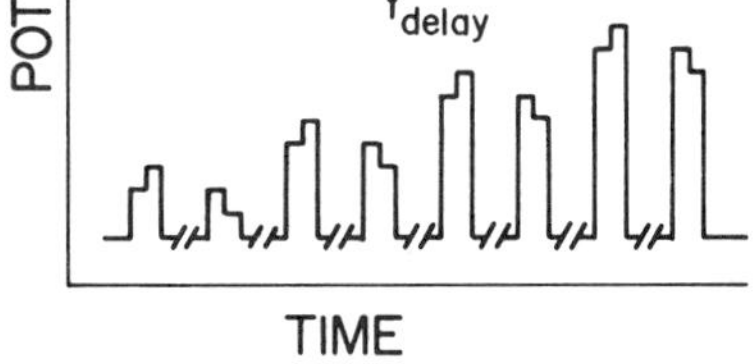


Figure 5.8 Excitation signal for differential "normal pulse" voltammetry (adapted from Ref. 12).

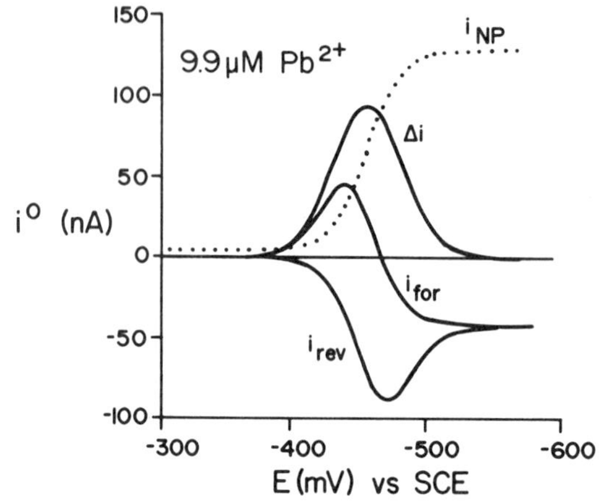


**Figure 5.9** Forward, reverse, and difference current DNP voltammograms of 9.9 μm (2.05 ppm) $Pb^{2+}$ in 0.5 M $CH_3COOH$/0.5 M $CH_3$ COONa. Conventional NP voltammogram is also shown (dotted line): $t_1 = t_p$ = 50 ms; $t_s$ = 16.67 ms; $t_{delay}$ = 0.4 s; $E_{delay}$ = −300 mV; ΔE = −25 mV; T = 20°C. Solid lines are experimental curves. Dots (•••) are theoretical response. (From Ref. 12. Reprinted with permission. Copyright 1981, American Chemical Society.)

for DNP in the alternate pulse mode [11-13]. The pulses are initiated from a potential at which no faradaic reaction takes place. Each pair of large pulses rise first to $E_1$ and then is incremented either above ($E_2$) or below ($E'_2$) $E_1$. $E_1$ is slowly increased for each subsequent pair of augmented pulses. A new mercury drop is initiated just after each large-amplitude pulse. If this is confusing, do not be alarmed, because this is one of the most complex electrochemical techniques and it will take time to master the details!

The current is sampled at $t_1$ and $t_2$. The forward current consists of that, $i(t_2) - i(t_1)$, monitored at $E_2$ on alternate pulses and the "reverse current" consists of that monitored at $E'_2$, $i(t_2) - i(t_1)$, on the even pulses. A differential current, $i = i(t_2,E_2) - i(t_2,E'_2)$, can also be plotted. All three responses are illustrated in Figure 5.9 for reduction of lead ion in an acetate buffer. There is a great deal of information in such a plot in which the forward and reverse currents combine to give a "cyclic voltammogram" and the differential current resembles a DPV curve with improved symmetry. The technique has the advantage of a relatively low duty cycle when compared to DPV since the action occurs during application of the large-amplitude pulse and not during the delay time between pulses. In DPV the potential applied between pulses is adequate to electrolyze the analyte continuously during most of the voltammogram. The DNP technique is a combination of a large-amplitude and a small-amplitude technique. The large-amplitude pulse sets up the concentration profiles which are the initial condition for the small-amplitude pulse riding on top.

## D. Square-Wave Voltammetry

The excitation signal for square-wave voltammetry (SWV) is illustrated in Figure 5.10 [14-17]. A symmetrical pulse train (the total amplitude is $2E_{SW}$) is added to a staircase (the step height is $\Delta E$) with a period of $\tau$. The response current is sampled at the end of both the forward and reverse half cycles (at 1 and 2 in Fig. 5.10). A difference current is determined by subtracting the current measured on the reverse cycle from that measured on the forward cycle. A dimensionless plot of the theoretical forward, reverse, and difference current is given in Figure 5.11 for a rapid electron transfer.

SWV is a very powerful technique. When a dropping mercury electrode is used one can apply the entire excitation waveform (Fig. 5.10) to a single mercury drop (allowing a delay time $T_d$ for the drop to grow to a predetermined size). In the previously described polarographic techniques as many as 1000 drops might be used in a typical experiment. Besides being fast, the technique is somewhat more sensitive than the very popular differential pulse technique because both

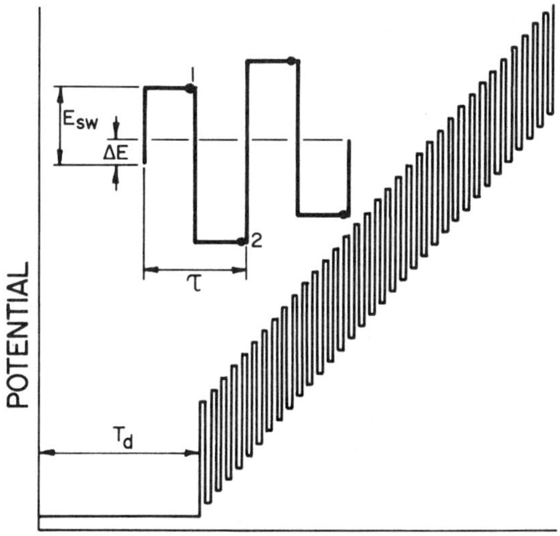


Figure 5.10 Excitation signal for square wave voltammetry (adapted from Ref. 16).

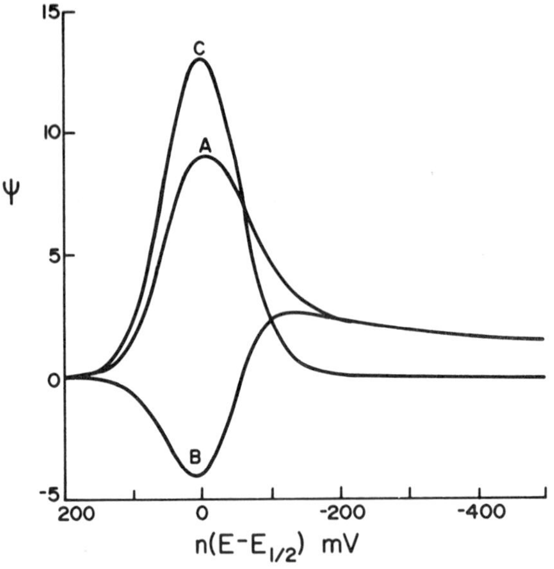


Figure 5.11 Calculated square wave voltammograms for reversible electron transfer: (A) forward current, (B) reverse current, and (C) net current, dimensionless units (adapted from Ref. 16).

forward and reverse currents are measured in SWV, whereas only forward currents are measured in DPV.

The speed of SWV makes it possible to monitor dynamic processes. For example, Osteryoung and co-workers have applied the technique as a detector for liquid chromatography [17]. Figure 5.12 illustrates a three-dimensional chromatogram obtained for two nitroso compounds.

## E. Staircase Voltammetry

Staircase voltammetry is a small-amplitude variation on single-scan or cyclic voltammetry and the technique was described briefly in Chapter 3 (Sec. III.B). The excitation signal is a staircase which closely approximates a ramp since the step height is typically 10 mV or less (Fig. 5.13). The idea is quite simple. A sample is taken at the end of the step (typically 50 ms wide or less), where charging current has decayed and the faradaic current is still quite significant. With the large number of samples acquired over the width of a given voltammetric wave, the waveshape closely resembles a voltammogram operated with a single ramp excitation (Chapter 3). The benefit results from a much reduced charging current contribution even at quite high scan rates [18-20].

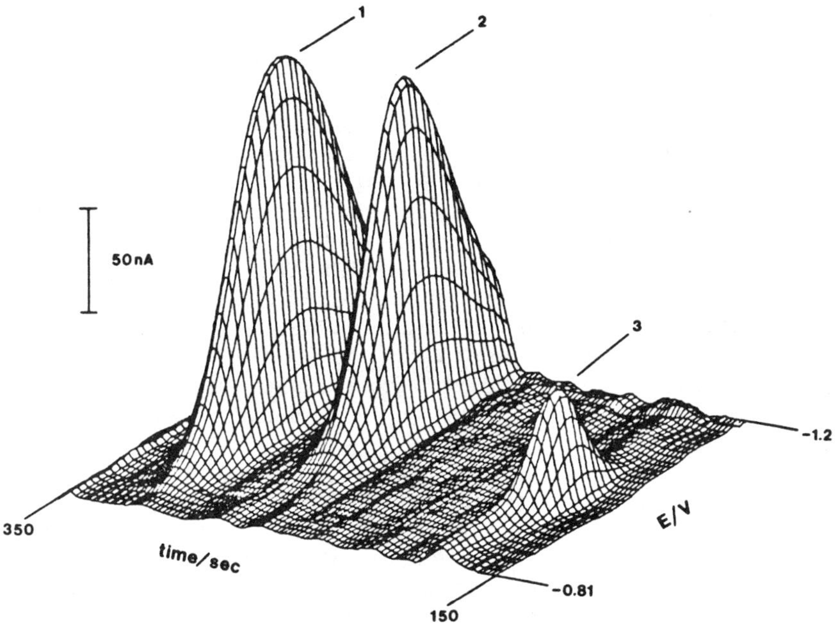


Figure 5.12 Three dimensional "chromatovoltammogram" for two nitrosamines (peaks 1 and 2) and an unknown impurity (peak 3). (From Ref. 17. Reproduced with permission. Copyright 1980, American Chemical Society.)

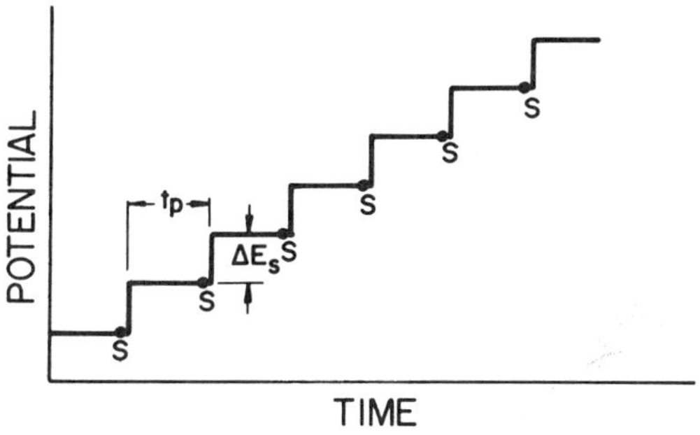


Figure 5.13 Excitation signal for staircase voltammetry.

## IV. CONCLUSION

Frequency-domain and time-domain small-amplitude techniques provide unique advantages for both analytical and fundamental studies. They tend to be more sensitive than large-amplitude techniques and are often more precise. These advantages result from the ease with which the signal can be distinguished from the background, the ease of minimizing charging current influences, and the fact that steady-state measurements can be made. In most cases, small-amplitude techniques are coupled to large-amplitude techniques, with the latter setting the surface concentration as an initial condition for the former. Understanding how to properly use small-amplitude techniques for fundamental purposes (e.g., determination of rates and transfer coefficients) requires a mathematical appreciation beyond the interest of most chemists. There are excellent sources, and Bard and Faulkner is the best place to start [1].

If nothing else has been accomplished, it should be very clear from Chapters 3 to 5 that there are an extraordinary number of finite current electroanalytical techniques. There is no doubt that this can cause considerable confusion for novices. Fortunately, all of these methods are based on relatively few fundamental concepts. It must be understood (1) that electron transfer rates and equilibrium constants vary with potential, (2) that mass transport to an electrode surface is precisely defined and reproducible, and (3) that the charge required to establish an electrode potential can be temporally distinguished from that utilized by a redox couple. These concepts are addressed in Chapter 2. Now that we have covered the more important electrochemical techniques, it is strongly recommended that Chapter 2 be reviewed with these techniques in mind.

## REFERENCES

1. A. J. Bard and L. R. Faulkner, *Electrochemical Methods*, Wiley, New York, 1980.
2. D. E. Smith, *Electroanalytical Chemistry*, Vol. 1, (A. J. Bard, ed.), Marcel Dekker, New York, 1966.
3. D. E. Smith, *Crit. Rev. Anal. Chem.* *2*:247 (1971).
4. D. D. Macdonald, *Transient Techniques in Electrochemistry*, Plenum, New York, 1977.
5. A. M. Bond, *Modern Polarographic Methods in Analytical Chemistry*, Marcel Dekker, New York, 1980.
6. A. M. Bond, R. J. O'Halloran, I. Ruzic, and D. E. Smith, *Anal. Chem.* *48*:872 (1976).
7. A. M. Bond, R. J. O'Halloran, I. Ruzic, and D. E. Smith, *Anal. Chem.* *50*:216 (1978).

8. J. Osteryoung and K. Hasebe, *Rev. Polarography (Japan)*, *22*:1 (1976).
9. B. H. Vassos and R. A. Osteryoung, *Chem. Instrum.* *5*:257 (1973-74).
10. J. H. Christie, L. L. Jackson, and R. A. Osteryoung, *Anal. Chem.* *48*:242 (1976).
11. K. Aoki, J. Osteryoung, and R. A. Osteryoung, *J. Electroanal. Chem.* *110*:1 (1980).
12. T. Brumleve, J. J. O'Dea, R. A. Osteryoung, and J. Osteryoung, *Anal. Chem.* *53*:702 (1981).
13. T. R. Brumleve, R. A. Osteryoung, and J. Osteryoung, *Anal. Chem.* *54*:782 (1982).
14. J. H. Christie, J. A. Turner, and R. A. Osteryoung, *Anal. Chem.* *49*:1899 (1977).
15. J. A. Turner, J. H. Christie, M. Vukovic, and R. A. Osteryoung, *Anal. Chem.* *49*:1904 (1977).
16. J. J. O'Dea, J. Osteryoung, and R. A. Osteryoung, *Anal. Chem.* *53*:695 (1981).
17. R. Samuelsson, J. J. O'Dea, and J. Osteryoung, *Anal. Chem.* *52*:2215 (1980).
18. H. L. Suprenant, T. H. Ridgway, and C. N. Reilley, *J. Electroanal. Chem.* *75*:125 (1977).
19. L. H. Miaw, P. A. Boudreau, M. A. Pichler, and S. P. Perone, *Anal. Chem.* *50*:1988 (1978).
20. S. Stefani and R. Steeber, *Anal. Chem.* *54*:2524 (1982).

# 6
# Introduction to Analog Instrumentation

PETER T. KISSINGER / Department of Chemistry, Purdue University, and Bioanalytical Systems, Inc., West Lafayette, Indiana



## I. CLASSICAL CONTROLLED-POTENTIAL INSTRUMENTATION

Instrumenting the controlled potential techniques described in Chapter 3 is not difficult. Nevertheless, beginners and old-timers alike often seem to have only the vaguest notion of how their instruments function and the logic implicit in various experimental designs. One reason for this is that there seems to be no good introduction to the subject. The extant literature is either too superficial or inordinately complex. Our goal here is the intermediate ground. We will attempt to provide a phenomenological basis for several fundamental instruments on a level satisfactory for utilizing them intelligently. The steps up the ladder to more complex situations (e.g., very fast experiments, alternating-current experiments, two working electrodes at different potentials in the same solution, iR compensation, digital computer control and analysis, etc.) will be relatively trivial ones. In our experience, most students encounter difficulties with electrochemical instrumentation at the outset primarily due to a lack of appreciation of basic electricity. The meaning of such terms as charge, voltage, current, and resistance must be intuitively clear.

Analytical techniques are conveniently discussed in terms of the excitation-system-response parlance described earlier. In most cases the system is some molecular entity in a specific chemical environment in some physical container (the cell). The cell is always an important consideration; however, its role is normally quite passive (e.g., in absorption spectroscopy, fluorescence, NMR, ESR) because the phenomena of interest are homogeneous throughout the medium. Edge or surface effects are most often negligible. On the other hand, interactions between phases are the central issue in chromatography and electrochemistry. In such heterogeneous techniques the physical characteristics of the sample container become of critical importance.

The potential difference which one wishes either to measure or control in an electrochemical experiment is that said to exist between the electrode and the solution. These tasks are not quite as easy to implement as it might appear. For one thing, the potential difference (strictly, the inner potential difference) $\Delta\phi$ between the two phases is not measurable. It quickly becomes obvious that a second electrode* in contact with the solution will be necessary and that this electrode will have its own metal-solution potential difference (again not measurable). It is important to recognize that in the media of interest (high dielectric constant, charge carriers present in great numbers) these potential differences exist only across a very small region contiguous with the metal (typically nanometers in width). There is no potential gradient across the bulk of the solution phase under zero-current conditions. This means that an ion (or molecule) in the bulk solution phase will have no way of knowing whether or not electrodes exist. The same ion will recognize that no potential *gradient* exists across this phase, and will have no way of ascertaining its own potential relative to that of either of the two electrodes.

The region across which the $\Delta\phi$ is impressed is broadly defined as the "electrical double layer," and it is in this region of very high potential gradient (typically $10^6$ V/cm) where all electrochemical reactions take place. The double layer has electrical characteristics similar to that of a conventional capacitor and one often hears about "double-layer capacitance" and "double-layer charging currents." The

---

*For our purposes an electrode is any solid conductor or semiconductor immersed in solution.

microscopic structural and kinetic details of the double layer are the subjects of continuing research. Charged interfaces (metal-solutions, membranes-solutions, solutions-solutions, etc.) are of paramount importance in numerous physical, chemical, and biological phenomena. It is, however, naive to assume that these things are well understood by anyone.

Figure 6.1 illustrates the classical apparatus for applying a potential difference to an electrochemical cell consisting of two electrodes. The electrode where the chemistry of interest occurs is called the working electrode (W) and the other electrode, for want of a better name, is dubbed the counter electrode. The applied potential $E_a$ is measured between the counter and working electrodes and the resulting current (if any) is measured in either the working-electrode lead (as shown) or the counter-electrode lead with equal facility. Notice that the working electrode is tied to ground potential and remains so irrespective of the value of $E_a$ selected with the potentiometer. It should be clear that to ions in the solution, it makes absolutely no difference whether or not this electrode is grounded.

In Figure 6.2a the situation in the cell is schematically illustrated in greater detail for the zero-current case. Note that the applied po-

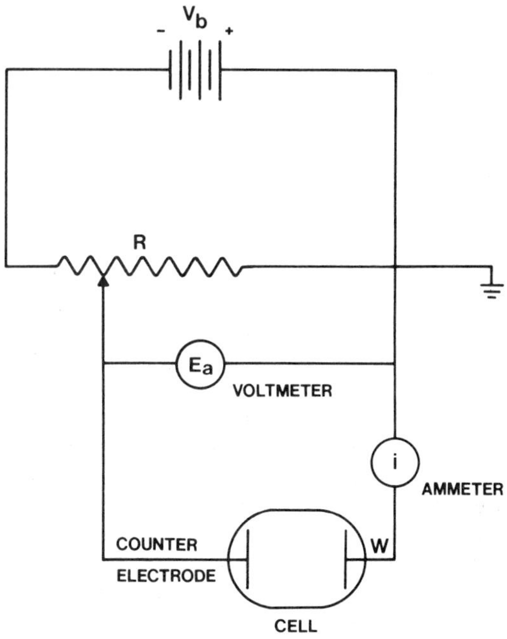


Figure 6.1 Classical two-electrode controlled-potential apparatus.

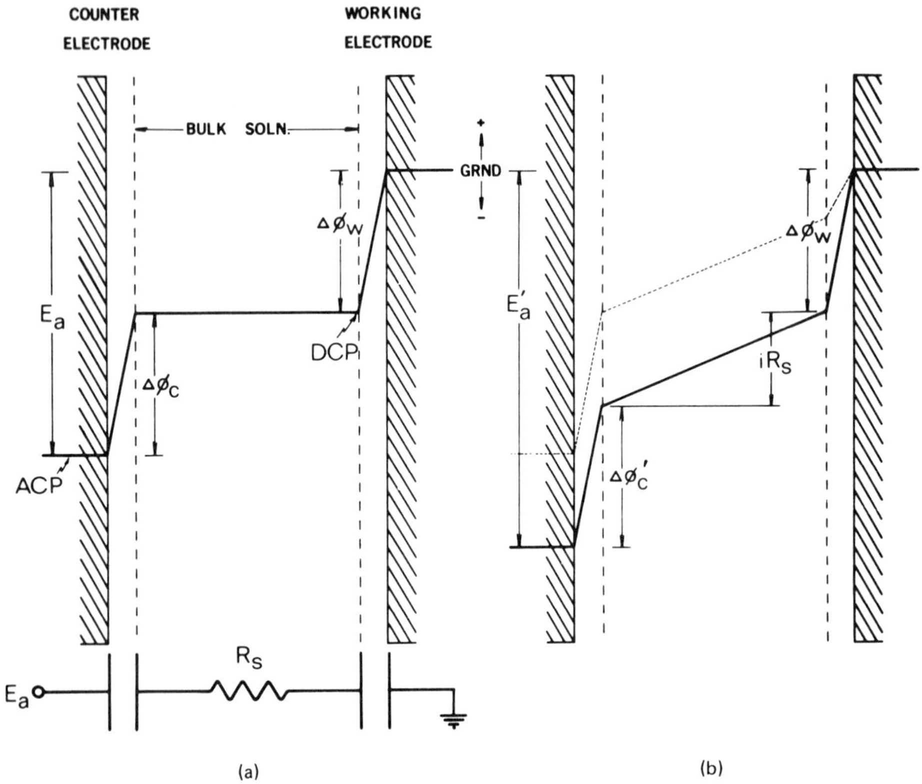


Figure 6.2 Schematic representation of potential gradients in a two-electrode cell.

tential is divided (not necessarily equally) between $\Delta\phi_c$ and $\Delta\phi_W$ and that no potential difference exists across the bulk solution. Although the working electrode is at ground as far as the outside world is concerned; it is *positive relative to the bulk solution* inner potential. The potential difference we wish to control is $\Delta\phi_W$ and the desired control point is indicated (DCP). The actual control point (ACP) as far as the circuit is concerned is on the metal side of the counter-electrode interface. Thus $\Delta\phi_W = E_a - \Delta\phi_c$ and one considers the working electrode to have a potential $E_a$ versus $\Delta\phi_c$. In general, the counter electrode is chosen to be one with a very rapid redox couple available, which means that $\Delta\phi_c$ does not change significantly as a function of current density if the latter is kept small (e.g., by using a large surface area). Such familiar electrodes as the calomel electrode or the silver-silver chloride electrode fit this bill quite well. Thus, although we never know $\Delta\phi_W$ in an absolute sense, we are perfectly capable of knowing it in a relative sense.

The two-electrode system is just fine under the zero-current condition, but this does not allow very much chemistry to be accomplished. Figure 6.2b represents the finite-current case, where $\Delta\phi_w$ is the same as in Figure 6.2a. *If* the current is small enough and the counter electrode is large enough (i.e., the current density at *this* electrode is kept small), $\Delta\phi_c$ may be assumed to be constant. *If* the solution resistance and the current are small enough, $E_a$ may be kept the same as in the zero-current case to maintain $\Delta\phi_w$ at the desired value. When the iR drop across the bulk solution phase is significant (as illustrated), $E_a$ must be increased by the value of iR or control will be lost. This undesirable situation derives from the remoteness of the ACP from the DCP. Fortunately, we can do much better. The problem just described is illustrated again in Figure 6.3. The importance of understanding this figure cannot be overestimated.

*The counter electrode in the two-electrode system serves two functions. First, it completes the circuit, allowing charge to flow through the cell; and second, it is assumed to maintain a constant interfacial potential difference regardless of the current.* These two functions are mutually exclusive only under very restricted circumstances. Both needs are better served by two electrodes, an auxiliary electrode to assume the first function and a reference electrode to assume the second.* This approach is illustrated schematically in Figures 6.4 and 6.5.

The reference electrode is used as a potentiometric (always zero current) probe to monitor $\Delta\phi_w$ relative to its own $\Delta\phi_r$. This value is compared with $E_a$ and if a difference (i.e., an error signal) exists, the potential impressed across the cell by the potentiometer is adjusted until balance is achieved (i.e., no error signal). A device that accomplishes this control function automatically is called a "potentiostat" for obvious reasons. Such behavior can be mimicked by the experimenter. Although this is assuredly almost never done these days, it is useful to think about a manual potentiostat as a pedagogical device.

Imagine the situation given in Figure 6.4 where the potential monitor is a high-impedance voltmeter connected to a saturated calomel electrode (SCE) as reference. The potentiometer is now adjusted to a value sufficient to initiate a reaction at the working electrode. The potential at which the reaction ensues ($E_a$ vs. SCE) is monitored with the meter. As the reactive species is consumed at W, the current through (and the iR drop across) the cell will decrease. This causes

---

**Nomenclature note*: "Counter electrode" is a term which is best applied in connection with two-electrode experiments and "auxiliary electrode" is a term reserved for the three-electrode experiments. Although this rule has not always been followed in the literature, there is a healthy trend in this direction.

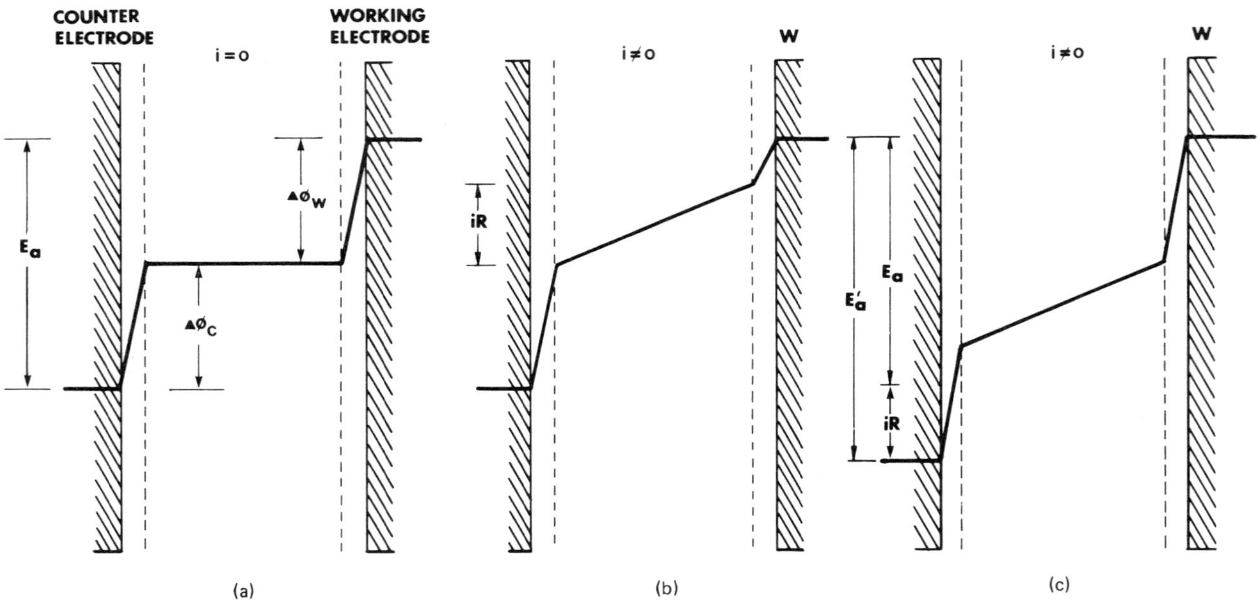


Figure 6.3 (a) An electrochemical cell at zero current. It is assumed that $\Delta\phi_C$ is fixed by a reversible redox couple which exists at the counter electrode and $\Delta\phi_W$ is adjustable by controlling the applied potential $E_a$. (b) If a solution component is electroactive at the given $\Delta\phi_W$, a reaction will occur resulting in a flow of charge and an iR drop across the bulk of the solution. $\Delta\phi_W$ necessarily will drop, but not so much that the reaction stops (because then $\Delta\phi_W$ would necessarily increase to its original value and the reaction would start again). (c) If (as is often the case) we wanted $\Delta\phi_W$ to be controlled at a given value regardless of whether or not a reaction takes place, it would be necessary to increase the applied potential to overcome any iR drop that results from such a reaction.

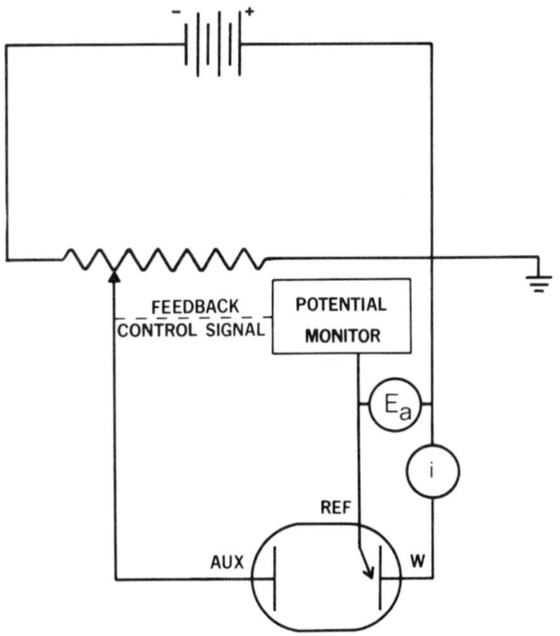


Figure 6.4 Schematic representation of a primitive three-electrode controlled-potential apparatus.

$E_a$ to *increase*. The experimenter observes this increase on the monitor and judiciously adjusts the potentiometer to *decrease* the voltage applied to the auxiliary electrode until $E_a$ returns to its initial value. Implicit in this process of monitoring and adjusting is *negative feedback*. The result is a constant $E_a$. How constant $E_a$ will be depends on the sensitivity of the meter and the response time of the experimenter.

Notice (Fig. 6.5) that although the DCP and ACP are much closer together in the three-electrode approach, they are still not one and the same. There will be some iR drop error, albeit small. The total cell resistance has been divided into two parts: a compensated resistance $R_c$ and an uncompensated resistance $R_u$. Only the ohmic loss $iR_u$ will be effective in moving the ACP from the DCP. The effectiveness of the potentiostat depends largely on the ratio of $R_c$ to $R_u$. The larger this ratio, the better the result. This suggests that the geometry of the reference electrode and its placement may be of critical importance, depending on the solution resistance. The efficacious design of electrochemical cells and the placement of electrodes will be discussed in Chapter 12.

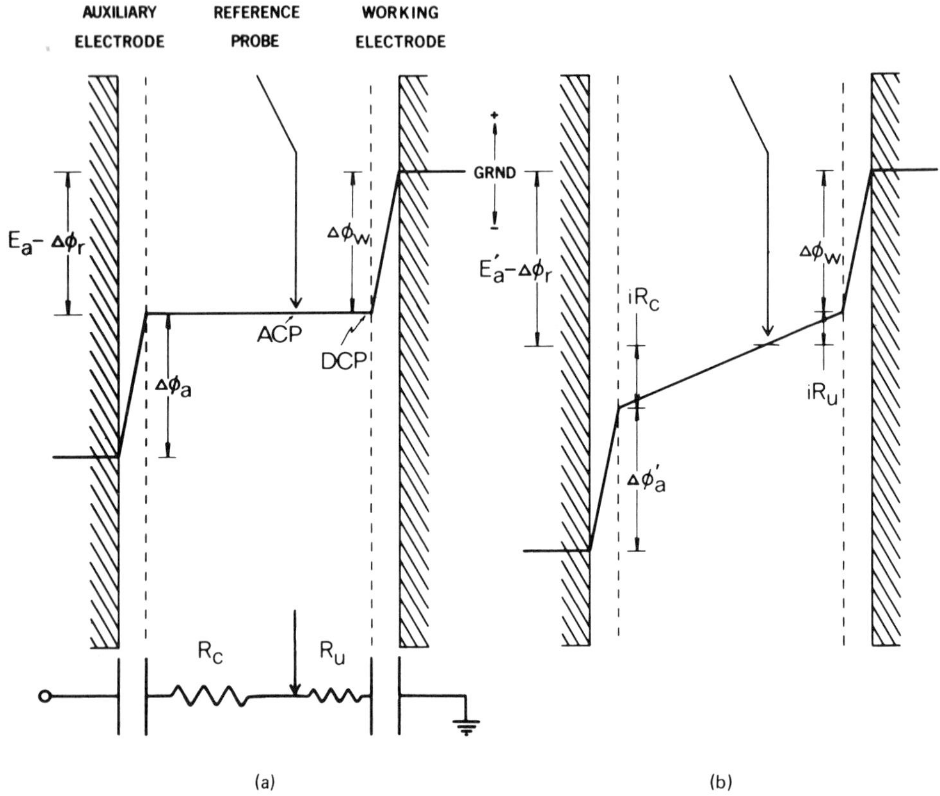


Figure 6.5 Schematic representation of potential gradients in a three-electrode cell: (a) $i = 0$; (b) $i \neq 0$.

We can summarize the virtues of the three-electrode potentiostat in two statements:

1. The iR drop across a cell is largely compensated by bringing actual experimental control of potential closer to that desired.
2. Since the monitoring circuit draws negligible current, the reference electrode redox couple need not be perfectly reversible (i.e., need not have a large exchange current density) and need not have a very large surface area.

These advantages are very much independent and either one may predominate in particular circumstances. The second advantage is quite often forgotten (particularly by those accustomed to two-electrode methodology). Limitations and experimental details will be described later.

## II. CONTROLLED-POTENTIAL INSTRUMENTATION BASED ON OPERATIONAL AMPLIFIERS

### A. Introduction

In order to assimilate the material that follows, the reader will have to be familiar with a few simple concepts in instrumentation electronics. These topics are more than adequately discussed in numerous texts and we will not belabor them in detail here. For the convenience of those who need to review, we have prepared a list of subjects *essential* to understanding these introductory sections. This list is all-inclusive, so study of topics not on the list (e.g., ac circuit theory, inductance, transformers, power supplies, digital electronics, transducers, transistors, etc.) will be of no immediate value.

*Essential Topics in Electronics*

- Resistors (Ohm's law) and capacitors
- Voltage dividers
- "Loaded" voltage dividers
- Passive differentiators
- Passive integrators
- Solid-state diode (including zener diodes)
- Ideal operational amplifier
- Real operational amplifier (compliance limitations, offset current and voltage)
  - Voltage follower
  - Follower with gain
  - Inverter
  - Integrator
  - Differentiator
  - Comparators
- Concept of input impedance and output impedance.

### B. The Voltage Follower as a Potentiostat

In the next few pages we will illustrate the evolution of modern controlled-potential instrumentation. The voltage follower (Fig. 6.6a) is a device that satisfies the need for an "impedance buffer" to isolate high-impedance sources (with no capability of supplying current) from low-impedance sinks (desirous of current). In other words, the follower transforms a *soft* voltage point, $E_i$, which is subject to *loading*, into a *hard* voltage point, $E_o$, which can supply considerable current. In this configuration the operational amplifier (OA) has no closed-loop voltage gain since $E_i = E_o$. The quality of the follower will depend on the impedance seen at the two inputs (the higher the better) and the current capability of the output stage of the OA. Of particular interest is the finite (though very small) "bias current"

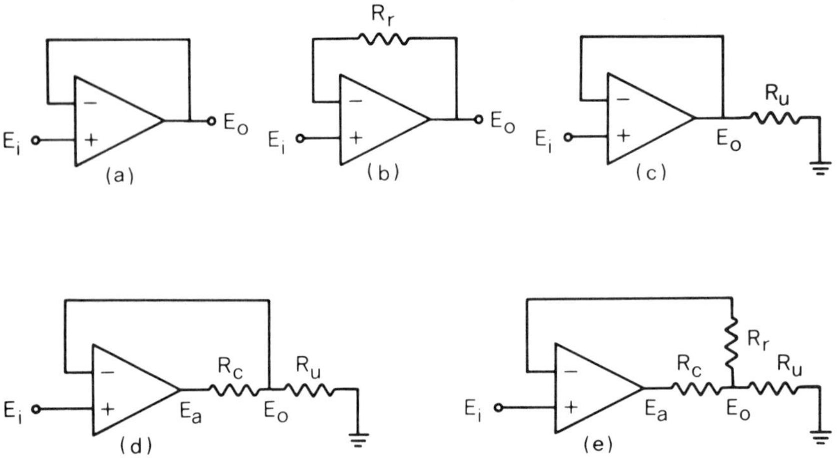


Figure 6.6 Evolution of a potentiostat based on the operational voltage follower.

which will flow across the input terminals and an "offset voltage" which will exist between the input terminals. The latter will, of course, cause a difference to exist between $E_i$ and $E_o$. For most electrochemical applications it is desirable that this difference be less than about 1 mV, a feat easily achieved in practice.

If a resistance is placed in the feedback loop (Fig. 6.6b), the bias current $i_b$ will also create a difference between $E_i$ and $E_o$ by an amount $i_b R_r$. Even very inexpensive (<$1) OAs can have bias currents of less than $10^{-9}$ A, which means that the value of $R_r$ will have to exceed 1 MΩ to create a 1-mV error. Amplifiers with bias currents of less than 0.1 pA ($10^{-13}$ A) are available. Using the same criterion, $R_r$ may then reach 1000 MΩ, a value well beyond any resistance commonly encountered in dynamic electroanalytical techniques. Such amplifiers are, however, eminently useful for constructing pH meters and pH stats and measuring potentials in electrophysiology, where very small high-impedance electrodes are often used.

Figure 6.6c depicts the voltage follower with a resistive load $R_u$. Limitations on $R_u$ are imposed only by the maximum output current of the OA, $i_o(\max)$. Quite obviously, $E_o/R_u$ cannot be allowed to exceed $i_o(\max)$. In other words, $R_u$ must be larger than $R_{min}$, where $R_{min} = E_o/i_o(\max)$.

In Figure 6.6d the change is much more dramatic. Here we have added an additional resistance $R_c$ between the OA output and the control point $E_o$. If readers have done their homework, they will immediately recognize this circuit as the familiar "follower with gain" drawn in an atypical way. The $E_o$ will still equal $E_i$ and it should be clear that $E_a = (1 + R_c/R_u)E_i$. If this expression is not clear, readers are advised to doodle on some scrap paper until it becomes obvious. Otherwise, there is no point in continuing with this material.

Notice that in Figure 6.6d, $i_o$(max) will still place a lower limit on $R_u$ (and only $R_u$), because the current through the two resistors is defined only by $E_o$ and $R_u$. The other resistance, $R_c$, is not controlling. Understanding this fact is also crucial to the reader's progress. There is, however, a limitation on $R_c$ imposed by the maximum *voltage* swing, $E_a$(max), permitted by the output stage of the OA. The limitation is defined by the fact that $i(R_c + R_u)$ cannot exceed $E_a$(max). Figure 6.6e incorporates the three resistors discussed above and all of the limitations will apply.

The symbology chosen for Figure 6.6 will become clear for the reader who reviews the three-electrode configuration introduced in Figures 6.4 and 6.5. To use the voltage follower as a potentiostat, the auxiliary electrode is placed at the OA output, the reference electrode is connected to the negative input, $R_r$ is the reference electrode impedance, $R_c$ is the compensated resistance, $R_u$ is the uncompensated resistance, and the working electrode is connected to ground.

The use of passive components (resistors and capacitors) to construct an equivalent circuit for the electrochemical cell is carried one step further in Figure 6.7a. In addition to the bulk phase resistances

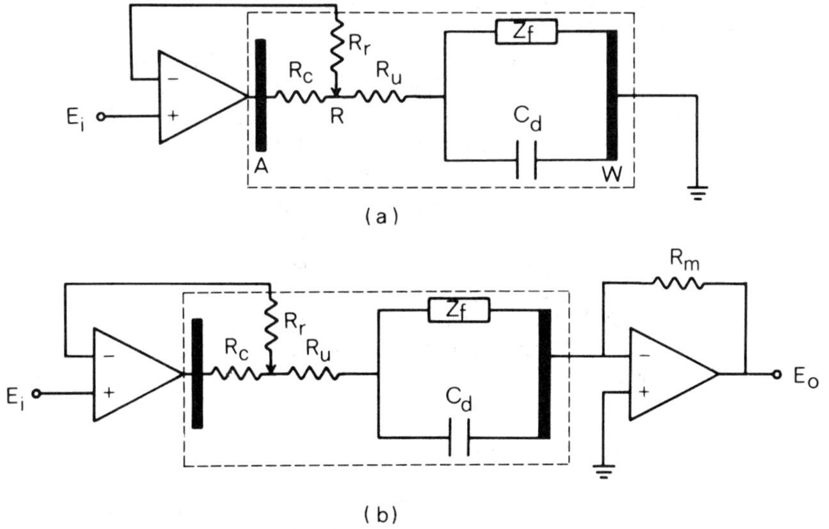


Figure 6.7 Evolution of a potentiostat (continued). (a) Equivalent circuit of a three-electrode cell. (b) Addition of current-to-voltage converter.

($R_c$ and $R_u$), the interphase components $C_d$ (double-layer capacitance) and $Z_f$ (faradaic impedance) have been added. When a faradaic reaction occurs, the electrode interface takes on the character of a leaky capacitor. Phenomenologically, one considers the charge stored in the double layer to leak away through some heterogeneous electron-transfer reaction represented by the impedance, $Z_f$. Naturally, in order to maintain the potential difference $\Delta\phi_w$, additional charge must flow to replenish that being lost through $Z_f$. This is, of course, the function of our potentiostat. Note that the potentiostat in this case consists of a single operational amplifier. At this point in time, amplifiers suitable for many experiments can be purchased for as little as \$0.50. A mathematical analysis of the equivalent circuit in Figure 6.7a will not be necessary until instrumentation for fast transient experiments is described in the following chapter.

Controlled-potential experiments (e.g., voltammetry, chronoamperometry) require some way of measuring the current without disturbing the controlled parameter. The most convenient way of accomplishing this is to connect an operational amplifier current-to-voltage converter (sometimes called a "current follower") to the working electrode as shown in Figure 6.7b. The current-to-voltage converter is a simplified version of the familiar operational amplifier inverter. One can think of the input resistor in an inverter as being a current source ($E_i/R_i$) for the feedback resistance, $R_f$. Replacing *this* current source with the working electrode is a trivial matter. The summing point of the OA will under most conditions remain at *virtual ground* and therefore connecting the current follower does not perturb the experiment.

The output of the converter, $E_o$, is a function of the current applied to it by the electrode reaction ($E_o = -iR_m$). Normally, current is measured by using a current-measuring resistor ($R_m$) to convert it into voltage. The circuit is most sensitive when $R_m$ is large. For any given current $R_m$ must not be so large that $iR_m$ exceeds the maximum output voltage of the OA (typically, ca. ±14 V). In addition, i (as determined by the concentration of electroactive species, area of the working electrode, convection of the solution, etc.) may not exceed the current capability of the amplifier (typically, ±10 mA). If the potentiostat (or "control amplifier") and current-to-voltage converter have different current capabilities, i must not exceed the $i_{max}$ of the less powerful OA.

### C. The Adder as a Potentiostat

Figure 6.8 depicts an alternative potentiostat design based on the operational amplifier adder rather than the voltage follower. This two-amplifier potentiostat or DeFord-type potentiostat is very widely used. As far as we can determine, the only advantage of this config-

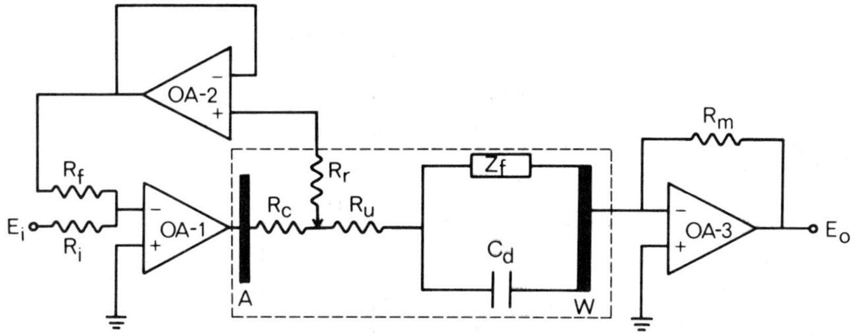


Figure 6.8 Two-amplifier adder-type potentiostat with current-to-voltage converter.

uration (developed before differential input OAs were widely available) is that excitation programs can be easily synthesized by adding a number of different waveforms via a sufficient number of input resistances $R_i$. For example, the basic ac voltammetry experiment can be performed by adding a sinusoid to a ramp via two input resistors. The contribution of each $E_i$ to the total excitation signal will depend on the relative magnitudes of the $R_i$ values, although in most applications these will all be the same.

The reader should study the adder-type potentiostat in detail and attempt to analyze it based on qualitative negative feedback principles (monitor a control point, compare with desired control, make adjustments as necessary). In your analysis first assume that $R_i$ and $R_f$ are equal. Why is the voltage follower required in the reference electrode lead? Can you envision circumstances where this might be eliminated? If a species oxidizes at +1.0 V vs. SCE, what should be the sign of $E_i$ to accomplish this chemistry in the two amplifier potentiostat? In the voltage follower potentiostat?

## D. Current Measurements

Figure 6.9 illustrates four methods of measuring the current which are equally applicable to either potentiostat design. It is *essential* that the reader be able to calculate $E_o$ as a function of the current and the various resistors. If this exercise is difficult, you are in *serious trouble* and will have to go back to the study topics listed earlier. Circuit 6.9a has already been described. The other three can be useful in high-current experiments when only the control amplifier (and not the current-to-voltage converter) is capable of carrying the maximum current anticipated. In circuit 6.9b, we cannot let $R_m$ get too large. Why not? In circuit 6.9c there is a limitation both on the value

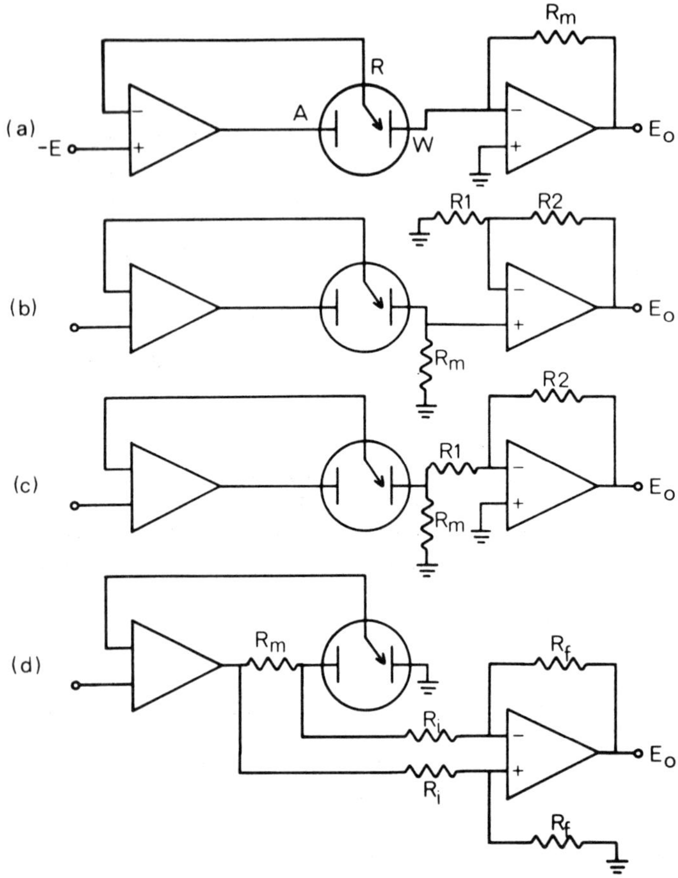


**Figure 6.9** Alternative modes of current measurement in three-electrode experiments.

of $R_m$ and the value of $R_m/R_1$. What is this limitation? In 6.9d there is also a limitation on the magnitude of $R_m$ and the ratio $R_m/R_i$. Can you figure this one out? If these questions are easy, you are doing just fine. There are several ways in which 6.9d may be improved if two additional amplifiers are available. Think about it.

## E. Cyclic Voltammetry Instrumentation

It is not the purpose of this volume to catalog the wide variety of circuits that have been used to generate excitation signals and measure the corresponding response of an electrode-solution interface.

The apparatus required for a given experiment may range from a battery, a resistor, and a microammeter to sophisticated systems operated under computer control. Since electrometric titrations and voltammetry are the most common electroanalytical experiments, we will make specific reference to the experimental requirements in several chapters. In the following section the operation of a popular cyclic voltammetry instrument (Bioanalytical Systems Model CV-1B, Fig. 6.10) will be described in some detail.

Most cyclic voltammetry instruments can be conveniently thought of as the combination of three circuits: a signal generator, a potentiostat, and a current-measuring amplifier. The latter two subunits consist of an operational amplifier voltage follower and current-to-voltage converter and are used in a conventional manner. On the other hand, the waveform generator is unique and requires careful study. We will examine a simple version of the circuit first and add refinements until the full design is explained.

A straightforward approach to triangular-wave generation is to integrate a square wave. *The amplitude of the square wave will determine the slope of the triangular wave and the period of the square wave will determine the amplitude of the triangular wave.*

Figure 6.11 illustrates the basic triangular-wave generator. Assume ideal amplifiers and that R1 = R2. In this configuration, OA-1 functions as a voltage comparator (square-wave generator) and OA-2 functions as the integrator with a time constant of R3·C seconds. For

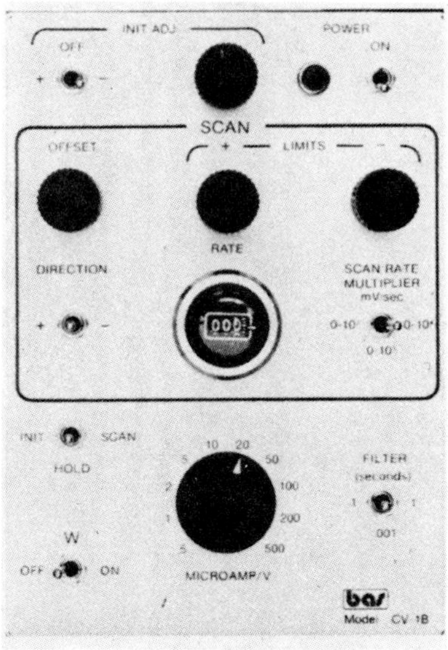


Figure 6.10 Top view of CV-1B.

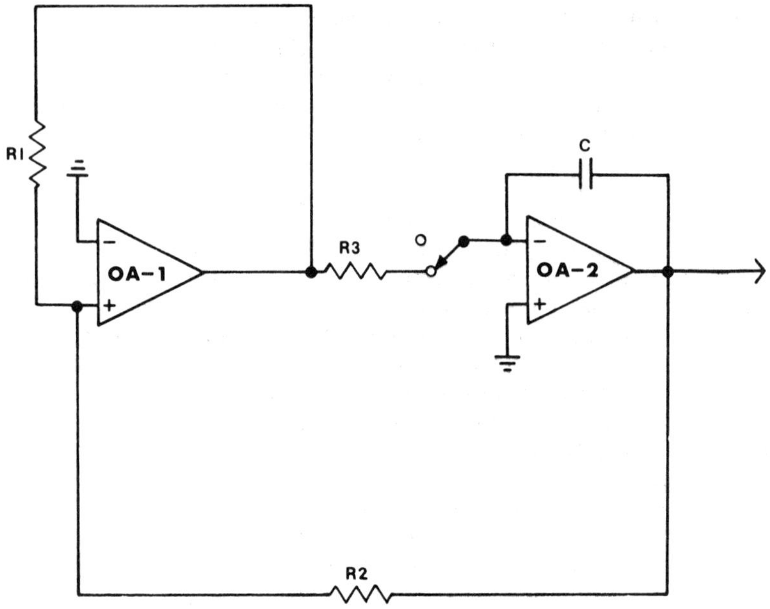


Figure 6.11 Comparator coupled to integrator.

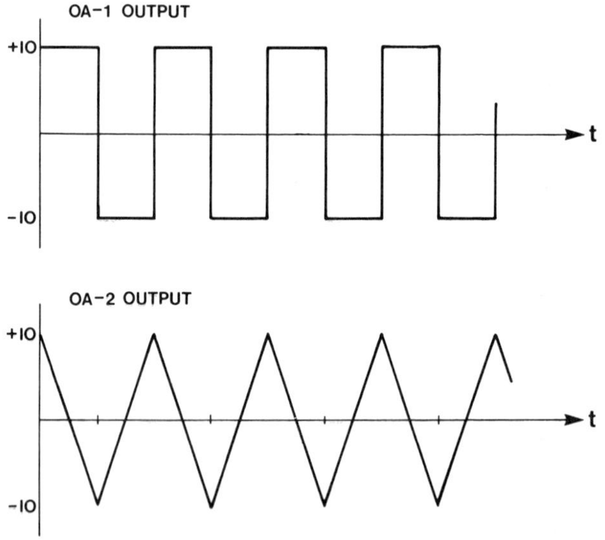


Figure 6.12 Integration of square wave to generate a triangle wave.

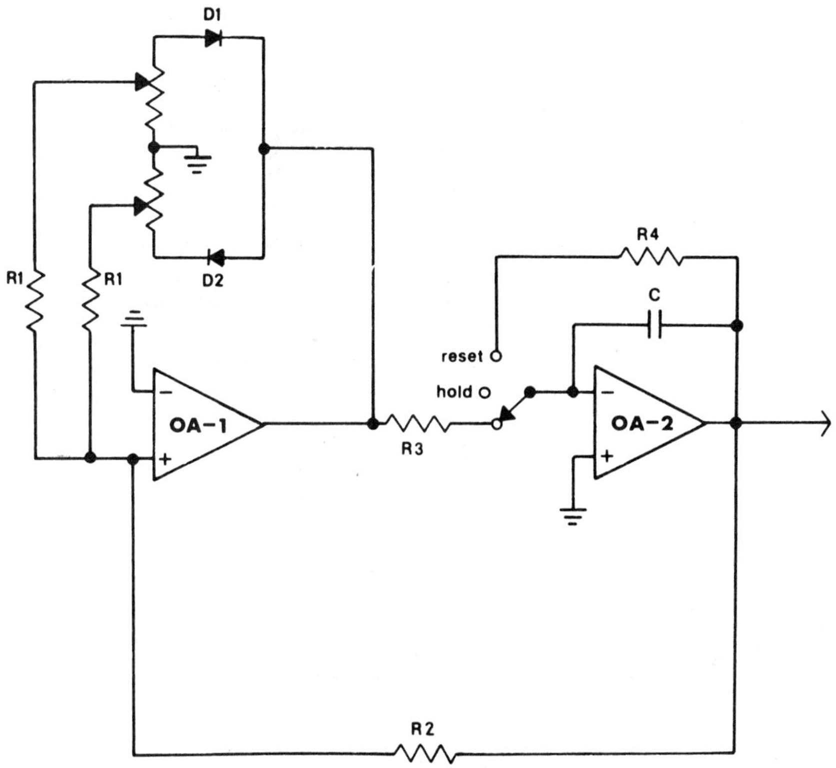


Figure 6.13 Adding circuitry to change comparator states asymmetrically.

the moment we will assume the output of OA-1 will be in either of two states (e.g., +10 or −10 V) depending on the polarity of the voltage at its positive input (i.e., a positive voltage at the + input results in a +10-V output). Note that the polarity at the positive input will depend on the voltage impressed across the series combination of R1 and R2.

If the output of OA-1 is +10 V, the output of OA-2 will be a negative-going ramp of slope −10/(R3·C). As the voltage of the ramp becomes more negative, the + input of OA-1 also moves in a negative direction. When the output of OA-2 becomes more negative than −10 V, the + input of OA-1 will cross zero and become negative. At this point OA-1 must necessarily change state and its output swings to −10 V. The integrator responds by changing the ramp to a positive direction. When the integrator output reaches +10 V, OA-1 changes state again (−10 V to +10 V) and the integrator again changes direction. This process continues indefinitely as long as the amplifiers are

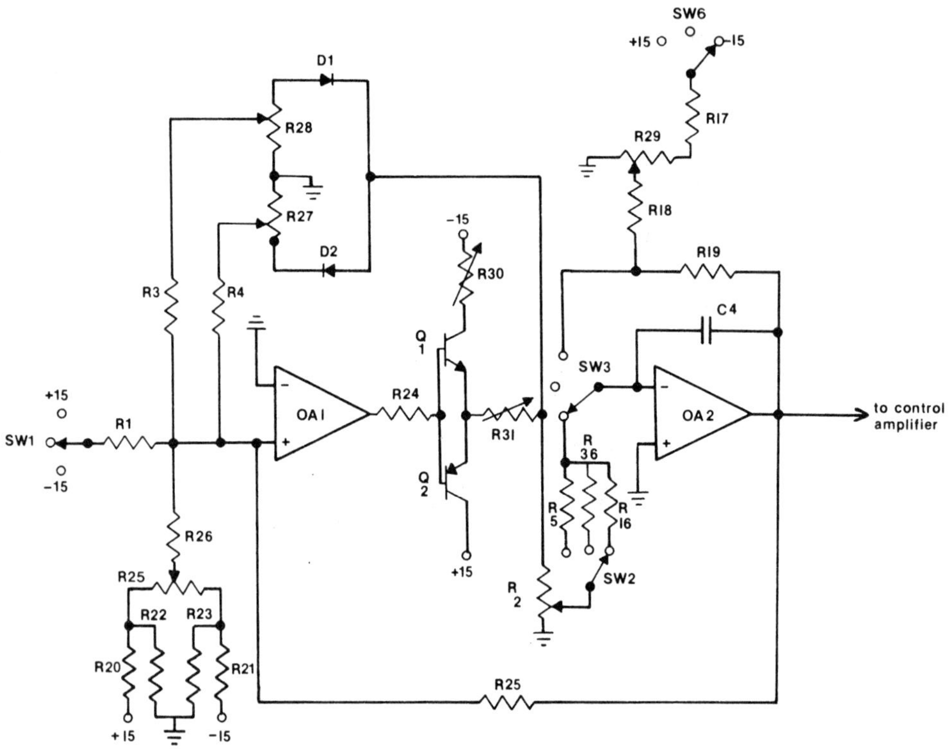


Figure 6.14 Signal generator for cyclic voltammetry experiments.

powered. The signal generator is said to be *free running*. To understand the events described above, it is useful to study the relation between the square wave (OA-1 output) and triangular wave (OA-2 output) as depicted in Figure 6.12.

The circuit of Figure 6.11 demonstrates the principles, but it is of little practical value for cyclic voltammetry. The embellishments added in Figure 6.13 improve utility by allowing the switching points of the triangular wave to be adjusted independently via two variable resistors. Now the voltage at which the sweep changes direction can be adjusted to values less than 10 V. In effect, only a portion of the OA-1 output is fed back via R1 as a reference to control the zero crossing for the + input of OA-1. The diodes, D1 and D2, permit independent control of the reference voltage for the two output states of OA-1.

The three position switch at the input to OA-2 enables the ramp voltage to be held at any value by opening the integrator input (center position). The third position leaves the integrator input open, but drains charge from the capacitor via R4, resetting the output of OA-2

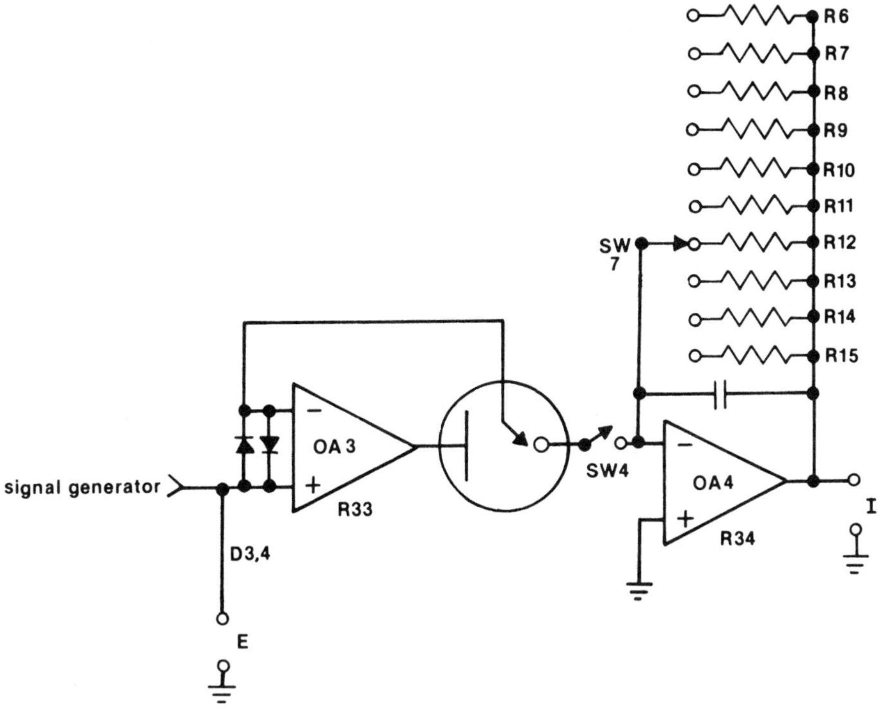


Figure 6.15 Potential control and current measuring circuitry.

to zero. The improvements illustrated in Figure 6.13 are not quite sufficient for a practical CV instrument and several additional features must be added. A completely satisfactory design is presented in Figure 6.14.

Momentary contact switch SW1 enables the state of OA-1 to be changed at will. A control (R23) has been added that permits the triangular wave to be moved (offset) along the voltage axis without influencing its time course. Note that the voltage set on R23 is not dependent on the square wave, unlike for R24 and R25. An initial voltage control (R26) has been added to enable a nonzero initial condition at the output of OA-2.

We have assumed ideality for OA-1 and OA-2. As long as the offset current for OA-2 is small (i.e., FET input stage) ideality is not a bad approximation of performance for OA-2 for the usual CV experiment. The nonideality of OA-1 does present problems. For example, its inherent output swing cannot be assumed to be symmetrical. Symmetry in the square wave is crucial because we must ensure that both legs of the triangular wave have the same absolute slope. Thus we must operate on the output of OA-1 to obtain a symmetrical swing around zero. A complementary pair (Q1 and Q2) is added to the output stage. Since the inherent + limit is always greater than the − limit, a trimmer (R27) is added to the collector Q1. To set a symmetrical ±10-V square wave at the top of R29, we adjust R27 to achieve symmetry and R28 to achieve accuracy. The other problems with OA-1 are finite slew rate and finite input current, which limit the rate and linearity of the triangular wave and the integrity of the apices.

For use in cyclic voltammetry the slope of the triangular wave must be known with precision and be calibrated in advance. This is accomplished with control R28, which calibrates the dial on R29 to read in units of mV/s even if the value of C4 is not precisely known. (Precision capacitators are extremely expensive and are fortunately not necessary here.) The precision resistors R5 and R13 provide a 10-fold change in time constant (and therefore scan rate) via SW2. Figure 6.15 shows the control and current conversion circuitry which when combined with Figure 6.14 give a complete cyclic voltammetry instrument.

## F. Test Circuits for Controlled-Potential Instrumentation

Figure 6.16 illustrates two common arrangements whereby simple performance checks may be implemented. Both experienced and novice electrochemists often encounter situations in which it is clear that there is some difficulty with their apparatus. This is certainly one of the

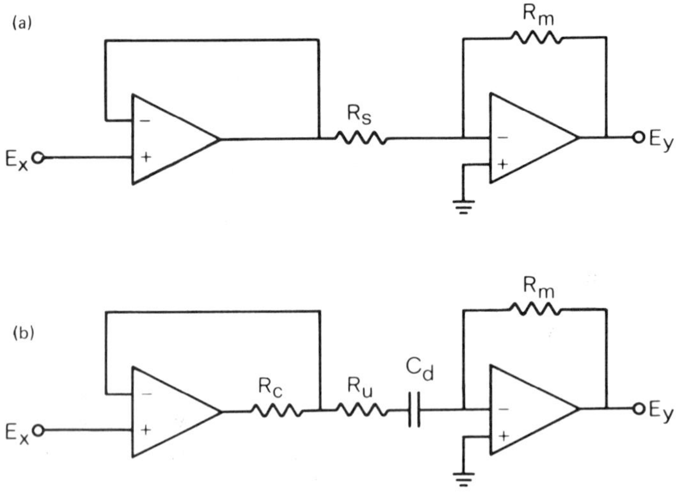


Figure 6.16 Dummy-cell test circuits for three-electrode instrumentation.

reasons electrochemical experiments have not been popular among chemists in general. An understanding of the apparatus, coupled with a little experience, should enable one to become operational without too much swearing and foot stomping. As is usual, the first job is to isolate the source of unhappiness and the second is to apply appropriate medication.

Almost always the problem resides in the vicinity of the cell and involves improper connection of the electrodes (one or more leads not connected or connected to the wrong electrode), shorting of the electrode leads together, an electrode not immersed in the solution, a gas bubble on an electrode or in a salt bridge, a broken lead, a broken electrode, no solution, no supporting electrolyte, no sample, solvent trapped under a mercury pool electrode, clogged capillary, poor platinum to glass seal, and so on. Careful observation of the cell will usually lead to a sound diagnosis and cure.

There is a tendency for beginners to connect their electrodes without regard for the electronics. This can not only ruin the best experimental intentions, but can also be *extremely dangerous* in the case of high-powered potentiostats. It is important to remember that a potentiostat will not function as such unless the feedback loop is connected. This loop will always include the auxiliary electrode, the reference electrode, and the solution between them. It is always best to set up the electronics first, and then connect the electrodes! Many students attempting a cyclic voltammetry experiment, for example, will connect their cell while the triangular-wave generator is sweeping willy-nilly all over the place. This approach can change the state of a working electrode (by exposing it to extremes in potential) and the desired initial condition will be lost. The initial potential and switching potentials should be chosen, then the cell connected, and only then should the sweep be initiated. If an electrode has been exposed to a potential extreme (e.g., an HMDE became too positive), it may have to be replaced before a meaningful experiment is attempted.

The undesirable response of the instrument will be a helpful datum. For example, if the reference electrode is not connected, then when the working electrode is turned on (i.e., connected), the current axis will be driven to high values (perhaps the saturation current *or* voltage of the current follower, whichever comes first) even at applied potentials where no reaction should occur. This results from breaking the control loop and thereby driving the control amplifier (and auxiliary electrode) into voltage saturation. Under such conditions the voltage applied to the control amplifier is meaningless with regard to what is actually going on at the working electrode interface.

The next most frequent problems involve the recording device. The sensitivity may be too high or too low or the leads may not be connected properly. Neglect of ground leads is extremely common among novitiates (particularly with regard to differential-input XY recorders where four of the six input terminals will normally *have to be* grounded by the user).

Malfunction of the instrumentation per se almost never occurs. To convince oneself that all is well with the electronics, it is useful to replace the cell with a passive analog frequently referred to as the "dummy cell." The simplest dummy cell consists of a single resistor as illustrated in Figure 6.16a. The reference and auxiliary leads are connected to one side and the working electrode lead to the other. The applied potential $E_x$ will thus be impressed across $R_s$ and the measured current, $-E_y/R_m$, will be equal to $E_x/R_s$. When a potential scan is applied, the response will take the form of an oblique straight line crossing the origin. The line will be independent of sweep rate. Typically, $R_s$ is a precision resistor of 10 kΩ. Although we have a tendency to avoid two-electrode experiments, this same configuration may be used for this purpose where the counter electrode is connected to *both* the auxiliary and reference leads. This further emphasizes the dual role of counter electrodes discussed earlier.

A more advanced dummy cell is employed in Figure 6.16b. In this case a time invariant $E_x$ will obviously result in no current response. A step change in $E_x$ will result in a current transient decaying away to zero. Note that the combination of $R_u$ and $C_d$ forms a passive integrator, the *current* through which will have a derivative format. If a sweep is applied, the responding current will have a steady nonzero value. If the sweep direction is reversed, a current value of the same magnitude but opposite in sign will be recorded. It is important to understand both test circuits because their operation is closely analogous to that with real cells. Circuit 6.16b will be discussed at length when fast transient experiments are described in detail. It is, for example, quite useful in determining the capabilities of an instrument to charge the double layer in as short a time as possible. For a step change in $E_x$ it will take a finite time for the solution side of $C_d$ to achieve $E_x$. How would you expect this time to depend on the value of Cd, $R_u$, Rc, and the ratio of $R_c$ to $R_u$?

## G. Multiple-Electrode Controlled-Potential Instrumentation

To carry out amperometric or voltammetric experiments simultaneously at different electrodes in the same solution is not difficult. In principle, any number of working electrodes could be studied, however, it is unlikely that more than two or three would ever be widely used in practice. The bulk of the solution can have only one *controlled* potential at a time (if there are significant iR drops, there will be severe control problems with multiple-electrode devices). It is necessary to use a single reference electrode to monitor the difference between this "inner" solution potential and the "inner" potential of W1 at the summing point of an operational amplifier current-to-voltage converter (this is the potential of the circuit common; see OA-2 in Fig. 6.17). The potential difference between the inner potentials of the solution and the working electrode determines the chemistry that will occur at the W1 interface. Nothing can be done to incorporate an independent *second* working electrode if this action would change the potential of the bulk solution. Otherwise, the interfacial potential of the first working electrode would be changed. The only recourse is to vary the interfacial potential difference of the second electrode by changing its potential *external to the electrochemical cell.*

If the external potential of W2 was at the potential of the circuit common, it would necessarily have the same interfacial potential as W1, since the bulk solution potential would also be the same for both electrodes. The solution is to float the circuit common for the second (third, etc.) current-to-voltage converter (OA-3) to a potential which is the difference between the desired electrode potentials for W1 and W2 (W1 and W3, etc.).

The circuit illustrated in Figure 6.17 is a simple schematic designed to introduce the principles of the system. Obviously, there is a great deal more involved in producing a practical device. One has to consider such factors as time constants, background current offset circuits, and a convenient means of controlling the two (or more) potentials without having to make inconvenient calculations. There are several different approaches to development of a practical circuit. While these details are beyond the scope of the present chapter, suffice

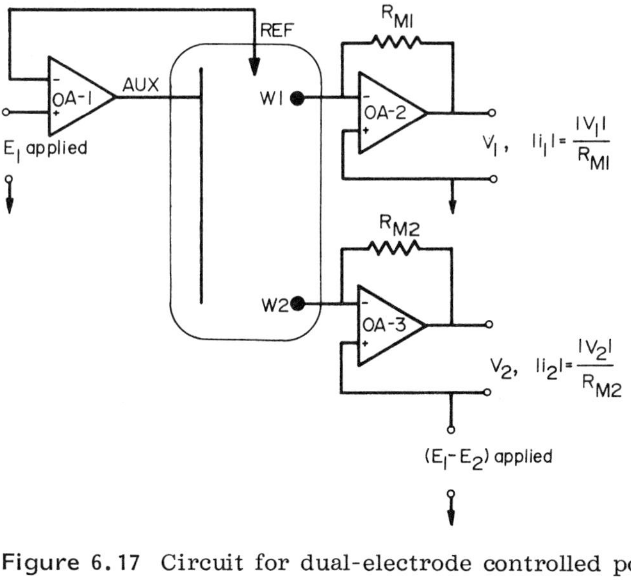


Figure 6.17 Circuit for dual-electrode controlled potential experiments.

it to say that such devices are now available commercially and have been used for many years by electrochemists, usually in connection with ring-disk electrodes).

As was hinted at above, if there is a significant potential drop across the bulk of the solution, problems will arise. This is especially significant with respect to thin-layer electrochemical transducers for liquid chromatography detection because the resistance of the bulk solution can be quite high. In our favor, however, are the very low currents (typically $10^{-10}$-$10^{-9}$ A) usually encountered in such cells. Nevertheless, these iR-drop problems can be very severe with multiple-electrode detectors, causing the events at one electrode to influence the events at the other electrodes. For example, if a reaction at one electrode results in a large current at that electrode, this current might cause an iR drop across the solution which would change the interfacial potential of all other electrodes.

Remember that the interfacial potential is a *difference* between the inner potential of the electrode material and the potential established in the bulk of the solution. An iR drop can cause the latter to change. One way to minimize this problem is to construct the cell with an auxiliary electrode(s) which is directly across from the working electrodes, resulting in a very short current path and, therefore, a very small iR drop. Multiple-electrode transducers constructed in this way minimize "crosstalk" between different electrodes.

## III. CLASSICAL CONTROLLED-CURRENT INSTRUMENTATION

The reader familiar with controlled-potential methodology will have no trouble understanding a controlled-current apparatus. Figure 6.18 illustrates classical approaches to two- and three-electrode constant-current chronopotentiometric experiments (see Chap. 4). The simplicity of these circuits was for many years an attractive feature of chronopotentiometry. Improvements in potentiostats have been largely responsible for a decline in the popularity of "chronopot" in recent years. Nevertheless, constant current experiments are even more important with respect to the coulometric titration.

The necessity for a constant-current source is satisfied (Fig. 6.18) simply by impressing a known voltage across a resistance large in comparison with the equivalent resistance of the cell. If the applied voltage is very much larger than that developed across the cell, the potential-time response of the cell will not significantly influence the applied current. Variation of V and/or R affords convenient adjustment of the excitation current. The interfacial potential of the counter electrode must be independent of the applied current for meaningful cell potentials to be measured using the two-electrode approach.

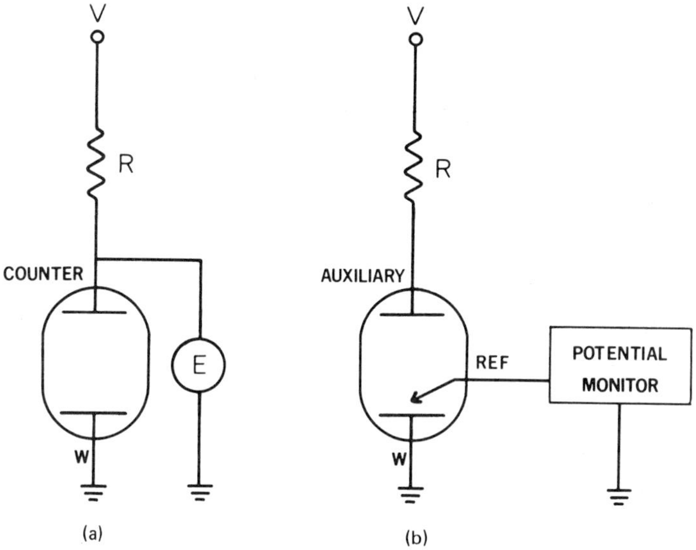


Figure 6.18 (a) Classical two-electrode controlled-current apparatus. (b) Primitive three-electrode controlled-current apparatus.

Figure 6.18b depicts the classical three-electrode constant current experiment. The counter electrode is now replaced by an auxiliary electrode and a reference electrode. The reference electrode enables one to monitor the working electrode potential as a function of time using a high-impedance (zero-current) measuring device. In many early experiments, carried out on a time scale of tens of seconds, a conventional pH meter served very nicely for this purpose. It is important to note that unlike the three-electrode potentiostat, the three-electrode amperostat shown here does not employ negative feedback principles to maintain control over the excitation signal.

## IV. CONTROLLED-CURRENT INSTRUMENTATION BASED ON OPERATIONAL AMPLIFIERS

The evolution of operational amplifier based amperostats is illustrated in Figure 6.19. The OA inverter is a satisfactory starting point. As mentioned earlier, $E_i$ and $R_i$ function together as a current source for $R_f$. Replacing $R_f$ with the electrochemical cell is all that is necessary. Notice (Figure 6.19b) that the working electrode is placed at the summing point and the auxiliary electrode at the output of the OA. Consider the electron flow in Figure 6.19b. What will the sign of $E_i$ have to be when a cathodic current is applied to W? The addition of

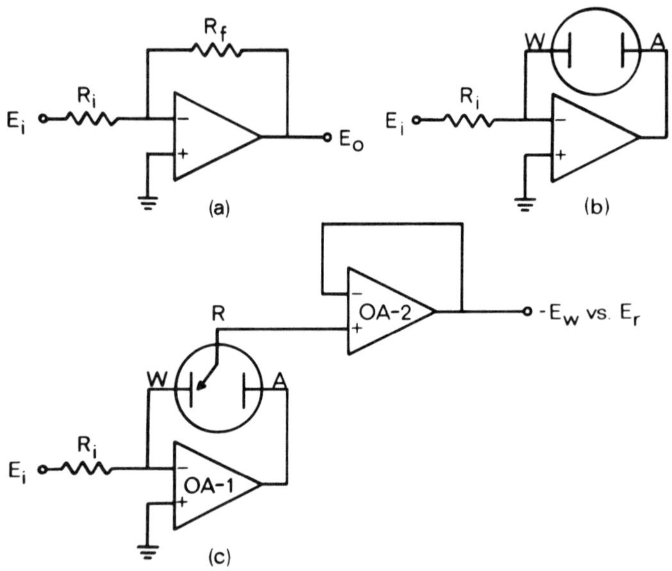


Figure 6.19 Evolution of an operational amplifier three-electrode amperostat.

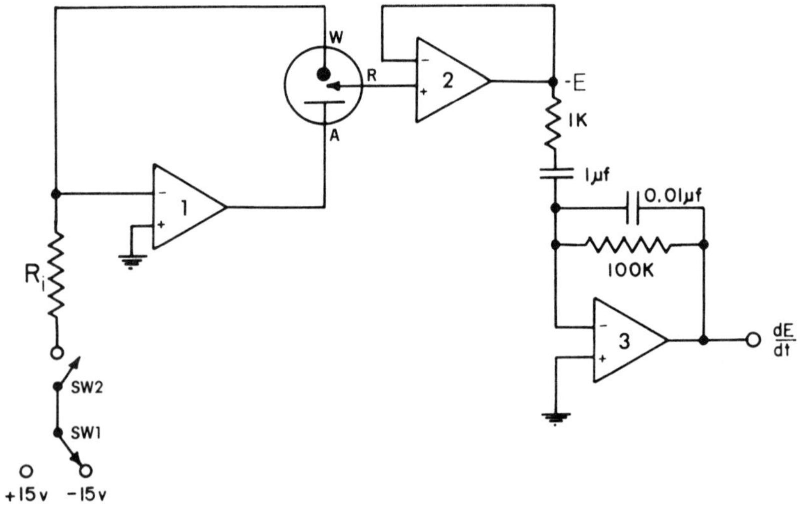


Figure 6.20 Current-reversal chronopotentiometry with derivative output.

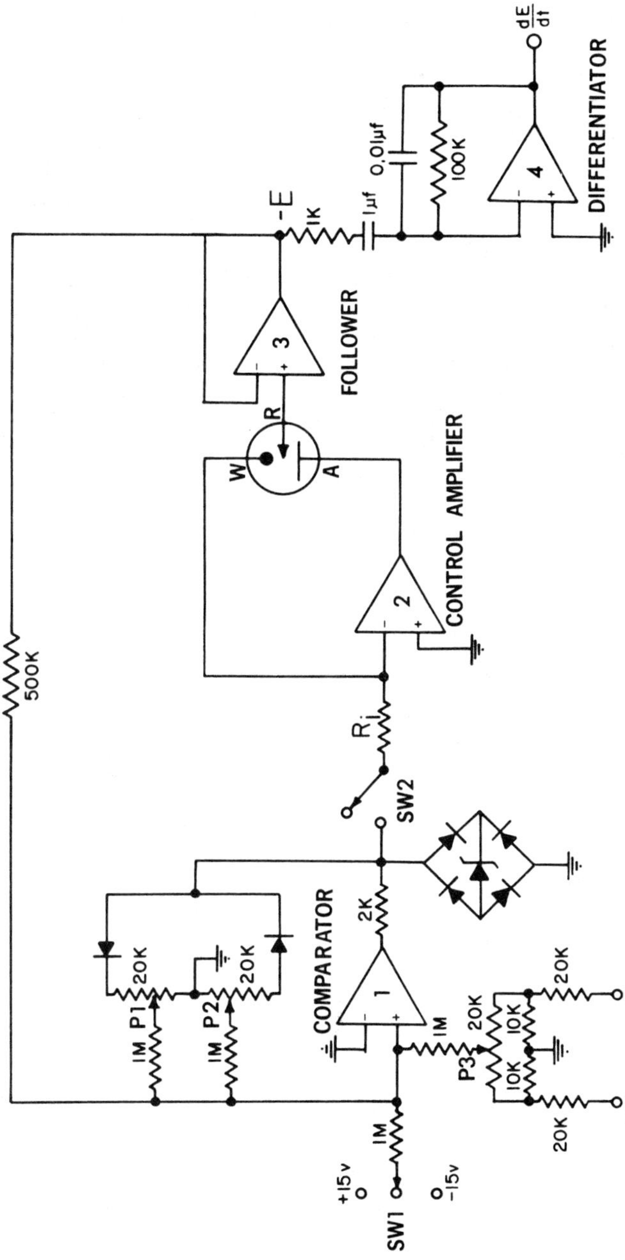


Figure 6.21 Cyclic chronopotentiometry.

a reference electrode and a voltage follower permits one to monitor the working electrode potential as illustrated in Figure 6.19c. The presence of the reference probe has no influence whatsoever on the processes occurring at the working electrode. This is contradistinct to potentiostatic experiments, where removing the reference probe destroys the control loop. There are, in fact, many constant-current experiments where a reference electrode is not needed (e.g., coulometric titrations, constant-current chronoabsorptometry) because the dependent variable of interest is not the potential.

Figure 6.20 illustrates a circuit that has been very widely used. SW1 affords choice of anodic or cathodic current, SW2 initiates the experiment, and then SW1 may be used for current reversal. OA-3 is available for differentiating E with respect to t. A more advanced circuit (Fig. 6.21) incorporates an additional feedback loop and a comparator to perform cyclic chronopotentiometry with automatic switching. Operation of this circuit is perfectly analogous to the cyclic voltammetry circuit discussed in Section II.E.

## V. MICROPROCESSOR-BASED ELECTROCHEMICAL INSTRUMENTATION

The arrival of large-scale integrated circuits in the last 10 years has revolutionized chemical instrumentation just as it has kitchens, automobiles, and television sets. With respect to electrochemistry, the microprocessor has been incorporated in signal generation and data processing, while the basic instrumentation (e.g., potentiostat and current-to-voltage converter) remains as described in earlier sections of this chapter. Microprocessor instruments provide *flexibility* in that different experiments can be implemented via software changes. Such instruments can also be *interactive* in the sense that decisions can be made from one experiment that define the conditions for subsequent experiments (with or without human intervention). Finally, *automated sample processing* and *data storage* can be incorporated more easily than for conventional instruments.

The major difficulty with processor-based instruments is the enormous investment in software development required to make all of these advantages a reality. This can take a great deal of time (person-years in some cases). While such units are flexible, it is necessary to limit flexibility to some degree to provide convenience of operation. This implies that some decisions have to be made early in the development phase that might be difficult to countermand later. In my laboratory, documentation of software by graduate students has been such a problem that it is sometimes easier to start from scratch rather than to modify existing programs!

Commercial processor-based electrochemical instruments are available in two forms. In the first configuration, a general-purpose laboratory or "personal" computer is interfaced to the analog instrumen-

Figure 6.22 Microprocessor-based electrochemical analyzer. (Courtesy of Bioanalytical Systems, Inc., West Lafayette, Ind.)

tation. In the other approach, the package is integrated in such a way that the processor is dedicated to the electrochemical experiments. Several dedicated, processor-based pulse polarographs have been available since the mid-1970s. The first units were developed by Princeton Applied Research at a time when microprocessors were just beginning to influence all chemical instrumentation.

In early 1983, Bioanalytical Systems introduced a new class of integrated processor-driven instrumentation based on a concept first developed by Faulkner and his co-workers [1] at the University of Illinois. This unit (Figs. 6.22 and 6.23) includes a repertoire of some 25 electrochemical techniques, including the most popular large-amplitude (Chap. 3) and small-amplitude (Chap. 5) controlled-potential methods. The unit also is capable of determining electrocapillary curves and can automatically measure and compensate for solution resistance ($R_u$ in Fig. 6.5). Thus in a *single instrument* it is possible to utilize virtually all of the diagnostic criteria introduced in Chapters 3 and 5 and also to explore quickly which technique is optimum for analytical purposes. Data can be digitally smoothed, signal averaged, and background subtracted, and such things as peak currents and peak potential are provided directly in numerical form. The excitation signal parameters and the electrochemical response can be displayed on a video terminal, printer, and/or digital plotter. This type of

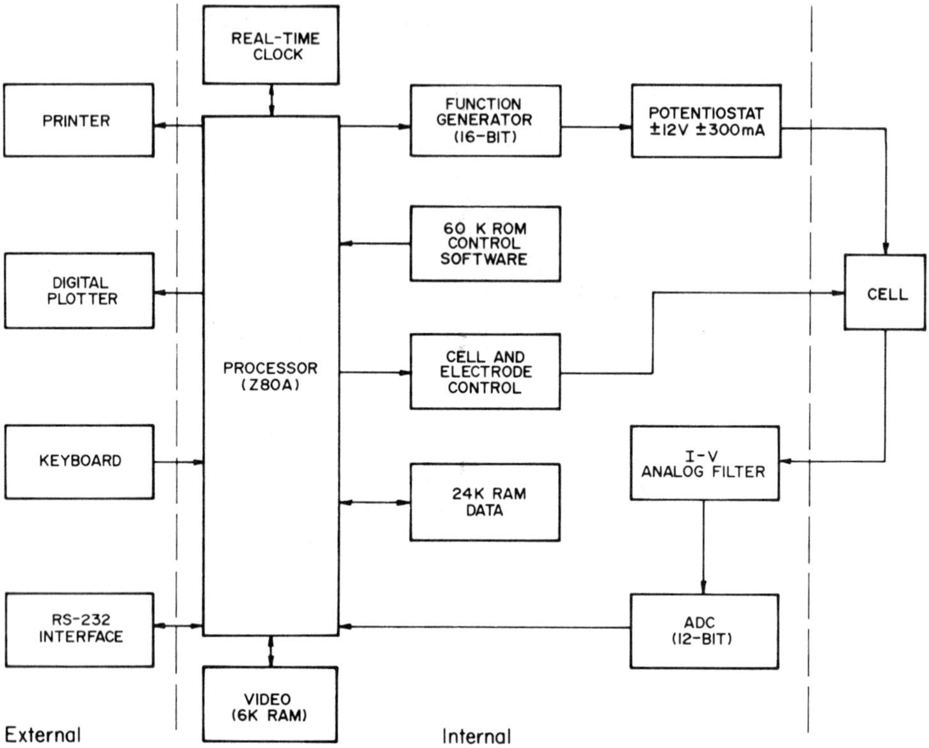


Figure 6.23 Block diagram for BAS electrochemical analyzer.

instrumentation provides unique opportunities which save a great deal of time in electrochemical investigations. A wide variety of experiments can be accomplished on the same solution without the need to change the electronics.

## VI. CONCLUSION

The goal of this chapter was to present an overview of electrochemical instrumentation, not to carry the reader to the point of instrument construction. Most universities have one or more courses in electronics at an adequate level to build rudimentary apparatus. The nature of electrochemistry is such that some knowledge of electronics is extremely useful to good science.

## REFERENCE

1. P. He, J. P. Avery, and L. R. Faulkner, *Anal. Chem.* *54*:1313A (1982).

# 7

# Overcoming Solution Resistance with Stability and Grace in Potentiostatic Circuits

DAVID K. ROE / Department of Chemistry, Portland State University, Portland, Oregon



## I. INTRODUCTION

Three advantages accrue from the use of operational amplifiers (OAs) in controlled-potential instrumentation, as described in Chapter 6. These are: (1) Many independent input signals can be accurately summed and applied to the cell; (2) potential control is relative to a reference electrode which is subject to negligible current; and (3) accuracy of potential control is essentially independent of the reactions at the auxiliary electrode and also of the solution resistance between the auxiliary and reference electrodes. The improvements over the traditional scheme (low-impedance voltage divider and two-electtrode cell) are especially significant with measurement techniques involving moderate to high frequencies or step changes in potential.

However, potential control is not yet perfect since some resistance remains between the reference and working electrodes; the value depends on solution conductivity, the distance between the two electrodes, and electrode geometry. Overcoming this last hurdle with poise is not an easy task, so it is worthwhile to first examine the magnitude of this resistance and its consequences.

In a cell with spherical symmetry, the resistance between the large auxiliary electrode (radius $r_a$) and a small working electrode (radius $r_w$) is given by

$$R_s = \frac{1}{4\pi K}\left(\frac{1}{r_w} - \frac{1}{r_a}\right) \tag{7.1}$$

The specific conductivity of the solution, K $(\text{ohm}\cdot\text{cm})^{-1}$, can be found in reference handbooks or estimated from limiting ionic conductances.

Between these two spheres, the reference electrode is subject to a part of the total resistance. If the distance between the surface of the working electrode and the reference tip is r, this resistance is

$$R_u = \frac{1}{4\pi K}\left(\frac{1}{r_w} - \frac{1}{r_w + r}\right) \tag{7.2}$$

It is $R_u$ that is not compensated by the feedback action of a potentiostat.

Many other electrode geometries in common use approximate parallel, circular planes, and these resistances can be adequately estimated from the following equations (which are approximate in that only straight-line current paths are considered):

$$R_s = \frac{\ell}{K\pi r_w(r_w + r_a)} \tag{7.3}$$

$$R_u = \frac{r}{K\pi r_w(r_w + r_a)} \tag{7.4}$$

Distance between the planes is $\ell$ and the reference electrode tip is located at distance r from the working electrode of radius $r_w$. Often the auxiliary electrode is isolated from the cell by a porous plug; in this case the radius of the plug should be used for $r_a$, instead of the radius of the auxiliary electrode. Additionally, the resistance between the plug and the auxiliary electrode should be included by another calculation based on the geometry of the auxiliary electrode and its compartment. Representative values of cell resistance are listed in Table 7.1 for several common electrolytes.

In addition to solution resistance, electrodes may have appreciable resistance in themselves. The common example is the mercury in a dropping mercury electrode (DME) capillary. Such resistances are summed with solutions resistances since they are in series and the treatment that follows is unchanged.

Table 7.1 Representative Cell Resistances (Ω)

| Cell geometry | 0.1 M KCl | | 1 M KCl | | 0.1 M HCl | | 1 M HCl | |
|---|---|---|---|---|---|---|---|---|
| | $R_s$ | $R_u$ | $R_s$ | $R_u$ | $R_s$ | $R_u$ | $R_s$ | $R_u$ |
| Spherical | | | | | | | | |
| $r_a$ = 10 cm | | | | | | | | |
| $r_w$ = 0.1 cm | | | | | | | | |
| r = 1 cm | 61 | 56 | 7.9 | 7.2 | 20 | 19 | 3.9 | 3.6 |
| r = 0.1 cm | 61 | 31 | 7.9 | 4.0 | 20 | 10 | 3.9 | 2.0 |
| $r_a$ = 1 cm | | | | | | | | |
| $r_w$ = 0.1 cm | | | | | | | | |
| r = 0.2 cm | 55 | 41 | 7.2 | 5.3 | 18 | 14 | 3.6 | 2.7 |
| r = 0.05 cm | 55 | 20 | 7.2 | 2.7 | 18 | 6.8 | 3.6 | 1.3 |
| Planar | | | | | | | | |
| $r_a$ = 10 cm | | | | | | | | |
| $r_w$ = 0.1 cm | | | | | | | | |
| $\ell$ = 10 cm | | | | | | | | |
| r = 0.2 cm | 242 | 4.8 | 32 | 0.64 | 81 | 1.6 | 16 | 0.32 |
| r = 0.05 cm | 242 | 1.2 | 32 | 0.16 | 81 | 0.40 | 16 | 0.08 |
| $r_a$ = 1 cm | | | | | | | | |
| $r_w$ = 0.1 cm | | | | | | | | |
| $\ell$ = 10 cm | | | | | | | | |
| r = 0.2 cm | 2200 | 44 | 2900 | 58 | 740 | 15 | 145 | 2.9 |
| r = 0.05 cm | 2200 | 11 | 2900 | 14 | 740 | 3.7 | 145 | 0.72 |

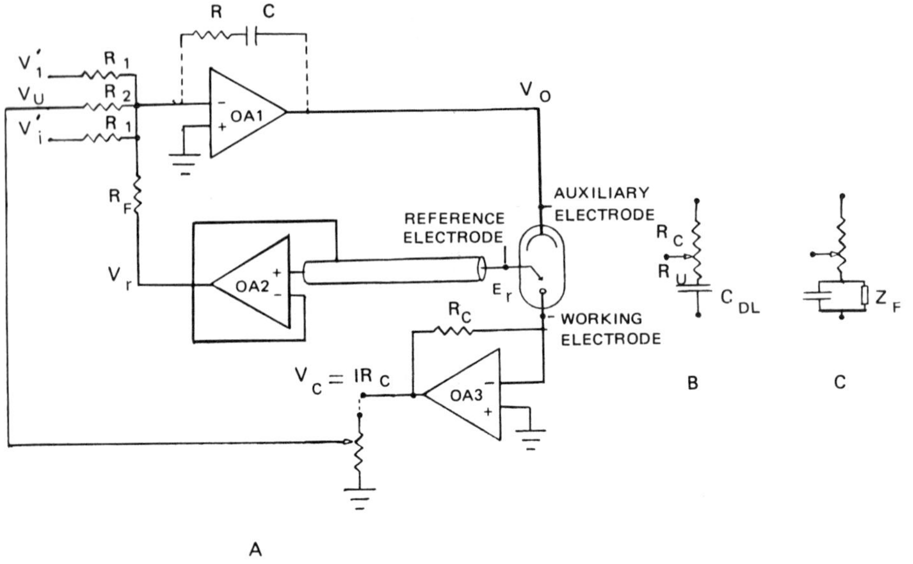


**Figure 7.1** (A) Typical controlled-potential circuit and cell: OAl, the control amplifier; OA2, the voltage follower ($V_r = E_r$); OA3, the current-to-voltage converter. (B) Equivalent circuit of cell: $R_c$, solution resistance between auxiliary and working electrodes; $R_u$, solution resistance between reference and working electrodes, $R_s = R_c + R_u$; and $C_{d\ell}$, capacitance of interface between solution and working electrode. (C) Equivalent circuit with the addition of faradaic impedance $Z_f$ due to charge transfer. Potentials are relative to circuit common and working electrode is effectively held at circuit common ($E_w = 0$) by OA3.

The control amplifier OAl in Figure 7.1A sums the input voltage and applies an output voltage, $V_o$, such that $V_r$ is equal in magnitude but opposite in sign to the scaled sum of input voltages, that is, $V_r = -\Sigma\, V_i' R_f/R_i$. Considering only one input signal and letting the symbol $V_1$ include the scale factor $R_f/R_1$, the reference electrode potential $E_r$ is equal to $-V_1$ since OA2 has a gain of exactly unity. The potential across the solution-working electrode interface is given by

$$E_e = -V_1 - iR_u \tag{7.5a}$$

or

$$E_e = E_r - iR_u \tag{7.5b}$$

The potential of the working electrode relative to the reference electrode, as usually measured, is opposite in sign to $E_r$. Here the current i at the working electrode is given a positive sign if anodic to agree with Ohm's law. The potential given by $iR_u$ is the error in potentiostatic control of the working electrode. Whatever the tolerable error is (e.g., $\leq 1$ mV), one can simply apply Ohm's law and arrive at acceptable trade-offs between i and $R_u$. The range of possible values of i and $R_u$ is so great that any given situation must be examined and decisions reached from experimental values. However, a few generalities can be proferred with qualifications. Conventional polarography and (slow) linear sweep voltammetry involving currents less than about 10 μA will usually be free of significant potential errors by proper positioning of the reference electrode. Coulometry and electrosynthesis experiments routinely involve much larger currents and the resulting $iR_u$ terms are also larger, but the acceptable error level also may be much larger (10-100 mV). Note that for a given current density, the term $iR_u$ is independent of electrode area for the planar arrangement, but for a spherical electrode the magnitude of $iR_u$ depends on $r_w$ and r. If $r \gg r_w$, $iR_u$ is proportional to $r_w$, but if $r_w \gg r$, this term is proportional to r.

Fast linear scans, step changes of potential, and ac methods may be subject to very significant errors because of large faradaic currents and large capacitive currents. Even if the currents are large for a moment, perturbations may be introduced that persist long enough to cause artifacts in the measurements. Some care toward minimization of $R_u$ is usually necessary with this group of techniques.

There are three methods for decreasing the magnitude of the error term $iR_u$: (1) use highly conducting solutions; (2) position the reference tip close to the working electrode; and (3) devise an electronic means to compensate for $iR_u$ by increasing the output of the control amplifier (positive feedback). Although the first method should not be overlooked, it is often impractical with organic solvents. While the second and third methods are feasible means for increasing the accuracy of potential control, they cannot be applied without some attention to stability of the feedback system. In the sections that follow, stability requirements for these two methods will be examined from a practical viewpoint and illustrated with well-defined examples. From this information, some general guidelines will be developed which allow approximate, but easy and useful, assessments of the most direct path to optimum performance. Detailed derivations of equations will be relegated to an appendix.

## II. INPUT-OUTPUT RELATIONS OF CELLS AND POTENTIOSTATS

The equivalent circuit of an electrochemical cell is shown in Figure 7.1B; it is adequate to illustrate the response of a potentiostat system to small (<100 mV) changes in potential. A potentiometer equal in value to $R_s(= R_c + R_u)$ allows $R_u$ to be changed without changing the auxiliary-to-working electrode resistance. This corresponds to changing the position of the reference electrode while the two other electrodes in the cell are fixed; thus only $R_u$ is changed and all other parameters of the system remain constant. To avoid nonlinear effects, the demands on the potentiostat must not exceed its capabilities in terms of output voltage, current, and slew rate. Small values of potential changes and some attention to choices of values of $R_s$, $R_u$, and $C_{d\ell}$ will ensure this condition.

A few comments are in order on the probable validity of conclusions based on this equivalent circuit to real cells. Quite simply stated, real cells that are properly designed will have the same properties as dummy cells of the same values of $R_s$, $R_u$, and $C_{d\ell}$. Important design features of a cell are: (1) equal resistance between all points on the surface of the working electrode and the auxiliary electrode; (2) low-impedance reference electrode; and (3) low stray capacitance between electrodes, between leads, and to shields. Spherical symmetry is a good, but somewhat inconvenient, method of meeting the first requirement; a parallel arrangement also works with planar electrodes. At the very least, two auxiliary electrodes on opposite sides of a spherical or cylindrical working electrode should be used. A low-impedance reference electrode is one that does not introduce an additional time constant into the loop. This means that electrodes designed for use with pH meters should be avoided and the shield of the lead should be connected to the output of the voltage follower, as shown in Figure 7.1. Common sense will help to minimize the third requirement. In particular, avoid long leads between the cell and the potentiostat. Chapter 12 contains many examples of and details of cell designs and further information on the high-frequency characteristics of cells and can be found in the bibliography at the end of this chapter.

Perhaps the most obvious difference between the equivalent circuit to be used and a real cell is the lack of a faradaic impedance in Figure 7.1B. The rationale for omitting $Z_f$ is that the most difficult stability problems occur when only a capacitative load is present. Placing $Z_f$ in parallel with $C_{d\ell}$, as in Figure 7.1C, actually increases the stability of the circuit when the reaction is either rate controlled or diffusion controlled. However, if $Z_f$ includes a capacitive component due to adsorption-desorption of a reactant or product, or if $C_{d\ell}$ changes markedly with potential, a sudden change in circuit stability may occur. As will be seen, it will then be important to assess the range of values of $C_{d\ell}$, or to allow extra margin for stability. So for the majority of situations, this equivalent circuit represents the worst case for design purposes.

From the viewpoint of the control OA, OAI in Figure 7.1A, the cell is simply a voltage divider, producing $E_r$ at the reference electrode terminal when $V_o$ is applied to the auxiliary electrode (note that $E_r$ and $V_o$ are small signal *changes*, not the dc signal). In terms of the equivalent cell of Figure 7.1B, the relation (or transfer function) between the two voltages as vectors, when $V_o$ is a sine wave of frequency f (hertz) is

$$\frac{\overline{E}_r}{\overline{V}_o} = \frac{R_u + X_c}{R_s + X_c} \tag{7.6}$$

Making use of the capacitive reactance equation, $X_c = 1/j2\pi fC_{d\ell}$, and solving for the ratio of the magnitudes of the peak values gives

$$\frac{E_r}{V_o} = \sqrt{\frac{1 + (2\pi fR_uC_{d\ell})^2}{1 + (2\pi fR_sC_{d\ell})^2}} \tag{7.7}$$

There is a phase lag between $E_r$ and $V_o$ because of the capacitance; the angle is

$$\phi = \tan^{-1}\left[-\frac{2\pi fC_{d\ell}(R_s - R_u)}{1 + R_uR_s(2\pi fC_{d\ell})^2}\right] \tag{7.8}$$

These equations express two facts: $E_r$ is less than $V_o$ and lags behind $V_o$ in time. Depending on the frequency, the effects may be very small or significant. It is quite easy to assess the frequency range over which the phase shift is significant: it occurs between $f_s = 1/2\pi R_sC_{d\ell}$ and $f_u = 1/2\pi R_uC_{d\ell}$, and decreases just below and above these points. In Figure 7.2A are plotted the ratio of voltages and the phase lag on linear scales and the frequency, as both f (hertz) and w (rad/s), on a log scale. At these two characteristic frequencies, the phase lag is 45° ($\phi = -45°$). As $R_u$ is decreased, the frequency range between $f_s$ and $f_u$ increases and the phase lag dips toward its limiting value of 90°. For example, if $R_u/R_s = 0.0001$, $\phi = -88.9°$ at the midpoint, regardless of all other parameters. Another characteristic of the graph is that $E_r/V_o = 0.707$ at $f_s$ and has a limiting value of $R_u/R_s$ at $f \gg f_u$.

In the case of potential steps, it is more direct to think in terms of characteristic time constants than in terms of impedances. Two time constants can be defined and used to describe the cell transfer function: $\tau_s = R_sC_{d\ell}$ and $\tau_u = R_uC_{d\ell}$. The units will be in seconds if

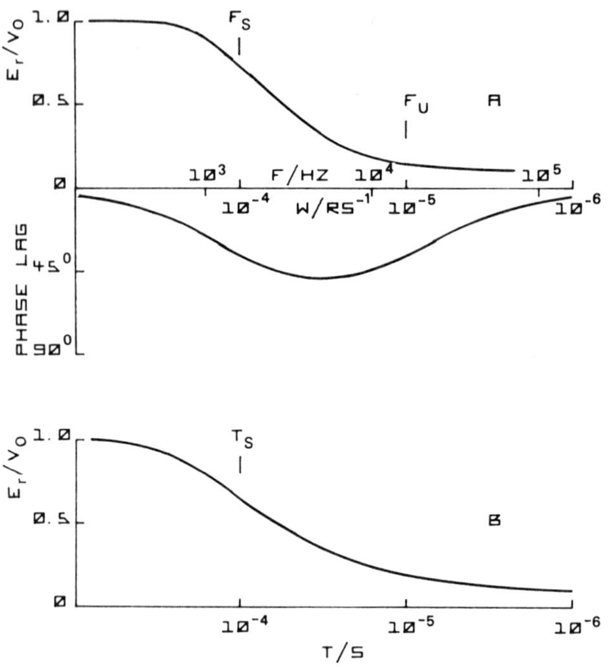


Figure 7.2 (A) Voltage-transfer function and phase lag of equivalent cell with $R_sC_{d\ell} = 1 \times 10^{-4}$ s and $R_uC_{d\ell} = 1 \times 10^{-5}$ s. $V_o$ is amplitude of sine wave of frequency f. (B) Transient response of same equivalent cell. $V_o$ is magnitude of potential step applied at t = 0.

ohms and farads are used. Note that Equations 7.7 and 7.8 could be cast in terms of these time constants. In these terms, the time dependence of $E_r$ of the equivalent cell after a step-change $V_o$ is

$$\frac{E_r}{V_o} = 1 - \left(1 - \frac{\tau_u}{\tau_s}\right) \exp\left(\frac{-t}{\tau_s}\right) \tag{7.9}$$

From the plot of this equation in Figure 7.2B, it is seen that immediately after a step change in $V_o$, the ratio $E_r/V_o$ has the value $\tau_u/\tau_s$, which is 0.1 in the example. Following this, the increase with time is exponential. Note that the abscissa is logarithmic, with time increasing to the left and that the frequency and time scales of Figure 7.2A and B are related by $f = 1/2\pi t$. This allows a direct comparison to be made between the response in the frequency domain (Fig. 7.2A) and the time domain (Fig. 7.2B). Now, the significant point is that over a certain frequency-time region there is a phase shift-time delay. The role of $R_u$, holding $R_s$ constant, in changing the shape of the curves in Figure 7.2 should be evident. The phase or time delay in $E_r$ relative to $V_o$ is an inherent property of electrolytic cells. The consequence of all this is that $E_r$ can fool the control amplifier which then overcompensates, leading to overshoot or even sustained oscillations. We should remember that $R_u$ is the adjustable parameter of interest and we are attempting to see how its value will influence circuit stability and accuracy.

Amplifier characteristics also enter into the list of factors that influence the accuracy of the circuit and its stability. The two parameters of particular interest to this analysis are the dc (or low-frequency limit) open-loop gain and the frequency at which the open-loop gain has decreased to $1/\sqrt{2}$ of the dc value. These parameters are specified by the manufacturer of all OAs and most potentiostats. They can also be easily measured. A common method of presenting performance characteristics of OAs is by means of a graph of log (gain) vs. log

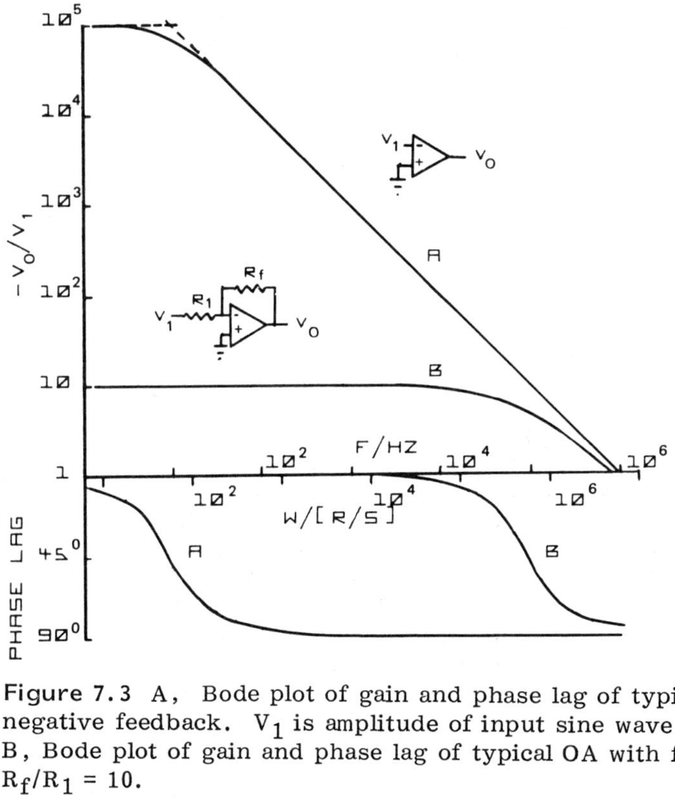


Figure 7.3 A, Bode plot of gain and phase lag of typical OA without negative feedback. $V_1$ is amplitude of input sine wave of frequency f. B, Bode plot of gain and phase lag of typical OA with feedback; $R_f/R_1 = 10$.

(frequency), often referred to as a Bode plot. An example is given in Figure 7.3, curve A, for which the open-loop dc gain, $A_o$, has a value of $1 \times 10^5$ V/V and the frequency at which the gain drops to $7.07 \times 10^4$, $f_a$, is 10 Hz. Alternative terms for this frequency are corner frequency, roll-off frequency, dominant pole frequency, and −3-dB point. Extrapolation of the two straight lines of Figure 7.3, curve A, results in their intersection at this frequency, as shown in dashed lines. If only $A_o$ and the upper bandwidth frequency ($f_o$ in Fig. 7.2A) are available, $f_a$ can be calculated from $f_a = f_o/A_o$; this is a consequence of the slope of gain vs. frequency being −1; at least it should be in a traditional feedback amplifier.

Phase shift always accompanies the decreasing gain of an OA; the plot in Figure 7.3 includes this information. In a manner similar to the equivalent cell circuit, the phase shift is −45° at $f_a$, relative to much lower frequencies, and increases beyond this point to −90° for the remaining, usable frequency region. This phase shift is in addition to the inverting action of the amplifier, so the total phase shift is ($\phi$ − 180°) between the inverting input and the output, but only $\phi$ between the noninverting input (which is not used in this circuit) and output. It is common to omit the −180° in discussing stability, since it is constant and understood to be present.

Such amplifiers are commonly used with a feedback connection; curve B of Figure 7.3 illustrates the relation between input and output for a ratio of feedback resistor $R_f$ to input resistor $R_1$ of Equation 7.10. Under this condition, a gain of −10 is obtained to about f = 10 kHz, beyond which the gain drops and the phase lag commences. The same results would be obtained with the control circuit of Figure 7.1 if $R_s/R_u = 10$ and if $C_{d\ell}$ is shorted. Curve B can be calculated from the equation

$$\frac{V_o}{V_1} = -\frac{R_f/R_1}{1 + (R_f + R_1)/R_1 A} \tag{7.10}$$

in which A is $A_o/(1 + jf/f_a)$, the gain in complex notation. The term $(R_f + R_1)/R_1 A$ must be much smaller than unity to ensure accuracy; this implies a high open-loop gain. The reciprocal, $AR_1/(R_f + R_1)$, is the loop gain under feedback and represents the fraction of $V_o$ that is mixed with the input signal at the inverting input.

A time constant $\tau_a$ can also be defined to represent the delay between output and input of an OA without feedback. It can be calculated from $f_a$ or $f_o$ using the relations $\tau_a = 1/2\pi f_a$ or $A_o/2\pi f_o$. A plot of the open-loop response to a step input of an OA having $\tau_a = 1.6 \times 10^{-2}$ s and $A_o = 1 \times 10^5$ V/V is given in Figure 7.4, curve A. This is the same OA as was used for Figure 7.3, as can be verified with the preceding relations. This plot shows the time lag between output

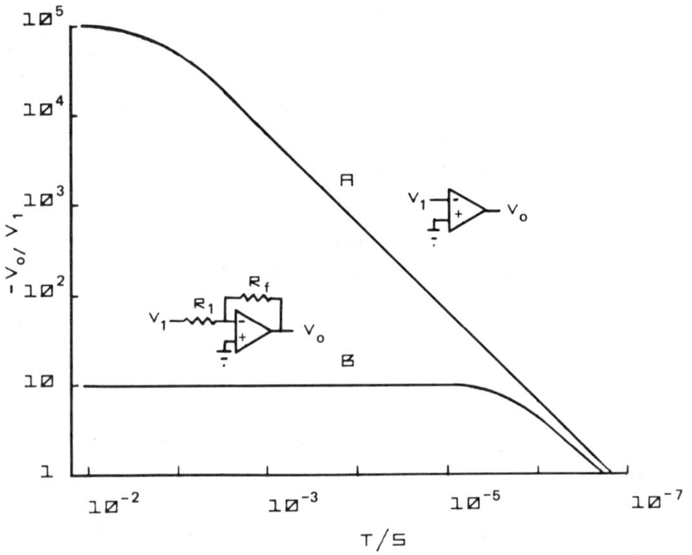


**Figure 7.4** A, Transient response of typical OA without negative feedback. $V_1$ is magnitude of potential step applied at $t = 0$. B, Transient response of typical OA with negative feedback; $R_f/R_1 = 10$.

and input and the time region over which it occurs. Again, a log scale was used for the time axis and time increases to the left to emphasize the similarity between frequency and time domains. The equation of the curve is

$$\frac{V_o}{V_{in}} = -A_o\left[1 - \exp\left(\frac{-t}{\tau_a}\right)\right] \tag{7.11}$$

and applies only if $V_o$ does not exceed the ratings of the amplifier, which is quite likely without feedback since $A_o$ is rather large. Normally, the step response of this type of amplifier is measured with feedback $R_f$ and input, $R_1$ resistors so that the ratio $V_o/V_{in}$ has a reasonable value. Curve B of Figure 7.4 shows the response for $R_f/R_1 = 10$. Under these conditions, the voltage-transfer ratio to a step input is

$$\frac{V_o}{V_{in}} = -\frac{R_f}{R_1 - (R_1 + R_f)/A_o}\left(1 - \exp\left\{-\frac{t[R_f + R_1(A_o - 1)]}{\tau_a(R_1 + R_f)}\right\}\right) \tag{7.12}$$

or more simply, but accurate to $\leqslant 0.1\%$ when $R_f/R_1 \leqslant 100$,

$$\frac{V_o}{V_{in}} = -\frac{R_f}{R_1}\left\{1 - \exp\left[-\frac{tA_oR_1}{\tau_a(R_f + R_1)}\right]\right\} \tag{7.13}$$

Now the effective time constant is decreased by a factor of $A_oR_1/(R_f + R_1)$, which is the loop gain in the long-time limit.

## III. STABILITY OF POTENTIOSTAT-CELL CIRCUITS AND THE ROLE OF $R_u$

So far, the individual transfer functions of cells and potentiostats have been described. It is time to put them together and seek answers to the question: How does the value of $R_u$ affect the stability of the circuit? An intuitive answer will be developed before delving into a quantitative relation.

Connecting an electrochemical cell, or its equivalent, to an OA introduces a phase or time lag between the output and the input (Fig. 7.2A and B). Depending on the amount of phase or time lag relative to the characteristics of the OA, we can anticipate that the output $V_o$ may become excessively large while the corrective feedback signal is delayed. Sinusoid signals in this frequency range will have a larger-than-expected amplitude and increased phase shift, leading to artifacts in ac voltammograms. Step and pulse response will have overshoot and perhaps damped oscillations. In both cases, sustained oscillations may result if the total phase or time lag is extreme.

With sinusoid input signals, the output signal, and therefore, $E_r$, will be larger than anticipated where the frequency is in the range where the phase shift from the cell adds to the phase shift of the amplifier. The magnitude of the error depends on the loop gain and it is at its worst near the upper frequency limit of the amplifier.

Figure 7.5, curve A, represents the combination of the equivalent cell of Figure 7.2A and the amplifier of Figure 7.3. The phase lag caused by $C_{d\ell}$ produced the rise in the ratio $E_r/V_1$ and also resulted in a rapid change in phase just beyond 10 kHz. The magnitude of both effects is related to $R_u$, as shown by curves B and C. In assessing these changes, it is useful to refer to Figure 7.2A; decreasing $R_u$ allows the phase lag to increase and also makes the loop gain decrease beyond $f_u$.

Before introducing further examples, the significance of phase shifts upon measurement error in ac voltammetry should be explored. Since the basic purpose is to measure an ac current due to a faradaic process, the current due to $C_{d\ell}$ must be rejected as completely as

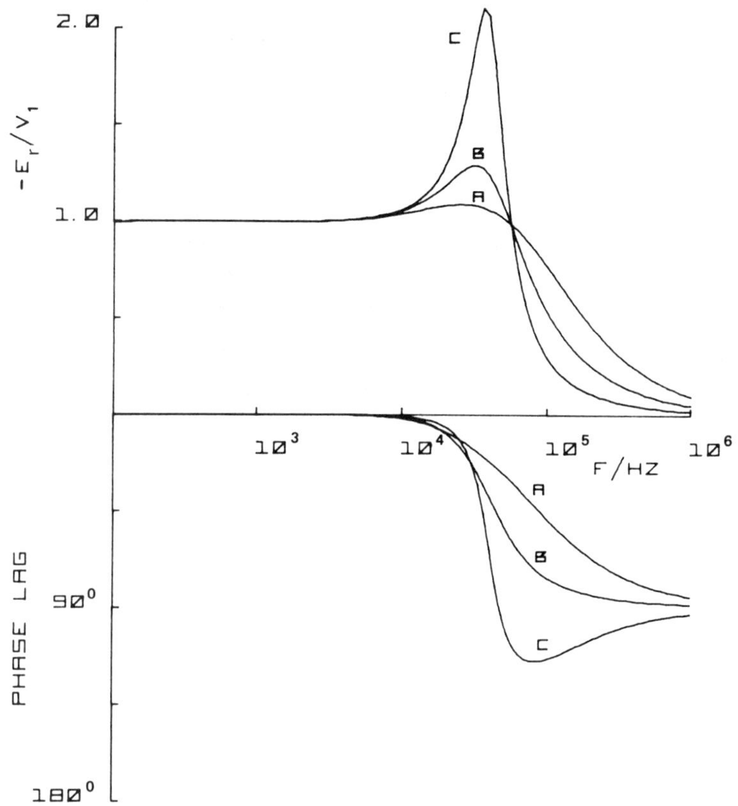


Figure 7.5 Transfer function and phase lag of potentiostat and equivalent cell at reference electrode. $A_o = 1 \times 10^5$, $f_a = 10$ Hz, $R_sC_{d\ell} = 1 \times 10^{-4}$ s with $R_uC_{d\ell} = 1 \times 10^{-5}$ s (A), $5 \times 10^{-6}$ s (B), and $2 \times 10^{-6}$ s (C).

possible. This is accomplished by making use of the fact that the capacitive current $i_c$ and the faradaic current $i_f$ are not in phase but differ by 45 to 90°, depending on the reversibility of the charge-transfer reaction. If the phase of the capacitive current relative to $V_1$ (or $E_r$) is precisely known, the rejection of $i_c$ is easily visualized if $i_f$ is measured the instant that $i_c$ passes through zero. Most instruments rely on a technique known variously as synchronous, lock-in, or phase-selective detection. If $E_e$ is the potential across the solution-electrode interface, it can be stated with certainty that $i_c$ leads $E_e$ by 90°. Unfortunately, $E_e$ is not experimentally available, but it is possible to illustrate the phase difference between $E_e$ and $V_1$. Then, using $V_1$ as the reference, the error can be predicted as a function of frequency and circuit parameters. Figure 7.6 gives the ratio $E_e/V_1$ and the phase lag for the same system as in the preceding figure. The main differences are the increase in phase lag and the decrease in the peak of voltage ratios. Both effects are due to the low-pass-filter action of $R_uC_{d\ell}$.

As long as $i_c$ does not change in phase relative to $V_1$, the measurement will be essentially free of this error after setting the synchronous detector's phase to 90°. Any change in the relative phase of $i_c$ introduces an error proportional to the cosine of the phase angle. For example, on curve A' of Figure 7.6 at f = $10^3$ Hz, the phase of $E_e$ has changed by about 4° and the same change must occur in the phase of $i_c$. Therefore, a fraction of the peak value of $i_c$ is measured and this fraction is cos 86° or 0.07. Depending on the circumstances, this may be an acceptable error in the baseline of an ac voltammogram.

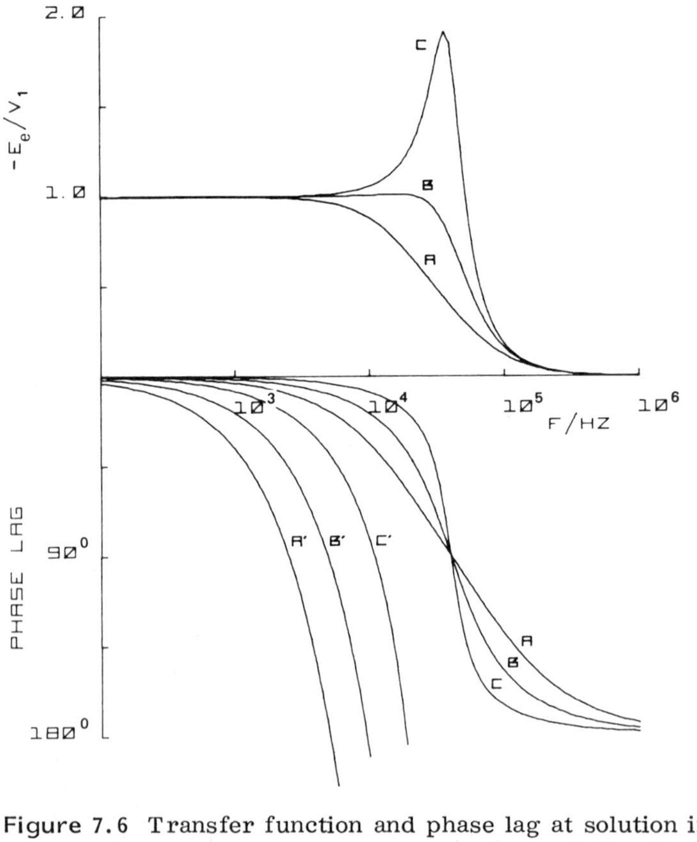


Figure 7.6 Transfer function and phase lag at solution interface of working electrode; same parameters as Figure 7.5. A', B', and C' are phase lag × 10.

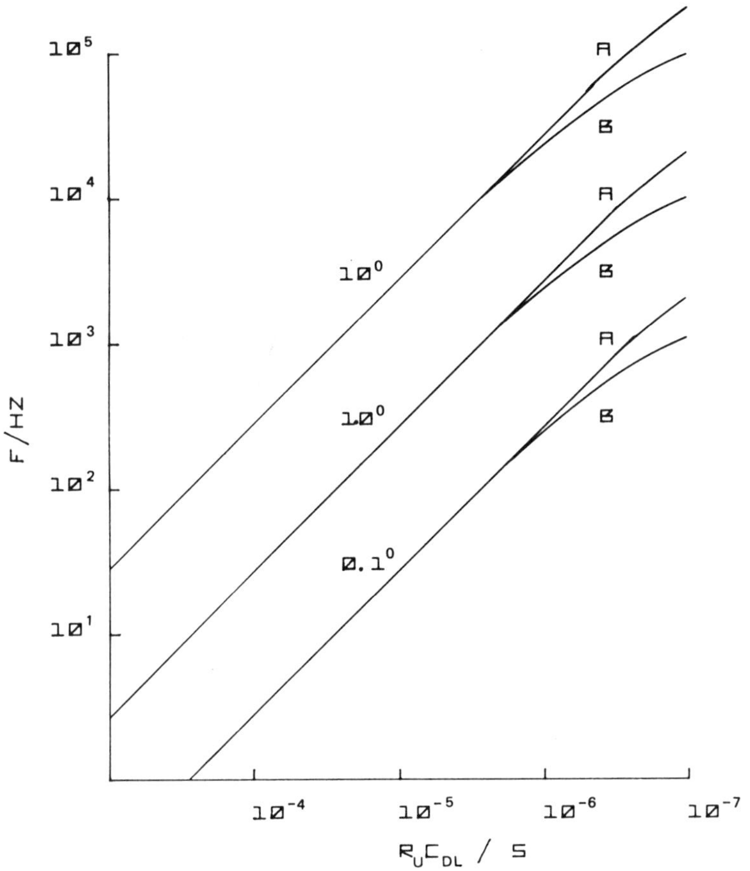


Figure 7.7 Frequency at which phase shift of capacitive current relative to $V_1$ equals 0.1°, 1.0°, and 10° as a function of $R_uC_{d\ell}$. Curve A applies if $f_a$ = 100 Hz, and curve B is $f_a$ = 10 Hz. $R_s/R_u \leqslant 50$ for all curves.

Figure 7.7 summarizes the frequency at which the phase error equals 0.1°, 1°, and 10° as a function of $\tau_u = R_uC_{d\ell}$. Over the range where the plots are straight lines, the phase error is due solely to $\tau_u$, but as this time constant is decreased, the control amplifier and $\tau_s$ cause the plots to curve.

Several further examples of response to ac signals are given in Figures 7.8 to 7.11 to show the changes caused by increasing amplifier bandwidth and decreasing $\tau_s$. It is evident that $R_u$ plays a pivotal role in all of these examples, both in determining total phase shift (stability) and maximum operating frequency for a given level of

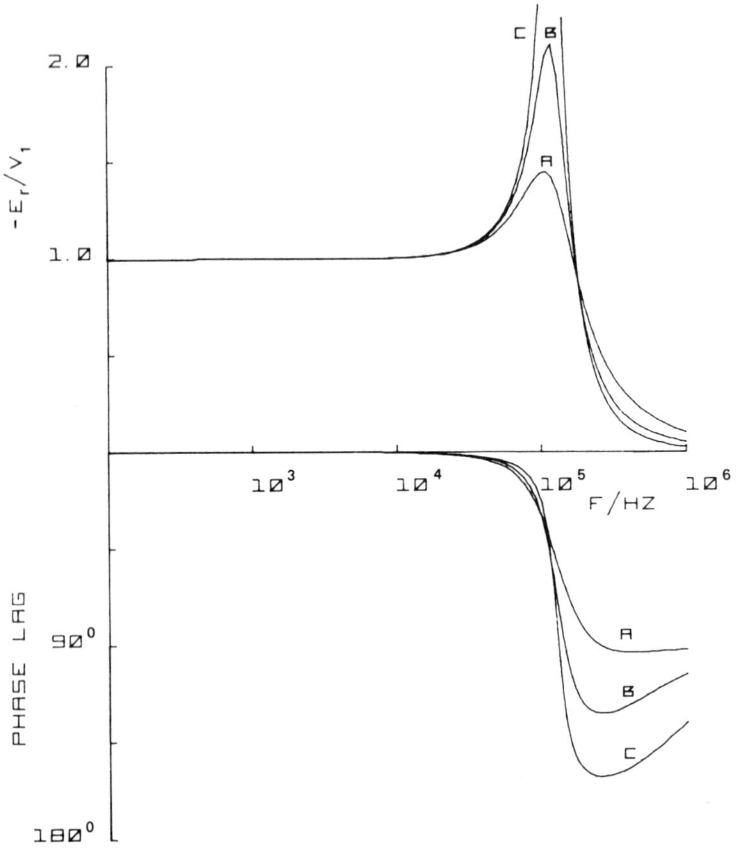


**Figure 7.8** Transfer function and phase lag at reference electrode of potentiostat and equivalent cell; $A_0 = 1 \times 10^5$, $f_a = 10$ Hz, $R_s C_{d\ell} = 1 \times 10^{-5}$ s with $R_u C_{d\ell} = 1 \times 10^{-7}$ s (B), and $2 \times 10^{-6}$ s (C).

measurement error. While the acceptable error can be selected to meet experimental requirements, the total phase shift between $E_r$ and $V_0$ should be less than $-135°$ due to the factors cited. Real systems introduce further phase errors (usually in the wrong direction) and some leeway must be retained. When the phase lag due to all causes reaches $-180°$, the potentiostat becomes an oscillator at the peak frequency.

Time-domain response of feedback amplifiers has been frequently correlated with frequency-domain behavior, and vice versa. Examples have usually been restricted to those situations in which only the amplifier contributes phase shift (single pole) or where a second source of phase shift (two poles) was included, such as from nonideal amplifier design or from the effects of stray capacitance at the input terminal. The system of interest in electrochemistry is more complicated than a two-pole system because there is also a decreasing phase shift caused by $R_uC_{d\ell}$. This is referred to as a "zero." Consequently, predicting rise times and overshoot in the time domain from transfer functions in the frequency domain may be unreliable. So exact solutions have been obtained and from them the detailed influence of $R_u$ on system stability and accuracy will be examined. From these results, a useful empirical relation was found to summarize the behavior for the control amplifier and equivalent cell of Figure 7.1.

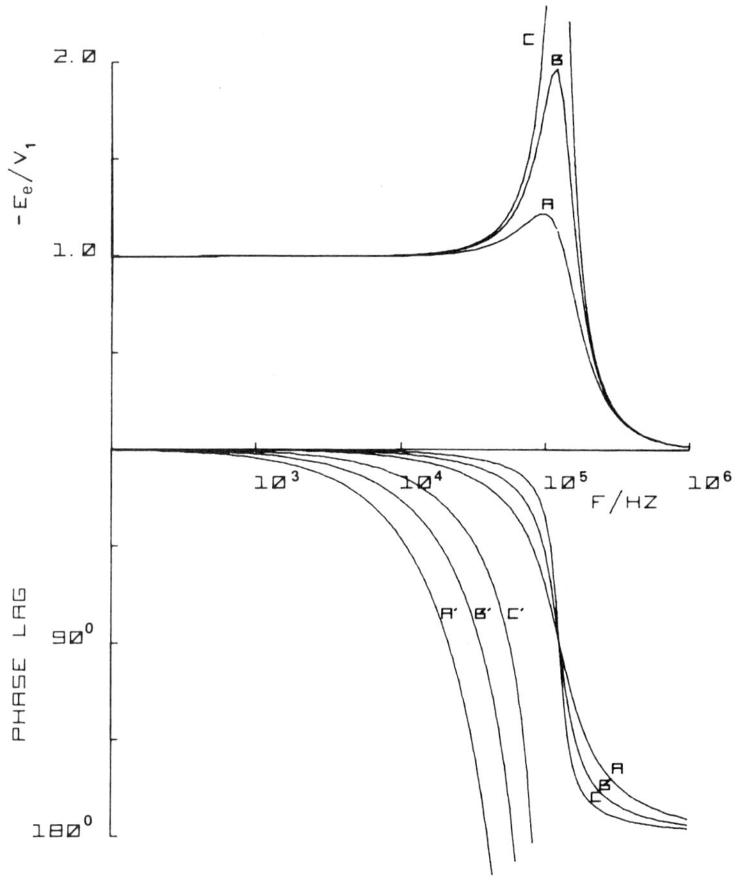


Figure 7.9 Transfer function and phase lag at solution interface of working electrode; same parameters as Figure 7.8. A', B', and C' are phase lag × 10.

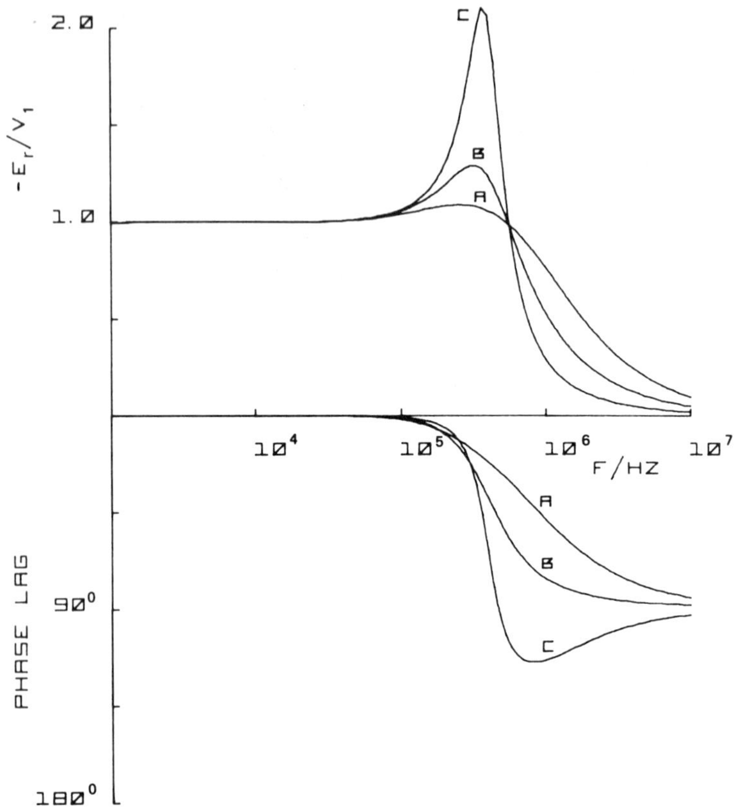


Figure 7.10 Same parameters as Figure 7.8 except that $f_a$ = 100 Hz.

Upon application of a potential step $V_1$ to the control amplifier, the potential $E_r$ rises with time at a rate determined by the combined effects of the amplifier gain $A_o$ and the three time constants $\tau_a$, $\tau_s$, and $\tau_u$. The potential across the solution-working electrode interface $E_e$ is delayed relative to $E_r$ by the time constant $\tau_u$. Calculated curves for a range of parameters are given in Figures 7.12 to 7.14.

Three general trends are evident from these graphs.

1. As $\tau_u$ is decreased ($\tau_s$ and $\tau_a$ held constant), the rise time of $E_r$ increases, the rise time of $E_e$ decreases, and overshoot increases.
2. As $\tau_s$ is decreased ($\tau_u/\tau_s$ and $\tau_a$ held constant), the rise time of $E_r$ and $E_e$ decrease and overshoot decreases.
3. As $\tau_a$ is decreased ($\tau_u$ and $\tau_s$ held constant), the rise time of $E_r$ and $E_e$ decrease and overshoot decreases.

A reflection on the curves of Figures 7.2B and 7.4 should reveal the same conclusions. The relative benefits of decreasing $R_u$, and therefore $\tau_u$, in order to decrease the $iR_u$ drop must be carefully compared with the disadvantage of marginal stability. In order to have a reasonable means of predicting instability, as measured by overshoot of $E_r/V_1$, a large number of curves were calculated and the percent overshoot was plotted vs. $(\tau_s\tau_a/A_o)^{1/2}\tau_u$, as shown in Figure 7.15. Three curves are plotted to account for the relative contributions of the time delays of the amplifier and of the cell. The curves are valid for any reasonable combination of cell and amplifier parameters of interest in electrochemistry. A real system will no doubt exhibit greater overshoot than predicted by those curves due to stray capacitance and lead inductance. However, results very close to these

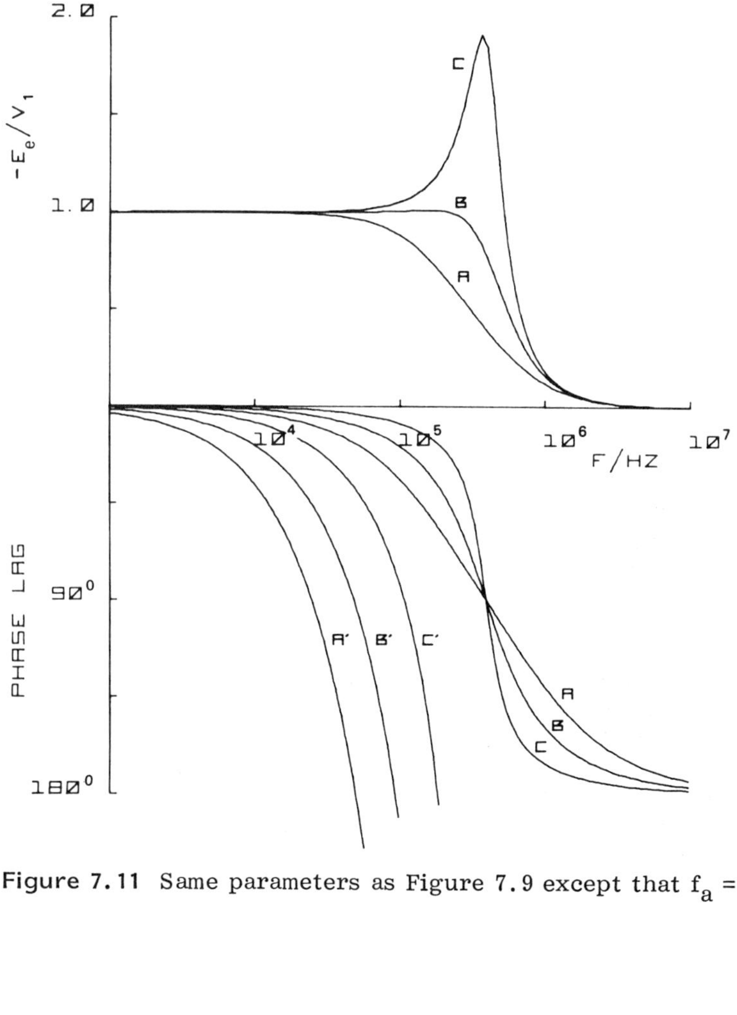


Figure 7.11 Same parameters as Figure 7.9 except that $f_a$ = 100 Hz.

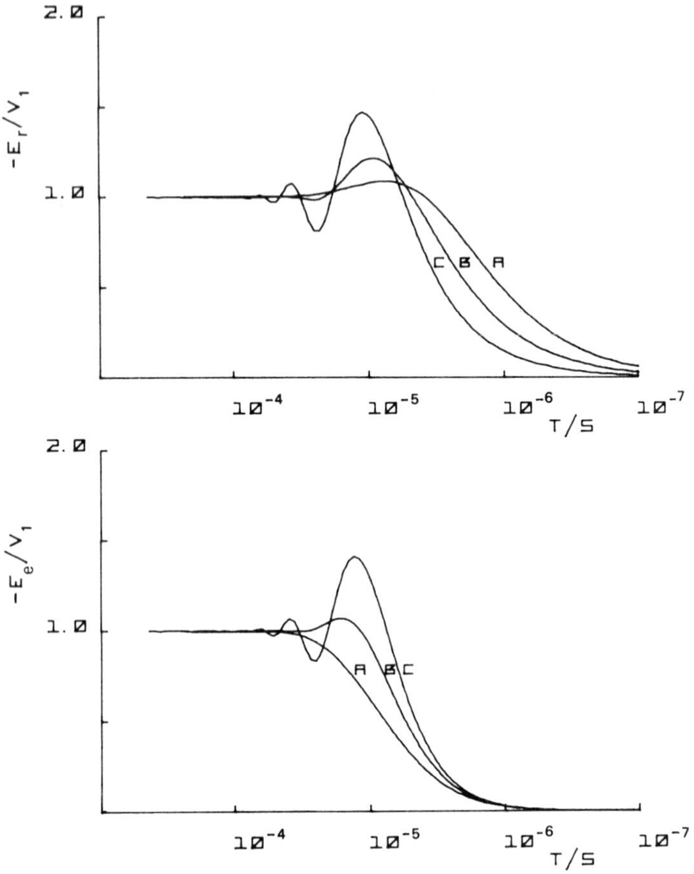


Figure 7.12 Transient response of potentiostat and equivalent cell at reference electrode (top) and at solution interface of working electrode (bottom); $A_o = 1 \times 10^5$, $f_a = 10$ Hz, $R_sC_{d\ell} = 1 \times 10^{-4}$ s with $R_uC_{d\ell} = 1 \times 10^{-5}$ s (A), $5 \times 10^{-6}$ s (B), and $2 \times 10^{-6}$ s (C).

are obtainable in practice and the graph at least provides a good starting point for system optimization.

A small amount of overshoot in $E_r$ is quite acceptable, as a comparison of curves in Figures 7.12 to 7.14, reveals. If $E_r$ has <10% overshoot, $E_e$ will have a negligible amount for a very wide range of cell parameters. Also, the presence of a faradaic process that does not alter $C_{d\ell}$ will tend to decrease overshoot in both $E_r$ and $E_e$ due to a decrease in phase and time lag, but the faradaic current increases the error due to $iR_u$.

In all of the numerical examples illustrated thus far, the potential changes across the interface were primarily governed by $\tau_u$ rather than by amplifier characteristics. Under these conditions, it is a simple matter to calculate the current due to double-layer charging if $R_u$ and $C_{d\ell}$ are known or can be estimated. The peak value of this current is equal to the change in potential divided by $R_u$ and it decays exponentially thereafter. Following a given number of time constants, the contribution of the charging current to the total current measured can be determined and thereby the extent of error in the quantity of interest, the faradaic current, can be estimated.

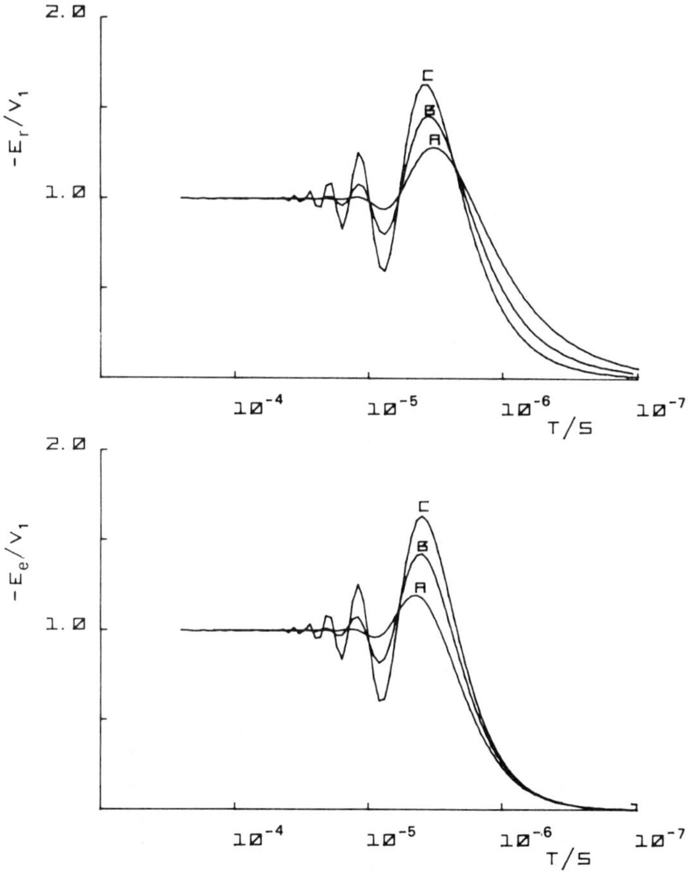


**Figure 7.13** Same as Figure 7.12 except that $R_sC_{d\ell} = 1 \times 10^{-5}$ s with $R_uC_{d\ell} = 1 \times 10^{-6}$ s (A), $5 \times 10^{-7}$ s (B), and $2 \times 10^{-7}$ s (C).

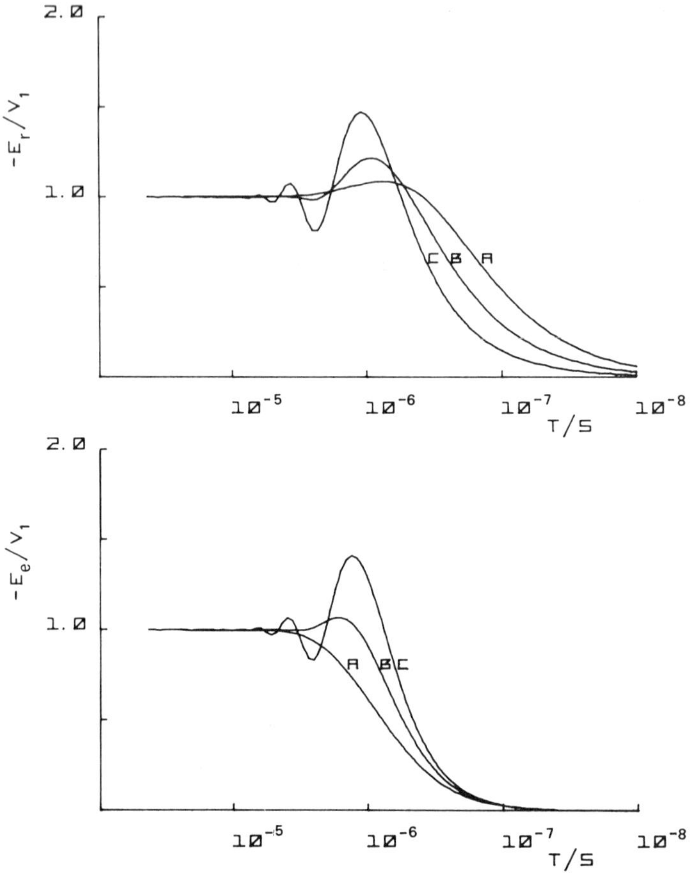


**Figure 7.14** Same as Figure 7.13 except that $f_a$ = 100 Hz.

The preceding suggests that a wide range of values of $\tau_s$ and $\tau_u$ might be worth examining. For example, if $\tau_s$ and $\tau_u$ are both less than $\tau_o$, the phase and time lag due to the cell may become negligible in the control loop and instability is decreased. Figures 7.16 and 7.17 show that this is indeed true; the overshoot becomes negligible as $\tau_s/\tau_o$ is made less than unity. There are two methods for meeting this condition: (1) decrease the cell resistance and/or electrode capacitance (area), and (2) increase $\tau_o$ by decreasing amplifier bandwidth. The first method is somewhat limited in practice, but calls attention to the advantages of low cell resistance and capacitance. The second approach should always work and has been extensively used. It is usually implemented by connecting a capacitor between the inverting input and output of the control amplifier; some potentiostats have this or an equivalent provision built in. There is a second benefit from this decrease in bandwidth; the system noise level decreases. Although this follows general rules of noise-bandwidth relations, it is particularly noticeable when changes in bandwidth eliminate a peak in $E_r/V_1$ because noise in the frequency region of the peak receives excess amplification. Further details on bandwidth tailoring will be given in a later section.

Finally, it should be noted that the condition $\tau_s \cong \tau_u$ also avoids instability regardless of the values of the parameters; at worst there will be a small amount of overshoot if $\tau_o$ is close to either time constant and the latter are separated by a small amount. When the loop gain is high, system response is smooth, as is illustrated for $\tau_s = 10^{-3}$ s, $\tau_u = 5 \times 10^{-4}$ s, and $\tau_o = 1.6 \times 10^{-7}$ s in Figures 7.18 and 7.19. The

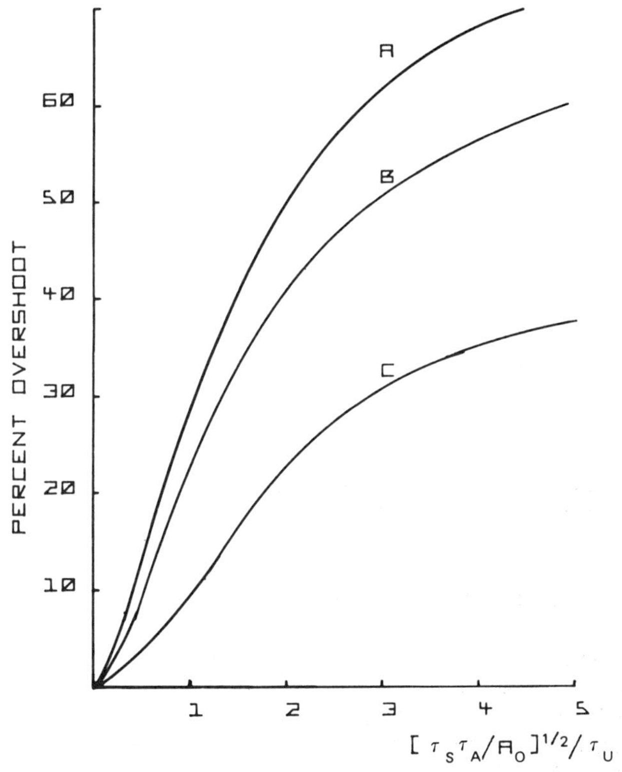


Figure 7.15 Overshoot of $E_r$ relative to $V_1$; A, $\tau_s = 1 \times 10^3\ (\tau_a/A_o)$; B, $\tau_s = 100\ (\tau_a/A_o)$; C, $\tau_s = 10\ (\tau_a/A_o)$.

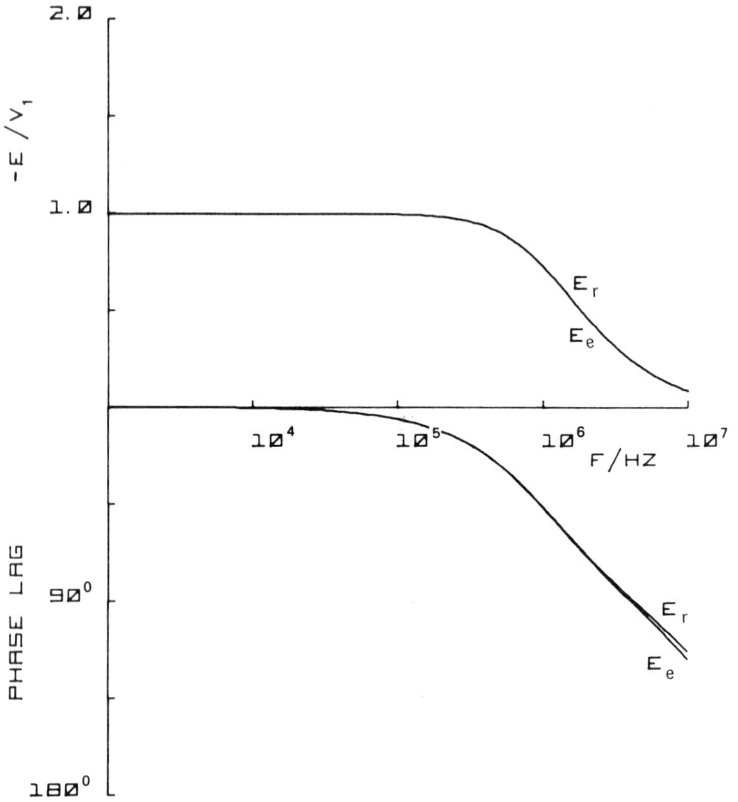


Figure 7.16 Transfer function and phase lag for $A_o = 1 \times 10^5$, $f_a = 10$ Hz, $R_sC_{d\ell} = 1 \times 10^{-8}$ s, and $R_uC_{d\ell} = 1 \times 10^{-9}$ s.

trade-off is a loss of potential control; $E_e$ will depend on cell current and the acceptability of the error must be determined in each case.

## IV. COMPENSATION FOR $iR_u$ BY POSITIVE FEEDBACK

Whenever cell parameters do not allow a reduction in $R_u$ commensurate with potentiostatic performance requirements, the technique of positive feedback can be implemented to achieve the goal. This method is nearly as old as the analog potentiostat and it has been applied in a variety of ways with highly variable results. In principle, the error due to $iR_u$ can be reduced to zero (or even made negative) through positive feedback. Fortunately, stability problems are unchanged in principle relative to the examples presented thus far, but they may be somewhat more demanding of careful analysis in practice. Through the illustrations of this section, one point will emerge: positive feedback of a signal proportional to cell current causes exactly the same changes in system response as moving the reference electrode closer to the working electrode. In the final section, methods will be given whereby graceful control may be achieved with or without positive feedback.

The signal $V_u$, for compensation of $iR_u$, must be proportional to this quantity and identical in phase. A convenient source of $V_u$ is the output of the current follower, OA3, of Figure 7.1A. Adjustment of the magnitude is provided by a potentiometer so that some fraction F of $iR_u$ is compensated. Because of the wide range of resistor values

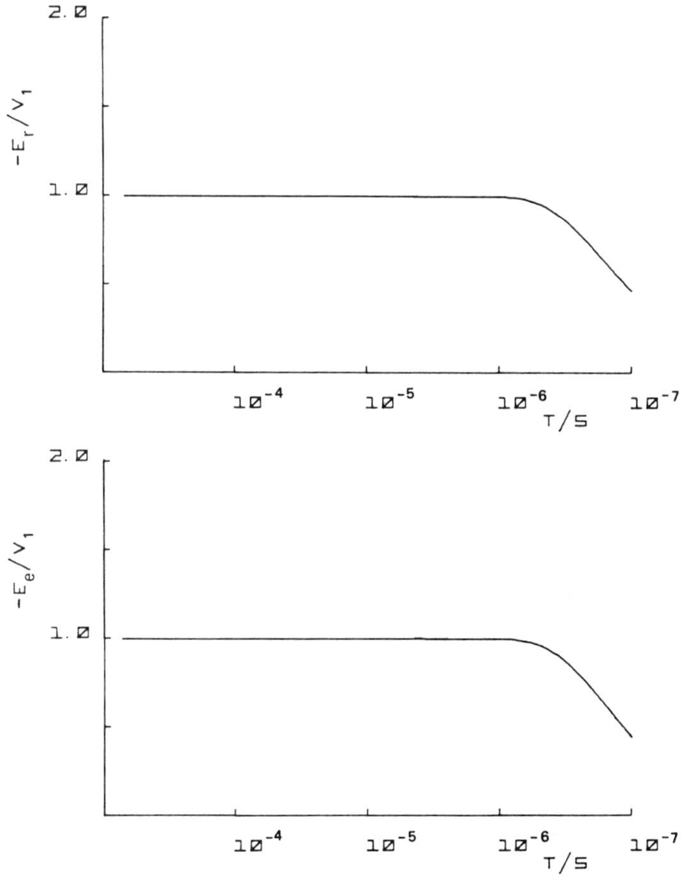


Figure 7.17 Transient response to step input. Same parameters as Figure 7.16.

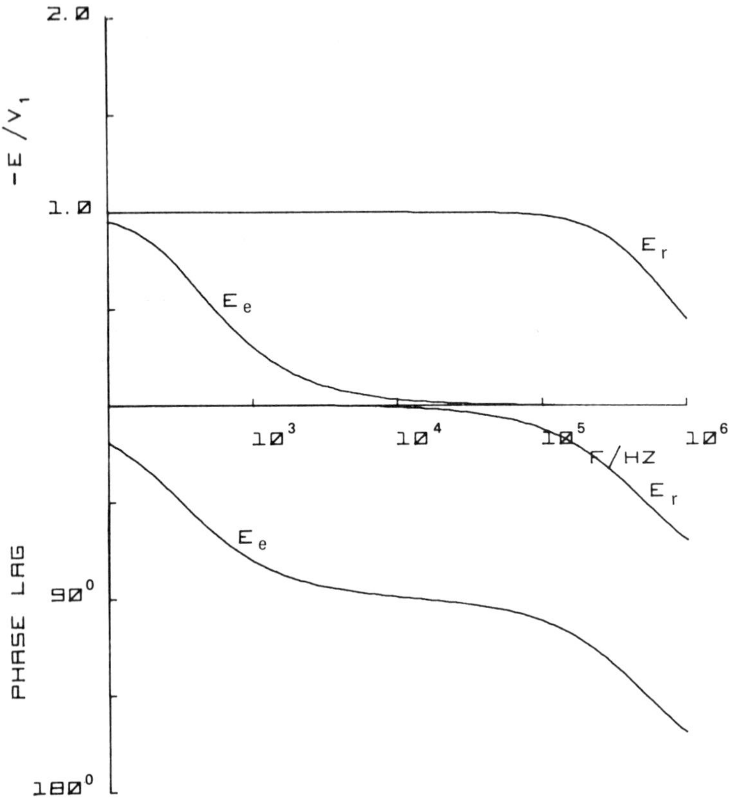


**Figure 7.18** Transfer function and phase lag for $A_o = 1 \times 10^5$, $f_a$ = 10 Hz, $R_sC_{d\ell} = 1 \times 10^{-3}$ s, and $R_uC_{d\ell} = 5 \times 10^{-4}$ s.

that may be used in the circuit, $V_u$ will not generally be equal to $F(iR_u)$, but will be scaled according to the equation

$$V_u = V_c \frac{FR_uR_2}{R_cR_f} \tag{7.14}$$

For any given set of resistors, the potentiometer may be calibrated in ohms of compensated resistance through the relation

$$R_u = \frac{\beta R_cR_f}{R_2} \tag{7.15}$$

where β is the fraction of $V_c$ applied to $R_2$. For clarity, the examples will refer to the fraction F of $R_u$ that is compensated, not to β.

Before attempting to complete the positive feedback loop, the signal $V_u$ (and therefore $V_c$) must meet the same requirements that were imposed on $V_f$: no phase-time lag in the frequency region of operation of the control loop around AO1. This is a demanding task and requires that OA3 be a high-performance device in the upper-frequency regions. The task is made easier if $R_c$ is as small as practical while allowing β to equal unity for the maximum value of $R_u$ anticipated. An additional OA in a noninverting configuration may be added to amplify $V_c$, allowing $R_c$ to be kept small. Two OAs in series

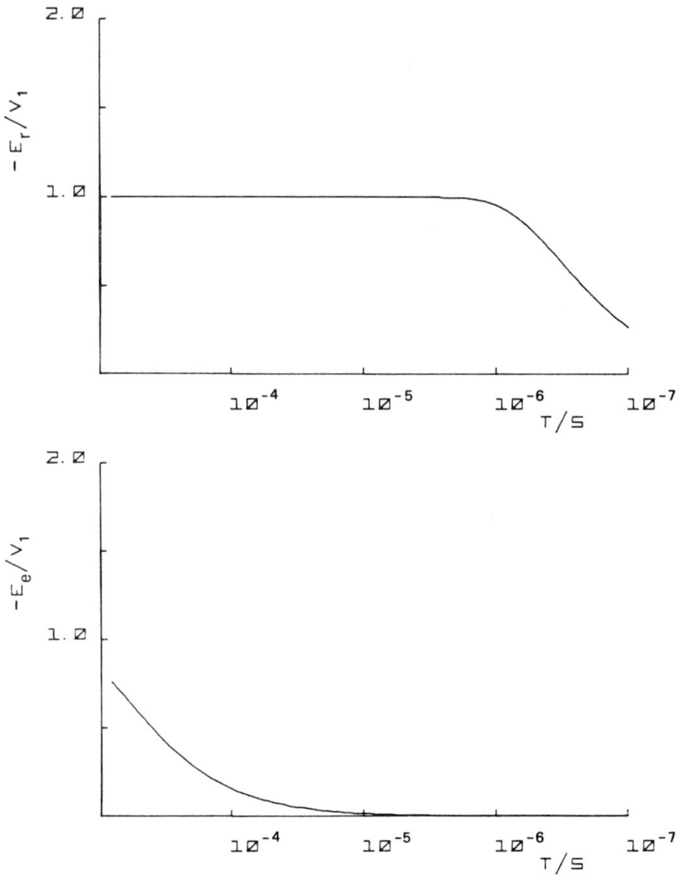


Figure 7.19 Transient response to step input. Same parameters as Figure 7.18.

will always outperform a single OA of the same type in terms of gain-bandwidth product and therefore phase/time lag will be reduced.

With the compensation signal, $V_u$, applied to $R_2$, the frequency-domain response for several values of F are plotted in Figure 7.20. The control OA and cell parameters are the same as those used for Figures 7.5 and 7.6, but here only $E_e/V_1$ is given. Curves A to C are identical with their counterparts of Figure 7.6, since values of F are 0, 0.5, and 0.8, corresponding to $C_{d\ell}R_u$ values of $10^{-5}$, $5 \times 10^{-6}$, and $2 \times 10^{-6}$ s. Complete compensation of the effects of $R_u$ results in curve D. Again it should be noted that a real system will undoubtedly have a small amount of additional phase lag in the high-frequency

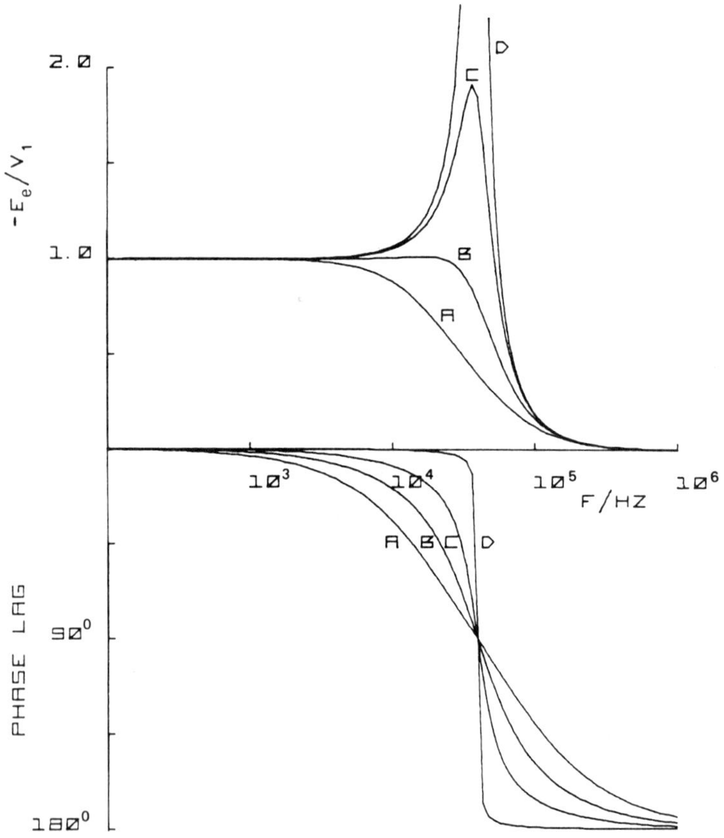


Figure 7.20 Transfer function and phase lag for $A_o = 1 \times 10^5$, $f_a$ = 10 Hz, $R_sC_{d\ell} = 1 \times 10^{-4}$ s, $R_uC_{d\ell} = 1 \times 10^{-5}$ s, and fractional compensation of $R_u$ equal to (A) 0, (B) 0.5, (C) 0.8, and (D) 1.0.

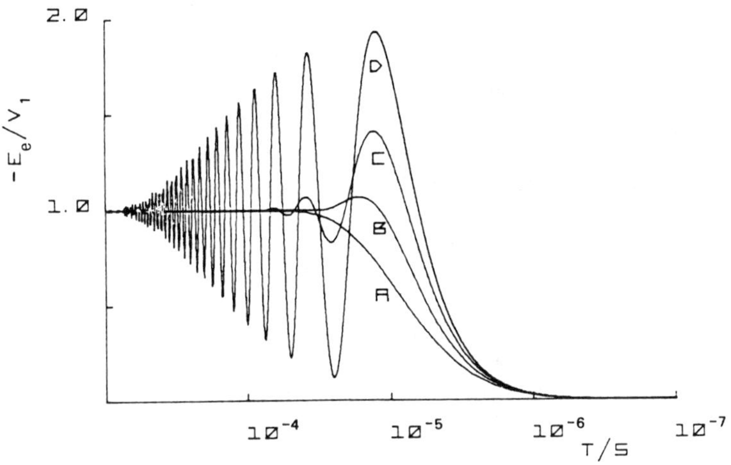


Figure 7.21 Transient response to step input. Same parameters as Figure 7.12.

region, which will add to the plotted values and produce a sustained oscillation when F = 1. The frequency of the oscillation will be about $5 \times 10^4$ Hz.

Time-domain response to a step change is given in Figure 7.21 for the same system. Again the first three curves are identical with $E_e/V_1$ curves of Figure 7.12 and the fourth results when F = 1. That the system is teetering on the edge of instability is very evident by curve D.

Accepting these results for an idealized system confirms the intuitive notion that the use of positive feedback from the current follower is identical to physically moving the reference electrode closer to the working electrode. There are no additional concepts to cope with in relating response to physical parameters. It will be seen evident that this situation allows a rather simple assessment of loop gain and a clear view of corrective measures which will lead to stability with $R_u$ compensation.

There is a possibility that concern over the new feedback loop and system instability has overshadowed the improvement in potentiostatic control when F = 1. This change is not readily apparent in the graphs, but the low-frequency, long-time value of $E_e/V_1$ is unity in the presence of faradaic current. After the instability is cured, the independence of potential on current will be more evident than it is now.

## V. ACHIEVING STABILITY THROUGH GAIN-FREQUENCY SHAPING

Reduction of the actual or effective value of $R_u$ is the basic reason for upsetting the feedback loop. The phase/time lag related to $C_{d\ell}R_s$ becomes extended as $R_u$ is decreased and this causes the feedback signal $V_r$ to be delayed over a wide frequency-time region. The effect is most apparent when the delay occurs over a frequency-time region where the loop gain is small but greater than unity. Attempts to change cell parameters will not lead to a general solution of the problem, but there is the possibility of altering the gain-frequency characteristics of the control OA. As has been usually implemented, the method involves trading off gain for stability, but in certain cases it is possible to arrive at stable control by increasing the gain. The principle involved is to decrease the phase-time lag of the control OA in a selected frequency-time region. A graphical method will be used to clarify the method.

Figure 7.22 is a straight-line approximation of the information presented in Figure 7.2 and 7.3. Amplifier gain A (inverting for the circuit of Fig. 7.1) represented by line segments A and B over the

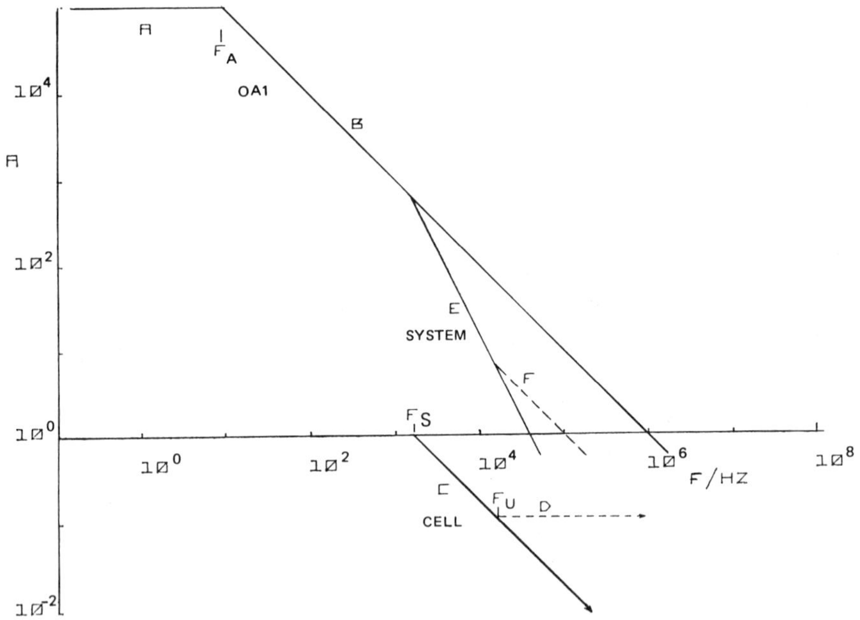


Figure 7.22 Bode plot of open-loop gain of OA (lines A and B), cell attenuation (lines C and D), and their product (lines E and F). See the text for details.

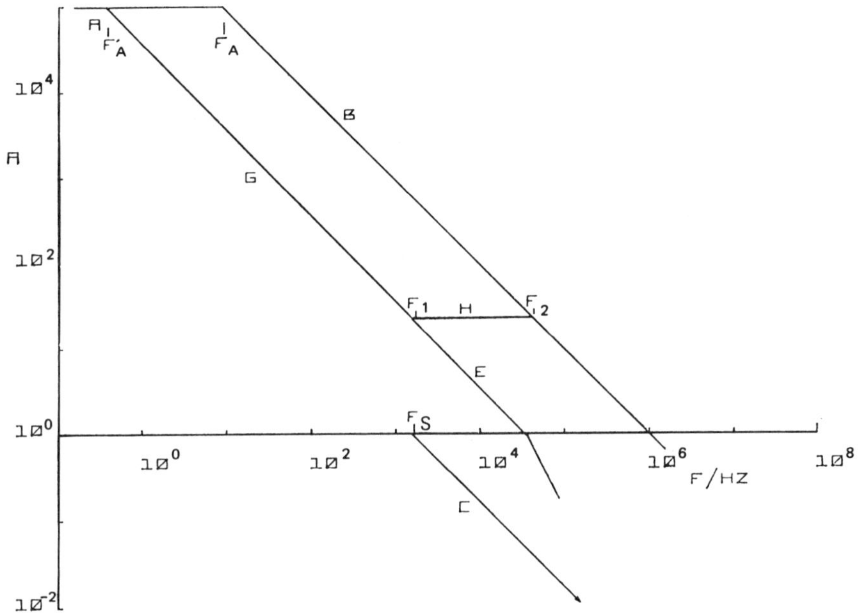


Figure 7.23 Bode plot of open-loop gain of OA (lines A and B), with gain-frequency shaping (lines G and H), cell attenuation (line C), and the product of H and C (line E). See the text for details.

frequency range of interest, and cell attenuation A (unity or less) is represented in the lower part of the graph by the abscissa below $f_s$ and then by lines C and D. These lines intersect at $f_u$ and decreasing $R_u$ ($C_{d\ell}$ constant) moves the intersection to the lower right, along C. For the present purpose, this can also be assumed to be the case when positive feedback of $V_u$ is used to compensate for $R_u$. Line C has been extended to show cell attenuation when $R_u$ has been effectively compensated; if F = 1, the slope of C is unchanged to infinite frequency. Note that phase lag was not plotted; it can be estimated by reference to previous figures. It is only necessary to remember that decreasing gain and increasing attenuation are accompanied by phase lags. Loop gain is the product of amplifier gain and cell attenuation, and this is obtained by addition of ordinate values on the log scale. Loop gain follows lines A, B, branching to E (and F if $R_u$ is uncompensated). The assumption is made that attenuation due to the resistor network on the input of OA1 is negligible, that is, $R_f \ll R_1 \parallel R_2$. A segment with a slope of $-2$ signals a phase lag increasing toward 180° and this condition leads to overshoot and oscillations. The situation becomes worse when this steep segment occurs in the region of low loop gain. All of the examples illustrated by frequency- or

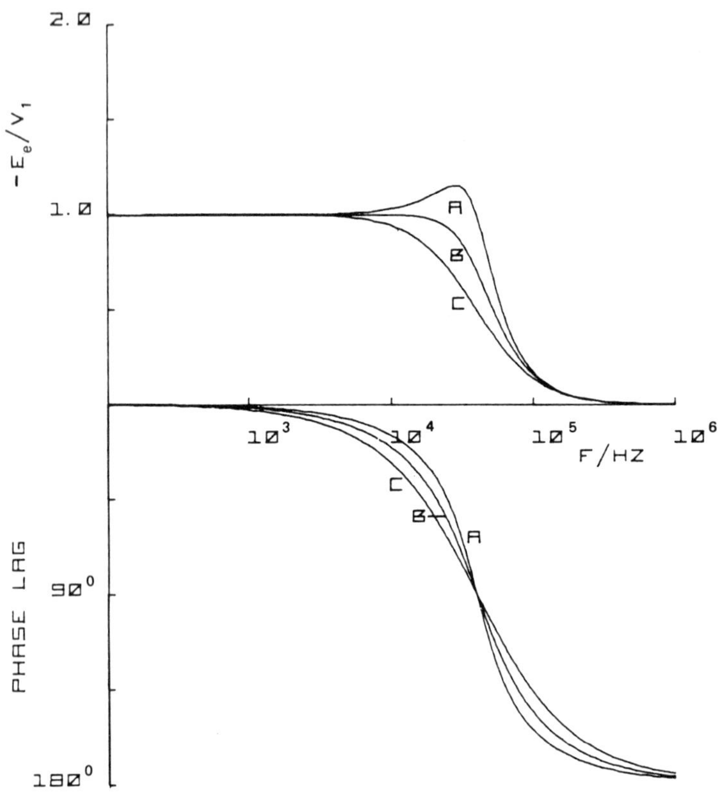


Figure 7.24 Transfer function and phase lag for potentiostat and cell as modified in Figure 7.23. A, $f_a = 0.4$ Hz, $f_1 = 1.6 \times 10^3$ Hz, $f_2 = 4 \times 10^4$ Hz; B, $f_a = 0.28$ Hz, $f_1 = 1.6 \times 10^3$ Hz, $f_2 = 5.6 \times 10^4$ Hz; C, $f_a = 0.20$ Hz, $f_1 = 1.6 \times 10^3$ Hz, $f_2 = 4.0 \times 10^4$ Hz. Other parameters as in Figure 7.20, curve D.

time-response curves can be quickly analyzed by constructing this type of straight-line Bode plots from values of $f_a$, $f_s$, $f_u$, and $f_o$ and comparing response curves in a relative fashion.

Graphically speaking, stability or first-order response can be achieved by elimination of the line segment with a slope of −2. One method depends on the introduction of a horizontal section in the amplifier's gain-frequency curve, as shown in Figure 7.23. A series capacitor and resistor between the inverting input and the output of OA1 as shown in Figure 7.1A will tailor the gain in this way. Segment G is due to the capacitor and segment H is due to the resistor; the normal behavior then dominates beyond the intersection of H and B at $f_2$. Loop gain now follows the segments A to G at $f_a'$ and to E at

$f_i$. Now the slope is −1 below frequency $f_1$, indicating that the phase lag is decreased in the critical region identified in Figure 7.22. The slope does increase to −2 to the right of $f_2$ where A < 1, and this change means the total phase lag is 135° at the frequency of unity loop gain.

Curves A of Figures 7.24 and 7.25 show the benefits of this technique of gain-frequency shaping; comparison with curves D of Figures 7.20 and 7.21 immediately reveals a dramatic improvement in stability, with potentiostatic conditions (F = 1) prevailing smoothly over the frequency region below about 1 kHz (phase lag < 1°) or at times longer than $5 \times 10^{-5}$ s ($E_e/V_1 > 0.99$). To the right of these points, the control OA becomes less able to maintain potentiostatic conditions because of limited loop gain. There is about 20% overshoot in $E_e/V_i$ because of the excess phase lag just before the loop gain reaches unity. By moving line G to the left and lowering line H by the same amount so that their intersection is unchanged on the abscissa ($f_1 = f_s$), the overshoot can be eliminated. Curves B and C of Figures 7.24 and 7.25 were plotted after shifting the line segments by factors of 0.71 and 0.5 in the directions noted. Taking the response represented by curves C as optimum, simple expressions can be obtained to guide the choice of $f'_a$, $f_1$, and $f_2$.

$$f'_a = \frac{1}{2}\left(\frac{f_a f_s}{A_o}\right)^{1/2} \tag{7.16}$$

$$f_1 = f_s \tag{7.17}$$

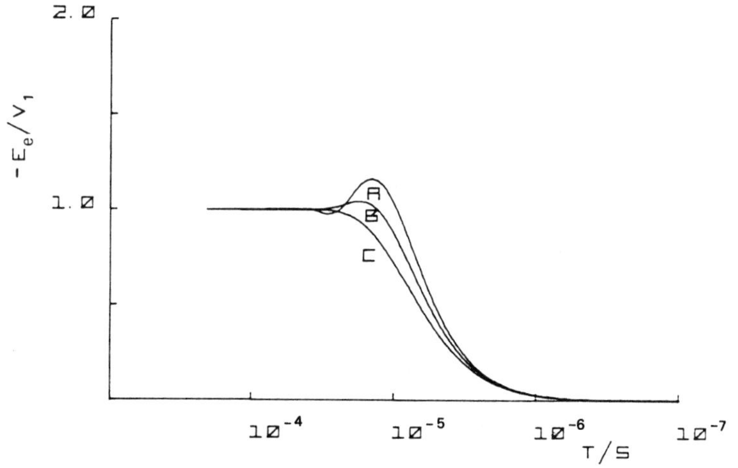


Figure 7.25 Transient response to step input. Same conditions as in Figure 7.24.

$$f_2 = 2(A_o f_a f_s)^{1/2} \tag{7.18}$$

Often it is easier to think in terms of time constants; the corresponding equations are

$$\tau'_a = 2(\tau_a \tau_s A_o)^{1/2} \tag{7.19}$$

$$\tau_1 = \tau_s \tag{7.20}$$

$$\tau_2 = \frac{1}{2}\left(\frac{\tau_a \tau_s}{A_o}\right)^{1/2} \tag{7.21}$$

Elimination of the numerical factors of 1/2 and 2 gives the conditions of curves A in Figures 7.24 and 7.25.

Shaping the gain-frequency curve of the control OA requires knowing the value of feedback resistor $R_f$ (Fig. 7.1) and an estimate of the cell roll-off frequency $f_s$. First-order approximations for the stabilizing elements are

$$R \simeq 0.1R_f \tag{7.22}$$

$$C \simeq \frac{1}{2\pi f_s} = \frac{\tau_s}{R} \tag{7.23}$$

The optimum value for R depends on the cell attenuation, so it is advantageous to use a potentiometer so that some adjustment is possible.

In practice, the one quantity that will be estimated is $f_s$ (or $\tau_s$) and it is important to have a rough idea of the consequence of uncertainty in this number. Again, using the same cell and control OA parameters as were used for Figures 7.24 and 7.25, responses to sine waves and a step input are presented in Figures 7.26 and 7.27 for variations of factors of 0.5 and 2 in $f_1$. That is, $f_s$ is unchanged but curves were calculated on the assumption that the estimated value was higher and lower than the actual cell break frequency and these values were set equal to $f_1$. Corresponding adjustments of $f'_a$ and $f_2$ were calculated to agree with Equations 7.16 and 7.18. One may conclude that an error of about a factor of 2 in the estimate of $f_s$ is just acceptable for this particular situation. However, the amount of overshoot or undershoot and the time constant for the final settling does depend on the loop gain in the vicinity of $f_1$. Increased loop gain

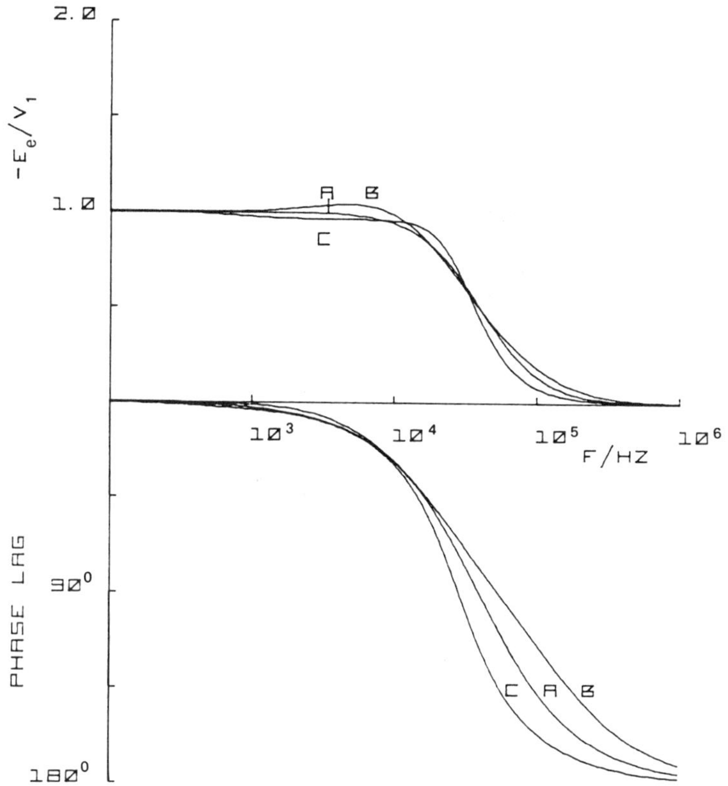


Figure 7.26 Transfer function and phase lag. A, $f_a$ = 0.20 Hz, $f_1$ = $1.6 \times 10^3$ Hz, $f_2 = 8 \times 10^4$ Hz; B, $f_a$ = 0.40 Hz, $f_1 = 3.2 \times 10^3$ Hz, $f_2$ = $1.6 \times 10^5$ Hz; C, $f_a$ = 0.10 Hz, $f_1 = 8 \times 10^2$ Hz, $f_2 = 4 \times 10^4$ Hz. Other parameters as in Figure 7.20, curve D.

produces less of a deviation, and such would result from a lower value of $f_s$ than was used in this example.

When experimental requirements are not met in terms of upper-frequency limit or rise time by the preceeding method, a second approach may be considered for tailoring amplifier response. This method greatly improves the gain-bandwidth limits but requires circuit modification of commercial instrumentation. The principle is obvious when the graphical problem of slopes discussed relative to Figure 7.22 is readdressed but with the view that line segment B is not a fixed boundary to gain-frequency capabilities of an amplifier. If line B can be moved toward the right, a horizontal section can be constructed that allows the loop gain line to maintain a slope of −1. This is the same stability objective used in the first method. Figure 7.28 shows

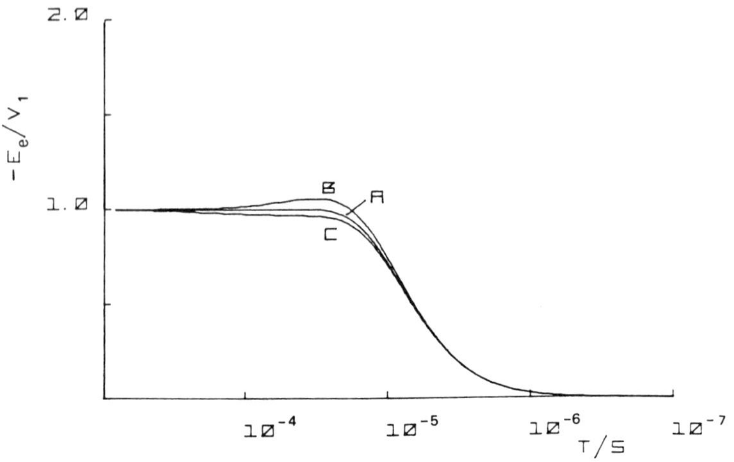


Figure 7.27 Transient response to step input. Same conditions as in Figure 7.26.

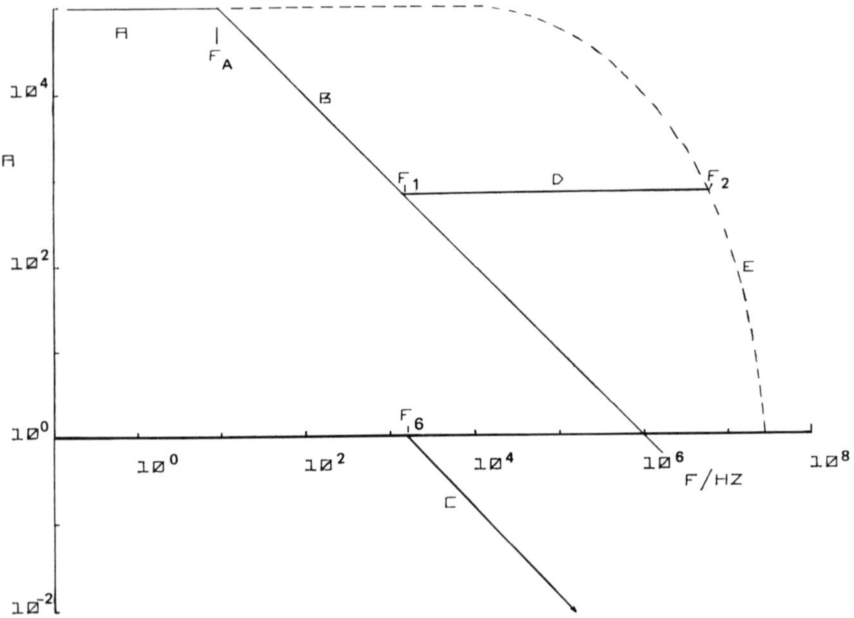


Figure 7.28 Bode plot of open-loop gain of OA (lines A and B), cell attenuation (line C), modified open-loop gain (curve D), and uncompensated open-loop gain (E). See the text for details.

the location of $f_1$ and $f_2$ that meet the basic requirement. The dashed line, E, is an estimated gain-frequency response of an OA without any internal compensation (which is usually provided by a single capacitor). This dashed line is an absolute limit for an OA. If the internal compensation network is modified to shape the gain-frequency curve successfully as shown by line segments A, B to D at $f_1$ and to E at $f_2$, it is to be expected that high-frequency performance and stability will be obtained. Figures 7.29 and 7.30 show the results, which are one to two orders of magnitude faster than those obtained with the same basic system used to calculate responses shown in Figures 7.24 and 7.25. There is another advantage and one disadvantage. The advantage is that this method always gives a higher loop gain in

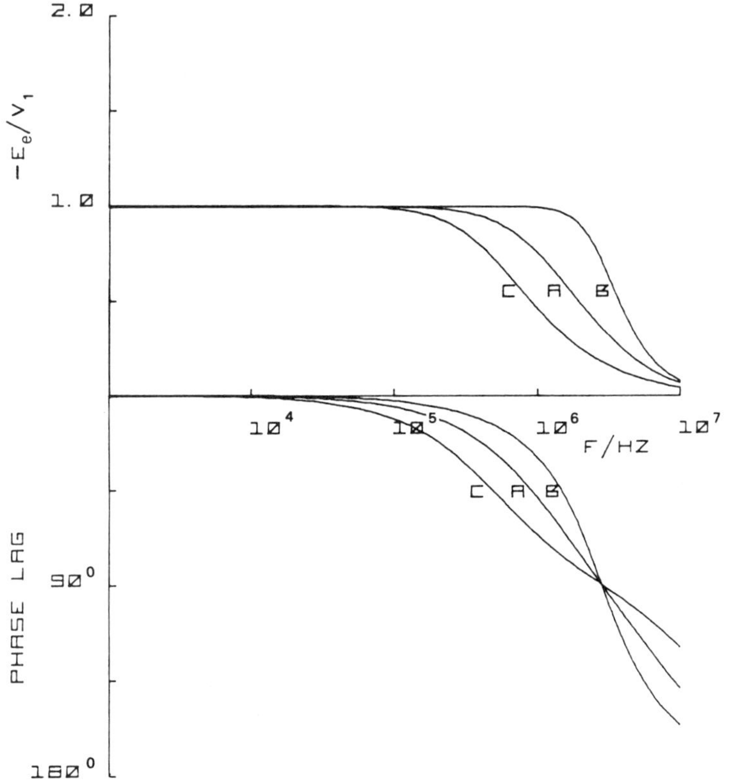


Figure 7.29 Transfer function and phase lag for potentiostat and cell as modified in Figure 7.28. A, $f_a = 10$ Hz, $f_1 = 1.6 \times 10^3$ Hz, $f_2 = 8 \times 10^6$ Hz; B, $f_1 = 8 \times 10^2$ Hz, $f_2 = 4 \times 10^6$ Hz; C, $f_1 = 3.2 \times 10^3$ Hz, $f_2 = 1.6 \times 10^7$ Hz. Other parameters as in Figure 7.20, curve D.

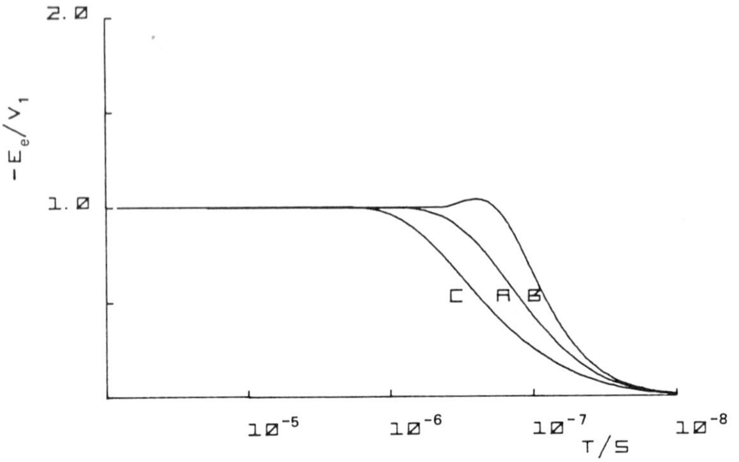


Figure 7.30 Transient response to step input. Some conditions as in Figure 7.29.

the vicinity of $f_1$ so that errors in the estimated value of $f_s$ will produce less distortion than with the first method (compare curves B and C of the figures). The disadvantage is that system noise will be increased by the increased bandwidth, so that the signal-to-noise ratio will suffer. In cases where fast response is needed, this is the best approach. The details of altering internal compensation circuits will be left to the reader (usually, the addition of a potentiometer will suffice) but it is advisable to know curve E and to then adjust $f_S$ so that OA phase lag at $f_2$ is less than 135°. In calculation of the curves, it was assumed that phase lag at $f_2$ was only 90°, although the Bode plot shows otherwise.

There are several other methods of achieving stability in potentiostatic circuits. A capacitor may be added between the counter and reference electrode to reduce phase shift in the critical frequency region. Some caution must be exercised since a low-resistance reference electrode then becomes the counter electrode at high frequencies. A particularly interesting method is known as input lead-lag compensation; a series RC is connected between the input terminals of the control amplifier and a second resistor is connected between the noninverting input and common. This form of compensation has minimum effect on the slew rate of the control amplifier. Further details can be found in the book by Stout and Kaufman listed in the bibliography.

It is reasonable to ask at this point: Are there other approaches to reach stability with grace in true potentiostatic circuits? The answer is indeed affirmative, but unfortunately with qualifications.

One technique is discontinuous control of cell potential. It is not a new approach; it was, in fact, the method used in the very first electronic potentiostat by Hickling in 1942. The principle is quite simple: Current pulses are applied to the counter electrode so that the desired potential is maintained between reference and working electrode. Since the potential can be measured between pulses, there is no iR drop. Cell potential is not steady; it depends on the sensitivity of the comparator circuit and the rate at which current demand can be met.

Another method is to apply feedback control as in Figure 7.1, but periodically interrupt the circuit for a brief moment so that the magnitude of $iR_u$ can be determined and a correction applied. Several variations of such a control system have been published and one version is available as an accessory for a potentiostat. The outstanding feature of these approaches is that $R_u$ does not have to be known and it can vary during an experiment with only minor decrease in control accuracy. In many situations, the brief interruption of the feedback loop is a small price to pay for such convenience. Microprocessors will foster further developments in this direction; the advent of digital potentiostats is imminent.

## VI. REAL SYSTEMS

It should be anticipated that there will not be a smooth transition from these idealized, simple systems into the real world. Some precautions and pitfalls have been cited, but usually in a parenthetical manner that lacked proper emphasis. The systems selected to illustrate the general principles of potentiostatic control have shown what can be expected under ideal conditions, but real systems have additional parameters that may tilt the balance from graceful control to chaos. Cell design is of paramount importance and a guide to transfer characteristics of cells is included in the bibliography. To bring the information in this chapter effectively into use, it is necessary to acknowledge the role that cells play and to design them to meet the measurement requirements as part of the system. A classical H cell and 10-ft coax cables is not likely to give satisfactory performance with even the best potentiostat available.

If one of the main points of this chapter is to be put in the form of a maxim, it would be: An operational amplifier is *not* a potentiostat; a potentiostat is a system comprised of operational amplifiers *and* a cell.

## APPENDIX

All of the system response curves in frequency and time domains were calculated numerically from equations that are much too involved to reproduce in detail here. Transfer functions in Laplace transform notation are easily defined for the potentiostat and cell of Figure 7.1. Appropriate combinations of these functions then yield system transfer functions that may be cast into time- or frequency-dependent equations by inverse Laplace transformation or by using complex number manipulation techniques. These methods have become rather common in electrochemical literature and will not be described here. The interested reader will find several citations in the bibliography to be helpful in clarifying details.

Transfer functions are as follows:

Control OA1:

$$\frac{V_o(s)}{V_-(s)} = -A(s) = -\frac{A_o(1 + s\tau_1)}{(1 + s\tau_a)(1 + s\tau_s)} \tag{A7.1}$$

where $V_o(s)$ is the output, $V_-(s)$ is the voltage at the inverting input, $A_o$ is the dc gain, and the time constants $\tau_a$, $\tau_1$, and $\tau_2$ are defined in the frequency domain by Figures 7.23 and 7.28. When gain-frequency shaping is not used, $\tau_1$ and $\tau_2$ are set equal to zero.

Auxiliary electrode to reference electrode of cell:

$$H(s) = \frac{E_r(s)}{V_o(s)} = \frac{1 + s\tau_u}{1 + s\tau_s} \tag{A7.2}$$

Auxiliary electrode to solution-working electrode interface:

$$K(s) = \frac{E_e(s)}{V_o(s)} = \frac{1}{1 + s\tau_s} \tag{A7.3}$$

where $\tau_u = R_uC_{d\ell}$ and $\tau_s = R_sC_{d\ell}$.

Relative value of voltage drop, $R_ui(s)$, to be compensated by positive feedback:

$$L(s) = -\frac{V_u(s)}{V_o(s)} = \frac{Fs\tau_u}{1 + s\tau_u} \tag{A7.4}$$

where F is the fraction of $R_u$ compensated.

In order to combine the transfer functions, the relation between $V_-(s)$, the summing resistors, and the other voltages is required. From Kirchhoff's current law, this equation is

$$V_-(s) = \left[ \frac{V_1'(s)}{R_1} + \frac{V_u'(s)}{R_2} + \frac{V_r(s)}{R_f} \right] R_p \tag{A7.5}$$

where

$$R_p = \left( \frac{1}{R_1} + \frac{1}{R_2} + \frac{1}{R_f} \right)^{-1} \tag{A7.6}$$

It is advantageous to make $R_1$, $R_2 >> R_f$ to avoid loss of amplifier gain, so $R_p \cong R_f$, and a simplified form results:

$$V_-(s) = V_1(s) + V_u(s) + V_r(s) \tag{A7.7}$$

where $V_1(s)$ and $V_u(s)$ are the scaled values of the actual signal voltages.

Combining transfer functions A7.1, A7.2, and A7.4 with Equation A7.7 gives the system transfer function between the signal at the reference electrode and the input, provided that $E_r(s) = V_r(s)$:

$$\frac{E_r(s)}{V_1(s)} = - \frac{A(s)\,H(s)}{1 + A(s)[H(s) - L(S)]} \tag{A7.8}$$

By using Equation A7.3 instead of Equation A7.4 the system transfer function at the solution-working electrode interface relative to the input signal is obtained:

$$\frac{E_e(s)}{V_1(s)} = - \frac{A(s)\,K(s)}{1 + A(s)[K(s) - L(s)]} \tag{A7.9}$$

Substitutions for the transfer functions in Equations A7.8 and A7.9, noting that $s = j\omega$ for steady-state sinuoid signals, give equations in complex notation. The magnitude ratio and phase relation follow by the usual methods. Transient response results by setting $V_1(s)$ equal to $V_1/s$, where $V_1$ is the magnitude of the voltage step. Inverse Laplace transformation then gives the time dependence of either signal of the cell relative to $V_1$.

The response curves were plotted by numerically solving the final equations at intervals of 0.01 or 0.04 or major divisions on the abscissa scale using a Tektronix 31 calculator and plotter.

## BIBLIOGRAPHY

The following is only a partial listing of significant literature on the design and use of potentiostats. References cited in these publications provide a thorough bibliography of chemical and control theory papers on the subject.


Bewick, A., "Analysis of the Use of IR Compensation in Potentiostatic Investigations," *Electrochim. Acta* *13*:825 (1968).

Booman, G. L., and W. B. Holbrook, "Electroanalytical Controlled-Potential Instrumentation," *Anal. Chem.* *35*:1793 (1963). (Related publications in same issue.)

Booman, G. L., and W. B. Holbrook, "Optimum Stabilization Networks for Potentiostats with Application to a Polarograph Using Transistor Operational Amplifiers," *Anal. Chem.* *37*:795 (1965).

Britz, D., "iR Elimination in Electro-chemical Cells," *J. Electroanal. Chem.* *88*:309 (1978).

Brown, E. R., H. L. Hung, T. G. McCord, D. E. Smith, and G. L. Booman, "A Study of Operational Amplifier Potentiostats Employing Positive Feedback for iR Compensation: II. Application to AC Polarography," *Anal. Chem.* *40*:1424 (1968).

Brown, E. R., D. E. Smith, and G. L. Booman, "A Study of Operational Amplifier Potentiostats Employing Positive Feedback for iR Compensation: I. Theoretical Analysis of Stability and Bandpass Characteristics," *Anal. Chem.* *40*:1411 (1968).

Garreau, D., and J. M. Saveant, "Linear Sweep Voltammetry. Compensation of Cell Resistance and Stability. Determination of Residual Uncompensated Resistance," *J. Electroanal. Chem.* *35*:309 (1972).

Harrar, J. E., "Techniques, Apparatus and Analytical Applications of Controlled-Potential Coulometry," *Electroanalytical Chemistry*, Vol. 8 (A. J. Bard, ed.), Marcel Dekker, New York, 1975.

Harrar, J. E., and C. L. Pomernacki, "Linear and Nonlinear System Characteristics of Controlled-Potential Electrolysis Cells," *Anal. Chem.* *45*:57 (1973).

Mumby, J. E., and S. P. Perone, "Potentiostat and Cell Design for the Study of Rapid Electrochemical Systems," *Chem. Instrum.* *3*:191 (1971).

Pilla, A. A., "Influence of the Faradaic Process on Nonfaradaic Resistance Compensation in Potentiostatic Techniques," *J. Electrochem. Soc.* *118*:702 (1971).

Schroeder, R. R., and I. Shain, "Application of Feedback Principles to Instrumentation for Potentiostatic Studies," *Chem. Instrum.* *1*:233 (1969).

Stout, D., and M. Kaufman, *Handbook of Operational Amplifier Circuit Design*, McGraw-Hill, New York, 1976.

# 8

# Conductivity and Conductometry

F. JAMES HOLLER* and CHRISTIE G. ENKE / Department of Chemistry, Michigan State University, East Lansing, Michigan



All electrochemical phenomena require a cell consisting of, at the minimum, two electrodes that are both in contact with an electrolyte. The ions of the electrolyte serve to transport charge (conduct current) from one electrode to another. In many electrochemical techniques the attention is focused on the diffusion or charge-transfer processes occurring at or near the electrode surface for particular ions. In such cases the ions of interest are generally in relatively low concentration and are assumed not to be significantly involved in the charge conduction process between electrodes. This condition is met in practice by adding a relatively large concentration of noninterfering ions to

*Present address: Department of Chemistry, University of Kentucky, Lexington, Kentucky.

provide the interelectrode conductance and minimize the electric field gradient between the electrodes. By this addition of "inert" electrolyte, it is often assumed that the interelectrode conductance part of the total electrochemical cell phenomenon can be ignored.

On the other hand, there are practical and important electrochemical techniques in which the ions of interest provide a significant portion of the interelectrode conductance and for which the cell solution conductivity is the major electrochemical parameter that is measured. As we shall see, the solution conductivity depends on the ion concentration and the characteristic mobility of the ions present. Therefore, conductivity measurements of simple, one-solute solutions can be interpreted to indicate the concentration of ions (as in the determination of solubility or the degree of dissociation) or the mobility of ions (as in the investigations of the degree of solvation, complexation, or association of ions). In multiple-solute solutions, the contribution of a single ionic solute to the total solution conductivity cannot be determined by conductance measurements alone. This lack of specificity or selectivity of the conductance parameter combined with the degree of tedium usually associated with electrolytic conductivity measurements have, in the past, discouraged the development of conductometry as a widespread electroanalytical technique. Today, there is a substantial reawakening of interest in the practical applications of conductometry. Recent electronic developments have resulted in automated precision conductometric instrumentation and applications are being developed in which conductometry is used to follow the course of highly specific chemical reactions.

In this chapter we take a careful look at the phenomenon of the electrical conductivity of materials, particularly electrolytic solutions. In the first section the nature of electrical conductivity and its relation to the electrolyte composition and temperature is developed. This section and the second (which deals with the direct-current contact methods for measuring conductance) introduce the basic considerations and techniques of conductance measurement. This introduction to conductance measurements is useful to the scientist, not only for electrolytic conductance, but also for understanding the applications of common resistive indicator devices such as thermistors for temperature, photoconductors for light, and strain gauges for mechanical distortion. The third section of this chapter describes the special techniques that are used to minimize the effects of electrode phenomena on the measurement of electrolytic conductance. In this section you will encounter the most recent solutions to the problems of conductometric measurements, the solutions that have sparked the resurgent interest in conductometry.

## I. SOME BASIC RELATIONSHIPS

A material exhibits the property of electrical conductivity when it contains charged particles that are free to move through the material. When an electrical potential is applied across such a material, the charged particles will experience a force along the field in a direction opposite to their charge. The resulting net motion of the charged particles is an electric current. The greater the current produced by a given electric field, the greater the conductivity of the material. Since conductivity is related to the rate of flow of charge in response to an electric field, the magnitude of the conductivity is dependent on the concentration, mobility, and charge of the charged particles. In this section the relationship between the conductivity and the mobility, charge, and concentration of the charged species is developed. Finally, the property of conductivity is related to the measured parameter, conductance.

### A. Conductivity

When a charged particle i (an electron or ion) in a liquid or solid material is subjected to an electric field, it quickly reaches a limiting average velocity of motion in the direction of the field opposite to the sign of its charge. The velocity $v_i$ (cm/s) is given by

$$v_i = u_i \mathcal{E} \tag{8.1}$$

where $u_i$ is the particle's mobility and $\mathcal{E}$ is the electric field strength (V/cm). The resulting electric current density ($J_i$ in C/s through 1 $cm^2$ of the material) is then

$$J_i = v_i N_i q_i \tag{8.2}$$

where $N_i$ is the number of i particles per cubic centimeter and $q_i$ is the coulombic charge on each particle. Note that the sign of $v_i$ will change with the sign of $q_i$ so that J is positive for particles of either sign. Substituting Equation 8.1 in Equation 8.2, we obtain

$$J_i = \mathcal{E} N_i |q_i| u_i \tag{8.3}$$

where the absolute value of $q_i$ is required since the vector quantity $v_i$ has been eliminated.

## B. Ionic Conductivity

The total current density J is the sum of the individual current density charge carriers in the substance.

$$J = \sum_{i=1}^{n} J_i = \mathcal{E} \sum_{i=1}^{n} N_i |q_i| u_i \tag{8.4}$$

The electrical conductivity K has been defined as the current density per unit electric field, that is,

$$K \equiv \frac{J}{\mathcal{E}} \tag{8.5}$$

Combining Equations 8.4 and 8.5, we have

$$K = \sum_{i=1}^{n} N_i |q_i| u_i \tag{8.6}$$

From Equation 8.6 it is clear that any change in the substance that affects the concentration or mobility of any of the charge carriers will affect the conductivity of the substance. The conductivity of virtually all substances is affected by temperature. In metallic conductors, electrons are the charge carriers. As the temperature increases, the mobility of the electrons decreases, decreasing the conductivity. On the other hand, the concentration of electrons and holes in semiconductors increases as the temperature increases. Thus the temperature dependence of conductivity for semiconductors is opposite that for metallic conductors. The effect of temperature on the conductivity of ionic solutions is more complex, as we shall see.

In chemical applications, the concentration units of mol/L are more useful than ions/cm$^3$. From Equation 8.3 and the relationship $N_i = C_i N_{av}/1000$, where $C_i$ is the molar concentration of i and $N_{av}$ is Avogadro's number,

$$J_i = \mathcal{E} \frac{C_i N_{av}}{1000} |q_i| u_i \tag{8.7}$$

The charge on an ion $q_i$ can be given by $q_i = Z_i q_e$, where $Z_i$ is the charge number of the ion and $q_e$ is the unit electron charge in coulombs.

Also, the quantity $N_{av}q_e$ is defined as the Faraday, F. Thus $|q_i| = |Z_i|F/N_{av}$. The individual ion current density can now be written

$$J_i = \mathcal{E} \frac{C_i F}{1000} |Z_i| u_i \tag{8.8}$$

and the conductivity, considering all species, is

$$K = \frac{F}{1000} \sum_{i=1}^{n} C_i |Z_i| u_i \tag{8.9}$$

In the case of a dissolved simple salt, the mobile charges are the cations and anions resulting from the solvation and dissociation of the salt. The conductance is thus

$$K = \frac{F}{1000} (C_+ |Z_+| u_+ + C_- |Z_-| u_-) \tag{8.10}$$

Electrical neutrality requires that the charge concentration for positive and negative charges be equal, that is,

$$C_+ |Z_+| = C_- |Z_-| \tag{8.11}$$

If the salt is *completely* dissociated, the normality of the salt solution, $C^*$, is equal to $C_+|Z_+| = C_-|Z_-|$. The equivalent conductance $\Lambda$ of a salt solution is defined as

$$\Lambda \equiv \frac{1000K}{C^*} \tag{8.12}$$

whether the salt is completely dissociated or not. In Equation 8.12, $C^*$ is the normality of the salt solution, *not* just the concentration of the ions. Clearly, $\Lambda$ depends on the degree of dissociation of the salt, having a lower value for a lower degree of dissociation. The dissociation reaction can be studied by measuring $\Lambda$ as a function of $C^*$ [1,2].

The definition in Equation 8.12 does not require complete dissociation, but *if the salt is completely dissociated,* from Equation 8.10,

$$K = \frac{C^*F}{1000}(u_+ + u_-) \qquad \text{complete dissociation} \tag{8.13}$$

Combining Equations 8.12 and 8.13 gives us

$$\Lambda = F(u_+ + u_-) \qquad \text{complete dissociation} \tag{8.14}$$

Now the ionic equivalent conductance $\lambda$ can be defined for each ion.

$$\lambda_+ \equiv Fu_+, \qquad \lambda_- \equiv Fu_-, \qquad \lambda_i \equiv Fu_i \tag{8.15}$$

Substituting the expressions for $\lambda_+$ and $\lambda_-$ into Equations 8.13 and 8.14, we have, for the case of complete dissociation,

$$K = \frac{C^*}{1000}(\lambda_+ + \lambda_-) \qquad \text{complete dissociation} \tag{8.16}$$

and

$$\Lambda = \lambda_+ + \lambda_- \qquad \text{complete dissociation} \tag{8.17}$$

Since $\lambda$ depends on the mobility of the ion, it is a function of the ion type and of the solution parameters (solvent, solute concentration, temperature, etc.). At very low solute concentrations, $\lambda$ reaches a steady "limiting" value, $\lambda°$, that is often tabulated for the common ions in water solvent at specific temperatures. A table of $\lambda°$ values is given in Table 8.1. The $\lambda°$ values from such a table can be used to estimate the conductivity of a completely dissociated salt solution by applying them in Equation 8.10. For finite salt concentrations the calculated values are indicative, but not exact. Onsager, Fuoss, and others [2, 3] have developed equations by which quite accurate values of $\lambda$ can be calculated from $\lambda°$ values under certain circumstances.

In analytical applications of conductometry, the sample often contains many species of ions, each contributing to the total conductivity. The total conductivity is then

$$K = \frac{1}{1000}\sum_{i=1}^{n} C_i^*\lambda_i = \frac{1}{1000}\sum_{i=1}^{n} C_i|Z_i|\lambda_i \qquad \text{complete dissociation or actual ion concentrations} \tag{8.18}$$

Table 8.1 Limiting Equivalent Conductance of Ions in Water at 25°C

| Cations | $\lambda_+^\circ$ | Anions | $\lambda_-^\circ$ |
|---|---|---|---|
| $H^+$ | 349.8 | $OH^-$ | 199.2 |
| $Li^+$ | 38.6 | $F^-$ | 55.4 |
| $Na^+$ | 50.1 | $Cl^-$ | 76.4 |
| $K^+$ | 73.5 | $Br^-$ | 78.1 |
| $Rb^+$ | 77.8 | $I^-$ | 76.8 |
| $Ag^+$ | 61.9 | $NO_3^-$ | 71.5 |
| $NH_4^+$ | 73.5 | $ClO_3^-$ | 64.6 |
| $(CH_3)_2NH_2^+$ | 51.8 | $ClO_4^-$ | 67.4 |
| $(CH_3)_3NH^+$ | 47.2 | $IO_4^-$ | 54.5 |
| $(CH_3)_4N^+$ | 44.9 | Formate | 54.6 |
| $Mg^{2+}$ | 53.0 | Acetate | 40.9 |
| $Ca^{2+}$ | 59.5 | Benzoate | 32.4 |
| $Ba^{2+}$ | 63.6 | $SO_4^{2-}$ | 80.0 |
| $Cu^{2+}$ | 53.6 | $CO_3^{2-}$ | 69.3 |
| $Zn^{2+}$ | 52.8 | $Fe(CN)_6^{4-}$ | 111.0 |
| $La^{3+}$ | 69.7 | — | — |
| $Ce^{3+}$ | 69.8 | — | — |

*Source*: Reprinted in part from Ref. 2, p. 463, with permission. Copyright 1968 Butterworths.

where $C_i^*$ and $C_i$ are the normality and molarity, respectively, of the ionic species i in the solution. Thus, in using Equation 8.18 to calculate (or estimate) the conductivity of a solution, *all* ionic species present must be included. Also, Equation 8.18 shows that the variation in concentration or mobility of any of the ionic species will result in a change in K. A change in temperature can affect both ion concentration and mobility by affecting ion dissociation, complexation, and solvation equilibria as well as solvent viscosity. The combined effects are often complex and rarely negligible.

## C. Conductance

When a voltage source of E volts is connected across the contacts of a conducting material as shown in Figure 8.1, the mobile charges move in the conducting material in response to the field. The electric field $\mathcal{E}$ applied to the conductor is $E/\ell$. From Equation 8.5 the resulting current density J is

$$J = \mathcal{E}K = \frac{EK}{\ell} \tag{8.19}$$

The total current I is the current density times the area a:

$$I = Ja = \frac{EKa}{\ell} \tag{8.20}$$

From Equation (8.20) we see that the current in a conductor is proportional to the voltage across it. The proportionality constant $Ka/\ell$ is defined as the conductance G:

$$G \equiv K\frac{a}{\ell} \tag{8.21}$$

The current-voltage relationship is now simply

$$I = EG \tag{8.22}$$

This relationship was first articulated by Ohm, who also defined the resistance R of a conductor as the reciprocal of the conductance,

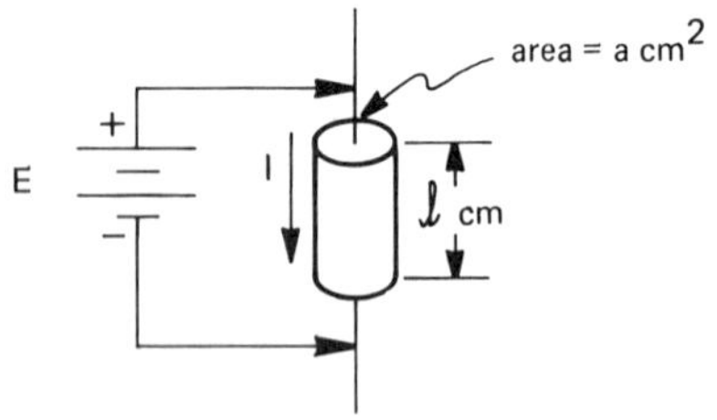


Figure 8.1 Application of a voltage source to a conductor.

$$R \equiv \frac{1}{G} \tag{8.23}$$

so that Ohm's law is also often written

$$I = \frac{E}{R} \qquad \text{or } E = IR \tag{8.24}$$

Another way to state Equation 8.24 is that the voltage across a resistor is proportional to the current through the resistor. These simple relationships are the basis of virtually all techniques of measuring conductance (or resistance); the measurement of the I/E ratio for a conductor yields G directly, while the E/I ratio is equal to R. The unit of R (volts/ampere) is the ohm (Ω) and the unit of G (amperes/volt) is the mho ($\Omega^{-1}$).

In the case of electrolytic solutions, it is G that is measured, that is, the conductance of a cell filled with the sample solution. The conductance of an actual cell is related to the conductivity of the solution and the geometry of the cell as given by Equation 8.21. The conductance of the solution is proportional to the area a of the electrodes and inversely proportional to the distance between them, so that K is calculated from a measurement of G by $K = G\ell/a$, where the factor $\ell/a$ is often called the cell constant. Actually, the exact cell constant is rarely obtained by geometric measurement. It is generally determined empirically by measuring the conductance of a cell filled with a solution for which the conductivity is exactly known.

## II. DC CONTACT MEASUREMENT OF CONDUCTANCE

The measurement of the conductance of a substance is simplified greatly if direct and effective contact can be made to that substance with ordinary metallic conductors. An ideal contact between the test substance and the measuring device has perfect conductance and introduces no additional voltages into the circuit such as the thermal or electrochemical potentials that are often part of circuits with dissimilar conductor composition. There are, fortunately, many devices, substances, and measurements for which metallic contacts are sufficiently ideal and for which the simple relationships of Ohm's law can be used as the basis of measurement. The ohmmeters and Wheatstone bridge described in the first two parts of this section can be used where the metallic contact is effective. The consideration of special contact problems concludes this section.

## A. Practical Ohmmeters

An ohmmeter is a device that provides a reading of the resistance of a conductor in ohms when the ohmmeter test probes are attached to the conductor. A simple ohmmeter circuit is shown in Figure 8.2. With the test probes short-circuited ($R_u = 0\ \Omega$), the "ohms adjust" control is turned so that the current $I_1$ through the total circuit resistance $R = R_m + R_f + R_a$ deflects the meter exactly full scale. Now, by connecting the test probes across the unknown resistance $R_u$, the current is decreased to a value $I_2$ which depends on the value of $R_u$:

$$I_2 = I_1 \frac{R}{R_u + R} \tag{8.25}$$

from which it follows that

$$R_u = \left(\frac{I_1}{I_2} - 1\right) R \tag{8.26}$$

For the 1-mA meter movement and 1.5-V battery shown in Figure 8.2, resistance $R = R_m + R_f + R_a$ will be 1500 $\Omega$ for full-scale deflection, so, from Equation 8.26,

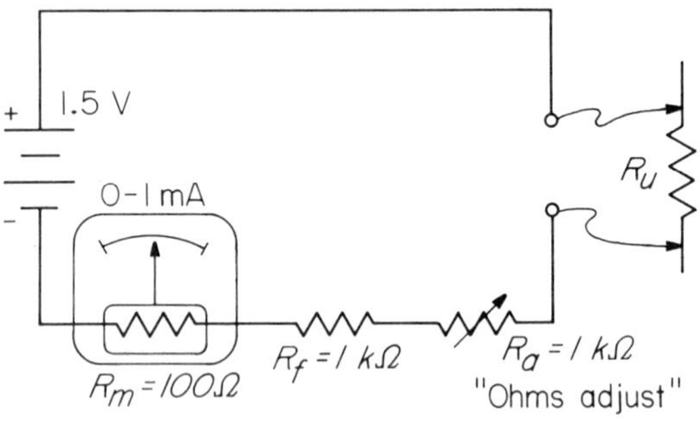


Figure 8.2 Ohmmeter of simple series type. (Reprinted with permission from *Electronics and Instrumentation for Scientists*, by H. V. Malmstadt, C. G. Enke, and S. R. Crouch, 1981. © Benjamin/Cummings Publishing Company.)

$$R_u = \left(\frac{I_1}{I_2} - 1\right) 1500 \tag{8.27}$$

With $I_2 = (1/2)I_1$ (midscale deflection), the unknown resistance $R_u$ = 1500 Ω. It is apparent from Equation 8.26 that the relationship between $R_u$ and $I_2$ or scale position is not linear, since $R_u$ goes from ∞ to 0 as $I_2$ goes from 0 to $I_1$. Values of $R_u$ much higher than 1500 Ω become crowded on the "infinite ohms" end of the scale, and values much lower become indistinguishable from zero. Different midscale resistance values are obtained by changing the meter sensitivity or the battery voltages.

### *Linear Resistance-to-Voltage Converters*

The nonlinear relationship between the unknown resistance and the current measured in the ohmmeter circuit of Figure 8.2 is inconvenient for many types of resistive transducer measurements. From Ohm's law, E = IR, the voltage across a resistor is proportional to the resistance for a constant value of current. Similarly, the conductance is proportional to the current through a resistor for a constant applied voltage (I = EG). To take advantage of these linear relationships, a constant-current source is required in the first case, and a current measurement with negligible input voltage is required in the second. The operational amplifier current follower circuit provides a convenient and accurate means to achieve either of these requirements [4]. In Figure 8.3a, $R_u$ is connected to the voltage source E, and the result is a current $i = E/R_u$ which is proportional to the conductance of $R_u$. The operational amplifier maintains an output voltage such that point S is at the common potential. Therefore, the input current, i, produces a voltage $e_o = iR_u$ at the current-follower input. Also, the connection of the G-to-I converter circuit to point S is equivalent to a connection to common; that is, essentially no additional voltage or current is added to the input circuit. Since $i = E/R_u$ and $e_o = -iR_u$,

$$e_o = \frac{-ER_r}{R_u} = -ER_rG_u \tag{8.28}$$

This results in an output voltage directly proportional to the conductance of $R_u$ within the output voltage and current capabilities of the operational amplifier.

In the circuit of Figure 8.3b, a fixed resistance $R_r$ is used in the input current-generating circuit. Since this same current passes through $R_u$ and the operational amplifier maintains point S at the common potential, the output voltage $e_o$ is equal to $-iR_u$. From

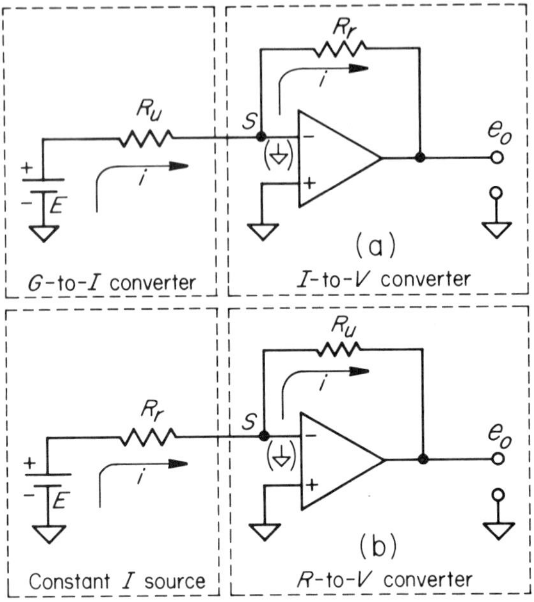


**Figure 8.3** Current-follower circuit for resistance measurement: (a) $e_o = -ER_r/R_u$; (b) $e_o = -ER_u/R_r$. (Reprinted with permission from *Electronics and Instrumentation for Scientists*, by H. V. Malmstadt, C. G. Enke, and S. R. Crouch, 1981. © Benjamin/Cummings Publishing Company.)

$$i = \frac{E}{R_r} \quad \text{and} \quad e_o = -iR_u \tag{8.29}$$

$$e_o = -\frac{E}{R_r} R_u \tag{8.30}$$

A very useful application of the circuit of Figure 8.3b is produced if we replace the feedback resistor $R_u$ with a resistive transducer such as a coated semiconductor thermistor. The commercial availability of small, rapid response, chemically inert thermistors which have conveniently measurable resistances at temperatures of 200-600 K makes them an excellent choice as the transducer in chemical applications that require the rapid and accurate measurement of temperature. Unfortunately, a continuous current in $R_u$ may produce undesirable Joule heating in the thermistor and its surroundings and thus cause an error in the measurement. This error may be reduced by utilizing a pulsed voltage source as illustrated in Figure 8.4. The logic signal A closes the electronic switch SW and produces a current

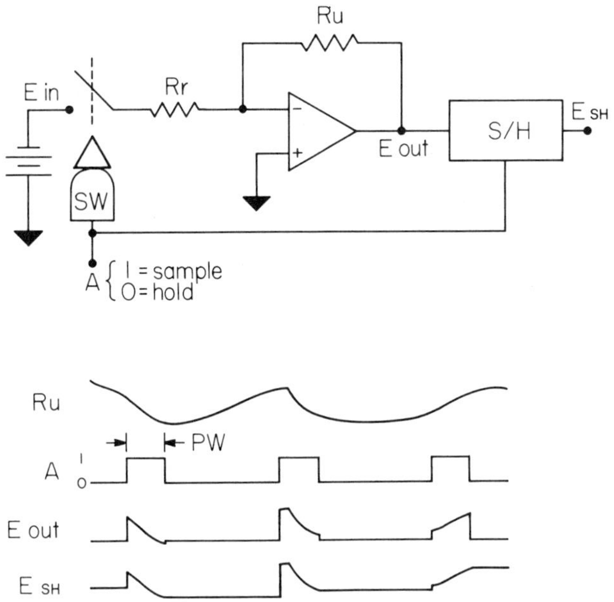


Figure 8.4 Pulsed dc measurement of resistance.

pulse in $R_u$ the duration of which is designated by PW in the figure. The logic signal also switches the sample-and-hold module (S/H) into the sample mode and causes the transient analog signal which is proportional to $R_u$ to be held at the S/H output between pulses. The pulses need only be applied at the frequency of the desired temperature sampling rate and the output of the S/H module may easily be connected to an analog-to-digital converter for computer data acquisition.

## B. Null Comparison Measurement

In a null comparison measurement of resistance, the effect of an unknown resistance must be compared with the effect of a variable standard resistance under nearly identical conditions as possible. Therefore, the unknown and standard resistances are placed in identical circuits in such a way that the resulting voltage or current in each circuit can be compared. Then the standard is varied until the difference in voltage or current between the two circuits is zero. Several methods for performing this comparison have been devised, of which the Wheatstone bridge is by far the most common. Comparison methods for resistance measurement offer great accuracy, resolution, and relative independence of the electronic sources and detectors used.

*The Wheatstone Bridge*

The Wheatstone bridge shown in Figure 8.5 provides the most direct and best known circuit for comparison of unknown resistances against standard resistances. Resistances $R_A$, $R_B$, and R are standard resistance values which are used in the measurement of the unknown resistance $R_u$. Resistance R is made variable and is adjusted until the null detector indicates that the bridge is balanced.

When the circuit is at balance, there is no current through the null detector and no potential difference between terminals x and y. At balance, four significant conditions exist:

1. The current through $R_A$ and $R_u$ is $I_1$.
2. The current through $R_B$ and R is $I_2$.
3. $I_2R_u = I_2R$.
4. $I_1R_A = I_2R_B$.

Therefore, $R_u/R = R_A/R_B$ and

$$R_u = R \frac{R_A}{R_B} \tag{8.31}$$

It can be seen from Equation 8.31 that the unknown resistance $R_u$ is determined from the values of the three standard resistances R, $R_A$, and $R_B$. It is common practice to make the ratio $R_A/R_B$ some exact

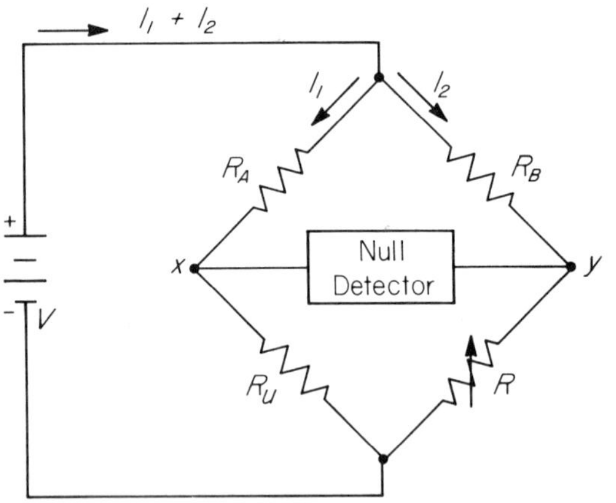


**Figure 8.5** Wheatstone bridge. (Reprinted with permission from *Electronics and Instrumentation for Scientists*, by H. V. Malmstadt, C. G. Enke, and S. R. Crouch, 1981. © Benjamin/Cummings Publishing Company.)

decimal fraction or multiple such as 0.01, 0.1, 10, or 100 and to refer to this ratio as the "multiplier." The resistance R is variable continuously or in small increments so that the dial reading of R times the multiplier setting equals the unknown resistance $R_u$. The main sources of error in a Wheatstone bridge are the inaccuracies of the three standard resistances R, $R_A$, and $R_B$ (as low as 0.001%), establishing the null point, and thermal electromotive force (EMF) values.

The magnitude of the off-balance indication of the null detector of a Wheatstone bridge is sometimes used as a measure of the change in resistance of the unknown resistor $R_u$. In this application, $R_u$ is frequently a light-, temperature-, or strain-dependent resistor. When matched devices are used for R and $R_u$, or $R_A$ and $R_u$, an off-balance output is obtained which is relative to the ratio of the resistances of the two devices. This "comparative" resistance measurement with matched devices can tend to cancel the effects of resistance changes due to environmental factors other than the measured quantity. For example, the resistance of strain gauge resistors depends on temperature and humidity as well as strain. If twin strain gauges are used in adjacent arms of a Wheatstone bridge with one gauge under strain but both in the same environment, the environmental effects will not affect the resistance ratio and only the strain effect will affect the off-balance output signal.

## C. Contact Considerations

In all the examples of conductance or resistance measurement above, it was assumed that the probe leads and the contacts to the measured conductor were ideal. An ideal lead and contact has zero resistance, and has no voltages whether generated by thermal, electrochemical, or any other effect. In a very wide range of measurements of electron-conducting devices, the ideal conditions are met within the desirable or practical error limits. However, the measurement of very low resistances can pose a problem in that the lead and contact resistance must be negligible compared to the resistance measured. Normal lead and contact resistance can be several tenths of an ohm, which limits 1% accuracy measurements to values greater than about 50 $\Omega$.

The best technique for lower resistances is the four-contact method shown in Figure 8.6. The leads and contacts that measure the voltage across the resistor that is caused by the current through the resistor are not the same leads and contacts that supply the current. Since the current does not appear in the voltmeter leads or contacts, no error due to an IR voltage in the voltmeter contacts will occur.

Direct contact techniques do not work with electrolytic solutions. One cannot, for example, measure the resistance of salt water in a cell with an ordinary ohmmeter. This is due to the fact that the

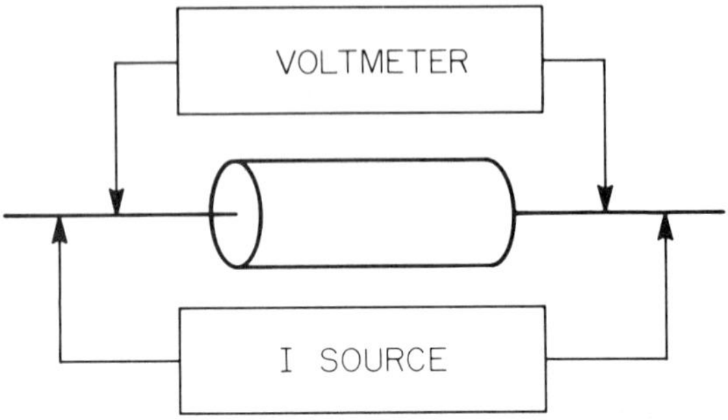


Figure 8.6 Four-contact technique for low-R measurement.

contact between the metallic electrode and the electrolytic solution is far from ideal. There is generally a rather large potential at the electrode-solution interface as a result of the electrochemical reaction(s) forced to occur by the measurement current. Normally, this problem is circumvented by using one of the capacitance contact techniques described in the next section. However, some investigators have used a four-contact approach successfully.

In the electrolytic conductance measurement application, the four-electrode method takes the form shown in Figure 8.7. The technique actually measures the conductance of the part of the solution that is in the capillary tubing between the two vials of the H cell. When a current is applied from the current source, this current must

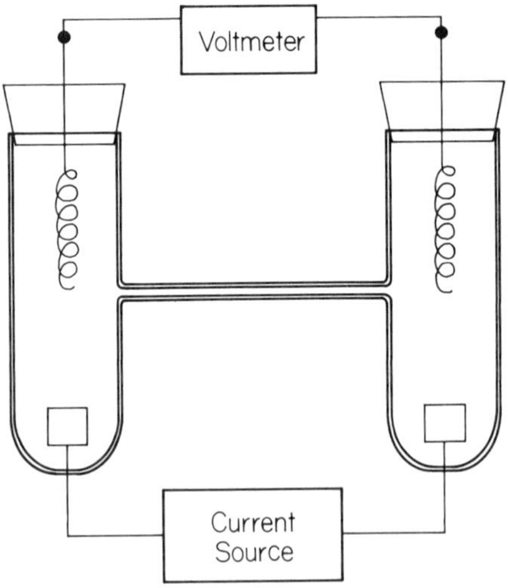


Figure 8.7 Four-contact measurement of electrolytic conductivity.

appear in the capillary section of the cell. The IR voltage in the capillary section is very much larger than in the vials because of the much smaller cross-sectional area of the capillary. Thus the voltmeter contacts to the solution in the vials measure the voltage (iR drop) developed in the capillary section. If the voltmeter probe contacts have stable and identical electrode-solution potentials, the interface voltages will cancel in the voltmeter circuit. Reference type electrodes are often used for the voltmeter probes.

## III. CAPACITIVE CONTACT MEASUREMENT OF CONDUCTANCE

A method of avoiding the effect of potential differences arising at the electrode-solution interface is to take advantage of the capacitive behavior of the double layer at the electrode surface to make ac (alternating current) contact with the solution. To understand how this may be accomplished, it is necessary to consider a simple model of a conductance cell and examine its behavior under the influence of ac excitation. A review of ac circuit principles at a level sufficient for understanding the behavior of conductance cells and the instrumentation for conductance measurement is presented. The reader who desires a more thorough study of this topic is directed to material contained in the references [4-7].

### A. Ac Excitation

At relatively low applied frequencies a conductance cell may be represented as the double-layer capacitance $C_S$, in series with the solution resistance R, as shown in Figure 8.8a. When a sinusoidal voltage $e_S$ is applied to the series RC circuit, the instantaneous current i is the same in every part of the circuit and is given by

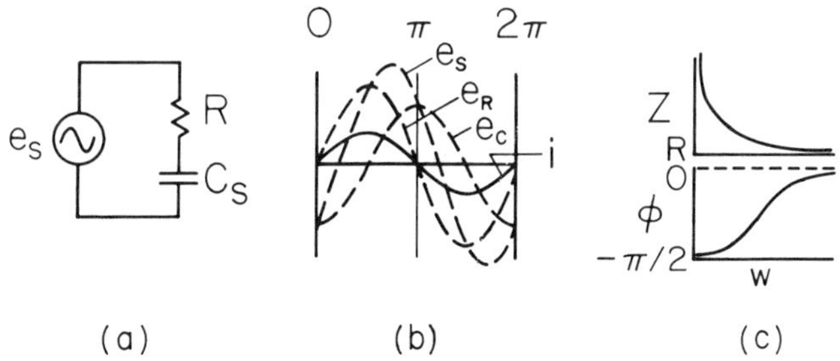


Figure 8.8 Series RC circuit: (a) circuit; (b) current-voltage relationships; (c) frequency dependence of impedance and phase angle.

$$i = I \sin \omega t \tag{8.32}$$

where I is the maximum current, $\omega$ is the angular frequency (rad/s) ($\omega = 2\pi f$, where f is the frequency in hertz) and t is time (s). From Ohm's law the instantaneous voltage $e_R$ across R is then

$$e_R = iR = IR \sin \omega t \tag{8.33}$$

Since the instantaneous current in a capacitor is given by $i = C_s(de_C/dt)$,

$$e_C = \frac{1}{C_s}\int i\, dt = \frac{1}{C_s}\int I \sin \omega t\, dt \tag{8.34}$$

$$= \frac{I}{\omega C_s}(-\cos \omega t) \tag{8.35}$$

It follows that the maximum voltage $E_C$ across $C_s$ occurs when $-\cos \omega t = 1$ and, therefore,

$$E_C = \frac{I}{\omega C_s} = I\frac{1}{\omega C_s} = IX_c \tag{8.36}$$

This expression has a form exactly analogous to Ohm's law in dc circuits. The quantity $1/\omega C_s$ is called the capacitive reactance $X_C$. It is a measure of the opposition to the flow of charge in a capacitor, and it is therefore measured in ohms. If we now substitute the definition of $X_C$ into the expression for $e_C$, it follows that

$$e_C = IX_C(-\cos \omega t) \tag{8.37}$$

Finally, applying the appropriate trigonometric identity, we obtain

$$e_C = IX_C \sin\left(\omega t - \frac{\pi}{2}\right) \tag{8.38}$$

A comparison of Equations 8.38 for $e_C$ and 8.33 for $e_R$ reveals two very important points: $e_C$ lags $e_R$ by $\pi/2$ (a quarter cycle or 90° of phase) and $e_C$ is directly proportional to $X_C$ while $e_R$ is proportional to R.

Capacitive reactance is a frequency-dependent quantity, decreasing with increasing frequency. Typically, double-layer capacitances for aqueous solutions are 10-100 μF/cm$^2$. Thus the capacitive reactance for a 1-cm$^2$ electrode with a 10-μF capacitance at an applied frequency of $10^4$ rad/s (1.6 kHz) is

$$X_C = \frac{1}{\omega C} = \frac{1}{(10^4)(10^{-5})} = 10\ \Omega \tag{8.39}$$

The relationships among $e_R$, $e_C$, $e_s$, and i for the series RC circuit are illustrated in Figure 8.8b. Since the instantaneous voltages across R and C are additive, $e_s$ may be obtained by graphically summing $e_R$ and $e_C$. Note that the current in the circuit is in phase with $e_R$ but is out of phase with $e_C$ by $\pi/2$ (90°), and therefore the signal voltage is always between 0 and 90° out of phase with the current. This phase difference is called the phase angle.

The total opposition to the current in an ac circuit is referred to as its impedance Z, and in a simple RC network is given by the vector sum of R and $X_C$; that is,

$$Z = \sqrt{X_C^2 + R^2} = \sqrt{\left(\frac{1}{\omega C_s}\right)^2 + R^2} \tag{8.40}$$

It should be clear from the expression above and the expressions for $e_C$ and $e_R$ that as the signal frequency is increased, $X_C$ decreases, Z approaches R, and G approaches 1/Z. The potential across $C_s$ also decreases gradually, and the phase angle $\phi$ between i and $e_s$ approaches zero, as is illustrated in the plot of $\phi$ vs. $\omega$ in Figure 8.8c. For an RC circuit, there is a frequency for which $X_C = R$. The reciprocal of this frequency is called the time constant $\tau$. When $R = 1/\omega C$, $1/\omega = RC$. Thus $\tau = RC$, in seconds.

It appears obvious that it is desirable to measure R (and thus G) at a high frequency in order to reduce $X_C$ to a negligibly small value compared to R. Unfortunately, other complications arise if the frequency is increased above a few kilohertz and therefore other means must be devised to decrease $X_C$. A commonly used remedy is to increase the surface area and thus the capacitance of the electrodes as much as possible. A 100-fold area increase is obtained by platinizing the electrodes, that is, electrodepositing a layer of platinum black onto the platinum electrodes, usually from a solution of chloroplatinic acid [8].

A high-frequency limit for the applied potential is encountered above several kilohertz where the impedance of the conductance cell again begins to deviate from the resistance R. Since the solution medium itself is a dielectric situated between two parallel charged surfaces, it can assume the characteristics of a capacitor placed in parallel across the solution resistance as shown in Figure 8.9a. The magnitude of this capacitance is given by

$$C_p = \frac{10^9 D}{4\pi c^2(\ell/a)} \tag{8.41}$$

where the quantities D, c, and $\ell/a$ are the dielectric constant of the medium, the speed of light, and the cell constant, respectively. The dilute aqueous solution dielectric constant D is approximately 80, and if we assume a cell constant of 1 $cm^{-1}$, the parallel capacitance is the approximately 10 pF. At an applied frequency of $10^4$ rad/s (1.6 kHz) this capacitance results in $X_C$ = 10 MΩ.

In order to assess the effect of $C_p$ on the conductance measurements, the analysis of the parallel RC network of Figure 8.9a may be carried out in a manner analogous to the series network above. The situation differs from the series circuit in that in the case, the *voltage* is the same across R and $C_p$, and the currents $i_C$ and $i_R$ in $C_p$ and R are different. The instantaneous currents are given by

$$i_C = \frac{e_s}{X_C} \quad \text{and} \quad i_R = \frac{e_s}{R} \tag{8.42}$$

The current in the resistor, $i_R$, is in phase with the applied voltage while the current existing in the capacitor leads the applied voltage by

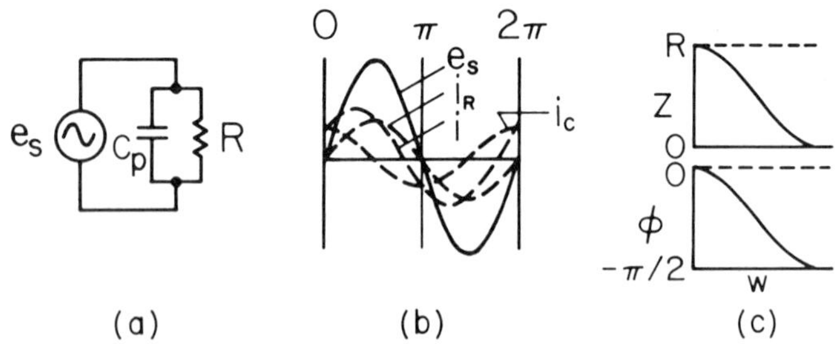


Figure 8.9 Parallel RC circuit: (a) circuit; (b) current-voltage relationships; (c) frequency dependence of impedance and phase angle.

90° as illustrated in Figure 8.9b. To obtain an expression for the total current i, we must find the vector sum of $i_C$ and $i_R$ as follows:

$$i = \sqrt{\left(\frac{C_s}{X_C}\right)^2 + \left(\frac{C_s}{R}\right)^2} = C_s \sqrt{\left(\frac{1}{X_C}\right)^2 + \left(\frac{1}{R}\right)^2} \tag{8.43}$$

The cell impedance is then

$$Z = \frac{e_s}{i} = \frac{1}{(1/X_C)^2 + (1/R)^2} \tag{8.44}$$

Thus if $X_C \simeq 10\ M\Omega$ as shown above, the current path through $C_p$ only becomes appreciable (for a 1% measurement) when the solution resistance approaches 100 kΩ. Such situations are usually avoided by a judicious choice of cell constant and concentration of electrolyte in the solution of interest.

Unfortunately, other experimental factors, such as contact capacitance at the junction of the cell leads and the measurement system, lead capacitance, and capacitance due to the dielectric properties of the thermostating medium, may contribute substantially to the parallel capacitance. These effects may be minimized by proper choice of cell design and use of oil rather than water in the thermostating bath. The art of making ac conductance measurements has been refined to a high degree of precision and accuracy, and detailed discussions of the rather elaborate procedures that are often necessary are available [9,10].

The simple high- and low-frequency models of a conductance cell may be combined in the network shown in Figure 8.10a. The response of this network to ac excitation is illustrated in Figure 8.10b, in which the logarithm of the ratio of the total impedance to the resistance is plotted as a function of the logarithm of the frequency for several values of series and parallel RC time constants. It is obvious from the curves that the solution resistance is best measured over a fairly narrow midfrequency range or plateau. For some solutions this plateau may not exist at all. However, if the plateau can be located, the solution resistance is easily determined by measuring the cell impedance as discussed in the following section.

## B. Impedance Measurements

It is apparent from the previous discussion that in order to measure electrolytic conductance we must devise means for compensating for

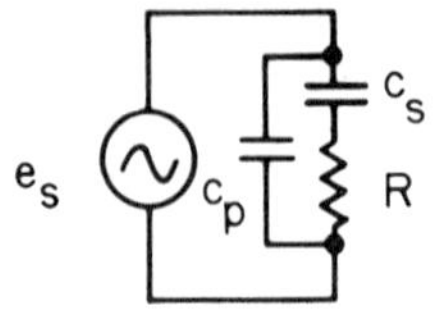


(a)

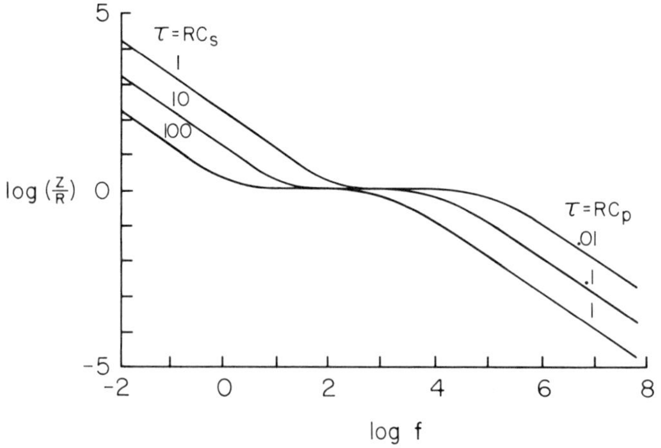


(b)

**Figure 8.10** Resistance with series and parallel capacitors: (a) circuit; (b) frequency dependence of impedance.

the effects of $C_p$ and $C_s$. Figure 8.11 illustrates in block diagram form the elements of the impedance measurement process. Basically, a sinusoid is impressed across the conductance cell and the resulting signal is amplified and demodulated in such a way as to produce a signal proportional to the resistive component of the cell impedance. In the last step of the process, the signal is filtered to provide a dc voltage suitable for driving a convenient readout device.

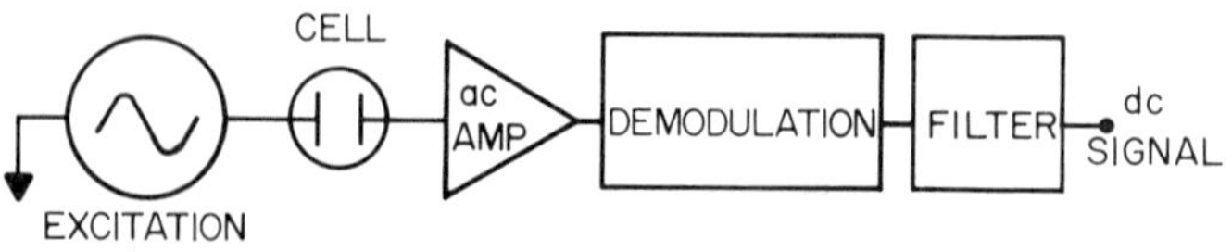


**Figure 8.11** Impedance measurement system.

The most straightforward method for measuring impedance is accomplished by choosing an excitation frequency for which Z = R (i.e., $X_{C_s} \gg R \gg X_{C_p}$), thus making the demodulation step trivial. One very simple circuit for measuring impedance at such frequencies is illustrated in Figure 8.12. This circuit functions in much the same way as the current-follower circuit of Figure 8.3 except that diodes $D_1$ and $D_2$ have been added to rectify (demodulate) the signal resulting from the ac excitation of the cell [12]. As the input signal A drops below zero volts, the summing point of the OA drops a very small amount below ground potential. This causes the output to swing positive, reverse biasing $D_1$ and forward biasing $D_2$, causing the selected feedback resistor, 10R, to be in the feedback loop. At this point, and for the remainder of the negative half-cycle, the gain of the OA is $10R/Z_{cell}$.

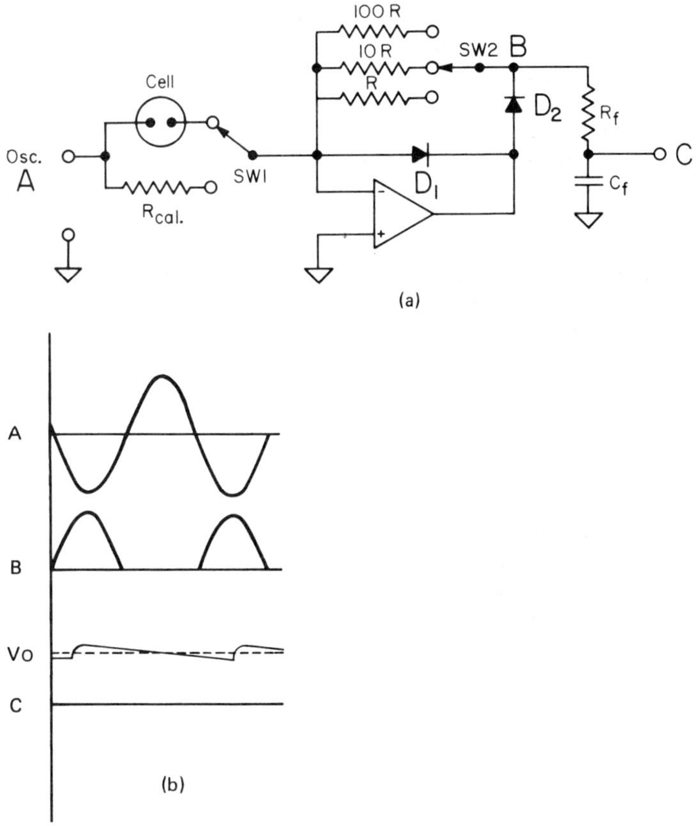


Figure 8.12 Impedance measurement by phase-selective demodulation: (a) circuit diagram; (b) waveforms. (From Ref. 11.)

As the input signal swings above 0 V, the output of the OA swings negative, causing $D_1$ to conduct and $D_2$ to block. This results in a zero-gain situation and the output of the circuit at point B becomes zero. The resulting rectified output is shown in waveform B. This waveform is filtered by the passive RC network at the output to provide a dc signal that may be fed to a strip-chart recorder, digital voltmeter, or other readout device. The switches SW1 and SW2 facilitate calibration of the output and selection of proper amplification for full-scale deflection of the readout device.

Since the output of the circuit is directly proportional to the conductance of the cell ($1/Z_{cell}$ in this case), the circuit may be used as a direct-reading detector for conductometric titrations, reaction kinetic studies of fairly slow reactions, liquid chromatography, or ion exchange chromatography.

The circuit may be easily constructed for a modest cost using inexpensive operational amplifiers and any of a number of integrated circuit oscillators which are commercially available. This rather simple piece of instrumentation is suitable for student laboratory experiments, conductometric monitoring of distilled water or other flowing streams, and other routine applications that do not require high precision or absolute accuracy. Obviously, the instrument is limited to systems in which the capacitive effects are negligible.

Another technique that is useful at frequencies of less than ~ 10 kHz is phase-selective demodulation. A distinct advantage of this technique is that it enables the separation of the resistive (in-phase or "real") and capacitive (90°-out-of-phase or "quadrature") components of the cell impedance. This is accomplished through the process of cross-correlation [13] (selecting that component of $e_s$ that correlates with the phase of i). The excitation voltage waveform is multiplied by a square wave that is in phase with the cell current waveform.

We can understand how this is carried out by considering the waveforms of Figure 8.13a. At frequencies for which parallel capacitive components of the conductance cell impedance are negligible, sinusoidal excitation of the cell produces the waveforms of A, where $e_s$, $e_R$, $e_c$, and i have the same significance as previously discussed. In order to measure the real component of the impedance, the magnitude of the correlation integral cc must be determined.

$$cc = \int r e_s \, dt \tag{8.45}$$

The function r is the bipolar square wave illustrated by waveform B, which has an amplitude of unity and is in phase with $e_R$. We shall demonstrate that the integral above is proportional to R over the first half-cycle. As an exercise, the reader should verify that this is true over the second half-cycle and that cross-correlation of $e_S$ with a quadrature square wave results in an integral that is proportional to $X_C$.

For f = 1 over the first half-cycle, the correlation integral is

$$cc = \int_0^{\pi} (+1)e_S \, dt \tag{8.46}$$

Since $e_S$ may be expressed as the sum of $e_R$ and $e_C$, we have the following:

$$cc = \int_0^{\pi} (+1)\left(IR \sin \omega t - \frac{I}{\omega C} \cos \omega t\right) dt \tag{8.47}$$

Separation of the sine and cosine terms gives

$$cc = IR \int_0^{\pi} \sin \omega t - \frac{I}{\omega C} \int_0^{\pi} \cos \omega t \tag{8.48}$$

$$= -\left(\frac{IR}{\omega} \cos \omega t\right)_0^{\pi} - \left(\frac{I}{\omega^2 C} \sin \omega t\right)_0^{\pi} \tag{8.49}$$

$$= \frac{2I}{\omega} R \tag{8.50}$$

Therefore, the magnitude of the correlation integral is proportional to the resistive component of the cell impedance at constant frequency. This is shown qualitatively in waveforms A to C of Figure 8.13a. Simple graphical multiplication of waveform B and $e_S$ in A gives the correlation waveform C. The shaded areas represent areas that will cancel upon integration and the unshaded area represents $\int re_S \, dt$.

Instrumentally, the cross-correlation and integration may be easily carried out by the circuit shown in Figure 8.13b. The cell is arranged so that it is the input impedance to the familiar current follower. If a sinusoidal voltage $e_S$ is impressed across the cell, the output of OA1 is proportional to and in phase with the current i (see waveform A). Since i is in phase with $e_R$, we may use the output of OA1 to generate the bipolar square wave of waveform B. This is accomplished by the comparator, which is adjusted so that its output

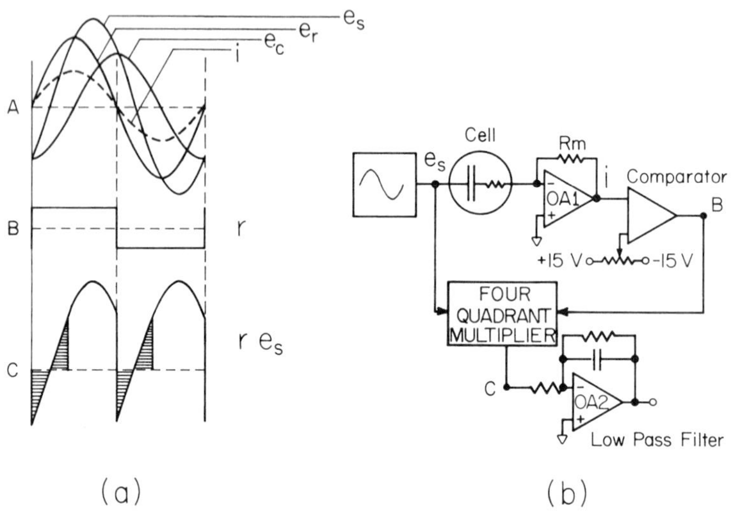


Figure 8.13 Conductance measurement by phase-selective demodulation: (a) waveforms; (b) circuit schematic.

alternates between the positive and negative limits of the OA as the current waveform changes sign. The output of the comparator is connected to the four-quadrant multiplier, which carries out the multiplication of the square wave with $e_s$. The output of the multiplier is then filtered by OA2 to provide a dc output voltage proportional to R.

Circuitry similar to that presented in Figure 8.13b has been used to analyze cells with impedances ranging from $10^2$ to $10^{11}$ Ω with 1% accuracy and resolution better than 1 part in $10^4$ over a frequency range of 0.005 Hz to 10 kHz [14]. The technique has been especially useful for studies of the reaction kinetics of moderately fast chemical reactions. Kadish et al. [14a] have recently used phase selective techniques to make ac impedance measurements to evaluate reference electrodes for use in nonaqueous solvents. Recent decreases in the cost of integrated function modules such as analog multipliers, oscillators, and phase-locked loops make this type of phase-selective instrumentation more accessible than ever before.

Traditionally, the instrument of choice for accurate conductance measurements which are relatively free of capacitance effects has been the ac Wheatstone bridge illustrated in Figure 8.14. The details of operation and the derivation of the balance condition of the ac bridge are presented in considerable detail elsewhere [15,16]. The balance condition is exactly analogous to that of the dc bridge except that impedance vectors must be substituted for resistances in the arms of the bridge when reactive circuit elements are present.

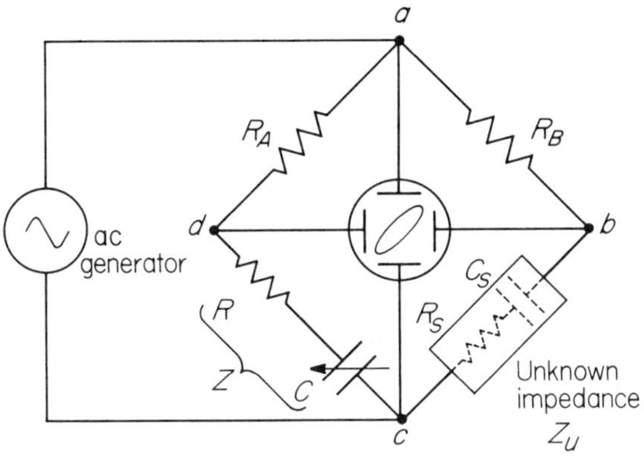


Figure 8.14 Impedance bridge.

At balance, the points at d and b on the ac bridge must be equal in magnitude as well as in phase. The simplest method for determining when this condition exists is to use an oscilloscope connected as shown in the figure as the null detector. If precautions are taken to ensure that the excitation and the null signal are electronically isolated, the balance condition is easily, albeit slowly, obtained [15].

Assuming that the bridge is operated at frequencies for which the parallel capacitance of the cell has negligible conductance, the bridge will balance when $R_s = R_{cell}$ and $C_s = C_{cell}$. Bridges that have been carefully calibrated against known standard resistors may achieve accuracies of 0.01%. With careful preparation of solutions and control of experimental conditions, the measurement of conductances by this method rate among the most accurate and precise of all electrochemical techniques.

## C. Bipolar Pulse Measurements

The bipolar pulse technique for measuring solution resistance minimizes the effects of both the series and parallel cell capacitances in a unique way. The instrumentation for this technique is illustrated in Figure 8.15. The technique consists of applying two consecutive voltage pulses of equal magnitude and pulse width but of opposite polarity to a cell and then measuring the cell current precisely at the end of the second pulse [17].

The pulses are provided by a precision bipolar voltage source which is switched into the input of the pulsing amplifier, OA1, by the switch at point A in the circuit. A very accurate crystal-controlled timing circuit (not shown) drives the switch to ensure that the pulses are symmetrical. The pulsing amplifier inverts the signal as shown by

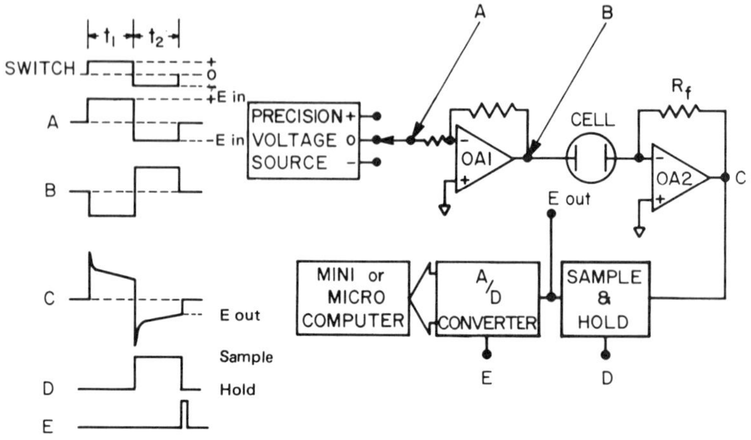


Figure 8.15 Bipolar pulse conductance measurement system.

waveform B and supplies current to the cell. The cell current is amplified by the current follower, OA2, the output of which is illustrated by waveform C.

At the beginning of the first pulse any parallel capacitance in the cell is rapidly charged, producing the small spike in waveform C. As long as $t_1 << R_s C_s$, the potential developed across $C_s$ will be small compared to the excitation potential $E_{in}$. During $t_1$, the current will decrease slightly as the potential across $C_s$ increases. At the end of the first pulse, the polarity is quickly reversed, again resulting in the small spike due to charging of the parallel capacitance. Assuming that $t_1 = t_2$, the double layer is exactly discharged during the second half of the cycle, and the cell current at the end of $t_2$ is due to the resistive component of the cell impedance alone.

The sample-and-hold amplifier samples the output of OA2 during $t_2$ as shown by waveform D and holds the signal $E_{out}$ at the exact end of the second pulse. The falling edge of signal D fires a monostable multivibrator generating a trigger pulse (E) for the analog-to-digital conversion of the sample-and-hold output. Between pulses the cell is at ground potential in order to eliminate any spurious currents.

The resistance of the cell is calculated from the following expressions. Since, at the end of $t_2$,

$$i_{cell} = \frac{E_{out}}{R_f} \quad \text{and} \quad R_s = \frac{E_{in}}{i_{cell}} \tag{8.51}$$

we have

$$R_s = \frac{E_{in}}{E_{out}} R_f \tag{8.52}$$

If the pulse widths of the two pulses are equal to within 1% and the double-layer voltage can be kept to less than 1% of the applied voltage by keeping the pulses short (i.e., 10 μs to 10 ms in duration), the conductance may be determined to within 0.01% [17].

The advantages of the bipolar pulse technique include speed (discrete measurements at a rate as high as 30 kHz), accuracy, and with computer control, facility for signal averaging for maximizing the signal-to-noise ratio. The system has been employed as a detector in automated conductometric titrations and in stopped-flow mixing systems with excellent results.

In another variation of the bipolar pulse technique, a bipolar current pulse is applied to a conductance cell, and the voltage is sampled at the end of the second pulse [18]. Analogously, the solution resistance is calculated from $R_s = E_{measured}/i_{applied}$. This technique has also been applied with good success to chemical problems similar to those mentioned above.

Recent applications of the bipolar pulse technique have demonstrated its utility in a variety of experimental situations. The technique has been shown to be useful in following reaction kinetics on the stopped-flow time scale by the investigation of the dehydration of carbonic acid [19]. The study of this widely used test reaction clearly demonstrates the accuracy and precision of the technique. A sample data set from a single experiment is shown in Figure 8.16, and the excellent precision obtainable in such experiments is evident. The standard error of estimate of the fitted first-order logarithmic plot (middle curve) amounts to 0.2% of the conductance change or 0.015% of the average conductance during the course of the reaction. Using a computer-controlled instrument with simultaneous measurement of conductance and temperature, it is also possible to correct conductance measurements to constant temperature.

A novel application of the bipolar pulse technique that introduces a degree of selectivity into conductometric methods is found in the measurement of the conductance of ion selective electrodes (ISEs) as a very convenient and rapid means of monitoring ion concentrations [20]. Working curves of conductance vs. log concentration for $Ca^{+2}$ are linear over at least four decades of concentration with a detection limit of $10^{-6}$ M $Ca^{+2}$. The most important advantage of monitoring conductance rather than electrode potential is clearly illustrated in the response-time curves of Figure 8.17. In the experiments shown,

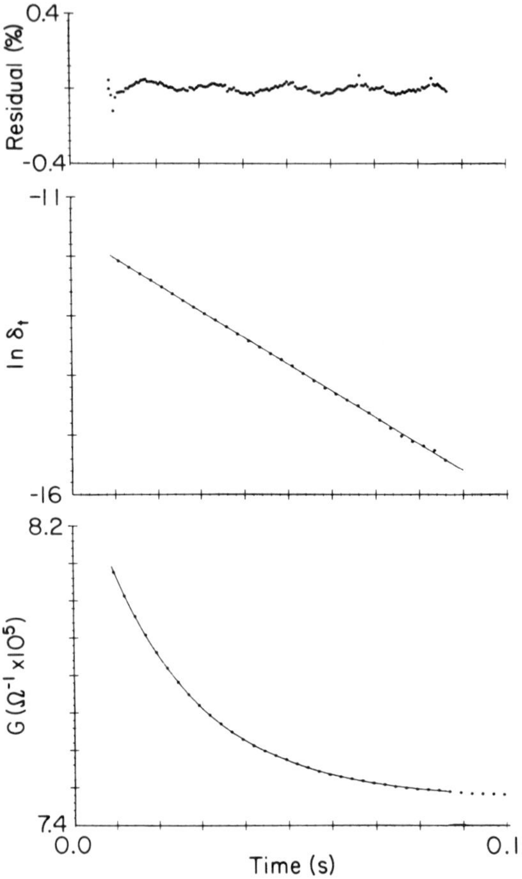


Figure 8.16 Reaction curve for the dehydration of carbonic acid by conductometric detection. Lower: dots, every tenth experimental point; solid line, fitted exponential curve. Middle: dots, logarithmic plot, every tenth point; solid line, least-squares line. Upper: residuals ($G - G_{calcd}$). [From Ref. 19, reprinted with permission from K. J. Caserta, F. J. Holler, S. R. Crouch, and C. G. Enke, *Anal. Chem.* *50*:1539 (1979). Copyright 1978 American Chemical Society.]

small volumes of concentrated solution of $Ca^{+2}$ were added to stirred solutions, and the resulting response curves were recorded conductometrically using both an ISE (open circles) and a pair of stainless steel electrodes (closed circles) and potentiometrically with an ISE (dashed curve). The 90% response time of the ISE was found to be 0.5 s by conductometry and 4.5 s with potentiometric detection.

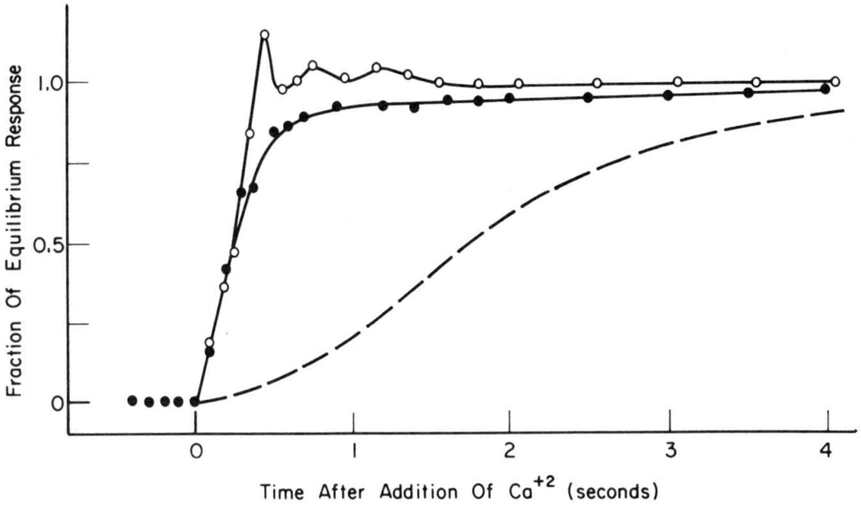


**Figure 8.17** Electrode response time. Conductometric with steel electrodes (open circles); conductometric with ISE (solid circles), potentiometric (dashed line). [From Ref. 20, reprinted with permission from C. R. Powley, R. F. Geiger, Jr., and T. A. Nieman, *Anal. Chem.* *52*:708 (1980). Copyright 1980 American Chemical Society.]

Plunge tests indicate an even faster response for conductometric detection (i.e., about 10 ms). This method also has the advantage of totally eliminating the need for a reference electrode, since a simple stainless steel or platinum electrode provides electrical contact with the analyte solution [20a].

Other applications of the bipolar pulse which have appeared in the recent literature include its use in the measurement of critical micelle concentrations [21] and as a detector in ion chromatography [22]. It is clear that the advent of the bipolar pulse conductance technique has generated renewed interest in one of the oldest instrumental methods.

## REFERENCES

1. J. E. Prue, *Ionic Equilibria*, Topic 15, Vol. 3 of *International Encyclopedia of Physical Chemistry and Chemical Physics*, Pergamon Press, New York, 1966.
2. R. A. Robinson and R. H. Stokes, *Electrolyte Solutions*, Second ed., revised, Butterworth, Sevenoaks, Kent, England, 1968.
3. R. M. Fuoss and L. Onsager, *J. Phys. Chem.* *66*:1722 (1962); *67*:621 (1963); *68*:1 (1964).

4. H. V. Malmstadt, C. G. Enke, and S. R. Crouch, *Electronics and Instrumentation for Scientists*, Benjamin-Cummings, Menlo Park, Calif., 1981, pp. 115-121.
5. H. V. Malmstadt, C. G. Enke, and E. C. Toren, Jr., *Electronics for Scientists*, Benjamin-Cummings, Menlo Park, Calif., 1962, pp. 545-581.
6. R. D. Sacks and H. B. Mark, Jr., *Simplified Circuit Analysis*, Marcel Dekker, New York, 1972, Chap. 1.
7. A. G. Diefenderfer, *Principles of Electronic Instrumentation*, Second ed., W. B. Saunders, Philadelphia, 1979, Chap. 2.
8. D. F. Evans and M. A. Matesich, in *Techniques of Electrochemistry*, (E. Yeager and A. J. Salkind, eds., Wiley-Interscience, New York, 1973, p. 49.
9. *Ibid.*, p. 30.
10. T. Shedlovsky, in *Technique of Organic Chemistry*, 2nd ed., Vol I, Part II (A. Weissberger, ed., Interscience, New York, 1959.
11. C. N. Reilley, *J. Chem. Educ.* *39*:A853 (1962).
12. J. I. Smith, *Modern Operational Circuit Design*, Wiley-Interscience, New York, 1971, Chap. II.
13. Ref. 4, p. 420.
14. A. J. Bentz, J. R. Sandifer, and R. P. Buck, *Anal. Chem.* *46*:543 (1974).
14a. S. Cai, T. Malinski, X. Lin, J. Ding, K. M. Kadish, *Anal. Chem.*, *55:*161 (1983); K. M. Kadish, S. Cai, T. Malinski, J. Ding, X. Lin, *Anal. Chem.*, *55:*163 (1983).
15. J. Braunstein and G. D. Robbins, *J. Chem. Educ.* *48*:52 (1971).
16. Ref. 1, p. 481.
17. D. E. Johnson and C. G. Enke, *Anal. Chem.* *42*:329 (1970).
18. P. H. Daum and D. F. Nelson, *Anal. Chem.* *45*:463 (1973).
19. K. J. Caserta, F. J. Holler, S. R. Crouch, and C. G. Enke, *Anal. Chem.* *50*:1534 (1978).
20. C. R. Powley, R. F. Geiger, Jr., and T. A. Nieman, *Anal. Chem.* *52*:705 (1980).
20a. C. R. Powley and T. A. Nieman, *Anal. Chim. Acta.*, *139:*61 (1982); *139:*83 (1982).
21. J. Baxter-Hammond, C. R. Powley, K. D. Cook, and T. A. Nieman, *J. Colloid Interface Sci.* *76*:434 (1980).
22. J. M. Keller, *Anal. Chem.* *53*:344 (1981).

# 9
# Mercury Electrodes

ZBIGNIEW GALUS / Department of Chemistry, University of Warsaw, Warsaw, Poland



## I. INTRODUCTION

Mercury is widely used in the practice of electroanalytical chemistry both for working electrodes and for reference electrodes, in the latter case usually as an electrode of the second kind.

The use of mercury is nearly an ideal choice for the construction of working electrodes for several reasons. Mercury has a large liquid range (−38.9 to 356.9°C at normal pressure) and therefore electrodes of various shapes may be easily prepared. The surface of such electrodes is highly uniform and reproducible if the mercury is clean.

The preparation of pure mercury is not very difficult. All metals with a standard potential more negative than that of mercury may be oxidized easily (with the exception of nickel, which forms a mercury intermetallic compound) by dispersing mercury into a solution of its salts acidified with $HNO_3$ and saturated with oxygen. The elimination from mercury of metals more noble than itself is accomplished by distillation under reduced pressure. Usually, such distillations are repeated several times.

One of the most important reasons for the application of mercury to the construction of working electrodes is the very high overvoltage for hydrogen evolution on such electrodes. Relative to a platinum electrode, the overvoltage of hydrogen evolution under comparable conditions on mercury will be −0.8 to −1.0 V. It is therefore possible in neutral or (better) alkaline aqueous solutions to reduce alkali metal cations at mercury electrodes, giving relatively well-defined polarographic waves at potentials more negative than −2.0 V vs. SEC.

Using supporting electrolytes such as tetraalkylammonium salts, one may apply potentials as high as −2.6 V vs. SCE in aqueous solutions while in some nonaqueous systems even −3.0 V vs. SCE (aqueous) is accessible. Unfortunately, mercury electrodes have serious limitations in applications at positive potentials (with the exception of passivated mercury electrodes, which are described in Section VI), and this has led to extensive research in the development of solid metal and carbon electrodes. Oxidation of mercury occurs at approximately +0.4 V vs. SCE in solutions of perchlorates or nitrates since these anions do not form insoluble salts or stable complexes with mercury cations. In all solutions containing anions which form such compounds, oxidation of the mercury proceeds at potentials less than +0.4 V vs. SCE. For example, in 0.1 M KCl this occurs at +0.1 V, in 1.0 M KI at −0.3 V, and so on.

The following types of mercury electrodes have been widely used for voltammetry: dropping mercury electrode (DME), hanging mercury drop electrode (HMDE), streaming mercury electrode (SME), and mercury film electrode (MFE). More recently a static mercury drop electrode (SMDE) has been developed. We will begin our discussion with a description of the construction and properties of the DME because this electrode has a long history and continues to be used extensively both for analytical and fundamental studies.

## II. DROPPING MERCURY ELECTRODE

### A. Introduction

The DME is the essential component of all polarographic experiments. It was introduced by Kucera in 1903 [1]. Later, in 1921, Heyrovsky applied this electrode in the original polarographic apparatus [2]. Usually, the DME is formed of glass capillary tubing connected to a standpipe of plastic or glass attached to a mercury reservoir. If the level of mercury in the standpipe is sufficiently high with respect to the tip of the capillary, mercury flows from the capillary at a steady rate and small drops form at the end and fall at a regular interval. These drops have a nearly spherical shape if their mass is not greater than about 15 mg and the radius of the capillary is on the order of 0.05 mm. Electrolysis is carried out on the surface of the mercury drops. One of the most important advantages of the DME is continuous renewal of the electrode surface. The periodicity of this renewal is governed by the life of an individual drop.

Our description of the DME will start with a discussion of basic concepts involved in the formation of mercury drops. These considerations will subsequently be used to define the capillary characteristics required for satisfactory electrodes. This discussion is based partially on the classical monograph of Kolthoff and Lingane [3].

## B. General Discussion of the Flow of Mercury from Capillaries

The rate of flow of a liquid through a capillary can be defined by the Poiseuille equation. The volume V of liquid that flows in time t is given by

$$V = \frac{\pi r_c^4 P t}{8 \ell \eta} \tag{9.1}$$

where $r_c$ and $\ell$ are the internal radius of a capillary and its length, $\eta$ is the viscosity of the flowing liquid, and P is the effective hydrostatic pressure (i.e., P is not exactly equal to the pressure created by the difference between the level of liquid in the reservoir and the lower end of the capillary). This equation may be strictly applied to the flow of a liquid which wets the walls of a capillary. This is not exactly true in the case of mercury flowing from glass capillaries; however, the resulting error is small.

Equation 9.1 may be easily rearranged to the form

$$m = \frac{Vd}{t} = \frac{\pi r_c^4 P d}{8 \ell \eta} \tag{9.2}$$

where d is the density of the liquid. In this rearrangement the following relationship was used:

$$mt = Vd \tag{9.3}$$

where m is the flow rate of mercury from the capillary (usually expressed as mg/s). Both mt and Vd express the total mass of mercury flowing for a given t and can easily be measured experimentally.

The effective pressure that acts on the mercury is not exactly equal to the total pressure $P_t$, which is given by the equation

$$P_t = h_t gd \tag{9.4}$$

where $h_t$ is the vertical distance between the end of the capillary and the level of the mercury in the reservoir and g is the gravitational constant.

Mercury flows from the capillary in small drops. The size of these drops is determined by the interfacial tension at the mercury-solution interface. The internal pressure in the drop acting against $P_t$ is, according to Kučera, given by the relation

$$P_b = \frac{2\gamma}{r} \tag{9.5}$$

This internal pressure $P_b$, often called back pressure, is proportional to the interfacial tension $\gamma$ at the mercury-solution interface and is inversely proportional to the radius of the drop.

It follows from Equation 9.5 that the amount of mercury flowing into the solution as drops should be somewhat less than the amount of mercury that would flow from the same capillary at the same $h_t$ when dipped into a larger volume of mercury. Equation 9.5 also shows that the back pressure in a given solution changes with the growth of a drop and reaches the smallest value at the moment of detachment of a drop from the capillary. If one considers, for example, a capillary with an internal diameter of 0.06 mm placed into 3 M KCl at a potential −0.6 V, the back pressure changes from 10.44 to 0.97 cm of mercury during the life of the drop. If the mercury column is lower than the first value, it is obvious that the outflow of mercury will not be observed.

The radius of the drop, r, at an instant $t_1$ following drop birth may be easily calculated using Equation 9.3 and the expression for the volume of a sphere,

$$V = \frac{mt_1}{d} = \frac{4}{3}\pi r^3 \tag{9.6}$$

It is important to know the mean value of the back pressure over the lifetime of the drop, $t_d$. To obtain this value one can calculate the mean radius of the DME, $\bar{r}$:

$$\bar{r} = \frac{1}{t_d}\int_0^{t_d} r\,dt = \frac{3}{4}\left(\frac{3mt_d}{4\pi d}\right)^{1/3} \tag{9.7}$$

By combining Equations 9.7 and 9.5 with elimination of r, one obtains

$$\bar{P}_b = 4.31\,\frac{\gamma d^{1/3}}{(mt_d)^{1/3}} \tag{9.8}$$

Now the effective pressure may be expressed as

$$P = P_t - \bar{P}_b = h_t gd - 4.31\,\frac{\gamma d^{1/3}}{(mt_d)^{1/3}} \tag{9.9}$$

By introducing P given by Equation 9.9 into 9.2, one obtains

$$m = \frac{\pi r_c^4 d}{8\ell\eta}\left[h_t dg - 4.31\,\frac{\gamma d^{1/3}}{(mt_d)^{1/3}}\right] \tag{9.10a}$$

Using the following values for the constants,

$d = 13.53\ g/cm^3$, at 25°C

$\eta = 0.0152\ dyn\cdot s/cm^2$, at 25°C

$g = 980.6\ cm/s^2$

and taking

$\gamma = 400\ dyn/cm$

one can evaluate m according to Equation 9.10b.

$$m = 4.64 \times 10^9 \frac{r_c^4}{\ell}\left[h_t - \frac{3.1}{(mt_d)^{1/3}}\right] \qquad (9.10b)$$

If $r_c$ and $\ell$ are given in centimeters and $t_d$ in seconds, one obtains m in milligrams per second.

One should remember that the product $mt_d$ is constant for a given capillary at a given potential. This constancy follows from the equation

$$mt_d g = 2\pi r_c \gamma \qquad (9.11)$$

The left-hand side represents the weight of the falling drop, which is dependent on the radius of the capillary and the interfacial tension.

The correction term in Equation 9.10a is constant independent of h; however, in view of equation 9.10b, this term is quite significant. Assuming that m = 2 mg/s and $t_d$ = 4 s (conditions that are very often met in practice), $h_t$ has to be corrected for 15.5 mm, which is quite significant in comparison to $h_t$, which sometimes may be as small as 150 mm.

The linear dependence between m and $h_t$ corrected for the back pressure has been found experimentally [4]. With a given capillary and the same $h_t$, m may be slightly different for different solutions because the back-pressure term will change from solution to solution due to variations in the interfacial tension. It follows from Equation 9.11 that the product $mt_d$ should be constant if $h_t$ is not changed. From Equations 9.10a and 9.11 it follows that $t_d$ should be inversely proportional to $h_t$.

## C. Construction of the Dropping Mercury Electrode

Capillaries suitable for construction of a DME are commercially available from many sources. They may also be prepared from ordinary capillary tubing with thick walls by heating and pulling to shrink the bore, but this is rarely worth the trouble.

In choosing the capillary one should remember that uniformity is a primary consideration. The capillary should have no irregularities or radial cracks extending from the orifice, because such capillaries often cause irregular limiting currents. Therefore, it is advisable to inspect each capillary with a magnifying glass before use. Although it is usual for analytical work to cut the tube normal to the capillary, this results in "shielding" the drop from the top side since the diameter of the glass is typically 5-7 mm; a conical tip can be made to avoid shielding to some extent.

The length of the capillary used depends on its internal diameter and also on the experimental objectives. Usually, the longest drop times should not exceed 7-8 s. In general, the capillary should be such as to give a drop time on the order of 3-4 s with $h_t$ equal to ca. 40 cm when the capillary is dipped into the solution. If drop times are too long, one should cut the capillary remembering that a linear dependence exists between the length of the capillary and $t_d$.

The appropriate capillary length may also be estimated on the basis of simple calculations. Neglecting the correction term in Equation 9.10b, one has for the mercury flow rate

$$m = 4.64 \times 10^9 \frac{r_c^4 h_t}{\ell} \tag{9.10c}$$

By combining Equations 9.10c and 9.11 and introducing numerical constants, we have

$$t_d = 5 \times 10^{-7} \frac{\ell}{r_c^3 h_t} \tag{9.12}$$

Using this equation, one may calculate the length of the capillary necessary to have proper drop times at reasonable $h_t$.

Let us assume that the radius of the capillary is 0.04 mm and the shortest $t_d$ that we would like to have is 2 s at $h_t$ equal to 80 cm. Introducing these data into Equation 9.12 gives $\ell = 18.6$ cm. For such a capillary if $h_t = 20$ cm, the drop time is about 8.5 s. These calculated values are only approximate, since we did not consider the influence of the back pressure.

When cutting the capillary one should remember that the end that enters the sample solution should be cut as close as possible to 90° to obtain the most reproducible currents [5,6]. The cut should also be smooth, although not mechanically polished. The other end may be sealed to a short glass tubing of several millimeters internal diameter or connected directly to flexible tubing. The connection to the mercury reservoir is normally made using a soft plastic tubing such as Tygon. The inside diameter of this tubing should be of the order of 5-7 mm and the connection to the capillary should be reinforced with wire. The plastic tube should not contain sulfur. Before use the tube interior should be washed with a dilute solution of sodium or potassium hydroxide, rinsed with distilled water, and thoroughly dried.

The mercury reservoir usually has a volume varying between 100 and 200 $cm^3$. The platinum wire used to make contact to the

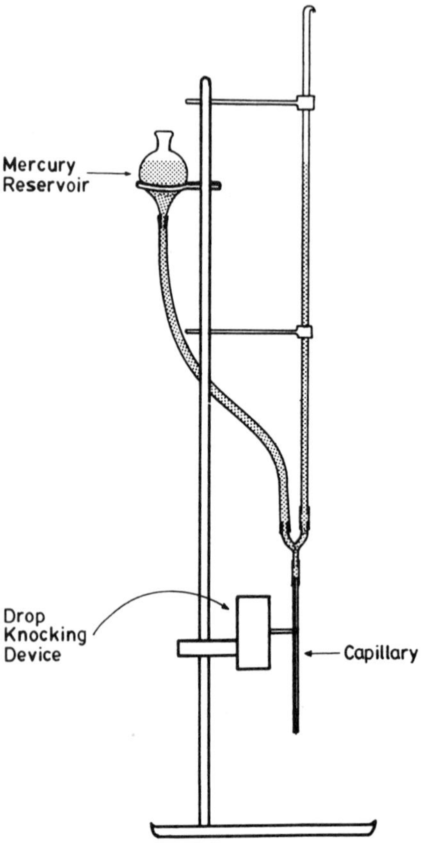


Figure 9.1 Stand tube and mercury reservoir assembly for use with dropping mercury electrodes. [From J. A. Wise, D. A. Roston, and W. R. Heineman, in *Methods and Techniques of Multi-element Determination* (O. Rose, ed.), Marcel Dekker, New York, to be published.]

DME may be dipped in the mercury reservoir or sealed in the stand-tube. A commonly used stand tube and mercury reservoir assembly is shown in Figure 9.1.

Once the capillary has been connected, the mercury reservoir should be filled with very pure mercury. For polarographic purposes the triply distilled mercury available commercially is usually satisfactory. One should be especially careful when filling the apparatus with mercury in order to minimize entrapped air bubbles.

The DME requires a stand that facilitates changing the position of the mercury reservoir. Such stands are produced by several manufacturers [7]. A centimeter scale is often included, making it convenient to determine $h_t$. Some authors recommend an all-glass apparatus. This seems inadvisable now that inert plastic materials are available. When a fixed glass reservoir is used, the drop time may be controlled by changing the gas pressure over the mercury. In any case, the polarographic apparatus should be protected against vibrations that might prematurely dislodge the drop.

## D. Maintenance of the DME and Determination of the Drop Time and Flow Rate of the Capillary

A capillary will work properly for a long time if special care is exercised both during experiments and when the electrode is not in use. An important rule of thumb is that solutions should never enter the capillary. The mercury reservoir should always be elevated and the flow of mercury observed, before the capillary is placed in solution. When the polarographic work is finished the DME should be withdrawn from the cell and then transferred to a beaker containing distilled water (or a nonaqueous solvent, if used). Mercury should drop there for several minutes to wash the end of the capillary. Finally, the capillary is carefully washed with distilled water.

There are several procedures for storing the DME after washing. Probably the best technique is to dry the capillary with a filter paper and then to dip it into a vial of mercury. Following this, the mercury reservoir is lowered to decrease the flow of mercury. The DME may also be stored in distilled water. In this case mercury should again continue to flow very slowly.

After washing and drying the capillary many workers simply lower the mercury reservoir sufficiently to stop the flow of mercury. This convenient procedure can occasionally cause problems. The inevitable metal impurities in mercury can be air oxidized at the capillary tip and the oxidation products may change the parameters of the electrode (namely, m and $t_d$). There is also a possibility that a small amount of the sample solution may remain in the capillary and then evaporate leaving a solid residue. The latter will also change the parameters of the electrode and may be very difficult to remove from the capillary.

If erratic behavior of the capillary is observed (this may be easily recognized from a decrease of the current or irregularities in the drop time), the tip may be cleaned with 1:1 nitric acid (ca. 6 M) by immersing it in a small beaker of this solution followed by thorough rinsing. Mercury should, of course, flow from the DME throughout the procedure. This simple treatment may be insufficient when the residue is at some distance from the end of the capillary. In these cases the capillary may be removed from the apparatus and carefully washed out with 1:1 hot nitric acid using a vacuum to suck this solution through the capillary. It may take quite a long time to remove the impurities. The capillary is then rinsed with distilled water, dried, and remounted. Sometimes even this procedure is not effective and it may be helpful to place the dry capillary in an oven for 24 h at approximately 450°C. For obvious reasons, such desperate measures are usually passed up in favor of replacing the capillary with a new one.

The parameters characterizing the DME may be determined in the following way. The capillary is placed in a solution of the supporting electrolyte. The mercury reservoir is adjusted to a given $h_t$ and a potential is applied, either a value related to experimental measurements or in the range −0.5 to −0.6 V vs. SCE. A stopwatch is used to measure the time for 10-20 drops to form and the droptime, $t_d$, is obtained by dividing the elapsed time by the number of drops collected. Mass flow rate of mercury, m, can be measured by collecting drops for a measured time and then weighing the washed and dried mercury. Usual units are milligrams per second.

For accurate calculation of the area of the drop, a correction must be made for the area of the capillary by use of the formula

$$A = 2\pi r[r + (r^2 - r_c^2)^{1/2}]$$

The radius of the drop is calculated from m and $t_d$ and $r_c$ is measured with a microscope and ocular micrometer.

## E. Unusual Dropping Mercury Electrodes

A variety of types of DMEs have been developed [8] and they are useful in some types of experiments. Blunt-ended DMEs are, for example, not useful in experiments carried out at higher frequencies. The shielding effect mentioned earlier creates a nonuniform distribution of current along the surface of the drop and spurious frequency dispersion effects appear.

To diminish this problem capillaries with a sharp tip are used [9], which may be prepared by drawing out a normal capillary. To overcome the shielding effect significantly, the overall diameter of the capillary at its end should be considerably smaller than the mean diameter of the mercury drop. It should, however, be mentioned that work with this type of capillary is more difficult and also the production of such capillaries with a uniform drop time requires considerable practice.

The deviation of the position of a capillary from the vertical changes its drop time considerably. The DME placed in a horizontal position was originally used by Smoler [10]. The shielding and depletion effects are diminished to some extent and also current oscillations are much smaller. Probably the best performance is observed when the electrode position deviates 45° from the vertical; however, other [11] opinions are not so positive about such electrodes. In some experiments, electrodes with small mercury flow rates are helpful. For these electrodes the probability of the appearance of polarographic maxima is diminished. When m is very low, the drop time is usually very long. In such cases artificial regulation of the drop time should be used (see Sec. II.F).

Capillaries with small m may be produced by restricting the internal diameter of a normal polarographic capillary at some distance from its end. Since the radius of the capillary at the end is not changed, the drop should be practically identical at $t_d$ with that of the original nonrestricted capillary. If the capillary were restricted at its end, the drop size would also be diminished. Electrodes of the latter type would also have the disadvantage of being easily blocked by small crystals of mercury salts or other impurities.

Very often anomalous effects are observed due to solution entering the capillary at the instant the mercury drop detaches from the thread. In ac measurements this phenomenon leads to anomalous frequency dispersion [12]. In addition, the drop time becomes irreproducible. These effects may be diminished to a large extent by coating the internal wall of the capillary with a film of silicone [13]. This procedure is also very commonly used for the hanging mercury drop electrode discussed in Section III.

### F. Mechanical Control of Drop Life

Very often polarographic experiments are best carried out with a prescribed drop time. If not controlled, the drop time will change slightly as a function of potential since the surface tension is dependent on double-layer structure. Drop time can be controlled, for example, by a simple arrangement consisting of a glass rod sealed to the end of the polarographic capillary [14]. For modern techniques using digital timing circuitry (e.g., sampled dc and pulse polarography) it is most convenient to achieve a constant $t_d$ by dislodging the drop prior to its natural fall. Many systems are based on a solenoid controlled hammer which periodically touches the capillary, leading to detachment of the mercury drop. Apparatus for mechanical control of the drop time are produced by several manufacturers [15].

## III. HANGING MERCURY DROP ELECTRODE

### A. HMDE Suspended on a Metallic Contact

A simple HMDE was developed by Gerischer [16]. Using his approach 1 or 2 mercury drops falling from the classical DME are collected on a

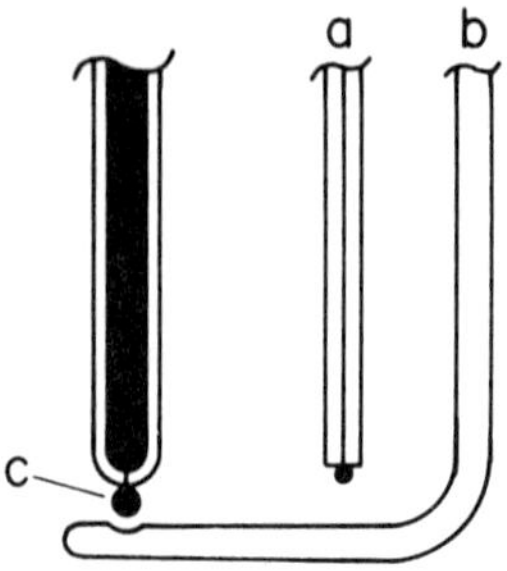


Figure 9.2 HMDE as developed by Gerischer: (a) DME; (b) transfer cup; (c) HMDE.

miniature spoon and transferred to a small metal contact sealed in glass or in pastic material (Fig. 9.2). Platinum or gold-plated platinum wires are used. The length of the exposed contact wire is usually on the order of 0.1-0.5 mm and its diameter is about 0.5 mm.

Electrodes of this sort have many different chemical properties from pure mercury electrodes, because of the formation of a gold or platinum amalgam [17]. Normally, a drop is suspended just prior to an experiment and this problem will therefore be of no consequence. Nevertheless, since the solubility of these noble metals in mercury is of the order of 0.05 M, the concentration of gold or platinum in mercury may be quite significant on a longer time scale. In such cases gold or platinum may form intermetallic compounds with several metals which are electrodeposited into the mercury [17].

## B. Capillary HMDE

This version of the HMDE consists of a small mercury drop with radius usually not exceeding 1 mm, which hangs on a thin mercury thread in a glass capillary. The inner diameter of this capillary is about 0.15-0.20 mm, considerably larger than for the DME. An electrode of this type was used for the first time by Antweiler [18]; however, the preparation of a reproducible HMDE using his original model was very difficult. As pointed out by Kemula and Kublik [19], the suspended HMDE is more reproducible in comparison with the early design capillary HMDEs.

In the most successful designs, the mercury is pushed out mechanically. This type of electrode was constructed and used in the experimental work of Randles and White [20]. Mercury was pushed from the capillary by turning a micrometer screw. The electrode which is now most popular was developed by Kemula and Kublik [21]. Many different variations exist and the model presented in Figure 9.3 has

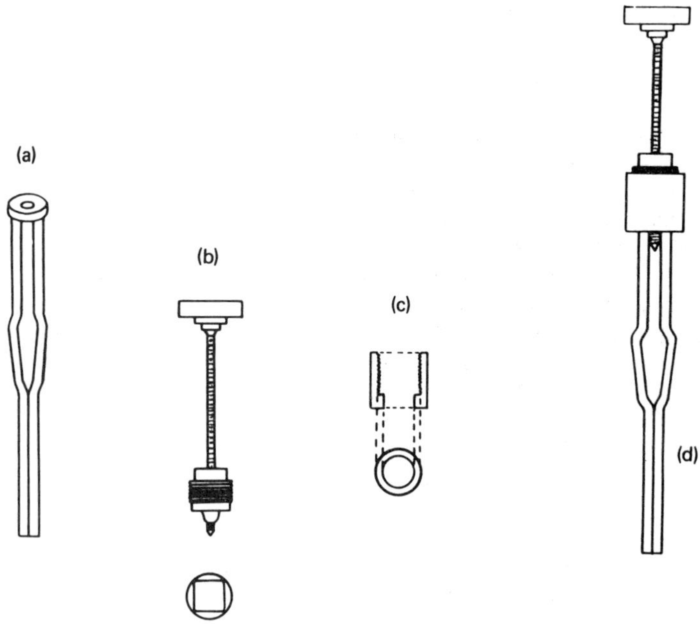


Figure 9.3 Capillary HMDE: (a) reservoir and capillary (glass), (b) threaded plunger, (c) plastic collar, (d) assembled unit.

been used for some time at the Department of Chemistry at Warsaw University.

A detail of the glass part of this electrode is shown in Figure 9.3a. It is composed of a thick-walled glass tubing with an internal diameter of 4 mm, approximately 5 cm long. This forms a mercury reservoir which is sealed to a thick wall capillary. The internal diameter of this capillary is commonly 0.18 mm, but may vary somewhat around this value.

The length of the capillary will depend on the size of the electrolytic cell. A ground-glass joint on the mercury reservoir can provide a convenient means of supporting the electrode in a cell. A Teflon thermometer adapter also works well. The top of the reservoir resembles a glass ring which is flat and very smoothly polished. This part of the electrode is connected to the plastic head, which is composed of two parts. These parts are shown in Figure 9.3b and c. After the capillary is inserted through part b, part c is screwed rather tightly into part b. The end of the mercury reservoir should then be tightly sealed by the bottom of part c.

To provide a tight seal a small plastic cylinder is extended from part c, being a bit thicker than the internal diameter of the glass tubing. This small cylinder, shown schematically in Figure 9.3c, should enter into the reservoir tubing with a pressed fit when all parts of the HMDE are put together.

A threaded steel piston of 2 mm in diameter moves in part c. By turning this screw by a fixed angle, for instance 90° or 180°, mercury will be pushed out from the capillary and will hang as a drop from its end. It is important that the piston screw be set tight in the polyethylene part shown in Figure 9.3b.

In Figure 9.3d the HMDE is shown ready for use. To ensure that the electrode works properly, special care should be taken to coat the internal part of the capillary with a hydrophobic silicone film. The capillary should be thoroughly cleaned before a new coating is applied. Sodium hydroxide (ca. 2 M) should be used to remove any residual silicone. The capillary should then be washed with 3 M nitric acid, rinsed with distilled water, and thoroughly dried. Then a 5% solution of dichlorodimethylsilane in $CCl_4$ is pulled through the capillary, followed by air. After several hours at room temperature or about 15 min at 110°C, the capillary is ready for use. The silicone coating should be renewed from time to time, especially when the hanging drop tends to fall during an experiment. This happens when the silicone coating on the capillary wall is partly destroyed and solution enters between the glass and the mercury thread.

The mercury reservoir and the capillary should be completely filled with mercury. There must be no entrapped air! A special vessel useful in filling the HMDE was described by Kemula [22]. This vessel should be equipped with a ground joint corresponding to that of the capillary and the end of the capillary should extend to within several millimeters of the bottom.

The vessel should be filled with mercury such that the end of the capillary is covered when the vessel is placed in a horizontal po-

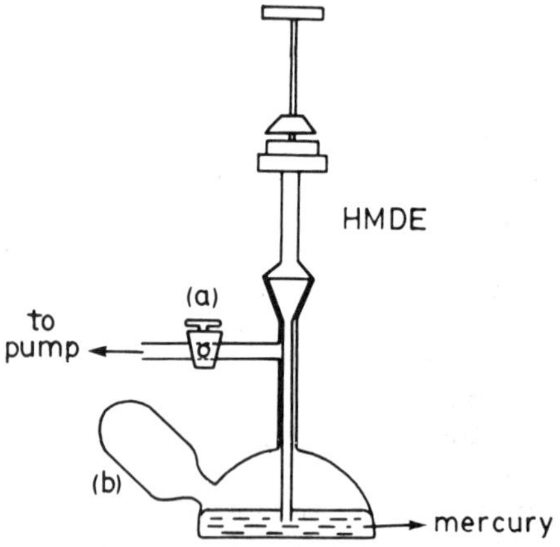


Figure 9.4 Arrangement for filling the HMDE.

sition, as depicted in Figure 9.4. With all of the mercury in part b, the vessel is connected to a vacuum pump, with a three-way stopcock to decrease the pressure in the vessel and in the electrode down to 0.01 mm.

After evacuation, the vessel is tilted into the horizontal position shown in Figure 9.4 and the stopcock is turned to disconnect the pump and then turned further to admit air to the vessel. The pressure of air acting on the mercury in the vessel pushes it into the reservoir and the capillary of the HMDE. If the pressure had been decreased to 0.01 mm, mercury fills reservoir and capillary completely. The HMDE is now ready for use. By turning the screw a fixed angle, a known amount of mercury is pushed out, forming a drop.

An electrode like that described above may be easily prepared in every laboratory. The connection of the mercury reservoir to the plastic cap may be made in other ways, but in every case the mercury reservoir must be closed tightly. Modifications of this type of HMDE have been described by other authors [23]. Variations on the HMDE described above are also available from several manufacturers. These electrodes are equipped with a micrometer screw for pressing out a well-defined amount of mercury.

### C. Sessile Mercury Drop Electrodes

These electrodes are obtained using J-shaped capillaries connected with the typical HMDE arrangement described above. The drop rests on the mercury thread rather than hanging from it. These electrodes are sometimes useful when working at very negative potentials. Then the classical HMDE may not be stable and the drop often falls off at the most inconvenient time. Sessile drops are more stable; however, they are not recommended for most other experiments.

Such electrodes may also be easily prepared by placing the mercury drop on a small contact made of metal which is wetted by mercury as was done by Barker [12]. This version of the electrode was also used by Neeb [24].

## IV. STATIC MERCURY DROP ELECTRODE

The static mercury drop electrode (SMDE) recently developed by EG&G Princeton Applied Research utilizes a method of drop formation in which the mercury drop is dispensed rapidly and then allowed to hang stationary at the capillary tip [25]. When used in a DME mode of operation, the drops can be repetitively formed and dislodged at desired time intervals for use in sampled dc polarography, pulse polarography, or differential pulse polarography. Alternatively, a drop can be maintained indefinitely for use as a HMDE.

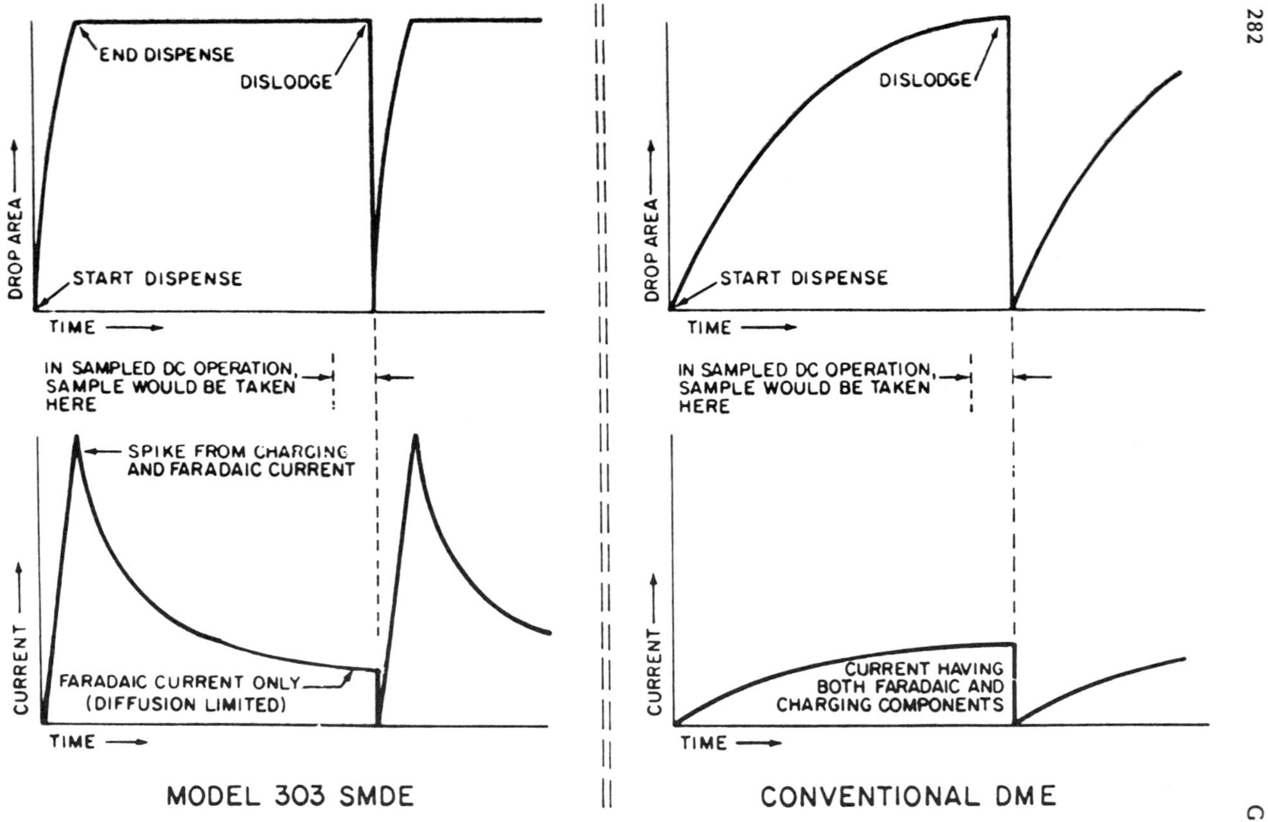


Figure 9.5 Drop growth and current waveforms for SMDE and conventional DME. [From Ref. 25, adapted from *Am. Lab.* *11*(12):69-78 (1979). Copyright 1979 International Scientific Communications, Inc.]

When used in the DME mode of operation, the SMDE exhibits a fundamental difference in its current-time behavior compared to the conventional DME. Figure 9.5 shows the growth of drop area and the accompanying current profiles for a SMDE and a conventional DME. In the SMDE the drop is rapidly expanded to a given area, which is then maintained constant. This feature enables the current measurement in various polarographic techniques to be made on an electrode with a static area, eliminating the contribution of nonfaradaic current from changing surface area. By comparison, the conventional DME exhibits a continuously increasing surface area until drop dislodgement. Thus a component of charging current due to increasing surface area always exists.

The mechanical details of the SMDE are illustrated in Figure 9.6. Activation of the solenoid lifts the plunger, allowing mercury to flow from the large reservoir on top through the capillary to form a drop.

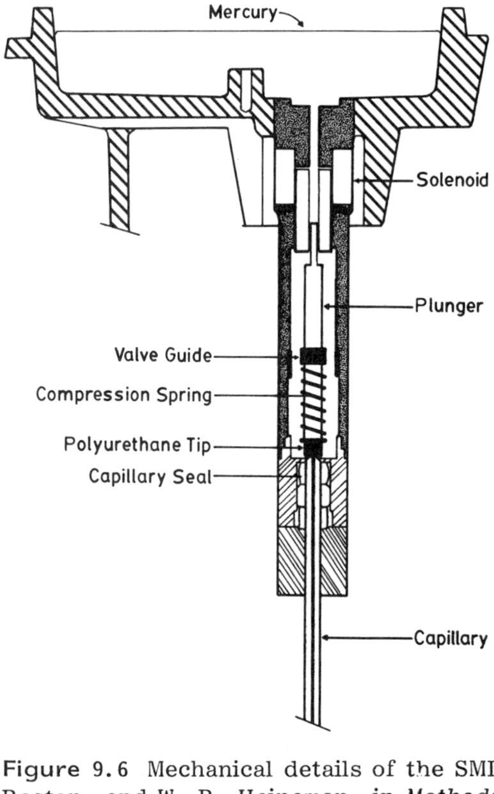


Figure 9.6 Mechanical details of the SMDE. [From J. A. Wise, D. A. Roston, and W. R. Heineman, in *Methods and Techniques of Multielement Determination* (O. Rose, ed.), Marcel Dekker, New York, to be published.]

The mercury flow rate is very rapid due to the relatively large bore of the capillary (0.006 in.). Drop size is controlled by the time of solenoid activation. Solenoid deactivation stops the flow of mercury, resulting in a static drop. At the drop time selected, the mercury drop is mechanically dislodged from the capillary; a new drop is then grown by activation of the solenoid. A SMDE with similar features but based on a different principle was developed by Tesla Laboratorni pristroje in Czechoslovakia.

## V. STREAMING MERCURY ELECTRODES

There are two types of streaming mercury electrodes (SMEs). The first type is used in the determination of the potential of zero charge, while the second one was introduced for voltammetry by Heyrovsky.

The rapid flow of mercury from a capillary will assume the potential of zero charge, which is then simply measured. This method was proposed at the end of the last century by Helmholtz and the first experiments were carried out by Ostwald [26]. Later a number of investigators adapted this approach [27-31].

In voltammetry it is occasionally necessary to have a mercury electrode with a constant but fresh mercury surface. Such an electrode was developed by Heyrovsky [32-34] for "oscillopolarography." It is usually made of a thick-walled capillary (6-7 mm o.d.) with an internal diameter of 1 to 2 mm, but at its end this diameter decreases to 0.1 mm. Mercury flows from the capillary as a stream which is usually directed upward, out of the solution. If the cylindrical shape of the mercury stream is maintained along its length, the surface of the electrode will be $A = 2\pi r_c \ell$ where $r_c$ is the radius of the capillary at its end and $\ell$ is the length of the continuous stream of mercury. The rate of mercury flow is approximately a hundredfold greater than for a classical DME (about 0.2 g/s or 0.7 kg/hr).

The streaming mercury electrode has been used in kinetic measurements [35,36]. Following introduction of the HMDE, streaming electrodes have become little more than historical curiosities. They are limited by the fact that the radius of the mercury stream and the flow rate are not exactly constant as a function of distance [37].

## VI. MERCURY FILM AND OTHER TYPES OF MERCURY ELECTRODES

In recent years mercury film electrodes (MFEs) are frequently used in electroanalytical practice. Using such electrodes, metal ions present in the solution in trace amounts may be determined with satisfactory accuracy by their reduction on the surface of MFEs, formation of relatively concentrated amalgam, and anodic oxidation from MFRs of the preconcentrated metal in a final step (see Chap. 19).

MFEs are also useful in hydrodynamic techniques such as rotating disk voltammetry (Chap. 3) and electrochemical detection for liquid chromatography (Chap. 22). MFEs are manufactured by the deposition of mercury either on carbon (polished glassy carbon is preferred [38] due to its inertness and low porosity) or on a noble metal support. In the former case, mercury film is produced in situ by electrodeposition on carbon of mercury ions introduced to the studied solution [38,39]. In fact, a mercury film is produced in situ support forms a layer composed of small droplets with a size dependent on the amount of mercury deposited and the deposition potential [40,41].

Since the distribution of mercury on a carbon support can not be exactly controlled, such MFEs should preferentially be used in analytical determinations, where experiments with a standard sample can be carried out under conditions identical with those used in the determination. The advantage of this type of the MFE is that mercury of the film is not contaminated by atoms of material on which it is deposited. This is not the case when a MFE is formed on platinum, gold, silver, or other metal support. However each of these metals dissolves to some extent in mercury and in consequence, properties of the MFE may be slightly changed.

For instance, the concentration of metals introduced in the electrochemical process to such contaminated mercury film may be changed by formation of intermetallic compounds between these metals with metal of a support dissolved in the film or with the surface atoms of the metal support-mercury film interface. Such effects were observed when zinc or tin was introduced to the MFE formed on gold or platinum support. Also when using such a MFE, the hydrogen evolution overvoltage may be decreased. The mercury film thickness can be easily regulated by the electrolytic deposition with a coulometric control. When film is relatively thick its thickness along the MFE surface may be not uniform. MFEs are not stable in time when the mercury film thickness is very low, due to the diffusion of mercury into the metallic support.

Some aspects of the MFEs are discussed in Chap. 19. Also, various types of mercury pool electrodes have been used for many purposes [42-49]. These electrodes are prepared by placing a suitable amount of mercury in containers of desired geometry equipped with metal contacts. Suitable holders may be produced from Teflon, or if made from glass, their surface should be made hydrophobic with a water-repellent material such as silicone. Such electrodes are usually used as working electrodes for slow experiments where the spherical effect on the HMDE is too great. Large mercury pools formed in the bottom of a cell are used for separation of some metals, purification of electrolytes, or for exhaustive electrolysis for coulometric or synthetic purposes.

An effort has been made to apply mercury electrodes at positive potentials. Such electrodes are constructed from mercury covered by a thin layer of ill-soluble mercury compound. Such an electrode was used for the first time by Kuwana and Adams [50,51]. In their experiments mercury was covered by a thin layer of calomel produced by the oxidation of mercury in chloride solutions. N,N'-dimethyl-p-phenylenediamine and ferrocyanide were oxidized at these passivated electrodes.

Using the HMDE covered by calomel, Kemula and co-workers studied the oxidation of several other substances [52]. Kublik studied the oxidation of various ions at the HMDE covered by a film of HgO [53]. These electrodes may be used at even more positive potentials than platinum electrodes in the same solutions. Although the mechanism of oxidation at such electrodes is very interesting, in practical work platinum or carbon electrodes are more useful since their properties and stability are not so dependent on the nature of the sample. When using such electrodes at positive potentials, some current due to oxidation of mercury is inevitable. This limits the application of passivated mercury electrodes to rather concentrated sample solutions.

## REFERENCES


1. B. Kucera, *Ann. Phys. 11*:529 (1903).
2. J. Heyrovsky, *Chem. Listy 16*:256 (1922); *Philos. Mag. J. Sci. 45*:303 (1923).
3. I. M. Kolthoff and J. J. Lingane, *Polarography*, Vol. 1, Interscience, New York, 1952.
4. L. Meites, *J. Am. Chem. Soc. 73*:2035 (1955).
5. I. M. Kolthoff and G. J. Kahan, *J. Am. Chem. Soc. 64*:2553 (1942).
6. O. H. Miiller, *J. Am. Chem. Soc. 66*:1019 (1944).
7. See, for example, Metrohm E 454: PAR 9337.
8. I. M. Kolthoff and Y. Okinaka, in *Progress in Polarography*, Vol. 2 (P. Zuman and I. M. Kolthoff, eds.), Wiley-Interscience, New York, 1962, p. 357.
9. D. C. Grahame, *J. Am. Chem. Soc. 68*:301 (1946).
10. I. Smoler, *Chem. Listy 47*:1667 (1953); I. Smoler, *Coll. Czech. Chem. Commun. 19*:238 (1954); I. Smoler, *J. Electroanal. Chem. 6*:465 (1963); E. Scarano, M. Forina, M. G. Bonicelli, and M. Mioscu, *ibid. 21*:279 (1969).
11. L. Meites, *Polarographic Techniques*, 2nd ed., Wiley-Interscience, New York, 1965, p. 75; L. Meites, *J. Am. Chem. Soc. 73*:3724 (1951).
12. G. C. Barker, *Anal. Chim. Acta. 18*:118 (1958).
13. G. Tessari, P. Delahay, and K. Holub, *J. Electroanal. Chem. 17*:69 (1968).
14. E. M. Skobets and N. S. Kavetskii, *Zavod. Lab. 15*:1299 (1949).
15. See, for instance, PAR 174/70; Radelkis OH-991.
16. H. Gerischer, *Z. Phys. Chem. 202*:302 (1953).

17. M. Kemula, Z. Galus, and Z. Kublik, *Bull. Acad. Pol. Sci. Cl. III* *7*:613 (1959); W. Kemula, Z. Galus, and Z. Kublik, *Rocz. Chem.* *33*:1431 (1959); W. Kemula, Z. Galus, and Z. Kublik, *Bull. Acad. Pol. Sci. Cl. III* *7*:723 (1959).
18. H. J. Antweiler, *Z. Electrochem.* *44*:831 (1938).
19. W. Kemula and Z. Kublik, in *Advances in Analytical Chemistry and Instrumentation*, Vol. 2 (C. N. Reilley, ed.), Wiley-Interscience, New York, 1963, p. 123.
20. J. E. B. Randles and W. White, *Z. Elektrochem.* *59*:666 (1955).
21. W. Kemula and Z. Kublik, *Rocz. Chem.* *30*:1005 (1956); *Anal. Chim. Acta* *18*:104 (1958).
22. W. Kemula, *Microchem. J.* *11*:54 (1966).
23. J. J. Vogel, *J. Electroanal. Chem.* *8*:82 (1964); H. Specker and H. Trub, *Z. Anal. Chem.* *186*:123 (1962); F. von Sturm and M. Ressel, *ibid.* *186*:63 (1962); Z. Kowalski, *Rocz. Chem.* *35*:365 (1961); Z. Kowalski, *J. Electroanal. Chem.* *21*:P9 (1969); see also R. Neeb, *Inverse Polarographie und Voltammetrie*, Akademie-Verlag, Berlin, 1969.
24. R. Neeb, *Z. Anal. Chem.* *171*:321 (1959).
25. W. M. Peterson, *Am. Lab.* *11*(12):69 (1979).
26. W. Ostwald, *Z. Phys. Chem.* *1*:583 (1887).
27. F. Paschen, *Ann. Phys.* *39*:43 (1890); *40*:36 (1890); *43*:568 (1891).
28. D. C. Grahame, E. M. Coffin, J. I. Cummings, and M. A. Poth, *J. Am. Chem. Soc.* *74*:1207 (1952).
29. D. C. Grahame, R. P. Larsen, and M. A. Poth, *J. Am. Chem. Soc.* *71*:2978 (1949).
30. B. B. Damaskin, N. B. Nikolaeva-Fedorovich, and A. N. Frumkin, *Dokl. Akad. Nauk SSSR* *121*:129 (1958).
31. R. Payne, *J. Phys. Chem.* *69*:4113 (1965); *70*:204 (1966).
32. J. Heyrovský and J. Forejt, *Z. Phys. Chem.* *193*:77 (1943).
33. J. Heyrovský, *Chem. Listy* *40*:222 (1946).
34. J. Heyrovský, F. Sorm, and J. Forejt, *Coll. Czech. Chem. Commun.* *12*:11 (1947).
35. J. Koryta, *Coll. Czech. Chem. Commun.*, *19*:433 (1954).
36. J. Koryta, *Coll. Czech. Chem. Commun.*, *20*:1125 (1955).
37. J. R. Weaver and R. W. Parry, *J. Am. Chem. Soc.* *76*:6258 (1954); *78*:5542 (1956).
38. T. M. Florence, *J. Electroanal. Chem.* *27*:273 (1970).
39. G. E. Batley and T. M. Florence, *J. Electroanal. Chem.* *55*:23 (1974).
40. D. N. Hume and J. M. Carter, *Chem. Anal. (Warsaw)* *17*:747 (1972).
41. M. Štulikova, *J. Electroanal. Chem.* *48*:33 (1973).
42. R. Kalvoda, *Coll. Czech. Chem. Commun.* *22*:1390 (1957).
43. R. Kalvoda, *Anal. Chim. Acta* *18*:132 (1958).
44. I. M. Kolthoff, J. Jordan, and S. Prager, *J. Am. Chem. Soc.* *76*:5221 (1954).

45. I. M. Kolthoff and J. Jordan, *J. Am. Chem. Soc.* *77*:3215 (1955).
46. A. G. Stromberg and V. E. Gorodovykh, *Zavod. Lab.* *26*:46 (1960).
47. C. A. Streuli and W. D. Cooke, *Anal. Chem.* *25*:1691 (1953).
48. C. A. Streuli and W. D. Cooke, *J. Phys. Chem.* *57*:824 (1953).
49. D. J. Rossie and W. D. Cooke, *Anal. Chem.* *27*:1360 (1955).
50. T. Kuwana and R. N. Adams, *J. Am. Chem. Soc.* *79*:3610 (1957).
51. T. Kuwana and R. N. Adams, *Anal. Chim. Acta* *20*:51, 60 (1959).
52. W. Kemula, Z. Kublik, and J. Taraszewska, *J. Electroanal. Chem.* *6*:119 (1963).
53. Z. Kublik, Habilitation thesis, University of Warsaw, 1968.

# 10

# Carbon Electrodes

GLENN DRYHURST and DAVID L. McALLISTER* / Department of Chemistry, University of Oklahoma, Norman, Oklahoma



## I. INTRODUCTION

Since about 1950 electrochemistry at solid electrodes has become increasingly popular, as evidenced by the number of publications in this area. This increased popularity can be attributed in part to the development of electrochemical techniques which employ solid or stationary electrodes, and in part to theoretical studies which have advanced the interpretation of reactions occurring at solid and other electrodes. Most important, of course, is the fact that many interesting reactions can be studied which are not accessible at a mercury surface. A wide range of materials have been employed as solid electrodes, but the noble metals, particularly platinum and gold, and carbon have emerged as the most popular.

The ideal properties of a material for use as a solid electrode are that it should be easy to incorporate into an electrode, it should have

*Present address: Great Lakes Chemical Corporation, West Lafayette, Indiana

a large useful potential range, a low electrical resistance, and an easily reproduced surface. Although these criteria are met to some degree by many materials, the properties of various types of carbon make it the material of choice for many applications.

In this chapter a description of the more useful types of carbon electrodes will be presented, giving brief construction details, advantages and disadvantages of each electrode, precautions that must be observed, sources of materials, and other information that might be useful to potential users of these electrodes.

## II. SPECTROSCOPIC GRAPHITE ELECTRODES

The first reported use of carbon as a working electrode in voltammetry was that of Rogers and Lord [1], who employed spectroscopic graphite. Spectroscopic graphite is most commonly used, in the form of rods about 6.4 mm in diameter, as electrodes in arc or spark emission spectroscopy. Many other diameters (e.g., 3.2 and 4.8 mm) and geometries are commercially available. Because of its purity and low electrical resistance it was an obvious and readily available material for early studies as a solid electrode. The outside of the spectroscopic graphite rods were coated with molten ceresin wax and the lower tip of the electrode (ca. 6 mm) was broken off to expose a fresh graphite surface before each voltammogram was run [2]. The ceresin wax was coated over the electrode so that only the broken tip would be conducting and hence behave as an electrode. It was also used to prevent excessive soaking of the test solution into the porous graphite structure. The rather large, nonreproducible residual currents obtained at spectroscopic graphite electrodes prepared in this way are caused primarily by the porosity of the electrode material [3]. The porosity and the very crude method of surface renewal (i.e., breaking off the tip) both contribute to the nonreproducibility of voltammetric currents.

In an effort to overcome these defects of the spectroscopic graphite electrode the wax-impregnated spectroscopic graphite electrode (WIGE) was introduced [3]. Wax impregnation, which fills the pores of the graphite, is accomplished by soaking the graphite rod in a suitable liquefied impregnating agent such as ceresin wax, castor wax, or opal wax for about 1 h. The wax is then allowed to solidify in the pores of the graphite. A short piece of rubber or plastic tubing forced over one end of the rod and containing a short column of mercury conveniently serves as a contact between the electrode and the instrument lead. Such impregnation reduces residual currents to much smaller values and considerably improves the reproducibility of the electrode. At this stage of development of the WIGE (ca. 1957) the electrode was resurfaced before each voltammetric run by sawing or grinding off the end of the rod.

Morris and Schempf [4] found that the performance of the WIGE could be further improved by impregnating the porous carbon rod under vacuum to eliminate entrained air pockets. They also designed a small manual lathe for resurfacing the electrode.

An interesting improvement in performance was introduced by Elving and Smith [5], who immersed the freshly resurfaced electrode (this was done by machining a new surface on the electrode in a small lathe) into a 0.003% solution of Triton X-100 for 1 min. This treatment with detergent ensured better wetting of the surface by aqueous solutions. Even with all of these precautions the WIGE often still does not compete favorably with other more recently developed types of graphite electrodes, especially in terms of ease of preparation and reproducibility. Accordingly, the use of this electrode per se in voltammetric studies has steadily declined in recent years. WIGEs remain very popular, however, because they are a common substrate for the mercury film electrodes (Chap. 9) used in anodic stripping voltammetry (Chap. 19). For this reason their properties continue to be under active investigation.

Probably the single most important piece of information about an electrode is the range of potentials over which the electrode can be used in a given medium. The positive and negative potential limits of an electrode are normally determined by the potential where the solvent, or a component of the supporting electrolyte, is oxidized or reduced or, on occasion, the potential, where the electrode material is itself oxidized. There is no generally agreed upon method for determining the potential limits for an electrode. Many workers use a technique where a voltammogram is obtained at the electrode of interest in a suitable supporting electrolyte system. The negative or positive potential limits are then taken to be at the point where the cathodic or anodic current, respectively, reaches, typically, 1 μA. Since the background or residual current, however, depends on electrode area, voltage sweep rate, and the rate of increase of current at potentials more negative or more positive than that where 1 μA flows, it is clear that this method of definition of potential limits is at best only a very rough qualitative guide. The importance of background current to a given experimental objective is very dependent on the concentration of material being electrolyzed. Some typical data on the negative and positive potential limits at various spectroscopic graphite electrodes are shown in Table 10.1. It can be concluded from these data that the WIGE in particular has very favorable positive and negative potential limits (i.e., it has a wide "potential window").

Exhaustive electrolysis of a few hundred milligrams of electroactive species can be carried out by using several wax-impregnated graphite rods as the working electrode. Struck and Elving [7] used several 7.6-cm lengths of rod, scraped on all sides with a knife to expose the maximum surface area.

Table 10.1 Potential Limits for Spectroscopic Graphite Electrodes

| Electrode (area, $cm^2$) | Solvent electrolyte | Potential limits volt vs. SCE: Positive | Potential limits volt vs. SCE: Negative | Current at potential limits (μA) | Sweep rate | Reference |
|---|---|---|---|---|---|---|
| Spectroscopic graphite (0.079) | 0.1 M HCl[a] | 1.19 | −0.46 | 1.0 | 1.83 mV/s | 6 |
| | 0.1 N $H_2SO_4$ | 1.46 | −0.33 | 1.0 | 1.83 mV/s | 6 |
| | 0.1 M NaOH | 0.85 | −0.39 | 1.0 | 1.83 mV/s | 6 |
| WIGE[b] (0.31) | 0.1 M KCl | 1.0 | −1.3 | NA[c] | 0.62 mV/s | 3 |
| WIGE (0.31) vacuum evacuated | Chloride pH 1.2 | 1.32 | −1.28 | Potential range obtained by extrapolation of residual current and background discharge | 2.47 mV/s | 4 |
| | Phthalate pH 3.95 | 1.33 | −1.30 | Potential range obtained by extrapolation of residual current and background discharge | 2.47 mV/s | 4 |

| | | | | | | |
|---|---|---|---|---|---|---|
| | Phosphate pH 7.02 | 1.35 | −1.38 | Potential range obtained by extrapolation of residual current and background discharge | 2.47 mV/s | 4 |
| | Borate pH 9.80 | 1.04 | −1.61 | Potential range obtained by extrapolation of residual current and background discharge | 2.47 mV/s | 4 |
| WIGE (0.31) (Triton-X treated) | "Neutral" | 1.2 | −1.2 | 1.0 | 3.33 mV/s | 5 |
| | "Acidic" | 1.2 | −1.2 | 1.0 | 3.33 mV/s | 5 |

[a]Aqueous solution.
[b]Wax-impregnated spectroscopic graphite electrode.
[c]Information not available.

## III. CARBON PASTE ELECTRODES

The carbon paste electrode (CPE) was introduced by Adams [8] and generally consists of a Teflon well into which is inserted a platinum, copper, steel, or graphite contact. The Teflon well may be mounted in the end of a suitable glass tube or consist simply of a piece of Teflon tubing as illustrated in Figures 10.1 and 10.2. The well is filled with a paste made by mixing powdered graphite with a suitable mulling liquid. Practical considerations require that the mulling liquid should have a low volatility, purity with respect to electroactive impurities, and very low solubility in the medium to be employed. These considerations generally narrow the choice of mulling liquids to bromoform, bromonaphthalene, or mineral oil (Nujol).

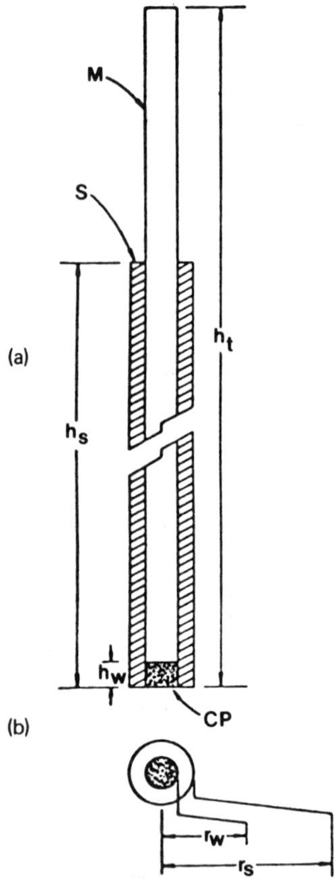


Figure 10.1 Semimicro carbon paste electrode. M, metal contact; S, Teflon sheath; CP, carbon paste; $h_s$, height of Teflon sheath; $h_t$, height of metal contact; $h_w$, well depth; $r_w$, well radius; $r_s$, sheath radius.

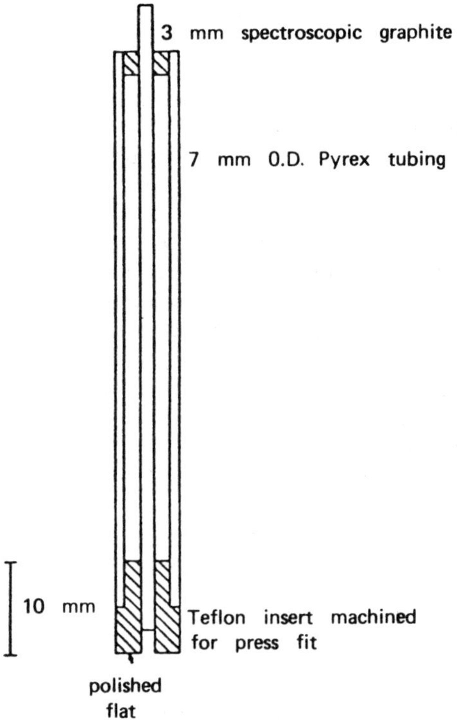


Figure 10.2 Conventional carbon paste electrode.

A typical paste for the CPE may be prepared by thoroughly mixing Nujol and graphite powder (Ultra Microcrystal Grade Graphite Powder UCP-1 M from Ultra Carbon Corporation) until the mixture is uniformly wetted. The resultant paste should have a consistency similar to that of peanut butter. The optimum formulation seems to consist of 3 g of carbon to 2 mL of oil. The choice of carbon for the paste is not critical, although gritty types such as Nuchar (Westvaco, New York, New York) should not be used since they give a rough surface to the electrode and hence somewhat poorer peak current reproducibility. Pastes that appear satisfactory for conventional experiments at millimolar concentrations may give rise to many disconcerting background peaks when pulse voltammetry at the micromolar level is attempted. High-purity graphite is then important and the graphite (and paste) should not be unnecessarily exposed to the laboratory atmosphere. It is good practice to divide batches of freshly prepared paste into lots of about 1 g and to store these in small wide-mouth vials.

Packing carbon paste electrodes to obtain a uniform reproducible surface remains an art. A small amount of paste is placed on a clean card (e.g., an unused IBM computer card); the electrode is gently pressed into this pile and then rubbed on the card to remove excess paste and to attain a flat surface. Care must be taken to fill the well completely, leaving no holes or channels. Any paste remaining on the mantle is carefully wiped away with a paper tissue. For the smallest electrodes, it is best to inspect the packed surface under a low-power microscope to ensure that gross imperfections are not extant. Electrodes may be assembled and packed in quantity and stored (e.g., inverted in Styrofoam) in a covered box. Electrodes may be renewed by removing a small amount of paste from the surface with a paper tissue and repacking with fresh paste. In the unusual event that an electrode must be completely restored, the well can be cleaned out by immersing the electrode in an ultrasonic bath. After drying, the electrode may be repacked as outlined above. With experience, large (1-5 mm in diameter) electrodes can be resurfaced in about 30 s with a peak current reproducibility (stationary electrode) of about ±2%. However, reproducibility is a function of the medium and the compound under study.

It is important when packing the well to avoid applying too much pressure. This may result in separation of the carbon and oil, with a resulting high-resistance contact between paste and metal. With the smallest electrodes there is some danger of a poor seal between the sheath and the paste material, resulting in a solution leak to the contact metal and hence a useless electrode.

Other pastes and methods of electrode preparation have been described [8-11]. Chambers and Chambers [12] prepared a CPE using a rather dry paste (graphite-to-Nujol ratio of 1.92) and smoothed the surface of the electrode on a piece of parchment paper covered with a thin layer of graphite paste. When the latter surface became slick, a fresh piece of parchment was used. It was reported [12] that this method gave the "most reversible current-potential curves which were reproducible from day to day."

McCreery and co-workers [13] have described a carbon paste electrode for use in vivo, particularly in brain tissue, which may be used to monitor quantitatively the fate of intracerebrally injected electroactive drugs. Conti and co-workers [14] have developed carbon paste microelectrodes having tip diameters of 75-200 μm which have been implanted in rat brains to monitor neurotransmitter substances.

The formulations and method of preparation of the CPE described above are satisfactory for use in aqueous solutions containing perhaps as much as 25% alcohol or other solvent. However, when carbon paste electrodes of this type are used in nonaqueous media (e.g., acetonitrile, nitromethane, propylene carbonate), the pool of carbon paste tends to disintegrate. The cause of this disintegration appears to be dissolution of the mulling liquid or the preferential wetting of the graphite by the solvent. This difficulty can be overcome by addition of a surface-active agent, such as sodium lauryl sulfate, to the paste [15]. The latter compound is thought to cause the graphite to be wetted equally well by both the mulling liquid and the nonaqueous solvent.

It is normal, in practice, to prepare a new electrode surface before running each voltammogram by removing the surface layer of paste and refilling the Teflon holder. This can be done quickly and reproducibly manually, although Lindquist [16] has described an electrode assembly which incorporates a piston arrangement that forces an appropriate amount of paste out of the holder when a threaded rod is turned (Figure 10.3). The surface is refinished by cutting off the excess paste with a stretched piano wire while rotating the electrode at 600 rpm in a lathe.

Lindquist has described the preparation of a CPE with an increased positive potential range (+1.7 V vs. SCE in 0.1 M $H_2SO_4$) [17]. The electrode was prepared from graphite powder which was heated under vacuum to remove adsorbed oxygen, then mixed with molten wax. Ten grams of graphite powder (Ringsdorff-Werke, RW-A) was placed in a 45 × 1.7 cm quartz tube. The tube was closed at one end, bent to a right angle, and connected to a 50-mL Erlenmeyer flask with a side arm for connection to a pump. In the Erlenmeyer flask were 5.6 g of ceresin wax and 1.4 g of paraffin oil (Marck's

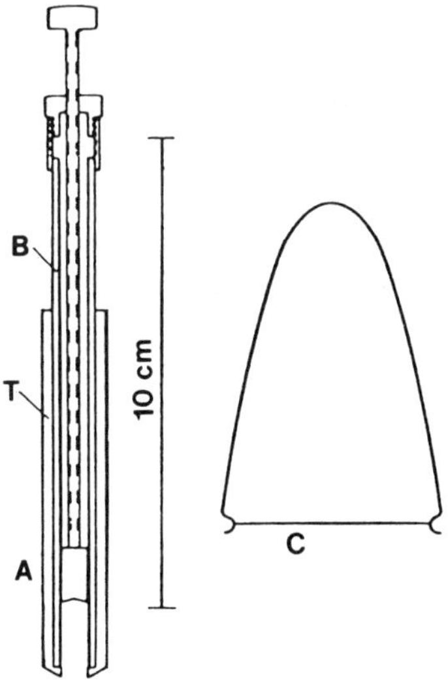


Figure 10.3 A, Carbon paste electrode holder: T, Teflon; B, nickel-plated brass; C, cutting device: a length of 0.2-mm piano wire stretched with a bow-shaped piece of 1.2-mm piano wire. (From Ref. 16.)

paraffin liquid for spectroscopy). The part of the quartz tube containing the graphite powder was placed in a tube furnace, the pressure reduced to $10^{-2}$ mmHg, and the temperature increased to about 1000°C. At the same time, the Erlenmeyer flask was heated for a while to melt the ceresin wax and drive off the dissolved oxygen. After 1 h the tube was removed from the furnace and the Erlenmeyer flask was placed on a steam bath. When the wax had melted, the graphite powder was poured into it, and after 30 min the mixture was cooled and air was admitted. The stiffened paste was pulverized and tamped into the electrode holder [16], which was then heated to 80°C to produce a homogeneous paste. The surface was prepared as before [16].

Atuma and Lindquist [18] have also developed a carbon paste electrode, impregnated with ceresin wax, which may be used in all of the common solvents used in electrochemistry. The carbon paste is prepared in the following way. Ceresin wax (0.5 g) is dissolved in 20 mL of warm (40-50°C) n-hexane in a beaker on a water bath and 9.5 g of powdered graphite is then added with stirring. The stirring is continued until all of the n-hexane has evaporated. The dry graphite powder, now containing 5% w/w of ceresin wax, is then carefully mixed with MS 510 silicone oil in the proportions 5:3 w/w to give a homogeneous paste. The paste is then tamped into an electrode holder with care being taken to ensure that it is not too firmly packed. The excess of paste is then cut off with a stretched length of 0.2-mm piano wire while the electrode is rotated.

Several authors have reported effects at the CPE characteristic of adsorption of electroactive species at the electrode surface [19-22]. It appears in most instances that adsorption of the electroactive species at the graphite particle surface is not of major importance, but rather the observed effects are due primarily to dissolution of the electroactive species in the organic phase of the CPE. One should be aware of this effect, which can be pronounced, especially when working with compounds that are highly soluble in the mulling liquid.

The range of potentials over which the CPE can be used is quite large (Table 10.2). In general, it appears that Nujol pastes have a very slightly larger positive potential range than that of bromonaphthalene pastes.

All carbon paste electrodes described in the literature have been designed for voltammetric studies. However, larger electrodes suitable for exhaustive electrolyses have been used. Alternatively, large sheets of the graphite of the same type as that used in the CPE electrode could be used. However, in this case it should be noted that the surface properties of the finely powdered graphite in the CPE and the sheets of graphite are probably quite different. Such differences could be reflected in differences in adsorption properties of the electroactive species at the two surfaces, alterations in electrode mechanisms, and possibly even formation of different reaction products.

Table 10.2 Potential Limits for Carbon Paste Electrodes

| Electrode (area, $cm^2$) | Mulling liquid | Solvent electrolyte | Potential limits volt vs. SCE | | Current at potential limits ($\mu A$) | Sweep rate | Reference |
|---|---|---|---|---|---|---|---|
| | | | Positive | Negative | | | |
| CPE[a] (0.31) | Bromonaphthalene | 1 M KCl[b] | 1.10 | NA[c] | 2.0 | NA[c] | 10 |
| | | 0.1 M KCl –HCl, pH 2.5 | 1.08 | NA | 2.0 | NA | 10 |
| | | 1 M HCl | 1.02 | NA | 2.0 | NA | 10 |
| | | 0.1 M NaOAc-HOAc | 1.27 | NA | 2.0 | NA | 10 |
| | | 1 M $Na_2SO_4$ | 1.28 | NA | 2.0 | NA | 10 |
| | | 0.2 N $H_2SO_4$ | 1.30 | NA | 2.0 | NA | 10 |
| | | BRB[d] + 0.2 M $Na_2SO_4$, pH 2.4 | 1.30 | NA | 2.0 | NA | 10 |
| | | BRB[d] + 0.2 M $Na_2SO_4$, pH 4.8 | 1.30 | NA | 2.0 | NA | 10 |
| | | 0.2 M NaOH | 0.87 | NA | 2.0 | NA | 10 |

Table 10.2 (continued)

| Electrode (area, $cm^2$) | Mulling liquid | Solvent electrolyte | Potential limits volt vs. SCE | | Current at potential limits (μA) | Sweep rate | Reference |
|---|---|---|---|---|---|---|---|
| | | | Positive | Negative | | | |
| CPE (0.17) | Nujol with Na lauryl sulfate[d] | Acetonitrile-0.1 M TEAP[f] | 1.10 | −0.7 | 3.0 | 33 mV/s | 15 |
| CPE (NA) | Nujol | 1 M $HClO_4$ | NA | −0.9 | NA | NA | 22 |
| | | 1 M HCl | NA | −0.9 | NA | NA | 22 |
| | | 1 M HOAc-NaOAc, pH 4.7 | NA | −1.0 | NA | NA | 22 |
| | | 1 M $KNO_3$ | NA | −1.1 | NA | NA | 22 |
| | | 1 M KCl | NA | −1.1 | NA | NA | 22 |
| | | 1 M $NaClO_4$ | NA | −1.1 | NA | NA | 22 |
| | | 1 M $NH_3$-$NH_4Cl$ pH 9.3 | NA | −1.2 | NA | NA | 22 |
| | | 1 M NaOH | NA | −1.4 | NA | NA | 22 |

| | | | | | | | |
|---|---|---|---|---|---|---|---|
| CPE (0.31) | Ceresin wax-paraffin oil | 0.1 M $H_2SO_4$ | 1.7 | −1.2 | 1.0 | 8.3 mV/s | 17 |
| CPE (0.31) | Ceresin wax-MS 510 silicone oil | 85% v/v ethanol[g] | 1.5 | NA | 1 | 8.3 mV/s | 18 |
| | | Acetonitrile[g] | 1.4 | NA | 1 | 8.3 mV/s | 18 |
| | | Acetic acid[g] | 1.3 | NA | 1 | 8.3 mV/s | 18 |
| | | DMSO[g,h] | 1.15 | NA | 1 | 8.3 mV/s | 18 |

[a]Carbon paste electrode.
[b]Aqueous solutions unless otherwise specified.
[c]Not available.
[d]Britton-Robinson buffer.
[e]Electrode composition: 3.3 g of powdered graphite (Acheson grade 38), 2.5 mL of Nujol, and 1.4 g of sodium lauryl sulfate (Matheson, Coleman and Bell, USP, powder).
[f]Tetraethylammonium perchlorate.
[g]The supporting electrolytes were sodium perchlorate, ammonium acetate, sodium acetate, and lithium chloride, but the correspondence between solvent and supporting electrolyte is not known.
[h]Dimethyl sulfoxide.

## IV. PYROLYTIC GRAPHITE ELECTRODES

Pyrolytic graphite has been known since 1880, but only in the last 20 years or so has the material become available in dimensions suitable for incorporation into electrodes. The recent availability of pyrolytic graphite is due primarily to the applications of this material in the space program.

The crystalline structure of pyrolytic graphite closely approximates that of the ideal graphite crystal. A good discussion of the preparation, properties, and structure of pyrolytic graphite has been prepared by Walker [23]. Because of its highly ordered structure, pyrolytic graphite is anisotropic and so must be properly oriented in the electrode. In describing the structure of pyrolytic graphite, the ab plane is that parallel to the layers of graphite, the ac and bc planes are those perpendicular to the graphite layers (Fig. 10.4).

The first reported applications of pyrolytic graphite as a material for an indicator electrode in electrochemistry were those of Miller and Zittel [24] and Beilby and Brooks [25]. Since that time the pyrolytic graphite electrode (PGE) has been used extensively and several different forms have been developed.

For voltammetric studies a PGE can be readily constructed by first cutting small rods (4 mm in diameter, 10 mm long) from a block of the material [26,27]. A tool suitable for machining these rods con-

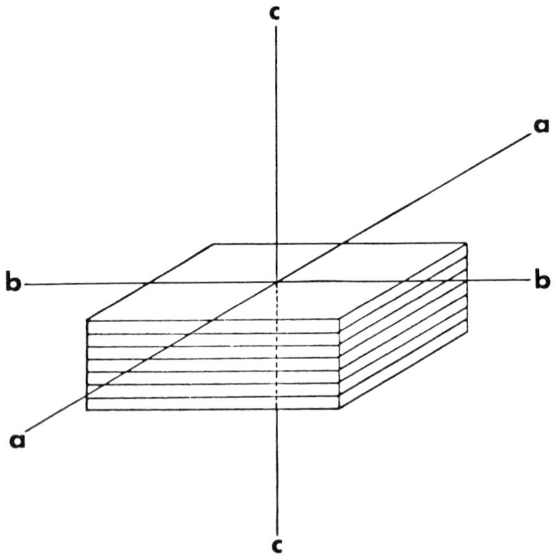


Figure 10.4 Schematic representation of the layered structure of pyrolytic graphite.

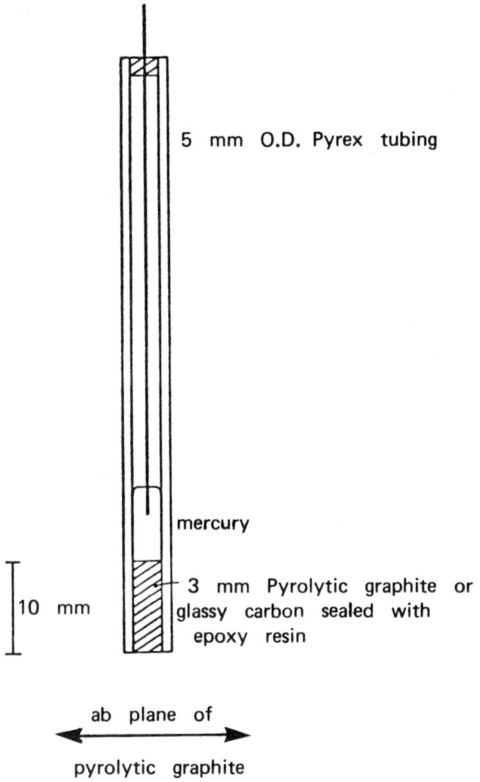


Figure 10.5 Typical pyrolytic graphite (or glassy carbon) electrode for voltammetric studies.

sists of a length of stainless steel tubing of appropriate inside diameter with serations on the cutting end. The tool is rotated in a lathe while the graphite block remains stationary. The graphite block must be positioned so that the tool enters exactly perpendicular to the ab plane. The small graphite rods so obtained are then cemented into a length (ca. 6 in.) of 4-mm bore glass tubing with epoxy cement (Hysol Epoxi-Patch) (Fig. 10.5). The pyrolytic graphite rods are thus mounted with only the parallel ab plane exposed to the solution. The latter epoxy cement is suitable for use in pyrolytic graphite electrodes which are to be used in aqueous and many nonaqueous solutions (e.g., alcohols, pyridine, acetonitrile). It is important to prevent the occurrence of air pockets between the glass tube and the pyrolytic graphite rod. This can usually be done by first coating the inside of the glass tubing with epoxy cement, coating the graphite rod with epoxy, and then sliding the rod carefully into the tube. Invariably, the backside of the graphite rod becomes coated with epoxy during this operation, which, after curing, prevents electrical contact to the graphite rod. The excess epoxy is easily removed after it has hardened by inserting a red-hot heavy wire through the top of the glass tube and melting away the epoxy. The end of the PGE is then ground on a 600-grit silicon carbide paper (Buehler Ltd.) mounted on a metallographic polishing wheel (e.g., Style R1138, Buehler Ltd.) so that the glass tube and pyrolytic graphite are flush. It is advisable to provide some type of support above the polishing wheel to hold the electrode perpendicular to the wheel during the polishing process. Electrical contact to the electrode is established by placing a small amount of mercury into the glass tube and inserting a suitable wire lead into the mercury. We have found it convenient to prepare batches of 10-20 electrodes at a time since approximately 50-75% of those constructed are not suitable for voltammetric work (i.e., the residual currents are too large or the useful potential range is too small). The exact causes of the latter effects are obscure. The decision as to whether a PGE is suitable for voltammetric work is based primarily on its useful potential range and residual currents observed in standard buffer systems.

The PGE is normally resurfaced by lightly polishing the dry electrode on a 600-grit silicon carbide paper before each voltammogram is run. The finely divided graphite powder that adheres to the surface of the PGE after polishing is removed with a powerful jet of distilled water. The electrode surface is then dried by touching *very gently* against a clean, absorbent paper tissue. Electrodes prepared and resurfaced in this fashion will last for several hundred runs. In the absence of pronounced adsorption of the electroactive species, peak currents at a stationary PGE are reproducible to ±5-10%. Because of the differences in surface roughness after each polish it is noticeable that the peak current reproducibility is considerably worse when adsorption of the electroactive species occurs, particularly at fast potential sweep rates. Peak potentials are normally reproducible to within ±5 mV.

Various other forms of the PGE have appeared. Laitinen and Rhodes [28], for example, have described an electrode composed of a film of vapor-deposited pyrolytic graphite on a graphite rod. This electrode was used in molten salt studies. The fabrication and use of a similar pyrolytic graphite film electrode for studies in aqueous solution has been described by Beilby et al. [29]. The properties of the electrode were compared to the WIGE on the ferrocyanide-ferricyanide system using chronopotentiometry and chronocoulometry. The system appeared to behave more reversibly at the film PGE than at the WIGE. The film PGE, however, does not appear to offer any significant advantages over PGEs.

When solvents are employed which attack the epoxy of the PGE described earlier, a KEL-F* holder as described by Miller et al. [30] may be employed. Owing to the arrangement of the electrode in the holder, it would appear that resurfacing of this PGE would be somewhat inconvenient.

Eisner and Mark [31] have used a ceresin wax-impregnated (Fisher) PGE for the anodic stripping determination of silver. Accurate reproducible peak currents as low as 0.03 $\mu A/cm^2$ could be measured.

Factors affecting the electrochemical performance of graphite have been studied by Panzer and Elving [32]. Pyrolytic graphite from several different sources was examined, as were different resurfacing methods. It was found that the most reproducible surface was produced by cleaving the electrode, rather than by grinding. It was also recommended that electrodes be resurfaced only when absolutely necessary, that is, when carryover of material from one run to another cannot be eliminated by simply washing the electrode.

For fairly large scale electrolyses (i.e., a few hundred milligrams of electroactive species) at pyrolytic graphite electrodes it is convenient [26] to employ plates of pyrolytic graphite approximately 1 × 5 cm and 2-3 mm thick. One to six (or more) such plates are inserted through slots in a suitable stopper which fits into the mouth of the electrolysis vessel and are connected in parallel with alligator clips. When necessary, the electrodes are cleaned with 600-grit silicon carbide paper. As with other types of electrode systems, products and n values of electrochemical reactions are sometimes different at a micro PGE under voltammetric conditions than at a massive PGE under controlled potential electrolysis conditions. It has also been our experience that gross changes in n values and hence products occasionally occur when massive electrodes of pyrolytic graphite from different sources are used. It is prudent, therefore, to select a particular type of pyrolytic (or other) graphite and use it exclusively.

Recently, a type of pyrolytic graphite known as stress-annealed pyrolytic graphite has become available. This material has a nearly perfect orientation along the c axis (i.e., perpendicular to the planes of graphite) and its structure is much more defined than that of ordinary pyrolytic graphite [33]. The surface of stress-annealed pyrolytic graphite (i.e., the ac planes) has a metallic, mirrorlike finish. The design of a suitable electrode is shown in Figure 10.6. The rods of graphite must be machined under compression. The resultant rods or disks are backed with nickel to increase their mechanical strength, and are then press-fitted into the Teflon holder. A fresh electrode surface is exposed for each experiment by placing a

---

*KEL-F is a fluorocarbon resin manufactured by the 3M Company.

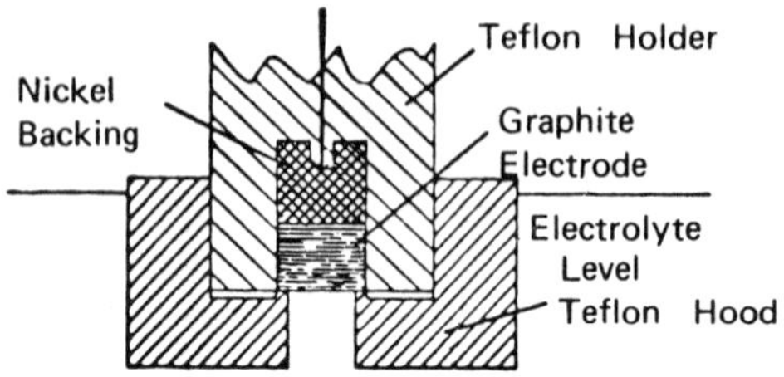


Figure 10.6 Cross section of stress-annealed pyrolytic graphite electrode holder and hood. (From Ref. 33, reprinted by permission of the publisher, The Electrochemical Society, Inc.)

piece of plastic adhesive tape, cut to the exact size, in contact with the graphite surface and then peeling off a layer of graphite with the tape. Care must be taken not to contaminate the Teflon surrounding the graphite disk with the adhesive on the plastic tape. The graphite surface so exposed has the appearance of a black mirror. This electrode has generally not been used for normal voltammetric studies but rather for differential capacitance studies, which indicated a high degree of surface reproducibility [33]. Nevertheless, electrodes of this material seem to potentially offer the highest degree of surface reproducibility of any type of graphite electrode. The potential limits reported for various pyrolytic graphite electrodes are shown in Table 10.3.

It is becoming increasingly apparent that pyrolytic graphite, and indeed other carbons, have a variety of oxides and other functional groups, such as —OH, —COOH, and quinones located on their surfaces [35,36]. The surface function types and population are probably governed by both the source and previous history (thermal, chemical) of the carbon specimens.

Cyclic voltammetry at pyrolytic graphite electrodes immersed into buffer solutions containing no electroactive species always gives rise to a small, but distinct quasi-reversible couple (e.g., at about 0.35 V vs. Ag/AgCl at pH 2.5) [37-39]. The potential of this couple is dependent on pH (ca. 60 mV per pH unit [39]), which is consistent with a 2e-$2H^+$ process. It has been concluded that this couple is due to a 1,2-naphthoquinone-like surface structure [39,40]. The presence of such surface redox couples on pyrolytic graphite and other carbon electrodes clearly raises the possibility of electrode reactions actually being mediated or catalyzed by such moieties. Indeed, some electrode reactions have been definitely shown to be catalyzed by such surface redox couples [39,41,42]. Other surface functionalities have been used to covalently attach other redox groups to pyrolytic graphite electrodes to produce "chemically modified" PGEs having unique catalytic and/or stereochemical properties [42-45].

Table 10.3 Potential Limits for Pyrolytic Graphite Electrodes

| Electrode (area, $cm^2$) | Solvent electrolyte | Potential limits volt vs. SCE | | Current at potential limits ($\mu A$) | Sweep rate | Reference |
|---|---|---|---|---|---|---|
| | | Positive | Negative | | | |
| 0.14 | 0.1 M HCl[a] | 1.0 | −0.8 | NA[b] | 1.66 mV/s | 34 |
| | 0.1 M $HNO_3$ | 1.0 | −0.8 | NA | 1.66 mV/s | 34 |

[a]Aqueous solutions.
[b]Not available.

## V. GLASSY CARBON ELECTRODES

The properties of glassy (vitreous) carbon have been described to some extent by Yamanda and Sato [46], although no details concerning the method of manufacture or the structure of the material were provided. The material is, however, isotropic, as opposed to pyrolytic graphite, which is anisotropic. It is also impermeable to gas, electrically conductive, and highly resistant to chemical attack [34]. The structure of glassy or vitreous carbon has been described as thin, tangled ribbons of graphitelike sheets which may be cross-linked to some extent [47].

Glassy carbon was first used as an electrode material by Zittel and Miller [34] and many workers continue to use the electrode; indeed, electrodes are commercially available. Several preparations of the electrode have been described, but one of the simplest is that of McBride and Evans [48]. For this electrode a length of 3-mm-diameter glassy carbon rod (Grade GC-20, International Carbon Company) is sealed into a length of 5-mm glass tubing with epoxy cement. A few millimeters of carbon are then exposed from the end of the glass tube using a glass cutting wheel. The end of the glass tube-carbon rod assembly is then polished with emery paper, followed

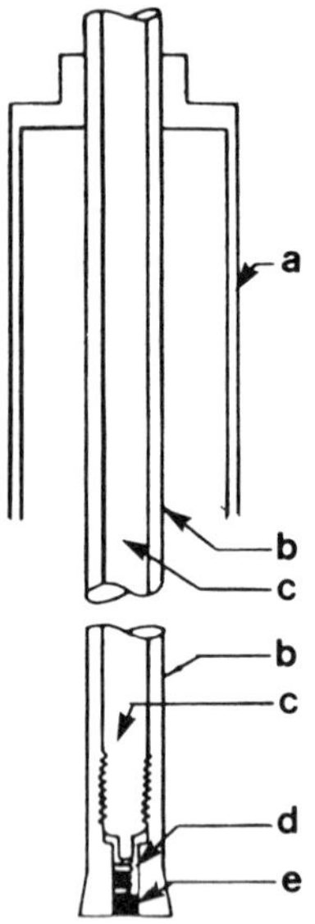


Figure 10.7 Schematic diagram of GCE. a, Teflon bell for liquid seal; b, Teflon; c, stainless steel shaft; d, contact spring; e, glassy carbon rod. (From Ref. 50, reprinted by permission of the publisher, The Electrochemical Society, Inc.)

Table 10.4 Potential Limits of Glassy Carbon Electrodes

| Electrode (area, $cm^2$) | Solvent electrolyte | Potential limits volt vs. SCE | | Current at potential limits (μA) | Sweep rate | Reference |
|---|---|---|---|---|---|---|
| | | Positive | Negative | | | |
| 0.11 | 0.1 M $HNO_3$ | 1.25 | −0.68 | 1.0 | 1.66 mV/s | 23 |
| | 0.1 M HCl | 1.05 | −0.80 | 1.0 | 1.66 mV/s | 23 |
| | 0.1 N $H_2SO_4$ | 1.32 | −0.80 | 1.0 | 1.66 mV/s | 23 |
| | 0.1 M $HClO_4$ | 1.22 | −0.88 | 1.0 | 1.66 mV/s | 23 |
| | 0.1 N $H_3PO_4$ | 1.36 | −0.80 | 1.0 | 1.66 mV/s | 23 |

[a]Aqueous solution.

by polishing alumina (Fisher) until a mirror finish is obtained. Before each voltammogram is run the electrode is buffed for about 5 s with 1-μm alumina, rinsed, and placed in the electrochemical cell.

Panzer and Elving [32] indicate that the glassy carbon electrode (GCE) is best resurfaced by polishing with 0.5-μm alumina, with only simple rinsing with a suitable solvent between voltammograms. This assumes, of course, that no contaminating products or reactants from the previous voltammogram remain on the electrode surface.

Kopanica and Vydra [49] have satisfactorily used a GCE for stripping analysis. The error in the determination of $5 \times 10^{-7}$ M silver in the presence of $5 \times 10^{-3}$ M copper was less than ±5%.

A GCE suitable for use in molten salts has been described by Holleck [50]. The electrode consists of a 3-mm glassy carbon rod press fitted into hot Teflon. Details of the electrode construction are shown in Figure 10.7. This electrode has been used in melts at temperatures up to 150°C.

Gupta [51] has also reported the preparation of a GCE for use in molten salts. A Pyrex glass tube is shrunk around a glassy carbon rod, making a gastight seal. The design incorporates a thermocouple to indicate the temperature of the electrode. Alder and coworkers [52] examined over 30 different electrode materials for anodic voltammetric studies and concluded that glassy carbon is generally the most useful and convenient material for anodic studies. Nevertheless, it is likely that many research groups will dispute this finding.

A glassy carbon crucible (30 mL in capacity) has been used both as an electrolysis vessel and as a working electrode by Plock and Vasquez [53]. Coulometric reductions of iron, neptunium, uranium, chromium, and copper were less precise with this GCE than with mercury or platinum electrodes. The loss of precision was attributed to longer electrolysis times and higher background currents with the GCE. Although massive glassy carbon electrodes have not apparently been used in large-scale oxidations, no problems should be encountered provided, of course, that material of suitable dimensions can be obtained.

Some typical data on the positive and negative potential limits at the GCE are shown in Table 10.4. The uses and applications of glassy carbon electrodes in electroanalytical chemistry have recently been reviewed by Van der Linden and Dieker [54].

Methods to modify glassy carbon electrodes chemically have begun to appear in the literature, whereby redox reagents are bonded to surface functionalities such as —OH or —COOH groups [41,45,55,56].

## VI. RETICULATED VITREOUS CARBON ELECTRODES

Reticulated vitreous carbon (RVC) is a glasslike carbon with a porous structure having a free void volume of about 97% and a surface area of about 65 $cm^2/cm^3$ [57]. Blaedel and Wang [58] have utilized a rotated RVC disk electrode as an indicator electrode in a flow-through electrochemical cell. These workers have evaluated the dependences of the limiting current and conversion efficiency on the rotation speed, flow rate, and disk thickness. Very high analytical currents were observed for low analyte concentrations.

An interesting application of RVC is an optically transparent electrode (OTE) for use in thin-layer spectroelectrochemistry. Thin-layer spectroelectrochemistry using OTEs is particularly useful for simultaneous spectroscopic and electrochemical studies [59,60]. Norvell and Mamantov [61] first proposed the use of RVC as an electrode material in thin-layer spectroelectrochemistry. The RVC-OTE has several characteristics which overcome some of the disadvantages of wire minigrid or metal oxide OTEs, while combining the advantages of both of these types of electrodes. For example, RVC can be used in strongly acidic or basic solutions or in the presence of strong oxidants or reductants. The electrode also has a usable potential range, extending, for example, from about 1.2 V to −1.0 V vs. SCE at pH 7.

A typical optically transparent thin-layer cell is constructed by cutting slices of RVC (ERG, Inc., Oakland, California) of approximate dimensions 2 cm × 4.5 cm with thicknesses ranging from ca. 0.5 to 1 mm. Norvell and Mamantov [61] recommended using RVC with a porosity of 100 pores per inch (ppi). However, we have found the 60-ppi material to be somewhat more useful since thinner, more transparent slices may be prepared with the result that faster electrolyses can be carried out and less diffusion from outside the electrode occurs. Electrical contact to the RVC is made by threading a fine copper wire through the corner of the electrode and folding it over to secure it. Liquid Organic Silver No. 9167 (Englehard Industries, East Newark, New Jersey) is then applied to the surrounding area and the organic portion burned away, leaving a silver coating. The latter procedure is normally repeated several times until a sturdy contact is achieved. The OTE is assembled by sealing the RVC between two clean glass or quartz (Esco Optics, Oak Ridge, New Jersey) microscope slides in a configuration similar to that suggested by DeAngelis and Heineman [62]. The RVC is positioned a few millimeters from the bottom of the slides and extended about 1 cm beyond the slides on each side. To prevent breakage of the RVC at the edge of the slides, the entire exposed portion of the electrode, including the silvered contact region, is coated with epoxy. A small-diameter Teflon tube is connected to the top of the cell to allow solution to be drawn into the cell with either a syringe or by connecting it to house vacuum. A major problem encountered with such an RVC thin-layer cell is that many air bubbles are often trapped within the electrode cavity. In this laboratory we typically draw 1-2 L of distilled water through a new cell followed by buffer or test solution. Bubbles may be effectively removed by alternately drawing solution, then air, through the cell. Repeating this procedure three to five times removes all trapped gas bubbles. On occasion, the RVC may become fouled by some electrode reaction product. This may usually be removed by drawing 0.5-1 L of water through the cell. Alternatively, 6 M $HNO_3$ followed by several hundred milliliters of water usually restores the electrode to its original activity.

Thin-layer cells containing an RVC electrode can be readily made with volumes within the electrode cavity of 200 μL or less. The volume of the electrode cavity may be determined by coulometry with a standard solution of $K_3Fe(CN)_6$ using the procedure given by DeAngelis and Heineman [62]. The electroanalytical applications of RVC electrodes have recently been reviewed and investigated by Strohl [63].

## VII. MISCELLANEOUS CARBON ELECTRODES

Although the CPE, PGE, and GCE are the most widely used carbon electrodes, other types have been reported. Pungor and co-workers [64-66] have employed an electrode consisting of graphite particles supported in a silicone rubber matrix. This electrode is very similar in many respects to the carbon paste electrode. However, it would appear that preparation and resurfacing of the silicone rubber electrode could be a rather serious problem. Perhaps in solutions that attack or dissolve the mulling liquid of the CPE, silicone-rubber-based electrodes could offer some advantages. Swofford and Carmen [67] have described a similar electrode but using 25% carbon powder in epoxy cement (Resiweld No. 1004, Fuller Adhesives, St. Paul, Minnesota). Potential limits of 0.8 to −0.6 V (vs. SCE) in 0.1 M $H_2SO_4$ were reported. The electrode surface was renewed by polishing on 600-grit emery paper. Some representative data on potential limits at silicone rubber and carbon-epoxy electrodes are shown in Table 10.5.

A number of electrodes used for purposes other than regular voltammetric studies have also been described. Flow-through electrodes, for example, have been used by several workers. Bamberger and Strohl [68] have used a Vycor tube packed with graphite powder as an electrode for the oxidation of hydroquinone and p-phenylenediamine, and later as a type of chromatography column for the separation of quinones [69]. The latter column separated quinones based on differing degrees of adsorption of various compounds at different potentials imposed on the graphite electrode. Tubular carbon electrodes, constructed of spectroscopic graphite rods through which a 1- or 2-mm hole has been made, have been described by Mason and Olson [70] and Sharma and Dutt [71]. The latter workers have derived working equations relating the current to flow rate of solution through the electrode, electrode area, and concentration of electroactive species. Such tubular electrodes might well also be successfully

Table 10.5 Potential Limits at Silicone Rubber and Epoxy-Carbon Electrodes

| Electrode (area, $cm^2$) | Solvent electrolyte[a] | Potential limits volt vs. SCE | | Current at potential limits ($\mu A$) | Sweep rate | Reference |
|---|---|---|---|---|---|---|
| | | Positive | Negative | | | |
| Silicone rubber (0.25) | 0.1 M KCl | 1.35 | ∽ −0.3 | 1.0 | 25 mV/s | 65 |
| | 1.0 M KCl | 1.32 | ∽ −0.3 | 1.0 | 25 mV/s | 65 |
| | 0.1 M $Na_2SO_4$ | 1.25 | ∽ −0.3 | 1.0 | 25 mV/s | 65 |
| | 1.0 M $NH_4NO_3$ | 1.30 | ∽ −0.3 | 1.0 | 25 mV/s | 65 |
| | 0.2 M HOAc | 1.40 | ∽ −0.3 | 1.0 | 25 mV/s | 65 |
| | 0.1 M HCl | 1.30 | ∽ −0.3 | 1.0 | 25 mV/s | 65 |
| | 0.1 M $H_2SO_4$ | 1.40 | ∽ −0.3 | 1.0 | 25 mV/s | 65 |
| | 0.1 M $NaNO_3$ | 0.65 | ∽ −0.3 | 1.0 | 25 mV/s | 65 |
| | 0.1 M NaOH | 0.70 | ∽ −0.3 | 1.0 | 25 mV/s | 65 |
| | BRB,[b] pH 8.1 | 1.20 | ∽ −0.3 | 1.0 | 25 mV/s | 65 |
| | BRB, pH 4.5 | 1.20 | ∽ −0.3 | 1.0 | 25 mV/s | 65 |
| | Phosphate, pH 7.0 | 1.25 | ∽ −0.3 | 1.0 | 25 mV/s | 65 |
| Epoxy (0.772) | 0.1 M $H_2SO_4$ | 0.8 | −0.6 | 2.0 | NA[c] | 67 |
| | 0.1 M KCl | 0.8 | −0.6 | 2.0 | NA | 67 |

[a]All in aqueous solution.
[b]Britton-Robinson buffer.
[c]Not available.

Table 10.6 Suppliers of Materials Used in the Construction of Graphite Electrodes

| Material | Trade name, type | Supplier and address |
|---|---|---|
| Spectroscopic graphite rods | 0.25-in. spectroscopic grade A graphite electrodes | National Spectrographic Laboratories<br>19500 S. Miles Rd.<br>Cleveland, OH 44128 |
| Spectroscopic graphite electrode | 9324 | Princeton Applied Research Corp.<br>P.O. Box 2565<br>Princeton, NJ 08540 |
| Paraffin-impregnated graphite electrode | 9319 | Princeton Applied Research Corp. |
| Graphite powder | Grade 38 | Acheson Colloids Co.<br>1635 Washington Ave.<br>Port Huron, MI 48060 |
| Graphite powder | Ringsdorff-Werke RW-A | Ringsdorff Corp.<br>P.O. Box 22<br>East McKeesport, PA 15035 |
| Graphite powder | Ultra Carbon UCP-1-M | Ultra Carbon Corp.<br>P.O. Box 747<br>Bay City, MI 48706 |
| Carbon paste electrode | 9326 | Princeton Applied Research Corp |
| | CP-O, CP-S, CP-W | Bioanalytical Systems, Inc.<br>2701 Kent Avenue<br>West Lafayette, IN 47906 |

| | | |
|---|---|---|
| Pyrolytic graphite | | Super-Temp Company<br>11120 S. Norwalk Blvd.<br>Santa Fe Springs, CA 90670 |
| | | Union Carbide Corp.<br>Carbon Products Divison<br>270 Park Ave.<br>New York, NY 10017 |
| | | Pfizer Minerals, Pigments and Metals Div.<br>Pfizer Inc.<br>640 N. 13th St.<br>P.O. Box 548<br>Easton, PA 18042 |
| Stress-annealed pyrolytic graphite | Highly ordered pyrolytic graphite | Union Carbide Corp.<br>Carbon Products Div.<br>270 Park Ave.<br>New York, NY 10017 |
| Glassy carbon rod | Grade GC-20 | International Carbon Co.<br>New York, NY |
| Glassy carbon crucible | | Tokai Electrode Mfg. Co.<br>Tokyo, Japan |
| Glassy carbon electrode | A-111 | Chemtrix, Inc.<br>P.O. Box 725<br>Beaverton, OR 97005 |
| | 9333 | Princeton Applied Research |
| | GCE | Bioanalytical Systems Inc. |

Table 10.6 (continued)

| Material | Trade name, type | Supplier and address |
|---|---|---|
| Epoxy cement | Hysol Epoxy-Patch | Hysol Corporation<br>Olean, NY 14760 |
| Conducting epoxy | Eccobond Solder 58C | Emmerson-Cummings, Inc.<br>Canton, MA 02021 |
| Carbon cloth | | Carbon Products Division<br>Union Carbide Corp.<br>270 Park Ave.<br>New York, NY 10017 |
| Silicon carbide paper | 600 grit | Buehler Ltd.<br>2120 Greenwood St.<br>Evanston, IL 60204 |
| Silicon carbide paper | Carborundum, Flexbac, Aloxite | Carborundum Co.<br>Abrasives Marketing Div.<br>P.O. Box 337<br>Niagara Falls, NY 14302 |
| Polishing wheel | Style R1138 | Buehler Ltd.<br>2120 Greenwood St.<br>Evanston, IL 60204 |
| Reticulated vitreous carbon | | ERG, Inc.<br>Lowell and 57th St.<br>Oakland, CA 94608 |
| Electrode polishing kits | | Bioanalytical Systems Inc. |

applied to the analysis of chromatographic column effluents or other monitoring processes, or to continuous electrochemical syntheses.

Voorhies and Davis [72] have used acetylene black powder (Shawinigan Products Corp.) upon which a compound of interest was adsorbed as an electrode for constant-current coulometry. Both organic and inorganic samples could be electrolyzed.

For large-scale electrolyses Nelson and co-workers [73] have used carbon cloth electrodes. This inexpensive material gave low residual currents and good n values in coulometric experiments.

## VIII. SOURCES OF MATERIALS

Frequently, a problem encountered in the construction of graphite electrodes is that of locating a suitable source of the required materials. In Table 10.6 a summary of suppliers of appropriate materials is presented.

## REFERENCES


1. L. B. Rogers and S. S. Lord, Pittsburgh Conference on Analytical Chemistry, March 1952 (source: Gaylor et al., *Anal. Chem. 25*:1078 (1953).
2. V. F. Gaylor, P. J. Elving, and A. L. Conrad, *Anal. Chem. 25*:1078 (1953).
3. V. F. Gaylor, A. L. Conrad, and J. H. Landerl, *Anal. Chem. 29*:224 (1957).
4. J. B. Morris and J. M. Schempf, *Anal. Chem. 31*:286 (1959).
5. P. J. Elving and D. L. Smith, *Anal. Chem. 32*:1849 (1960).
6. S. S. Lord and L. B. Rogers, *Anal. Chem. 26*:284 (1954).
7. W. A. Struck and P. J. Elving, *Biochemistry 4*:1343 (1965).
8. R. N. Adams, *Anal. Chem. 30*:1576 (1958).
9. R. N. Adams, *Electrochemistry at Solid Electrodes*, Marcel Dekker, New York, 1969, pp. 280-283.
10. C. Olsen and R. N. Adams, *Anal. Chim. Acta 22*:582 (1960).
11. S. Piekarski and R. N. Adams, in *Techniques of Chemistry*, Vol. 1: *Physical Methods of Chemistry* Part IIA (A. Weissberger and W. B. Rossiter, eds.), Wiley, New York, 1971, p. 531.
12. C. A. Chambers and J. Q. Chambers, *J. Am. Chem. Soc. 88*:2922 (1966).
13. R. L. McCreery, R. Dreiling, and R. N. Adams, *Brain Res. 73*:23 (1974).
14. J. C. Conti, E. Strope, R. N. Adams, and C. A. Marsden, *Life Sci. 23*:2705 (1978).

15. L. S. Marcoux, K. B. Prater, B. G. Prater, and R. N. Adams, *Anal. Chem.* *37*:1446 (1965).
16. J. Lindquist, *J. Electroanal. Chem.* *18*:204 (1968).
17. J. Lindquist, *Anal. Chem.* *45*:1006 (1973).
18. S. S. Atuma and J. Lindquist, *Analyst* *98*:886 (1973).
19. C. A. Chambers and J. K. Lee, *J. Electroanal. Chem.* *14*:309 (1967).
20. E. P. Meier and J. Q. Chambers, *Anal. Chem.* *41*:914 (1969).
21. T. Kuwana and W. G. French, *Anal. Chem.* *36*:241 (1964).
22. D. G. Davis and M. E. Everhart, *Anal. Chem.* *36*:38 (1964).
23. P. L. Walker, Jr., *Am. Sci.* *50*:259 (1962).
24. F. J. Miller and H. E. Zittel, *Anal. Chem.* *35*:1866 (1963).
25. A. L. Beilby and W. Brooks, 144th Meeting, American Chemical Society, Los Angeles, Calif., April 1963 (source: A. L. Beilby, W. Brooks, and G. L. Lawrence, *Anal. Chem.* *36*:22 (1964).
26. G. Dryhurst and P. J. Elving, *J. Electrochem. Soc.* *115*:1014 (1968).
27. G. Dryhurst, *Talanta* *19*:769 (1972).
28. H. A. Laitinen and D. R. Rhodes, *J. Electrochem. Soc.* *109*:413 (1962).
29. A. L. Beilby, W. Brooks, and G. L. Lawrence, *Anal. Chem.* *36*:22 (1964).
30. F. Miller, H. Zittel, G. Mamantov, and D. Freeman, *J. Electroanal. Chem.* *9*:305 (1965).
31. U. Eisner and H. B. Mark, *J. Electroanal. Chem.* *24*:345 (1970).
32. R. E. Panzer and P. J. Elving, *J. Electrochem. Soc.* *119*:864 (1972).
33. J. Randin and E. Yeager, *J. Electrochem. Soc.* *118*:711 (1971).
34. H. E. Zittel and F. J. Miller, *Anal. Chem.* *37*:200 (1965).
35. J. S. Mattson and H. B. Mark, *Activated Carbon*, Marcel Dekker, New York, 1971.
36. B. R. Puri, "Surface Complexes on Carbon," in *Chemistry and Physics of Carbon*, Vol. 6 (P. L. Walker, ed.), Marcel Dekker, New York, 1970.
37. H. A. Kozlowska, J. Klinger, and B. E. Conway, *J. Electroanal. Chem.* *75*:45 (1977).
38. H. A. Kozlowska, J. Klinger, and B. E. Conway, *J. Electroanal. Chem.* *75*:61 (1977).
39. J. F. Evans and T. Kuwana, *Anal. Chem.* *49*:1632 (1977).
40. J. P. Randin and E. Yeager, *J. Electroanal. Chem.* *58*:313 (1975).
41. D. C. S. Tse and T. Kuwana, *Anal. Chem.* *50*:1315 (1978).
42. J. F. Evans, T. Kuwana, M. T. Henne, and G. P. Royer, *J. Electroanal. Chem.* *80*:409 (1977).
43. A. W. C. Lin, P. Yeh, A. M. Yacynych, and T. Kuwana, *J. Electroanal. Chem.* *84*:411 (1977).

44. M. F. Dautartas, J. F. Evans, and T. Kuwana, *Anal. Chem.* *51*:104 (1979).
45. J. F. Evans and T. Kuwana, *Anal. Chem.* *51*:358 (1979).
46. S. Yamada and H. Sato, *Nature* *193*:261 (1962).
47. G. M. Jenkins and K. Kawamura, *Nature* *231*:175 (1971).
48. H. D. McBride and D. H. Evans, *Anal. Chem.* *45*:446 (1973).
49. M. Kopanica and F. Vydra, *J. Electroanal. Chem.* *31*:175 (1971).
50. G. L. Holleck, *J. Electrochem. Soc.* *119*:1158 (1972).
51. N. K. Gupta, *Rev. Sci. Instrum.* *42*:1368 (1971).
52. J. F. Adler, B. Fleet, and P. O. Kane, *J. Electroanal. Chem.* *30*:427 (1971).
53. C. Plock and S. Vasquez, *Talanta* *16*:1490 (1969).
54. W. E. Van der Linden and J. W. Dieker, *Anal. Chim. Acta* *119*:1 (1980).
55. C. M. Elliot and R. W. Murray, *Anal. Chem.* *48*:1247 (1976).
56. A. Bettelheim, R. J. H. Chan, and T. Kuwana, *J. Electroanal. Chem.* *99*:391 (1979).
57. *Bull. No. 176,* Chemotronics International, Inc., Ann Arbor, Mich.
58. W. J. Blaedel and J. Wang, *Anal. Chem.* *52*:1697 (1980).
59. T. Kuwana, *Ber. Bunsenges. Phys. Chem.* *77*:858 (1973).
60. T. Kuwana and W. R. Heineman, *Acc. Chem. Res.* *9*:241 (1976).
61. V. E. Norvell and G. Mamantov, *Anal. Chem.* *49*:1470 (1977).
62. T. P. DeAngelis and W. R. Heineman, *J. Chem. Educ.* *53*:594 (1976).
63. A. N. Strohl, *Diss. Abstr. Int. B* *40*:1149 (1979).
64. E. Pungor, E. Szepesvary, and J. Havas, *Anal. Lett.* *1*:213 (1968).
65. E. Punger and E. Szepesvary, *Anal. Chim. Acta* *43*:289 (1968).
66. E. Siska and E. Pungor, *Z. Anal. Chem.* *257*:7 (1971).
67. H. S. Swofford and R. L. Carmen, *Anal. Chem.* *38*:966 (1966).
68. R. L. Bamberger and J. H. Strohl, *Anal. Chem.* *41*:1450 (1969).
69. J. H. Strohl and K. L. Dunlap, *Anal. Chem.* *44*:2166 (1972).
70. W. D. Mason and C. L. Olson, *Anal. Chem.* *42*:548 (1970).
71. L. R. Sharma and J. Dutt, *Indian J. Chem.* *6*:593, 597 (1968); *7*:485 (1969).
72. J. D. Voorhies and S. M. Davis, *Anal. Chem.* *32*:1855 (1960).
73. R. Hand, A. K. Carpenter, C. J. O'Brien, and R. F. Nelson, *J. Electrochem. Soc.* *119*: 74 (1972).

# 11

# Film Electrodes

NICHOLAS WINOGRAD / Department of Chemistry, The Pennsylvania State University, University Park, Pennsylvania



## I. INTRODUCTION

The incorporation of metal or semiconducting film electrodes into electrochemical experiments often offers a unique advantageous approach to electroanalysis. This type of electrode was originally developed for use in spectroelectrochemical studies where a light beam could be directed through a film which is optically transparent [1]. The convenience of being able to prepare electrodes on virtually any type of substrate with high purity and chemical reproducibility has spurred the development of many other types of electrodes with diverse applications. For example, aside from the optically transparent electrode (OTE) application, the planarity of film electrodes deposited on optically flat glass substrates has proven valuable for optical reflection spectroscopy [2] and for electrodes in thin-layer cells, where it is necessary to obtain parallel electrode surfaces to within tolerances of several micrometers [3]. These well-defined surfaces are also finding applications in many photoelectrochemical experiments, particularly those involving water splitting [4]. The synthesis of chemically modified electrodes has usually begun with a film electrode which has been previously characterized by ESCA or some other surface analysis technique [5]. A final, pressing motive for utilizing film electrodes is their cost. Considering the current market value for Pt of \$50 per gram, a 1000-Å, 1-cm$^2$ electrode consists of 15 cents worth of Pt. In this chapter, some of the details of the preparation and properties of both semiconducting and conducting films will be presented. Hopefully, the material will provide enough of a guide so that, through the help of relevant literature articles, it would be possible to construct and utilize the desired electrode.

## II. PROPERTIES OF FILM ELECTRODES

### A. Electrical Conduction

Film electrodes are generally prepared from conducting or semiconducting materials deposited either chemically or in vacuo onto an appropriate substrate, such as a glass microscope slide. The thickness of the deposited film is generally determined by either the desired electrical resistance or optical transparency needed in the experiment. For most metals, a resistivity $\rho$ approaching that of the bulk metal can be obtained for rather thin layers, say on the order of 1000 or 2000 Å. The conduction mechanisms are then essentially "bulklike" in that the electrical properties of the film are indistinguishable from conventional foils or wires.

When the thickness of the film becomes less than several hundred angstroms, however, the film may have many characteristics that are not found in a bulk metal. For example, during the initial stages of film formation the metal atoms form many small separated "puddles" or clusters resulting in a discontinuous metal surface [6]. The deposition conditions of film formation can influence the cluster distribution and affect film continuity, surface roughness, grain size, lattice defects, crystal orientations, crystal composition, and film purity. Variations in these properties lead to variations in film resistance that would not be predicted solely from its thickness.

Electrical conduction through discontinuous films probably proceeds by transfer of an electron from one puddle to the next. There are several theories for this process, but a simple one [7] considers that a potential energy barrier exists between the puddles, and that electron hopping occurs by tunneling through this barrier. The presence of an applied field may also decrease the magnitude of the activation energy required for puddle jumping, making the development of a rigorous theory difficult. In fact, for discontinuous films, Ohm's

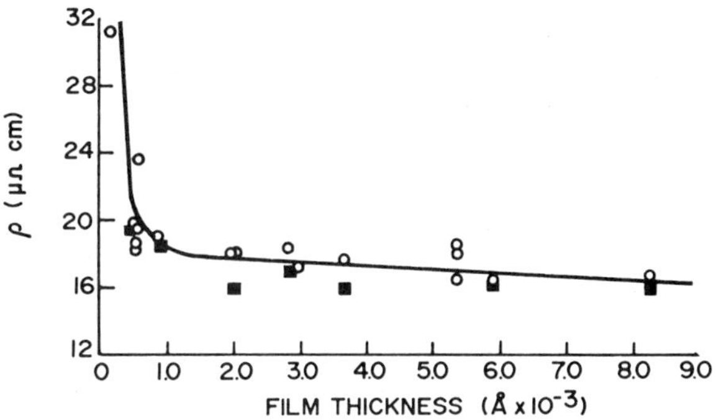


**Figure 11.1** Resistivity ρ in μΩ-cm of sputtered platinum as a function of film thickness, in angstroms. The circle and square designations refer to the different substrates of glass and GaAs, respectively, that were used for the substrate. (From Ref. 8.)

law does not generally apply. The variation of ρ with film thickness is shown in Figure 11.1. Remember that ρ is a property of the metal and, in principle, should not vary with the metal geometry. The point where the value of ρ becomes independent of thickness is the point where the film has become continuous.

In general, semiconducting films are much thicker than metal films due to their higher resistivity. Thicknesses to several micrometers are common, so that the puddle problem is not a concern for these systems. The conduction mechanism can be thought of as being essentially the same as for polycrystalline semiconductors.

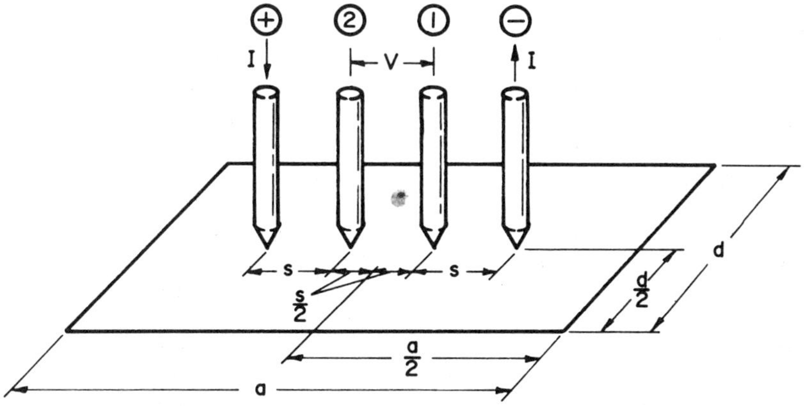


**Figure 11.2** Arrangement of a four-point probe on a rectangular sample. Note that the dimensions a, d, and s must be specified to calculate a specific value for the resistance. (From Ref. 9.)

The measurement of electrode resistance presents a tricky but necessary chore since any resulting iR losses present in the electrochemical experiments can then be anticipated. The process is not a simple one for a sheet since the measured resistance depends on the film geometry. The four-point-probe depicted in Figure 11.2 has been most commonly employed for measuring the resistivity of a film [9]. The device operates by applying a known current between the outer probes and measuring the voltage drop between the inner probes. The measured resistance must be related to the geometry of the sheet material (i.e., a and d) as well as to the spacings between the probes. Several tabulations are available for specific film geometries. The most common configuration is a square sheet and the measured resistance is reported in $\Omega/\square$ (ohms per square), independent of the actual size of the square.

## B. Optical Transparency

The key properties of an OTE to the electrochemist are its optical transmission properties vs. wavelength as well as its resistance, say in $\Omega/\square$, for a given optical transparency. The major factors governing these properties are the concentration of free carriers in a metal and the band gap of a semiconductor material. For metals, intense optical absorption by the free carriers (e.g., electrons) is observed in the visible and IR region of the spectrum. For example, a Pt film of 200 Å has an absorbance of approximately 0.5 at 500 nm. Absorption generally begins in the ultraviolet (UV) at a sharp edge defined by the plasma frequency of the metal. The value of this frequency depends on the concentration or density of free electrons. Sodium, for example, has a free electron concentration of $10^{22}$ electrons/cm$^3$ and is transparent below 210 nm [10]. Interband transitions can also influence the optical properties of metal film electrodes. These transitions arise from accessible optical excitations within the valence band of the metal. In gold, for example, the color is believed to be due to the 5d → 6s interband transition at 470 nm.

For semiconducting materials the concentration of charge carriers is very low, $10^{13}$-$10^{17}$ electrons/cm$^3$, so that the intrinsic optical absorption from the free carriers is negligible. When the incident light energy is greater than the band gap of the semiconductor, however, excitations from the valence band to the conduction band are initiated and intense light absorption occurs. Many common semiconductor electrodes such as $SnO_2$ and ZnO have band gaps of about 3 eV (415 nm), resulting in transparent optical properties in the visible and strong absorption in the UV. The semiconductors Si and Ge have a very low energy band gap (1140 and 1900 nm, respectively), so that these materials absorb very strongly in the UV and visible region. They are suitable only as transparent electrodes in the IR.

The intrinsic semiconductors can be doped with n-type (electron donor) or p-type (electron acceptor) impurities, which effectively increase the concentration of free carriers, electrons or holes, by several orders of magnitude. A common example is the heavy doping of $SnO_2$ with antimony to generate an n-type semiconducting OTE. Hall measurements indicate that the concentration of the carriers is high, of the order of $10^{20}$-$10^{21}$ $cm^{-3}$, for the n-type tin oxide with resistances of 5-10 Ω/□. This compares to a carrier density of ca $10^{19}$ $cm^{-3}$ for tin oxide coatings made by the spraying of tin chloride without Sb doping. The transparency of the doped tin oxide in the visible region is explained by the fact that the optical absorption by the free carriers begins in the infrared region at about 2000 nm for the foregoing carrier density. The intrinsic optical absorption edge occurs in the UV at an energy of 3.7-3.8 eV (300 nm). On glass, of course, the transmittance decreases sharply at 360 nm due to absorption by the glass itself.

## III. SEMICONDUCTING FILM ELECTRODES

A vast number of semiconductor materials have been tested for use as electrodes. Some of these are elucidated in Chapter 23, particularly with respect to their unique properties toward photochemically excited molecules, although most have been used as bulk single crystals and not as polycrystalline films. The two most common semiconducting film electrodes are $SnO_2$ and $InO_x$.

### A. Tin Oxide OTE

The optical and semiconducting properties of tin oxide coatings on glass or quartz plates have been known for a long time. These coatings have been available commercially under the trade names of Nesa and Electropane glass and were reputedly used in the early days of aviation for deicing windshields. More recently, tin oxide has been deposited on plastic substrates such as polyester sheets [11]. Antimony has been the most commonly used doping material and the resulting tin oxide is an n-type semiconductor. Various references are available in journal articles [12] and in the patent literature [13], but most involve the spraying of the substrate with $SnCl_4$ at high (ca. 500°C) temperatures followed by hydrolysis of the surface. Commercial preparations are available from Libbey, Owen Ford Glass Co. (Brackenridge, Pennsylvania). The p-type surfaces doped with fluoride are also available.

Electron microscopic examination indicates a surface without any defined grain boundaries but having microcrystalline particles with mean size of ca. 100 Å. The commercial coatings are usually between

600 and 1000 nm thick, depending on the source and sample. The refractive index is 2.0 in the visible region of the spectrum. Because of the refractive index and the thickness of these coatings, it is not surprising that the absorption spectrum, taken with the light beam passing normal to the surface, exhibits an interference pattern. Typical spectra of n-type tin oxide on transparent substrates are shown in Figure 11.3. Coating thicknesses can be easily calculated from the interference maxima and a knowledge of the refractive index [14].

The doped tin oxide surface can be described best as a poor conductor in its behavior as an electrode. It can be used in aqueous solutions over a potential range quite similar to that of platinum. The linear sweep cyclic voltammetric scans for $SnO_2$ in comparison to other film electrodes are shown in Figure 11.4. It is not recommended for use in highly alkaline pH values since some surface degradation occurs. According to Laitinen and co-workers [16], oxygen or chlorine (in HCl media) evolution occurs without disruption of the surface. The positive potential extreme is quite high, on the order of 2 V, presumably since the Sn exists in its highest oxidation state, so that further oxidation, as is evident for Pt and Au, does not occur. At the negative limit, hydrogen ion discharge is accompanied by the concurrent reduction of the tin oxide to metallic tin, resulting in destruction of the electrode. In an intermediate region of potential, a charg-

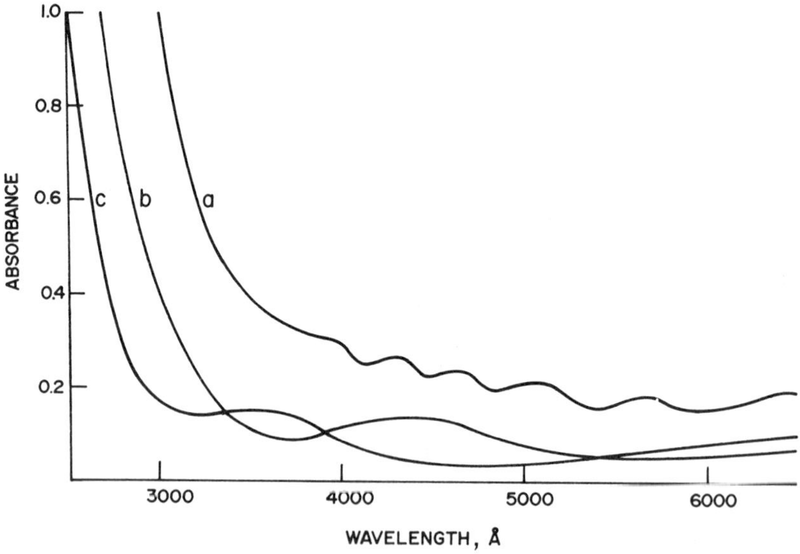


Figure 11.3 Transmission spectra of $SnO_2$ coatings on various substrates. a, glass, 3 Ω/□; b, vicor, 6 Ω/□; c, quartz, 20 Ω/□.

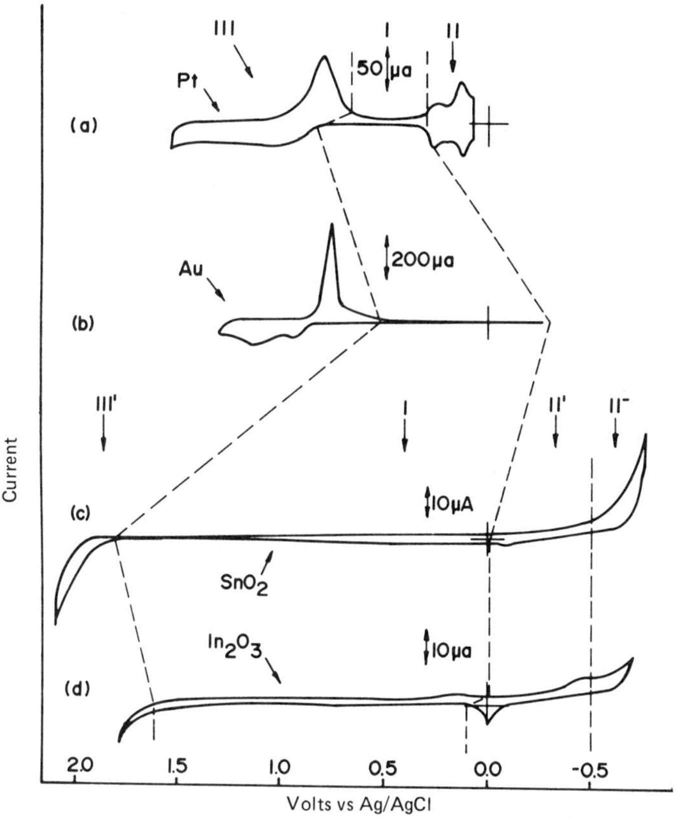


**Figure 11.4** Comparison of current-voltage curves for Pt, Au, $SnO_2$, and $In_2O_3$ electrodes in 1 N $H_2SO_4$ media. Potential scan rate was 87 mV/s. All electrodes were 1 $cm^2$ in geometric area. (a) Platinum thin film, (b) gold thin film, (c) $SnO_2$ thin film, (d) $In_2O_3$ thin film. I corresponds to the double-layer region; II indicates hydrogen reduction; III represents oxygen evolution. [From Ref. 15, N. R. Armstrong, A. Lin, M. Fujihira, and T. Kuwana, *Anal. Chem.* *48*:741 (1976), with permission. Copyright 1976 American Chemical Society.]

ing current is observed by linear sweep cyclic voltammetry. This current is attributed to the charging and discharging of positive surface states and is most pronounced in chloride media. The amount of charge is several times greater than the estimated surface population of tin atoms. The surface chemistry may also be affected by pH through groups which can be influenced by possible acid-base equilibria:

Table 11.1 Background Potential Limits in Various Nonaqueous Solvents

| | Potential limits[a] | | |
|---|---|---|---|
| Solvents | Tin oxide-OTE | Pt-OTE | Au-OTE |
| Acetonitrile | +2.47 to −2.77 | +2.07 to −1.88 | −2.68 |
| DMF | +1.72 to −2.70 | +1.65 to −2.21 | +1.61 to −2.39 |
| Nitromethane | +2.48 to −1.08 | +1.42 to −1.15 | — |
| Dichloroethane | +2.36 to −2.17[b] | — | — |
| Propylene carbonate | +2.52 to −2.66 | — | — |
| Monoglyme | +2.37 to −2.82[b] | — | — |
| Diglyme | +2.40 to −2.93[c] | — | — |

[a]Potentials reported vs. SCE; 0.1 M $Et_4NClO_4$ supporting electrolyte unless otherwise noted; scan rate 0.053 V/s, temperature 25°C, potentials are for current levels of 100 μA/cm$^2$.
[b]0.1 M n-$Bu_4NClO_4$.
[c]0.05 M n-$Bu_4NClO_4$.
*Source*: Ref. 17.

$$\sim SnOH \rightleftharpoons SnO^- + H^+$$

or

$$\sim SnOH \rightleftharpoons Sn^+ + OH^-$$

In nonaqueous solutions, the usable potential range is large, the order of 5 V. This potential range allows a wide variety of oxidations and reductions to be examined. Some typical values on different types of film electrodes are reported in Table 11.1. The electrolysis of several organic compounds in these solvents has been characterized. A general observation is that the tin oxide electrode appears more suited to reduction studies than to oxidation ones in nonaqueous media. For example, the reversible oxidation of ferrocene at Pt electrode is quite irreversible at tin oxide. Most reductions, on the other hand, give similar results for Pt, with the added advantage of a larger accessible negative range.

## B. Others

Although $SnO_2$ has found the widest electroanalytical utility as a semiconducting electrode, other types have been investigated over the years. One promising type is the tin-"doped" In oxide coated glass electrode manufactured by PPG (Barberton, Ohio) under the trade name of Nesatron. The electrochemical characteristics are similar to $SnO_2$ as shown in Figure 11.4, although the available potential range is somewhat less. The optical transmission is only slightly improved over $SnO_2$, so that Nesatron does not appear to possess any marked advantage over $SnO_2$.

Several other varieties of semiconducting electrodes have been produced by chemical vapor deposition. For example, films of $TiO_2$ on Ti of 10,000 Å in thickness can be prepared by heating $Ti(OC_3H_7)_4$ at 80-85°C into a flowing $N_2$ carrier gas [18]. This gas jet is then combined with a similar water stream which hydrolyzes the $Ti(OC_3H_7)_4$. The resulting films are then annealed at 600°C at $10^{-5}$ torr to introduce oxygen vacancies in the n-type material which considerably lowers its resistance. $TiO_2$ has also been deposited on plastic substrates [11]. Other materials, such as $Fe_2O_3$ and $SnO_2$ films, have been prepared in a similar fashion [19]. Most of these other electrodes have been tested as electrodes for use in photoelectrochemically induced water splitting and have not found conventional electroanalytical utility.

Germanium plates have been employed as OTEs in the infrared region, although films have not been examined due to the high resistance. The surface also exhibits a limited usable potential range [20]. The electrochemical properties of germanium are substantially improved by vapor-depositing a thin carbon film on the surface [21].

A carbon OTE can be prepared by vapor-depositing a thin carbon film on glass or quartz [22]. The electrochemical properties are similar to those of bulk graphite. Optical transparency is good in the UV-visible range.

# IV. METAL FILM ELECTRODES

Metal films can be prepared either by chemical deposition of the metal onto the desired substrate or by vapor deposition and sputtering, generally in high vacuum. Chemical methods, although quick and convenient, do not provide pure, reproducible films and therefore have not been seriously used except for Hg films. Vapor deposition, on the other hand, has been extensively explored not only by electrochemists, but by a large body of engineers in the coatings and electronics industries. A complete description of this technique may be found in the classic text by Holland entitled *Vacuum Deposition of Thin Films* [23]. Although films of most metals can be prepared by vapor deposition, Pt and Au films have been most extensively employed as electrodes, so preparation of these materials as thin films will be discussed to illustrate some of the recommended procedures.

### A. Substrate Preparation

The most common substrate for thin-film electrodes is, by far, the glass microscope slide or quartz disk. These substrates are flat, readily cleaned, and are well suited for spectroelectrochemical studies. Rigorous cleaning is required if the metal films are to adhere strongly to the glass. Many common procedures have been attempted, although boiling in a detergent, rinsing in triply distilled water, followed by an isopropanol vapor degreasing is quite successful. The substrate should then immediately be placed in the vacuum system used for deposition and pumped to $10^{-6}$ torr. A glow discharge with about 50 torr of Ar or residual air for about 10-20 min can provide a final cleaning in the vacuum [24]. Typically, a dc voltage of 1-5 kV at 50-200 mA is applied between an aluminum block holding the glass and an aluminum discharge electrode.

Adherence of metal films on glass substrates can often be improved by first depositing a thin layer of a different material on the glass. Evaporation or sputtering of only several layers of Ti, Cr, or Ta has improved the adherence of Pt on glass without degrading the optical transparency or the electrochemical properties of the electrode [25]. Deposition of other oxides, such as $Bi_2O_3$ and $PbO_2$, have improved the sticking of Au films to glass [24]. Other combinations of undercoats have been suggested. These undercoats are critical for metal film electrodes deposited on glass or quartz, especially when operating in aqueous solution. The films invariably strip off during $H_2$ or $O_2$ evolution unless they are well bonded to the substrate.

Polyester sheet covered with a thin film of gold has also proved suitable as an OTE [11].

### B. Film Preparation

Metal films are best prepared within the confines of a high-vacuum system ($10^{-6}$-$10^{-8}$ torr) by vapor deposition or sputtering. Evaporation is initiated either by melting the metal onto a heater filament, usually tungsten or tantalum, followed by further electrical heating of the metal to its boiling point. For higher-melting metals such as Pt, electron beam evaporation has been successfully employed. Here, the metal is placed in a water-cooled copper bucket and bombarded with high-energy electrons. The metal surface is rapidly heated and evaporates quickly, so the electron beam method is most commonly used to prepare very thick films. Sputtering is also employed on high-melting metals. A large foil of the desired material is negatively

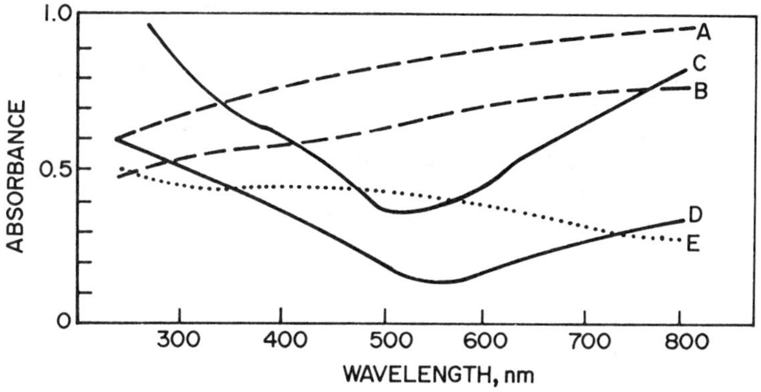


Figure 11.5 Spectra of Pt and Au films. A, Pt on quartz, 11 Ω/□; B, Pt on quartz, 20 Ω/□; C, Au-bismuth oxide on quartz, 2.5 Ω/□; D, Au-bismuth oxide on quartz, 11 Ω/□, (film of bismuth oxide on quartz). [From Ref. 24, W. Von Benken and T. Kuwana, *Anal. Chem.* *42*:1114 (1970), with permission. Copyright 1970 American Chemical Society.]

charged to about 1-3 kV and bombarded with $Ar^+$ ions. The ions can conveniently be generated by backfilling the evaporation chamber with 10-100 mtorr of Ar. The ion impact on the metal then kicks out metal atoms or ions, which can be deposited on the glass substrate. Adherence to the substrate by sputtering is usually excellent and the process occurs at very low temperatures. The presence of ions all over the vacuum system makes high-purity deposition difficult and the rate of deposition is generally quite low (ca. 25 Å/min).

The electrical and optical properties and the adherence of Pt and Au evaporated films can be markedly improved by annealing at temperatures up to 600°C in air [24]. These changes are probably manifestations of the coalescing of individual islands into a more continuous film. The effects are particularly pronounced for gold, where the resistance drops by 30-50% and the optical transparency as shown in Figure 11.5 improves considerably. Similar effects are observed for Pt [24,26].

## C. Optical and Electrochemical Properties

Pt acts like a neutral density filter through a wide wavelength range, and can be used in the UV when the film is on quartz. Au has an optical window with maximum transmission at 5400 Å. On quartz, Au can also be used in the UV when the conductivity is comparable to Pt. For the visible region, Au films with good transmission and with resistance as low as a few $\Omega/\square$ can be prepared. Typical optical properties are shown in Figure 11.5. The electrochemical properties are similar to bulk Pt and Au, Figure 11.4, except that the overpotential for hydrogen ion discharge is slightly less on a Pt film than for bulk Pt. Whether this activity is due to the "cleaner" surface or to the crystal structure of the Pt film is presently unknown. Characteristics of these films in nonaqueous solutions have been evaluated by Osa and Kuwana [17]. The background potential limits are summarized in Table 11.1.

## V. TECHNIQUES OF CELL DESIGN BASED ON FILM ELECTRODES

A few special technical problems arise when attempting to mount film electrodes into conventional electrochemical cells due to difficulties in defining the electrode area and in making electrical connection to the active surface. These constraints usually result in cells that have been modified somewhat, although the same principles that are discussed in Chap. 12 still apply. In this final section we address some of these practical difficulties.

The film electrode should usually be partially masked to the minimum desired electrochemical area to reduce power requirements of the electronic equipment and to minimize any iR drop that may occur across the electrode surface. These problems are especially

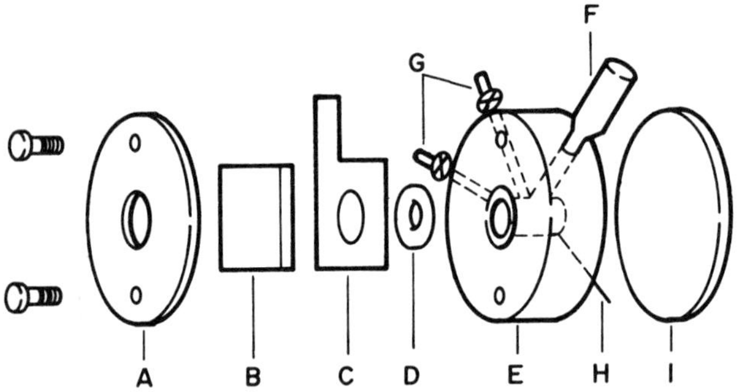


Figure 11.6 Cell for use of thin-film electrodes. A, Aluminum retaining plate; B, OTE; C, copper foil; D, O ring; E, Lucite body; F, glass salt bridge for reference electrode; G, Hamilton valves; H, auxiliary electrode; I, quartz disk. [From Ref. 27, W. R. Heineman and T. Kuwana, *Anal. Chem.* 44:1972 (1972), with permission. Copyright 1972 American Chemical Society.]

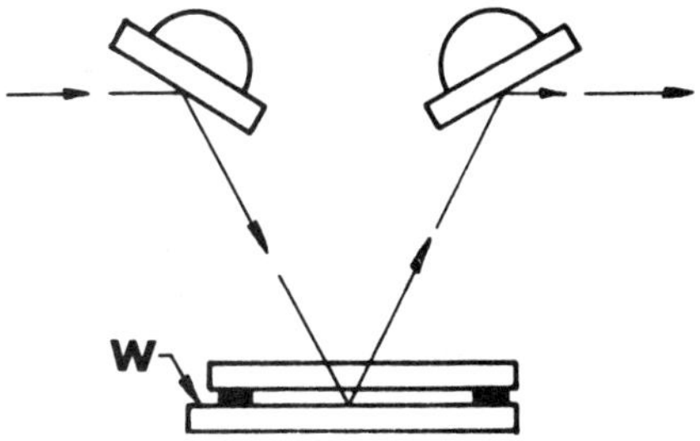


Figure 11.7 Basic configuration for the specular reflectance thin-layer electrochemical cell. [From Ref. 3, P. T. Kissinger and C. N. Reilley, *Anal. Chem.* *42*:12 (1970), with permission. Copyright 1970 American Chemical Society.]

critical if short-time pulse experiments using electrodes with a significant resistance are being performed. Masking is conveniently accomplished using a rubber O ring which serves to define the electrode area and to provide a leaktight seal to the body of the cell. Electrical contact to the film electrode could be made using conducting silver paint or epoxy, although high resistance often results. A mechanical contact to a thin copper sheet held tightly against the electrode by the O ring provides an excellent solution to this problem. A typical cell arrangement is shown in Figure 11.6.

A more sophisticated approach to this problem involves the use of integrated circuit technology, allowing the electrode to be directly press-fit onto the cell with an O ring [28]. The electrical contact can be provided by evaporation of a thick film of gold onto the film electrode. The active area can then be defined by applying a photoresist which is exposed using a contact photographic mask of the shape of the desired electrode area. Subsequent removal of the exposed photoresist then allows etching of the gold by a concentrated KI solution, exposing the electrode area. Any gold that may be exposed to the electrochemical solution can then be covered by evaporation of SiO or $MgF_2$ in vacuo or simply by coating the gold with an epoxy impervious to the solvent of interest.

A variety of cell designs have been proposed to accommodate film electrodes. Of special interest here is the development of a cell compatible with a vacuum manifold for use in studies of oxygen-sensitive systems [29]. In the design shown in Figure 12.13 the reference and auxiliary electrodes can be attached to vacuum valves on the cell body. The vacuum line attachment can be made to the remaining valve. This cell features a small volume of less than 2 mL and can keep the oxygen concentration below $5 \times 10^{-7}$ M for several days. Thin-layer electrochemical cells have relied on a very simple scheme

[3] where a thick and a thin-film electrode are separated by a spacer of a given thickness, perhaps several hundred micrometers. As shown in Figure 11.7, a light beam can then be utilized for spectroscopic studies and the thin-layer cell can be operated in the conventional manner. A myriad of other cell designs exist in the literature—a list too long to describe—with some as simple as to utilize disposable cells with a film electrode held in place and sealed liquid-tight with epoxy. In general, the same rules apply for the design of these cells as those utilizing foil electrodes.

## REFERENCES


1. For a review, see T. Kuwana and N. Winograd, in *Electroanalytical Chemistry*, Vol. 7 (A. J. Bard, ed.), Marcel Dekker, New York, 1974.
2. J. D. E. McIntyre, in *Advances in Electrochemistry and Electrochemical Engineering*, Vol. 9 (R. H. Muller, ed.), Wiley, New York, 1973.
3. P. T. Kissinger and C. N. Reilley, *Anal. Chem. 42*:12 (1970).
4. K. L. Hardee and A. J. Bard, *J. Electrochem. Soc. 123*:1024 (1976).
5. R. W. Murray, *Acc. Chem. Res. 13*:135 (1980).
6. K. L. Chopre, L. C. Bobb, and M. H. Francombs, *J. Appl. Phys. 34*:1669 (1963).
7. C. A. Mengebauer and M. B. Webb, *J. Appl. Phys. 33*:74 (1962).
8. S. P. Muraka, *Thin Solid Films, 23*:323 (1974).
9. F. M. Smits, *Bell Syst. Tech. J. 37*:711 (1958).
10. For a more detailed explanation, see C. Kittel, *Introduction to Solid State Physics*, Wiley, New York, 1971, p. 276.
11. R. Cieslinski and N. R. Armstrong, *Anal. Chem. 51*:565 (1979).
12. H. Kim and H. A. Laitinen, *J. Electrochem. Soc. 122*:53 (1975).
13. J. M. Mochel, U.S. Patent, 2,564,707 (1951).
14. S. Tolansky, *Multiple Beam Interferometry of Surfaces and Films*, Clarendon Press, Oxford, 1948.
15. N. R. Armstrong, A. Lin, M. Fujihira, and T. Kuwana, *Anal. Chem. 48*:741 (1976).
16. H. A. Laitinen, C. A. Vincent, and T. M. Bednarski, *J. Electrochem. Soc. 115*:1024 (1968).
17. T. Osa and T. Kuwana, *J. Electroanal. Chem. 22*:389 (1969).
18. K. L. Hardee and A. J. Bard, *J. Electrochem. Soc. 122*:739 (1975).
19. K. L. Hardee and A. J. Bard, *J. Electrochem. Soc. 123*:1024 (1976).
20. D. E. Tallant and D. H. Evans, *Anal. Chem. 41*:835 (1969).

21. J. S. Mattson and C. A. Smith, *Anal. Chem.* *47*:1122 (1975).
22. T. P. DeAngelis, R. W. Hurst, A. M. Yacynych, H. B. Mark, Jr., W. R. Heineman, and J. S. Mattson, *Anal. Chem.* *49*:1395 (1977).
23. L. Holland, *Vacuum Deposition of Thin Films*, Wiley, New York, 1956.
24. W. Von Benken and T. Kuwana, *Anal. Chem.* *42*:1114 (1970).
25. J. O'M. Bockris and B. D. Cahan, *J. Chem. Phys.* *50*:1307 (1969).
26. R. B. Belser, *J. Appl. Phys.* *28*:109 (1957).
27. W. R. Heineman and T. Kuwana, *Anal. Chem.* *44*:1972 (1972).
28. J. Holland, Ph.D. Thesis, University Microfilms, Ann Arbor, Michigan, 1979.
29. F. M. Hawkridge and T. Kuwana, *Anal. Chem.* *45*:1021 (1973).

# 12

# Electrochemical Cells

FRED M. HAWKRIDGE / Department of Chemistry, Virginia Commonwealth University, Richmond, Virginia



## I. DESIGN CONCEPTS

### A. Introduction

During a nonexhaustive "semi-infinite" electroanalytical experiment the initial concentration of the electroactive material remains unchanged at some distance into the solution. Under these conditions the solution might just as well extend an infinite distance from the working electrode surface, since the chemical changes that occur are confined to

---

*Written by Ray G. Clem, Lawrence Livermore Laboratory, University of California, Berkeley, California

†Written by Peter T. Kissinger, Department of Chemistry, Purdue University and Bioanalytical Systems, Inc., West Lafayette, Indiana

a small volume adjacent to the interface. Nevertheless, the electrochemical cell is an important part of the total experimental system (the control instrumentation, cell, and sample) and its characteristics must be understood to ensure that meaningful data are obtained. The physical design of the cell and the materials used in its construction must be chosen with both the nature of the sample and the experimental objectives in mind. A configuration that yields valid results under one set of conditions might cause experimental artifacts when the objectives are changed.

The investigator must define the experimental goals and select the technique(s) most appropriate to satisfy them. Viewing the system as a whole, one can then optimize the characteristics of the electronics, the cell, and the sample. The geometry of the cell body and each electrode, as well as the placement of the electrodes within the cell are important factors contributing to overall cell performance. The transfer function of the cell must be compatible both with the electronics and the electrochemical technique used.

The controlled-potential three-electrode apparatus can be conveniently used to discuss the matching of the cell, sample, and the instrument. Controlled-potential experiments are common and the following discussion is relevant to many electroanalytical techniques. The operation of a potentiostat is discussed in Chapters 6 and 7. Operational amplifiers are the major building block of modern electrochemical instruments and the reader should be familiar with their operation and characteristics. In short, feedback of an error signal to the input of the potentiostat maintains control of the potential difference between the working and reference electrodes (Chap. 6).

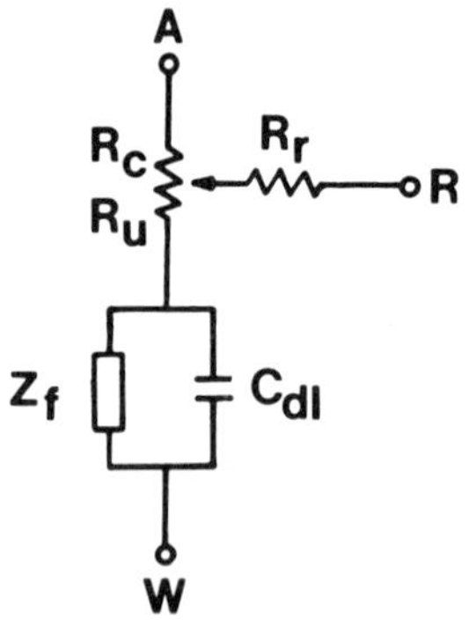


Figure 12.1 Equivalent circuit of an electrochemical cell. A, Auxiliary electrode; R, reference electrode; W, working electrode; $R_c$, compensated resistance; $R_u$, uncompensated resistance; $R_r$, reference electrode impedance; $Z_f$, faradaic impedance; $C_{d\ell}$, double-layer capacitance.

An equivalent circuit of the three electrode cell discussed in Chapters 6 and 7 is illustrated in Figure 12.1. In this simple model $R_r$ is the resistance of the reference electrode (including the resistance of a reference electrode probe, i.e., salt bridge), $R_c$ is the resistance between the reference probe tip and the auxiliary electrode (which is compensated for by the potentiostat), $R_u$ is the uncompensated resistance between the reference probe and the working-electrode interphase ($R_t$ is the total cell resistance between the auxiliary and working electrodes and is equal to the sum of $R_c$ and $R_u$), $C_{d\ell}$ is the double-layer capacitance of the working-electrode interface, and $Z_f$ is the faradaic impedance of the electrode reaction.

Electrochemical techniques can be broadly divided into two categories based on whether the time is long and the current is low or the time is short and the current is high. It is the latter category that presents major challenges in the design of an electrochemical cell.

## B. Low-Current Conditions

Low-current experiments present few problems with regard to cell construction. Unless there are chemical complications due to substances generated at the auxiliary electrode or released from the reference electrode probe, all three electrodes can be placed in a single compartment. Design of the "ideal" cell for accurate potential control is explained using the equivalent circuit shown in Figure 12.1. The total cell resistance $R_t$ is essentially the solution resistance between the working and auxiliary electrodes and is obviously dependent on the distance between the two electrodes, the electrode areas, and the solution conductivity. When a current i exists between the auxiliary and working electrodes, the potential at any point just outside of the working-electrode interface will be affected by the potential drop between that point and the auxiliary electrode (see Chap. 6). A symmetrical arrangement of the auxiliary and the working electrodes should be used, where possible, to keep the potential and current density the same at all points along the working electrode surface.

Single-surface planar electrodes placed parallel to each other provide an ideal arrangement. Cylindrical (Pt wire) or spherical (HMDE) working electrodes placed in the center of a larger concentric auxiliary electrode (usually a metal screen or a wire loop) will be equally satisfactory for low-current applications in solutions of high conductivity. The $iR_t$ is small along any line from A to W. Variations from point to point along the working-electrode surface are therefore negligible. As indicated in Figure 12.1, the placement of the tip of the reference probe divides $R_t$ into a compensated and an uncompensated resistance. The reference probe tip *should* be placed close to the working electrode on a line *between* the working and auxiliary electrodes. Nevertheless, in very low current experiments (common in analytical electrochemistry) the effect of electrode placement is usually negligible because the iR drop throughout the bulk solution may be less than 1 mV.

The potential drop in solution between the reference electrode probe tip and the auxiliary electrode, $iR_c$, is corrected for by the operation of the potentiostat (see Chap. 6 if this is not understood). The potential drop existing between the reference probe tip and the working electrode, $iR_u$, remains as an error in the accuracy of potential control. If the probe tip is too large or is placed too close to the working-electrode surface, it will interfere with mass transfer of the electroactive material to the working electrode, an effect called "shielding" of the working electrode [1,2]. It is possible to correct for the uncompensated iR drop mathematically or by positive feedback electronic techniques, but the correction is rarely significant in low-current measurements (see Chap. 7).

Often the chemistry of an experiment dictates that the cell have special features. For example, *products generated at the auxiliary electrode or materials leaking from the reference probe will occasionally interfere with the reaction at the working electrode.* In such cases the auxiliary electrode and the reference electrode must be isolated from the working-electrode compartment. The auxiliary electrode and the reference electrode are often isolated by a semipermeable barrier (i.e., a glass frit, an ion exchange membrane, "thirsty" glass, an asbestos fiber junction, a platinum glass seal) from the working electrode compartment. This restricts mass transfer between the electrode compartments and yet permits electrical contact. For long-term experiments where interference is most likely, an intermediate solution compartment or salt bridge can be employed [3]. Use of strong electrolyte solutions and placement of the reference probe tip on the line of closest approach between the other two electrodes minimizes potential gradients along the working electrode surface.

The accuracy of working-electrode potential control can be established by introducing an independent reference electrode into the cell with the probe tip positioned about the same distance from the working-electrode surface as the original reference electrode probe and also on a line between working the auxiliary electrodes. The experimentally applied potential may then be compared with that measured between the working electrode (usually at real or virtual ground) and the independent reference electrode. This check is usually adequate for cases where the product $iR_u$ is small, such as is true in low-current experiments in strong electrolyte solutions. Correction of the working-electrode potential error can be made by determining the magnitude of $iR_u$ experimentally or by instrumental compensation for $R_u$ (see Chap. 7).

## C. High-Current Conditions

High-current conditions can exist under constant-potential control when the working electrode is large, the concentration is high, the solution is in motion, or when the potential is rapidly changed from one value to another (i.e., usually requiring high nonfaradaic current levels for double-layer charging). In order for the potentiostat to control the working electrode potential, the current between the auxiliary and working electrodes must not overload the capability of the control amplifier at the applied potential.

A typical example would be a coulometric controlled-potential experiment where, initially, a large current exists between the auxiliary and the working electrode (the load of the control amplifier). If a potentiostat is rated to have a maximum output of ±20 V at 1 A, it cannot supply more than 20 W of power [power (watts) = current (amperes) × potential (volts)]. If $R_t$ were 100 Ω, the potentiostat would *not* be able to control the potential of the working electrode at −2.0 V (or any other potential for that matter) if 0.5 A were demanded. At least 25 W ($I^2R_t$) of power would be required of the potentiostat in order for potential control to be maintained. As a result of our 5-W deficiency, the potential of the working electrode would be uncontrolled at a value less than the −2.0 V and less than 0.5 A would, in fact, pass through the cell.

Accurate potential control in an electrolysis is often necessary to limit the electrode reaction(s) to the one(s) of interest. Failure of the potentiostat in the case above could be solved by using a lower sample concentration or a smaller working-electrode area in order to decrease the current demand. Decreasing the resistance between the auxiliary and working electrodes, $R_t$, by increasing the conductivity of the medium might also lower the power ($I^2R_t$) required to operate the cell to a value within the potentiostat's capability. If a diffusion barrier is used to isolate the auxiliary electrode, it will usually contribute significantly to $R_t$. It might then be desirable to use a larger barrier or a material with a larger pore size. As should now be obvious, there usually are a number of ways to get around the problem of too much power demand.

The example above points out that *it is the potentiostat's voltage swing, not its current capability, which often limits its performance.* If we had used a 40-W potentiostat (±20 V at 2 A) the experiment might still have failed, whereas a 25-W model (±50 V at 0.5 A) could do the job.

Some attention must be paid to the speed at which a potentiostat can change the controlled potential from one value to another when using any of a variety of fast-sweep or potential-step experiments. The rise time of a potentiostat intended for high-current applications should be experimentally determined by measuring the time required to charge a realistic resistive-capacitive "dummy cell" following a known potential step. *Statement of an experimentally determined rise time without reference to the specific load driven is meaningless.* The experimental factors that limit the rise time of a potentiostat are complicated and have been examined in some detail in Chapter 7. The power of the potentiostat, the phase shift of the feedback signal with respect to the output signal of the control amplifier, the characteristics of the cell and sample, and the magnitude of $C_{d\ell}$ all influence the actual rise time of the interfacial potential excursion.

When the potential of the working electrode is stepped from one value to another, large amounts of current must be supplied in a very short time to charge the double-layer capacitance, $C_{d\ell}$ in Figure 12.1, before the final potential difference across $C_{d\ell}$ can be controlled and the faradaic current can commence. The apparatus and the experimental design must be matched to make maximum use of the rise time of the potentiostat. Experimentally, there is an advantage in using smaller potential steps, since large potential steps may be accompanied by large changes in the double-layer capacitance. Using the voltammogram in Figure 12.2 as an example, one would prefer to perform a potential step from an initial value B to a final value of C rather than from A to C. Small working-electrode areas are often desirable, since $C_{d\ell}$ is directly proportional to the area of the working electrode. For fast potential-step experiments a working-electrode area of about 0.01 $cm^2$ is often a reasonable compromise.

It is important to realize that a potentiostat's rise time using a dummy cell cannot be used as evidence for accepting a measurement made on a real sample. Moreover, establishing the optimum placement

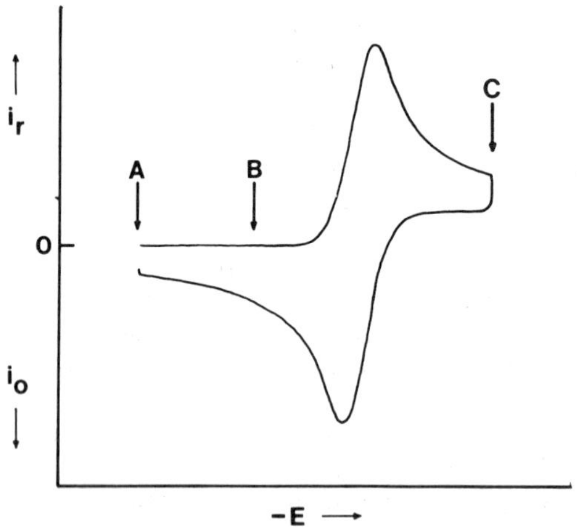


Figure 12.2 Model cyclic voltammogram. A, B, Initial applied potentials; C, step potential.

of the electrodes and their geometry ought to be performed while monitoring the effect these changes have on the response for the real sample. The time required for double-layer charging and therefore control of the working-electrode potential should be determined by monitoring both the current and the potential in a potential-step experiment on a real sample. Reference electrode probe-tip geometry is particularly critical in the optimization of cell design for very fast potential-step experiments [1,2]; however, these experiments are rarely performed by analytical electrochemists.

## II. STATIONARY-SOLUTION EXPERIMENTS

Numerous electroanalytical techniques are performed on "quiet" or stationary solutions. The ideal stationary solution has no convective motion due to vibrations, mechanical stirring, motion of the electrode(s), temperature gradients, or density gradients. At long times only "natural convection" effects arising from the electrode reaction itself will contribute to nonideality.

Several popular cell designs are shown in Figure 12.3. These cells can be scaled to fit the available sample volume and the size and type of electrodes to be used. The cell may include a stationary semimicro working electrode [i.e., platinum wire, bead, or button; DME (shown in Fig. 12.3C), HMDE, carbon paste (shown in Fig. 12.3A), graphite, etc.], although simple modification could allow for rotating electrodes. The auxiliary electrode shown in Figure 12.3A and C is a platinum wire loop concentric with the working electrode. A platinum wire or foil, a mercury pool, or carbon can also be used for this purpose. The graphite rods used for emission spectroscopy are useful as auxiliary electrodes. For optimum performance at short times the position of the reference probe might have to be changed from that shown when using different auxiliary electrodes.

Physical isolation of the auxiliary electrode can be essential, unnecessary, or even detrimental to the experimental goal. It is important to recognize which situation exists for a given experiment. In general, isolation is essential for exhaustive coulometric experiments, unnecessary for nonexhaustive experiments (i.e., voltammetry), and detrimental to fast experiments (i.e., chronoamperometry). In the normal stationary-electrode voltammetry experiment, so little material is actually converted at the electrodes that the possibility for interference is very remote. When necessary the auxiliary electrode can be easily isolated by using a glass frit as shown in Figure 12.3B. The sample and the auxiliary-electrode filling solution should be adjusted to the same level to minimize convection between the two compartments. The isolation of the auxiliary electrode historically was achieved with an agar gel, but there is no advantage to this more tedious approach. Porous Vycor and ion exchange membranes have

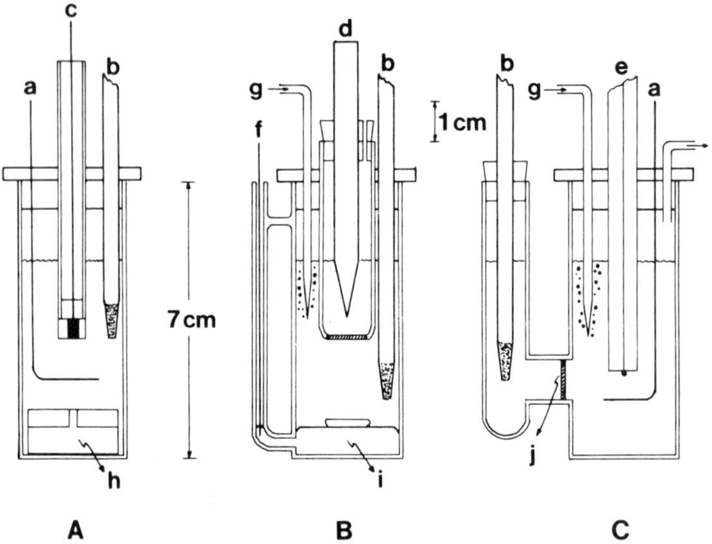


Figure 12.3 Stationary solution voltammetry cells. (a) Platinum wire loop auxiliary electrode, (b) reference electrode or reference electrode probe tip, (c) carbon paste working electrode, (d) graphite auxiliary electrode, (e) dropping mercury electrode, (f) platinum wire contact to mercury pool working electrode, (g) nitrogen gas inlet tube, (h) magnetic stirrer, (i) mercury pool working electrode, (j) glass frit isolation barrier.

been widely substituted for the glass frit with satisfactory results [2,4,5 and references therein]. Another type of auxiliary electrode can be used in cases where interferences from the product(s) of this electrode reaction are of concern in reductive experiments. Silver metal in contact with chloride electrolyte results in the formation of silver chloride on the surface of the electrode. This type of auxiliary electrode further ensures that contamination from the auxiliary compartment into the working-electrode compartment does not occur. The silver auxiliary electrode is polished at the conclusion of an experiment to provide a clean surface for subsequent experiments.

The reference-electrode probe shown in Figure 12.3 may be the reference electrode itself or a salt bridge connecting and isolating the sample solution from the reference electrode. As mentioned earlier, the primary reference-electrode filling solution must sometimes be isolated from the sample. Contamination of nonaqueous samples by water and denaturing of biological samples by metal ions in the filling solution are examples of experiments requiring an intermediate bridge.

It is important to consider the stability of reference electrodes and salt bridges in work done in nonaqueous solvents. Fouling of either the reference electrode or the salt bridge junction can lead to potential errors and distorted voltammetric responses. A procedure for testing aqueous saturated calomel electrodes used in nonaqueous solvents employing impedance measurements has recently been described [6].

Several types of reference electrodes are convenient for use in analytical electrochemistry. The use of high-input-impedance operational amplifiers in the reference electrode inputs of potentiostats ensures that very low levels of current are drawn from the reference electrode (see Chap. 6). This permits the use of reference electrodes that do not have to contain a large number of redox equivalents in order to ensure a constant reference potential and that therefore are very small. Several reference-electrode designs that are convenient for use in analytical electrochemistry are shown in Figure 12.4. Saturated calomel and silver-silver chloride (of various concentrations of chloride) are among the most common reference electrodes that are commercially available or that are conveniently fabricated.

Provision for nitrogen purging is shown in Figure 12.3B and C, although for many experiments employing the application of positive potentials, removal of dissolved oxygen is unnecessary. Techniques such as differential pulse polarography can be used without removal of oxygen when the redox potential of the reaction of interest is suf-

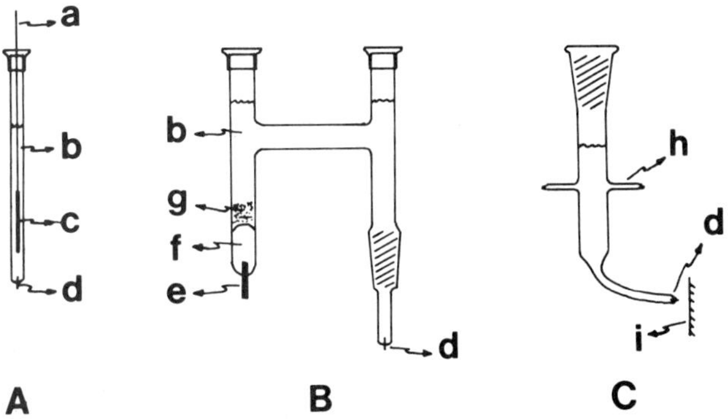


Figure 12.4 Reference electrodes and reference electrode probe. (A) Silver-silver chloride reference electrode, (B) calomel reference electrode, (C) reference electrode probe. a, Silver wire, ca. 22 gauge; b, reference electrode filling solution, saturated or known concentration of KCl; c, silver chloride electrochemically precipitated on silver wire by anodization; d, reference electrode junction; e, Pt/glass seal; f, mercury; g, calomel; h, electrochemical cell body wall; i, working electrode surface.

ficiently positive or negative of the reduction potential of oxygen. More efficient removal of oxygen is achieved if a fritted glass dispersion tube (available from laboratory glass suppliers) is used; however, a small-diameter glass tube drawn out to a fine tip providing many small bubbles is usually adequate. In cases where removal and exclusion of dissolved oxygen is required, the bubbling of gas through the solution must be stopped prior to experiments employing quiet solutions. A slight positive pressure of inert gas is then maintained over the solution. This can be done by including another gas delivery tube in the cell cap positioned above the solution at a distance that does not physically agitate the solution surface. The inert gas is alternatively passed through the dispersion or the delivery tubes. In some cases this can simply be accomplished by raising and lowering the inert-gas delivery tube above and into the solution.

The cell cap and the cell body can be constructed of any suitable material that is inert to the sample. A snug, but not a tight, press fit of the cap to the cell is desirable if glass is used for the cell since the diameter of ordinary glass tubing is imprecise and may break at the most inopportune moment. The cell body may be a beaker, test tube, glass tubing, any of a number of plastic containers, or may be machined out of Lucite, Delrin, Teflon, MACOR (machinable glass), or similar materials. The cell cap may be made of Lucite, Delrin, Teflon, rubber, neoprene, and the like. Cork borers may be used to make the necessary holes in the soft rubberlike cell caps, but freezing in liquid nitrogen or an acetone-dry ice bath allows the convenient use of an electric drill to bore necessary holes. Adjustment of the depth to which each part can be positioned in the sample solution is simple. A short length of the appropriate-diameter rubber tubing or an O ring may be slipped around each part, providing a collar that can be adjusted for positioning. Micro cells based on this design [7] can be used for sample volumes of from 10 to 0.5 μL. Other micro cells using similar design principles are useful for anodic stripping voltammetry on sample volumes down to as little as 5 μL [4,5] as well as for other voltammetric techniques [8,9].

Many variations of the design of the cells shown in Figure 12.3 have been reported. Reasonable cost, simple construction, flexibility, and ease of correct use have led to widespread acceptance of these general designs. These cell designs fit most aqueous and nonaqueous sample requirements where the presence of oxygen or water is not critical. When the removal and exclusion of these contaminants is required, special care must be taken to work in an inert atmosphere. A dry box (Chap. 15) or a cell that can be interfaced to a vacuum line (Chap. 14) may be required.

## III. CONVECTED-SOLUTION EXPERIMENTS

### A. Stirred Solutions

Exhaustive coulometric techniques can be used both as an analytical tool and as a preparative tool. These two applications often require different cell designs. Present discussion will be restricted to analytical coulometric cells. The design of electrochemical cells for preparative electrolysis has been treated elsewhere [10].

Figure 12.5 (a more elaborate version of Fig. 12.3B) shows a versatile design for analytical controlled-potential coulometry [11]. The working electrode is a stirred mercury pool with provision for rapid and convenient exchange of spent sample and mercury through the stopcock. This same cell could be used with a platinum gauze working electrode, with the only change being provision for electrical contact through the cell cap. The auxiliary electrode is isolated by a porous glass tube and is located far enough above the surface of the

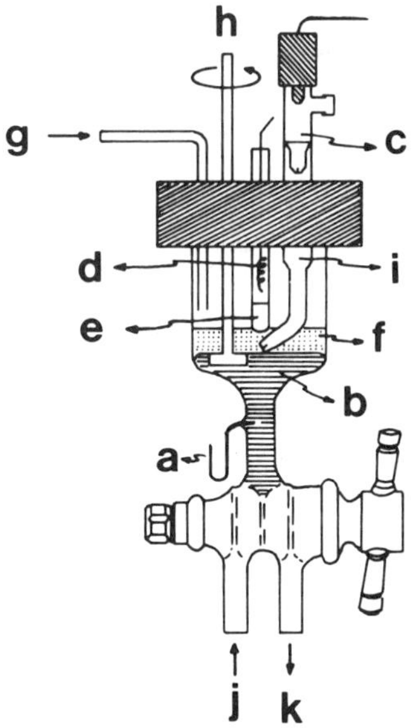


Figure 12.5 Controlled-potential coulometry cell with a mercury pool working electrode. a, Platinum wire contact to mercury pool working electrode; b, mercury pool working electrode; c, reference electrode; d, auxiliary electrode; e, porous Vycor; f, sample solution; g, inert gas inlet; h, stirrer; i, reference electrode salt bridge; j, clean mercury; k, waste. [From Ref. 11, adapted with permission from J. E. Harrar and E. L. Pomernacki, *Anal. Chem.* 45:57 (1973). Copyright 1973 American Chemical Society.]

working electrode to minimize potential gradients. Positioning of the reference probe tip is shown on the line of closest approach between the working and auxiliary electrodes but would have to be moved if a cylindrical platinum gauze electrode were used. A mechanical stirring paddle is included. The stirring rate should be adjusted to achieve maximum agitation of the mercury surface while keeping the agitation of the sample surface at a minimum to prevent splattering. The dimensions of the cell can be designed to hold sample volumes from ca. 20 mL to several hundred milliliters. Efficient cells (minimum volume, maximum electrode area) are desirable in controlled-potential coulometry to keep the time of the experiment short. A nitrogen gas dispersion tube is included and the rate of nitrogen flow should be adjusted so that splattering is minimized. Several other controlled-potential electrolysis cell designs are given in the same reference [11]. A discussion of cell design philosophy is also included in this reference and in a previous paper [12].

### B. Rotating Electrodes

Rotating electrodes have been used to provide some degree of control of the rate of mass transfer to the working electrode through control of the rate of electrode rotation. Wire, wire gauze, ring, and ring-disk electrodes are among the types of rotating electrodes that have found wide use in analytical and kinetic applications. Details of construction of some of these electrodes are found in Chapters 9 to 11.

The cells shown in Figures 12.3 and 12.5 can be readily adapted for use in rotating electrode experiments. Clamping the cell and top into position and drilling the cap to allow free rotation of the electrode shaft is required. Constant-speed rotating electrode experiments usually have analytical objectives (e.g., amperometric titrations) and the cell volume may be any convenient size ranging upward from about 0.5 mL, as demonstrated in a study of sample volume requirements in rotating electrode voltammetry [13].

Kinetic studies employing rotating disk and rotating ring-disk electrodes require convenient and precise adjustment of the rotation speed from below 10 rps to above 1000 rps. The high rates of rotation make wobble of the electrode a point of concern and mating of the electrode shaft to the cell cap can require the use of a bearing joint. Commercial systems are available so that this problem can be avoided. The rotating electrode should be positioned in the cell just below the solution surface to minimize convection in the solution due to electrode vibration. Kinetic experiments can require care with respect to both cell dimensions and electrode placement [14,15]. It should be recognized that the larger currents encountered in rotating electrode experiments can result in depletion of the sample concentration if too many voltammetric scans are performed on a single sample. The use of cells containing large sample volumes (i.e., 50-500 mL) can minimize this problem.

Rotating platinum screens have been used for electrogravimetric analyses. Commercial instruments employ two concentric cylindrical platinum screen electrodes with one or both electrodes rotating to increase convection. The cell itself is usually a beaker with a sample volume of about 150 mL. No cell top is generally used and when running multiple analyses there should be adequate ventilation to prevent accumulation of hydrogen. A simple operational amplifier circuit can be constructed for the instrument. The commercial instruments available for this experiment are expensive given the simplicity of this type of experiment.

## C. Rotating Cells

Rotated mercury [16] and rotated platinum [17] coulometry cells were originally developed for the analysis of small samples, or samples of high value that must be recovered after the analysis, such as separated isotopes. These cells have proved to have such exceptional, timesaving characteristics that they are now used for general analytical purposes.

The mercury cell (Fig. 12.6) is quite simple in design and operation. The cell consists of a plastic cylinder, which is partially open at the tip and closed at the bottom, mounted on a copper turntable attached to the end of the shaft of a synchronous 1800-rpm motor. Sample and mercury (only 2 mL of each are required) are aliquoted into the cell and the cell is started rotating. The mercury is held rigidly against the wall of the cylinder in a thin film by the centrifugal force and the solution, owing to the density difference, is held in a thin film against the mercury. The mercury is contacted through a threaded platinum wire that is screwed through the cell wall and is connected to the copper turntable. A brush contact of bronze wool that touches the turntable is used to make connection to the potentiostat.

The solution phase is contacted and very efficiently stirred with a unique, stationary, coaxial probe that contains both the reference bridge and auxiliary electrode bridge. The reference is terminated in fritted glass and is partially filled with a fumed-silica gel [18]. In this configuration the reference electrode is positioned in the region of the highest current density; this assures that the control potential will not be exceeded [12]. Fulfilling this requirement is essential to the proper functioning of the cell. Failure to do this will result in the loss of resolution between adjacent redox couples if the second couple is irreversible or, in the limiting case, a positive but nonlinear

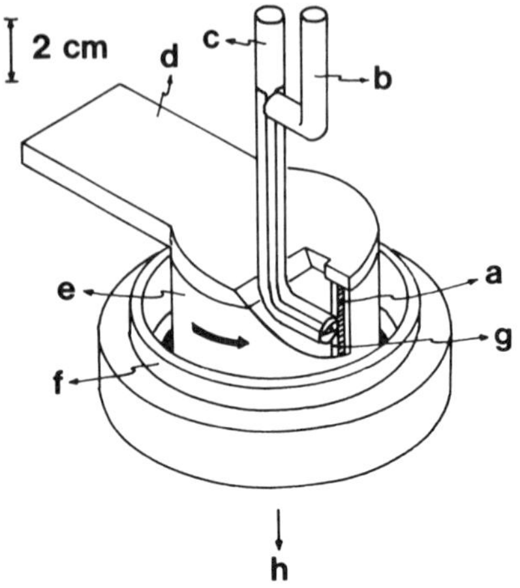


Figure 12.6 Rotated cell for high-speed controlled-potential coulometry. a, Mercury or platinum working electrode; b, auxiliary electrode compartment; c, reference electrode compartment, stationary; d, cell cover, stationary; e, rotated cell; f, positioning collar; g, sample solution; h, motor. [From Ref. 16, adapted with permission from R. G. Clem, F. Jakob, D. H. Anderberg, and L. D. Ornelas, *Anal. Chem.* 43:1398 (1971). Copyright 1971 American Chemical Society.]

bias will result when electrolyzing a sample at a potential near the limiting potential of the solvent system.

The electrolytic rate constants for the rotated mercury cell are on the order of 0.01-0.02 $s^{-1}$; thus the cell is at least as fast as the fastest conventional cell. The stirring noise level is perhaps the lowest of any cell ever built. Low-noise cell design is at a premium when performing predictive coulometry [19].

The most outstanding feature of the cell is that the sample can be deoxygenated in as little as 20 s because the ratio of solution area to solution volume is quite large (see Fig. 12.7). In addition, since the solution is rigidly held by centrifugal force, very much higher gas flow rates can be employed. The sparging gas is admitted to the cell through a stationary, hinged, cell cover positioned only 250 $\mu$m above the rotating cell. In its open position, it prevents the ingress of air.

The sample solution is virtually deoxygenated during the aliquoting step and before the addition of mercury to the cell. This is an important consideration when the sample or supporting electrolyte is acidic, especially if it contains chloride. In the latter case, the mercury is converted to calomel, and the calomel is only slowly reduced, thus extending the preelectrolysis time and possible contributing to a positive bias in the coulometric results.

The rotated cell is most useful for samples in the range 0.2 to 20 μeq. Samples greater than 20 μeq cause current or potential limiting by the iR drop between the rotated working and stationary reference electrodes, thus increasing the electrolysis time. The precision obtained for samples smaller than 0.2 μeq suffers because of difficulties in resolving the faradaic current from the continuous background current. This problem can be ameliorated by the use of a digital computer and is described briefly below.

The rotated cells exhibit a very large cell potential gradient over the working electrode surface. This has never been found to be of particular trouble for inorganic analysis; however, it may be of concern in the coulometry of some organic substances [20]. This large gradient may be of use in scanning coulometry—perhaps the last underdeveloped area in the electroanalytical field. The scanning coulometer would increment or "step and hold" the cell potential by sensing cell current demand [21]. A high current demand would manifest itself by a large potential gradient sensed with a second reference electrode stationed 180° from the coaxial electrode. In potential ranges in which there is no sample current, the instrument would scan rapidly. In potential ranges in which the sample could be reduced, the instrument would increment the control potential only after the sample in that potential time frame was exhaustively electrolyzed.

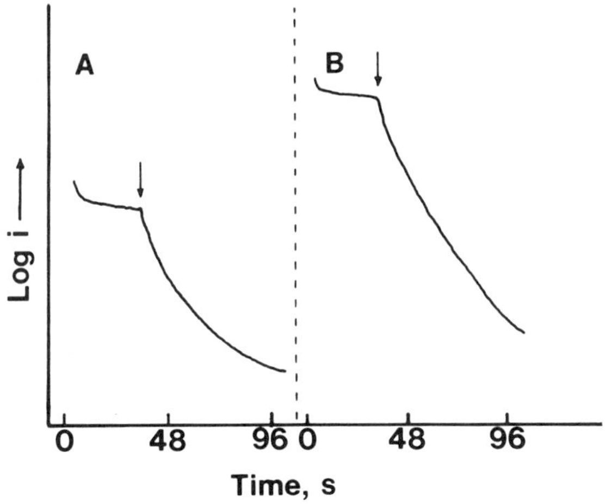


Figure 12.7 Sparging behavior of the rotated cell. (A) Current-time data as recorded. (B) Curve A expanded by a factor of 10 and normalized. Arrows denote the point in time at which the nitrogen was turned on. [From Ref. 16, adapted with permission from R. G. Clem, R. Jakob, D. H. Anderberg, and L. D. Ornelas, *Anal. Chem.* *43*:1398 (1971). Copyright 1971 American Chemical Society.]

It appears that the small electrolytic rate constants result from the lack of efficient stirring in the mercury phase. To destroy any mercury-phase diffusion gradient and thus to promote efficient mercury phase mixing, a pulse-rotated cell was developed [22]. In operation, the synchronous cell motor is alternatively driven for 0.1-s intervals with 60-Hz, then 50-Hz ac current. Consequently, the rotational rate of the cell is constantly changing between 1800 and 1500 rpm, respectively. The cell wall is fitted with shallow scallops which, owing to the high inertia of the mercury, cause riffling, and subsequently, very efficient stirring in the mercury phase. Electrolysis rate constants on the order of 0.03-0.04 $s^{-1}$ have been obtained in the controlled-potential electrolysis of lead in 1 M KCl. High-speed photography of the mercury cell reveals a condition approaching ultrasonic cavitation in the solution phase, upon suddenly increasing or decreasing the cell rate of rotation. The shear between the mercury and solution phase is so great that vacuum pockets are ripped in the sample solution.

The fact that these cells were designed for small samples, and hence require little current for the electrolysis, minimized electronic considerations and costs. The cells can be driven with a simple two-operational-amplifier potentiostatic circuit capable of delivering 20 mA at 25 V [23]. Placement of the reference electrode bridge in the highest cell current location and near the working electrode assures very good potentiostatic stability [11].

A digital normalization technique was developed to permit the assessment of the background current for controlled-potential electrolysis data on an individual basis [17]. The result is improved precision (see Fig. 12.8). To effect this background correction procedure required only primitive digital data processing capabilities. While viewing an oscilloscope display of the recorded log i-time data, the same number of counts are subtracted from each data channel until all the data points fall on a straight line. The total number of data counts subtracted is equal to the background current contribution. Precision approaching 0.01% is easily attained for samples at the 10-to 20-μeq level. The normalization technique becomes progressively less applicable as the sample available becomes less than 0.2 μeq. The results suffer from a progressively greater positive bias. This occurs because the background current is never truly constant; rather, it decreases very gradually but continually. A computer program that fits the sum of two exponentials to the data yields much better results. Results accurate to 1% or better have been obtained in cases in which the background current equals the initial electrolysis current.

The rotated platinum cell appears to offer little opportunity for further improvement [17]. Like the mercury cell, it can be deoxygenated in as little as 20 s, but unlike the mercury cell, the electrolysis

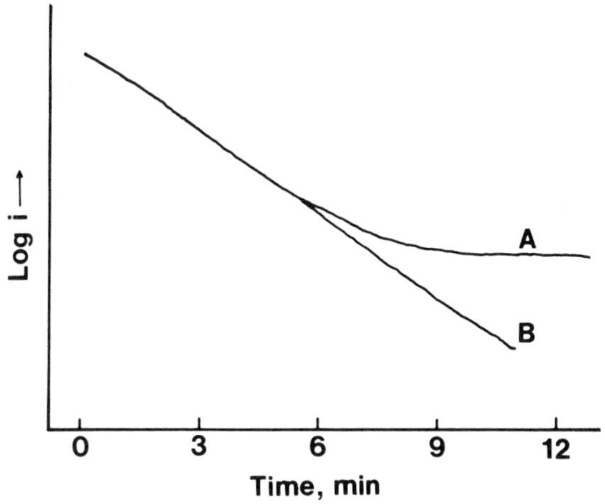


Figure 12.8 Digital normalization of a controlled-potential electrolysis of $U^{6+}$. A, Raw data: 569.8 μg of $U^{6+}$ found. B, normalized data: 561.6 μg of $U^{6+}$ found. Uranium present, 561.8 μg.

rate constants are >0.1 $s^{-1}$ for many substances. A determination can be completed within 5 min, aliquoting and cell cleaning included.

The platinum cell is made entirely of platinum. Three layers of fine-mesh platinum gauze are spot welded to the cylinder to increase the active surface area. The total surface area approximates that of conventional cells, but the rate constant is about 10 times as large. This results in a large increase in real sensitivity and is due to the very favorable electrode surface area/solution volume ratio. Data obtained with the platinum cell are also amenable to the digital normalization techniques described above for the mercury cell.

## D. Flow-Through Cells

Flowing solution methods are particularly useful in on-line determinations and in providing a ready means of controlling mass transfer in analytical and kinetic measurements. Cell designs for flowing solution analyses have used a wide variety of electrode materials, often seeking to take advantage of a particular electrode material's electrochemical characteristics or its ease of use in certain geometric configurations. Platinum wire [24-26], microconical platinum [27], platinum grid [26, 28], hanging mercury drop [29], gold micromesh [30], graphite packing [31], reticulated vitreous carbon [32], tubular platinum [33-35], carbon [36-38], stainless steel [39], gold [40], mercury-covered

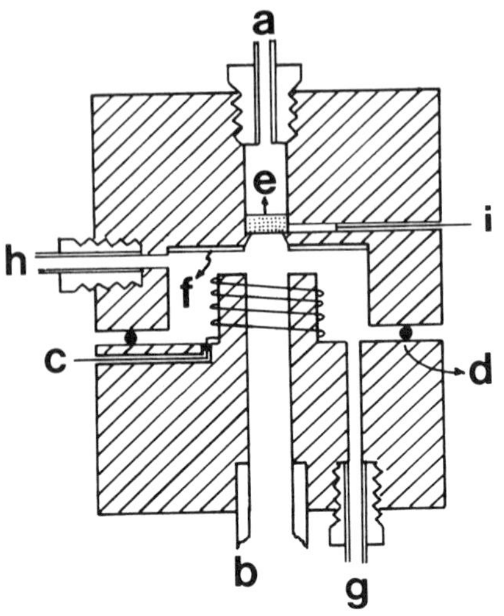


Figure 12.9 Flow-through cell with RVC electrode and reference electrode. a, Sample solution inlet; b, sample solution outlet (Tygon); c, lead to reference electrode; d, O ring; e, RVC electrode, one disk shown; f, cation exchange membranes; g, reference solution inlet; h, reference solution outlet; i, lead to working electrode. [From Ref. 32, adapted with permission from W. J. Blaedel and J. Wang, *Anal. Chem.* *51*:799 (1979). Copyright 1979 American Chemical Society.]

platinum [41,42], and mercury-covered graphite [43] were among the electrodes studied in flowing solution analyses. Also, flow cells have been employed as reagent generators for titrations [44,45] and for electrosynthesis [46]. Flow electrochemical cells that have been used as detectors in liquid chromatography with excellent success are described in Chapter 22.

A flow cell design, shown in Figure 12.9, is one of the many configurations that incorporate various types of working electrodes described above. In this example, reticulated vitreous carbon (RVC) is used as the working electrode, providing a large surface area and sample flow through the electrode itself. The reference electrode is isolated from the sample solution by a cation exchange resin and the internal filling solution of the reference electrode is continually replaced by flow. The cell body is machined conveniently from Plexiglas and conventional connectors serve to mate the necessary input and output tubes to the cell body. The detection limits for this cell design are below $10^{-9}$ M [32]. Another flow cell design provides for the use of rotated electrodes [47-49]. Increased sample convection is achieved leading to high levels of current efficiency.

One major advantage of flowing solution methods is that the measurement is of steady-state currents. This simplifies the instrumentation and sllows for very accurate and precise measurements at very low sample concentrations.

## IV. THIN-LAYER-CELL DESIGN

Problems in designing cells for semi-infinite diffusion techniques are well known to electrochemists. When the same considerations must be applied to a layer of solution less than 100 μm thick, the difficulties are significantly enhanced. Like most real-world components, thin-layer cells do not always conform to theory (Chap. 3), and design trade-offs are necessary to optimize performance in a particular application.

The thickness L of the layer ought to be uniform; thus the electrode must have a minimum surface roughness and be parallel with the opposing wall. Normally, the interstitial volume will have a juncture with bulk solution to accommodate the reference and/or auxiliary electrodes, or to permit filling. This is accomplished either at the edge of the thin layer or via a small orifice in the barrier wall. This cross-sectional area of this edge or orifice must be small relative to the electrode area in order to minimize diffusion of material in and out of the cell volume proper. The edge effects are minimized when the thin solution layer extends beyond the working electrode area. In general, it is best to keep the solution film horizontal to mitigate enhancement of edge effects by natural convection phenomena.

The rapid response theoretically inherent with thin-layer experiments is difficult to realize in practice because of the problem of achieving a uniform current-potential distribution. This is, of course, especially true for nonaqueous media. Except in steady-state twin-electrode experiments (see Chap. 3), current paths usually run parallel to the electrode surface. This most unfavorable situation can be rectified by use of a barrier wall that will effectively transmit current without significant transport of electroactive material during the experiment. This is a difficult property to achieve in practice while still maintaining the integrity of the thin-layer geometry. Porous glass ("thirsty" glass or porous Vycor) barriers have been used with some success; however, these frequently are more of a nuisance than the iR drop they purport to correct. The poor current distribution can make placement of the reference electrode more important than usual. Although "internal" reference probes are possible, they can result in overcompensation of the iR drop if placed in certain reaches of the cavity (e.g., above the working electrode in the sandwich cells discussed below).

The final cell design, as in semi-infinite electrochemistry, depends very much on the goals of the experimenter. For example, when one is interested in single-electrode coulometry for n value, concentration, or spectral measurements, the cell requirements are minimal. Potential scan experiments (voltammetry, potential scan coulometry, steady-state voltammetry) or experiments where transient response is important (e.g., kinetic studies) require the most grandiose thin-layer cells.

Cells with solid electrodes have incorporated a thin-layer volume either in the form of a slab or cylindrical shell. The latter have all been of fixed L value, whereas the former have been constructed with either fixed or variable thickness. A comprehensive review of thin-layer cells and applications has been published [50]. Since many of these are quite complex and will only be of interest to the specialist, only one approach will be described in detail.

Sandwich-type cells with fixed L values are easily constructed and have been widely employed for coulometric and spectroelectro-

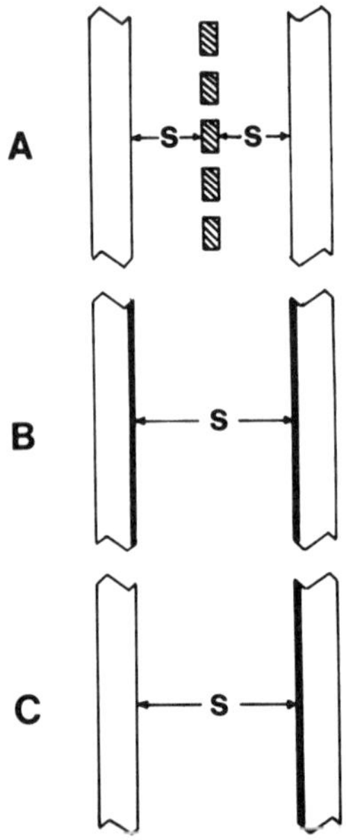


Figure 12.10 Three configurations for sandwich-type thin-layer cells. (A) Minigrid suspended between two spacers. (B) Twin-electrode cell using metal films on glass. (C) Single-electrode cell, barrier plate and electrode plate. s, Sample solution.

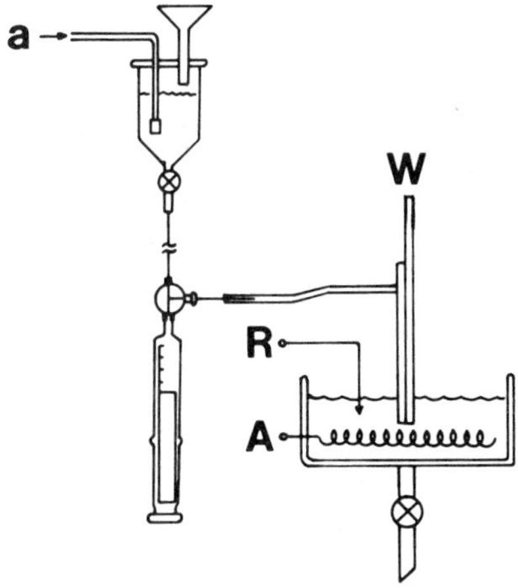


Figure 12.11 Experimental arrangement for use of thin-layer electrode constructed with glass or quartz plates. W, Working electrode; R, reference electrode; A, auxiliary electrode; a, inert-gas inlet.

chemical experiments. These designs have incorporated metal foils, a metal minigrid, deposited metal films, and semiconducting metal oxide electrodes [50-53]. A popular configuration uses glass slides and an inert spacer (e.g., Teflon) to form the sandwich. Cross sections of three models are shown in Figure 12.10. This specific approach originated with the minigrid cell of Murray et al. [54].

Figure 12.11 illustrates one practical scheme for using sandwich-type cells. The sample is introduced by capillary action from below or through a side arm from above. Solution contact to the reference and auxiliary electrodes is made via the bottom edge of the cell in a small cup filled with supporting electrolyte. The thin-layer electrode itself (Fig. 12.12) consists of two glass microscope slides (an electrode plate and a barrier plate, two electrode plates, or two barrier plates with a central grid electrode) pressed to a thin (ca. 20-75 μm) spacer. Either epoxy cement or a clamp has been commonly used to hold the cell together. A spacer(s) of suitable geometry is carefully cut from stock film with a razor blade. The sandwich is then assembled and clamped tightly together with a large spring-loaded paper clip, making sure that the spacer has no folds and the plates are properly oriented. Epoxy is lightly applied to three edges and the unit is placed in an oven or under an infrared lamp to facilitate curing of the resin. A large number of cells can be prepared in short order and at very little expense. Should convenient disassembly be necessary, a clamp will have to be devised. Two steel plates backed by rubber gaskets can be screwed together around the cell or the entire cell can be machined from an engineering plastic [55].

Table 12.1 Commercial Sources of Electrochemical Cell Components[a]

| Component | Source | Comments |
|---|---|---|
| Cation exchange washers | American Machine & Foundry<br>689 Hope St.<br>Springdale, CO 06879 | Diffusion barriers for isolating reference and auxiliary electrode compartments |
| Electrodes | Buckbee-Mears Co.<br>245 E. 6th St.<br>St. Paul, MN 55101 | Minigrids: nickel, copper, silver, gold |
| | Bioanalytical Systems, Inc.<br>Purdue Research Park<br>2701 Kent Ave.<br>West Lafayette, IN 47906 | Wide variety of working electrodes, reference electrodes, components |
| | Princeton Applied Research Corp.<br>P.O. Box 2565<br>Princeton, NJ 08540 | Static mercury drop electrode, hanging mercury drop electrode, dropping mercury electrode, reference electrode |
| | Brinkman Instruments, Inc.<br>Cantiague Road<br>Westbury, NY 11590 | Hanging mercury drop electrode |
| | Pine Instrument Company<br>101 Industrial Drive<br>Grove City, PA 16127 | Rotating disk and ring-disk electrodes made of several materials |
| | PPG Industries<br>1 Gateway Center<br>Pittsburgh, PA 15222 | Semiconductor optically transparent electrodes |

| | | |
|---|---|---|
| | ERG, Inc.<br>Lowell and 57th St.<br>Oakland, CA 94608 | Reticulated vitreous carbon, various porosities |
| O rings | Fournier Rubber & Supply Co.<br>459 N. High St.<br>Columbus, OH 43215 | O rings of all sizes and materials |
| Plastics | Cadillac Plastic and Chemical Co.<br>15841 Second Ave.<br>P.O. Box 801<br>Detroit, MI 48232 | Wide variety of types and shapes of plastics for cell construction |
| Porous Vycor | Corning Glass Works<br>Corning, NY 14830 | Diffusion barrier for auxiliary and reference electrode compartments |
| Quartz | ESCO Products<br>171 Oak Ridge Rd.<br>Oak Ridge, NJ 07438 | Quartz in various shapes and qualities |
| Reference junctions | Corning Glass Works<br>Scientific Products Div.<br>Medfield, MA 02052 | Ceramic rod for making reference electrode junctions, No. 210035 (not listed in catalog) |
| Teflon tape | Dilectrix Division<br>465 Ohio Street<br>Lockport, NY 14094 | Adhesive Teflon tape for thin-layer electrochemical cell spacers |
| Valves | Hamilton Company<br>P.O. Box 17500<br>Reno, NV 89510 | Inert stainless steel valves |

[a] A comprehensive listing of electrochemical equipment, cells, and accessories is given in the August issue of *Analytical Chemistry*, Lab Guide.

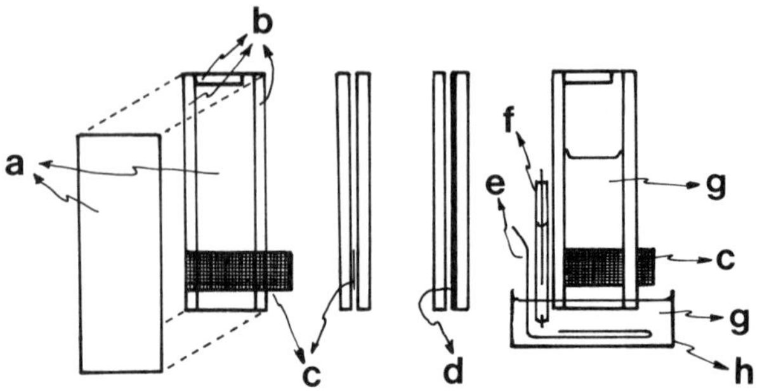


Figure 12.12 Assembly of sandwich-type optically transparent thin-layer electrochemical cell. a, Glass or quartz plates; b, adhesive Teflon tape spacers; c, minigrid working electrode; d, metal-thin-film working electrode which may be used in place of (c); e, platinum wire auxiliary electrode; f, silver-silver chloride reference electrode; g, sample solution; h, sample cup. [Adapted with permission from T. P. DeAngelis and W. R. Heineman, *J. Chem. Educ.* 53:594 (1976). Copyright 1976 American Chemical Society.]

Some attention must be paid to the electrode dimensions (see Fig. 12.12). The working electrode's lower edge should be close to the bottom of the cell plates to minimize iR-drop problems. The width of the working electrode in contact with the thin layer of solution should be small to minimize edge diffusion. As noted earlier, a vertical orientation is not desirable; however, it is convenient and compatible with the horizontal optical path of virtually all commercial spectrophotometers. Recommended sources of cell components (including minigrids) are listed in Table 12.1. Thin-layer cells for chromatographic detection and electron spin resonance spectroscopy are discussed in Chapters 22 and 24, and their application in optical studies is described in Chapter 3.

## V. CELLS FOR SPECTROELECTROCHEMISTRY

The coupling of electrochemical and spectroscopic techniques is finding increased use in analytical, kinetic, and mechanistic studies. The product of an electrode reaction or the product of a homogeneous reaction between the electrogenerated species and some material (e.g., an enzyme) present in solution can be examined using optical absorption spectroscopy. Several approaches to joining electrochemical and optical absorption techniques have been taken. In one approach the sample solution is continuously circulated from the working electrode to an optical cell where the absorbance is measured. A variety of

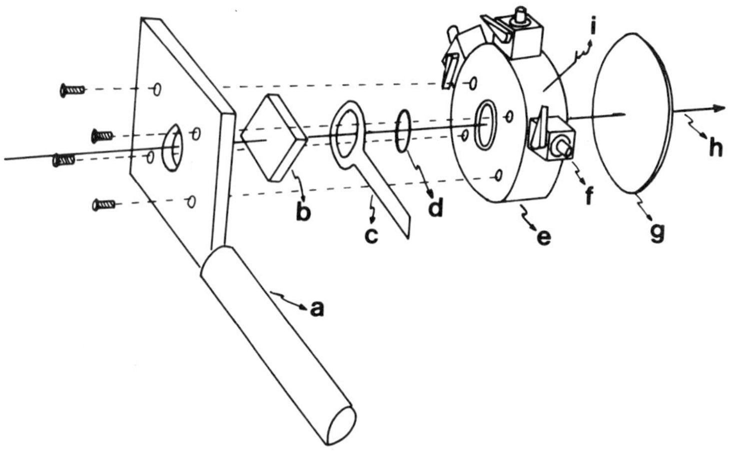


Figure 12.13 Assembly of sandwich-type optically transparent electrochemical cell. a, Face plate and cell holder; b, optically transparent working electrode; c, brass working-electrode contact; d, Viton O ring; e, Lucite cell body; f, inert valve; g, quartz back plate; h, path of optical beam; i, platinum indicating or auxiliary electrode. [From Ref. 62, adapted with permission from F. M. Hawkridge and T. Kuwana, *Anal. Chem.* 45:1021 (1973). Copyright 1973 American Chemical Society.]

designs have been described [56-60]. Possibly the simplest design employs a conventional cuvette as the electrochemical cell [58]. Classical kinetic experiments may also be performed on the electrogenerated product as long as the rate of solution mass transport is fast compared with the kinetic process under study, ensuring a uniform concentration gradient throughout solution. Determinations employing electrochemically generated titrants are also well suited to this approach.

Another approach is to employ a metal minigrid or a thin metal or semiconductor film for the working electrode (Chap. 11). These electrodes permit transmittance of the optical beam through the electrode and are readily adapted for use in thin-layer cells. Fast heterogeneous or homogeneous electron transfer reactions or homogeneous chemical reactions initiated by the electrode reaction product can best be followed via spectroelectrochemistry under semi-infinite diffusion conditions. The popular sandwich cell design that has evolved over many years in Kuwana's laboratory remains flexible and relatively simple to assemble [61]. Figure 12.13 illustrates a sandwich cell employing an optically transparent electrode (OTE). Some OTEs are available commercially with varying film conductivities and optical

Table 12.2 Commercially Available Electrochemical Cells

| Type | Source | Comments |
|---|---|---|
| General designs | Bioanalytical Systems, Inc.<br>Purdue Research Park<br>2701 Kent Ave.<br>West Lafayette, IN 47906 | Flow cells, reference electrodes, working electrodes, accessories |
| | IBM Instrument, Inc.<br>475 Main St.<br>P.O. Box 3020<br>Wallingford, CT 06492 | Voltammetry cells, electrodes, accessories |
| | Princeton Applied Research Corp.<br>P.O. Box 2565<br>Princeton, NJ 08540 | Voltammetry cells, electrolyte cleaning cells, electrodes, accessories |
| | Sargent-Welch Co.<br>7300 N. Linden Ave.<br>Skokie, IL 60076 | Wide selection of voltammetry cells |

| | | |
|---|---|---|
| Rotating cells | Pacific Precision Inst.<br>1040 Shary Court<br>Concord, CA 94518 | Rotating cells as described in this chapter |
| Jacketed cells for constant temperature | Bolab Inc.<br>359 Main St.<br>Reading, MA 01867 | Cell volumes for 2-500 mL |
| Optical cells | Helma Cells, Inc.<br>Box 544<br>Borough Hall Station<br>Jamaica, NY 11424 | Cell employing 1-cm cuvette bodies, cells amenable to sandwich design |
| | Harrick Scientific Corp.<br>Box 867<br>Ossining, NY 10562 | Platinum film optically transparent electrodes |

properties (see Table 12.2). In general OTEs are fabricated by vapor deposition on the appropriate substrate as described by Winograd in Chapter 11. For the cell shown in Figure 12.13, a brass or copper foil cut into an appropriate shape serves to make electrical contact to the conducting OTE face, minimizing the effect of film resistance. The Viton O ring seals the OTE to the cell body when the face plate is screwed into place. The deoxygenation bulb permits removal of oxygen by vacuum and nitrogen pressure. The cell body itself is made of Lucite. This cell and variations on its design have been used in the study of biological redox molecules but can also be used for other applications. Nonaqueous solvents can be used in this cell when the cell body is made from MACOR, a machinable glass-ceramic material [63].

The design principles described in the preceding sections must also be considered in configuring an electrochemical cell for spectroscopic measurements. The need to minimize variations in current density, and therefore potential, along the surface of a disk electrode suggests the use of small-diameter OTEs. Physical placement of the auxiliary electrode and the reference electrode probe must be outside of the optical beam path. Thin-metal-film OTEs are often not mechanically stable to large applied potentials and good experimental technique must be followed to prevent destruction of the OTE. The working-electrode lead should be connected last and disconnected first when starting and finishing experiments, respectively.

Internal reflection spectroscopy has been coupled with electrochemical experiments at thin-metal-film OTEs and several cells based on the sandwich design have been reported [61]. Specular reflection techniques have been similarly employed with solid electrodes to probe fundamental electrode-solution interface phenomena via vibrational as well as electronic spectroscopy. Even transparent rotating disk electrodes or opaque disk electrodes have been used for spectroscopy or photoelectrochemistry. Light diffracted by a beam of light passing parallel to a planar electrode has also been shown to increase significantly the effective path length in spectroelectrochemistry [64].

A variety of optically coupled electrochemical techniques have been recently advanced for addressing analytical and kinetic problems. The advantages and limitations of these different spectroscopic techniques as applied to particular problems have been described in other chapters in this book and in recent reviews [51-53,65].

## REFERENCES

1. J. E. Mumby and S. P. Perone, *Chem. Instrum.* 3:191 (1971).
2. B. D. Cahan, Z. Nagy, and M. A. Grenshaw, *J. Electrochem. Soc.* *119*:64 (1972).
3. J. Lindberg, *J. Electroanal. Chem.* *40*:265 (1972).
4. L. Huderova and K. Stulik, *Talanta 19*:1285 (1972).

5. K. Stulik and M. Stulikova, *Anal. Lett. 6*:441 (1973).
6. K. M. Kadish, S.-M. Cai, T. Malinski, J.-Q. Ding, and X.-Q. Lin, *Anal. Chem. 55*:163 (1983).
7. G. Park, R. N. Adams, and W. R. White, *Anal. Lett. 5*:887 (1972).
8. M. Karolczak, R. Dreiling, R. N. Adams, L. J. Felice, and P. T. Kissinger, *Anal. Lett. 9*:783 (1976).
9. J. Wang and B. A. Frelha, *Anal. Chem. 54*:334 (1982).
10. H. Lund and P. Iverson, in *Organic Electrochemistry* (M. M. Baizer, ed.), Marcel Dekker, New York, 1973, pp. 165-249, and references therein.
11. J. E. Harrar and E. L. Pomernacki, *Anal. Chem. 45*:57 (1973).
12. J. E. Harrar and I. Shain, *Anal. Chem. 38*:1148 (1966).
13. B. Miller and S. Bruckenstein, *Anal. Chem. 46*:2033 (1974).
14. R. N. Adams, *Electrochemistry at Solid Electrodes*, Marcel Dekker, New York, 1969, pp. 80-92, and references therein.
15. W. J. Albery and M. L. Hitchman, *Ring-Disc Electrodes*, Clarendon Press, Oxford, 1971.
16. R. G. Clem, F. Jakob, D. H. Anderberg, and L. D. Ornelas, *Anal. Chem. 43*:1398 (1971).
17. R. G. Clem, *Anal. Chem. 43*:1853 (1971).
18. R. G. Clem, F. Jakob, and D. H. Anderberg, *Anal. Chem. 43*:918 (1971).
19. F. L. Stephens, F. Jakob, L. P. Rigdon, and J. E. Harrar, *Anal. Chem. 42*:765 (1970).
20. J. E. Harrar, *Anal. Chem. 35*:893 (1963).
21. R. G. Clem, *Ind. Res. 15*:50 (1973).
22. R. G. Clem, *LBL Nucl. Chem. Div. Annu. Rep.* 386 (1972).
23. R. Jakob and R. G. Clem, *MPI Appl. Notes 6*:9 (1971).
24. O. H. Muller, *J. Am. Chem. Soc. 69*:2992 (1947).
25. J. Jordan, *Anal. Chem. 27*:1708 (1955).
26. R. E. Sioda, *Electrochim. Acta 15*:783 (1970).
27. J. Jordan, R. A. Javich, and W. E. Ranz, *J. Am. Chem. Soc. 80*:3846 (1958).
28. R. E. Sioda, *Electrochim. Acta 13*:375 (1968).
29. G. Koster and M. Ariel, *J. Electroanal. Chem. 33*:339 (1971).
30. W. J. Blaedel and S. L. Boyer, *Anal. Chem. 45*:258 (1973).
31. R. E. Sioda, *Electrochim. Acta 13*:1559 (1968).
32. W. J. Blaedel and J. Wang, *Anal. Chem. 51*:799 (1979).
33. W. J. Blaedel, C. L. Olsen, and L. R. Sharma, *Anal. Chem. 35*:2100 (1963).
34. W. J. Blaedel and L. N. Klatt, *Anal. Chem. 38*:879 (1966).
35. W. J. Blaedel and S. L. Boyer, *Anal. Chem. 43*:1538 (1971).
36. W. D. Mason and C. L. Olson, *Anal. Chem. 42*:488 (1970).
37. W. D. Mason and C. L. Olson, *Anal. Chem. 42*:548 (1970).
38. W. D. Mason, T. D. Gardner, and J. T. Stewart, *J. Pharm. Sci. 61*:1301 (1972).
39. J. C. Bazan and A. J. Ariva, *Electrochim. Acta 9*:17 (1964).

40. D. B. Easty, W. J. Blaedel, and L. Anderson, *Anal. Chem.* *43*:509 (1971).
41. T. O. Oesterling and C. L. Olson, *Anal.Chem.* *39*:1543 (1967).
42. T. O. Oesterling and C. L. Olson, *Anal. Chem.* *39*:1547 (1967).
43. W. R. Seitz, R. Jones, L. N. Klatt, and W. D. Mason, *Anal. Chem.* *45*:480 (1973).
44. R. E. Meyer, M. C. Banta, P. M. Lantz, and F. A. Posey, *J. Electroanal. Chem.* *30*:345 (1971).
45. H. S. Wroblowa and A. Saunders, *J. Electroanal. Chem.* *42*:329 (1973).
46. D. Danly, *Organic Electrochemistry* (M. M. Baizer, ed.), Marcel Dekker, New York, 1973, pp. 911-924, and references therein.
47. J. Wang and M. Ariel, *Anal. Chim. Acta* *99*:89 (1978).
48. B. Oosterhuis, K. Brunt, B. H. C. Westerink, and D. A. Doornbos, *Anal. Chem.* *52*:203 (1980).
49. J. L. Anderson, *Anal. Chem.* *51*:2312 (1979).
50. A. T. Hubbard and F. C. Anson, *Electroanalytical Chemistry*, Vol. 4 (A. J. Bard, ed.), Marcel Dekker, New York, 1970, pp. 129-214, and references therein.
51. T. Kuwana and W. R. Heineman, *Acc. Chem. Res.* *9*:241 (1976), and references therein.
52. W. R. Heineman, *Anal. Chem.* *50*:390A (1978), and references therein.
53. W. R. Heineman, F. M. Hawkridge, and H. N. Blount, *Electroanalytical Chemistry*, Vol. 13 (A. J. Bard, ed.), Marcel Dekker, New York, 1984, pp. 1-113, and references therein.
54. R. W. Murray, W. R. Heineman, and G. W. O'Dom, *Anal. Chem.* *39*:1666 (1967).
55. C. W. Anderson, H. B. Halsall, and W. R. Heineman, *Anal. Biochem.* *93*:366 (1979).
56. J. M. Sosonkin, A. N. Domarev, V. A. Subbotin, F. F. Lakomov, V. I. Kumantsov, and A. Ya. Kaminskii, *Elektrokhimiya* *8*:1168 (1972).
57. A. M. Gary, E. Pietmont, J. Roynette, and J. P. Schwing, *Anal. Chem.* *44*:198 (1972).
58. P. Hobza, I. Nykl, and K. Uhlir, *Coll. Czech. Chem. Commun.* *38*:994 (1973).
59. M. T. Stankovich, *Anal. Biochem.* *109*:295 (1980).
60. J. L. Anderson and J. R. Kincaid, *Appl. Spectrosc.* *32*:356 (1978).
61. T. Kuwana and N. Winograd, *Electroanalytical Chemistry*, Vol. 7 (A. J. Bard, ed.), Marcel Dekker, New York, 1974, pp. 1-78, and references therein.
62. F. W. Hawkridge and T. Kuwana, *Anal. Chem.* *45*:1021 (1973).
63. F. M. Hawkridge, J. E. Pemberton, and H. N. Blount, *Anal. Chem.* *49*:1646 (1977).
64. P. Rossi, C. W. McCurdy, and R. L. McCreery, *J. Am. Chem. Soc.* *103*:2524 (1981).
65. M. D. Ryan and G. S. Wilson, *Anal. Chem.* *54*:20R (1982).

# 13

# Solvents and Supporting Electrolytes

ALBERT J. FRY / Department of Chemistry, Wesleyan University, Middletown, Connecticut

WAYNE E. BRITTON / Programs in Chemistry, University of Texas at Dallas, Richardson, Texas



## I. INTRODUCTION

All electrochemical phenomena occur in a medium, which generally consists of a solvent containing a supporting electrolyte. An enormous number of such combinations have been reported by electrochemical experimenters; as a glance at the Appendix will show, selection of a solvent and electrolyte and purification of these has been the subject of much discussion in the literature. We consider that the inexperienced electrochemical practitioner will be best served by our recommending a few good solvent systems (for brevity's sake, we will use the term "solvent system" to describe the medium consisting of both solvent and supporting electrolyte) from the many that have been reported, rather than by an extended discussion of all such systems. We do include in the Appendix, however, a guide to further readings of a more comprehensive nature. We also believe that the novice will want to develop some feeling for those characteristics to be desired in an electrochemical solvent, as well as the ways in which electrochemical results may depend on the nature of the particular solvent and electrolyte employed in an investigation. We have for this reason examined each of these matters at some length.

## A. Solvent Selection Criteria

There is no universal solvent. Generally, a system is used whose merits outweigh its disadvantages *for a particular application.* A good solvent system for one type of experiment or compound may be totally unacceptable for other applications. One must therefore factor into one's choice of solvent (and electrolyte) some informed guesswork. Consideration must be given to how the chemical and electrochemical properties of the electrode reaction in which one is interested may be affected by the solvent system, as well as to how conditions may be optimized for the particular experiment to be conducted.

One could construct a long list of properties which the ideal electrochemical solvent system ought to exhibit. The most important are *electrochemical inertness, electrical conductivity, good solvent power, chemical inertness,* and *convenient liquid range,* although a number of others must also be considered. Let us now examine each of these in more detail.

### *Electrochemical Inertness*

In order to permit the widest latitude in application, the solvent system should not undergo any electrochemical reaction over a range of potentials from very positive (strongly oxidizing) to very negative (strongly reducing). The potential at which electrochemical reaction of the solvent system commences is known variously as the *background limit, solvent breakdown,* or *decomposition potential.* Certain systems exhibit a very wide accessible potential range; one of the best for this purpose is a solution of tetrabutylammonium hexafluorophosphate in acetonitrile, which exhibits positive and negative decomposition potentials of +3.4 and −2.9 V (SCE), respectively [1,2]. In practice, however, one rarely needs both extended positive and negative ranges in a single set of experiments, and solvents are often used which exhibit either a very negative or very positive decomposition potential, but not both. Thus both nitrobenzene [3] and methylene chloride [4] have been used for studies of oxidation processes because they are oxidized with difficulty, notwithstanding the fact that they are actually quite easily reduced: conversely, liquid ammonia [5] and methylamine [6] are good solvents for electrochemical reduction, although poor for oxidations.

Frequently, the potential limit in a given solvent system is set not by the solvent itself but by the electrochemical behavior of the supporting electrolyte. It has been known for many years, for example, that the reduction of alkali metal ions to the corresponding metals (at ca. −2 V vs. SCE) is generally the potential-limiting process in aprotic solvents, and that by use of tetraalkylammonium salts one can reach considerably more negative potentials. The decomposition potentials of tetraalkylammonium ions at mercury become more negative as the alkyl groups become larger ($Me_4N^+$: −2.65 V; $Bu_4N^+$: −2.88 V) [7] (but double-layer and adsorption effects make it unwise to draw the tempting conclusion that it is always preferable to use a large tetraalkylammonium ion because of the greater cathodic range it affords) [7,8]. At very positive potentials oxidation of the anion of the supporting electrolyte is often the limiting process; at such potentials tetrafluoroborate and hexafluorophosphate salts are preferred over the more readily oxidized ions (e.g., perchlorates or nitrates) [1].

Electrochemical reaction of the solvent or supporting electrolyte need not always be undesirable. Sometimes one can put this to good use by employing the product of such a reaction as a reagent. For example, one can generate bromine by anodic oxidation of bromide ion or lithium metal by cathodic reduction of lithium ion. There can be real advantages in this procedure as opposed to external addition of the reagents [9]. In the present context suffice it to point out the fact that in such applications the solvent decomposition potential is not critical, as long as it is not reached before electrolysis of the electrolyte begins.

### *Electrical Conductivity*

In order to support passage of an electrical current, the solvent system should have low electrical resistance, and hence a moderately high dielectric constant ($\geqslant 10$) since the prevalence of ion pairing and even multiple ionic associations in less polar solvents results in low ionic mobility and conductance in such solutions. A number of the most common solvents used in organic electrochemistry have satisfactory dielectric constants (dimethylformamide [DMF], 36.7; acetonitrile [AN], 37.5; dimethyl sulfoxide [DMSO], 46.7), and their electrolytic solutions have acceptably high conductances [10]. Solvents of substantially lower polarity than these have, however, been employed successfully in some electrochemical investigations despite their high electrical resistance (e.g., tetrahydrofuran [11] and methylene chloride [4], dielectric constants 7.3 and 8.9, respectively) [12]. It should be noted that uncompensated iR errors can become especially troublesome in high-resistance solvents.

### *Good Solvent Power*

Needless to say, an electrochemical solvent must be able to dissolve a wide range of substances at acceptable concentrations. This generally means that electrolytes must be soluble at least to the extent of 0.1 M, while the electroactive material must form approximately millimolar solutions (for application of the various electroanalytical techniques) or greater (for preparative electrolyses). Tetraalkylammonium perchlorates, hexafluorophosphates, and tetrafluoroborates generally exhibit the most satisfactory solubility as electrolytes in organic solvents; the latter two are becoming more popular than perchlorates because of safety hazards involved in the use of the latter.

### *Chemical Inertness*

The solvent should not react with the electroactive material, nor with intermediates or products of the electrode reaction under investigation. No one solvent could reasonably be expected to be unreactive toward all of the many kinds of reactive species which can be generated electrochemically. Nevertheless, one can use certain generalizations based on chemical experience. For example, it is known that many anodic organic reactions generate cationic species [13]; one would therefore not wish to use a nucleophilic solvent when it is hoped to extend the lifetime of such species sufficiently for adequate characterization. Acetonitrile is commonly used in anodic studies (vide infra) but it is moderately nucleophilic; hence attempts to characterize electrochemically generated cations may require a less reactive solvent (e.g., methylene chloride) or one of the "superacid" solvents to be discussed in Section II.A. On the other hand, reductions of organic species generally form anions, and here one must avoid the use of protic solvents to avoid acid-base reactions between the solvent and cathodically generated intermediates. Finally, one would like the solvent to be reasonably stable, so that purification, preparation, and storage of standard solutions will present no major problems.

### *Liquid Range*

The solvent should have a convenient liquid range; this again depends on the experiment involved, for example, whether it is to be carried out at temperatures well removed from room temperature, whether (and how) the solvent must be removed and solutes recovered following the experiment, the solvent purification scheme to be employed, and so on.

### *Miscellaneous Considerations*

For obvious reasons solvents and electrolytes should also be inexpensive, nontoxic, and nonflammable. The latter two characteristics are not well satisfied by most organic solvents, but with reasonable ventilation and safety precautions they can be used routinely without incident. A final solvent property, viscosity, may on occasion be of importance. High viscosities are useful in experiments in which one wishes to extend the time interval over which mass transfer is occurring solely by diffusion (e.g., potential-step experiments) but would be a disadvantage where efficient mass transfer is required, as in preparative-scale electrolyses.

## II. RECOMMENDED SOLVENTS AND ELECTROLYTES

### A. Anodic Electrochemistry

Many of the preceding desirable characteristics (low electrical resistance, low cost, ease of purification, etc.) are easily met by solutions of inorganic salts in water, to which the vast polarographic literature bears ample testimony. Much inorganic oxidative and reductive electrochemistry has been carried out in water with little difficulty. Of course, organic and organometallic compounds generally either are not very soluble in water or are reactive toward it; such considerations (and others) have stimulated interest in the use of nonaqueous solvents in electrochemistry. Ethanol and methanol are good organic solvents, inexpensive, readily available in high purity, dissolve salts readily, and in fact generally have much to commend them for electrochemical purposes. However, because alcohols, like water, are rather easily oxidized and are highly nucleophilic, it has been more common to use aprotic organic solvents for electrochemical oxidations. Acetonitrile (AN) is the most widely used organic solvent for anodic electrochemistry at the present time. *We believe it to be the preferred solvent for most purposes.* It can be purified (Sec. V), is a good solvent for organic compounds and many inorganic and organic salts, has a high dielectric constant, is transparent in the ultraviolet region, and has a convenient liquid range (−45.7-81.6°C). Recommended electrolytes include the tetrabutylammonium hexafluorophosphate (TBAHFP) or tetrafluoroborate (TBATFB) salts, both of which are readily obtained in pure form (Sec. V), have high solubilities in AN, and are stable to very positive and negative potentials. It is noteworthy that these salts, particularly the hexafluorophosphate, are rather insoluble in water, which means that they are nonhygroscopic, and can often be separated from reaction products by evaporation of the AN, followed by appropriate treatment (e.g., extraction with a nonpolar solvent, chromatography, or volatilization of the desired product), although sometimes separation of the electrolyte from an organic product presents difficulties. An inorganic electrolyte (e.g., $NaBF_4$ or $NaClO_4$) may be necessary in such cases. Very dry AN can be made by addition of a small amount of trifluoroacetic anhydride [14] or an alumina suspension [15] to the solvent.

The nucleophilicity of AN sometimes militates against its use when one wishes to extend the lifetime of anodically generated cations; several choices present themselves for this application: TBAHFP or TBATFB in methylene chloride [4,16] or liquid sulfur dioxide [17], trifluoroacetic acid (no supporting electrolyte necessary) [18], or an aluminum chloride-alkali halide melt [19]. Such intermediates may also be stabilized by operation at low temperatures [20] or in very dry AN [14,15]. Methylene chloride is useful in organometallic electrochemistry in situations where acetonitrile is found to displace ligands coordinated to a metal center.

### B. Cathodic Electrochemistry

Dimethylformamide, acetonitrile, and dimethyl sulfoxide are all excellent solvents for reductive electrochemistry of organic compounds; again, water and alcohols are useful for inorganic electrochemistry. *TBAHFP and TBATFB* are good electrolytes for use in organic solvents. *Dimethyl sulfoxide is a particularly useful solvent for cathodic electrochemistry,* although it has a somewhat inconvenient liquid range (18-189°C). It is quite pure as received, and is an excellent solvent for both organic and inorganic compounds. Water (which is always present in dipolar aprotic solvents) is a poor proton donor in DMSO. For most applications the small amount of water in DMSO as received has no observable effect on voltammetric measurements. The best commercial-grade DMSO need not be further purified for most electrochemical studies, but purification is not difficult. DMSO is synthesized by oxidation of dimethylsulfide; probably the only serious contaminants are dimethylsulfide and water, which can be removed by vacuum distillaton and drying over molecular sieve or activated alumina. *TBAHFP and TBATFB are two of the many satisfactory electrolytes for use in DMSO.* We have observed irregular drop-time behavior during polarography in DMSO; it is possible that this is due to strong adsorption of DMSO or dimethylsulfide on the mercury surface.

Probably the most widely used organic solvent for cathodic chemistry has been DMF. It has a number of desirable properties; at the same time it is quite toxic and undergoes undesirable reactions under fairly mild conditions (e.g., storage). (Thus it can react with traces of base to liberate formic acid and dimethylamine, which catalyze further decomposition.) *We feel that AN and DMSO are preferable to DMF as solvents for most applications.*

## III. SOME OTHER SOLVENTS

Many solvents (and supporting electrolytes) have been used in electrochemical applications. In the preceding sections we suggested a few solvents for general use. It may happen that none of these is adequate for a given application. Confronted with this situation, one may turn to the many extended discussions of solvent systems already available in the literature. We shall not attempt to duplicate those discussions here (but see the Appendix). A few solvents deserve special mention here, however, because each has a property likely to be useful in certain applications.

*Hexamethylphosphoramide* (HMPA) [21], *ammonia* [5,22] and *methylamine* [6] are excellent solvents for electrochemical generation of solvated electrons. Lithium chloride is recommended as the supporting electrolyte for this application. The properties of HMPA are otherwise generally similar to those of DMF and DMSO.

*Tetrahydrofuran* (THF) and other ether solvents are useful aprotic solvents for cathodic electrochemistry in spite of their low dielectric constants, because they can be dried more efficiently than the more polar solvents. Anderson has claimed that, in fact, water levels less than $10^{-5}$ M can be attained in THF [23]. THF and glyme (1,2-dimethoxyethane) [24] both dissolve salts more readily than does, for example, diethyl ether. Lithium salts are generally very soluble in ether solvents.

*Low-temperature applications* present special problems. With most solvents, solvating power diminishes and viscosity increases as the temperature is lowered. The solvent may even solidify—DMSO is, in fact, useless for low-temperature applications because of its rather high melting point (18.6°C). Reilley and Van Duyne surveyed a series of solvents in light of these problems [20]. They concluded that *butyronitrile* suffers the least degradation of solvent properties with decreasing temperature and recommended it for use at temperatures as low as −105°C.

## IV. SOLVENT AND ELECTROLYTE-DEPENDENT PHENOMENA

Electrochemical reactions, occurring in solution as they do, are necessarily influenced by the nature of their surroundings, that is, the solvent and supporting electrolyte (and other components of the medium too, of course, but these are not our province here). We have already alluded at several points to the possibility of chemical reaction of electrochemically generated intermediates with the solvent. One must always be on guard against this possibility, which sometimes may be unavoidable due to the solvent's being in large excess in the medium. The nature of the problem with highly reactive intermediates is exemplified by a recent study of the electrochemical reduction of iodobenzene in DMF containing tetraethylammonium bromide [25]. A variety of experiments demonstrated that the highly basic and nucleophilic phenyl anion was generated in this reaction. Benzene was the major product, formed by attack on traces of water in the solvent. When the water content was lowered, benzene was still the major product, this time by proton abstraction (Hoffmann elimination) from the supporting electrolyte. When the electrolyte was changed to lithium bromide, to prevent the Hoffmann elimination, the phenyl anion was found to react with the solvent by a combination of proton abstraction and nucleophilic addition. Clearly, electrochemical generation of long-lived phenyl anions would require changing to a more felicitous nonelectrophilic, nonacidic solvent (perhaps an ethereal solvent containing a metal salt as electrolyte). The anodic counterpart of this problem has already been mentioned: the fact that anodically generated cations (electrophiles) will attack nucleophilic solvents such as acetonitrile, requiring selection of electrophilic solvents when it is desired to extend the lifetime of such cations. All of the special solvent systems mentioned at the end of Section II meet this criterion.

There are a variety of other ways in which the solvent and electrolyte can affect an electrode process other than simply by direct chemical reaction. Rates of heterogeneous electron transfer from the electrode to the electroactive material can be affected by the structure of the electrical double layer at the electrode surface, which, in turn, is affected by, among other things, the nature of the solvent and supporting electrolyte [26]. The electrolyte can play other roles when incorporated into the double layer. It appears, for example, that some weakly acidic cations, such as triethylammonium ion and guanidinium ion, when incorporated into the double layer, can protonate intermediates generated at the electrode surface before these intermediates can escape into bulk solution [27]. Similarly, preferential solvation of small ions by one constituent of a mixture can increase the effective concentration of this constituent at the electrode surface if the ion is in the double layer [anions at potentials positive of the potential of zero charge (PZC), cations at potentials negative of PZC]. Examples of both cathodic [27a] and anodic [28,40] electrochemistry thought to be affected in this way have been reported. In recent years a third type of surface electrolyte effect has been discovered: optically active products have been produced in a number of reactions where an optically active supporting electrolyte has been employed [29]. Clearly, a chiral environment has been generated at the electrode surface in such reactions. It is a challenging task to develop a picture of what is happening in such reactions on the molecular level. Horner has begun to assemble data that may lead eventually to such a picture [29].

Probably the most common electrochemical effects exerted by solvents and electrolytes in bulk solution are related to association (solvation, ion pairing, complex formation, etc.) with the electroactive species or electrochemically generated intermediates [30]. The importance of solvation is exemplified by comparison of calculated and measured values of the parameter $\Delta E_{1/2}$, defined as the difference in volts between the half-wave potentials of the first and second polarographic waves exhibited by aromatic hydrocarbons in dipolar aprotic solvents. It can be shown [31] that $\Delta E_{1/2}$ is related to the equilibrium constant for the disproportionation of the aromatic radical anion into neutral species and dianion, that is,

$$2ArH^{\overline{\cdot}} \underset{}{\overset{K}{\rightleftharpoons}} ArH + ArH^{2-} \tag{13.1}$$

The computed value of $\Delta E_{1/2}$ by a reliable computational method [32] is ca. 5 V [33]; measured values are generally only about one-tenth of this value (0.4-0.6 V) in a variety of dipolar aprotic solvents such as DMF, AN, and DMSO, indicating substantial perturbation of the disproportionation equilibrium by solvation. (The computed value is essentially a gas-phase, solvent-free value). It also happens that the disproportionation equilibrium constant is dependent on the nature of the supporting electrolyte [31,33], demonstrating that ion pairing is operative even in such highly polar solvents. In solvents of low dielectric constant, where ion pairs and even higher ion aggregates are very prevalent, or with electrolytes (e.g., lithium) which form strong covalent bonds, association between the electrolyte and charged species plays an even greater role in affecting measured potentials. In the specific case of radical anion disproportionation under discussion here, $\Delta E_{1/2}$ can even become negative; that is, a single two-electron polarographic wave is observed [34].

## V. EXPERIMENTAL PROCEDURES

### A. General

Water is the most ubiquitous impurity in organic solvents. Water is removed rather efficiently from most solvents by allowing the solvent to stand over activated 4A molecular sieve. A series of experiments in which the water content of a sample of DMF over molecular sieve was monitored by gas chromatography as a function of time demonstrated that several days may be necessary to reach the lowest water levels [35]. Presumably, other viscous solvents, such as DMSO, will take similarly long times for complete drying. Dipolar aprotic solvents are generally very hygroscopic, and contact with the atmosphere should be avoided following drying. The best procedures for obtaining very dry solvents involve distillation into the electrochemical cell in vacuo [36] (Chap. 14), or in a dry box (Chap. 15); a simpler and almost as satisfactory alternative involves drying the solvent over a molecular sieve in a dropping funnel attached to the cell, followed by direct addition to the cell [35]. For work at very negative potentials, a few cautions should be noted: alkali metal ions in the sieve can be exchanged for the electrolyte cation [37], and residual currents can be higher in solvent dried over molecular sieve [35]. The latter problem seems to be due to fine particles adhering to the sieve, since it can be alleviated by distillation of the solvent from the sieve or by washing the sieve with solvent and then activating at high temperature before use. Water in organic solvents can be determined by a number of methods. One of the fastest and most sensitive involves gas chromatographic analysis on columns packed with Porapak Q [38a] or N [38b]. Addition of activated alumina directly to the voltammetric cell is a convenient in situ final purification and drying procedure for solvents such as AN, DMSO, and DMF [15]. Experimental procedures have been devised [39] to do this, yet avoid direct contact between the electroactive substance and alumina.

Some general advice may be in order here concerning the nature and use of drying agents with organic solvents:

1. Drying agents differ in their mode of action. Water levels are lowest after drying with those agents that destroy water chemically [e.g., lithium aluminum hydride (LAH), calcium hydride, phosphorus pentoxide, sodium, etc.] rather than by physical action (molecular sieves and metal salts such as magnesium or copper sulfates). LAH affords very dry solvents, but its use is restricted to chemically inert solvents such as ethers. Calcium hydride is a good drying agent, although not as good as LAH, and is widely used for producing quite dry solvents (<1 mM in water). In our experience water levels in solvents dried over molecular sieve are 1-10 mM unless long contact times and well-activated sieve are employed. Eberson and Olofsson have reached a similar conclusion [40].
2. For safety reasons, solvents should receive a preliminary treatment with a less efficient drying agent before coming in contact with the more reactive chemical drying agents, unless one is assured that the initial water levels are low. (Commercial DMF is ca. 0.1 M in water; other solvents can be as wet.)
3. Many solvents, especially but not only ethers, contain peroxides. These can be removed by treatment with LAH (caution!) or passage through a column of activated alumina.

## B. Solvent Purification Procedures

In view of the more extended discussions presented in the references cited in the Appendix, we present here only selected procedures for purification of materials for general use, and only for those which we advocated in Section II, since the large majority of applications will be satisfied by one or another of these.

*Acetonitrile (AN)*

Many procedures for the purification of AN have been reported [37-41]. Problems in purification of this solvent appear to stem at least in part from variability in composition of the material as received from commercial suppliers. Common impurities are water and acrylonitrile. Kuwana and co-workers [41a] surveyed a variety of procedures for purification of AN, and concluded that the best procedure involved simply passing the solvent twice through a column packed with alumina previously dried at 400°C. We have used this procedure and have found it quite satisfactory for general use as long as the alumina is replaced regularly (when voltammetric background currents begin to increase). Anderson et al. recommend that the alumina treatment be followed by distillation from calcium hydride to obtain highest-purity material [41b]. A recent four-step procedure has been proposed and claimed to produce very high quality acetonitrile [41d].

*Dimethyl Sulfoxide (DMSO)*

This solvent is fairly pure as received. Water can be removed by allowing the solvent to stand over 4A or 5A molecular sieve (activation of the sieve at 500°C for 16 h in an inert atmosphere has been recommended [42]). Vacuum distillation removes the other likely impurity, dimethyl sulfide. This is also removed by bubbling nitrogen through the solvent for some time. The polarographic behavior of aromatic hydrocarbons in DMSO was found to be identical, in commercial DMSO used as received, with that in the purified material; in fact, intentional addition of substantial amounts of water to the solvent had no effect [33].

*Dimethylformamide (DMF)*

The principal impurities in commercial DMF are water, dimethylamine, and formic acid. Distillation from calcium hydride in vacuo (aspirator pressure) affords material of quality adequate for most purposes. For special applications, the residual water content may be substantially lowered by treatment with activated alumina [15,39].

*Tetrahydrofuran (THF)*

THF is chemically inert, hence can be obtained in a high state of purity by treatment with potent reagents which will destroy both peroxides and water. Distillation from sodium benzophenone [43] or lithium aluminum hydride [10] produces high-quality solvent, which must then be stored under nitrogen because of the facile formation of peroxides when the solvent is in contact with air. Similar purification procedures can be used with other ether solvents.

*Methylene Chloride*

This solvent is commercially available in a high state of purity. It is adequately purified by passage through a column of neutral alumina immediately before use [4].

*Ethanol*

This solvent is also available commercially in pure form, and can be used as received for most purposes. Distillation from magnesium ethoxide affords very dry material [44].

*Water*

Double-distilled water is adequate for almost all purposes, although more extensive purification might be necessary in rigorous studies because of the possible presence of organic impurities at very low concentration.

## C. Preparation and Purification of Electrolytes

*Tetrabutylammonium hexafluorophosphate (TBAHFP)* is a good general-purpose electrolyte: it exhibits good solubility in a variety of solvents, including methylene chloride [16] and liquid sulfur dioxide [17], is nonhygroscopic, has a wide accessible potential range, and is readily prepared in a high state of purity by the following procedure [45]. Tetrabutylammonium iodide, 100 g (0.27 mol) is dissolved in the minimum quantity of reagent acetone. To this is added, with stirring, an acetone solution containing ammonium hexafluorophosphate, 50 g (0.31 mol). It is important to maintain at least a 5% excess of hexafluorophosphate ion. The acetone mixture is filtered to remove the precipitate of ammonium iodide. Water is then added slowly to precipitate the TBAHFP, which is collected by filtration and washed several times with water to remove all ammonium salts. The crude product is again precipitated from a 5% solution of ammonium hexafluorophosphate to ensure complete removal of iodide. The white solid is recrystallized from ethanol-water and dried in vacuo at 100°C for 10 h. The yield is ca. 80%.

*Tetrabutylammonium tetrafluoroborate (TBATFB)* is another excellent electrolyte for general use. It is somewhat more soluble in both organic solvents and water than the corresponding hexafluorophosphate, a characteristic it exhibits in common with other tetrafluoroborate salts. House and co-workers have described procedures for the synthesis of tetraalkylammonium tetrafluoroborates [10]; the following procedure for the synthesis of TBATFB is representative. A solution of 8.4 g (0.025 mol) of tetrabutylammonium bromide (Eastman) in a minimum volume of water (ca. 18 mL) is treated with 3.6 mL (ca. 0.026 mol) of aqueous 48-50% tetrafluoroboric acid (Allied Chemical).

The resulting mixture is stirred at 25°C for 1 min; the crystalline salt is then collected by filtration, washed with water until the washings are neutral, and dried. The crude salt, 6.3 g (79%), m.p. 155-157°C, can be recrystallized three times from ethyl acetate-pentane to afford 6.0 g (75%) of TBATFB, m.p. 162-162.5°C. A similar procedure is employed for the synthesis of tetramethylammonium tetrafluoroborate [58]. This compound ought to be useful when the NMR spectra of electrolysis solutions are to be examined directly, because of its simple spectrum.

## APPENDIX: OTHER LITERATURE

One of the most comprehensive reviews on solvent use and purification was prepared by C. K. Mann [12]. Many solvents are discussed, and for each particular solvent suitable supporting electrolytes and reference electrodes are presented. This is certainly an excellent starting point for an in-depth study of the use of solvents. Lund and Iversen [46], among others [47], have also surveyed the use, purification, and properties of solvents. Although there is much overlap among these articles, each has its own flavor and the different viewpoints make for instructive comparisons. There are also a number of articles devoted to the use of specific solvents (e.g., DMSO [48], DMF [49], methylene chloride [50], and pyridine [51]).

As we have suggested earlier, tetraalkylammonium hexafluorophosphates and tetrafluoroborates are displacing the corresponding perchlorates as the most useful supporting electrolytes for general use. There does not appear to be any reason to use the perchlorates, in view of the potential safety (explosion) hazard associated with their use [27b,52,53]. A number of methods for preparation of tetraalkylammonium tetrafluoroborates and hexafluorophosphates have been reported [10,11,16,39,54]; tributylethylammonium tetrafluoroborate appears particularly attractive for use in tetrahydrofuran because of its good solubility in that solvent [11].

Tetraalkylammonium tosylates [55] and trifluoromethanesulfonates [53] also exhibit excellent properties as supporting electrolytes. Tetraalkylammonium salts are most often employed of these anions and the anions mentioned above, because such cations have excellent solubility properties and extended potential limits. For special applications other cations are sometimes used [e.g., methyltriphenylphosphonium (whose tosylate happens to be freely soluble in methylene chloride) [55] or metal ions], where undesirable reactions of the tetraalkylammonium ion might otherwise occur [12,56]. The properties of many electrolytes suitable for nonaqueous use have been surveyed [57].

A barely explored idea in organic electrochemistry has been that of incorporating the functions of solvent and supporting electrolyte into a single substance by use of a low-melting organic salt. Tetrahexylammonium benzoate, a liquid at room temperature, has indeed been found suitable for electrochemical use [58]. It is likely that better candidates for this purpose could be found among the many low-melting organic salts now available [59].

## ACKNOWLEDGMENTS

AJF acknowledges the financial support of the National Science Foundation, and WEB, the financial support of the Robert A. Welch Foundation.

## REFERENCES

1. M. Fleischmann and D. Pletcher, *Tetrahedron Lett.*, 6255 (1968).
2. R. L. Krieger, unpublished observation, 1974.
3. L. Marcoux, P. Malachesky, and R. N. Adams, *J. Am. Chem. Soc. 89*:5766 (1967).
4. O. Hammerich and V. D. Parker, *J. Am. Chem. Soc. 96*:4289 (1974).
5. (a) A. J. Bard and A. Demortier, *J. Am. Chem. Soc. 95*:3495 (1973); (b) W. H. Smith and A. J. Bard, *ibid.*, *97*:5203 (1975); (c) O. R. Brown, *Specialist Periodical Reports in Electrochemistry*, Vol. 4, Chemical Society, London, 1974, p. 55ff.
6. R. A. Benkeser and S. J. Mels, *J. Org. Chem. 35*:361 (1970).
7. A. J. Fry and R. L. Krieger, *J. Org. Chem. 41*:54 (1976).
8. S. R. Missan, E. I. Becker, and L. Meites, *J. Am. Chem. Soc. 83*:58 (1961).
9. A. J. Fry, *Synthetic Organic Electrochemistry*, Harper & Row, New York, 1972, p. 149ff.
10. H. O. House, E. Feng, and N. P. Peet, *J. Org. Chem. 36*: 2371 (1971).
11. T. J. Curphey, C. W. Amelotti, T. P. Layloff, R. L. McCartney, and J. H. Williams, *J. Am. Chem. Soc. 91*:2817 (1969).
12. C. K. Mann, in *Electroanalytical Chemistry*, Vol. 3 (A. J. Bard, ed.), Marcel Dekker, New York, 1969, pp. 57-134.
13. Ref. 9, Chap. 8.
14. U. Svanholm and V. D. Parker, *J. Am. Chem. Soc. 96*:1234 (1974).
15. (a) O. Hammerich and V. D. Parker, *Electrochim. Acta 18*:537 (1973); (b) B. S. Jensen and V. D. Parker, *J. Am. Chem. Soc. 97*:5211 (1975); (c) D. M. LaPerriere, W. F. Carroll, Jr., B. C. Willett, E. C. Torp, and D. G. Peters, *ibid. 101*:7561 (1979).

16. L. Byrd, L. L. Miller, and D. Pletcher, *Tetrahedron Lett.*, 2419 (1972).
17. L. L. Miller and E. A. Mayeda, *J. Am. Chem. Soc. 92*:5818 (1970).
18. (a) Hammerich, N. S. Moe, and V. D. Parker, *J. Chem. Soc., Chem. Commun.*, 156 (1972); (b) U. Svanholm and V. D. Parker, *Tetrahedron Lett.*, 471 (1972).
19. V. R. Koch, L. L. Miller, and R. A. Osteryoung, *J. Org. Chem. 39*:2416 (1974).
20. R. P. Van Duyne and C. N. Reilley, *Anal. Chem. 44*:142, 153, 158 (1972).
21. L. A. Avaca and A. Bewick, *J. Chem. Soc. Perkin Trans. 2*, 1712 (1972).
22. O. R. Brown and P. O. Stokes, *J. Electroanal. Chem. 57*:425 (1974).
23. J. L. Mills, R. Nelson, S. G. Shore, and L. B. Anderson, *Anal. Chem. 43*:157 (1971).
24. R. E. Dessy and R. L. Pohl, *J. Am. Chem. Soc. 90*:1995 (1968).
25. (a) J. W. Sease and R. A. de la Torre, *J. Am. Chem. Soc. 101*:1687 (1979); (b) R. Alvarado, M. A. thesis, Wesleyan University, 1975.
26. D. M. Mohilner, in *Electroanalytical Chemistry*, Vol. 1 (A. J. Bard, ed.), Marcel Dekker, New York, 1966, pp. 241-409.
27. (a) A. J. Fry and R. G. Reed, *J. Am. Chem. Soc. 94*:8475 (1972); (b) R. Breslow and R. F. Drury, *ibid. 96*:4702 (1974). *Caution*: The electrolyte guanidinium perchlorate, employed by the latter workers, is explosive and was in fact surveyed for possible military use by Nernst in 1915: K. Mendelssohn, *The World of Walther Nernst*, University of Pittsburgh Press, Pittsburgh, Pa., 1973, p. 90. Substitution of a more stable anion (see the Appendix) is advised.
28. K. Nyberg, *J. Chem. Soc., Chem. Commun.*, 774 (1969).
29. L. Horner and D. Degner, *Electrochim. Acta 19*:611 (1974), and references therein.
30. L. A. Avaca and A. Bewick, *J. Electroanal. Chem. 41*:405 (1973), and references therein.
31. A. J. Fry, L. L. Chung, and V. Boekelheide, *Tetrahedron Lett.*, 445, 1354 (1974).
32. M. J. S. Dewar and C. de Llano, *J. Am. Chem. Soc. 91*:789 (1969).
33. A. J. Fry, C. S. Hutchins, and L. L. Chung, *J. Am. Chem. Soc. 97*:591 (1975).
34. H. Lehmkuhl, S. Kintopf, and E. Janssen, *J. Organomet. Chem. 56*:41 (1973).
35. J. W. Sease and R. C. Reed, personal communication, 1971.
36. A. J. Bard, *Pure Appl. Chem. 25*:345 (1971).

37. J. P. Billon, *J. Electroanal. Chem. 1*:486 (1960).
38. (a) M. Walter and L. Ramaley, *Anal. Chem. 45*:165 (1973); (b) L. C. Portis, J. C. Roberson, and C. K. Mann, *ibid. 44*:294 (1972).
39. (a) H. Kiesele, *Anal. Chem.* 53:1952 (1981); (b) R. Lines, B. S. Jensen, and V. D. Parker, *Acta Chem. Scand. 32B*: 510 (1978).
40. L. Eberson and B. Olofsson, *Acta Chem. Scand. 23*:2355 (1969).
41. (a) T. Osa and T. Kuwana, *J. Electroanal. Chem. 22*:389 (1969); (b) J. L. Mills, R. Nelson, S. G. Shore, and L. B. Anderson, *Anal. Chem. 43*:157 (1971); (c) J. F. Coetzee and C. D. Ritchie, *Recommended Purification Methods for Selected Solvents*, Marcel Dekker, New York, 1969, p. 607; (d) H. Kiesele, *Anal. Chem. 52*:2230 (1980).
42. J. N. Butler, *J. Electroanal. Chem. 14*:89 (1967).
43. See, for example, A. J. Gordon and R. A. Ford, *The Chemist's Companion*, Wiley, New York, 1972, p. 439.
44. L. F. Fieser, *Experiments in Organic Chemistry*, D. C. Heath, Lexington, Mass., 1955, p. 286.
45. J. A. Ferguson, *Interface 6*(2) (1970).
46. H. Lund and P. Iversen, in *Organic Electrochemistry* (M. Baizer, ed.), Marcel Dekker, New York, 1973, pp. 208-230.
47. (a) J. A. Riddick and W. B. Bunger, *Organic Solvents*, 3rd ed., Wiley, New York, 1970; (b) Ref. 43, p. 429ff.
48. (a) J. N. Butler, *J. Electroanal. Chem. 14*:89 (1967); (b) T. R. Koch and W. C. Purdy, *Talanta 19*:989 (1972); (c) T. B. Reddy, *Pure Appl. Chem. 25*:459 (1972).
49. (a) D. S. Reid and C. A. Vincent, *J. Electroanal. Chem. 18*: 427 (1968); (b) C. D. Ritchie and G. H. Megerle, *J. Am. Chem. Soc. 89*:1447 (1967); (c) M. Bréant and G. Demange-Guérin, *Bull Soc. Chim. Fr.*, 2935 (1969).
50. D. Coutagne, *Bull. Soc. Chim. Fr.*, 1940 (1971).
51. D. A. Hall and P. J. Elving, *Anal. Chim. Acta 39*:141 (1967).
52. *Handbook of Laboratory Safety*, 2nd ed., CRC Press, Cleveland, Ohio, 1971, pp. 268-269.
53. K. Rousseau, G. C. Farrington, and D. Dolphin, *J. Org. Chem. 37*:3968 (1972).
54. N. S. Moe, *Acta Chem. Scand. 19*:1023 (1965).
55. M. M. Baizer, *J. Electrochem. Soc. 111*:215 (1964).
56. J. L. Webb, C. K. Mann, and H. M. Walborsky, *J. Am. Chem. Soc. 92*:2042 (1970).
57. C. J. Janz and R. P. T. Tomkins, *Nonaqueous Electrolytes Handbook*, Academic Press, New York, Vol. I, 1972; Vol. II, 1974.
58. C. G. Swain, A. Ohno, D. K. Roe, R. Brown, and T. Maugh II, *J. Am. Chem. Soc. 89*:2648 (1967).
59. (a) J. E. Gordon, *J. Am. Chem. Soc. 87*:1499, 4347 (1965); (b) J. E. Gordon, *J. Org. Chem. 30*:2760 (1965); (c) W. T. Ford, *ibid. 38*:3916 (1973).

# 14

# Vacuum Line Techniques

CSABA P. KESZTHELYI / Department of Chemistry, Louisiana State University, Baton Rouge, Louisiana



## I. INTRODUCTION

The proper use of nonaqueous solvents in electrochemistry often requires a vacuum line either in the preparative stages or throughout an experiment. This chapter is devoted mainly to how rather than why a vacuum line is used. Accordingly, we may dismiss the why by pointing out that oxygen and water can participate in many electrochemical processes and may also contribute to solvent decomposition.

The subject of vacuum lines is approached here from the vantage point of the novice user who may wish to construct a unit in close association with a glassblower. The literature in this area is vast [1] and there are many commercial firms that specialize in vacuum technology.

Electrochemists remove dissolved oxygen from their cells by *degassing, deaerating, deoxygenating, or sparging*. These terms, although they appear to be self-explanatory, are frequently used improperly. Deaerating often implies only bubbling (sparging) some gas such as pure $N_2$ or Ar through the solution via a fine capillary or glass frit. In effect, such a procedure merely serves to deoxygenate the medium. Atmospheric pressure is maintained but the partial pressure of oxygen in the system is reduced to near zero. Deoxygenation can also be accomplished by adding a reducing agent to a sealed system, but this is rarely done in practice. In contrast to these methods, degassing utilizes a vacuum technique such as "freeze-pump-thaw" which serves to remove all gases. The principal advantages of vacuum line methods are thorough degassing (or "outgassing") and effective long-term isolation from the atmosphere. Relatively few electrochemists work with compounds which are themselves sensitive to oxygen and moisture. On the other hand, the intermediates and products of electrode reactions are often highly reactive and are usually present at concentrations low enough to make small amounts of water and oxygen of critical importance.

## II. VACUUM LINE

Figure 14.1 shows a vacuum line that is relatively simple and has been successfully used in nonaqueous electrochemical studies. When high-quality commercially available $N_2$ is used, the sections labeled "$Mg(ClO_4)_2$" and "hot Cu" can be omitted in most situations. The number of stopcocks should be kept to a minimum—they are not only expensive, but all leak to some extent. If some special experiment requires additional outlets, it is best to extend the vacuum line with a small portable segment. When ground-glass stopcocks are used, they should be regularly greased with "high-vacuum" lubricant (e.g., Dow-Corning silicone or Apiezon hydrocarbon, the latter being somewhat less resistant to organic solvents). After experiments where good lubricant solvents such as tetrahydrofuran (THF) or methylene chloride (MC) are vapor transferred (cf. Sec. III), the stopcocks should always be checked and regreased, if necessary. "Sticky" stopcocks must never be tolerated, since the uneven friction can either break them or cause them to develop permanent leaks.

Cleaning a vacuum line is facilitated if the line is sloping down toward the open end. By keeping the first 4-mm stopcock (the one that isolates the line from the gauge and pumps) closed, and replacing the other 4-mm stopcocks with appropriate corks, soap solution or cleaning solvents such as ethanol may be introduced through the space of the first 2-mm stopcock. Several such rinsings will usually clean the line sufficiently. A clean line minimizes the opportunity for sample contamination and decreases the time required to pump the system down. Once the line has been opened to the atmosphere, a significant amount of water will be adsorbed onto the glass surface. Removal of

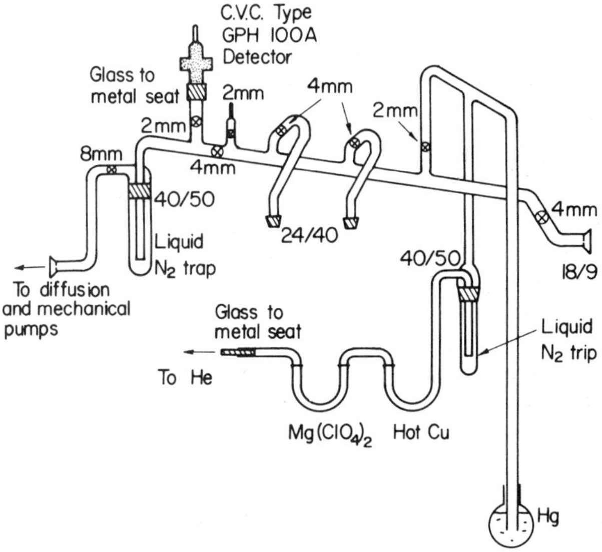


Figure 14.1 Vacuum line for electrochemical studies. The electrochemical cell attaches to the ball joint ($\frac{S}{J}$) at the right or at one of the tapered joints ($\frac{T}{S}$) near center. Purified helium can be introduced by the metal line at the bottom left. (From Ref. 2.)

adsorbed water is greatly accelerated if the line is carefully warmed with a heat gun. In the following sections the components of the vacuum line are discussed in more detail.

## A. Pumps

A mechanical forepump (or "roughing pump") and a different pump present a suitable combination for the modest vacuum ($10^{-5}$-$10^{-6}$ torr) usually needed by an electrochemist. The two pumps should be linked by special rubber hose (2 in. o.d., 3/4 in. i.d.) fastened by chain clamps. An oil diffusion pump is safer than a Hg diffusion pump, and it can be conveniently linked to the line by a ball-and-socket joint fastened by clamps. The pumping rate of the diffusion pump is normally adjusted with a variable transformer controlling the heater. For vacuum distillation of large volumes of solvents from drying agents and the like, a separate "prep line," without the diffusion pump, is most convenient.

The oil should be changed in the mechanical pump at regular intervals, and also whenever it becomes contaminated with solvents. The latter should be an infrequent occurrence if the cold trap(s) are faithfully attended. It is very bad practice to pump on solvents in an unfrozen condition, because, for example, DMF attacks the seals of inexpensive mechanical pumps. The drive belt should be periodically checked for cracks and replaced before a failure occurs.

In the case of the diffusion pump, if circulating water is used for cooling, accumulated deposits in the cooling section should be removed periodically by disconnecting the water and flushing with a suitable liquid (dilute $HNO_3$ seems to work). A small copper wire placed in the cooling water will keep down algae growth and deposits. If the mechanical forepump is shut off while the system is under vacuum, pump oil may slowly back up and flood the diffusion pump. To avoid this, air or an inert gas should be let in immediately after the mechanical pump is shut off. Most modern pumps have a built-in positive antisuckback valve which seals the pump from the atmosphere in the event of a power failure, thus eliminating the possibility of oil being drawn into the system. To avoid contact of the hot fluid in the diffusion pump with air, the heater should be turned off well in advance to allow it to cool. When starting up the pumps, the cold diffusion pump should be evacuated for several hours before its heaters are turned on.

Once the vacuum line has been brought up to atmospheric pressure and the pumps turned off, there is no longer a need to keep the traps cold and dewars should be removed. A liquid-nitrogen trap can condense oxygen from the air. The liquid oxygen thus collected (usually unknowingly) can result in an explosive mixture with other condensates.

To avoid unnecessary contamination of the diffusion pump, the arrangement shown in Figure 14.1 has been modified in the author's present laboratory; the difference is that preliminary evacuation and freeze-pump-thaw takes place through a bypass connection. In relation to Figure 14.1, the bypass starts below the 2-mm "vent" stopcock, has its own 8-mm stopcock and cold trap, and leads to the mechanical pump while the vacuum gauge and diffusion pump are closed off temporarily; this operating procedure has no adverse effect on the oil diffusion pump itself. In order to minimize the weight of the connections to the mechanical pump, 3/4-in.-o.d. copper tubing, secured to the supporting structure, can be used for most of the length, with rubber hose providing the connections to the pumps and vacuum line.

## B. Vacuum Gauge

The vacuum gauge is an important part of the vacuum line insofar as numbers are quoted from it in describing experimental conditions. There are many types of vacuum gauges described in the literature. In the following discussion several common vacuum gauges are briefly mentioned, followed by a somewhat more detailed description of the gauge used in conjunction with the vacuum line of Figure 14.1.

There are various mechanical gauges (e.g., Bourdon) which are sensitive to a pressure difference between two chambers, but these do not provide the type of information usually needed by experimental electrochemists. The McLeod [3] and Knudsen [4] gauges are absolute manometers when properly constructed. The former utilizes the compression of a definite volume of residual gases to produce sufficient pressure to be measured with a hydrostatic head of mercury and an ordinary scale, whereas the latter measures the momentum transferred by gas molecules from a hot to a cold surface. The major advantage of the Knudsen over the McLeod gauge is that it responds uniformly, whereas water vapor, ammonia, carbon dioxide, and pump oil vapors can cause erratic response of the McLeod gauge. The sensitivity limits of both types is ca. $10^{-6}$ mmHg, readings becoming qualitative and increasingly difficult as this limit is approached.

Thermocouple gauges employ a heating element and a fixed current to maintain a temperature that is a function of the vapor pressure. A single metal filament rather than a thermocouple is used in the Pirani [5] gauge. The filament should have a high-temperature coefficient of electrical resistance, as it will comprise one arm of a Wheatstone bridge. The reference arm is usually a similar filament permanently sealed at a very low pressure. The typical useful range of a Pirani gauge is $0.1\text{-}10^{-4}$ torr.

The operating range of the Langmuir gauge [6] is similarly restricted, and it is used only for special applications. Since the operating principle involves the measurement of viscosity, this gauge must be calibrated for different gases or vapors.

Ionization gauges have been vastly improved since their appearance in 1916 [7]. They are essentially vacuum tubes, the plate current being proportional to the amount of residual gas present. It would be an unwise investment for most electrochemists to install one of these tubes, for although pressure in the vicinity of $10^{-10}$ torr can be monitored, they will burn out at $10^{-3}$ torr or higher pressure; hence a sudden leak of the electrochemical glassware on the vacuum line could prove to be expensive.

Although somewhat less sensitive, the Penning ionization discharge gauge (affectionately known as "PIG") is a durable and versatile unit, having no delicate grid and filament. A high voltage is applied between anode and cathode, and the path of the electron is greatly extended by a ring magnet surrounding the unit. Unlike the ionization gauges, which are readily contaminated by organic matter, the PIG will satisfactorily operate between 25 and $10^{-7}$ torr. The PIG referred to in Figure 14.1 is from Consolidated Vacuum Gauge Corp.

The position of the vacuum gauge on the line is one of the least appreciated factors in vacuum line design. For example, if the position of the GPH 100A in Figure 14.1 were to be changed to the "end" of the line, the vacuum readings would show an increase in pressure. A particularly misleading arrangement would put the vacuum gauge outlet on the line between two cold traps; such an arrangement is feasible only for special applications where the pressure of a noncondensable material is sought, and undesirable in many electrochemical experiments. For the perfectionist with sufficient funds, there is no substitute for having two vacuum gauges, one at the end of the line most distant from the pump and one in the position shown in Figure 14.1. This arrangement allows convenient leak checks by shutting off the stopcocks leading to the cold traps and the first vacuum gauge.

When several different sensitivity scales are available on the gauge electronics, unattended monitoring should always be on the least-sensitive scale. Protective circuitry that shuts off the gauge when a certain preset pressure is expected, or sounds an alarm, is available at moderate extra cost.

The vacuum gauge tube directly attached to the line should require no maintenance, although it can be dismantled and cleaned manually if the less radical steps outlined below fail to correct a malfunction. Two difficulties commonly occur if the unit is exposed to high pressure while in operation:

1. No reading at all (needle stays off scale, indicating a "perfect vacuum" regardless of actual conditions) is almost always caused by malfunction of the principal vacuum tube in the meter housing unit. This usually occurs when the vacuum line is exposed to atmospheric pressures while the vacuum gauge tube is "on" [e.g., due to a sudden leak caused by breakage of an electrochemical cell during freeze-pump-thaw cycles (cf. Sec. III.C)]. After replacing the vacuum tube in the meter housing unit, the power should not be turned on until the unit is well evacuated (see item 2).
2. After the vacuum gauge tube has become exposed to air or solvent vapors for a considerable length of time, erroneous vacuum readings (this time indicating too high pressures in the order of several torr) may prevail for a number of hours. There is no remedy except patience, although flushing the unit with helium periodically helps somewhat (the power to the vacuum gauge is turned off during flushing, of course, to protect the vacuum tube; see item 1).

In an active research laboratory, once the gauge is pumped down, the power should be left on indefinitely. After closing the stopcock between the vacuum gauge tube and the rest of the line, the gauge will maintain the vacuum through ionization discharge in its own section indefinitely.

## C. Stopcocks

The larger the bore of the stopcocks serving the main evacuation path, the faster the system will pump down. Very large stopcocks are, however, difficult to turn and are also rather expensive. The choices indicated in Figure 14.1 illustrate a typical compromise. Stopcocks should be numbered permanently so that in cleaning procedures there can be no confusion as to which came from where. Although nominally No. 2, 4, and so on, no two stopcocks are alike. To obtain a best fit, they should be "lapped in" using a suitable grinding compound in place of lubricant. In some cases, Teflon stopcocks are preferable to all-glass ones (cf. Sec. III.B). Many other varieties are available, including all-metallic stopcocks.

Stopcocks should be periodically greased after they are cleaned with solvents such as ligroin or methylene chloride. The latter is an excellent solvent for removing silicone lubricants. Both the stopcock and the housing should be whitish at the ground sections before fresh lubricant is *sparingly* applied. Novices almost invariably use several times the amount necessary, plugging the evacuation hole of high-vacuum stopcocks. The lubricant should be applied with a small glass rod instead of fingers and to the stopcock only, not to the housing. After greasing, the stopcock should be turned several times under considerable pressure using both hands until all striations are absent. After assembling, vacuum stopcocks should be pumped down before other work is resumed. It is good practice to put a small rubber ring or some fingernail polish on the stopcock handle to indicate the direction in which vacuum is applied; for example, when the labeled handle points toward the pump, the stopcock is closed.

## D. Connections Between the Vacuum Line and Auxiliary Glassware

Auxiliary glassware is usually connected to the vacuum line by either of two types of ground-glass connections: ball-and-socket joints and tapered joints. If flexibility beyond that provided by a ball-and-socket joint is needed, flexible stainless steel connections can be used rather than rubber hose. These two are supplied with glass already securely attached to one or both ends (Cajon Company, Cleveland, Ohio). There are many types of couplings useful in vacuum technology, including the familiar Swagelok and Cajon tube fittings. Although originally designed for metal tubing, many of these fittings can be used with glass, polyethylene, or Teflon tubing as well. If it is desired to make a connection between two glass tubes, all one need do is select the appropriate fitting with Teflon ferrules and the job is done in a few minutes. Such couplings can hold a high vacuum. No glassblowing is necessary, but the fittings are quite expensive.

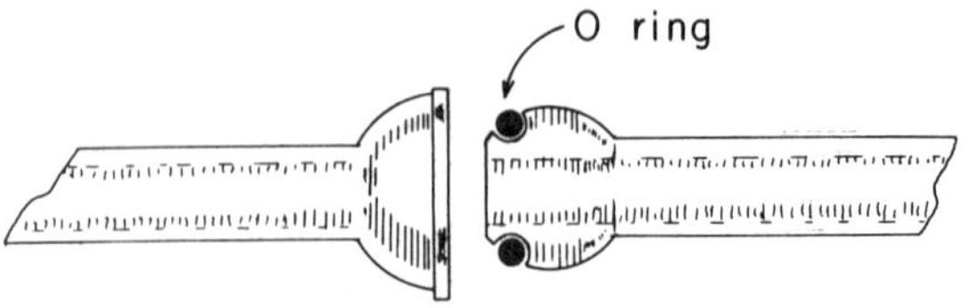


Figure 14.2 O-ring grooved ball joint that is preferable in certain applications to the ordinary type; note that physical contact between the socket and ball end is minimized by the use of the O ring, flexibility of the connection is increased, and lubricant is not essential to hold a vacuum.

There is one kind of ball joint (Fig. 14.2) that is often preferable to the ordinary type. Physical contact between the socket (normally the permanent outlet from the vacuum line) and the ball end is minimized by the use of a small O ring. It should be noted that a special socket is not needed. If a special-purpose outlet can be accommodated, nonground, precision-tooled flanged joints [8] can be used.

An adapter, such as that shown in Figure 14.3, forms the interface between the vacuum line and attached glassware (electrochemical cells, solvent storage containers, etc.). The adapter may be attached to the vacuum line and pumped down by itself. Later in the experiment it might be removed together with the auxiliary glassware.

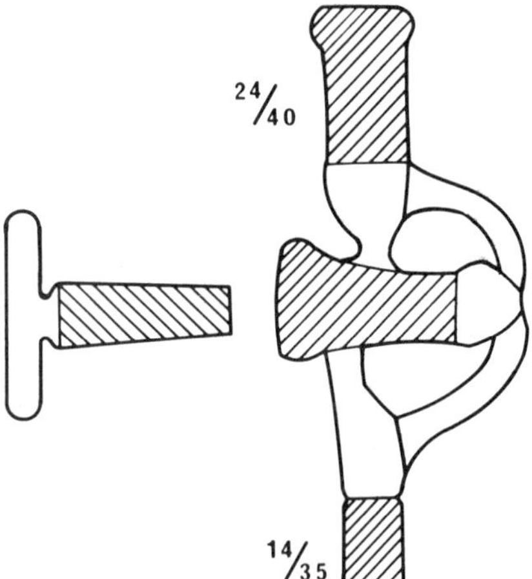


Figure 14.3 Adapter to connect electrochemical cells or storage vessels to the vacuum line (cf. $\frac{T}{S}$ joints in Fig. 14.1). [Reprinted with permission from N. E. Tokel, C. P. Keszthelyi, and A. J. Bard, *J. Am. Chem. Soc.* *94*:4872 (1972). Copyright 1972 American Chemical Society.]

An adapter should always be used when a large volume is to be evacuated, for example, when molecular sieves are being activated. If there is a chance that some solid material will be swept from the sample to the vacuum line during the initial stages of evacuation, a clean plug of glass wool should be placed in the adapter.

## III. ELECTROCHEMICAL GLASSWARE FOR THE VACUUM LINE

### A. Storage Containers

The purification of solvents on the "prep line" by distillation provides the experimenter with a batch of precious liquid gained by many hours or several days work. Subsequent storage should preserve the quality. Two principal types of storage containers are used in vacuum line work, depending on the volatility of the solvent.

*Storage Containers for Low-Volatility Solvents*

Low-volatility solvents such as propylene carbonate (PC) or N,N-dimethylformamide (DMF) can be transferred under gravity. A suitable arrangement is shown in Figure 14.4. These solvents usually do

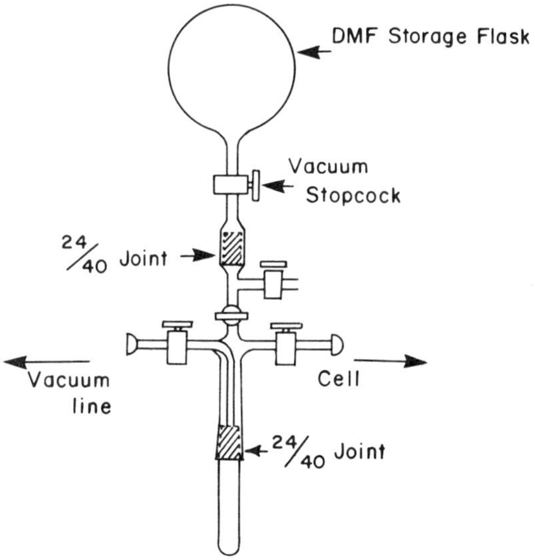


Figure 14.4 Storage container for vapor transfer of low-volatility solvents under vacuum. [Reprinted with permission from L. R. Faulkner, H. Tachikawa, and A. J. Bard, *J. Am. Chem. Soc. 94*:691 (1972). Copyright 1972 American Chemical Society.]

not attack lubricant, and ground-glass high-vacuum stopcocks are satisfactory. Under no circumstances should a specially purified solvent of a hygroscopic nature be stored in a container with nonvacuum glass stopcocks.

### *Storage Containers for Volatile Solvents*

Solvent such as tetrahydrofuran (THF), acetonitrile (AN), or methylene chloride (MC), on the other hand, can be conveniently vapor transferred from containers that have the shape of a large test tube with a ground-glass tapered joint on top to fit the adapter (cf. Sec. II.D). In order that solvent may be poured into the container conveniently, its top should be the male joint, and the bottom of the adapter should be the female joint. An alternative, more elaborate design is shown in Figure 14.5. In this case a trap is included to prevent solvent from "bumping" into the vacuum line.

The appropriate drying agent (NaK alloy for THF, activated molecular sieves for MC) can still be in the container when the final vapor transfer is made, although long-term storage of some solvents

Figure 14.5 Storage container for vapor transfer of volatile solvents under a temperature gradient. To freeze-pump-thaw such containers is unadvisable. (From John T. Bowman, Ph.D. thesis, The University of Texas at Austin, 1970; courtesy of John T. Bowman, Dean McGee Eye Institute, Oklahoma City, Okla.)

over a drying agent is not advisable. It is customary to freeze-pump-thaw (FPT) the solvents, as described below, before the vapor transfer of solvent to the cell. FPT cycles should be carried out as soon as the solvent is brought into contact with the drying agent unless degassed solvent is transferred to the evacuated drying material.

The FPT method, as suggested by the name, involves three basic steps which are repeated in cycles:

1. With the container closed (adapter-to-container stopcock cavity being evacuated in the meantime), the contents are thoroughly frozen by placing a proper-size dewar holding liquid $N_2$ or a slush bath around the storage vessel. The liquid-nitrogen level should not reach the ground-glass connection!
2. With the cooling bath left in place, the stopcock to the vacuum line is carefully opened and the container is evacuated for 10-30 min. Because it is more advantageous to go through several cycles, this step should not be unnecessarily prolonged for any one cycle. During the first cycle, the vacuum gauge should not be opened. Once the gauge is contaminated with solvent, it may require considerable time to pump down again (cf. Sec. II.B).
3. After evacuation, the stopcock is closed and the frozen solvent is thawed. This is accomplished by quickly placing tepid water in a large beaker in contact with the container. Letting the frozen solvent thaw by itself, or using too small a container of water that melts only the bottom of the frozen solvent is an invitation to disaster, for the vessel may well break. The storage flask shown in Figure 14.5 is prone to break even under the greatest care when the frozen solvent has a large thermal expansion coefficient. Thus large containers of AN should be subjected to FPT cycles only if they have a round bottom and are less than half full (this allows the expanding solid to detach itself from the container walls and visibly pop up in the initial stage of thawing).

The FPT cycles can be repeated until the pressure over the frozen solvent is $\simeq 10^{-5}$ torr. Because some of the volatile solvents attack stopcock grease, greaseless high-vacuum stopcocks are useful in constructing these containers.

## B. Vessels for Preparing Solutions

Electrochemical cells with elaborate interior construction are ill suited for freeze-pump-thaw, and when the solution needs to be degassed, a convenient experimental arrangement is to use a special vessel for the FPT cycles, and subsequently gravity transfer the degassed so-

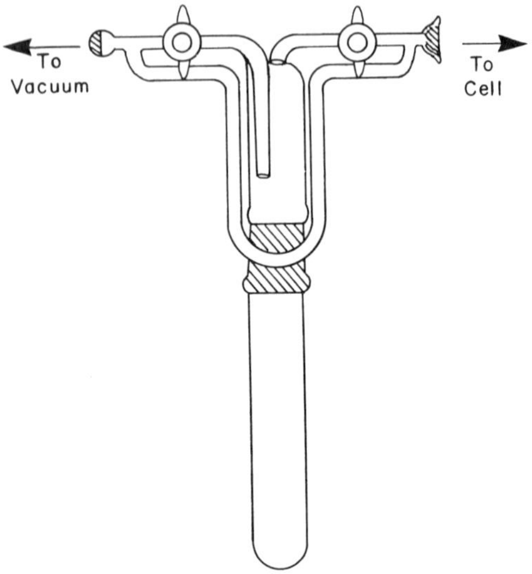


Figure 14.6 Vessel for preparing solutions. This "flipchucker" is suitable for freeze-pump-thaw and relatively accurate volume measurements. The final degassed solution is gravity transferred to the electrochemical cell by flipping the container up 180°. (From John T. Bowman, Ph.D. thesis, The University of Texas at Austin, 1970; courtesy of John T. Bowman, Dean McGee Eye Institute, Oklahoma City, Okla.)

lution into the cell. Figure 14.6 shows suitable glassware [9]. After the solution is degassed, the vessel can be rotated into the upright position and the fluid gravity-transferred. The physical position of such a preparation vessel would be immediately at the end of the vacuum line, followed by the electrochemical cell itself.

When solvent is to be vapor transferred from a large storage container to make up the solution, it would seem feasible to omit the FPT step in the final preparation. However, to have an exact volume, it is still convenient to use a calibrated preparation vessel (scratching with a file must be avoided, for glassware so marked will readily break during FPT cycles). A major drawback of the use of such an intermediate vessel is that the final solution must flow a long distance. Contamination by stopcock grease can be avoided by using the O-ring ball joints and nonground glass stopcocks mentioned earlier. An alternative arrangement is discussed in Sec. IV.

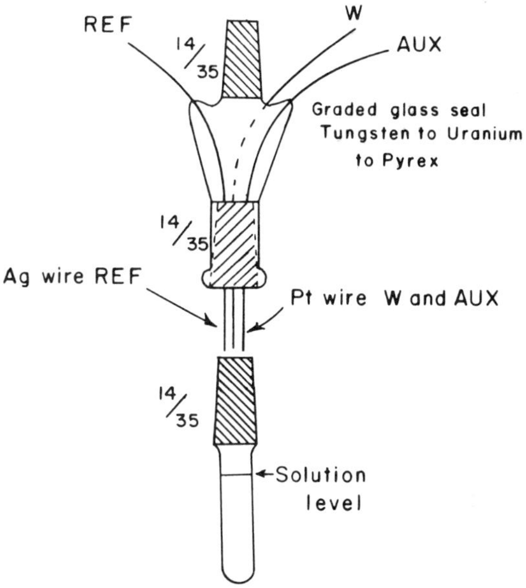


Figure 14.7 Direct freeze-pump-thaw of solutions is feasible in simple electrochemical cells designed for quick exploratory experiments under vacuum. [Reprinted with permission from N. E. Tokel, C. P. Keszthelyi, and A. J. Bard, *J. Am. Chem. Soc.* *94*:4872 (1972). Copyright 1972 American Chemical Society.]

## C. Cell Design and Freeze-Pump-Thaw

It is possible to degas solutions in simple cells such as the three-electrode cell shown in Figure 14.7. Cells of this type have also been found to be suitable for carrying out elevated temperature electrochemical experiments using molten scintillator dyes as solvents [10]. We will not dwell on the shortcomings of such cells in this chapter. They are essentially for preliminary work, and great time-savers at that (cf. Sec. IV).

A larger volume cell is shown in Figure 14.8. This cell incorporates an FPT chamber and micro working electrode as well as a macroelectrode for electrolysis.

# IV. EXPERIMENTAL METHODS

Final drying of solids (e.g., tetraalkylammonium supporting electrolytes) in the cell just prior to making up the final solution is advisable. Boiling water may be placed under the cell to facilitate the

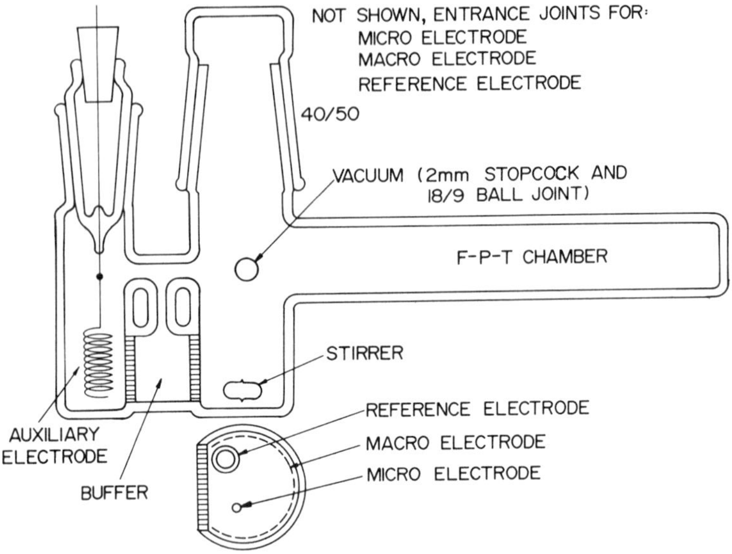


Figure 14.8 Electrochemical cell having a freeze-pump-thaw chamber. (From Ref. 2.)

process. Together with pumping down the adapter and cell, this will take approximately 2 h when a small cell (Fig. 14.7) is used. On the other hand, for a large cell (Fig. 14.8) the time required can exceed 20 h [11].

The vacuum line is a very time consuming apparatus. A cyclic voltammogram may, for example, be obtained in just a few minutes, but the preparation leading up to that step may take a good number of hours. The only poor investment of time spent on the vacuum line is when the rigorous conditions are foresaken in one of the many experimental steps. An example would be the addition of solids to a solution through the top of the cell, even under a positive inert gas pressure. Once such means of addition are incorporated, the purpose of the vacuum line is compromised, and one might just as well use a glove bag or box for the entire preparation. It is sometimes convenient to introduce liquids or gases to the cell by means of syringe injection through an inert septum. Although not in keeping with a totally sealed system, this procedure can be carried out with little risk of sample contamination.

Naturally, some means for controlling concentration is needed, and one possibility is indicated in Figure 14.9. Here the solution is first thoroughly degassed in the vessel described in Section III.A; then it is gravity transferred to the thoroughly evacuated auxiliary sample reservoir. Subsequently, the auxiliary sample reservoir is

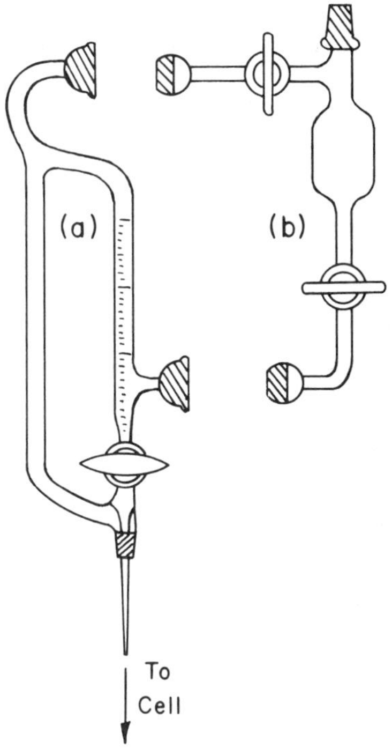


Figure 14.9 Concentration-dependence study under vacuum line conditions is possible with the aid of an auxiliary sample buret (a) and auxiliary sample reservoir (b). Degassed solution is initially transferred from a preparative vessel (cf. Fig. 14.6) to reservoir (b). (From John T. Bowman, Ph.D. thesis, The University of Texas at Austin, 1970; courtesy of John T. Bowman, Dean McGee Eye Institute, Oklahoma City, Okla.)

brought into contact with the auxiliary sample buret, which is already attached to the electrochemical cell. In a typical application a different solution, which is prepared in a different auxiliary vessel, would be gravity transferred into the cell, allowing a concentration dependence study under vacuum line conditions.

This chapter must also make mention of that "twilight zone" of paraphernalia which are not a part of traditional glassblowing and vacuum line technique, yet are gaining in popularity. Metal connectors with "swage" action (Swagelok, Gyrolok, or Cajon fittings) are often used if it is desired to make a connection between two glass tubes of less than 1 in. o.d., since all that needs to be done is to select the right size fittings, preferably with Teflon ferrules, and connect them. When a special coating has been applied to the glass tubing

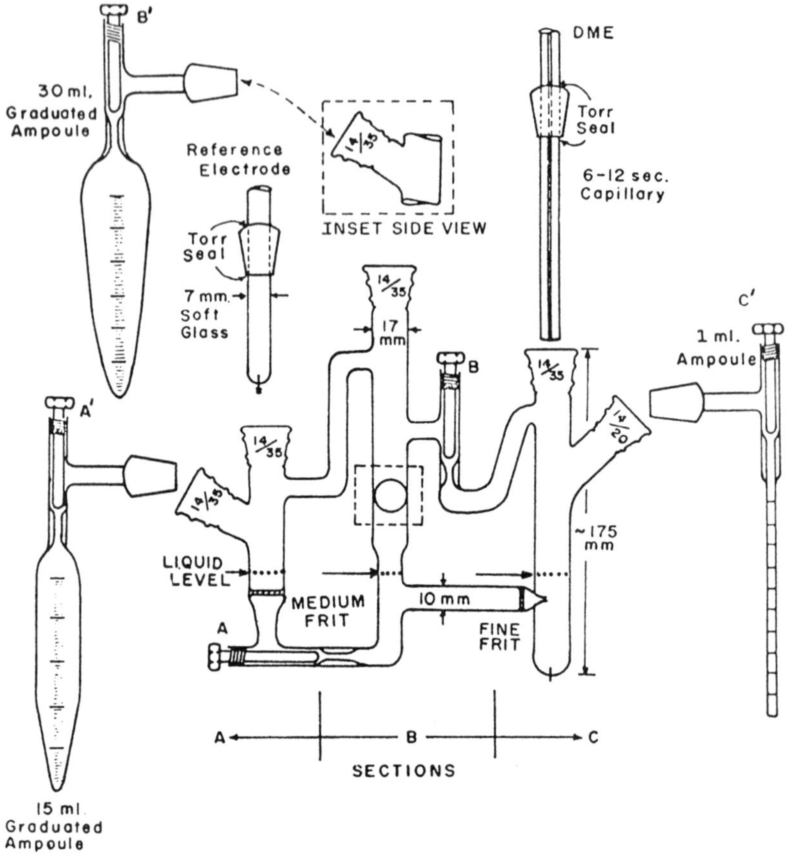


Figure 14.10 Electrochemical cell and assorted glassware allowing concentration dependence studies under vacuum line conditions. [From Ref. 13, reprinted with permission from J. L. Mills, R. Nelson, S. G. Shore, and L. B. Anderson, *Anal. Chem.* 43:157 (1971). Copyright 1971 American Chemical Society.]

that would be destroyed by heat, such connections were indeed found to be the best available means to carry out the experiment [12]. Rubber septa, part of the standard technique in gas chromatography, have also been employed in reduced pressure studies to facilitate concentration dependent measurements.

Figure 14.10 represents a cell capable of a variety of electrochemical experiments under conditions of isolation from the atmosphere. The working electrode (Hg pool, DME, Pt bead) compartment (section

C) is rigidly connected to a separate compartment containing a reference electrode (section A). Semiquantitative concentration studies may be accomplished by addition of concentrated reagents from ampoule C'. Sample may also be added by gastight syringe (e.g., Hamilton Gas Tight Model 1705 NCH); this method results in greater precision of addition but introduces small quantities of $O_2$ and $H_2O$.

A further feature of the system in Figure 14.10 is the capacity for removal of samples during electrolysis. Microliter samples, removed by syringe through a rubber septum in side arm C', have been analyzed by GCMS to separate, identify, and measure products of electrolysis of an olefin at $1 \times 10^{-3}$ M initial concentration. A Solv-Seal connector can be positioned so that samples of the contents of the electrolysis vessel may be removed for UV-visible, EPR, or NMR analysis and reintroduced into the electrolysis compartment.

This brief introduction to vacuum line techniques has, we hope, shown that we are dealing with a rather extensive area of experimental science [14]. It should be clear that this is a worthy area of competence for the electrochemist. Those who have invested in the systems described above have been rewarded with higher-quality media and more meaningful data than would otherwise be possible.

## REFERENCES

1. Books dealing with properties of glass and glassblowing can be of help: (a) R. H. Wright, *Manual of Laboratory Glass-Blowing*, Chemical Publishing Company, Brooklyn, New York, 1943; (b) J. D. Heldman, *Techniques of Glass Manipulation in Scientific Research*, Prentice-Hall, Englewood Cliffs, N.J., 1946; (c) H. McKearin and George S. McKearin, *Two Hundred Years of American Blown Glass*, Crown, New York, 1950; (d) *Symposium on the Art of Glassblowing*, American Scientific Glassblower Society, Corning, N.Y., 1956; (e) A. J. B. Robertson, *Laboratory Glass Working for Scientists*, Academic Press, New York, 1957; (f) W. E. Barr and V. J. Ahorn, *Scientific and Industrial Glass Blowing and Laboratory Techniques*, Instruments Publishing Company, Pittsburgh, Pa., 1949, 1959; (g) F. L. Wheeler, *Scientific Glassblowing*, Interscience, New York, 1958, 1963; (h) R. Barbour, *Glassblowing for Laboratory Technicians*, Pergamon Press, Oxford, 1968.
2. A. J. Bard, *Pure Appl. Chem.* *25*:379 (1971).
3. W. Gaede, *Ann. Phys.* *41*:289 (1913).
4. M. Knudsen, *Ann. Phys.* *28*:75 (1909).
5. M. von Pirani, *Dtsch. Phys. Gesell.* *8*:24 (1906).
6. I. Langmuir, *J. Am. Chem. Soc.* 35:107 (1913).
7. D. E. Buckley, *Proc. Natl. Acad. Sci. USA* *2*:683 (1916).
8. W. R. Doty, *Rev. Sci. Instrum.* *30*:1053 (1959).

9. J. T. Bowman, Ph.D. dissertation, University of Texas at Austin, 1970.
10. C. P. Keszthelyi and A. J. Bard, *J. Electrochem. Soc. 120*: 241 (1973).
11. A. Demortier and A. J. Bard, *J. Am. Chem. Soc. 95*:3495 (1973).
12. C. P. Keszthelyi and A. J. Bard, *J. Electrochem. Soc. 120*: 1726 (1973).
13. J. L. Mills, R. Nelson, S. G. Shore, and L. B. Anderson, *Anal. Chem. 43*:157 (1971).
14. (a) R. R. Lapelle, *Practical Vacuum Systems*, McGraw-Hill, New York, 1972; (b) D. F. Shriver, *The Manipulation of Air-Sensitive Compounds*, McGraw-Hill, New York, 1969; (c) N. T. M. Dennis and T. A. Heppell, *Vacuum System Design*, Chapman & Hall, London, 1968; (d) P. A. Redhead, J. P. Hobson, and E. V. Kornelsen, *The Physical Basis of Ultrahigh Vacuum*, Chapman & Hall, London, 1968; (e) J. Yarwood, *High Vacuum Technique; Theory, Practice, Industrial Application and Properties of Materials*, Chapman & Hall, London, 1943, 1945, 1955, 1967; (f) K. Diels and R. Jaeckel, *Leybold Vacuum Handbook*, Pergamon Press, Oxford, 1966; (g) A. Roth, *Vacuum Sealing Techniques*, Pergamon Press, Oxford, 1966; (h) W. Espe, *Materials of High Vacuum Technology*, Pergamon Press, Oxford, 1966; (i) W. F. Brunner, Jr., and T. H. Batzer, *Practical Vacuum Techniques*, Reinhold, New York, 1965; (j) G. S. Lewin, *Fundamentals of Vacuum Science and Technology*, McGraw-Hill, New York, 1965; (k) C. M. Van Atta, *Vacuum Science and Engineering; Properties of Gases at Low Pressure, Vacuum Measurements, Design and Operating Features of Vacuum Pumps and Systems*, McGraw-Hill, New York, 1965; (l) R. W. Roberts and T. A. Vanderslice, *Ultrahigh Vacuum and Its Applications*, Prentice-Hall, Englewood Cliffs, N.J., 1963; (m) A. H. Turnbull, R. S. Barton, and J. C. Riviere, *An Introduction to Vacuum Technique*, Wiley, New York, 1963; (n) A. Guthrie, *Vacuum Technology*, Wiley, New York, 1963; (o) H. A. Steinherz, *Handbook of High Vacuum Engineering*, Reinhold, New York, 1963; (p) S. Dushman, *Scientific Foundations of Vacuum Line Techniques* (J. M. Lafferty, revising ed.), Wiley, New York, 1962; (q) M. S. von Pirani and J. Yarwood, *Principles of Vacuum Engineering*, Chapman & Hall, London, 1961; (r) *Advances in Vacuum Science and Technology—Proceedings of the 1958 International Congress on Vacuum Techniques* (E. Thomas, ed.), Pergamon Press, Elmsford, N.Y., 1960; (s) *Transactions of the 1954 National Symposium on Vacuum Technology*, American Vacuum Society, Pergamon Press, Elmsford, N.Y., 1959; (t) J. H. Leck, *Pressure Measurement in Vacuum Systems*, Institute of Physics

Press, London, 1957; (u) *Vacuum—A Review of Developments in Vacuum Design and Engineering*, W. Edwards, London, 1951; (v) *Vacuum Equipment and Techniques—National Nuclear Energy Series, Manhattan Project Technical Sections*, Div. I, Vol. 1 (A. Guthrie and R. K. Wakerling, eds.), McGraw-Hill, New York, 1949; (w) L. H. Martin and R. D. Hill, *A Manual of Vacuum Practice*, Melbourne University Press, Melbourne, 1949; (x) R. T. Sanderson, *Vacuum Manipulation of Volatile Compounds: A Laboratory Manual Describing the Application of High Vacuum Technique in Experimental Chemistry*, Wiley, New York, 1948; (y) S. Jnanananda, *High Vacua: Principles, Production, and Measurement*, D. Van Nostrand, New York, 1947; (z) L. Dunoyer, *Vacuum Practice*, G. Bell and Sons, London, 1926; (aa) *The Journal of Vacuum Science and Technology*, American Institute of Physics, New York, 1964 to date.

# 15

# Electrochemistry in the Dry Box

STEVEN N. FRANK* / Department of Chemistry, University of Texas at Austin, Austin, Texas

SU-MOON PARK / Department of Chemistry, University of New Mexico, Albuquerque, New Mexico



## I. INTRODUCTION

Controlled atmospheres are employed in electrochemistry whenever oxygen, water, or other constituents of the air may interfere with the reaction under study. The problem is basically twofold: to supply and contain an atmosphere of suitable composition and to exclude air. For our present purposes it is sufficient to confine the discussion to inert atmospheres [i.e., gases devoid of oxygen and water (and sometimes nitrogen)]. The use of high-purity argon, helium, or nitrogen is most common.

Much progress has been made in the last 25 years in making available high-purity inert gases at reasonable cost. The design of enclosed working spaces to contain these inert atmospheres and exclude air have resulted from the demands of high-technology areas such as metallurgy, semiconductor physics, and atomic energy [1]. The enclosed working space most often used in the research lab is the dry (or glove) box.

The relatively recent interest in nonaqueous solvents [2,3] has presented the electrochemist with a number of new problems in solvent manipulation. Since most anhydrous solvents are quite hygroscopic,

*Present address: Texas Instruments, Inc., Dallas, Texas

they should not be exposed to the laboratory atmosphere unnecessarily. The time it takes to pour or pipette from one container to another can be sufficient to pick up significant quantities of water. In many cases even minute quantities of water cannot be tolerated since it will often participate in the reaction mechanism. Small amounts can also significantly reduce the available potential range. Nonaqueous solvents often also suffer from the disadvantage of being too volatile to be conveniently deoxygenated by the usual method of bubbling an inert gas through the solvent. Electrochemists have typically circumvented these problems by using the vacuum line techniques discussed in Chapter 14. Vacuum line techniques, however, suffer from a number of disadvantages, not the least of which is that they are often very time consuming. Using a dry box, on the other hand, can often shorten the time required to prepare for an experiment to approximately 1 h. There is also much less chance of breaking very expensive cells such as are often used on a vacuum line.

The advantages of using a dry box over vacuum techniques may be summarized as follows:

1. There is a considerable saving of time.
2. Dry box procedures are often simpler than vacuum line techniques.
3. A high quality of oxygen and water removal is more easily maintained. Attainment of $10^{-4}$ torr vacuum may require several hours for large cells, but it is a simple matter to maintain routinely less than a 1 ppm level of oxygen (we have successfully maintained less than 0.1 ppm oxygen as determined with a Chemtrix Model 30 Oxygen Meter) in the inert atmosphere of the dry box.
4. There is less danger of breaking important and expensive cells.
5. Solvent volume can be conveniently measured to any degree of accuracy desired. A high accuracy in weighing materials may be obtained by placing a balance in the dry box.
6. The handling of sublimable materials is easier than with vacuum techniques.

There are, however, several disadvantages of using a dry box for electrochemistry:

1. Every liquid has to be thoroughly deoxygenated before transferring to the dry box.
2. An SCE reference electrode cannot be used for nonaqueous work. This disadvantage can be overcome by using a nonaqueous Ag-$AgNO_3$ electrode and calibrating it against an SCE or by using a silver wire "pseudo"-reference.
3. It is difficult to carry out temperature studies.
4. The work space is rather limited.

5. So-called "inert" gases are not always inert. A vacuum may be better.
6. A good dry box can be very expensive to install and maintain.

## II. CHOOSING AN INERT-ATMOSPHERE SYSTEM

In choosing an inert-atmosphere system one should select a system commensurate with the job at hand. It makes no sense to invest thousands of dollars in a sophisticated dry box when a $3 glove bag might serve just as well.

The cheapest and simplest system is the glove bag, which is basically a plastic bag open at one end with glove inserts. The materials to be used are placed in the bag in the collapsed condition. Holes are cut in the plastic for making electrical connections, and so on, and then sealed with tape or other suitable material. The bag is then purged with inert gas for several minutes and the open end is sealed with the gas continuing to flow.

The glove bag has several advantages. It is economical, $3 to $12, and available in a variety of sizes. Although a glove bag is also versatile, it is difficult to maintain the quality of the atmosphere. At best the atmosphere can be no better than the inert gas used. The highest-quality tanked gases usually have in excess of 10 ppm impurity and this establishes a lower limit on the impurity level. Glove bags also tend to leak and there is a large diffusional area (most plastics are somewhat permeable to air). The purging procedure is usually not very efficient at removing all of the air initially trapped in the bag. The impurity level of the atmosphere in the glove bag is therefore usually much higher than that in the gas supply. When a large number of experiments are anticipated that require an inert atmosphere, the glove bag is not very convenient. It is ideal for the occasional experiment when the systems being investigated are only moderately sensitive to water and oxygen. There are a number of systems somewhere between the glove bag and the conventional dry box. Nevertheless, in our experience these systems offer no significant advantages over the glove bag.

Dry boxes vary widely in design and the materials used in construction. Prices range from about $300 for economy models to over $10,000 for more sophisticated units. All dry boxes contain a working space that incorporates glove ports, gas connections, electrical connections and lighting, and a viewing window. In addition, most boxes have an antechamber or transfer box for moving material in and out of the working area without disturbing the integrity of the dry box atmosphere. Two kinds of transfer chambers are available: an evacuation chamber and a gas purging chamber. In an evacuation chamber the samples are placed in the chamber and the chamber evacuated. Gas of the same composition and purity as in the working area is then introduced and the dry box opened. With a purging chamber, gas, usually from the same cylinders as the supply gas for the box, purges air from the chamber. The evacuation transfer chamber is preferred since it does a more complete job of removing air and adsorbed water and saves the cost of the purging gas.

The inert atmosphere for the box may be supplied continually with a bleed-off valve, or recirculated through purification trains that remove oxygen and water. Since all boxes leak to some extent (1 $f^3$/day is typical), recirculation through a purification train is desirable. Figure 15.1 shows a block diagram for a typical dry box with the recirculation option. Most trains can maintain oxygen and water to less than 1 ppm. If economics is a consideration, one can construct his own purification trains [1], although they would be much less convenient than the commercial product.

Systems that permit automatic regulation of the gas pressure in the box within present limits are preferable to those that do not. This

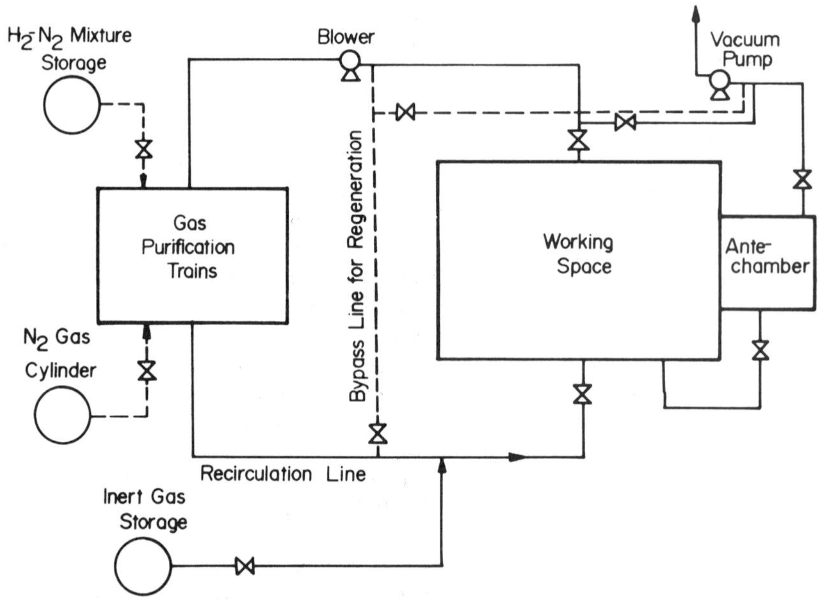


Figure 15.1 Block diagram of dry box with inert-gas recirculation through purification train. The $N_2$ and $N_2$-$H_2$ mixtures are used during regeneration of the purification train.

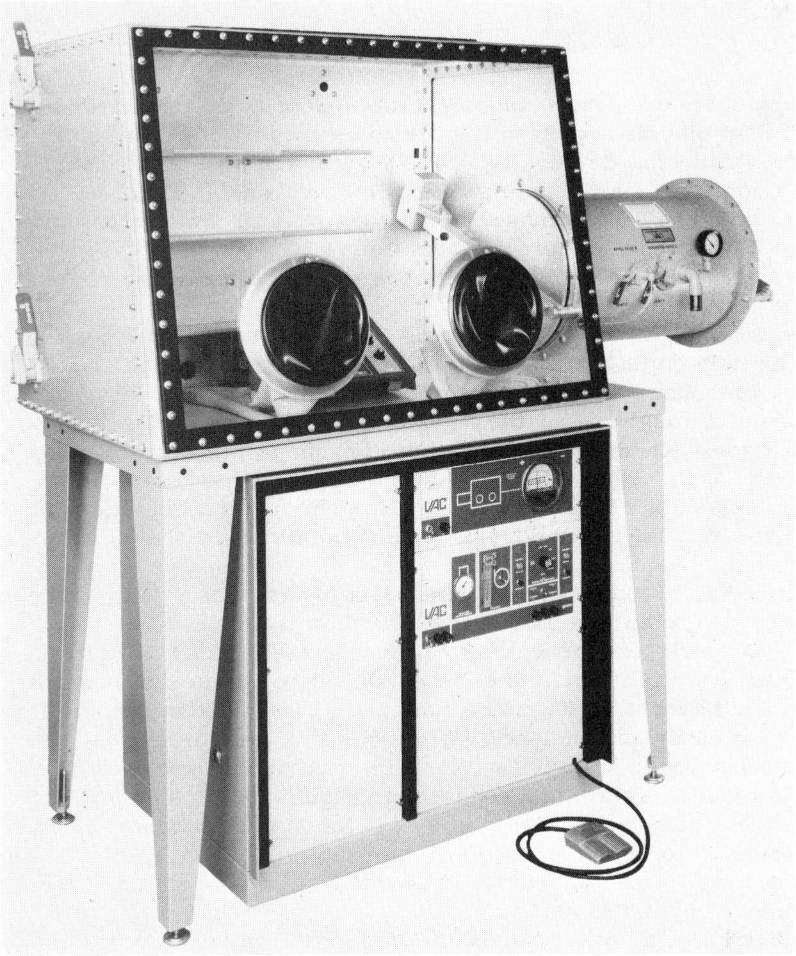


Figure 15.2 Typical dry box suitable for high-quality electrochemistry. The purification train, recirculation unit, and automatic pressure control system are housed in the unit beneath the dry box. (Courtesy of Vacuum/Atmospheres Corporation, Hawthorne, California.)

feature can prevent nasty accidents that may result from over- or underpressurizing the box. All-metal boxes with glass or plastic windows (Fig. 15.2) are generally better than all-plastic boxes. Metal boxes offer the advantage of durability, resistance to chemical attack, and increased ease in modification.

## III. EXPERIMENTAL

### A. Basic Dry Box Procedures and Considerations

Although there are several classes of dry boxes as described above, we will limit our discussion to the considerations involved in using a fairly sophisticated dry box equipped for gas recirculation through a purification train and including an evacuation transfer chamber. The dry box will also have provision for automatic control of the gas pressure within the box. These features make a dry box especially well suited for carrying out high-quality nonaqueous electrochemistry with a minimum of effort.

Initial purging of the dry box is readily carried out by passing approximately eight box volumes of the inert gas through the box at positive pressure (the equivalent of about 2 in. of water) with the gas flowing out through one of the service ports of the box. During the purge the box should be recirculating in order to purge the purification train of air. While the box is still under positive pressure the service port should be sealed. After three or four regenerations of the purification train, a high-quality inert atmosphere should be established.

Immediately following the initial assembly (and at any other time leaks are suspected) the box and purification train should be leak tested. A quick test for leaks is to pressurize the dry box until the gloves stick straight out. The gloves should remain in this position for several hours if no leaks are present. If leaks are indicated, testing is most easily accomplished while the box is pressurized. If the inert atmosphere is helium, the preferred method is the helium "sniff" test. All joints, welds, and connections should be checked. In the absence of a helium-sensitive probe or if the inert atmosphere is other than helium, the bubble method may be used. Here a small amount of soapy water is placed on leak-prone welds and joints and the appearance of any bubbles is noted.

Glassware and other material to be introduced into the box should be clean and as dry as possible. This saves in the time required to pump down the antechamber, minimizes the possibility of contaminating the dry box, and increases the time between regenerations of the purification train. Simple glassware that is clean and dry should be pumped down for only 20-30 min. Large electrochemical cells containing many frits and compartments should be pumped down for several hours. By removing glassware directly from a drying oven and placing it immediately in the antechamber, the efficiency of the pump-down procedure is considerably enhanced since desorption of adsorbed material is greatly enhanced at elevated temperatures. The efficiency of oxygen and water removal is also increased if the antechamber is evaculated a second time after filling it with inert gas from the working area of the dry box (via the appropriate plumbing). Before pumping down the transfer box, one should make sure that the door between the transfer box and the working area is well sealed or a rapid depletion of the supply gas for the dry box will result. When possible we have found it convenient to pump down complicated equipment such as stirring motors or cells overnight.

Most solid chemicals can be introduced without any difficulty after they have been thoroughly dried. Supporting electrolytes such as tetra-n-butylammonium perchlorate (TBAP) are usually vacuum dried, at elevated temperatures if possible, for 24-48 h and can be pumped down in the antechamber in the bottle with the top removed. Sublimable compounds should not be pumped down in the antechamber, for obvious reasons. These materials should be placed in specially designed sample containers having two small stopcocks. The containers can be purged with a dry inert gas for several minutes (Fig. 15.3). The design should be such as to permit evacuation of the surrounding atmosphere. Screw caps are especially convenient for sealing the top opening of the vials. As an alternative, metal pipes or cylinders sealed at one end with screw caps and O rings and stainless or brass valves for purging make excellent containers for introducing sublimable compounds.

Solvents present special problems in handling since they must be dried and deoxygenated before they are introduced into the dry box. Solvent manipulation best illustrates how vacuum line techniques complement dry box techniques. For this reason and because the published procedures [3] seem unduly complicated and time consuming, we present the procedure for purifying (drying) the widely used nonaqueous solvent acetonitrile and the introduction of the thus purified solvent (less than 0.1 m*M* in $H_2O$) into the dry box.

The design of the equipment used in this laboratory for the purification is shown in Figure 15.4. Spectrograde acetonitrile is successively distilled three times from $P_2O_5$ under vacuum at room temperature with the final distillation being into the receiving flask. The pur-

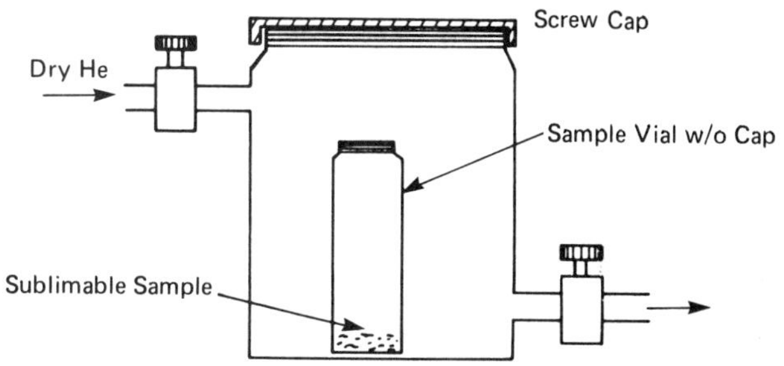


Figure 15.3 Design of apparatus for introducing sublimable materials into a dry box equipped with an evacuation transfer chamber.

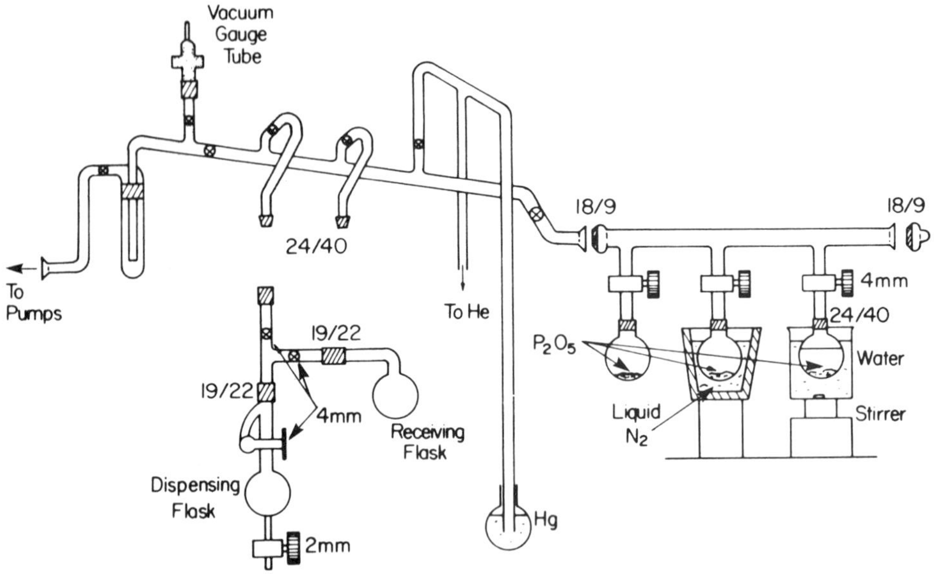


Figure 15.4 Apparatus for vacuum distillation of acetonitrile. For more details, see Figure 14.1.

ified and dry acetonitrile is degassed in this flask by performing one or two freeze-pump-thaw (FPT) cycles. Only one or two FPT cycles are necessary since the solvent has already been pumped down to $10^{-3}$ torr three times during the distillation. The acetonitrile is now transferred to the final holding and dispensing flask with the aid of a little positive helium pressure in the receiving flask, obtained by passing some helium onto the vacuum line (see Chap. 14 for more details). Before placing the dispensing flask in the antechamber and evacuating, the stopcocks should be tied to the flask to prevent them from popping out. It is important to consider similar possibilities any time any container that is not itself to be evacuated is placed in the transfer chamber.

Enough solvent should always be prepared to last for some time, since solvent preparation may be time consuming and it is always inconvenient. Any liquid, be it solvent or solute, should be degassed and dried (*not* always with $P_2O_5$) before transferring it to the working area.

What equipment should always be in the dry box? We have found that placing a large plastic sheet on the bottom of the box facilitates cleaning. A small brush and adhesive tape for cleaning up solid chemical spills may prove useful. The arrangement of the equipment is important since the working area is rather limited. Usually, several ring stands, a magnetic stirring motor, a buret stand, a pair of tongs, several clamps, and an automatic balance are necessary. A large plastic syringe attached to a rubber tube is useful for pipetting. Tongs with rubber tubing over the tips are especially convenient for handling glassware when using the balance. A small spatula is necessary for transferring solid materials. A box of wipes (these should be pumped down in the vacuum chamber for at least 12 h to avoid introducing oxygen and water into the dry box) and weighing paper are also useful, as is silicone vacuum grease should one wish to remove apparatus or solutions containing air-sensitive compounds from the dry box.

For electrochemical experiments one should be certain that there are enough electrical connections and plugs. Few if any of the commercial dry boxes will have enough of these, and they will have to be installed. Connections to the outside world are a source of leaks, and considerable care should be exercised to ensure that the connectors are well sealed. By having enough connections, two or three people can work simultaneously. For example, we have installed two connectors with six contacts on each. Long color-coded wires to the connectors are recommended to avoid confusion.

When working in the box, a slightly lower pressure inside the box than atmospheric pressure is preferred since it makes handling objects with the gloves easier. When not in use, however, the pressure in the box should be higher than atmospheric to minimize leaking into the dry box.

Sprinkling a little talc or baby powder in the gloves facilitates putting the gloves on and off. It also helps absorb moisture, as the hands and arms tend to perspire when they have been in the gloves for a few minutes. Better yet, wearing a pair of disposable surgical gloves used with a small amount of talc will greatly aid insertion and removal of hands from the dry box gloves. The gloves should be inspected periodically for leaks or holes and replaced when they are found.

Always double check to make sure the inner door between the working area and the antechamber is closed before opening the outer door to the antechamber. It is surprising how often this simple task is forgotten.

The presence of oxygen can be detected by exposing the filament of a broken light bulb to the inert atmosphere. Under ideal conditions the filament should not burn out for several weeks. Unfortunately, the oxygen transients that sometimes result on introducing materials into the box can exceed 15 ppm. Although the quality of the atmosphere is rapidly reestablished within a few minutes, the filament may have already burned out.

Oxygen can also be detected chemically. Tetrakis(dimethylamino) ethylene forms a luminescent compound with oxygen and is very sensitive to trace amounts of oxygen [4]. One drawback of this compound is that it has a high affinity for silicone stopcock grease, which it deteriorates quite rapidly. A small bottle of diethylzinc mixed with heptane is also useful. The absence of a vapor cloud indicates less than 5 ppm of oxygen. (Caution: This material is pyrophoric, use extreme care in handling.)

Probably the best method of oxygen measurement is an oxygen meter. We have successfully used a Chemtrix Type 30 Oxygen Meter (Hillsboro, Oregon). The amperometric oxygen electrode is placed in the box and connected to the external meter via the aforementioned electrical connections. When placing the oxygen probe in the dry box it should be placed in a vacuum-tight container when pumping down the antechamber. Otherwise, the pressure differential will rupture the electrode membrane. Periodically, the electrode will have to be removed and the filling solution replenished. No significant water contamination results from using this electrode.

Titanium tetrachloride will indicate 10 ppm or more of water by the appearance of a vapor cloud. Hygrometers are also available, but the expense hardly seems justified. If significant quantities of oxygen are detected, it is safe to assume that the dry box is also contaminated with water. When oxygen is detected, the gas purification train should be regenerated.

Since the gas purification trains consist of chemicals such as manganous oxide, calcium, or copper to remove oxygen and molecular sieves to remove water, they eventually become saturated and deactivated and must be regenerated. It is best to regenerate the train on a regular basis before its effectiveness decreases. Regeneration usually consists of heating the train while purging with nitrogen and nitrogen-hydrogen mixture for 4-12 h.

The vacuum pump should be serviced regularly. Whenever the pump oil seems contaminated, it should be replaced with new oil.

### B. Electrochemistry in the Dry Box

As we have pointed out previously, oxygen and water concentrations can be kept at extremely low levels with a properly maintained purification train. In fact, contamination by water is much more easily controlled in a dry box than on a vacuum line. This may result in part from the number of operations and manipulations necessary to use a vacuum line. The oxidation of the cation radical of thianthrene to the dication illustrates this point. The dication is very electrophilic and is rapidly attacked by any nucleophiles (e.g., water). The electrochemistry of the dication in solutions prepared in the dry box (with acetonitrile as the solvent purified as described above) is reversible if a little care is taken in preparing the solvent and in drying the glassware [5]. It is more difficult to obtain such reversible behavior when solutions are prepared on a vacuum line.

Cyclic voltammetric measurements are well suited for the dry box. Solvents, supporting electrolytes, and any compounds to be used can be kept in the box. Only the cell and electrodes need be introduced for experiments. Solution preparation usually takes only 20-30 min. The electrodes can be connected to external apparatus as described above. Measurements of voltammetric current functions should be avoided at slow sweep rates because vibrations due to the various motors and pumps result in some convection. In fact, theoretical measurements using any technique that assumes mass transfer by diffusion should not be attempted in the dry box if the time of data acquisition is greater than 2 or 3 s (e.g., sometimes chronopotentiometry and chronocoulometry). In these cases the solutions may be prepared in the dry box and the cells made airtight (we recommend using cells with ground-glass joints) with silicone grease and removed from the dry box. A considerable saving of time is realized by preparing the solutions in the dry box instead of on a vacuum line. The dry box offers similar advantages when performing controlled-potential electrolysis.

Conventional polarography with a dropping mercury electrode can conveniently and routinely be carried out in the dry box, although due to height limitations it may be necessary to use a lower mercury head than is possible outside the box.

Dry boxes are especially advantageous for rotating ring-disk electrode (RRDE) experiments [6]. Because the motor can be placed in the dry box, one is saved the expense of a rather expensive RRDE cell and sealed motor and the attendant problems of working outside the dry box.

Electrogenerated chemiluminescence (ECL) experiments [7] are difficult to perform in the dry box due to the need to exclude light from all external sources. Placing a black cloth over the dry box window is usually not sufficient unless the intensity of the ECL is very high. It may be necessary to use a monochromator in these experiments. This is especially unfortunate in the case of ECL at a RRDE since the complexity of cell and motor assembly is greatly increased. Nevertheless, the dry box is very useful for performing initial ECL experiments, perhaps to establish whether or not ECL occurs.

Dry boxes also offer significant advantages over vacuum line techniques in the study of mixed-solvent systems. It is difficult to mix two different solvents accurately on a vacuum line. In a dry box, the procedure is rendered almost trivial. Definite advantages of a dry box can also be envisioned for electrochemistry in molten salts, which are often very hygroscopic.

For those who wish to work with the old model Kemula hanging mercury drop electrode, some problems may be encountered when evacuating the antechamber (separation of the mercury column in the capillary and retraction of the mercury column up the capillary after the drop is dislodged). To prevent these problems, the electrode should be clamped in an upright position with the tip of the electrode submerged in a small mercury pool. This electrode also suffers the disadvantage that a vacuum is necessary for filling and it therefore cannot be refilled in the dry box. The new model Kemula electrode (available from Brinkmann Instruments and Princeton Applied Research) and the hanging mercury drop electrode available from Beckman Instruments can be evacuated in the antechamber and filled in the dry box without any problems.

Aqueous reference electrodes should not be used since they cannot be evacuated in the antechamber. In any case there would be a serious possibility of contaminating the dry box with water. A silver wire in contact with a solution of silver nitrate makes a satisfactory reference electrode. The silver nitrate can be dissolved in the solvent being used in the dry box. The electrode can later be calibrated against an aqueous SCE. For many experiments a bare silver wire will serve satisfactorily, although the potential of this electrode will not be constant.

With the high solution resistance that may be present with nonaqueous solvents and the long electrode connections recommended above, some impedance problems may result. We have found the severity of these problems to be highly dependent on the potentiostat being used [e.g., we have had far fewer problems when using the PAR (Princeton Applied Research) Model 173 than the more expensive PAR 170 electrochemical system]. To remedy this difficulty, one might consider placing a voltage follower in the dry box to minimize the lead length to the reference electrode.

A note of caution is needed if one plans to use open electrochemical cells in the dry box. Because the rate of recirculation of the atmosphere is rather high in most dry boxes (one complete change of atmosphere every few minutes), evaporation of most organic solvents is rapid. Open containers such as beakers are recommended only during solution preparation.

## APPENDIX

A reasonable place to begin one's search for a dry box is the Lab Guide issue of *Analytical Chemistry*, which appears yearly. When selecting a box the criteria set forth in the text should be considered. Consideration should also be given to the physical dimensions of the box that are needed. For those on a tight budget, one of the plastic or fiberglass dry boxes may be satisfactory. If purification trains or a vacuum transfer chamber are not listed as options, the manufacturer should be consulted to see if the dry box can easily be modified to accommodate them. We recommend buying a complete system (which would include all pumps and motors) from one manufacturer when possible. Two manufacturers that make especially good systems are Kewaunee Scientific Corp., Ardian, Michigan, and Vacuum/Atmospheres Company, Hawthorne, California.

Glove bags may be purchased from Instruments for Research and Industry ($I^2R$), 108 Franklin Avenue, Cheltenham, PA 19012.

## REFERENCES

1. P. A. F. White and S. E. Smith, *Inert Atmospheres in Chemical, Metallurgical, and Atomic Energy Industries*, Butterworth, Washington, D.C., 1962.
2. C. K. Mann and K. N. Barnes, *Electrochemical Reactions in Nonaqueous Systems*, Marcel Dekker, New York, 1970.
3. C. K. Mann, in *Electroanalytical Chemistry*, Vol. 3 (A. J. Bard, ed.), Marcel Dekker, New York, 1969.
4. (a) R. L. Purett, J. T. Barr, K. E. Rapp, C. T. Bahner, J. D. Gibson, and R. H. Lafferty, Jr., *J. Am. Chem. Soc. 72*:3646 (1950). (b) J. P. Paris, Symposium on Chemiluminescence, Durham, N.C., March 31-April 2, 1965.
5. S.-M. Park and A. J. Bard, unpublished results, 1975.
6. (a) W. J. Albery, *Trans. Faraday Soc. 62*:1915 (1966); (b) W. J. Albery and S. Bruckenstein, *ibid. 62*:1920 (1966); (c) W. J. Albery and S. Bruckenstein, *ibid. 62*:1932 (1966); (d) W. J. Albery, S. Bruckenstein, and D. C. Johnson, *ibid. 62*:1938 (1966); (e) W. J. Albery and S. Bruckenstein, *ibid. 62*:1964 (1966); (f) W. J. Albery and S. Bruckenstein, *ibid. 62*:2584 (1966); (g) W. J. Albery and S. Bruckenstein, *ibid. 62*:2596 (1966); (h) W. J. Albery, *ibid. 63*:1771 (1967).
7. (a) L. R. Faulkner and A. J. Bard, in *Electroanalytical Chemistry*, Vol. 10 (A. J. Bard, ed.), Marcel Dekker, New York, 1977; (b) S.-M. Park and D. A. Tryk, *Rev. Chem. Intermed. 4*:43 (1981).

# 16

# Digital Simulation of Electrochemical Problems

J. T. MALOY* / Department of Chemistry, West Virginia University, Morgantown, West Virginia



## I. THE FINITE DIFFERENCE REPRESENTATION OF FICK'S LAWS

Any discussion of the digital simulation of problems involving diffusion begins with a consideration of the combined form of Fick's laws [1],

$$\frac{\partial}{\partial t} C(x,t) = D \frac{\partial^2}{\partial x^2} C(x,t) \qquad (16.1)$$

*Present address: Department of Chemistry, Seton Hall University, South Orange, New Jersey for the concentration C(x,t), a function of position and time, where D is the diffusion coefficient. In terms of its finite difference definition, this law may be written

$$\lim_{t\to 0} \frac{C(x, t + \Delta t) - C(x, t)}{\Delta t} = \lim_{\Delta x\to 0} \frac{D}{(\Delta x)^2}\{[C(x + \Delta x, t) - C(x, t)] - [C(x, t) - C(x - \Delta x, t)]\} \quad (16.2)$$

This may be readily rearranged to yield

$$C(x, t + \Delta t) - C(x, t) = \Delta C(x, t) = \lim_{\substack{\Delta x\to 0 \\ \Delta t\to 0}} \frac{D\ \Delta t}{(\Delta x)^2}[C(x + \Delta x, t) - 2C(x, t) + C(x - \Delta x, t)] \quad (16.3)$$

Two substitutions are necessary to obtain the diffusion algorithm. First, division of both members of Equation 16.3 by C, the bulk concentration, transforms all concentration to fractional values:

$$f(x,t) \equiv \frac{C(x,t)}{C} \quad (16.4)$$

Second, the definition of $D_M$, the model diffusion coefficient, relates $\Delta x$, $\Delta t$, and D:

$$D_M = \lim_{\substack{\Delta x\to 0 \\ \Delta t\to 0}} \frac{D\ \Delta t}{(\Delta x)^2} \quad (16.5)$$

Thus the fundamental equation used in the simulation of diffusion is obtained by substituting Equations 16.4 and 16.5 into Equation 16.3:

$$f(x,t) = D_M[f(x + \Delta x, t) - 2f(x,t) + f(x - \Delta x, t)] \quad (16.6)$$

To use Equation 16.6 to calculate the fractional concentration change resulting from a diffusion process, one partitions the medium (the solution) into an as yet unspecified number of volume elements of thickness $\Delta x$ and requires that f(x,t) be uniform within each element. In the simplest case, these elements would form a linear array

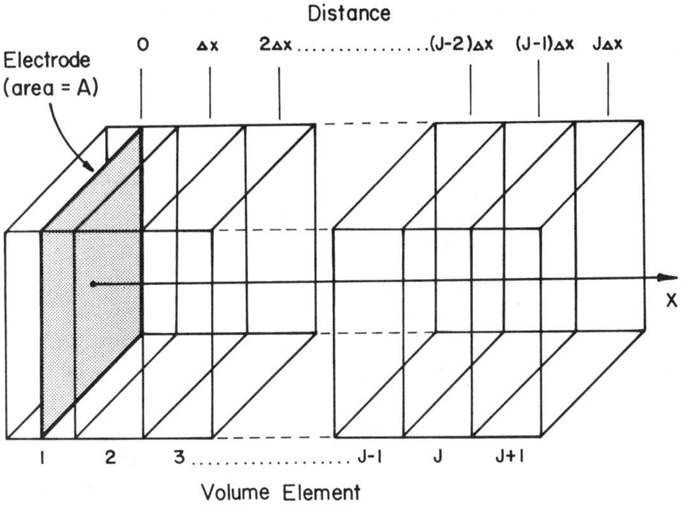


**Figure 16.1** Model volume element array used in elementary digital simulations of electrochemical problems. Note that the planar electrode has been placed in the middle of the first volume element in this model.

originating at a planar surface (the electrode) of area A located at $x = 0$. Therefore, the volume of each element will be $A\,\Delta x$, as represented in Figure 16.1. Each element may be assigned a serial number, J, as indicated in this figure, so that any distance x is given as

$$x = (J - 1)\,\Delta x \tag{16.7}$$

provided that the electrode is placed in the center of the first volume element. [This placement is particularly convenient in modeling the potential-step experiment; other techniques (e.g., chronopotentiometry) are sometimes more easily treated by placing the electrode at the exterior face of the first volume element. This choice is arbitrary but not without consequences, and is discussed further below.]

With this model in mind, one may abbreviate Equation 16.6 as

$$\Delta f(J) = D_M[f(J + 1) - 2f(J) + f(J - 1)] \tag{16.8}$$

to obtain an expression for the change in fractional concentration occurring in the Jth element at any time due to diffusion. In this form, $D_M$ is seen as the proportionality constant governing flux between adjacent elements containing different fractional concentrations.

Applied successively to each element in the array, Equation 16.8 predicts the new fractional concentration,

$$f'(J) = f(J) + \Delta f(J) \tag{16.9}$$

that will occupy each element after diffusion has occurred during the interval $\Delta t$.

Utilization of this basic method of treating diffusion requires

1. The assignment of a value to $D_M$
2. The assurance that $\Delta t$ and $\Delta x$ approach zero
3. The establishment of initial conditions ($t = 0$) and boundary conditions ($x = 0$, $x \to \infty$) so that concentration gradients may be obtained
4. The expression of time, distance, and material flux in experimentally accessible dimensionless parameters instead of the more ambiguous computer parameters (e.g., volume element J) indicated above

## II. DEFINING THE MODEL DIFFUSION COEFFICIENT

Although the assignment of a value to $D_M$ is arbitrary, it is by no means unrestricted. From its definition (Equation 16.5) it is clear that $D_M$ must be positive. To find the upper limit for $D_M$, one need only substitute the Einstein definition [2] of the diffusion coefficient,

$$D = \lim_{\substack{\Delta y \to 0 \\ \Delta t \to 0}} \frac{(\Delta y)^2}{2\Delta t} \tag{16.10}$$

(where $\Delta y$ is the distance a diffusing particle travels during $\Delta t$) into Equation 16.5 to obtain

$$D_M = \lim_{\substack{\Delta x \to 0 \\ \Delta y \to 0}} \frac{(\Delta y)^2}{2(\Delta x)^2} \tag{16.11}$$

Since $\Delta y$ may approach zero in the limit but $\Delta x$ may only approach some small finite value in the finite difference method, it is clear that $(\Delta y)^2 \leqslant (\Delta x)^2$. In other words, particle diffusion between nonadjacent volume elements is forbidden in this model because $\Delta y$ may never exceed $\Delta x$. Thus the maximum acceptable value for $D_M$ is obtained when $\Delta y$ equals $\Delta x$ and the condition

$$\frac{1}{2} \geqslant D_M > 0 \tag{16.12}$$

is a requirement of the technique.

Since the assignment of a value is arbitrary, it would appear that setting $D_M$ equal to its maximum value would tend to minimize $\Delta x$ with respect to $\Delta t$. This is true, but setting $D_M$ equal to 0.50 in the simulation of a simple potential step to diffusion limiting conditions results in simulated currents that oscillate between accurate and very inaccurate representations of the Cottrell equation [3]. These oscillations are especially severe during the early stages of the simulation when relatively large currents are predicted, but they become insignificant sources of error in the predicted current as the number of iterations (each of duration $\Delta t$) is increased. Because the extent of oscillation decreases with decreasing $D_M$, Feldberg has used $D_M = 0.45$ to obtain reliable simulations of current-time behavior [4]. Unfortunately, the prediction of any behavior requiring the numerical integration of simulated current-time data may not be accurate when $D_M$ is this large because of the accumulation of large errors in current during the first few iterations. Marked improvements in the accuracy of simulated charge-time [5,6] or current semi-integral-time [7] behavior is obtained if smaller values of $D_M$ are used; a satisfactory value that has been used in these simulations of potential-step behavior is $D_M =$ 0.2184.

Differences in real diffusion coefficients may also be taken into account using these methods. Because $\Delta x$ and $\Delta t$ would be the same for two different diffusing species A and B, it follows from Equation 16.5 that

$$\frac{D_{MA}}{D_{MB}} = \frac{D_A}{D_B} \tag{16.13}$$

Thus different rates of diffusion may be simulated by defining $D_{MA}$ and $D_{MB}$ in the same proportion as the actual coefficients $D_A$ and $D_B$. The conditions given in Equation 16.12 hold for both $D_{MA}$ and $D_{MB}$, however.

## III. DEFINING $\Delta t$ AND $\Delta x$

As indicated above, Equation 16.8 is applied successively to each element in the array of volume elements to simulate material transfer during the interval $\Delta t$. To define a $\Delta t$ unit one must select some known time ($t_k$) in the physical experiment and partition that known time into L equal intervals for the purpose of the simulation. Thus

$$\Delta t \equiv \frac{t_k}{L} \tag{16.14}$$

and increasing L, the number of iterations representing a known time in the physical experiment, causes $\Delta t$ to approach zero. (The known time parameter $t_k$ may change as different electrochemical techniques are simulated. For example, in the simulation of a double-potential-step experiment, $t_k$ is usually taken to be $t_f$, the interval between the two potential steps. In the simulation of a constant-current experiment, $t_k$ may be taken as $\tau$, the chronopotentiometric transition time; given a known current, the value of L may be calculated during the execution of the simulation rather than being defined. In some simulations, time-related quantities may be used to represent $t_k$. Thus $\Delta E/\nu$ may be used as the known time in potential sweep simulations, while $\omega^{-1}$ may be taken as $t_k$ in the simulation of rotating disk or ring-disk behavior.)

Real times may be related to $t_k$ through Equation 16.14. Thus, if Equation 16.8 has been applied to all the volume elements in the array K times so that diffusion has occurred for a time

$$t = \left(K - \frac{1}{2}\right) \Delta t \tag{16.15}$$

(1/2 is subtracted from K because time is measured midway through each iteration), the real time t may be related to the known time $t_k$ by substituting Equation 16.14 into Equation 16.15:

$$\frac{t}{t_k} = \frac{K - 1/2}{L} \tag{16.16}$$

This provides a dimensionless representation of time throughout the simulation.

Because $\Delta x$, $\Delta t$, and $D_M$ are related by Equation 16.5, the specification of $D_M$, L, and $t_k$ results in the definition of $\Delta x$.

$$\Delta x \equiv \left(\frac{Dt_k}{D_M L}\right)^{1/2} \tag{16.17}$$

Thus, increasing L causes both $\Delta t$ and $\Delta x$ to approach zero. It is important to note, however, that $\Delta x$ decreases with $L^{1/2}$; therefore, once reasonable accuracy has been obtained by increasing L to a large value (ca. 1000), little is to be gained by increasing L any further.

## IV. ESTABLISHING INITIAL AND BOUNDARY CONDITIONS

The establishment of initial conditions is straightforward: usually, an initial relative concentration equal to 1.0 is set in each theoretical volume element. It is possible that the concentration array might be initialized with some preestablished concentration profile other than bulk concentration throughout, but these cases do not arise frequently.

In the simulation of diffusion-limited semi-infinite linear diffusion, the establishment of the electrode boundary condition is equally straightforward: the concentration at the electrode surface is set equal to zero. If the electrode is placed in the middle of the first volume element in the model envisioned, the concentration in that element is set equal to zero to achieve the diffusion-limited condition. If the electrode is placed at the exterior face of the first volume element in the model under consideration, the concentration in the first element must be adjusted to one-third of that in the second element to maintain zero electrode surface concentration (see Fig. 16.2).

The simulation of other electrochemical experiments will require different electrode boundary conditions. The simulation of potential-step Nernstian behavior will require that ratio of reactant and product concentrations at the electrode surface be a fixed function of electrode potential. In the simulation of voltammetry, this ratio is no longer fixed; it is a function of time. Chronopotentiometry may be simulated by fixing the slope of the concentration profile in the vicinity of the electrode surface according to the magnitude of the constant current passed. These other techniques will be discussed below; a model for diffusion-limited semi-infinite linear diffusion will be developed immediately.

In the case of semi-infinite linear diffusion, the concentration of electroactive species A will be maintained at zero at the electrode surface if the diffusion-limited conditions exist. Thus the electrode boundary condition for species A may be written

$$f_A'(1) = f_A(1) = 0 \tag{16.18}$$

For the electrode reaction

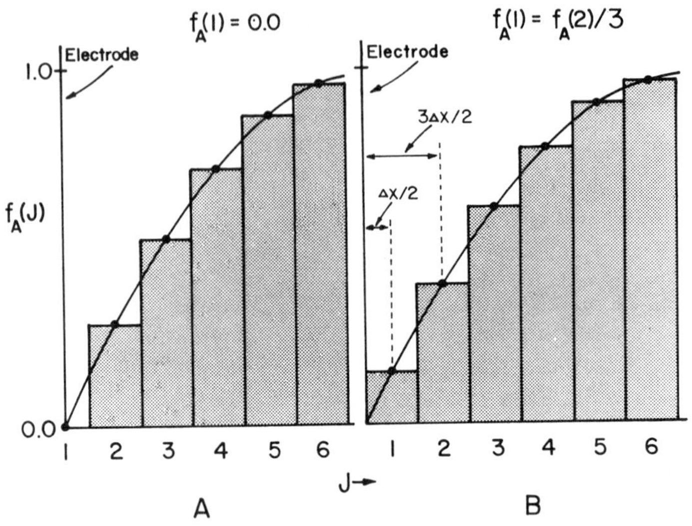


Figure 16.2 Effect of electrode position in establishing the diffusion-limited boundary condition. (A) The electrode has been positioned in the middle of the first volume element so that zero concentration at the electrode surface may be readily maintained. (B) The electrode has been placed at the exterior edge of the first volume element so that zero surface concentration can be obtained only by maintaining the concentration in the first element at one-third of that in the second element.

$$A \rightarrow B$$

the electrode boundary condition for the product B is not as straightforward. It must be written

$$f_B'(1) = f_B(1) + D_{MA} f_A(2) - D_{MB}[f_B(1) - f_B(2)] \tag{16.19}$$

where the second term of the right member accounts for the A converted to B by the electrode reaction while the third term accounts for the diffusion of B out of the first volume element.

At the other boundary, bulk concentration of A must be maintained at some finite distance from the electrode, while the concentration of B will be zero at the same point. This distance may be regarded as the diffusion layer thickness. In terms of the simulation, the establishment of the semi-infinite boundary condition requires the determination of the number of volume elements making up the diffusion layer. This will be a function of the number of time iterations that have taken place up to that point in the simulation. At any time in the physical experiment, the diffusion layer thickness is given by $6(Dt)^{1/2}$. This rule of thumb may be combined with Equation 16.7 to calculate $J_d$, the number of volume elements in the diffusion layer:

$$J_d = \frac{6(Dt)^{1/2}}{\Delta x} + 1 \tag{16.20}$$

Direct substitution of Equations 16.16 and 16.17 into Equation 16.20 yields

$$J_d = 6\sqrt{D_M\left(K - \frac{1}{2}\right)} + 1 \simeq 3\sqrt{2K} \tag{16.21}$$

Thus the number of volume elements required to model the diffusion layer increases with the square root of the iteration number, and the maximum number of volume elements required in any simulation of this sort is given by

$$J_{max} = 3\sqrt{2K_{max}} \tag{16.22}$$

where $K_{max}$ is the serial number of the last time iteration attempted. To maintain the semi-infinite boundary condition, then, one needs only to maintain bulk concentration of electroactive species A in the $J_{max}$th volume element, while holding the product concentration of B at zero at the same point.

These observations bring up an interesting situation that occurs when the diffusion layer is of some fixed thickness d. This occurs when one models thin-layer cell behavior or membrane diffusion. In this case, $\Delta x$ is defined

$$\Delta x = \frac{d}{J_{max} - 1} \tag{16.23}$$

because one partitions the known thickness d into $(J_{max} - 1)$ volume elements. Since d is constant, it is possible to define a known time in terms of the diffusion coefficient

$$t_k = \frac{d^2}{D} \tag{16.24}$$

These definitions may be combined with Equations 16.5 and 16.14 to yield

$$D_M = \frac{(J_{max} - 1)^2}{L} \tag{16.25}$$

This result is generally applicable to any situation in which a fixed-width diffusion medium is partitioned into ($J_{max}$ − 1) volume elements. In this case, the selection of $J_{max}$ fixes the magnitude of the model diffusion coefficient $D_M$. The constraints of Equation 16.12 still apply, however. Therefore, one must resist the temptation to increase the size of $J_{max}$ (without limit) to reduce the size of $\Delta x$. The maximum value of $J_{max}$ selected must be such that $D_M$ is still less than one-half.

## V. DIMENSIONLESS PARAMETERS

The concept of a dimensionless representation of time has been introduced above in Equation 16.16. This was obtained by substituting the definition of $\Delta t$ into Equation 16.15 and rearranging the result so that physical quantities appeared on one side of the equation and model parameters appeared on the other. Since the model parameters are all dimensionless, this method assures that the arrangement of physical quantities obtained will also be dimensionless. This procedure (of substituting in the definitions of $\Delta t$, $\Delta x$, and $D_M$ and rearranging the result) will be followed below in all assignments of dimensionless parameters.

The dimensionless distance parameter may be found by substituting the definition of $\Delta x$ into Equation 16.7. In the case of semi-infinite linear diffusion, Equation 16.17 is used to define $\Delta x$ and this substitution yields

$$x = (J - 1)\left(\frac{Dt_k}{D_M L}\right)^{1/2}$$

which may be rearranged to obtain a dimensionless distance parameter

$$\frac{x}{(Dt_k)^{1/2}} = \frac{J - 1}{(D_M L)^{1/2}} \tag{16.26}$$

Thus the ambiguous "box number" may be transformed to a meaningful dimensionless distance $[x/(Dt_k)^{1/2}]$ by applying Equation 16.26. In the case of thin-layer diffusion, a different expression for $\Delta x$ is used (Eq. 6.23). When this is substituted into Equation 16.7 and rearranged, a different dimensionless distance parameter is obtained:

$$\frac{x}{d} = \frac{J - 1}{J_{max} - 1} \tag{16.27}$$

Dimensionless current representations also exist. If the electrode is placed at the center of the first volume element in the model, this current will be proportional to the material flux into the first element from the second. For electroactive species A, this flux is given by $D_{MA}[f_A(2) - f_A(1)]$, where $f_A(J)$ is the fractional concentration of A in the Jth element. This fractional flux may be converted to moles of flux through multiplication by C (bulk concentration) and A $\Delta x$ (volume element volume, assuming a planar electrode of area A). Appropriate electrochemical conversion and the recognition that this material flux occurs during the interval $\Delta t$ yields the current expression

$$i(t) = \frac{nFAC\ \Delta x D_{MA}[f_A(2) - f_A(1)]}{\Delta t}$$

Substitution of the definition of $\Delta x$ and $\Delta t$ (Eqs. 16.14 and 16.17) and rearrangement yields the appropriate dimensionless current parameter

$$\frac{i(t)t_k^{1/2}}{nFACD_A^{1/2}} = \left(\frac{L}{D_{MA}}\right)^{1/2} D_{MA}[f_A(2) - f_A(1)] \tag{16.28}$$

Thus $(L/D_{MA})^{1/2}$ is the appropriate factor used to convert an ambiguous material flux to a meaningful current term. In the simulations reported below, the dimensionless current parameter is given by Z(K). If diffusion-limited conditions are obtained the relative concentration of electroactive species in the first element is zero. Thus the form most frequently used is

$$Z(K) = \frac{i(t)t_k^{1/2}}{nFACD_A^{1/2}} = \left(\frac{L}{D_{MA}}\right)^{1/2} D_{MA}f_A(2) \tag{16.29}$$

In the case of thin-layer diffusion,

$$Z(K) = \frac{i(t)d}{nFACD_A} = \frac{L}{J_{max} - 1} D_{MA} f_A(2) \qquad (16.30)$$

by substituting Equations 16.24 and 16.25 into Equation 16.29.

Other dimensionless parameters, particularly those related to the study of kinetics, are obtained in a similar manner. These will be discussed below.

## VI. A SAMPLE FORTRAN PROGRAM

It is now possible to write a computer program that will simulate the current-time behavior that occurs following a single potential step to diffusion-limiting conditions. The solution to this problem (the Cottrell equation) is well known and there is no need to perform this simulation to obtain any electrochemical information. However, it is instructive to see how the principles developed above are put to use in an actual program. Further examples of programming will be left as an exercise to the reader.

The annotated FORTRAN program appears in Figure 16.3. This program is suitable for up to 1000 time iterations; therefore, 1001 elements are reserved in the DIMENSION statement for Z and T, the dimensionless current and time parameters. This fixes the maximum number of volume elements that will be necessary (Eq. 16.22). Two volume element arrays (old and new) are reserved for each of two diffusing species (A and B). New concentrations are calculated using the diffusion algorithm (Eq. 16.8), and these become old concentrations just before the next time iteration. In this example, L is set at 1000, while $D_{MA}$ and $D_{MB}$ are each set at 0.45. The old concentration arrays are then initialized [$f_A(J) = 1.0$ and $f_B(J) = 0.0$ throughout] and the first time iteration takes place. All the A in the first element is transformed to B at the midpoint of the time iteration. This exhaustive electrolysis results in "unit" current ($\sqrt{L/D_{MA}}$) flow. Each subsequent iteration begins at statement number 69. Electrode boundary conditions are established and the maximum number of volume elements necessary is calculated for each iteration in order to save computer time. The diffusion algorithm is applied to each element in the array and the dimensionless time and current are calculated at the end of each iteration. The new arrays become old arrays and the process is repeated until a maximum number of time iterations (in this case, L) have transpired. The dimensionless current-time curve is then printed out point by point.

```
      DIMENSION FAOLD(136), FANEW(136), FBOLD(136), FBNEW(136),
     1Z(1001), T(1001)
  200 FORMAT(2F20.9)
```

Establishes Concentration and Current Arrays. Sets Format for Print-Out.

```
      L=1000
      XL=L
      DMA=0.45
      DMB=0.45
```

Designates the Defined Variables.

```
      DO 10 J=1,136
      FAOLD(J)=1.0
   10 FBOLD(J)=0.0
```

Initializes the Volume Element Array (t=o).

```
      K=1
      FAOLD(1)=0.0
      FBOLD(1)=1.0
      T(1)=0.5/XL
      Z(1)=SQRT(XL/DMA)
```

Sets up First Iteration Conditions. Calculates First Current-Time Point.

```
   69 K=K+1
      XK=K
```

All Subsequent Iterations Begin Here.

```
      FANEW(1)=0.0
      FBNEW(1)=FBOLD(1)+DMA*FAOLD(2)-DMB*(FBOLD(1)-FBOLD(2))
```

Defines Electrode Boundary Conditions for All Subsequent Iterations.

```
      JMAX=3.0*SQRT(2.0*XK)+1.0
```

Calculates Maximum Value of J.

```
      DO 20 J=2,JMAX
      FANEW(J)=FAOLD(J)+DMA*(FAOLD(J+1)-2.0*FAOLD(J)+FAOLD(J-1))
   20 FBNEW(J)=FBOLD(J)+DMB*(FBOLD(J+1)-2.0*FBOLD(J)+FBOLD(J-1))
```

Employs Diffusion Equation 8 to Calculate New Concentrations in All Other Significant Elements.

```
      T(K)=(XK-0.5)/XL
      Z(K)=DMA*FAOLD(2)*SQRT(XL/DMA)
```

Calculates Each Subsequent Current-Time Point by Equation 29.

```
      DO 30 J=1,JMAX
      FAOLD(J)=FANEW(J)
   30 FBOLD(J)=FBNEW(J)
```

Transforms New Arrays to Old Arrays for Next Iteration.

```
      IF(L-K)68,68,69
```

Starts Next Iteration Unless K≥L.

```
   68 DO 40 K=1,L
   40 PRINT 200,T(K),Z(K)
```

Prints Calculated Current-Time Curve.

```
      STOP
      END
```

Figure 16.3 Sample FORTRAN program.

Selected results obtained with this program appear in Table 16.1. The exact value for Z(K) as obtained from the Cottrell equation is shown here also. [The Cottrell equation may be written

$$i(t) = \frac{nFACD^{1/2}}{\pi^{1/2} t^{1/2}} = \frac{nFACD^{1/2}}{\pi^{1/2} t_k^{1/2}} \left( \frac{L}{K - \frac{1}{2}} \right)^{1/2}$$

by employing Equation 16.16. Rearrangement of this expression results in the determination of an exact expression for Z(K):

$$Z_c(K) = \frac{i(t) t_k^{1/2}}{nFACD^{1/2}} = \frac{1}{\pi^{1/2}} \left( \frac{L}{K - \frac{1}{2}} \right)^{1/2} \qquad (16.31)$$

Table 16.1 Comparison of Simulated and Exact Current-Time Data Obtained from the Sample Program

| K | $T(K) = t/t_k$ | $Z(K) = \dfrac{i(t)t_k^{1/2}}{nFACD^{1/2}}$ (simulated) | $Z_c(K) = \dfrac{i(t)t_k^{1/2}}{nFACD^{1/2}}$ (exact) | Percent error |
|---|---|---|---|---|
| 1 | 0.0005 | 47.14 | 25.23 | 86.8 |
| 5 | 0.0045 | 8.684 | 8.410 | 3.26 |
| 10 | 0.0095 | 5.962 | 5.788 | 3.01 |
| 50 | 0.0495 | 2.548 | 2.536 | 0.47 |
| 100 | 0.0995 | 1.793 | 1.789 | 0.22 |
| 500 | 0.4995 | 0.7987 | 0.7983 | 0.05 |
| 1000 | 0.9995 | 0.5645 | 0.5643 | 0.04 |

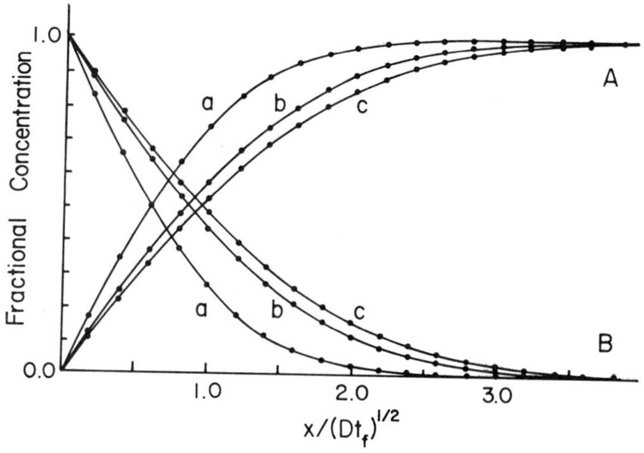


Figure 16.4 Typical concentration profiles obtained using digital simulation. The conditions are identical to those employed in the sample program where the electrode reaction is A ± ne→B. The lines represent the results of the simulation while the points show the exact solution obtained using partial differential equations. For a, b, and c the values of $t/t_f$ employed were 0.40, 0.80, and 1.00, respectively.

which may be compared with Z(K) obtained in the simulation to give some idea of the reliability of the simulation at any point.]

A comparison of the simulated and exact values of Z(K) in Table 16.1 reveals that most of the error associated with the simulated current occurs during the first few iterations. Once K exceeds 50, however, the error in the simulated current is less than 1% and if K exceeds 500, the error drops below 0.1%. This is well within the uncertainty associated with typical electrochemical measurements. The worst error occurs during the first iteration when Z(1) is set equal to $\sqrt{L/D_{MA}}$. Since this choice is somewhat arbitrary anyhow, it may be eliminated by setting K = 1 in Equation 16.31 and using

$$Z(1) = \sqrt{\frac{2L}{\pi}} \tag{16.32}$$

in its place.

Typical concentration profiles obtained using a program similar to that given above are shown in Figure 16.4. The agreement between the results of the simulation (solid lines) and the exact solution (points) is obvious.

## VII. CHRONOCOULOMETRY

The dimensionless current parameter Z(K) may be shown as a function of the dimensionless time parameter $t/t_k$. This is illustrated in Figure 16.5. The mathematical form of this function is not obtained; instead, the function is evaluated at $t/t_k = 0.5/L$ and $L^{-1}$ intervals thereafter. This immediately suggests a method for obtaining the integral of the dimensionless current by summing the areas of the rectangular strips that make up the area under the curve. While $Z(t/t_k)$ is evaluated at the midpoint of each strip, the integral of $Z(t/t_k)$ will always be obtained at a time corresponding to the right edge of the strip (at integer numbers of time iterations).

Formally, then

$$\sum_{K=1}^{K=K_t} Z\left(\frac{K - \frac{1}{2}}{L}\right) L^{-1} = \int_0^{K_t/L} Z\left(\frac{t}{t_k}\right) d\left(\frac{t}{t_k}\right)$$

$$= \int_0^{t/t_k} \frac{i(t/t_k)t_k^{1/2}}{nFACD^{1/2}} d\left(\frac{t}{t_k}\right) \qquad (16.33)$$

where $K_t$ is the iteration number corresponding to time t. Since

$$Q(t) = \int_0^t i(t)\ dt$$

an appropriate change in the variable of integration results in

$$\sum_{K=1}^{K=K_t} Z\left(\frac{K - \frac{1}{2}}{L}\right) L^{-1} = \frac{Q(t)}{nFACD^{1/2} t_k^{1/2}} \qquad (16.34)$$

Thus the sum of the areas of the rectangular strips making up $Z(t/t_k)$ form a dimensionless representation of the total charge passed at that time. In the discussions that follow, the dimensionless charge parameter that makes up the right member of Equation 16.34 is referred to (albeit ambiguously) as $Q(t/t_k)$ or $Q(K)$.

Dimensionless current data generated by the sample program given above may be integrated to obtain $Q(t/t_k)$. Some results are

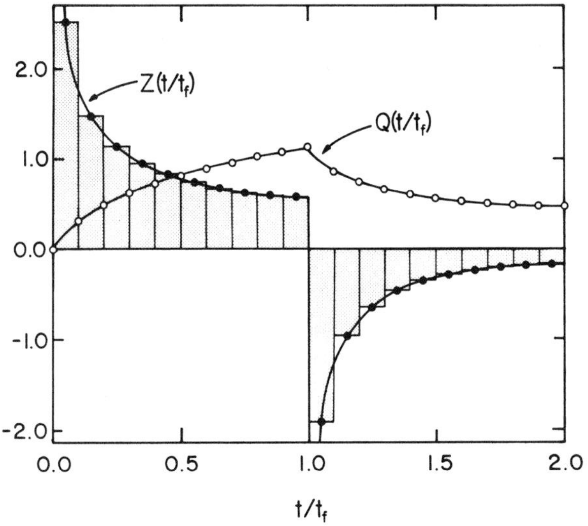


Figure 16.5 Model for the integration of the dimensionless current. Note that while the dimensionless current $[Z(t/t_f) = i(t)_f^{1/2}/nFACD^{1/2}]$ is sampled at the midpoint of each item iteration, the dimensionless charge $[Q(t/t_f) = Q(t)/nFACD^{1/2}t_f^{1/2}]$ is evaluated at the end of each iteration.

shown in Table 16.2 for selected values of $t/t_k$. These results may be compared with the expected value of $Q_c(K)$ obtained from the integration of the Cottrell equation:

$$Q_c(K) = \frac{\int_0^t (nFACD^{1/2}/\pi^{1/2}t^{1/2})\,dt}{nFACD^{1/2}t_k^{1/2}} = \frac{2}{\pi^{1/2}}\left(\frac{t}{t_k}\right)^{1/2} = \left(\frac{4K}{\pi L}\right)^{1/2} \tag{16.35}$$

This comparison is not nearly as encouraging as that in Table 16.1; even after 1000 iterations the simulated chronocoulometry is more than 2% in error. It is encouraging, however, that the absolute error is essentially constant after 10 iterations. This indicates that most of the error is introduced during the early stages of the simulation when the absolute error in the current is quite large. Therefore, some sort of fix-up at these early stages of the simulation is necessary if the accuracy of simulated charge-time curves is to approach that of simulated current-time curves. An immediate improve-

Table 16.2 Comparison of Simulated and Exact Charge-Time Data Obtained from the Results of the Sample Program

| K | $T(K) = t/t_k$ | $Q(K) = \dfrac{Q(t)}{nFACD^{1/2}t_k^{1/2}}$ (simulated) | $Q_c(K) = \dfrac{Q(t)}{nFACD^{1/2}t_k^{1/2}}$ (exact) | Error | Percent error |
|---|---|---|---|---|---|
| 1 | 0.001 | 0.047 | 0.036 | 0.011 | 32.0 |
| 5 | 0.005 | 0.099 | 0.080 | 0.019 | 24.6 |
| 10 | 0.010 | 0.134 | 0.113 | 0.021 | 18.8 |
| 50 | 0.050 | 0.275 | 0.252 | 0.023 | 8.99 |
| 100 | 0.100 | 0.380 | 0.357 | 0.023 | 6.49 |
| 500 | 0.500 | 0.821 | 0.798 | 0.023 | 2.90 |
| 1000 | 1.000 | 1.152 | 1.128 | 0.023 | 2.09 |

Table 16.3 Simulated Current-Time and Charge-Time Data Obtained from the Modified Sample Program[a]

| K | T(K) = $t/t_k$ [defined for Q(K)] | Z(K) | Percent error in Z(K) | Q(K) | Percent error in Q(K) |
|---|---|---|---|---|---|
| 1 | 0.001 | 25.23 | 0.00 | 0.036 | 0.00 |
| 5 | 0.005 | 8.555 | 1.72 | 0.080 | 0.63 |
| 10 | 0.010 | 5.840 | 0.90 | 0.114 | 0.77 |
| 50 | 0.050 | 2.540 | 0.16 | 0.253 | 0.54 |
| 100 | 0.100 | 1.790 | 0.06 | 0.358 | 0.43 |
| 500 | 0.500 | 0.7984 | 0.02 | 0.800 | 0.21 |
| 1000 | 1.000 | 0.5644 | 0.01 | 1.130 | 0.14 |

[a] $Z(1) = Z_c(1)$; $Q(1) = Q_c(1)$; $D_{MA} = 0.2184 \ldots$

ment may be obtained by substituting the exact value for Q(1) as determined from Equation 16.35:

$$Q(1) = \sqrt{\frac{4}{\pi L}} \tag{16.36}$$

Additional improvement may be noted if the simulated current-time curve is rendered more exact through the judicious selection of $D_{MA}$. This may be done by noting that unit relative concentration of A exists in the second volume element at the start of the second time iteration. Therefore, Z(2) as calculated in the simulation will be $\sqrt{LD_{MA}}$. This may be set equal to $L[Q_c(2) - Q_c(1)]$, as obtained from Equation 16.35 and solved for $D_{MA}$. Thus

$$\sqrt{LD_{MA}} = L\left[\left(\frac{8}{\pi L}\right)^{1/2} - \left(\frac{4}{\pi L}\right)^{1/2}\right]$$

and

$$D_{MA} = \left[\frac{2}{\pi^{1/2}}(\sqrt{2} - 1)\right]^2 = 0.2184 \ldots \tag{16.37}$$

This value of $D_{MA}$ will render Q(2) to be exact, at no loss in accuracy for subsequent values of Z(K). This may be seen in Table 16.3, which shows values of Z(K) and Q(K) obtained using this value of $D_{MA}$. With these two most serious sources of error in Q(K) eliminated, errors no greater than 0.3% can be expected in the simulation of chronocoulometry if values of L in the vicinity of 1000 are employed.

## VIII. OTHER NERNSTIAN ELECTRODE BOUNDARY CONDITIONS

One of the more common electrochemical methods treated by digital simulation techniques is double-potential-step chronoamperometry [5,7]. In this method a product B is electrogenerated from a reactant A under diffusion-limited conditions for a period $t_f$ (the conditions of the sample program above). The electrode potential is then reversed to a point such that B is converted back to A under diffusion-limited conditions also. The current so obtained gives a measure of the amount of B remaining in the vicinity of the electrode following electrogeneration and provides a very good measure of the stability of B to homogeneous reaction within the diffusion layer. Therefore, this method is suited to the study of homogeneous kinetics.

To model this method using digital simulation techniques one need only change the electrode boundary conditions after some predetermined number of time iterations (representing $t_f$) have taken place. The electrode boundary conditions become

$$f'_A(1) = f_A(1) + D_{MB}f_B(2) - D_{MA}[f_A(2) - f_A(1)] \tag{16.38}$$

and

$$f'_B(1) = f_B(1) = 0 \tag{16.39}$$

which may be compared with Equations 16.18 and 16.19. The dimensionless current expression becomes

$$Z(K) = -\left(\frac{L}{D_{MA}}\right)^{1/2} D_{MB}f_B(2) \tag{16.40}$$

by comparison with Equation 16.29; note that currents generated under these conditions are opposite in sign to those obtained following the initial potential step. Current passed during this second half-cycle may be compared with the current at some known time during the first half-cycle to get some measure of the stability of B. Methods for simulating the instability of B will be discussed below.

A different set of boundary conditions is obtained during the second half-cycle if the potential of the electrode is stepped to a value such that both B and A are converted to a third oxidation state (D) of the initial substance under diffusion-limited conditions. (A situation like this is encountered in the study of electrogenerated chemiluminescence when an aromatic hydrocarbon is first reduced to its anion radical, and then both the parent and the anion radical are oxidized to the cation radical at the same electrode.) In this case the electrode boundary conditions are

$$f'_A(1) = f_A(1) = f'_B(1) = f_B(1) = 0 \tag{16.41}$$

and

$$f'_D(1) = f_D(1) + D_{MA}f_A(2) + D_{MB}f_B(2) - D_{MD}[f_D(2) - f_D(1)] \tag{16.42}$$

The dimensionless current now results from the flux of two different materials,

$$Z(K) = -\left(\frac{L}{D_{MA}}\right)^{1/2}[n_A D_{MA} f_A(2) + n_B D_{MB} f_B(2)] \tag{16.43}$$

where $n_A$ and $n_B$ now take into account the differing number of faradays transferred per mole of A and B. In general, the application of the principles developed above to other multicomponent systems is equally straightforward.

In the event that the electrode potential is stepped to a level that does not bring about diffusion-limiting conditions, the relative concentration of A and B at the electrode surface must be calculated. If the reaction

$$A + ne \rightleftarrows B$$

is reversible, the ratio of the electrode surface concentration of A to that of B is given by the Nernst equation

$$\theta = \frac{[A]_{x=0}}{[B]_{x=0}} = \frac{f_A(1)}{f_B(1)} = \exp\left[\frac{nF}{RT}(E - E^{\circ\prime})\right] \tag{16.44}$$

where all quantities have their usual electrochemical significance. Given the necessary potential parameters, one may calculate $\theta$. This may be used in the simulation to calculate a new electrode surface concentration ratio from any set of surface concentrations

$$\theta = \frac{f'_A(1)}{f'_B(1)} = \frac{f_A(1) - \Delta f_A(1)}{f_B(1) + \Delta f_A(1)} \tag{16.45}$$

The change of fractional surface A concentration (which is proportional to the current) may be obtained by rearranging Equation 16.45:

$$\Delta f_A(1) = \frac{f_A(1) - \theta f_B(1)}{\theta + 1} \tag{16.46}$$

In the simulation it is probably best to allow diffusion to occur at the electrode surface before using Equation 16.46 to adjust the surface concentrations:

$$\begin{aligned} f'_A(1) &= f_A(1) + D_{MA}[f_A(2) - f_A(1)] \\ f'_B(1) &= f_B(1) - D_{MB}[f_B(1) - f_B(2)] \end{aligned} \tag{16.47}$$

These new surface concentrations of A and B may then be used to calculate $\Delta f_A(1)$, which may be used to achieve the desired concentration ratio

$$\begin{aligned} f''_A(1) &= f'_A(1) - \Delta f_A(1) \\ f''_B(1) &= f'_B(1) + \Delta f_A(1) \end{aligned} \tag{16.48}$$

The current is calculated from the quantity $\Delta f_A(1)$ also

$$Z(K) = \left(\frac{L}{D_{MA}}\right)^{1/2} \Delta f_A(1) \tag{16.49}$$

rather than from the flux into the first element. This same general procedure is used to simulate linear sweep voltammetry (when $\theta$ is a functional of time).

## IX. HOMOGENEOUS KINETICS

The real power of digital simulation techniques lies in their ability to predict current-potential-time relationships when the reactants or products of an electrode reaction participate in some intervening chemical reaction. These "kinetic complications" often result in a fairly difficult differential equation (when combined with the conditions for diffusion or convection encountered in electrochemical problems) that resists solution by ordinary means. Through simulation, however, the effect of any number of chemical steps may be predicted. In practice, it is best to limit these predictions to cases where the reactants and products participate in one or two rate-determining steps; each independent step adds another dimensionless kinetics parameter that must be varied over the range of experimental significance in order to predict the effect of that step. The method for obtaining these dimensionless kinetic parameters will be discussed in this section; their use in the simulation will also be illustrated.

Consider the electrode reaction

$$A \pm ne \rightleftharpoons B \tag{16.50}$$

where the product B undergoes subsequent chemical reaction. This may be a simple first-order (or pseudo-first-order) decomposition

$$B \xrightarrow{k_1} C \tag{16.51}$$

where C may be electroactive (as in the ECE mechanism) or may not be electroactive (as in the EC mechanism) at the potential at which B is formed. Equation 16.50 represents a reaction taking place at the electrode surface and its effect on electrode boundary conditions in the simulation has already been discussed; Equation 16.51 represents a homogeneous reaction that occurs in each volume element of the diffusion layer array. In order to account for the effect of Equation 16.51 on each element of the array, one must find an expression for the fractional change in the concentration of B due to the chemical reaction. This is obtained from the rate law for Equation 16.51:

$$-\frac{d[B]}{dt} = \frac{d[C]}{dt} = k_1[B]$$

by dividing through by the bulk concentration and rewriting the expression in finite difference form

$$-\frac{\Delta f_B(J)}{\Delta t} = \frac{\Delta f_C(J)}{\Delta t} = k_1 f_B(J)$$

Rearrangement and combination with Equation 16.14 yields

$$-\Delta f_B(J) = \Delta f_C(J) = k_1 \, \Delta t f_B(J) = \frac{[k_1 t_k]}{L} f_B(J) \qquad (16.52)$$

Introduced as a factor in the last term of Equation 16.52 is the dimensionless first-order kinetics parameter $k_1 t_k$. Variations in this parameter will allow the effect of any first-order reaction to be simulated.

In the simulation of the EC mechanism, a double potential step may be employed; these boundary conditions have been discussed previously. Diffusion would be modeled in the usual way. Following the diffusion step, however, the concentration in each element of the resulting array is adjusted according to Equation 16.52 using the input parameter $k_1 t_k$:

$$f'_B(J) = f_B(J)\left(1 - \frac{[k_1 t_k]}{L}\right)$$
$$f'_C(J) = f_C(J) + \frac{[k_1 t_k]}{L} f_B(J) \qquad (16.53)$$

This process is repeated throughout the simulation and the current obtained at some known time during the first half-cycle is compared with that obtained at some time during the second. The simulation is then repeated and a new value is used for the kinetics parameter $k_1 t_k$; another comparison of currents is made. The variation of the current ratio with values of $k_1 t_f$ when displayed graphically is known as a working curve. Typical working curves are shown in Figure 16.6. These working curves may be compared with experimental data in an attempt to elucidate the mechanism and evaluate the rate constant $k_1$.

The simulation of the ECE mechanism may also employ the double-potential-step technique, but a working curve can be constructed from single-potential step data also. This is because some of the current that passes as A is converted to B is due to the electrolysis of C, the decomposition product of B. The greater the decomposition rate of B, the more current flows, approaching the rate of

$$A \pm (n_A + n_c)e \rightleftarrows D$$

in the limit of fast kinetics (where D is the product of the electrode reaction of C). The treatment of the electrode boundary condition in

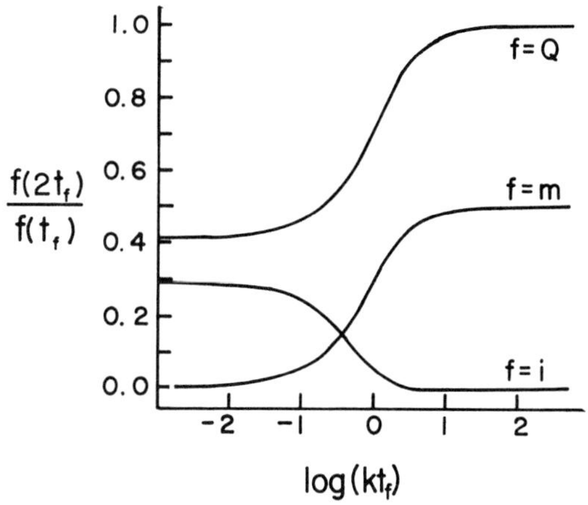


Figure 16.6 Typical working curves for the first-order EC mechanism. The curve shown for f = i shows the behavior of the current ratio $[i(2t_f)/i(t_f)]$ as a function of $kt_f$. That given for f = Q shows the charge ratio, while that given for f = m shows the current semi-integral ratio over the same range. (From Ref. 7.)

this case is along the same lines as the treatment of multicomponent systems given above:

$$f'_A(1) = f_A(1) = 0$$

$$f'_B(1) = f_B(1) + D_{MA} f_A(2) - D_{MB}[f_B(1) - f_B(2)]$$

$$f''_B(1) = f'_B(1)\left(1 - \frac{[k_1 t_k]}{L}\right) \tag{16.54}$$

$$f'_C(1) = 0$$

$$f'_D(1) = f_D(1) + D_{MC} f_C(2) - D_{MD}[f_D(1) - f_D(2)]$$

$$f''_D(1) = f'_D(1) + \frac{[k_1 t_k] f'_B(1)}{L} \tag{16.55}$$

Equations 16.54 and 16.55 result from the fact that the ongoing chemical reaction occurs at the electrode surface also; since this does take place at the electrode, however, the decomposition product of B is immediately converted to D. This flux must be accounted for in the determination of the dimensionless current parameter for that iteration

$$Z(K) = \left(\frac{L}{D_{MA}}\right)^{1/2} \left[D_{MA} f_A(2) + D_{MC} f_C(2) + \frac{[k_1 t_k] f'_B(1)}{L}\right] \tag{16.56}$$

The treatment of second-order kinetics is equally facile. Given the chemical reaction

$$B + A \xrightarrow{k_2} D$$

one may write the rate law

$$-\frac{d[B]}{dt} = -\frac{d[A]}{dt} = \frac{d[D]}{dt} = k_2[A][B]$$

and the corresponding finite difference representation

$$-\frac{\Delta f_B(J)}{\Delta t} = -\frac{\Delta f_A(J)}{\Delta t} = \frac{\Delta f_D(J)}{\Delta t} = k_2 C f_A(J) f_B(J) \tag{16.57}$$

where C represents bulk concentration of A. This may be rearranged with Equation 16.14 to yield

$$-\Delta f_B(J) = -\Delta f_A(J) = \Delta f_D(J) = (k_2 C \, \Delta t) f_A(J) f_B(J)$$
$$= \frac{[k_2 C t_k]}{L} f_A(J) f_B(J) \tag{16.58}$$

which shows $k_2Ct_k$ as the proper dimensionless kinetics parameter for a second-order reaction. This would be supplied as an input parameter and kinetically adjusted concentrations would be determined as indicated previously:

$$f'_A(J) = f_A(J)\left(1 - \frac{[k_2 C t_k] f_B(J)}{L}\right)$$
$$f'_B(J) = f_B(J)\left(1 - \frac{[k_2 C t_k] f_A(J)}{L}\right) \tag{16.59}$$

$$f'_D(J) = f_D(J) + \frac{[k_2Ct_k]f_A(J)f_B(J)}{L}$$

No special treatment like Equation 16.54 is necessary in the calculation of the surface condition, even if D is assumed electroactive, if diffusion-limited conditions are obtained; since $f_A(1) = 0$ under these circumstances, no conversion of B to D is possible in the first volume element.

Dimerizations, which yield nonlinear partial differential equations if treated rigorously, may be handled with ease using digital simulation techniques. In this case the reaction

$$2B \xrightarrow{k_2} D$$

yields the rate law

$$-\frac{1}{2}\frac{d[B]}{dt} = \frac{d[D]}{dt} = k_2[B]^2$$

which provides the finite difference form used in the simulation

$$-\Delta f_B(J) = 2\Delta f_D(J) = 2\frac{[k_2Ct_k]}{L} f_B(J)^2 \qquad (16.60)$$

This furnishes the method for determining the kinetically adjusted concentrations of B and D:

$$\begin{aligned} f'_B(J) &= f_B(J)\left(1 - \frac{2[k_2Ct_k]f_B(J)}{L}\right) \\ f'_D(J) &= f_D(J) + \frac{[k_2Ct_k]f_B(J)^2}{L} \end{aligned} \qquad (16.61)$$

One admonition is necessary concerning the maximum kinetics parameter that may be employed in any simulation. As is evident in Equations 16.53, 16.59, and 16.61, if too large a value is used for $k_1t_k$ or $k_2Ct_k$, a negative concentration may result. Since this is impossible, any values of these kinetics parameters that yield negative concentrations may be regarded as too large for treatment by these methods. In the case of first-order kinetics, the maximum rate that can be handled occurs when

$$k_1 t_k = L \tag{16.62}$$

Any values of $k_1 t_k$ in excess of this will produce erroneous results. Similarly, the upper limit of $k_2 C t_k$ is approached when

$$k_2 C t_k = L \tag{16.63}$$

(or L/2 in the case of Eq. 16.61). Values higher than this should be attempted only when some check for negative concentrations is built into the simulation. It is of interest to note that in the case of a second-order reaction, this restriction defines the rate of diffusion control for the purpose of the simulation

$$k_d = \frac{L}{C t_k} \tag{16.64}$$

Substituting in "typical" electrochemical values for L, C, and $t_k$, one obtains $k_d \simeq 10^8$ L/(mol s). Comparison of this result with the equation of Osborne and Porter [8] shows that diffusion control in the simulation occurs at a rate two orders of magnitude less than predicted by accepted theory. In other words, one cannot distinguish any difference between the effects of second-order rate constants in the vicinity of $10^8$ L/(mol s) from those in the vicinity of $10^{10}$ L/(mol s) by using digital simulation.

## X. PARAMETRIC SUBSTITUTIONS

The definition of a second time-dependent variable (such as $k_1 t_k$ or $k_2 C t_k$) within a simulation permits one to have some flexibility in the way other time dependent quantities are displayed. For example, in the case where no kinetic complications are introduced, one has no choice but to reference all times with respect to some known time in the physical experiment (Eq. 16.16). With the introduction of $k_1 t_k$ or $k_2 C t_k$, however, experimental times may be referenced with respect to $k_1^{-1}$ or $(k_2 C)^{-1}$. That is,

$$\frac{t}{t_k} k_1 t_k = k_1 t = k_1 t_k \frac{K - \frac{1}{2}}{L} \tag{16.65}$$

Thus one may obtain kt by multiplying the quantity previously referred to as the dimensionless time by $k_1 t_f$, the dimensionless rate constant. This is particularly useful in constructing the single-potential-step working curve for the ECE mechanism mentioned above. This parametric substitution allows the experimental time to be rendered dimensionless by the inverse of the rate constant instead of by some known time $t_k$.

Similarly, the dimensionless current may be expressed in terms of $k_1^{-1}$ instead of $t_k$. That is,

$$\frac{i(t)t_k^{1/2}}{nFACD^{1/2}}(k_1 t_k)^{-1/2} = \frac{i(t)}{nFACk_1^{1/2}D^{1/2}} = \frac{Z(K)}{\sqrt{k_1 t_k}} \qquad (16.66)$$

This parametric substitution allows the current to be rendered dimensionless by a quantity that is independent of the time of the physical experiment (all the factors of $nFACk_1^{1/2}D^{1/2}$ are constant for a given

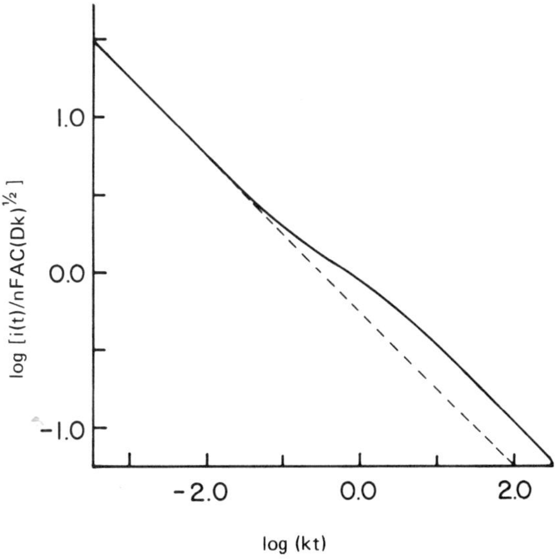


Figure 16.7 Working curve for the ECE mechanism obtained by parametric substitution. Note that this represents the current-time behavior over five time decades. The slope of the dashed line is negative one-half, and the separation between it and the parallel working curve immediately above it is 0.301 logarithmic units.

physical system). A typical working curve for the ECE mechanism obtained using these parametric substitutions is illustrated in Figure 16.7.

It must be emphasized that no new information is obtained through these parametric substitutions. Rather, the information that comes from the simulation is presented in an alternative form that may be more understandable or more readily compared with experimental data. Neither are the parametric substitutions discussed herein the only ones possible. The variety of possible substitutions is greatly increased when two different kinetic processes are considered in the model. The consideration of these possibilities is left as an exercise to the reader.

## XI. ELECTROGENERATED CHEMILUMINESCENCE

No discussion of digital simulation would be complete without some mention of the methods employed in treating electrogenerated chemiluminescence (ECL). In the most common statement of the problem, an aromatic hydrocarbon (A) is alternately oxidized and reduced at a single electrode to produce its radical anion and cation species (B and D). These species, in turn, react within the diffusion layer to regenerate the hydrocarbon.

$$\mathrm{B} + \mathrm{D} \xrightarrow{k_2} 2\mathrm{A} \tag{16.67}$$

The boundary conditions and the necessary kinetic considerations have been introduced previously and will not be considered here. However, some fraction ($\phi_{ECL}$) of the radical ions reacting according to Equation 16.67 produces a molecule of A in an excited state that goes on to produce a photon. The sum of this photon flux originating in the diffusion layer may be related to I(t), the intensity of ECL emission.

To define an ECL intensity parameter one first notes that the photon flux from an individual volume element is given by $\phi_{ECL}(A\ \Delta x)$ $C(k_2Ct_k)f_B(J)f_D(J)/L$. The total ECL intensity during any iteration is given by

$$I(t) = \frac{\phi_{ECL} A\ \Delta x\ C(k_2Ct_k)}{\Delta t\ L} \sum_J f_B(J)f_D(J) \tag{16.68}$$

where the sum is taken over all volume elements. Substitution of Equations 16.14 and 16.17 into Equation 16.68 and rearrangement yields a dimensionless representation of ECL intensity

$$\frac{I(t)t_k^{1/2}}{\phi_{ECL}ACD^{1/2}} = \left(\frac{L}{D_M}\right)^{1/2} \frac{[k_2Ct_k]}{L} \sum_J f_B(J)f_D(J) \tag{16.69}$$

where I(t) is expressed in einsteins/s. This dimensionless intensity factor may be rendered independent of $t_k$ through an appropriate parametric substitution [division of both members of Eq. 16.69 by $(k_2Ct_k)^{1/2}$]. Numerous examples of digital simulations of ECL problems exist in the literature [9-13].

## XII. CHRONOPOTENTIOMETRY

As indicated previously, in the simulation of chronopotentiometric behavior one writes the computer program to calculate L, the number of iterations associated with $\tau$, the transition time. This is achieved by defining a constant current $i_c$ such that

$$Z(K) = \frac{i_c\tau^{1/2}}{nFACD^{1/2}} = \left(\frac{L}{D_M}\right)^{1/2} \delta = \text{constant} \tag{16.70}$$

by analogy with Equation 16.28. The fractional concentration $\delta$, a defined constant, is simply subtracted from $f_A(1)$ and added to $f_B(1)$ during each iteration to simulate constant-current flow:

$$\begin{aligned} f'_A(1) &= f_A(1) + D_{MA}[f_A(2) - f_A(1)] - \delta \\ f'_B(1) &= f_B(1) - D_{MB}[f_B(1) - f_B(2)] + \delta \end{aligned} \tag{16.71}$$

Because the transition time corresponds to that time at which the concentration of A reaches zero at the electrode surface, a record is kept of the number of iterations required to achieve this end. This represents L in Equation 16.70 and combined with input parameters $D_M$ and $\delta$, this allows the computation of Z(K) at the conclusion of the simulation.

During the course of the simulation the most important variables are the electrode surface concentrations of A and B because they determine E(t). These may be calculated directly by placing the electrode in the center of the first volume element in the model as in previous simulations. In this case, however, it turns out to be more straightforward to place the electrode at the exterior edge of the first

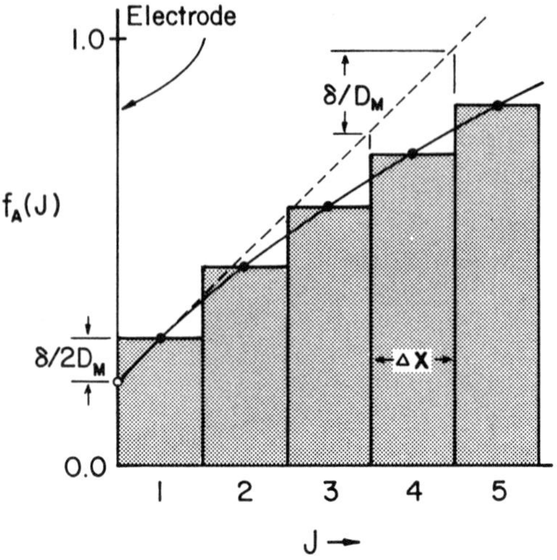


Figure 16.8 Model used in the digital simulation of chronopotentiometric behavior. A constant amount ($\delta$) is removed from the contents of the first element following the diffusion step. The electrode surface concentration is determined by extrapolating the line of known slope through the point representing first element concentration to the electrode surface.

volume and to calculate the electrode surface concentrations by extrapolating the concentration profiles to x = 0. This is illustrated in Figure 16.8. The extrapolation is made easier by the fact that one boundary condition in constant-current electrolysis requires that the concentration gradient at the electrode surface be constant:

$$\frac{\partial C(0,t)}{\partial x} = \frac{i_c}{nFAD} = \text{constant}$$

Written in finite difference form this becomes

$$\frac{\Delta f_A}{\Delta x} = \frac{i_c}{nFACD} = \text{constant} \qquad (16.72)$$

Equation 16.72 may be solved for $\Delta f_A$, the change in fractional A concentration over one volume element, and combined with Equation 16.17

$$\Delta f_A = \frac{i_c \Delta x}{nFACD} = \frac{i_c \tau^{1/2}}{nFACD^{1/2}} (D_M L)^{-1/2} = \frac{\delta}{D_M} \tag{16.73}$$

Inspection of Figure 16.8 reveals that the distance of extrapolation is one-half of a volume element. Therefore, the electrode surface concentration is given by

$$f_A(x = 0) = f_A(1) - \frac{\delta}{2D_M} \tag{16.74}$$

$$f_B(x = 0) = f_B(1) + \frac{\delta}{2D_M}$$

These surface concentrations may be used in the determination of L or in the calculation of E(t) by means of the Nernst equation.

Some comments about the size of the input parameter $\delta$ are necessary. If $\delta$ is too large, not enough iterations will take place before the transition time to make the simulation accurate. If $\delta$ is too small, a great deal of computer time will be wasted reaching the transition point. A useful technique is to estimate the number of iterations that one would like to represent the transition time. One may then use this estimate to calculate $\delta$ by means of the Sand equation:

$$\frac{i_c \tau^{1/2}}{nFACD^{1/2}} = \frac{\pi^{1/2}}{2} = \left(\frac{L'}{D_M}\right)^{1/2} \delta \tag{16.75}$$

where L' represents the estimated value for L. Rearrangement yields

$$\delta = \sqrt{\frac{\pi D_M}{4L'}} \tag{16.76}$$

In kinetically uncomplicated simulations of chronopotentiometric behavior the value of L that is calculated during the simulation is found to agree with L' within a fraction of one iteration out of a thousand. The corresponding simulated potential-time behavior is generally accurate within 0.1 mV.

## XIII. LINEAR SWEEP VOLTAMMETRY

One can also use these methods to model linear sweep voltammetric behavior. To do this, one first selects an appropriate potential scan range

$$\Delta E = E_1 - E_2 \tag{16.77}$$

where $E_1$ and $E_2$ are the initial and final potentials (vs. $E^{\circ\prime}$ for the electrode reaction) to be achieved with $E_1$ the more positive of the two. The combination of $\Delta E$ with $\nu$, the potential scan rate, yields the known time parameter for the simulation

$$t_k = \frac{\Delta E}{\nu} \tag{16.78}$$

For a potential scan in the negative direction, then

$$E = E_1 - \nu t = E_1 - \nu \frac{K - \frac{1}{2}}{L} t_k = E_1 - \frac{K - \frac{1}{2}}{L} \Delta E \tag{16.79}$$

so that the potential scan range is seen to have been partitioned into L equal parts.

Equation 16.79 may be combined with Equation 16.44 to calculate

$$\theta = \frac{f_A(1)}{f_B(1)} = \exp\left[\frac{nF}{RT}\left(E_1 - \frac{K - \frac{1}{2}}{L} \Delta E - E^{\circ\prime}\right)\right] \tag{16.80}$$

where $\theta$ now depends on K. This time-dependent $\theta$ may be used to calculate $\Delta f_A(1)$ according to Equation 16.46, which in turn may be used to calculate the dimensionless current according to Equation 16.49. In this case, however, the dimensionless current is given by

$$Z(K) = \frac{i(t)\nu^{1/2}}{nFACD^{1/2}(\Delta E)^{1/2}} \tag{16.81}$$

in view of Equation 16.78. Examples of simulated linear sweep voltammetry have been reported in the literature [14].

## XIV. SIMULATION OF ROTATING DISK HYDRODYNAMICS

The study of rotating disk electrode behavior provides a unique opportunity to develop a model that predicts the effect of diffusion and convection on the current. This is one of the few convective systems that have simple hydrodynamic equations that may be combined with the diffusion model developed herein to produce meaningful results. The effect of diffusion is modeled exactly as it has been done previously. The effect of convection is treated by integrating an approximate velocity equation to determine the extent of convective flow during a given $\Delta t$ interval. Matter, then, is simply transferred from volume element to volume element in accord with this result to simulate convection. The whole process repeated results in a steady-state concentration profile and a steady-state representation of the current (the Levich equation).

In rotating disk hydrodynamics, an approximate equation describing the velocity of the fluid in the vicinity of the electrode is given by

$$v_x = \frac{dx}{dt} = -0.51\sqrt{\frac{\omega^3}{\nu}}\,x^2 \tag{16.82}$$

where $\omega$ is the angular velocity of the rotating electrode, $\nu$ is the kinetic viscosity of the solution, and x is measured on an axis perpendicular to the rotating disk. Integration of Equation 16.82 yields

$$\int_{x_2}^{x_1} \frac{dx}{x^2} = \frac{1}{x_2} - \frac{1}{x_1} = -0.51\sqrt{\frac{\omega^3}{\nu}}\int_{t_1}^{t_2} dt = -0.51\sqrt{\frac{\omega^3}{\nu}}\,\Delta t \tag{16.83}$$

where $x_2$ represents the position of a given volume element of fluid initially (at $t_1$) and $x_1$ represents the position of that same fluid after the interval $\Delta t$. Equation 16.83 may be solved for $x_2$ to obtain an expression for the initial position of the fluid volume element that resides at $x_1$ at the conclusion of the $\Delta t$ interval

$$x_2 = \frac{x_1}{1 - 0.51\sqrt{\omega^3/\nu}\,\Delta t\, x_1} \tag{16.84}$$

Substitution of Equations 16.14 and 16.26 into Equation 16.84 results in

$$J_2 - 1 = (J_1 - 1) \Bigg/ \left(1 - 0.51\sqrt{\frac{\omega^3 D t_k^3}{\nu}}\ \frac{J_1 - 1}{\sqrt{D_M L^3}}\right) \qquad (16.85)$$

where $J_1$ and $J_2$ are volume element numbers corresponding to $x_1$ and $x_2$. Examination of Equation 16.85 reveals a simplification that occurs if

$$t_k \equiv \left(\frac{\nu}{D}\right)^{1/3} \omega^{-1} \qquad (16.86)$$

for this simulation; Equation 16.85 then becomes

$$J_2 - 1 = \frac{J_1 - 1}{1 - 0.51(J_1 - 1)/\sqrt{D_M L^3}} \qquad (16.87)$$

Thus given the volume element $J_1$, one may calculate the serial number of that volume element $J_2$ that supplies material to element $J_1$ during one iteration. In theory, one then simply transfers the contents of element $J_2$ to element $J_1$ in order to simulate convection.

Of course, it is not quite that simple. Given an integer value for $J_1$, it is quite unlikely that an integer will be obtained for $J_2$. Thus some sort of interpolation scheme must be devised to obtain the relative concentrations of the species of interest at noninteger values of $J_2$; these concentrations are then transferred to element $J_1$. In a typical program, this process would take place after the diffusion step in any iteration but before any kinetic perturbations are applied in that iteration. Care must be exercised to transfer to interior elements first so that one does not write over a meaningful value in the array.

Finally, it should be noted that the choice of $t_k$ indicated in Equation 16.86 provides the proper hydrodynamic expression for Z(K). The steady-state value of Z(K) obtained by this method (with L = 1000) is 0.61. That is,

$$\lim_{K\to\infty} Z(K) = \frac{i_{lim}\nu^{1/6}}{nFACD^{5/6}\omega^{1/2}} = 0.61 \tag{16.88}$$

This expression is identical to the Levich equation within the reliability of the simulation.

## XV. SIMULATION OF ROTATING RING-DISK BEHAVIOR

The methods developed for rotating disk hydrodynamics are easily extended to ring-disk electrodes. This extension is facilitated by the fact that one needs only to consider radial convection in addition to linear convection and diffusion to extend disk methods to ring-disk problems. This does require some model development, however, because mass transfer is now permitted in two directions, x and r. Thus a linear volume element array is no longer sufficient, and a two-dimensional volume element array is required. The linear axis of this array is identical to that developed above for linear diffusion and convection. Thus linear distances are still expressed in volume element units

$$x = (J_x - 1)\,\Delta x \tag{16.89}$$

Radial distances are expressed in a similar fashion

$$r = (J_r - 1)\,\Delta r \tag{16.90}$$

where $\Delta r$ is defined in terms of $r_c$, the radial distance to the outermost edge of the ring electrode

$$\Delta r = \frac{r_c}{J_c - \frac{1}{2}} \tag{16.91}$$

where $J_c$ is the serial number associated with the outermost ring volume element (ca. 100). Other radial dimensions such as $r_a$, the radius of the disk, and $r_b$, the radial distance to the innermost edge of the ring, may be expressed in terms of Equations 16.90 and 16.91.

The approximate equation describing radial convection in rotating disk and ring-disk hydrodynamics is

$$v_r = \frac{dr}{dt} = 0.51 \sqrt{\frac{\omega^3}{\nu}}\, x\, r \qquad (16.92)$$

This may be integrated just as Equation 16.82 was to obtain an expression for the position of a radial fluid volume element at the end of a $\Delta t$ interval with respect to its position at the start of the interval:

$$\ln \frac{r_2}{r_1} = 0.51 \sqrt{\frac{\omega^3}{\nu}}\, x \Delta t \qquad (16.93)$$

where $r_1$ represents the position at the start of the interval and $r_2$ the position at the end. Substitution of Equations 16.14, 16.17, 16.86, 16.89, and 16.90 into Equation 16.93 yields

$$\ln \frac{J_{r_2} - 1}{J_{r_1} - 1} = 0.51 \frac{J_x - 1}{\sqrt{D_M L^3}} \qquad (16.94)$$

This may be rearranged to permit one to calculate the radial position of a given fluid volume element at the start of an iteration given the linear and radial position of that element at the conclusion of the iteration:

$$J_{r_1} - 1 = \frac{J_{r_2} - 1}{\exp[0.51(J_x - 1)/\sqrt{D_M L^3}]} \qquad (16.95)$$

Equation 16.95 represents the algorithm for simulating radial convection. Fractional concentrations, now a function of $J_x$ and $J_r$, found in the element $(J_x, J_{r_1})$ are transferred to the element $(J_x, J_{r_2})$ according to Equation 16.95. Since $J_{r_2}$ need not be an integer, some interpolation scheme must be developed to calculate $f_A(J_x, J_{r_2})$ and $f_B(J_x, J_{r_2})$, but this is accomplished easily. Some care must be exercised to transfer concentrations to outermost radial elements first to maintain the proper radial concentration gradient. And, of course, all volume elements must be subjected to the same treatment for linear convection and linear diffusion that was developed in the preceding section. The digital simulation technique does permit one to handle this fairly complicated problem of linear diffusion, linear convection, and radial convection with relative ease. Of course, the effects of homogeneous kinetics may be superimposed on this problem of mass transfer by using the methods developed previously. Examples of simulations that predict the effect of kinetic perturbation on rotating ring-disk electrode behavior have appeared in the literature [10, 15-19].

## XVI. BEYOND THE BASIC TECHNIQUES

Basic digital simulation techniques have been demonstrated to be most useful in predicting the effects of diffusion, convection, and simple chemical reactions on a wide variety of current-potential-time experiments. The methods outlined above have been employed to handle most of the elementary electrochemical problems that have developed over the past two decades, and it seems reasonable to assume that digital simulation methods will continue to be useful in solving the problems associated with mass transfer and chemical reaction. It is obvious, however, that the supply of simple kinetic problems is probably now exhausted. Future applications will require the simulation of more complicated kinetic schemes which require multiparameter fits. Many of these problems will be handled more readily by making the steady-state assumption in the development of the algorithm that predicts the effect of the chemical reaction on the results of the experiment. This has already been done in at least two cases [20-25] and similar contributions are anticipated in the future.

The use of the steady-state assumption in digital simulations is illustrated in the first work cited above, the development of a model for enzyme catalysis in the amperometric enzyme electrode. In this model, the enzyme E is considered to be immobilized in a diffusion medium covering an electrode that is operated at a potential such that the product (P) of enzyme catalysis is electroactive. The substrate (S) diffuses through the medium that contains the immobilized enzyme and is catalyzed to form P by the usual enzyme kinetics

$$S + E \underset{k_2}{\overset{k_1}{\rightleftharpoons}} ES \xrightarrow{k_3} P \qquad (16.96)$$

The membrane containing the immobilized enzyme is handled by partitioning it into a specified number of volume elements so that Equations 16.23 and 16.24 are valid in this model. While the concentration of each species may vary from element to element, the steady-state assumption (d[ES]/dt = 0) may be invoked independently for each volume element. This results in the definition of the Michaelis-Menton constant, $K_M$:

$$K_M = \frac{k_2 + k_3}{k_1} = \frac{[E][S]}{[ES]} \tag{16.97}$$

The rate of product formation (and substrate loss) is given by

$$\frac{d[P]}{dt} = k_3[ES] \tag{16.98}$$

If $C_E$ is taken as the total enzyme concentration in each volume element so that

$$C_E = [E] + [ES] \tag{16.99}$$

a differential rate law may be formulated by combining Equations 16.96-16.98,

$$\frac{d[P]}{dt} = -\frac{d[S]}{dt} = \frac{k_3 C_E}{(K_M/[S]) + 1} \tag{16.100}$$

In finite difference form this becomes

$$\frac{\Delta f_P(J)}{\Delta t} = -\frac{\Delta f_S(J)}{\Delta t} = \frac{k_3 C_E/C}{(K_M/C)/f_S(J) + 1} \tag{16.101}$$

Combination of Equation 16.101 with Equations 16.14 and 16.24 and rearrangement yields

$$\Delta f_P(J) = \Delta f_S(J) = \frac{k_3 C_E d^2/DK_M}{1 + (C/K_M)f_S(J)} \times \frac{f_S(J)}{L} \tag{16.102}$$

Thus the invocation of the steady-state assumption results in a rate law involving two independent dimensionless parameters: $C/K_M$, which compares the bulk substrate concentration with the Michaelis constant for the enzyme, and $k_3C_Ed^2/DK_M$, which, in effect, compares the rate of catalysis with the rate of diffusion. These two parameters may be varied to show their effect on the detected current, and the reader is directed to the cited papers to view the results of this simulation. What is intended here, however, is a specific example of the use of the steady-state assumption in the development of an atypical rate expression (Equation 16.102).

The derivation of a similarly atypical rate expression is required for the simulation of the electrochemical behavior encountered in electrohydrodimerization studies. In these studies, the variation of the bulk concentration of the olefin (e.g., ethyl cinnamate, diethyl furmarate) reveals that there is a concentration dependence to the reaction order associated with the dimerization of the electrogenerated radical ion [24]. This variation in apparent reaction order with concentration can only be attributed to a two-step mechanism [16] involving two independent rate or equilibrium processes. A mechanism that meets this criterion and appears to fit the electrochemical data is the "preequilibrium" mechanism [25] in which the electrogenerated radical ions first engage in an equilibrium dimerization before the rate-determining ring closure of the dimer takes place. Symbolically, this mechanism may be written:

At $E_1$:

$$R + e \longrightarrow R^{\overline{\cdot}} \tag{16.103}$$

Chemical equilibrium:

$$2R^{\overline{\cdot}} \underset{}{\overset{K}{\rightleftharpoons}} R_2^{2-} \tag{16.104}$$

Chemical reaction:

$$R_2^{2-} \xrightarrow{k} X \tag{16.105}$$

At $E_2$:

$$R^{\overline{\cdot}} \longrightarrow R + e \tag{16.106}$$

where X is the cyclic product formed and $E_1$ and $E_2$ are potentials necessary to bring about the diffusion-controlled reaction of R and $R^{\overline{\cdot}}$, respectively. This mechanism would appear to be first order if the cyclization is rate determining or second order if the radical ion equilibrium is rate determining. Intermediate reaction orders may be obtained depending on the thermodynamic extent of Equation 16.104. The rate law for this mechanism may be derived by combining the equilibrium constant expression for Equation 16.104:

$$K = \frac{[R_2^{2-}]}{[R\dot{^-}]^2} \tag{16.107}$$

with the material balance equation for $R\dot{^-}$ and $R_2^{2-}$:

$$C_{R\dot{^-}} = [R\dot{^-}] + 2[R_2^{2-}] \tag{16.108}$$

(where $C_{R\dot{^-}}$ is the total concentration of the product of the electrode reaction exclusive of any product of the homogeneous chemical reaction) to yield

$$K = \frac{[R_2^{2-}]}{(C_{R\dot{^-}} - 2[R_2^{2-}])^2} \tag{16.109}$$

Solution of Equation 16.109 yields

$$[R_2^{2-}] = \frac{(\sqrt{8KC_{R\dot{^-}} + 1} - 1)^2}{16K} \tag{16.110}$$

which may be substituted into the first-order rate law equation for the reaction given in Equation 16.105 to obtain a rate law for the mixed mechanism:

$$\frac{d[X]}{dt} = k[R_2^{2-}] = \frac{k(\sqrt{8KC_{R\dot{^-}} + 1} - 1)^2}{16K} \tag{16.111}$$

The finite difference form of this rate law may be represented by

$$\frac{\Delta f_X(J)C}{\Delta t} = \frac{k[\sqrt{8KCf_{R\dot{^-}}(J) + 1} - 1]^2}{16K} \tag{16.112}$$

or

$$\Delta f_X = \frac{(kt_f)[\sqrt{8(KC)f_{R\dot{^-}}(J) + 1} - 1]^2}{16(KC)L} \tag{16.113}$$

where C is the bulk concentration of parent olefin, L is the number of time iterations representing step time $t_f$, and $f_X(J)$ and $f_{R\cdot}^{-}(J)$ are the fractional concentrations of product X and R· at any time or position within the diffusion layer.

The application of this rate law to the simulation of electrochemical behavior requires two dimensionless input parameters $kt_f$ and KC. When these are supplied, three-dimensional chronoamperometric or chronocoulometric working surfaces [25] are generated. These working surfaces both indicate first-order behavior when KC is large and second-order behavior when KC is small. Intermediate values of KC produce the variable reaction orders between one and two which are observed experimentally when the bulk olefin concentration is varied. Appropriate curve fitting of the experimental i(t, C) data to the simulation results in the evaluation of k and K; details appear in the referenced work.

These two advanced techniques will obviously be useful in the treatment of kinetic perturbations on the laws of diffusion. Another technique appears to be particularly attractive when one wishes to model more complicated transfer phenomena. This method involves the variation of the volume element size ($\Delta x$) within the space array according to the position of the volume element within the array. This causes a local variation in $D_M$ in accord with Equation 16.5. This method has been used by Joslin and Pletcher [27] and Feldberg [28] to achieve rapid simulations of electrochemical problems by varying element size in an exponential manner with distance from the electrode; this describes the events taking place in the vicinity of the electrode in greater detail than those taking place within the bulk. Accurate representations of the current may be achieved within only a few time iterations. Thus the technique may be suited for "stiff" problems—those in which two events are taking place on vastly different time scales. For example, if one wishes to model the complete electrochemical transfer problem (involving diffusion, convection, and migration) through the solution of the Nernst-Planck equation [29] to obtain a transient expression for the total current due to double-layer charging and a faradaic process, one must address the fact that migration in the vicinity of the electrode takes place on a much more rapid time scale than diffusion from the bulk. Because of this, those simulations of this process which have been attempted [30,31] have resulted in steady-state solutions without achieving good transient results. Variations in volume element size in the vicinity of the electrode permit the simulation of the combined effects of diffusion and migration, so that it appears to be possible to treat the transient problem in this manner [32]. Another example of the utility of this method lies in the simulation of the complex diffusion problems encountered in polymer film electrodes [33]. These problems are complicated by the fact that diffusion coefficients within the polymer film are vastly different than those within the solution phase; thus transfer takes place at significantly different rates on either side of the solution-membrane interface. By employing different volume element sizes on either side of the membrane, we [34] have been able to model the combined effects of solution and membrane transfer in the polymer-film-covered rotating disk electrode. Without resorting to this method, the simulation of this problem is quite time consuming.

Another development that appears to be forthcoming is an attempt to eliminate "working curves" by fitting the results of simulations to specific functional forms. Examples of this have already appeared in the literature [10,12,13,20,35]. Equations obtained in this manner permit one to analyze experimental data functionally, even though these functions are not rigorously derived. Even if the functions so obtained only fit the simulation results over a limited range, they may still be quite useful in predicting experimental results falling within that region. More developments along these lines are anticipated both in the simulation of electrochemical problems and in the simulation of other kinds of problems involving the competitive effects of mass transfer and chemical reaction [36].

## REFERENCES

1. A. Fick, *Ann. Phys. (Pogg.) 94*:59 (1855).
2. A. Einstein, *Z. Electrochem. 14*:235 (1908).
3. F. G. Cottrell, *Z. Phys. Chem. 42*:385 (1902).
4. S. W. Feldberg, in *Electroanalytical Chemistry*, Vol. 3 (A. J. Bard, ed.), Marcel Dekker, New York, 1969, p. 271.
5. W. V. Childs, J. T. Maloy, C. P. Keszthelyi, and A. J. Bard, *J. Electrochem. Soc. 118*:874 (1971).
6. R. D. Grypa, M. S. thesis, West Virginia University, 1974.
7. R. J. Lawson and J. T. Maloy, *Anal. Chem. 46*:559 (1974).
8. A. D. Osborne and G. Porter, *Proc. R. Soc., Ser. A, 284*:9 (1965).
9. S. W. Feldberg, *J. Am. Chem. Soc. 88*:390 (1966).
10. S. W. Feldberg, *J. Phys. Chem. 70*:3928 (1966).
11. S. A. Cruser and A. J. Bard, *J. Am. Chem. Soc. 91*:267 (1969).
12. J. T. Maloy, K. B. Prater, and A. J. Bard, *J. Am. Chem. Soc. 93*:5959 (1971).
13. L. R. Faulkner, *J. Electrochem. Soc. 122*:1190 (1975).
14. R. P. Van Duyne and C. N. Reilley, *Anal. Chem. 44*:153 (1972).
15. K. B. Prater and A. J. Bard, *J. Electrochem. Soc. 117*:207 (1970).
16. K. B. Prater and A. J. Bard, *J. Electrochem. Soc. 117*:335 (1970).

17. K. B. Prater and A. J. Bard, *J. Electrochem. Soc. 117*:1517 (1970).
18. K. B. Prater, in *Computers in Chemistry and Instrumentation*, Vol. 2 (J. S. Mattson, H. B. Mark, Jr., and H. C. MacDonald, Jr., eds.), Marcel Dekker, New York, 1972, Chap. 8.
19. J. T. Maloy, *ibid.*, Chap. 9.
20. L. D. Mell and J. T. Maloy, *Anal. Chem. 47*:299 (1975).
21. L. D. Mell and J. T. Maloy, *Anal. Chem. 48*:1597 (1976).
22. F. R. Shu and G. S. Wilson, *Anal. Chem. 48*:1679 (1976).
23. R. D. Grypa and J. T. Maloy, *J. Electrochem. Soc. 122*:377 (1975).
24. R. D. Grypa and J. T. Maloy, *J. Electrochem. Soc. 122*:509 (1975).
25. B. M. Bezilla, Jr. and J. T. Maloy, *J. Electrochem. Soc. 126*: 579 (1979).
26. E. L. King, *J. Chem. Educ. 51*:186 (1974).
27. T. Joslin and D. Pletcher, *J. Electroanal. Chem. 49*:171 (1974).
28. S. W. Feldberg, *J. Electroanal. Chem. 127*:1 (1981).
29. L. R. Faulkner and A. J. Bard, *Electrochemical Methods*, Wiley, New York, 1980.
30. S. W. Feldberg, *J. Phys. Chem. 74*:87 (1970).
31. D. Laser and A. J. Bard, *J. Electrochem. Soc. 123*:1828, 1837 (1976).
32. S. W. Feldberg, personal communication, 1982.
33. P. J. Peerce and A. J. Bard, *J. Electroanal. Chem. 112*:97 (1980).
34. Work done in collaboration with J. Leddy and A. J. Bard, 1983.
35. R. Bezman and L. R. Faulkner, *J. Am. Chem. Soc. 94*:3699 (1972).
36. B. M. Bezilla, L. Pyles, and J. T. Maloy, 1981 Pittsburgh Conference on Analytical Chemistry and Applied Spectroscopy, Atlantic City, N.J., Abstract 197.

# 17

# Electrochemical Evaluation of the Mechanisms of Organic Reactions Through Examples

M. DALE HAWLEY / Department of Chemistry, Kansas State University, Manhattan, Kansas



## I. INTRODUCTION

In this chapter we examine the applications of various electrochemical methods to the study of several organic electrode processes. Since space limitations permit only the salient features of each electrochemical system to be presented, the discussion of much of the original data, the experimental procedures, and the instrumentation is often very abbreviated. The reader is encouraged to consult the primary literature if more information is desired.

The goal of the electrochemical studies will be the elucidation of the sequence of electron transfer and chemical reactions that occur at or near the electrode surface. Although no single experimental plan can be devised that will suffice for all studies, answers to the following questions will provide much of the information needed to describe a redox pathway:

1. Is the redox pathway chemically reversible, or are solution reactions coupled to electron transfer?
2. How many electrons are involved in each of the observed redox processes?
3. Is each redox process kinetic or diffusion controlled?
4. What are the reaction kinetics of each of the kinetically controlled processes?
5. Is the redox behavior affected by a change in the concentration of the electroactive species, a component of the solvent system, or the electrode surface?
6. What are the intermediates in the electrode reaction?
7. What are the products and their yields?

The selection of the experimental techniques to be used in mechanistic studies is dictated most frequently by the instrumentation available to the investigator and less frequently by the chemical system itself. A potentiostat and a waveform generator providing cyclic voltammetric, chronoamperometric, and coulometric capabilities are essential to the performance of most electrochemical studies. Such instrumentation is discussed in Chapter 6. In addition, kinetic studies usually require a rapid recording device, such as an oscilloscope or, preferably, a laboratory digital computer. Although spectroscopic studies at optically transparent electrodes were not used in the present examples, spectroscopic methods have considerable utility and, no doubt, will be used increasingly in future studies of kinetically controlled processes (Chap. 3).

## II. REDUCTION OF p-CHLOROBENZONITRILE [1]

Although the one-electron reduction of nitrobenzene to its radical anion in dipolar aprotic solvents is a classical example of a chemically reversible redox couple, the reductions of many organic compounds are chemically irreversible. The redox behavior of p-chlorobenzonitrile is typical of those systems in which the initial electrode product undergoes rapid, irreversible chemical reaction to give another reducible species.

### A. Cyclic Voltammetry

Nearly all electrochemical studies of organic systems are initiated by a cyclic voltammetric examination of the redox behavior (Chap. 3, Sec. III.B). From several cyclic voltammograms obtained at widely

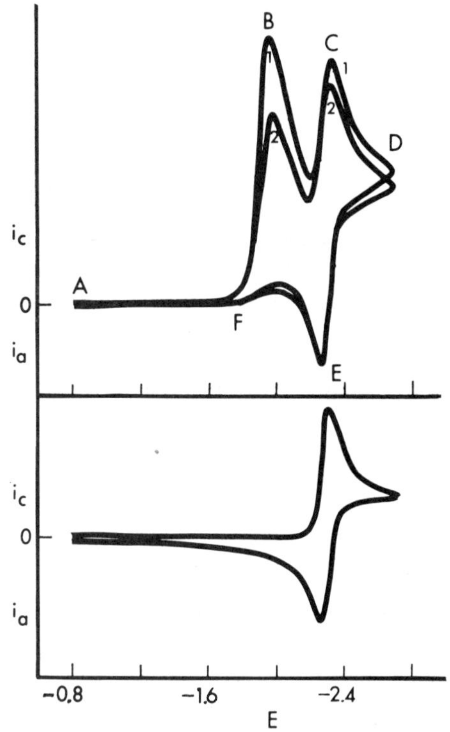


Figure 17.1 Cyclic voltammograms in DMF-0.1 M $Et_4NClO_4$ at a scan rate of 80.6 mV/s on a planar platinum electrode: (Top) $2.15 \times 10^{-3}$ M p-chlorobenzonitrile; (Bottom) $1.44 \times 10^{-3}$ M benzonitrile. Numbers 1 and 2 represent cycles 1 and 2, respectively. [From Ref. 1, reprinted with permission from D. E. Bartak, K. J. Houser, B. C. Rudy, and M. D. Hawley, *J. Am. Chem. Soc.* *94*:7526 (1972). Copyright 1972 American Chemical Society.]

separated scan rates, one can determine not only which redox processes are chemically reversible, but one can often identify intermediates and final products and ascertain the relationships among them.

In the case of p-chlorobenzonitrile (refer to Fig. 17.1),

A. The potential scan is initiated at −0.8 V in the negative-going direction.

B. The one-electron reduction of p-chlorobenzonitrile

$$p\text{-}ClC_6H_4CN + e \rightleftharpoons p\text{-}ClC_6H_4CN^{-} \qquad (17.1)$$

is followed by the rapid loss of chloride ion from the unstable radical anion,

$$p\text{-ClC}_6\text{H}_4\text{CN}^{\overline{\cdot}} \xrightarrow{\text{fast}} p\text{-NCC}_6\text{H}_4^{\cdot} + \text{Cl}^- \qquad (17.2)$$

the reduction of the p-cyanophenyl radical at the electrode surface at the applied potential,

$$p\text{-NCC}_6\text{H}_4^{\cdot} + e \longrightarrow p\text{-NCC}_6\text{H}_4^- \qquad (17.3)$$

and the abstraction of a proton from a component of the solvent system, HA, by the p-cyanophenyl anion,

$$p\text{-NCC}_6\text{H}_4^- + \text{HA} \longrightarrow \text{C}_6\text{H}_5\text{CN} + \text{A}^- \qquad (17.4)$$

C. Benzonitrile, the product of these coupled chemical reactions, is reduced reversibly to its radical anion

$$\text{C}_6\text{H}_5\text{CN} + e \rightleftharpoons \text{C}_6\text{H}_5\text{CN}^{\overline{\cdot}} \qquad (17.5)$$

D. The direction of the potential scan is reversed.
E. And on the positive-going sweep, benzonitrile is regenerated.

$$\text{C}_6\text{H}_5\text{CN}^{\overline{\cdot}} \rightleftharpoons \text{C}_6\text{H}_5\text{CN} + e^- \qquad (17.6)$$

F. The absence of an anodic peak here indicates that $p\text{-ClC}_6\text{H}_4\text{CN}^{\overline{\cdot}}$ decomposes rapidly on the time scale of this experiment.

The interpretation of the cyclic voltammetric behavior is relatively straightforward, since only one stable electroactive product arises from the reduction of p-chlorobenzonitrile. The fact that electron attachment to halogen-containing compounds frequently results in the cleavage of the carbon-halogen bond immediately suggests benzonitrile as the possible product. The similarity between the cyclic voltammetric behavior of benzonitrile and that of the p-chlorobenzonitrile reduction product supports this prediction (see Fig. 17.1).

## B. Chronoamperometry

Although cyclic voltammetry is eminently suited for the qualitative investigation of complex electrode processes, it is less satisfactory for the quantitative study of reaction kinetics. The shape of the current-voltage curve, and hence its interpretation, may be seriously affected by the rate of heterogeneous electron transfer, the rate of the follow-up chemical reaction, the charging of the electrical double layer, uncompensated resistance, and inadequate potential control. These problems are much less important with potential-step techniques, such as chronoamperometry, chronocoulometry, and chronoabsorptometry, and accordingly, these methods are usually preferred for the quantitative study of reaction rates (Chap. 3, Sec. II). Moreover, whereas each current-voltage curve gives only one value of the rate constant, the single-potential-step experiment gives as many values as the number of times the variable is sampled. With the potentiostats and data acquisition equipment which are presently available at rather moderate cost, studies of electrochemically generated intermediates which have half-lives ranging from a millisecond or less to tens of seconds can be performed readily.

The single-potential-step chronoamperometric technique was selected here for the study of diffusion vs. kinetic control for the formation of benzonitrile.

A. In this experiment (refer to Fig. 17.2), the potential is stepped abruptly from (A), where p-chlorobenzonitrile is electroinactive, to B, where p-chlorobenzonitrile and bezonitrile are reduced concurrently and the concentration of each species is effectively zero at the electrode surface.

C. Note that the chronoamperometric $it^{1/2}/CA$ value (C) is constant throughout the accessible time range. These data indicate that the rate of benzonitrile formation is controlled only by the diffusion rate of p-chlorobenzonitrile, a conclusion which further requires that the half-life of any intermediate in the electrochemical dechlorination reaction be less than the shortest observation time, which in this case is 1 ms.

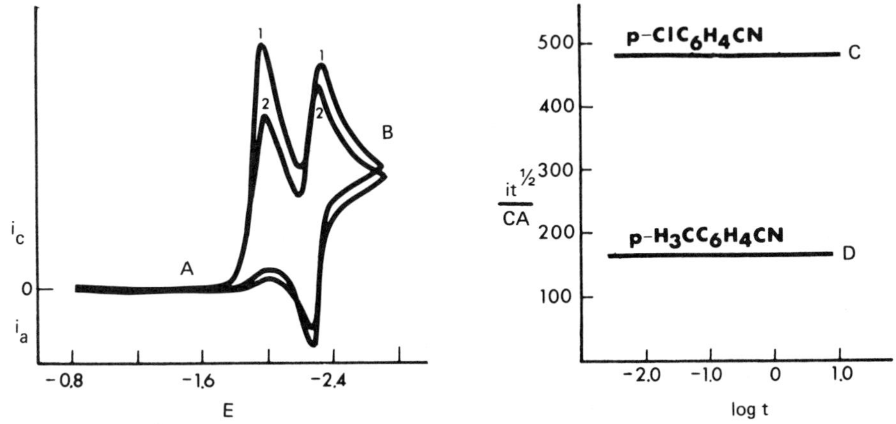


Figure 17.2 (Left) Cyclic voltammogram of p-chlorobenzonitrile. (Right) Chronoamperometric data for the reduction of p-chlorobenzonitrile (top curve) and p-tolunitrile (bottom curve). [From Ref. 1, reprinted with permission from D. E. Bartak, K. J. Houser, B. C. Rudy, and M. D. Hawley, *J. Am. Chem. Soc.* *94*:7526 (1972). Copyright 1972 American Chemical Society.]

D. The chronoamperometric results can also be used to ascertain the number of electrons involved in the formation of benzonitrile from p-chlorobenzonitrile. In order to translate the chronoamperometric data into a meaningful n value, a compound is selected which has a diffusion coefficient very similar to that of p-chlorobenzonitrile and which gives a stable, known product upon electroreduction. Tolunitrile, which satisfies these criteria, is known to be reduced to its radical anion at a diffusion-controlled rate. Since this one-electron process gives a value of 168 $\mu A \cdot s^{1/2}/M \cdot cm^2$ for $it^{1/2}/CA$, the corresponding value of 480 $\mu A \cdot s^{1/2}/M \cdot cm^2$ for the reduction of p-chlorobenzonitrile to benzonitrile anion radical must represent an overall three-electron process. When we subtract the one electron that is required to reduce benzonitrile to its radical anion from this total, we immediately conclude that two electrons are required to cleave the carbon-chlorine bond in p-chlorobenzonitrile. A scheme that is consistent with these data is described by Equations 17.1 to 17.5.

## C. Capture of the Intermediate p-Cyanophenyl Radical

An intermediate in an electrode process often can be studied directly by electrochemical and spectroscopic methods, or its existence can be inferred from product and kinetic analyses. In the present example, the intermediacy of the p-cyanophenyl radical can be demonstrated by a trapping experiment (refer to Fig. 17.3).

When the cathodic switching limit is adjusted to −2.2 V, the

A. Only cathodic peak that is observed on the first negative-going sweep is due to p-chlorobenzonitrile.

B. The sweep is interrupted and the potential held for approximately 30 s at the cathodic switching limit. A small fraction of the short-lived p-cyanophenyl radical intermediate is captured by cyanide ion, the trapping agent, to give

C. Terephthalonitrile radical anion (Eq. 17.7), a species that is oxidized reversibly (Eq. 17.8) at this potential when the potential scan is resumed

$$\text{p-NCC}_6\text{H}_4\cdot + \text{CN}^- \longrightarrow \text{p-NCC}_6\text{H}_4\text{CN}^{\overline{\cdot}} \qquad (17.7)$$

$$\text{p-NCC}_6\text{H}_4\text{CN}^{\overline{\cdot}} \rightleftharpoons \text{p-NCC}_6\text{H}_4\text{CN} + e \qquad (17.8)$$

in the positive-going direction. Note that the reduction of terephthalonitrile to its radical anion is not seen until the second negative-going potential sweep is made.

D. An authentic sample of terephthalonitrile gives this cyclic voltammetric behavior.

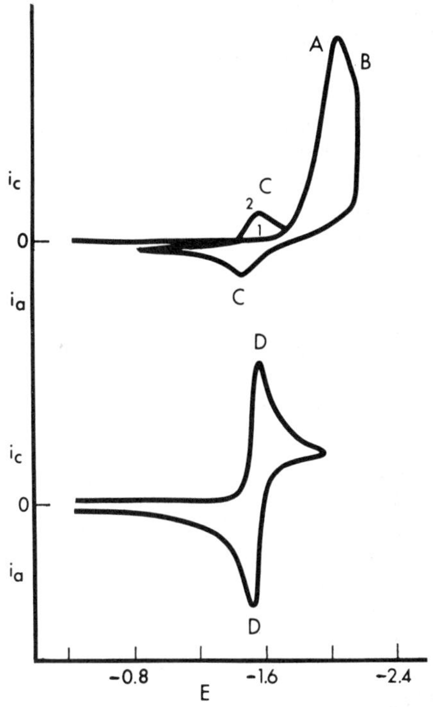


Figure 17.3 (Top) Cyclic voltammogram of $3.75 \times 10^{-3}$ M p-chlorobenzonitrile in DMF-0.2 M $Et_4NClO_4$. (Bottom) Cyclic voltammogram of $1.80 \times 10^{-3}$ M terephthalonitrile in DMF-0.1 M $Et_4NClO_4$. The working electrode in both experiments is platinum; the scan rate is 80.6 mV/s. [From Ref. 1, reprinted with permission from D. E. Bartak, K. J. Houser, B. C. Rudy, and M. D. Hawley, *J. Am. Chem. Soc.* *94*:7526 (1972). Copyright 1972 American Chemical Society.]

### D. Coulometry and Product Identification

Numerous examples could be cited in which two or more suspected products of an electrode process exhibit similar electrochemical behavior. In other instances, the species that is stable during the time required to complete the cyclic voltammetric experiment may undergo a slow chemical reaction to give the product that is isolated. These problems arise sufficiently frequently that the identification of products and the determination of the product distribution are required.

The procedure employed in many laboratories involves exhaustive controlled-potential electrolysis (Chap. 3, Sec. IV.F) of the parent species and the subsequent identification of the products and their yields by standard analytical procedures. In the case of p-chlorobenzonitrile, exhaustive electrolysis at a potential between the first and second cathodic waves (−2.08 V; see Fig. 17.1) gave the anticipated n value ($n_{exptl} = 2.02 \pm 0.05$). Cyclic voltammetric examination of the electrolyzed solution immediately after electrolysis indicated a nearly quantitative yield of benzonitrile (95%); a gas-chromatographic analysis of the same solution showed that benzonitrile was formed in 85% yield. No other product was detected by either of these methods.

## III. OXIDATION OF ADRENALINE

p-Chlorobenzonitrile and adrenaline, our second example, both give electrode products that are unstable with respect to subsequent chemical reaction. Since the products of these homogeneous chemical reactions are also electroactive in the potential range of interest, the overall electrode reaction is referred to as an ECE process (i.e., a chemical reaction is interposed between electron transfers). Adrenaline differs from p-chlorobenzonitrile in that (1) the product of the chemical reactions, leucoadrenochrome, is now more readily oxidized than the parent species, and (2) the overall rate of the chemical reactions is sufficiently slow so as to permit kinetic studies by electrochemical methods. As a final note before the experimental results are presented, the enzymic oxidation of adrenaline was known to give adrenochrome. Accordingly, the emphasis in the work described by Adams and coworkers [2] was on the preparation and study of the intermediates.

### A. Cyclic Voltammetry

The cyclic voltammetric behavior of adrenaline and several other catecholamines was studied as a function of pH. In strongly acidic media (refer to Fig. 17.4), extensive protonation of the amine precludes cyclization of the adrenalinequinone on the time scale of this experiment.

A. As a result, the oxidation of adrenaline to adrenalinequinone is a chemically reversible process (Eq. 17.9).

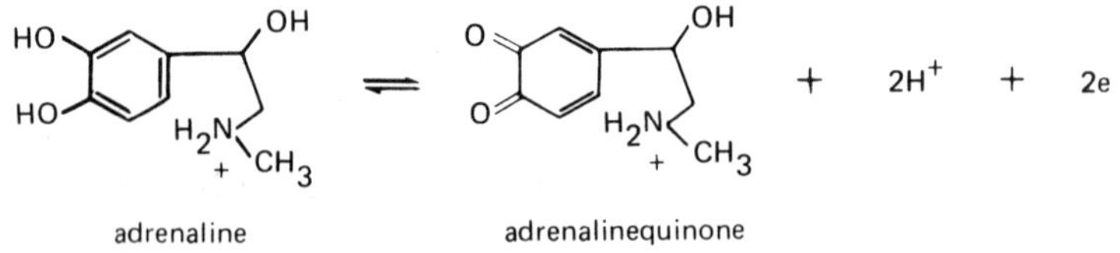


(1) (17.9)

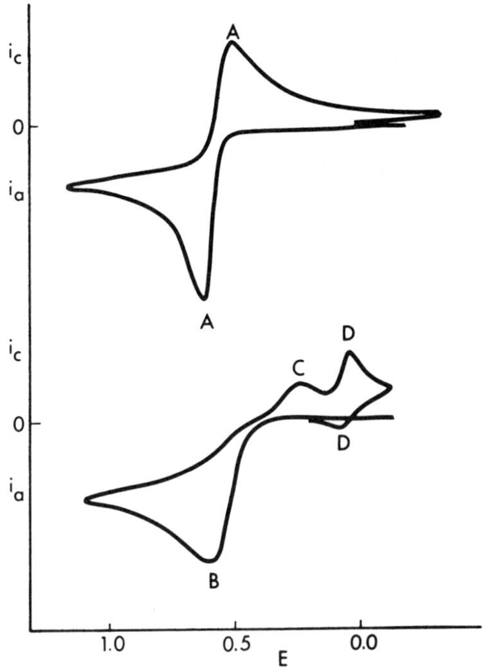


**Figure 17.4** (Top) Cyclic voltammogram of adrenaline in 0.29 M $HClO_4$ at a scan rate of 55 mV/s. (Bottom) Cyclic voltammogram of adrenaline at pH 3.0 at a scan rate of 278 mV/s. [From Ref. 2, reprinted with permission from M. D. Hawley, S. V. Tatawawadi, S. Piekarski, and R. N. Adams, *J. Am. Chem. Soc. 89*:447 (1967). Copyright 1967 American Chemical Society.]

B. At pH 3.0, however, this cyclization reaction involving the unprotonated form of adrenalinequinone,

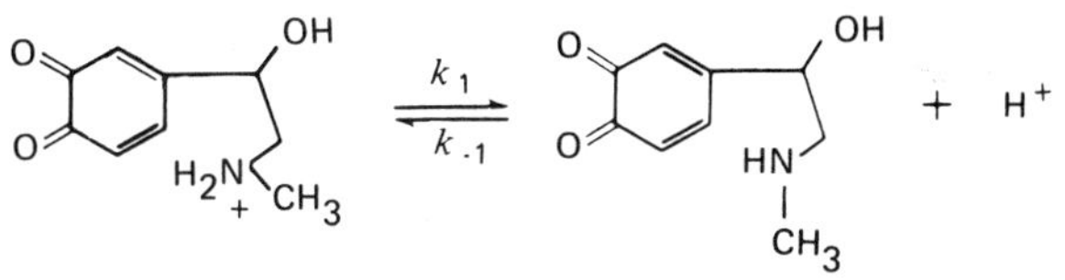


(17.10)

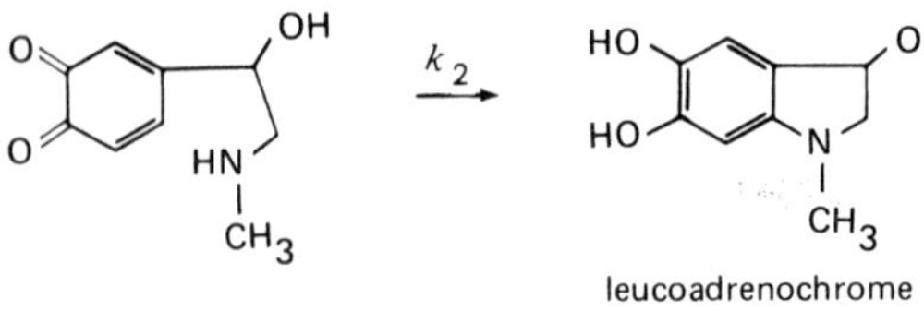


(17.11)

(2)

and this solution redox reaction,

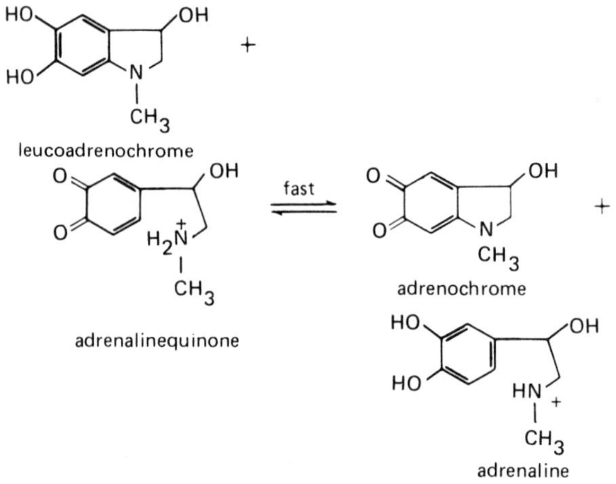


(17.12)

(3)

cause the oxidation of the adrenaline to be kinetically controlled, the peak height for

C. The reduction of unreacted adrenalinequinone on the reverse, negative-going sweep to be reduced, and

D. A redox couple to emerge at a less positive potential for the chemically reversible reduction of adrenochrome to leucoadrenochrome (Eq. 17.13). Note that the anodic peak due to the oxidation

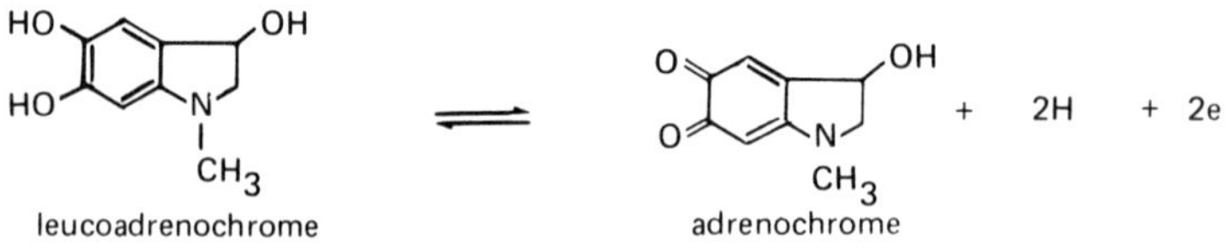


(17.13)

of leucoadrenochrome to adrenochrome near 0 V is not seen until the second positive-going potential sweep is made. The voltage separation between the anodic and cathodic peaks for the oxidation of adrenaline (peak B, Fig. 17.4, bottom) and the reduction of adrenalinequinone (peak C) is large compared to most chemically reversible redox couples. However, this behavior is typical of many quinone-hydroquinone systems on a carbon paste surface at intermediate values of pH.

## B. Kinetics of the Solution Reactions

Kinetic studies of ECE (sometimes called a DISP mechanism when the second electron transfer occurs in bulk solution) [3] processes are often best performed using a constant-potential technique such as chronoamperometry. The advantages of this method include (1) relative freedom from double-layer and uncompensated iR effects, and (2) a new value of the rate constant is obtained each time the current is sampled. However, unlike certain large-amplitude techniques, an accurately known, diffusion-controlled value of $it^{1/2}/CA$ is required of each solution before a determination of the rate constant can be made. In the present case, diffusion-controlled values of $it^{1/2}/CA$ corresponding to n = 2 and n = 4 are obtained in strongly acidic media (i.e., when kt is small) and in solutions of intermediate pH (i.e., when kt is large), respectively. The experimental rate constant is then determined from a dimensionless working curve for the proposed reaction scheme in which the apparent value of n ($n_{app}$) is plotted as a function of kt. Unfortunately, there is no single working curve which suffices for all first-order ECE processes. The equilibrium constant of the solution redox reaction has a marked effect on the current-time behavior, and accordingly, an important influence on the shape of the working curve. Experimentally, the value of the equilibrium constant is calculated using the peak potentials determined by cyclic voltammetry. The actual working curve is then obtained by the digital simulation of the diffusion, kinetics, and electron-transfer processes according to Feldberg's procedure [4] (see Chap. 16).

Once the appropriate working curve has been determined, either of two methods may be used for the calculation of the rate constant. In the method that was employed here, each experimental value of $n_{app}$ gives a corresponding value of kt from the working curve. If the limiting values of $it^{1/2}/CA$ have been determined correctly, a plot of kt vs. t should yield a straight line with a zero intercept (Fig. 17.5). The second and preferred method involves fitting the experimental $n_{app}$ vs. log t plot to the working curve ($n_{app}$ vs. log kt) until the best match is obtained. A value of k is then calculated directly.

Distinction among the several working curves for ECE processes is sometimes difficult, especially if data are obtained over a relatively restricted time range. Although the use of pen-and-ink recording limited the dynamic range in the present example, modern data acquisi-

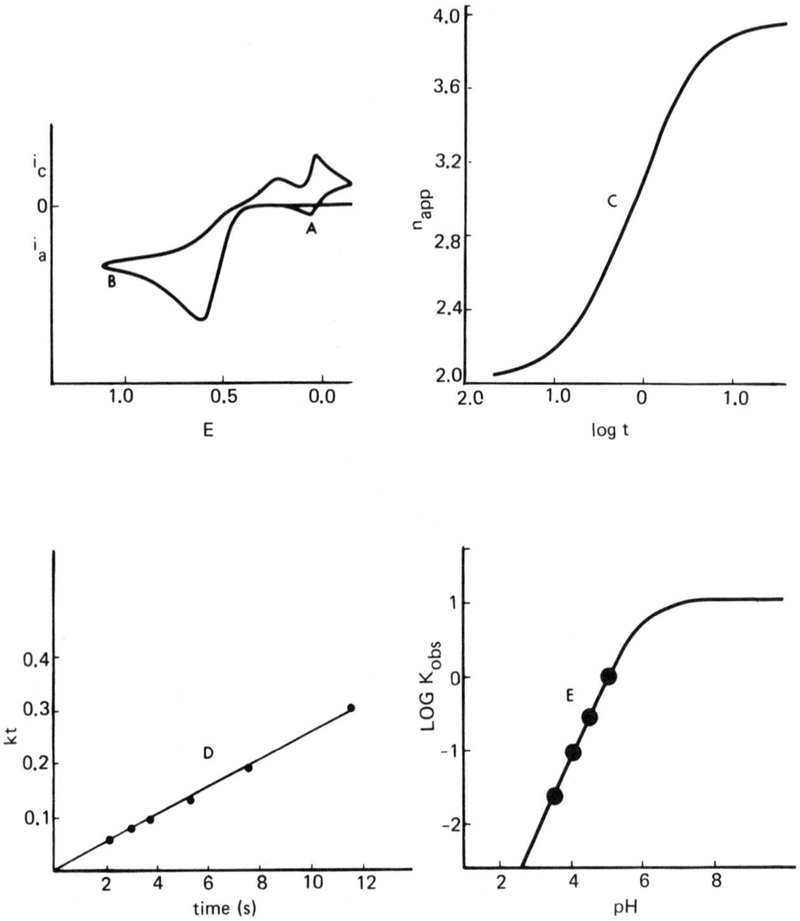


Figure 17.5 (Upper left) Cyclic voltammogram of adrenaline at pH 3.0. The scan rate is 0.278 V/s. (Upper right) Working curve for this ECE process. (Lower left) Chronoamperometric data for the cyclization of adrenalinequinone at pH 3.5. (Lower right) Observed rate constant for the cyclization of adrenalinequinone as a function of pH. [From Ref. 2, reprinted with permission from M. D. Hawley, S. V. Tatawawadi, S. Piekarski, and R. N. Adams, *J. Am. Chem. Soc.* *89*:447 (1967). Copyright 1967 American Chemical Society.]

tion devices and potentiostats usually permit acquisition of data over several decades of time. The full capability of the instrumentation should be utilized to minimize errors in interpretation.

The method for determining the observed rate constant for cyclization of adrenalinequinone is illustrated in Figure 17.5. The potential is stepped

A. From where adrenaline is electroinactive, to

B. Where the concentrations of adrenaline and the cyclization product, leucoadrenochrome, are effectively zero at the electrode surface. Kinetically controlled values of $it^{1/2}$ are divided by the diffusion-controlled value of $it^{1/2}$ when kt = 0. (The diffusion-controlled value of $it^{1/2}$ may be obtained either at very small values of t at this pH or in a strongly acidic medium where k is very small.) This ratio, multiplied by 2, gives $n_{app}$. Each experimental value of $n_{app}$ gives a

C. Corresponding value of kt from this working curve. Values of kt are then plotted

D. As a function of t to give the observed rate constant. Measurement of the rate constant at several values of pH suggest

E. A rate law of the form

$$k = \frac{k_1 k_2}{k_{-1}[H^+] + k_2}$$

The solid curve is calculated from this equation using $k_{-1} = 10^{10}\ M^{-1}\ s^{-1}$ $k_2 = 10^4 s^{-1}$ and $K = k_1/k_{-1} = 10^{-9}$.

## IV. OXIDATION OF α-TOCOPHEROL [5–7]

The elucidation of organic electrode processes frequently requires studies in which the chemical environment for the electrochemical and accompanying chemical reactions is varied extensively. The studies involving the oxidation of α-tocopherol illustrate how the effects of pH, compositions of the solvent and the buffer system, and the variation of an electroinactive reagent concentration might be used to deduce the principal electrochemical and chemical steps in an electrode process.

### A. Kinetics of the Hydrolysis Reaction (Fig. 17.6)

A. The cyclic voltammetric oxidations of α-tocopherol (Ia) and its model compound, 2,2,5,7,8-pentamethyl-6-hydroxychroman (Ib), are chemically reversible in acetonitrile-1.0.1 M $Et_4NClO_4$ (eq. 17.4). However, when water

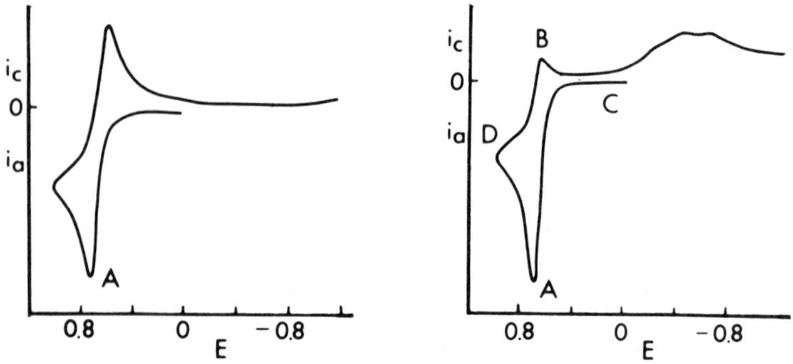


Figure 17.6 (Left) Cyclic voltammogram of α-tocopherol in acetonitrile with 0.1 M $Et_4NClO_4$ as the supporting electrolyte. The scan rate is 83.5 mV/s. (Right) Cyclic voltammogram in acetonitrile-0.1 M $Et_4NClO_4$ to which 200 mM water has been added. The scan rate is 83.5 mV/s. (From Ref. 5.)

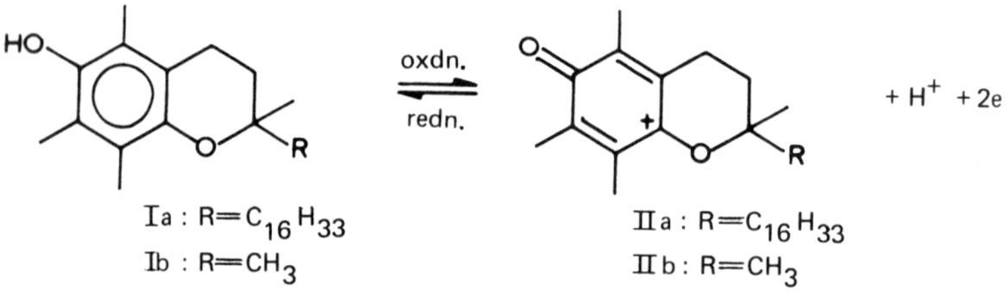


(17.14)

(2)

B. Is added as a nucleophile, hydrolysis of the carbonium ion II occurs to give III, which is electroinactive (Eq. 17.15).

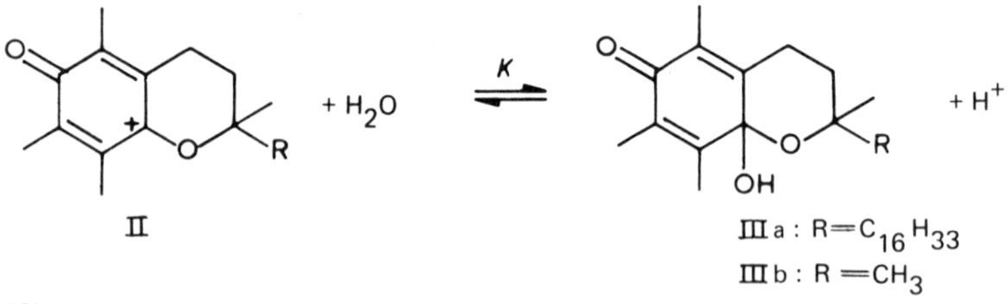


(17.15)

(3)

C. The kinetics of the hydrolysis reaction were studied by double-potential-step chronoamperometry. The potential is stepped from here, where I is electroinactive,

D. To here, where the oxidation of I occurs at a diffusion-controlled rate. After a prescribed electrolysis time t, the potential is stepped back to (C). Only the reduction of the unreacted carbonium ion II occurs at this potential. The value of the pseudo-first-order rate constant $k_1$ is then determined from a dimensionless working curve which relates the ratio of the cathodic and anodic currents to kt. Details for the construction of the working curves (each ratio of $t_r/t$ requires a different working curve) and their subsequent use may be found in the literature [8].

Results of the electrochemical measurements of the hydrolysis rate constant with 60 mM added water are shown in Figure 17.7 (left). The pseudo-first-order rate constant is determined from the slope of the plot of $k_1t$ as a function of the forward electrolysis time, t. From a series of these measurements at different concentrations of water, a second plot (Fig. 17.7, right) is obtained from which a value of 2.5 $M^{-1}s^{-1}$ is calculated for a second-order rate constant, k. Note that "dry" acetonitrile may contain approximately 25 mM water.

## B. Ring Opening of III and the Formation of α-Tocopherylquinone

Inspection of Figure 17.6 (right) will show that three small cathodic peaks are present near −0.2, −0.4, and −0.6 V. The first of these

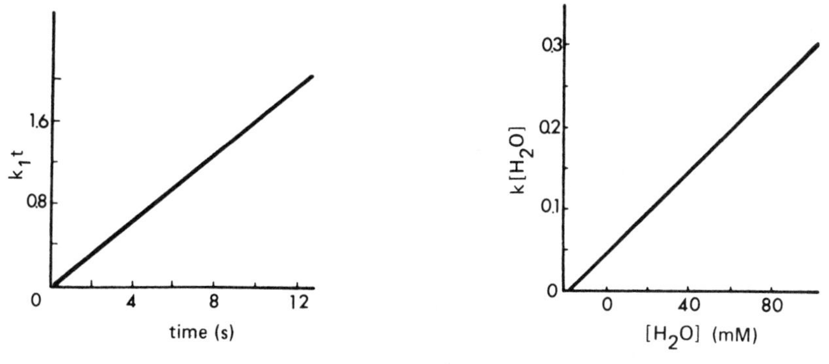


Figure 17.7 (Left) Plot of $k_1t$ vs. t for the hydrolysis of the carbonium ion prepared by the electrochemical oxidation of 2,2,5,7,8-pentamethyl-6-hydroxychroman. The solution contains 60 mM $H_2O$ and 0.1 M $Et_4NClO_4$ supporting electrolyte. (Right) Plot of $k[H_2O]$ for the hydrolysis of the carbonium ion prepared by the electrochemical oxidation of 2,2,5,6,7-pentamethyl-6-hydroxychroman. The solution is 1.1 mM in the chromanol and 0.1 M in $Et_4NClO_4$ supporting electrolyte. (From Ref. 5.)

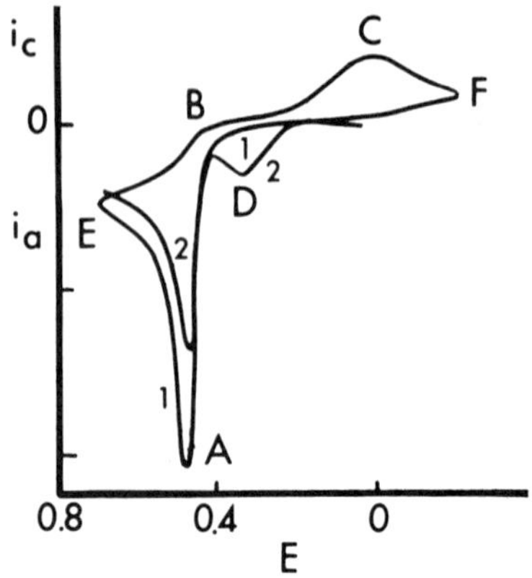


Figure 17.8 Cyclic voltammogram of 2,2,5,7,8-pentamethyl-6-hydroxy-chroman in water-acetonitrile (72:25 v/v). The solution is 0.95 mM in the chromanol and 0.87 M in perchloric acid. The scan rate is 0.167 V/s. The numbers 1 and 2 denote the first and second cycles, respectively. (From Ref. 5.)

peaks arises from the reduction of the protons which are formed during the electrooxidation of α-tocopherol, while the two more negative peaks arise from the stepwise reduction of α-tocopherylquinone to its hydroquinone. The formation of the quinone involves ring opening of the electroinactive III, a reaction that may be studied conveniently in a water-acetonitrile (75 vol % – 25 vol %) mixture (Fig. 17.8).

A. The reaction of the intermediate carbonium ion II with water in this solvent system is rapid, causing the oxidation of I to be

B. Chemically irreversible at this scan rate. In addition, the ring-opening reaction (Eq. 17.16), which occurs relatively rapidly in a strongly

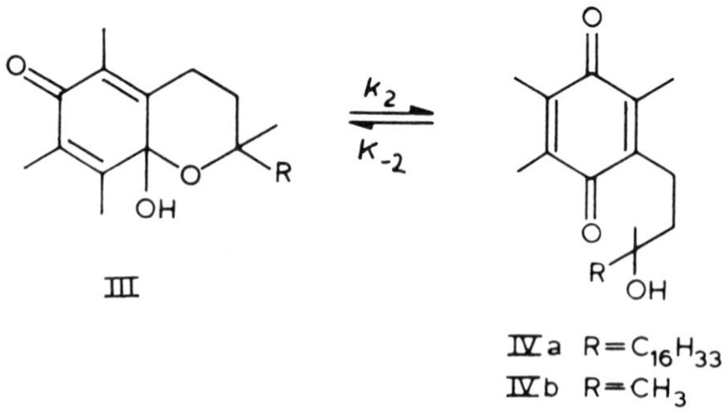


(17.16)

acidic medium, affords the quinone IV, which is reduced

C. At this potential to its hydroquinone (Eq. (17.17). Note that the

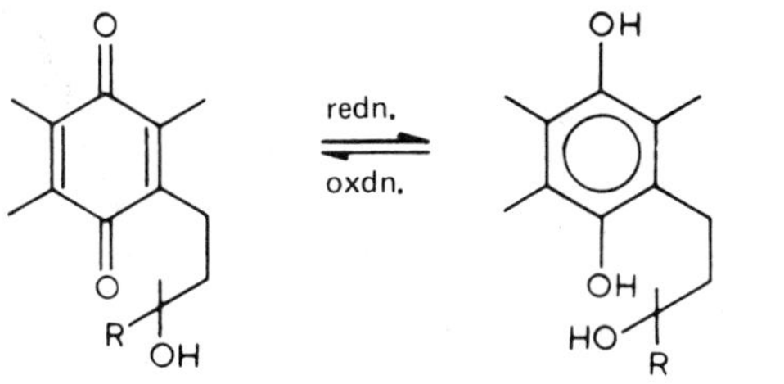


(17.17)

IVa $R=C_{16}H_{33}$ Va $R=C_{16}H_{33}$
IVb $R=CH_3$ Vb $R=CH_3$

(5)

oxidation of hydroquinone V near

D. 0.3 V is not observed until the second positive-going sweep is made. Note also the rather large separation in peak potentials of the quinone-hydroquinone redox couple.

The ring-opening reaction (Eq. 17.16) is relatively slow in the pH range 2–7. Since the hydrolysis of carbonium ion II remains rapid in this pH range in this solvent system, the kinetics of the ring-opening reaction can be ascertained by following the rate at which quinone IV is formed. Again, double-potential-step chronoamperometry was used for the study. The potential is stepped from here

E. To here, where to oxidation of I occurs to a diffusion-controlled rate. After time $\tau_f$, the potential is jumped to

F. Here. Any quinone IV formed by the ring-opening reaction is reduced to its hydroquinone at this potential. Each ratio of the cathodic to anodic current gives a corresponding value of $k\tau_f$ from appropriate, dimensionless working curves. The rate constant is then determined from a plot of $k\tau_f$ vs. $\tau_f$ (Fig. 17.9). The working curves were obtained by the digital simulation (Chap. 16) of the processes described by Equations 17.14 to 17.17.

## C. Effect of pH on the Ring-Opening Reaction

Numerous measurements of the rate constant in the pH range 2-7 and various buffer concentrations give the results shown in Figure 17.10. Catalytic constants for the ring opening of IIIb at 23° are: $H^+$, $4 \times 10^2\ M^{-1}\ s^{-1}$ $OH^-$, $1 \times 10^7\ M^{-1}\ s^{-1}$; $CH_3COO^-$, $1.6 \times 10^{-1}\ M^{-1}\ s^{-1}$; and intrinsic, $4 \times 10^{-3} s^{-1}$. The solid curve in Figure 17.10 is constructed using these data.

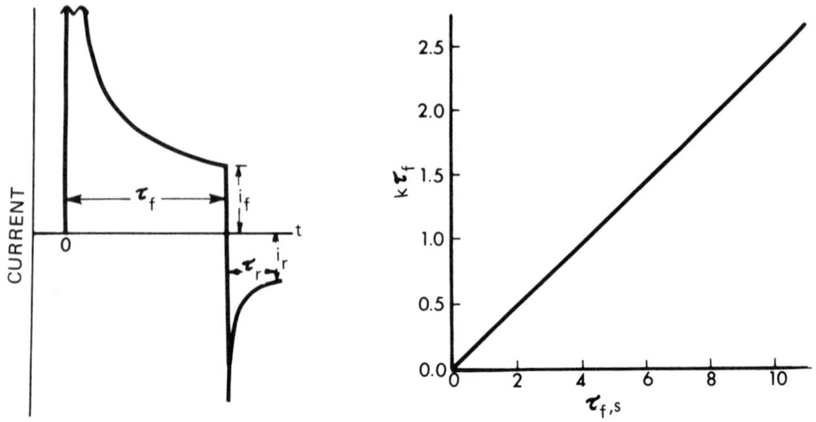


Figure 17.9 (Left) Anodic-cathodic current-time curve for 2,2,5,7,8-pentamethyl-6-hydroxychroman (Ib) showing the method of current and time measurements for the double-potential-step chronoamperometric method. The $\tau_r/\tau_f$ ratio shown here is 0.3. (Right) Plot of $k\tau_f$ vs. $\tau_f$ for the ring opening of IIb in 75 vol % water-25 vol % acetonitrile: pH 3.99; 0.5 M chloroacetic acid, 0.5 M sodium chloroacetate. [From Ref. 6, reprinted with permission from M. F. Marcus and M. D. Hawley, *J. Org. Chem.* 35:2185 (1970). Copyright 1970 American Chemical Society.]

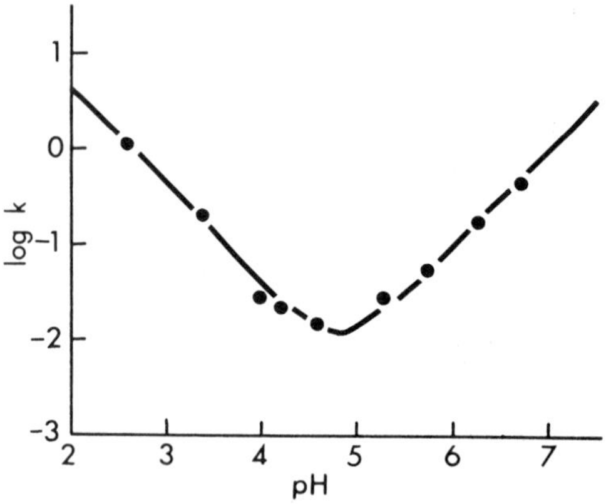


Figure 17.10 Plot of log k vs. pH for the ring opening of 8a-hydroxy-2,2,5,7,8-pentamethyl-6-chromanone (IIIb) at zero ionic strength and zero buffer concentration. [From Ref. 6, reprinted with permission from M. F. Marcus and M. D. Hawley, *J. Org. Chem.* 35:2185 (1970). Copyright 1970 American Chemical Society.]

## V. CONCLUDING REMARKS

Errors in the formulation of the reaction pathway will invariably arise. Their numbers can be minimized, however, if product studies are performed carefully and the electrochemical data are analyzed critically. Predicted changes in electrochemical behavior when a proton donor or acceptor is added, the concentration of the electroactive species is altered, and so on, should be examined and confirmed. It should also be clear from the examples above that one electrochemical technique is seldom capable of yielding all information that is needed to specify the electrode process. A variety of techniques, including nonelectrochemical methods, should be standard fare for the electroanalytical chemist.

## REFERENCES

1. D. E. Bartak, H. J. Houser, B. C. Rudy, and M. D. Hawley, *J. Am. Chem. Soc. 94*:7526 (1972).
2. M. D. Hawley, S. V. Tatawawadi, S. Perkarski, and R. N. Adams, *J. Am. Chem. Soc. 89*:447 (1967).
3. C. Amatore and J. M. Saveant, *J. Electroanal. Chem. 85*:27 (1977).
4. M. D. Hawley and S. W. Feldberg, *J. Phys. Chem. 70*:3459 (1966).
5. M. F. Marcus and M. D. Hawley, *Biochim. Biophys. Acta 201*:1 (1970).
6. M. F. Marcus and M. D. Hawley, *J. Org. Chem. 35*:2185 (1970).
7. M. F. Marcus and M. D. Hawley, *Biochim. Biophys. Acta 222*:163 (1970).
8. W. M. Schwarz and I. Shain, *J. Phys. Chem. 69*:30 (1965).

# 18

# Selected Aspects of Transition Metal Electrochemistry

WILLIAM E. GEIGER / Department of Chemistry, University of Vermont, Burlington, Vermont



## I. INTRODUCTION

Just a few short years ago it might have been appropriate to take a polemical point of view when discussing the use of electrochemical techniques in transition metal and organometallic chemistry, but it is quite apparent now that these methods will be fully and properly exploited by workers in these fields. It is an indisputable fact, however, that the field of metal complex electrochemistry has developed much more slowly and has been of much more limited scope than organic electrochemistry. The striking successes of more recent practitioners in metal complex electrochemistry have proven that this fact is due to historical reasons, and is not due to a limited applicability of these techniques to solution of inorganic problems.

Electroanalytical chemists of the mid-1950s to late 1960s who were mainly responsible for the rapid growth in sophistication of the methodology tended to have mainly organic interests, and the emphasis on study of mechanisms of organic reactions fit in perfectly well with the development of mechanistic electroanalytical chemistry (radical ions, ECEs, etc.). On the other hand, mechanisms were less widely studied in inorganic chemistry; instead, oxidation state changes were of overriding concern, and except for work of the type performed by Vlcek [1], most transition metal electrochemical studies had very limited goals, usually looking only for reversible vs. irreversible behavior of a voltammetric wave. More sophisticated problems are being approached now, and an almost unlimited future for transition metal electrochemistry is predicted, especially when broad application is made of mating electrochemical generation to physical techniques such as electronic and vibrational spectroscopy and electron spin resonance.

The intention of this chapter is to review and to preview some of the important areas which the author feels have gotten less attention than deserved in metal complex electrochemistry, but at the same time promise interesting and important future applications. Neither the range of topics (chosen on the basis of personal prejudice) nor individual treatments is intended to be comprehensive, and the reader will be referred to the literature for more extensive discussions. Only nonlabile complexes (i.e., metal complexes with essentially zero concentration of free ligand at equilibrium) will be discussed. For a thorough discussion of the electrochemical properties of labile complexes the reader is referred to the monograph by Crow [2].

## II. SIGNIFICANCE OF ELECTROCHEMICAL POTENTIALS

### A. Guide to Synthesis

Consider a situation in which it is desired to synthesize a compound which differs from the starting material only by one or more electrons (e.g., ferrocinium ion from ferrocene), which in principle can be effected through either chemical or electrochemical methods. If a "clean" chemical route to the reduced or oxidized form is found, this method may be preferable to an electrochemical method because it will not be necessary to separate the product from the supporting electrolyte in order to effect isolation of the desired material. An electrochemical experiment can complement chemical redox synthesis, however, by establishing (1) the reversibility of a redox process and hence the feasibility of chemical synthesis, and (2) the potential of the redox process. By knowing the latter, the investigator can choose the chemical redox agent of the appropriate strength to effect the synthesis. This approach has been used with success in metal dithiolene chemistry, and Davison and Holm [3] have listed reagents to use for certain easily oxidized and reduced compounds.

Headridge [4] has commented on these empirical observations within the context of the known standard potentials of the oxidizing or reducing agents. Among the compounds employed within the last few years for synthetic one-electron redox chemistry are: (for oxidations) bromine and iodine [3,5-7], silver salts [8], bis(cis-trifluoromethyl-ethylene-1,2-dithiolato)nickel [3,9], $XeF_2$ [5], $Ce^{4+}$ [10]; (for reductions) alkali metals [11], metal amalgams [12,13], and hydrazine [3]. The question of whether to use electrochemical or chemical synthesis must always be resolved according to the particular problem at hand, and general guidelines are difficult to present. Electrochemical redox power is obviously more versatile (e.g., excellent control of potential over 5 V in media such as $CH_3CN$-$R_4NPF_6$) and usually cleaner than chemical agents, and in experiments in which one of these factors is critical (e.g., partial reduction of a complex exhibiting two closely spaced reduction processes), electrochemical methods will obviously be the ones of choice. Often, however, actual isolation of salts from an electrolysis solution is made difficult by the presence of excess supporting electrolyte necessary to perform the electrolysis. Therefore, in facile redox processes involving only a clean, single, route of electron transfer, chemical redox agents are often preferred.

## B. Question of Metal vs. Ligand Participation in the Redox Process

Increasingly during the last decade coordination chemists have wrestled with the very important question of whether or not the metal oxidation state changes in a particular redox process involving a metal-ligand complex. This question of metal vs. ligand participation in the redox process is pivotal in understanding the properties of many important chemical and biochemical systems. Although this would seem to be a question more properly answered by electron spin resonance (ESR) measurements, many systems are not amenable to easy ESR study because of diamagnetism of the electrode product, difficult-to-observe or difficult-to-interpret spectra (fast relaxation effects, complex ligand hyperfine, lack of magnetic nuclei), or the degree of expertise required of the investigators. Hence an electrochemical approach has been tried and has been qualitatively successful in certain well-defined cases. Some misconceptions appear to have arisen, however, and we intend to review those here and discuss several different approaches which have yielded results of nonuniform success. Our coverage is broken down into two approaches: the first employs data from only one redox process per molecule and the second attempts to make use of potential separations of two or more redox processes for the same compound.

### *Variations of Single Redox Processes*

Given the propensity of electrochemists to make plots of half-wave potentials, it is no surprise that several studies have been made in which $E_{1/2}$ values have been measured either for a series of metals with the same ligand or a particular metal with various related ligands, and the resulting values plotted versus some parameter such as a metal ionization potential or ligand substituent constants. The success of such treatments has varied. Consider first an example in which metal and ligand orbitals are well defined and undergo little mixing: metalloporphyrins. Wolberg and Manassen [14] used data on oxidation of metal complexes of tetraphenylporphine [M(TPP)] (Eq. 18.1) to test whether metal or ligand participation in the redox process could be distinguished on the basis of electrochemical data. The redox processes involved were of the general type (Eq. 18.2) in which some oxidations not involving metal valency change occurred at the TPP ligand. The advantage of this model system was that extensive physical data (ESR, optical spectroscopy) were available which qualitatively defined the metal vs. ligand character of the various orbitals involved. For several metals (Fe, Ni, Co, Cu) with the TPP ligand, plots were made of potentials of three oxidation waves vs. the third metal ionization potential.

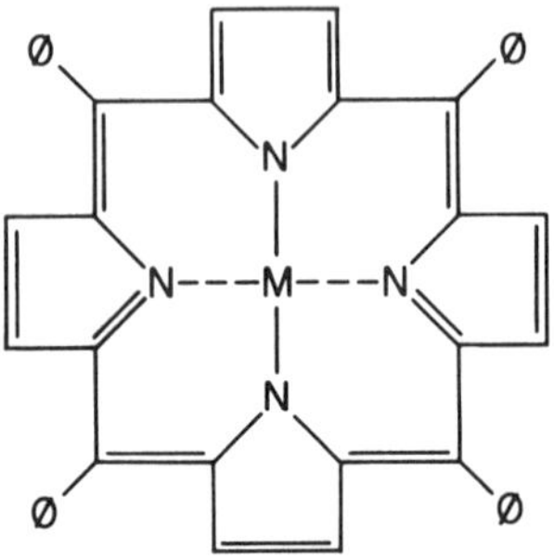


(18.1)

$$M(II)TPP \rightleftharpoons M(III)TPP^{+} \rightleftharpoons M(III)TPP^{2+} \rightleftharpoons M(III)TPP^{3+} \qquad (18.2)$$

The waves known to be metal-based gave a good linear relationship, but the ligand-based oxidations did not (see Fig. 18.1). Unfortunately, the success of this treatment could not be extended to other systems with more complex metal-ligand bonding. In fact, a similar linear plot for metal 1,2-dithiolene complexes led the authors to conclude that metal-based redox also occurs in bisdithiolene complexes of Ni, Co, Fe, and Cu, whereas it is well known that these species give predominantly ligand-based redox behavior [15].

In this regard, as long as a metal-ligand system exhibits covalency, dependence of the $E_{1/2}$ on both metal and ligand properties is to be anticipated. What is presently lacking is a theory to account quantitatively for the degree of metal-ligand participation on the basis of the degree of dependence (i.e., slopes of plots) of $E_{1/2}$

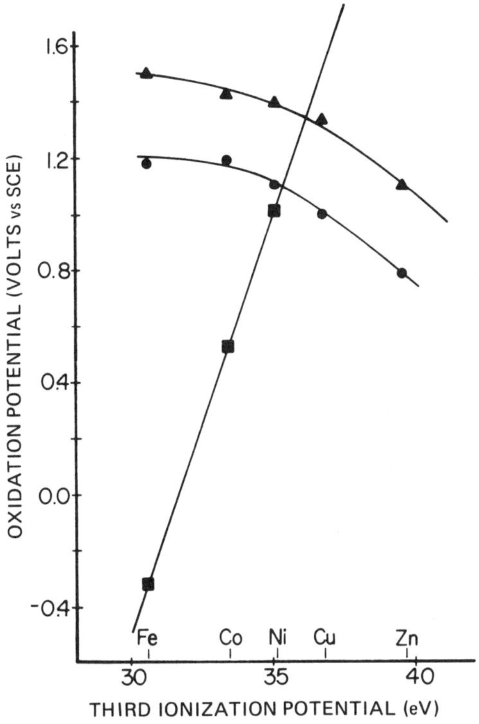


Figure 18.1 Oxidation potentials of M(TPP) complexes plotted against third ionization potential of the metal: solid square, metal valency change; solid circle, first ligand oxidation; and solid triangle, second ligand oxidation. [From Ref. 14, reprinted with permission from A. Wolberg and J. Manassen, *J. Am. Chem. Soc. 92*:2982 (1970). Copyright 1970 American Chemical Society.]

on parameters such as ionization potentials. It is already clear in a qualitative sense that these slopes can be related to metal vs. ligand contributions, for the data of Ref. 14 indicate a 250-mV/eV slope for a plot of $E_{1/2}$ vs. metal ionization potential for the *metal-based* M(TPP) oxidations, but only a 50-mV/eV slope for the predominantly *ligand-based* metal dithiolene redox.

In earlier work, Olson et al. [16] commented on the slopes observed in plots of $E_{1/2}$ for metal dithiolenes (M = Ni, Pd, Pt) with various substituents, R, versus a Taft substituent constant appropriate for the R moiety. The straight lines observed were taken to mean extensive *ligand* participation in the metal complex pi system; additionally, the lower slopes observed for Pd and Pt (0.95 and 0.89 of the Ni slope, respectively) were used to argue qualitatively for greater

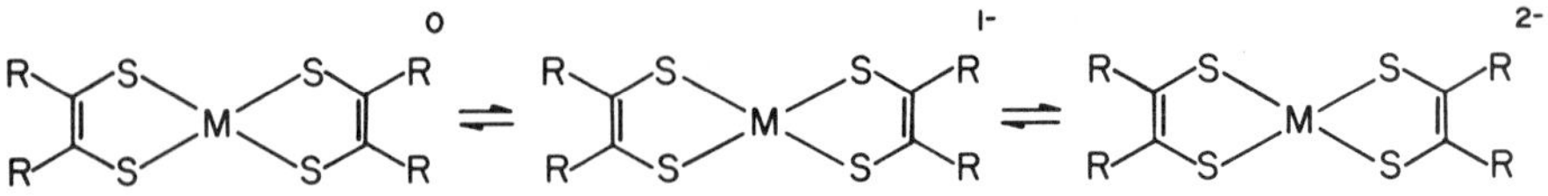


(18.3)

metal participation in the heavier metals. These conclusions have been brought into question by the ESR study of Schlupp [17] on complexes of the type $M(S_2C_2R_2)_2^{-1}$, in which the metal contribution to the half-filled orbital has been estimated at 20-30% (M = Ni), 10-20% (M = Pd), and 50-70% (M = Pt).

A variation of this approach has been to investigate a series of complexes of the same metal with a succession of similar ligands and to plot $E_{1/2}$ of the complex vs. $E_{1/2}$ of the free ligand. Performing this procedure on the reduction of a series of complexes of bipyridine and methyl-substituted bipyridines, Saji and Aoyagui [18] observed a straight line of approximately unit slope for M = Fe, Ru, and Os, concluding that occupancy of weakly antibonding ligand orbitals was involved in these reductions. This approach would appear to have validity, but needs to be tested for systems in which known amounts of metal-ligand orbital mixing is present. An additional limiting feature is that the free ligands must be stable and also undergo reversible electron transfers if ligand E° values are to be obtained.

### *Potential Differences Between Successive Electron Transfers*

Apparently, the first attempt to use electrochemical data to address the question of metal vs. ligand redox was made by Vlcek [19] (and reviewed by Gubin [20]). Vlcek attempted a calculation of the metal and ligand contributions to the redox molecular orbital of certain pi complexes of chromium by comparing differences in reduction potentials of structurally similar noncoordinated ligands and their chromium complexes. Gubin has employed differences between successive reductions in complexed and noncomplexed ligands for the same purpose [21], but the success of this approach has yet to be confirmed by independent probes to the electron density such as ESR.

In systems that undergo more than one reversible electron transfer step, it is possible to derive a great deal of information out of the *differences* in potential between successive steps. A textbook-grade example of this was found by Davis and co-workers [22] on metal octaethylporphyrin, M(OEP), complexes. It had been established through various physical studies that a wide range of M(OEP) complexes underwent as many as five reversible electron transfer steps, giving (1) one metal oxidation, (2) two ligand oxidations, and (3) two ligand reductions. Davis and colleagues measured the potential separations between the various ligand-based redox processes and found that $\Delta E_{1/2}$, the modulus of the potential differences, fell into the following classes:

$\Delta E_{1/2}$ ligand (OEP) *oxidations* = 0.29 V ± 0.05 V

$\Delta E_{1/2}$ ligand (OEP) *reductions* = 0.42 V ± 0.05 V

$\Delta E_{1/2}$ *first* OEP *oxdn.* − *first* OEP *redn.* = 2.25 V ± 0.15 V

Redox processes that did not fit these criteria could be assigned to *metal* redox processes. These authors also pointed out the danger of using the traditional $E_{1/2}$ vs. ionization potential or electron affinity plots to try to reach similar conclusions. The authors plotted the first OEP reduction and oxidation vs. the electronegativity of central metal ion and found a linear relationship to exist. As pointed out above, this is not unexpected for systems showing metal-ligand covalency and means that even predominantly ligand-based redox processes may show a linear dependence of $E_{1/2}$ on some property of the metal (e.g., electronegativity or ionization potential), and that such a fit should not be construed as proof of a metal-based redox process. Using the data of Ref. 22, it can be shown that the slope of a plot of either the first ligand oxidation or reduction vs. metal ionization potential is approximately 30 mV/eV, close to the value of 50 mV/eV observed in the metal dithiolene reductions (which have about 1/4 metal character), but much lower than the 250 mV/eV observed for the M(TPP) metal-based redox reactions.

In summary, measurement of reversible $E_{1/2}$ potentials for a series of metal complexes potentially holds useful information about the metal-ligand distribution of the orbital involved in the redox process. Simple plots of $E_{1/2}$ vs. properties like metal ionization potentials are unlikely to give unambiguous answers on the question of metal vs. ligand involvement, but more sophisticated treatments based on differences between successive redox potentials are more reliable for the systems in which they are applicable. In any case, development of procedures based on sound theoretical grounds, and carefully compared with independent experimental data, would provide a valuable contribution to this area.

## III. MECHANISMS AND REACTIONS OF ELECTROGENERATED COMPLEXES

### A. General Remarks

Mechanistic aspects of organic electrochemistry have been actively pursued for many years, and quite sophisticated information is now available on broad classes of compounds [23]. One of the outgrowths of these efforts has been a boom in the use of electrogenerated reagents, most specifically in organic electrosynthesis. By comparison, mechanistic electrochemistry on metal complexes is relatively unexplored, especially in cases where the electrode process is not a simple, reversible, electron transfer uncomplicated by following chemical reactions. For a long time, the only extensive systematic studies of mechanisms of organometallic electrochemical reactions were carried out by Vlcek [1] and by Dessy and co-workers. The latter investigated the mechanisms of reactions of electrogenerated metal complex anions with various electrophiles, and the reader is referred to Dessy's review of these efforts for further details [24]. While a literature review is not an objective of this chapter, the point we wish to make is that inorganic and organometallic electrosynthesis will become an important and versatile weapon in the chemist's arsenal only as more information on the nature and reactivity of electrogenerated reagents becomes available. This is particularly true in the case of metal complex oxidations.

What kinds of reactions can generally be expected of electrogenerated species? If we assume that the electrochemical process involves an initial electron transfer to or from a neutral starting material, it can be anticipated that a high reactivity of the resulting charged complex to either electrophiles or nucleophiles, respectively, will occur. One of the questions to be answered for each chemical system under study is whether such attack will occur at the metal or the ligand, and what the consequences of each might be. It is anticipated that some general routes to ligand modification might be available by this process; subsequent freeing of the ligand from the metal by electrochemical or chemical techniques could result in useful routes to organic compounds in cases where a complexed ligand has different properties (e.g., stability and reactivity) from the free ligand. In this sense, metal template electrosynthesis might be performed. On the other hand, the electrogenerated reagent might possess a metal in a reactive oxidation state and undergo reactions such as oxidative addition or reductive elimination at the metal. In systems wherein the metal reactivity and coordinating ability is a function of oxidation state, electrocatalytic phenomena may be anticipated.

Our immediate purposes here are less ambitious. In the subsequent section we shall indicate how a mechanistic study of the electroreduction of bis(cyclopentadienyl)cobalt, cobaltocene, in nonaqueous solvents, was carried out in our laboratory.

## B. An ECE Process: Reduction of Cobaltocene in Acetonitrile

The bis($\pi$-cyclopentadienyl)cobalt (III) cation, $Cp_2Co^+$, cobaltocinium ion, is reducible in two separate steps, the first reduction at about

−0.9 V vs. the SCE being a highly reversible, one-electron step [25, 26] resulting in formation of the well-characterized neutral cobaltocene. The second reduction, at about −1.9 V, involves the reduction of neutral $Cp_2Co$ and is of interest because of the rarity of anions derived from metallocenes. It is this process, the reduction of $Cp_2Co$, which is discussed in this section. The experimental study was aided greatly by comparison of data for the $Cp_2Co$ reduction with that obtained for the $Cp_2Co^+ \rightleftarrows Cp_2Co$ couple, which was used as an internal "calibrant" for a reversible, uncomplicated electron transfer. However, for discussion purposes, the reader must be careful not to confuse aspects of the multielectron reduction of $Cp_2Co$ with the wave at −0.9 V due to the $Cp_2Co^+ \rightleftarrows Cp_2Co$ couple.

The first report of the reduction of neutral $Cp_2Co$ used a comparison of the first and second $Cp_2Co^+$ polarographic plateau currents in $CH_3CN$ to conclude that one electron was also involved in the reduction of $Cp_2Co$ [27]. These authors suggested that the reduction was irreversible, ending up in the formation of $(C_5H_5)Co(C_5H_6)$. Shortly thereafter, a cyclic voltammogram (CV) of the reduction was reported in basic ethanol solution which showed the process to be electrochemically reversible [28]. However, macro-scale-controlled potential electrolyses in our laboratory in $CH_3CN$ yielded neither $Cp_2Co^-$ nor $CpCoC_5H_6$ as the product. The following is a detailed account of how the reduction mechanism was diagnosed as an ECE process in $CH_3CN$ and how the rate constant for the intervening chemical reaction was found.

### *Polarography*

The $Cp_2Co^+PF_6^-$ was very convenient in that the $Cp_2Co^+/Cp_2Co$ couple provided a one-electron "internal standard" against which to measure currents for the $Cp_2Co \rightleftarrows Cp_2Co^-$ couple. After corrections for changes in drop time, our polarographic data showed that the second plateau current was slightly greater than expected for a one-electron process. The ratio of diffusion current constants for the two waves varied from 2.08 to 2.12, depending on drop life. This measurement was only slightly outside of experimental error of the 2.00 expected for a simple one-electron change in $Cp_2Co \rightleftarrows Cp_2Co^-$, but provided the first suspicion of a more complicated mechanism.

### *Cyclic Voltammetry*

For an uncomplicated reduction the cathodic current function ($i_{pc}/\nu^{1/2}$) is independent of $\nu$ (the scan rate), and the ratio $i_{pa}/i_{pc}$ is unity. For an EC mechanism, the current function is still practically independent of $\nu$, but $i_{pa}/i_{pc}$ is less than unity unless the scan rate is much larger than the rate of the chemical reaction. For an ECE mechanism in which the second reduction occurs at a potential more positive than the first, at appropriately slow scan rates the current function

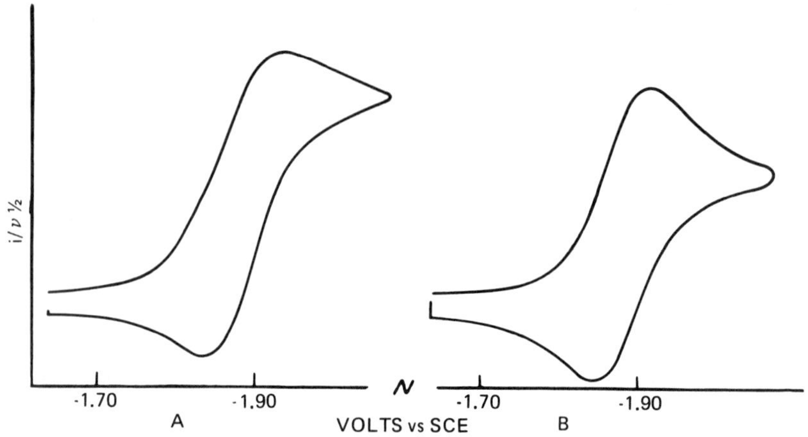


Figure 18.2 Cyclic voltammetry of the $Cp_2Co^{0/-}$ wave in $CH_3CN/0.1$ M $Bu_4NPF_6$ at a mercury electrode, with the current normalized for changes in scan rate. The scan rate for (A) was 0.05 V/s, and for (B), 0.5 V/s. Note that at the higher scan rate of (B) the current function ($i_{pc}/v^{1/2}$) is lower than at the slower scan rate of (A).

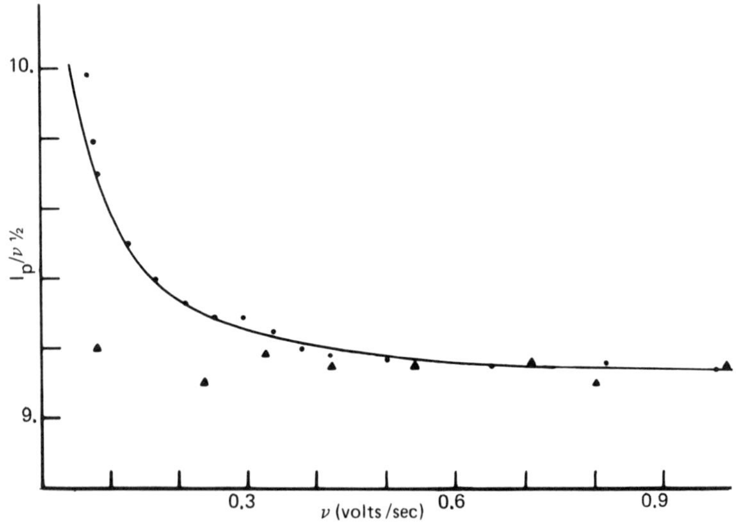


Figure 18.3 Plots of cathodic current function vs. scan rate for the reduction of $Cp_2Co^+$ (solid triangles) and $Cp_2Co$ (solid circles) in acetonitrile. The solid line gives the theoretical values assuming $k = 0.16\ s^{-1}$ for the follow-up reaction rate of an ECE process.

**Table 18.1** Anodic to Cathodic Current Ratios ($i_{pa}/i_{pc}$) for the Reduction of Cobaltocene in $CH_3CN$ as a Function of Scan Rate

| Scan rate (mV/s) | $i_a/i_c$ | $\tau^a$(s) | $k\tau^b$ |
|---|---|---|---|
| 58 | 0.75 | 4.17 | 0.31 |
| 83 | 0.81 | 2.92 | 0.22 |
| 166 | 0.89 | 1.46 | 0.11 |
| 249 | 0.92 | 0.97 | 0.083 |
| 332 | 0.94 | 0.73 | 0.065 |
| 415 | 0.95 | 0.58 | 0.055 |

[a] $\tau$, time elapsed from $E_{1/2}$ to $E_\lambda$.
[b] $k\tau$ determined from working curve of Nicholson and Shain.
(Adapted from Ref. 30.)

will be greater than the E or EC case and $i_{pa}/i_{pc}$ will be less than unity. The observed cathodic current function for both the $Cp_2Co^+$ reduction (E) and the $Cp_2Co$ reduction (ECE) are shown in Figures 18.2 and 18.3 as a function of scan rate, and ratios of $i_{pa}/i_{pc}$ are given in Table 18.1. These data establish that (1) a follow-up reaction occurs for $Cp_2Co^-$ in this medium, and (2) the product is electroactive, undergoing reduction at a more positive potential than $Cp_2Co$ itself. Controlled-potential coulometry was used to establish the total number of electrons transferred per mole of $Cp_2Co$. An n value of 2.0 was measured for the $Cp_2Co$ reduction, meaning that an ECE process with one electron per "E" was involved.

Our attention then turned toward measurement of the rate of the chemical reaction in the ECE process. For this, both CV and polarography were employed, using the procedures of Nicholson and Shain [29,30], and Nicholson et al. [31], respectively. First, let us consider the CV measurements. For an ECE process, the rate constant, k, of the follow-up reaction can be measured either by (1) monitoring the rate of disappearance of $Cp_2Co^-$ by $i_{pa}/i_{pc}$ values, or (2) monitoring the appearance of Z by measuring the cathodic current enhancement due to the $Z \rightarrow Z^-$ reduction. The latter requires careful measurement of the cathodic current function over an appropriate range of scan rates.

$$\text{E:}\quad Cp_2Co \underset{}{\overset{e}{\rightleftharpoons}} Cp_2Co^-$$

$$\text{C:}\quad Cp_2Co^- \xrightarrow{k} Z$$

$$E: \; Z \underset{}{\overset{e}{\rightleftharpoons}} Z^-$$

To measure k, the rate of the follow-up reaction, from the rate of disappearance of $Cp_2Co^-$, the ratio $i_{pa}/i_{pc}$ is measured at different scan rates using Nicholson's method [32]. The rate constant is then computed by using the theoretical working curve [30] of $i_{pa}/i_{pc}$ vs. $k\tau$ (where $\tau$ is the time in seconds it takes to scan from the $E_{1/2}$ potential to the switching potential) and dividing the experimental $k\tau$ by $\tau$. In practice, it is best to vary the scan rate over a range in which $i_{pa}/i_{pc}$ ranges from about 0.6 to 0.9 and then to plot $k\tau$ for each data point vs. $\tau$ for that scan rate. The slope of the resulting plot gives k. Our data (Fig. 18.4) yielded $k = 0.09\ s^{-1}$, assuming pseudo-first-order kinetics.

The rate constant k can also be measured from the enhancement of the cathodic peak current which arises when the chemical reaction is fast enough, compared to $\nu$, that the electroactive intermediate Z is formed in sufficient quantity during the scan. To make quantitative use of this current enhancement, one must know the value of the cur-

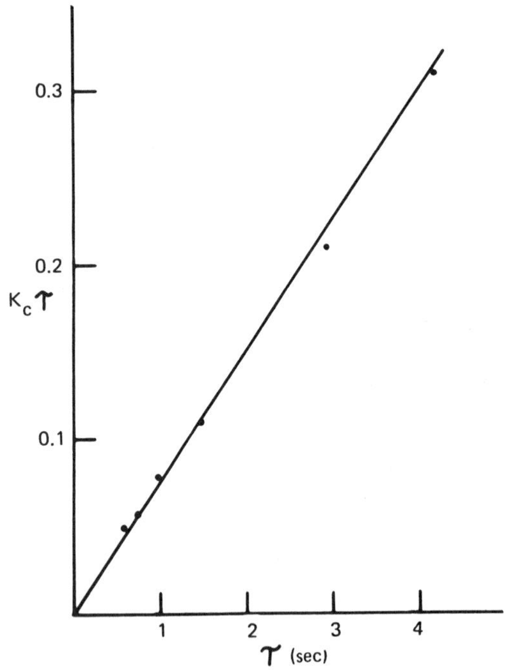


Figure 18.4 Plot of $k\tau$ vs. $\tau$ (time lapsed from $E_{1/2}$ to switching potential) for the $Cp_2Co$ reduction.

rent function, denoted $\chi_d$, which would occur in the *absence* of the complicating follow-up reaction. $\chi_d$ can be obtained by measuring $i_{pc}/\nu^{1/2}$ at scan rates that outrun the rate of the chemical reaction. In the present case (Fig. 18.3), this occurred at $\nu$ above about 1.5 V/s. Current functions measured at slower scan rates exceeded this value. Nicholson has shown that if the number of electrons transferred in the first and second "E's" are equal, these current functions can be related to k and to the scan rate parameter, a ($a = nF\nu/RT$), by the equation

$$\frac{\chi_k}{\chi_d} = \frac{0.400 + k/a}{0.396 + 0.469(k/a)}$$

One simply "plugs in" the value of $\chi_k/\chi_d$ observed at each scan rate to obtain a value of k/a, then plots the resulting k/a vs. 1/a and the slope determines k (Fig. 18.5). This treatment yielded k = 0.16 $s^{-1}$ in our case, compared to the value of 0.09 $s^{-1}$ measured from $i_{pa}/i_{pc}$ values. The probable reason for the discrepancy in results lies in this case in the rather low value of k, which resulted in relatively small current enhancements down to the lowest scan rate at which we consider quantitative results to be obtainable. For example, at $\nu$ = 58 mV/$s^{-1}$, $\chi_k/\chi_d$ was 1.14.

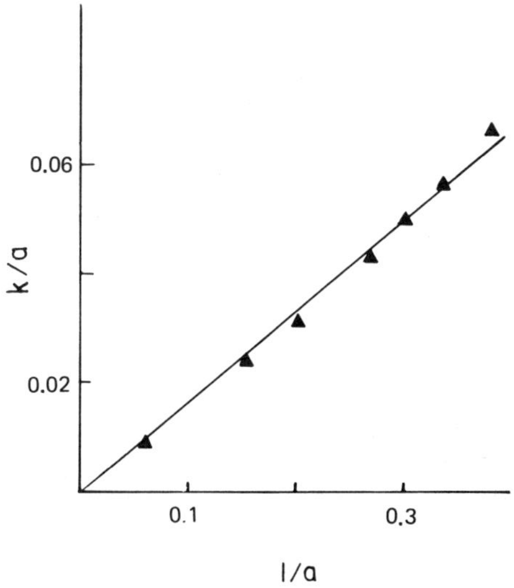


Figure 18.5 Plot of k/a vs. 1/a (the scan rate parameter) for the cobaltocene reduction.

Finally, the rate constant of the chemical follow-up reaction can be measured by polarography, using the theory of Nicholson et al. [31], for an ECE process with pseudo-first-order chemical kinetics. Use of this procedure is limited to systems in which the diffusion-controlled current, $i_d$, is known in the absence of the follow-up reaction. The colbaltocene problem lends itself very nicely to this analysis, because the $Cp_2Co^+ \rightleftarrows Cp_2Co$ wave acted as an internal standard against which to measure the current enhancements on the second wave. The ratio of the second wave height (corrected for drop time) to the first served as the measure of $i_k/i_d$ necessary to use the working curve of Ref. 31 to find $k\tau$, the product of the chemical rate constant and the capillary drop time.

To illustrate these calculations, when the drop time at the crest of the second wave was 4.77 s, the I value (diffusion current constant) of the second wave was 22% higher than that of the first ($Cp_2Co^{+/0}$) wave. This represented a current enhancement, over the diffusion-controlled value, of 22% due to the electroactive product of the ECE reaction. The value of 1.22 was used for $i_k/i_d$ and from Table I of of Ref. 31, and a value of $k\tau = 0.54$ was found. Taking the measured drop time into account, a rate constant of 0.11 $s^{-1}$ was calculated. These measurements were repeated at several drop times and an average of 0.10 $s^{-1}$ was obtained. This value can be compared with that measured from enhancement of the CV cathodic current function (0.16 $s^{-1}$) and from diminution of the CV anodic current (0.09 $s^{-1}$).

### *The Overall Reduction Mechanism*

This kinetic study did not identify the intermediate, Z, or the final product of the reduction of $Cp_2Co$ in acetonitrile. In view of the fact that cobaltocene can be reduced to cyclopentadienylcobalt cyclopentadiene, $CpCoC_5H_6$, by $BH_4^-$ [33] or by electrolysis in aqueous dimethylformamide [34], we need to address the possibility that $CpCoC_5H_6$ is formed at some point in $CH_3CN$ also. Bulk electrolysis of $Cp_2Co^+$ in $CH_3CN$ at −2.0 V consumes three electrons (one for $Cp_2Co^{+/0}$) and the resulting solution shows an oxidation wave at about −0.7 V and a reduction wave at −2.38 V (irreversible; cathodic peak potential reported). The former (oxidation) wave is unidentified; the latter matches up exactly with the reduction of $CpCo_5H_5CH_2CN$, in which addition of $CH_2CN$ to the Cp ring has taken place.

The compound $CpCoC_5H_6$ gives a reduction wave at −2.2 V. Since it does not seem to be the final product, we can still ask if it is formed as a transient intermediate by the following ECE mechanism, involving protonation of $Cp_2Co$:

$$\text{E:}\quad Cp_2Co + e^- \rightleftarrows Cp_2Co^- \qquad E_{1/2} = -1.88\ \text{V} \tag{18.4}$$

$$\text{C:}\quad Cp_2Co^- + H^+ \longrightarrow CpCoC_5H_6 \tag{18.5}$$

$$\text{E: } CpCoC_5H_6 + e^- \rightleftharpoons CpCoC_6H_6^- \qquad E_{1/2} > -1.88\text{ V} \qquad (18.6)$$

$$CpCoC_5H_6^- \longrightarrow \text{fast reaction to products} \qquad (18.7)$$

The source of protons in this scheme would be from adventitious water or from $CH_3CN$ itself. But this mechanism can be *eliminated* because of the requirement that our intermediate in the ECE process has to have an E° *positive* of the reduction $Cp_2Co^{0/-}$. Since the observed $E_{1/2}$ for the reduction of $CpCoC_5H_6$ (−2.28 V) is negative of −1.88 V ($Cp_2Co^{0/-}$), $CpCoC_5H_6$ must not be formed in this solvent. Since the rate of the follow-up reaction is not altered even in highly purified $CH_3CN$ under high-vacuum conditions, the reaction appears to be between the cobaltocene anion and acetonitrile.

## REFERENCES


1. A. A. Vlcek, *Prog. Inorg. Chem.* 5:211 (1963).
2. D. R. Crow, *Polarography of Metal Complexes,* Academic, New York, 1969.
3. A. Davison and R. Holm, *Inorganic Synthesis 10*:8 (1967).
4. J. B. Headridge, *Electrochemical Techniques for Inorganic Chemists,* Academic, New York, 1969.
5. J. Fajer, D. Borg, A. Forman, D. Dolphin, and R. H. Felton, *J. Am. Chem. Soc. 92*:3451 (1970).
6. R. J. Haines and A. L. du Preez, *Inorg. Chem. 11*:330 (1972).
7. J. A. de Beer, R. J. Haines, and R. Greatrex, *J. Chem. Soc., Chem. Commun.,* 1094 (1972).
8. N. G. Connelly and L. F. Dahl, *J. Am. Chem. Soc. 92*:7470, 7472 (1970).
9. W. E. Geiger, Jr. and A. H. Maki, *J. Phys. Chem. 75*:2387 (1971).
10. R. B. King, D. M. Braitsch, and P. N. Kapoor, *J. Am. Chem. Soc. 97*:60 (1975).
11. Alkali metal reductions have been widely employed. Some representative papers are: J. W. Eastes and W. M. Burgess, *J. Am. Chem. Soc. 64*:1187 (1942) (K in liquid $NH_3$); G. Bigam, et al., *Can. J. Chem. 50*:1825 (1972); C. Elschenbroich, F. Gerson, and F. Stohler, *J. Am. Chem. Soc. 95*:6956 (1973).
12. M. Hawthorne, D. Young, T. Andrews, D. Howe, R. Pilling, A. Pitts, M. Reintjes, L. Warren, and P. Wegner, *J. Am. Chem. Soc. 90*:879 (1968).
13. J. Ferguson and T. J. Meyer, *J. Am. Chem. Soc. 94*:3409 (1972).

14. A. Wolberg and J. Manassen, *J. Am. Chem. Soc. 92*:2982 (1970).
15. J. A. McCleverty, in *Reactions of Molecules at Electrodes* (N. S. Hush, ed.), Wiley-Interscience, New York, 1971, Chap. 9.
16. D. C. Olson, V. P. Mayweg, and G. N. Schrauzer, *J. Am. Chem. Soc. 88*:4876 (1966).
17. R. L. Schlupp, Ph.D. thesis, University of California, 1973.
18. T. Saji and S. Aoyagui, *J. Electroanal. Chem. 58*:401 (1975).
19. A. A. Vlcek, *Z. Anorg. Allg. Chem. 304*:109 (1960).
20. S. P. Gubin, *Pure Appl. Chem. 23*:463 (1970).
21. S. P. Gubin and V. S. Khandkarova, *J. Organomet. Chem. 12*:523 (1968).
22. J. H. Fuhrhop, K. Kadish, and D. Davis, *J. Am. Chem. Soc. 95*:5140 (1973).
23. See, for example, M. Baizer (ed.), *Organic Electrochemistry*, Marcel Dekker, New York, 1973.
24. R. Dessy and L. Bares, *Acc. Chem. Res. 5*:415 (1972).
25. A. Vlcek, *Coll. Czech. Chem. Commun. 30*:952 (1965).
26. W. E. Geiger, Jr. and D. Smith, *J. Electroanal. Chem. 50*:31 (1974).
27. S. P. Gubin, S. Smirnova, and L. Denisovich, *J. Organomet. Chem. 30*:257 (1971).
28. N. El Murr, R. Dabard, and E. Laviron, *J. Organomet. Chem. 47*:C13 (1973).
29. R. Nicholson and I. Shain, *Anal. Chem. 36*:706 (1964).
30. R. Nicholson and I. Shain, *Anal. Chem. 37*:179, 190 (1965).
31. R. Nicholson, J. Wilson, and M. Olmstead, *Anal. Chem. 38*:542 (1966).
32. R. Nicholson, *Anal. Chem. 38*:1406 (1966).
33. M. Green, L. Pratt, and G. Wilkinson, *J. Chem. Soc.*, 3753 (1959).
34. W. E. Geiger, Jr., W. L. Bowden, and N. El Murr, *Inorg. Chem. 18*:2358 (1979).

# 19

# Electrochemical Preconcentration

WILLIAM R. HEINEMAN, HARRY B. MARK, JR., and JOHN A. WISE* / Department of Chemistry, University of Cincinnati, Cincinnati, Ohio

DARYL A. ROSTON / Department of Chemistry, Northern Illinois University, DeKalb, Illinois



## I. INTRODUCTION

An important aspect of electrochemical methodology is the ease with which solution species can be extracted from a sample solution either into or onto an electrode by, for instance, a controlled-potential deposition. Such an extraction step enables material to be preconcentrated at the electrode, which can then be analyzed by any number of electrochemical or spectroscopic techniques. The electrochemical preconcentration step provides the means of substantially improving the detection limit of the technique used for the analytical step. This chapter describes certain aspects of electrochemical preconcentration with subsequent analysis by electrochemical or spectroscopic tech-

*Present address: The Goodyear Tire & Rubber Company, Akron, Ohio

niques. The term "stripping" is used in conjunction with the electroanalytical techniques, since the analysis typically involves electrochemical removal of the preconcentrated analyte from the electrode. The purpose of this chapter is to describe some of the principles and methodology of electrochemical preconcentration techniques, rather than to give detailed procedures for the determination of various analytes.

## II. ELECTROCHEMICAL STRIPPING ANALYSIS

### A. Introduction

Electrochemical stripping analysis has the lowest detection limit of any commonly used electroanalytical technique [1-3]. Analyses can be performed at the "trace" level, concentrations of $10^{-10}$ M and lower having been determined. Other advantageous features of stripping voltammetry include the capability for simultaneous multielement determination and instrumentation that is relatively inexpensive compared to that required for many of the spectroscopic techniques for trace metal analysis.

An electrochemical stripping analysis experiment consists of two parts. During the first step, analyte is deposited (preconcentrated) into or onto the surface of an electrode (usually mercury) by controlled-potential electrolysis. In the second step the deposited analyte is removed ("stripped") from the electrode, and the accompanying electrochemical signal is used to determine the concentration of each analyte species in the sample. In stripping voltammetry the stripping step is accomplished by a potential scan and the resulting current peaks are measured to determine concentration. In potentiometric stripping analysis stripping is accomplished by a chemical oxidant in solution and the potential of the electrode is recorded. The time intervals between potential breaks are then measured to determine concentration.

Anodic stripping voltammetry (ASV) is the stripping variation used primarily for the determination of heavy metals. In this technique preconcentration is caused by reduction of the metal ions to the elemental state, and the stripping step is accomplished by a positive potential scan that gives an anodic current when the preconcentrated metals are oxidized. The technique is applicable to metal ions that can be preconcentrated at an electrode by reduction and is especially effective for metals that dissolve in mercury. Examples of metal ions which have been determined by ASV at a mercury electrode are $Bi^{3+}$, $Cd^{2+}$, $Cu^{2+}$, $Ga^{3+}$, $Ge^{4+}$, $In^{3+}$, $Ni^{2+}$, $Pb^{2+}$, $Sb^{3+}$, $Sn^{2+}$, $Tl^{+}$, and $Zn^{2+}$ [1,2]. Other metals ions, such as $Hg^{2+}$, $Au^{3+}$, $Ag^{+}$, and $Pt^{4+}$, have been determined by ASV at solid electrodes such as wax-impregnated graphite and glassy carbon [1,4].

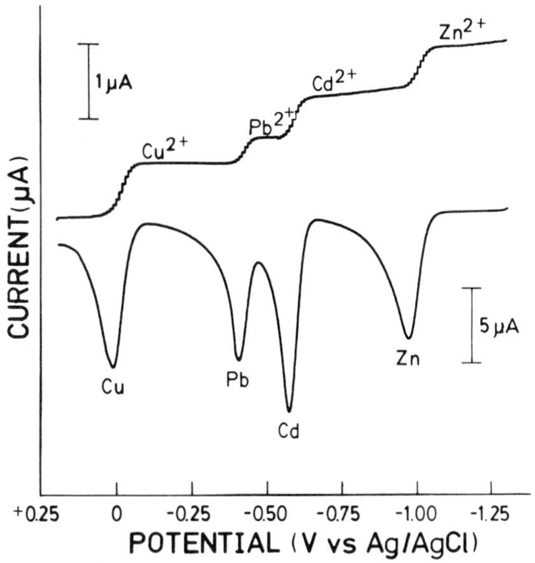


**Figure 19.1** Current-sampled polarogram (top) and anodic stripping voltammogram (bottom) of 2.5 ppm $Cu^{2+}$, $Zn^{2+}$ and 5 ppm $Pb^{2+}$, $Cd^{2+}$ in 0.1 M sodium acetate.

The lower curve in Figure 19.1 is a typical multielement stripping voltammogram obtained at a hanging mercury drop electrode. Positioned above it on the same potential axis is a current-sampled polarogram (see Chap. 3) of the same solution. The waves on the polarogram show the potential regions in which particular metal ions are reduced at the mercury electrode. For example, $Cu^{2+}$ can be selectively reduced in the potential range 0 to −0.3 V, whereas $Cu^{2+}$, $Pb^{2+}$, $Cd^{2+}$, and $Zn^{2+}$ are reduced simultaneously at a potential of −1.3 V. The stripping voltammogram is obtained by first applying a potential of −1.3 V for 1 min while stirring the solution. During this step Cu, Pb, Cd, and Zn are extracted into the mercury drop by reduction of the corresponding ions:

$$M^{2+}_{\substack{\text{sample}\\ \text{solution}}} + 2e \longrightarrow M_{Hg} \qquad (19.1)$$

Following the preconcentration period, stirring is stopped, and the potential is scanned in the positive direction. The dissolved metal atoms are stripped from the mercury electrode back into the sample solution by oxidation to the ionic form:

$$M_{Hg} \longrightarrow M^{2+}_{\substack{\text{sample}\\\text{solution}}} + 2e \qquad (19.2)$$

The different metals strip back into solution in sequence according to their reduction potentials. The resulting current peaks can then be used to determine the concentrations of the analyte species. Note the correspondence between polarographic $E_{1/2}$ and the peak potentials in the stripping voltammogram of the four metals.

Cathodic stripping voltammetry (CSV) involves preconcentration by oxidation with subsequent stripping via a negative potential scan. Anions may be determined by deposition as insoluble mercury salts on the electrode surface according to the following general reaction:

$$2Hg + 2X^{-} \longrightarrow Hg_2X_2 + 2e \qquad (19.3)$$

where the stoichiometry depends on what $X^{-}$ is. The negative potential scan causes the reduction of the salt to Hg and $X^{-}$ (the reverse of Eq. 19.3), giving a cathodic current peak. The height of this reduction peak is proportional to the concentration of the anion in solution. Thus anions may be determined by CSV in a similar way that metal ions are determined with ASV. Silver can be used as the electrode for the determination of anions that form insoluble silver salts. CSV has been applied to the determination of $Cl^{-}$, $Br^{-}$, $I^{-}$, $S^{2-}$, $Se^{2-}$, $CrO_4^{2-}$, $WO_4^{2-}$, $MoO_4^{2-}$, $VO_3^{-}$, $SO_4^{2-}$, oxalate, succinate, dithizonate, and diethylthiophosphate [1,4,5]. The determination of $Pb^{2+}$ is possible by cathodically stripping a film of $PbO_2$ deposited on a $SnO_2$ electrode [6]. Analogous procedures have been used for metals such as $Ce^{3+}$, $Tl^{+}$, $Mn^{2+}$, and $Fe^{2+}$ in which the sample is concentrated as an insoluble hydroxide by oxidation at a graphite electrode [4].

The following discussion is written in terms of determining metal ions by ASV. Many of the points, however, also apply to CSV and potentiometric stripping analysis.

## B. Preconcentration Step

The remarkable sensitivity of stripping voltammetry is attributable to the preconcentration step. In essence, the analyte ions are electrochemically extracted from the sample solution as metal atoms into a mercury electrode (or as a metal film on a solid electrode or as a salt onto the electrode surface). Since the volume of the mercury electrode is considerably less than the volume of sample solution in the cell, the resulting "solution" of metal atoms in the liquid mercury is more concentrated than the solution of metal ions being determined by a factor of up to $10^6$ if the solubility in mercury is not exceeded.

The deposition potential should be adjusted so that ions undergo electron transfer as rapidly as they are transported to the electrode surface. Correct selection of deposition potentials can be made by considering polarograms such as the one in Figure 19.1 or hydrodynamic voltammograms which can be obtained for the metal ions of interest in solutions of higher concentration. The deposition potential should be in (or beyond) the limiting current region of the metal ion(s) being deposited. In the example shown in Figure 19.1, Cu can be selectively deposited from the sample by preconcentration at a potential that is too positive for reduction of the other metal ions. Similarly, Cu and Pb could be selectively deposited. However, selective deposition of Pb without Cu is not feasible, since the potential required to deposit Pb will also cause reduction of $Cu^{2+}$ in this particular supporting electrolyte.

The extraction of sample into (or onto) the electrode is by an electrochemical redox process that occurs only at the electrode-solution interphase. Consequently, in order to minimize the time required to deposit a sufficient amount of material at the electrode for analysis, the mass transfer of electroactive species to the electrode surface should be as efficient as possible. For this reason, the solution is stirred, or the electrode is rotated, or the solution is passed through a flow cell during the deposition step.

The concentration of metal in a mercury electrode after a given deposition time (i.e., after the preconcentration step) is described by the following equation, assuming a constant limiting current [3]:

$$C_{Hg} = \frac{i_{\ell} t}{VnF} \tag{19.4}$$

where $i_{\ell}$ is the limiting current during the deposition step in amperes, t is the length of the deposition period in seconds, and V is the volume of the electrode in $cm^3$. For a given volume of mercury, the concentration of metal in the mercury electrode is directly proportional to $i_{\ell}$ and t. Accordingly, a given concentration of metal in the mercury electrode can be achieved by an appropriately long deposition time. Or, for a given deposition time, a greater concentration can be achieved by maximizing $i_{\ell}$ (i.e., improving the efficiency of the preconcentration process). Since lengthening t constitutes an increase in the time required per analysis, it is sometimes advisable to consider increasing the efficiency of the preconcentration process by maximizing $i_{\ell}$.

An empirical equation that delineates $i_{\ell}$ is

$$i_{\ell} = \frac{nFAD}{\delta} C^{\circ} \tag{19.5}$$

where A is the surface area of the electrode in $cm^2$, D is the diffusion coefficient of the electroactive species in $cm^2/s$, $C^o$ is the concentration of the electroactive species in the solution in $mol/cm^3$, and $\delta$ is the thickness of the Nernst diffusion layer. (The reader is referred to the discussion on hydrodynamic voltammetry in Chap. 3, Sec. IV, for further discussion of the Nernst diffusion layer and Eq. 3.44.) Experimentally controllable parameters in Equation 19.5 are $\delta$, the thickness of the Nernst diffusion layer, and A, the area of the electrode. The Nernst diffusion layer is the thin layer of stagnant solution next to the electrode across which metal ions must diffuse to reach the electrode. The thickness of the Nernst diffusion layer can be decreased by more efficient stirring of the solution, faster rotation of the electrode, or increasing flow rate in a flow cell. Hence, by decreasing the thickness of the Nernst diffusion layer, forced convection increases $i_\ell$ and the efficiency of the deposition process.

Electrode surface area represents another exploitable parameter in terms of improving the efficiency of the preconcentration process. The number of metal ions reduced during the preconcentration process is directly proportional to the surface area of the electrode, as shown in Equation 19.5. If two electrodes have the same surface area, higher metal concentrations are achieved at the electrode with the smaller volume. This is reflected in the relative sensitivities of the two electrodes most commonly employed for ASV, the hanging mercury drop electrode (HMDE) and the mercury film electrode (MFE). The MFE has a higher surface-to-volume ratio than the HMDE and consequently exhibits better sensitivity for a given deposition time [7].

The electrolysis of sample ions during the deposition step is generally not exhaustive. As the concentration of metal ion in solution decreases while electrolysis proceeds, the rate of deposition drops proportionately. Consequently, the benefit of extremely long depositions is not always commensurate with the expended time. Usually, only a fraction of the metal ions in the electrolyte solution needs to be deposited at the electrode to obtain a measurable current during the stripping step. For analytical purposes it is important that the same fraction of metal ion be removed during each ASV experiment. Therefore, it is essential for the electrode surface area, electrode volume, rate of solution stirring (or electrode rotation, or flow rate), and length of the preconcentration period to be as reproducible as possible to ensure that good precision is obtained.

Although much stripping analysis is done with an HMDE that is immersed in a solution with a stirring bar or an MFE that is immersed in a solution with a stirring bar or is rapidly rotated, flowing solution across an MFE in a manner analogous to flow injection analysis shows much promise, especially for the analysis of small-volume samples. Notable examples of flow methodologies include anodic stripping voltammetry with collection (ASVWC) at tubular electrodes [8,9] potentiometric stripping analysis in a thin-layer flow cell [10], electrodeposition and stripping at graphite cloth electrodes in a flow-through cell [11], differential pulse anodic stripping voltammetry in a thin-layer flow cell [12], and a flow cell for use with the static mercury drop electrode (see Chap. 9) [13]. The preconcentration, or more important, the electrochemical separation portion, of stripping analysis has also been utilized with flow cells for the purification of electrolytes [14] and eluents for liquid chromatography with electrochemical detection [15].

## C. Electroanalytical Stripping Techniques

Once some fraction of the analyte has been deposited at the electrode during the preconcentration step, the electrode is then "analyzed" by the stripping step. The resulting electrochemical response signal is proportional to the concentration (or amount) of analyte at the electrode, which is in turn proportional to its concentration in the sample solution. Numerous electroanalytical techniques have been used for this step in the analytical procedure.

### *Linear Potential Sweep Voltammetry*

An effective stripping technique that requires simple instrumentation is linear potential sweep voltammetry (Chap. 3, Sec. III). After preconcentration of the metal atoms at the electrode, the potential of the electrode is scanned in the positive direction. At characteristic potentials the metal atoms are oxidized (stripped) from the electrode. The potentials of the stripping peaks serve to identify the respective metals (see Fig. 19.1), while the magnitude of the stripping current reflects concentration. Theory for the stripping waves produced by a linear potential sweep centers on three parameters which characterize the stripping signal: the peak stripping current, $i_p$; the position of the peak stripping current on the potential axis, $E_p$; and the peak width at half-height, $W_{1/2}$ [16-19]. The measurement of these parameters is illustrated in Figure 19.2. The equations that describe these parameters for stripping voltammetry at the HMDE and the MFE differ due to the difference in the geometry of these electrodes.

For an MFE, Equation 19.6 relates the peak stripping current in amperes ($i_p$) to the number of electrons transferred per metal ion (n), the surface area of the electrode in cm$^2$ (A), the concentration of the metal in the electrode in mol/cm$^3$ ($C_R$), the thickness of the Hg film comprising the MFE in centimeters ($\ell$), and the scan rate in V/s ($\nu$).

$$i_p = 1.1157 \times 10^6 n^2 A C_R \ell \nu \qquad (19.6)$$

This expression can be simplified if it is considered that the charge ($Q_m$) required to electrolyze all of the metal in the MFE is [3]

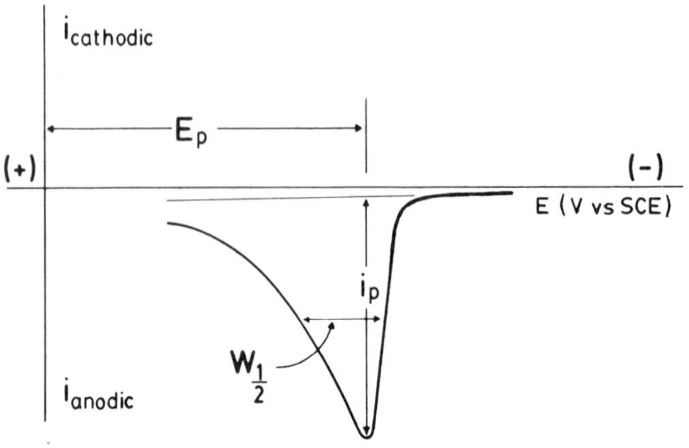


Figure 19.2 Parameters employed to characterize individual stripping waves: $E_p$, $W_{1/2}$, and $i_p$.

$$Q_m = nFAC_R\ell \tag{19.7}$$

Therefore, Expression 19.6 simplifies to

$$i_p = 11.157nQ_m\nu \tag{19.8}$$

which indicates that $i_p$ is proportional to the total amount of metal deposited in the MFE. The value of $i_p$ is also directly proportional to the scan rate used for the stripping step.

Equation 19.9 references the position of peak stripping current $E_p$ to the polarographic half-wave potential (mV):

$$n(E_p - E_{1/2}) = -1.43 + 29.58 \log \frac{\ell^2 \nu nFR}{RTD_R} \tag{19.9}$$

where T is the absolute temperature and $D_R$ is the diffusion coefficient of the metal in mercury in $cm^2/s$. This expression predicts that $E_p$ is dependent on the thickness of the MFE and on the scan rate, but is independent of concentration. An additional characteristic of a stripping wave at an MFE is predicted by Equation 19.10:

$$n(W_{1/2}) = 75.53 \text{ mV} \tag{19.10}$$

Namely, the peak width at half-height is independent of experimental parameters.

The validity of Equations 19.6, 19.9, and 19.10 is dependent on the value of the dimensionless parameter H [18]:

$$H = \ell^2 \nu \frac{nF}{RTD_R} \tag{19.11}$$

It has been shown that the behavior observed at MFEs is generally consistent with the behavior predicted by Expressions 19.6, 19.9, and 19.10 if the value of H is less than $1.6 \times 10^{-3}$ [20-22]. Examination of Equation 19.11 reveals that the value of H is directly proportional to the product of the square of the film thickness and the scan rate. Therefore, these two parameters can be manipulated to ensure that the value of H is appropriate for use of these equations. Slow scan rates must be used with relatively thick films or, conversely, faster scan rates can be used with thinner films.

Stripping waves obtained at the HMDE exhibit values of $i_p$ and $E_p$ (V) as described by Equations 19.12 and 19.13 [2].

$$i_p = 2.72 \times 10^5 n^{3/2} A D_R^{1/2} C_R \nu^{1/2} \tag{19.12}$$

$$E_p = E_{1/2} - \frac{1.1\,RT}{nF} \tag{19.13}$$

At the HMDE $i_p$ is proportional to $\nu^{1/2}$ and $E_p$ is invariant with scan rate.

Equations 19.6 to 19.13 describe the faradaic portion of the stripping wave observed during an ASV experiment at a mercury electrode. The capacitive or nonfaradaic portion of the signal during the potential scan is estimated by the following expression (see Eq. 3.22):

$$i_c = AC_{d\ell}\nu \tag{19.14}$$

where $i_c$ is the capacitive current in amperes, A is the area of the electrode in $cm^2$, $C_{d\ell}$ is the capacitance of the double layer in $F/cm^2$, and $\nu$ is the scan rate in V/s. Equation 19.14 indicates that the magnigude of the nonfaradaic component of the signal is directly proportional to the area of the electrode and the scan rate. At an MFE (as indicated by Eq. 19.6) the magnitude of $i_p$ is also proportional to the scan rate and electrode surface area, whereas at the HMDE $i_p$ is proportional to the electrode area and the square root of the scan rate

(Eq. 19.12). Hence an increase in scan rate or surface area at either the HMDE or the MFE does not improve the faradaic signal of interest relative to the interfering nonfaradaic signal for a given concentration of metal dissolved in the mercury electrode.

*Differential Pulse Stripping Voltammetry*

Like differential pulse polarography (see Chap. 5), differential pulse anodic stripping voltammetry (DPASV) has become very popular. The major background component of the stripping signal is due to capacitive effects of the electrode. DPASV imparts discrimination against the capacitive component of the stripping signal, just as it does in differential pulse polarography, by judicious current sampling during the differential pulse waveform. As discussed in Chapter 5, the charging current can be minimized by sampling the current before the pulse and during the section of the pulse where the faradaic to nonfaradaic ratio is enhanced. The difference in the two sampled signals is then recorded as the current response.

The peak stripping current is related to the concentration of the metal in the MFE [23] by

$$i_p = 0.138 \frac{Q_m}{t} \tag{19.15}$$

where $Q_m$ is the charge in coulombs produced by the total metal in the electrode and t is the pulse width in seconds. Equation 19.15 indicates that the signal can be enhanced by decreasing the pulse width. However, if the pulse width is too narrow, insufficient time for the decay of the nonfaradaic component is allowed and the needed discrimination against the capacitive component of the current is reduced.

Another feature of the differential pulse waveform contributes to increased sensitivity. Only a portion of the metal atoms oxidized as a result of the potential step ending a pulse have a chance to diffuse away from the surface of the electrode before the potential step beginning the next pulse returns the potential to a value at which metal is redeposited. Consequently, metal atoms make multiple contributions to the analytical signal. The combination of discrimination against capacitive current and "recycling" of analyte metal atoms renders the detection limits of DPASV one to several orders of magnitude lower than those of ASV with linear potential sweep.

Figure 19.3 shows a differential pulse anodic stripping voltammogram for a seawater sample containing $Zn^{2+}$, $Cd^{2+}$, $Pb^{2+}$, $Cu^{2+}$, $Sb^{3+}$, and $Bi^{3+}$. This example is indicative of the multielement capabilities of this technique. The calibration plot in Figure 19.4 was used to determine the linearity of $i_p$ with concentration [24].

In addition to the differential pulse waveform, a number of other waveforms have been shown to discriminate against the capacitive

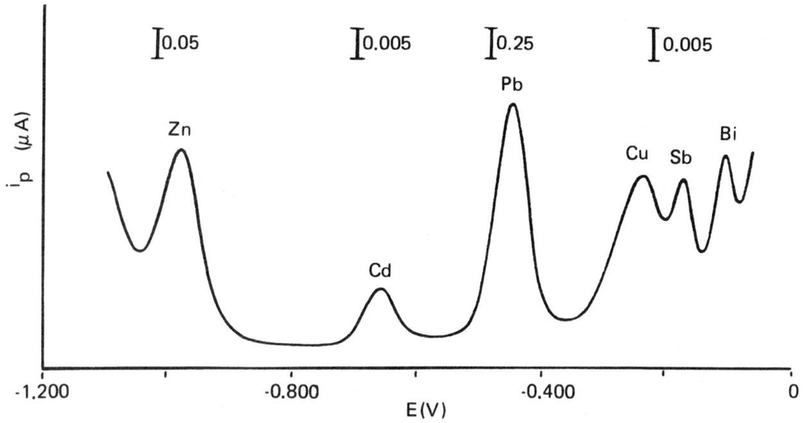


Figure 19.3 Differential pulse stripping voltammogram for the simultaneous determination of $Zn^{2+}$, $Cd^{2+}$, $Pb^{2+}$, $Cu^{2+}$, $Sb^{3+}$, and $Bi^{3+}$ in seawater at pH 1 containing 2 M chloride. Deposition timing, 20 min (Zn, Cd, Pb) and 40 min (Cu, Sb, Bi); pulse amplitude, 35 mV (Zn, Cd, Pb) and 10 mV (Cu, Sb, Bi); scan rate, 2 mV/s (Zn, Cd, Pb) and 0.5 mV/s (Cu, Sb, Bi). (From Ref. 24.)

current contribution during the anodic scan: derivative linear potential sweep ASV [25], differential linear potential sweep ASV [26], alternating current voltammetry [27], and staircase stripping voltammetry [28]. Often the detection limits of these techniques are comparable to or better than those obtained with DPASV.

*Potentiometric Stripping Analysis*

The last of the three stripping techniques to be discussed is also the most recently developed. First proposed by Bruckenstein and Bixler in 1965 [29] and further developed by Jagner and Graneli [30], potentiometric stripping analysis (PSA) shows promise as a technique for multielement analysis. The previously mentioned electrochemical methods are all based on the current-potential relationship. PSA, however, relies on the variation of potential with respect to time.

In a typical PSA experiment, a known amount of a suitable oxidizing agent (usually $Hg^{2+}$) is added to the sample. The analyte species are then preconcentrated at the working electrode via potentiostatic control of the electrode potential in the same manner as discussed in Section II.B. During this preconcentration period there are two electrochemical reactions taking place: $Hg^{2+}$ ions are being reduced to Hg metal, forming an MFE; and the analyte metal ions are being reduced to their metallic state dissolved in mercury. When the potentiostat is disconnected, the oxidizing species ($Hg^{2+}$) continues

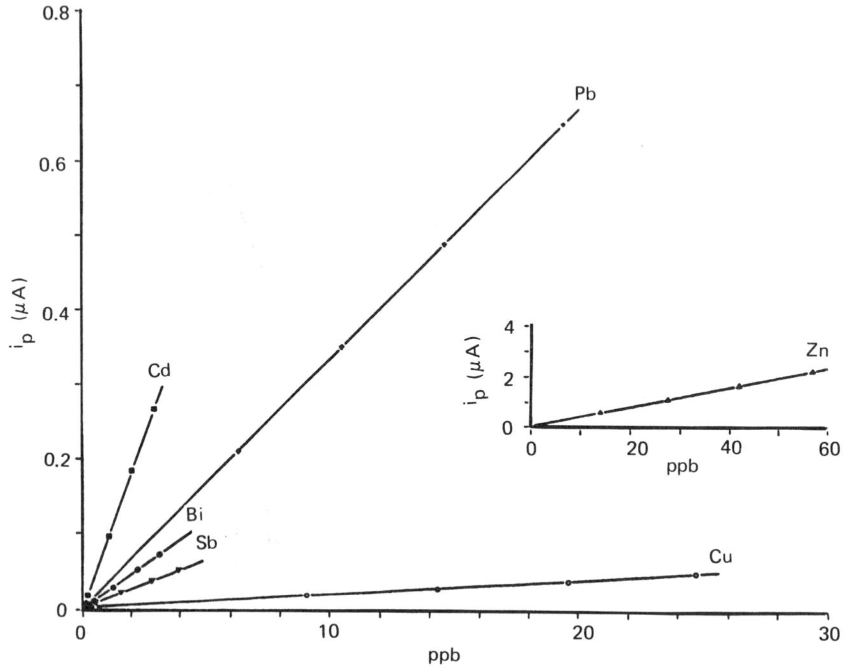


Figure 19.4 Calibration curves for $Zn^{2+}$, $Cd^{2+}$, $Pb^{2+}$, $Cu^{2+}$, $Sb^{3+}$, and $Bi^{3+}$ in a seawater sample at pH 1 containing 2 M chloride. Deposition timing, 15 min (Zn, Cd, Pb) and 30 min (Cu, Sb, Bi). Pulse amplitude and scan rate as for Figure 19.3. (From Ref. 24.)

to diffuse to the electrode surface, where it oxidizes the analyte metal dissolved in the MFE as described by the following redox reaction:

$$M_{Hg} + Hg^{2+} \longrightarrow M^{2+} + Hg \qquad (19.16)$$

In a multielement analysis the metals are oxidized in the order of their redox potentials. The time required for oxidation of each metal is proportional to its concentration in the MFE. Thus the length of time required for chemical oxidation of each species is the measured parameter.

Typical stripping potentiograms are shown in Figure 19.5. They were recorded by simply monitoring potential as a function of time with a time base recorder after the deposition step. Of the four deposited metals, Cd is most easily oxidized and is therefore stripped

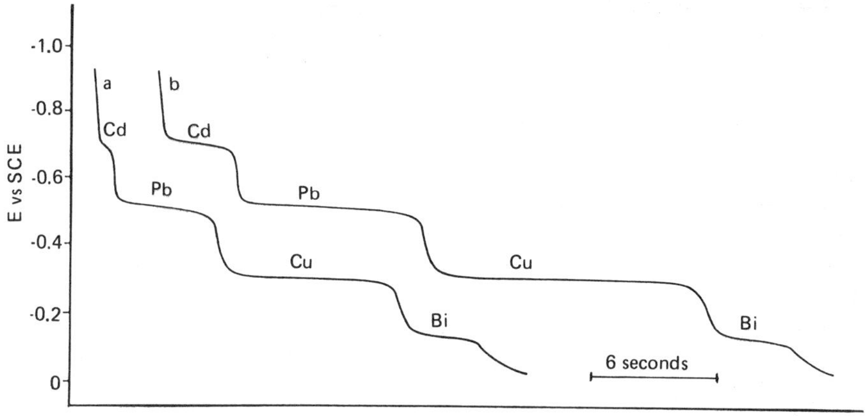


**Figure 19.5** Potentiometric stripping curves obtained in a sample containing 0.05 M HCl, 0.5 M NaCl, and 1 mg/L of $Hg^{2+}$ before (curve a) and after (curve b) the addition of 0.5 g/L of $Cd^{2+}$, $Pb^{2+}$, and $Cu^{2+}$. [From Ref. 31, reproduced with permission from *Anal. Chem.* *50*:1924 (1978). Copyright 1978 American Chemical Society.]

from the electrode first at a potential of ca. −0.7 V. Pb, Cu, and Bi are then stripped in that order at the potentials shown in the figure. Curve a is the original potentiogram, and curve b shows the effect of standard additions of $Cd^{2+}$, $Pb^{2+}$, and $Cu^{2+}$.

## D. Electrodes

Mercury is the electrode of choice for many ASV, CSV, and potentiometric stripping analyses. In the case of ASV, the analyte is deposited into the mercury as an amalgam. The solubilities of various metals in mercury are shown in Table 19.1. The liquid environment enables the various metals deposited to diffuse freely to the electrode-solution interphase for oxidation during the stripping step. This is a highly advantageous situation compared to deposition of analyte as a solid film on the surface of a solid electrode such as platinum or graphite. For solid films thicker than a few monolayers, the underlying material can be restrained from diffusing to the interphase during the stripping step until the overlying deposit has been oxidized. This can cause poorly defined stripping peaks, particularly with mixtures where the various components strip at different potentials. Consequently, although good results can be obtained at solid electrodes such as graphite under certain conditions [4], mercury is the most frequently used electrode. Important exceptions are the determination of $Hg^{2+}$, $Ag^{+}$, and $Au^{3+}$, which are not measurable at a mercury electrode since

Table 19.1 Solubilities of Various Metals in Mercury

| Metal | Solubility (wt %) |
|---|---|
| Indium | 68.3 |
| Thallium | 42.4 |
| Cadmium | 5.0 |
| Zinc | 5.6 |
| Tin | 1.3 |
| Lead | 1.2 |
| Bismuth | 1.2 |
| Copper | $8 \times 10^{-3}$ |
| Manganese | $6.6 \times 10^{-3}$ |
| Antimony | $3.8 \times 10^{-4}$ |
| Nickel | $2.1 \times 10^{-3}$ |
| Cobalt | $3 \times 10^{-4}$ |

*Source*: Refs. 32 and 33. (Also see Ref. 3, p. 59.)

the electrode itself oxidizes simultaneously with or before sample oxidation [4]. Mercury is a very useful electrode for CSV since it forms insoluble salts with many anions. Silver is also useful in this respect.

### *Hanging Mercury Drop Electrode*

The hanging mercury drop electrode (HMDE) is frequently used for stripping voltammetry. It usually consists of a mercury drop suspended from a mercury thread in a glass capillary. (Details are found in Chap. 9.) The main advantage of this electrode is the excellent negative potential range with the low residual current of pure mercury. Compared to the mercury film electrode (vide infra), it is less susceptible to problems associated with intermetallic compound formation and with exceeding the solubility of metals in mercury. Of practical importance is the ease with which reproducible drops can be formed with a micrometer syringe or a static mercury drop electrode. Disadvantages include a restriction to rather slow stirring rates with no electrode rotation during the deposition step in order to avoid dislodging the mercury drop and a relatively low surface area-to-volume ratio. Both of these factors contribute to lengthened deposition times. The low surface area-to-volume ratio also leads to broader stripping peaks, causing loss of peak resolution in multielement analysis.

### *Mercury Film Electrodes*

The mercury film electrode (MFE) consists of a thin mercury coating (ca. 1-1000 nm thick) deposited on an electrically conductive substrate, which provides physical stability to the liquid film [34]. This electrode is becoming more popular in ASV because of (1) higher sensitivity and (2) higher resolution compared to the HMDE.

The usual procedure for preparing the MFE is by electrochemical reduction of $Hg^{2+}$ to Hg on a conductive substrate. Procedures using both preformed films and in situ deposited films have been used extensively. Preformed films with thicknesses in the range 1-100 nm are typically made by deposition of mercury at $-0.4$ to $-1.0$ V vs. SCE for 2-5 min from an acidic 5-100 × $10^{-6}$ M $Hg^{2+}$ solution [7,35,36]. The electrode is then removed from the plating solution, rinsed, and used for sample analysis. In situ-deposited films are formed simultaneously with deposition of sample metals by adding a small amount of $Hg^{2+}$ (typically 5-100 × $10^{-6}$ M) to all sample solutions [7,34,37,38]. The mercury film is removed after each analysis either by wiping it off or by anodically stripping it to a soluble mercury species (in the absence of anions that may cause precipitation).

Many materials have been evaluated for use as MFE substrates. Glassy carbon and wax-impregnated graphite (WIG) electrodes have been most widely used. Both of these materials are rather inert substrates toward mercury, so that contamination of the MFE by a mercury-soluble substrate is avoided. Excellent results have been obtained with both substrates. Wax impregnation to fill pores in the graphite is important for preventing analyte creepage into the substrate. A variety of graphite types has been evaluated [35], several of which perform quite well. Wax types have also been evaluated [38]. Microcrystalline waxes fail rapidly due to cracking and separating from the graphite. Waxes such as Sonneborn and Ultraflex-Amber, which crystallize very slowly, are best. Very careful polishing procedures are absolutely essential for good results with both the WIG and glassy carbon substrates. The reader is referred to Refs. 35 and 38 for preparation procedures for the WIG MFE and to Refs. 7 and 37 for the glassy carbon MFE. Deposition of mercury on WIG and glassy carbon substrates results in a coating of mercury droplets rather than a continuous liquid film [7,38,39]. Nonetheless, ASV results of good accuracy and precision are obtainable at these "droplet" electrodes.

A variety of metals, including Pt, Ni, Au, and Ag, have been used as substrates for mercury film electrodes. Continuous mercury films can be formed by procedures including electrodeposition by reduction of $Hg_2^{2+}$ [40] or $Hg^{2+}$ [41], plunging a cathodized metal electrode which is evolving hydrogen into a mercury pool [42], or abrading a metal surface under mercury [43]. Mercury droplets formed during the electrodeposition procedure can be coalesced into a continuous mercury film by vigorous evolution of hydrogen [40]. Most metal substrates slowly contaminate the overlying mercury film by amalgamation, causing loss of hydrogen overvoltage and possible problems with intermetallic compound formation between the dissolved substrate and metals that might be deposited into the film. For this reason, metal substrates are rarely used for ASV.

A problem with the MFE is its deterioration in highly acidic media. Although an MFE on a glassy carbon or WIG substrate is reasonably long-lived in neutral or weakly acidic media (20-40 days), failure can occur within a few hours when exposed to pH 1-2 solutions [44]. Failure is indicated by a rapid decrease in hydrogen overvoltage. Evidence suggests that carboxyl groups on the graphite surface are responsible [44]. Some improvement on longevity in acid solution has resulted from substrate treatments such as chemical bonding of the impregnator by cross-linking impregnated styrene monomer to the graphite substrate by gamma radiation [45] and impregnation with epoxy [46]. However, the problem is yet to really be solved.

The MFE offers two significant advantages over the HMDE in the preconcentration step. One advantage of the MFE results from the greater electrode surface area-to-volume ratio of a film as compared to the spherical HMDE. The larger surface area enables more sample to be concentrated into a given amount of mercury during a specified deposition time. This means greater analytical sensitivity for the MFE compared to the HMDE (or shorter deposition times for the MFE for equal sensitivity with the HMDE). Another advantage of the MFE is its compatibility with stirring rates that would dislodge an HMDE and with flow systems. This ability to use greater stirring rates improves mass transfer during the concentration step, again reducing the time required to deposit a given quantity of metal. Efficiency is further improved by rotating the MFE. Problems associated with exceeding the solubility of analyte in mercury and with the formation of intermetallic compounds (vide infra) are more likely to be encountered with the MFE because of this more rapid rate of concentration during deposition.

The MFE is also superior to the HMDE in resolution of the stripping peaks. In a 50-nm-thick film, the electrodeposited sample is constrained to within 50 nm of the interphase, to which it must diffuse for oxidation during the stripping step. By comparison, metals deposited into an 0.5-mm-diameter HMDE can diffuse to a depth of 0.25 mm into the electrode, which greatly lengthens the return trip to the interphase for oxidation. The practical consequence of this is much sharper stripping peaks for an MFE than an HMDE, as shown in Figure 19.6. This minimizes the overlap of stripping peaks in multielement analysis, reducing the uncertainties in baseline estimation for the measurement of peak current.

Two disadvantages of the MFE are a slightly reduced negative potential range (perhaps due to formation of a droplet coating on the substrate rather than a continuous film) and some difficulty in preparing reproducible, stable films.

### *Amalgam Considerations*

An important aspect of the determination of metals at mercury electrodes is the solubility of the analyte metals in mercury. A list of the mercury

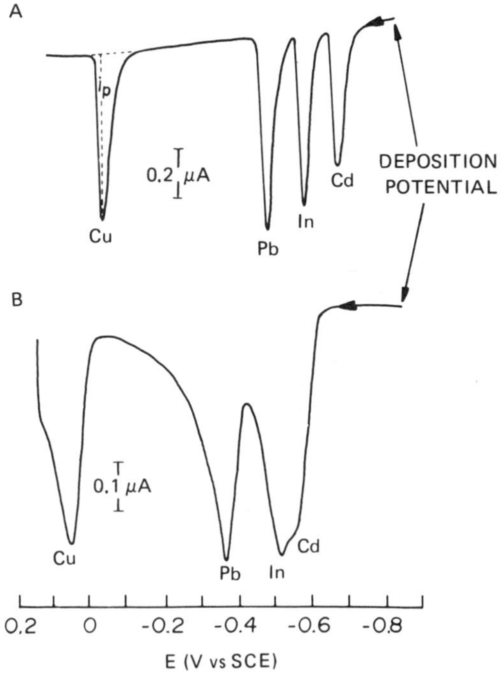


Figure 19.6 Anodic stripping voltammograms for $2 \times 10^{-7}$ M $Cd^{2+}$, $In^{3+}$, $Pb^{2+}$, and $Cu^{2+}$ in 0.1 M $KNO_3$. Linear sweep voltammetry, scan rate 0.3 V/min. (A) Mercury film electrode on polished glassy carbon, 2000-rpm rotation speed, 5-min deposition. Solution also contained $2 \times 10^{-5}$ M $Hg^{2+}$. (B) Hanging mercury drop electrode, 30-min deposition. (Adapted from Ref. 37.)

solubilities of several metals that are commonly determined by ASV is shown in Table 19.1. Consideration of the solubility of analyte metals in mercury is important because exceeding the solubility can result in the occurrence of phenomena that have a detrimental effect on the subsequent stripping determination [47]. Usually, a separate crystalline phase of pure metal results when the solubility is exceeded. However, some metals, such as Cu and Mn, form intermetallic compounds with mercury when the solubility is surpassed [32]. In either case, the stripping of the metal is affected because the electrode is no longer homogeneous. The effect of such changes is documented in a series of studies that provided experimental verification of the stripping theory at the MFE [20,22]. It was shown that good agreement between theory and experiment results were obtained when metal-mercury solubilities were not exceeded. However, theory and experiment deviated when nonideal amalgams were formed. The most severe of these deviations is the depression of the stripping current and shifts in peak potential due to changes in the nature of the electrode itself. Many metal ion determinations with ASV are accomplished with the standard additions method. If metal-mercury solubility phenomena change the behavior of the system when the concentration of the metal ion is increased by standard addition, erroneous results can be obtained [3].

Another amalgam phenomenon that can affect analysis with ASV is the formation of intermetallic compounds by metals deposited in the mercury electrode. Table 19.2 lists a number of intermetallic compounds that have been reported. Intermetallic compound formation is detrimental to analysis with ASV if it changes the stripping currents of the constituent metals. Usually, the intermetallic compound(s) oxidize at a potential between the oxidation potentials of the individual metals [32]. It has also been reported that the intermetallic compound(s) strip at the same potential as one of the constituent metals [48]. The most important ramification of intermetallic compound formation is that it can render the results of standard addition analysis incorrect [49,50].

Problems with exceeding mercury solubility and with intermetallic compound formation are more common with the MFE than with the HMDE because of the efficiency with which metal ions can be preconcentrated into the small volume of mercury that constitutes the MFE.

## E. Experimental Techniques

As with all analytical methods that are capable of analysis at the trace level, contamination of the sample must be carefully guarded against. Stringent cleaning of all glassware used for sample preparation and of the cell components is necessary. Sample aliquotes are typically added to supporting electrolyte, which serves a variety of functions including the adjustment of pH to an optimum value for the species being determined, the improvement of solution conductance to eliminate

Table 19.2 Intermetallic Compounds

| | | |
|---|---|---|
| Ag-Cu | Cu-Mn | Ni-Zn |
| Ag-Zn | Cu-Zn | Pt-Sb |
| Au-Cd | Fe-Mn | Pt-Sn |
| Au-Mn | Cu-Zn | Pt-Zn |
| Au-Sn | Mn-Ni | Cu-Ga |
| Au-Zn | Ni-Sb | Pt-Hg |
| Cu-Cd | Ni-Sn | Ag-Cu |

*Source*: Refs. 1 and 32.

Table 19.3 Stripping Voltammetry Detection Limits (ng/mL)

| Element | Differential pulse | Linear sweep |
|---|---|---|
| Bi | — | 0.01 |
| Cd | 0.005 | 0.01 |
| Cu | 0.005 | 0.01 |
| Ga | 0.4 | — |
| In | 0.1 | — |
| Pb | 0.01 | 0.02 |
| Rh | — | 10.0 |
| Sn | 2.0 | — |
| Tl | 0.01 | 0.04 |
| Zn | 0.04 | 0.04 |

*Source*: Ref. 2.

mass transfer by migration during the deposition step and to minimize peak distortion by iR drop during the stripping step, and provision of a complexing agent if necessary to aid in the separation of stripping peaks of different metals. Contamination of the sample by metal impurities in the chemicals and solvent(s) used for preparation of the supporting electrolyte is a serious problem when concentrations in the ppb range are sought. Even high-purity reagents and most house-distilled water supplies have a detectable level of electroactive impurities. These impurities can severely limit the analysis of low-level samples. Stripping voltammetry is especially susceptible to these limitations. Supporting electrolytes can be cleansed of trace elements by preelectrolysis. Figure 19.7 shows a cell for the preelectrolysis of supporting electrolytes. The large-volume Hg pool electrode can remove a majority of the electroactive impurities by reducing them. The cell can be operated by a very simple potentiostat which is constructed from a high-current operational amplifier. Large volumes of supporting electrolyte can be cleansed continuously with aliquots taken as needed. For most determinations the supporting electrolyte can be sufficiently purified in 48-72 h. For extremely low level determinations, several days may be necessary, depending on the purity of the reagents used. For low-level work, further purification of the triply distilled mercury usually used for the HMDE may be necessary. (Technical bulletins by Princeton Applied Research Corporation contain useful information for purifying nitrogen, mercury, and electrolyte.)

Once the sample with supporting electrolyte has been added to the cell, a procedure of the following general type is used for analysis by stripping voltammetry. The solution is first deoxygenated by vigorous bubbling with nitrogen gas (purified and saturated with solvent) for

Table 19.4 Typical Applications of Electrochemical Stripping Analysis

| Metals determined | Sample type | Electrode type[a] | Stripping technique[a] | Reference |
|---|---|---|---|---|
| Cu, Bi, Sb, Pb, In, Cd, Tl, Zn | Seawater and marine samples | Rotating MFE deposited in situ | LPSASV | 51 |
| Cu, Bi, Sb, Pb, Cd, Zn | Seawater | HMDE | DPASV | 25 |
| Cu, Pb, Cd, Zn | Seawater | HMDE | DPASV | 52 |
| Pb, Cd | Seawater | HMDE, GCMFE | DPASV | 53 |
| Cu, Pb, Cd | Seawater | Rotating MFE deposited in situ | LPSASV | 54 |
| Cu, Pb, Cd, Zn | River water sewage effluent | HMDE | DPASV | 55 |
| Cu, Pb | EDTA extracts of soil | GCMFE | DPASV | 56 |
| Cd | Fulvic acid solutions | SMDE | DPASV | 57 |
| Cu, Bi, Pb, Tl, Cd, Zn | Urine, blood, hair, teeth, feces, bone, tissue | HMDE | DPASV | 58 |
| Cu, Pb, Cd, Zn | Human teeth | HMDE | DPASV | 59 |
| Cu, Pb, Cd | Urine | HMDE | DPASV | 60 |
| Cu, Pb, Cd, Zn | Human hair | HMDE | DPASV | 61 |

| | | | | |
|---|---|---|---|---|
| Sb, Pb | Firearm discharge residues | WIGMFE | DPASV | 62 |
| Ge | Acidic pyrogallol, phosphate, borate, and carbonate buffers | HMDE | LPSASV | 63 |
| Ag | $H_2SO_4$ | RRDE | ASVWC | 64 |
| Sn | Pb | RMFRDE | ASVWC | 65 |
| Cu, Pb, Cd | Tap water | GCMFE (tubular) | ASVWC | 9 |
| As | $ZnSO_4$ solutions | PGMFE | CSV | 66 |
| Se | Soils and plants | HMDE | DPCSV | 67 |
| S | NaOH | RAgDE | LPSCSV, DPCSV | 68 |
| As | Drinking water | SMDE | DPSCV | 69 |
| Pb | Wine | GCMFE | PSA | 10 |
| Pb, Cd | Whole blood and serum | GCMFE | PSA | 70 |
| Pb, Cd, Tl | Fly ash | GCMFE | PSA | 71 |
| Na, K, Ca, Sr, Ba, Mg | Blood serum synthetic seawater | GCMFE | PSA | 72 |
| Cd, Pb, Tl, Sn, Bi, Cu | Groundwater | (in situ) MFGCE | PSA | 73 |

[a]LPSASV, linear potential sweep anodic stripping voltammetry; DPASV, differential pulse anodic stripping voltammetry; ASVWC, anodic stripping voltammetry with collection; CSV, cathodic stripping voltammetry; PSA, potentiometric stripping analysis; MFE, mercury film electrode; GC, glassy carbon; WIG, wax-impregnated graphite; PG, pyrolytic graphite; HMDE, hanging mercury drop electrode; SMDE, static mercury drop electrode; RRDE, rotating ring-disk electrodes; RMFRDE, rotating mercury film ring-disk electrode; RAgDE, rotating silver disk electrode.

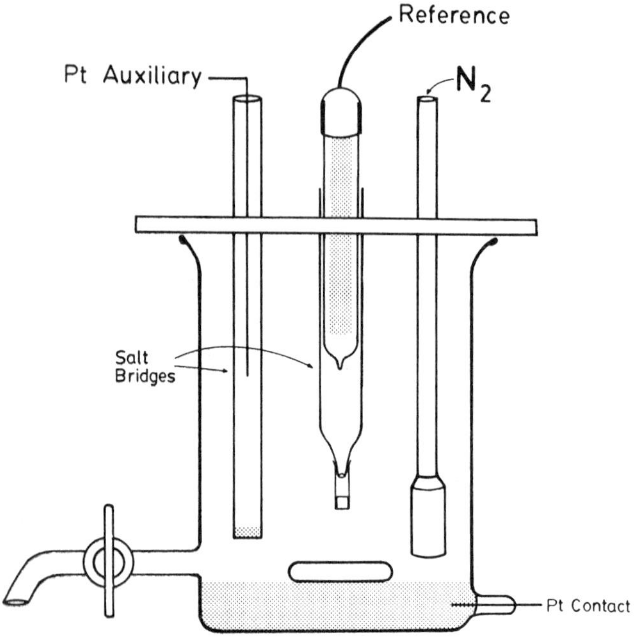


Figure 19.7 Cell for large-scale preelectrolysis of supporting electrolyte.

5-15 min. The nitrogen stream is then diverted to pass over the solution, maintaining a nitrogen blanket over the solution and a positive pressure of nitrogen in the cell. Stirring is activated (or the electrode is rotated) and the deposition potential is applied for a carefully measured time interval. This varies from 1 to 30 min depending on the sample concentration, the type of electrode, and the stripping technique. The use of the rotating MFE and differential pulse voltammetry minimize the deposition time required for adequate stripping peaks. The stirring is switched off (or electrode rotation stopped) at the prescribed time and convection is allowed to cease during a 10- to 60-s rest period, depending on the cell volume. The potential scan for stripping is then activated and the voltammogram recorded.

Since the technique of stripping voltammetry assumes that the same fraction of sample is deposited for standards and unknowns alike, it is very important that all parameters, such as electrode area, deposition time, rest time, stirring rate, and stripping parameters, be reproduced as carefully as possible during a series of measurements.

Determinations are usually based on the measurement of the peak current $i_p$. Figures 19.2 and 19.6 show typical procedures for measuring $i_p$ from the background current. Conversion of $i_p$ to a concentration value is done by either a calibration curve or by the method of standard additions. The latter technique is preferred for most natural samples since it minimizes matrix effects. In this technique a small aliquot of standard is added directly to the solution in the cell after the initial voltammogram is obtained; the analysis procedure is then repeated. The concentration of the original solution is calculated by the following equation from the increase in $i_p$ caused by addition of the standard:

$$C_u = \frac{i_1 v C_s}{i_2 v + (i_2 - i_1)V} \tag{19.17}$$

where $i_1$ is the sample peak height, $i_2$ the sample + standard peak height, v the volume of standard added, V the original volume of sample, $C_s$ the concentration of standard solution added, and $C_u$ the concentration of original sample solution. To improve precision, a second (or more) addition can be made to the same solution and a plot of $i_p$ vs. concentration of standard prepared.

## F. Applications

Anodic stripping voltammetry has been used extensively for the determination of metal ions in a variety of sample types, including biological fluids (especially lead in blood) and tissues, food products, and environmental samples. The technique is usually employed when analyte concentrations are very low. Some typical detection limits are shown in Table 19.3. Representative applications with references to the original papers are shown in Table 19.4.

# III. SPECTROSCOPIC ANALYSIS

## A. Introduction

As discussed in Section I, there are certain limitations in the use of electrochemical stripping of the film produced in the electrochemical deposition or preconcentration step. Such effects as intermetallic compound formation, less than 100% stripping efficiency, slow electrodissolution kinetics, multiple interactions between deposited films and surface sites of solid electrodes which lead to multiple dissolution peaks for the same species, overlapping dissolution peaks, and so on, result in complications in the quantitative interpretation of the analytical results obtained. However, as electrodeposition has been shown to have a high degree of precision for extremely dilute solutions and is instrumentally very simple, it is extremely useful as a preconcentration step for other methods, in particular, spectroscopy. Numerous examples of different spectroscopic techniques coupled with electrochemical preconcentration have been developed and are discussed below. An important factor is that the characteristics and limitations of the spectroscopic technique and/or sample matrix often impose strict specifications on the design of the electrochemical deposition step. The major portion of this section is designed to illustrate this point. Thus the great potential as well as limitations of this combination of techniques can be appreciated.

## B. Application to Neutron Activation Analysis

One of the first methods developed combined an electrodeposition step with neutron activation analysis (NAA) [74]. The activation of nuclides by thermal neutrons to yield unstable gamma-ray emitting-nuclides is a very sensitive trace method for many elements [75]. The resulting gamma emission spectra obtained by a multichannel analyzer exhibits gamma photopeaks whose energies are characteristic of a particular nuclide. Furthermore, the gamma spectrum is time dependent, as the concentration of the unstable nuclide decays with time. This is convenient in analysis, as often in complex samples there is considerable overlap of photopeaks. However, as the fast-decaying nuclides decay away, new peaks for other elements become resolved. In addition, the half-life of a photopeak's decay can be used for additional qualitative identification information of a particular nuclide. Thus nuclides which have similar photopeak energies can be identified by their characteristic decay half-life. Also, different nuclides having identical photopeak energies can be identified and analyzed using differential kinetic methods [76].

The qualitative determination of each nuclide present is also possible from a $\gamma$ spectrum as the number of counts (decays of a nuclide) of a given photopeak is directly proportional to the amount of that nuclide present in the sample, as given by the activation equation

$$A = N\sigma\Phi(1 - e^{\lambda t}) \qquad (19.18)$$

where A is the decay rate per second, N the number of atoms present in the sample, $\sigma$ the capture cross section for the particular nuclide, $\Phi$ the thermal neutron flux through the sample, $\lambda$ the decay constant, and t the irradiation time. This equation demonstrates one of the inherent difficulties in doing quantitative measurement by neutron activation analysis. For quantitative analyses it is necessary to know the integrated neutron flux Φ in each sample in order to make an intercomparison between samples (or correlate samples with a series of knowns as when a calibration curve technique is employed). This comparison can be a major problem, however, as flux distribution within or around the reactor core generally varies over a large range as a function of position and time. For example, a typical flux distribution is shown in Figure 19.8 within the "rabbit" pneumatic system, which carries the sample to the reactor core of a reaction. It can be seen that the flux density varied by as much as 50% over this small volume (ca. 2 in.$^2$). Also, the position of a rabbit within a core varies slightly from run to run.

Thus for quantitative determinations it is necessary to employ flux monitors (pure standards of known weights) which are placed in *very close proximity* to the irradiated samples to compensate for flux variations. These variations frequently occur from one sample to another. In general, these monitors or standards are fine wire, thin foils, or standard solutions of a pure element of a known weight. When trace analyses are performed, long irradiation times are necessary. This necessitates that the quantity of material employed as the monitor be in small quantities (usually milligrams or less); otherwise, extremely long waiting times are necessary to allow the activity of the monitor to decay to a reasonable counting level. Obviously, the weighing and handling problems arising from the use of such small samples are very troublesome. With the particular sample system employed in this method, special electrodeposited metal film standards, as described in a later section, circumvent this monitor problem.

It should be pointed out that the lack of availability of a reactor and its expense is also an unavoidable drawback in NAA. However, the technique described here partially avoids this problem by making

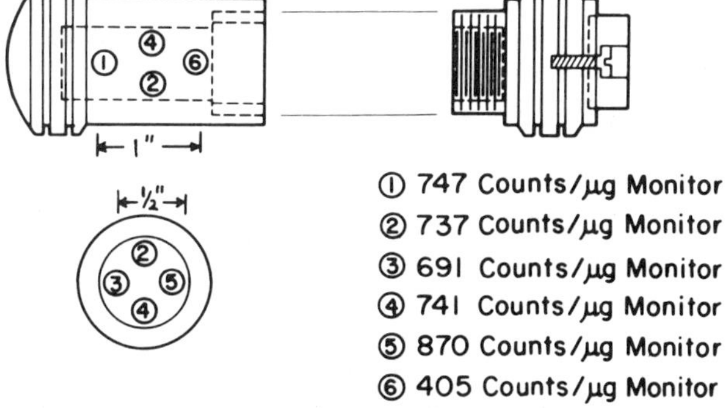


Figure 19.8 Thermal neutron flux distribution within a sample container (pneumatic tube "rabbit").

the sample chemically and physically inert and very small. Thus the preconcentrated form of the sample can be sent to commercial NAA labs.

In the ideal situation, neutron activation analysis can be used for the direct nondestructive determination of a species of interest within a matrix of other material. In actual practice, however, the matrix itself undergoes activation, and in many cases, its own photon emissions obscure the gamma spectrum of the species of interest,

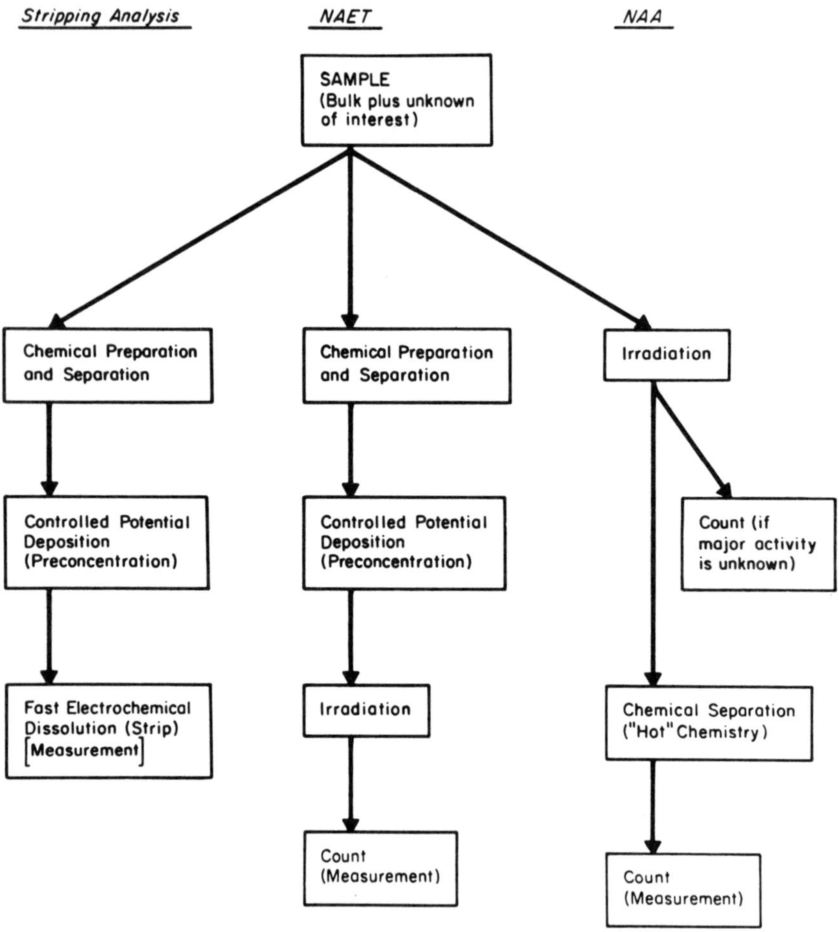


Figure 19.9 Flow diagram of the procedures of stripping analysis, neutron activation electroseparation technique (NAET), and neutron activation analysis (NAA).

necessitating a "hot" chemical separation (with a carrier if the species of interest is in trace amounts) before the activity of the unknown can be measured. This is especially true in the case of seawater and biological samples. The flow diagram for conventional neutron activation analysis is shown in the right-hand column of Figure 19.9, which also contains the flow diagrams for neutron activation electrodeposition technique (NEAT) and anodic stripping voltammetry.

When "hot" chemical separations are necessary, numerous problems arise: considerably elaborate, expensive facilities and equipment are required for safe handling of "hot" solutions; radioactive waste disposal must be considered; and chemical separation times can preclude the use of short-lived nuclides and result in a general decrease in sensitivity in some cases. The problems, however, could be completely circumvented by a quantitative micro or submicro separation step prior to irradiation which simultaneously prepares the species of interst as a thin film (preconcentration step), which suffices for easy handling, irradiation, and counting. This separation preconcentration step requires that no appreciable quantity of the species to be determined (or interfering species) is introduced as an impurity from the chemicals used in the preparation of the sample solution for the separation step (irradiation prior to separation eliminates this possible source of error, of course).

### *General Experimental Approach*

A combination technique would take advantage of the micro and submicro selective separation ability of controlled-potential electrodeposition as a preconcentration step and both the sensitivity and ability to determine mixtures simultaneously of neutron activation analysis as the measurement step (as shown by the center column of Fig. 19.9) and would circumvent all the problems inherent in each individual method. In the combination neutron activation-electroseparations technique (NAET), the small electrode with the deposited film of suitable geometry for counting (disk) is irradiated in a thermal neutron reactor to activate the species of the film, and the gamma spectrum is determined. As the electrode disk, on which the film is deposited, is irradiated also, this combination technique is feasible only if the electrode material employed does not undergo any reaction on thermal neutron bombardment, which forms an appreciable gamma-emitting product. If it did, the measured activity of the irradiated disk electrode plus deposited film would consist mainly of the electrode gamma spectrum as a background which would obscure the gamma spectrum of the trace species of interest in many cases.

Lux had actually previously devised a somewhat similar combination method for the determination of traces of silver in the presence of very large amounts of cadmium [77]. He employed electrodeposition on a platinum electrode to separate the trace quantity of silver from the cadmium, and irradiated and then dissolved the film with aqua regia. The silver was determined from the gamma spectrum of the solution. This technique suffers from a disadvantage in that the background count of the solution itself is relatively high. If the activated silver is stripped chemically, a considerable amount of activated platinum plus activated impurities in the platinum also dissolve and are, therefore, present in the solution. It is possible to strip the silver film electrochemically to reduce this source of contamination, but it is very difficult, experimentally, to obtain 100% (or even a reproducible percent) dissolution of silver from platinum electrochemically [78]. Direct counting of the silver film on the platinum electrode is not feasible, as this would result in a very high background from the platinum itself.

The discovery of pyrolytic graphite, developed initially as a nose-cone material for rocket reentry because of its unusual heat conduction properties [79], makes the combination NAET method possible, however. Pyrolytic graphite, which is a good electrical conductor, is a human-made and extremely pure form of graphite [79] which has also proved to be an excellent electrode material [83]. Because of its high purity, the amount of activity induced in the pyrolytic graphite disk electrode itself is quite low (primarily from the impurities, see below). The activation of $^{12}C$ (small cross section) on thermal neutron bombardment yields stable $^{13}C$, and the subsequent activation of $^{13}C$ also has a low cross section to form $^{14}C$ which is a $\beta$ emitter only [75]. Thus gamma emission from the product formed from carbon itself is zero for the activation times employed in any analysis. The gamma spectrum of a thermal neutron irradiated film of the species of interest electrodeposited on a pyrolytic graphite disk will be relatively free of a background spectrum from the electrode material itself, and from other species in the solution from which it was deposited. With this small background from the pyrolytic graphite, the sensitivity of determinations by this method is limited only by the sensitivity of the element for activation analysis (cross section) and the flux of the reactor.

Virtually any commercial three-electrode potentiostat and cell systems can be employed to carry out the controlled (constant potential electrodeposition and the experimental conditions for the deposition are identical to those in a stripping analysis experiment described in Section I.

The unique feature of this NAA method is the nature of the pyrolytic graphite used as the working electrode. Pyrolytic graphite as sold is almost a perfect single crystal of graphite. Figure 19.10 shows diagrammatically a block of pyrolytic graphite defining the crystallographic axes. The graphite plane (two-dimensional crystal lattice of fused rings) is the ab plane. These crystal layers are stacked along the c axis. These layers are very easily separated by a razor blade (like mica sheets). The electrical conductivity is greatest

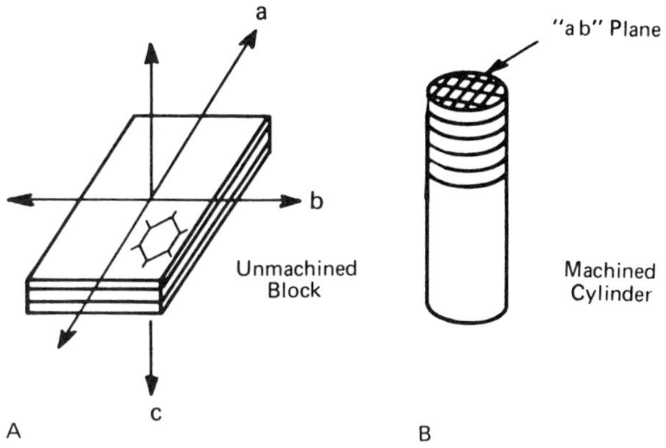


Figure 19.10 Pyrolytic graphite. (A) Crystallographic designation. (B) Machined pyrolytic graphite electrode.

along the ab direction; however, it is still very good along the c direction [79]. One potential problem of pyrolytic graphite electrodes is that the solution in contact with the c plane of the graphite enters into the graphite matrix between ab planes, which causes the electrode cylinder to swell along the c axis. When this happens, the electrode behavior of the material becomes erratic. To prevent this swelling, the cylinders are dropped in molten paraffin (a pure hydrocarbon which exhibits essentially no activity on short irradiation), which seals up the c axis edges of the ab planes and prevents the diffusion of the solution into the lattice. Both ends (ab planes) are cleaved off prior to insertion into the Teflon holder so that there would be no paraffin at the Pt electrical contact of the electrode assembly (see Fig. 19.11) and at the electrolysis solution interface.

The Teflon holder containing the pyrolytic graphite cylinder is then attached to a modified microburet assembly. Thus, after the electrodeposition, the calibrated micrometer head can be turned any reproducible amount which causes the graphite cylinder to emerge from the Teflon holder a reproducible amount. The graphite is then cleaved along the ab plane, flush with the Teflon holder. Thus the size (or weight) of the irradiated disk of graphite (on which the species of interest was deposited) is quite reproducible. This enables any background of activity from the graphite to be maintained at a constant level for many samples (provided that all the disks are from the same manufacturer's batch), which improves the accuracy of very trace determinations.

After electrodeposition of the film, graphite electrodes are gently rinsed in water to remove any adsorbed nondepositable species, allowed to dry, and then sprayed with a clear plastic coating.

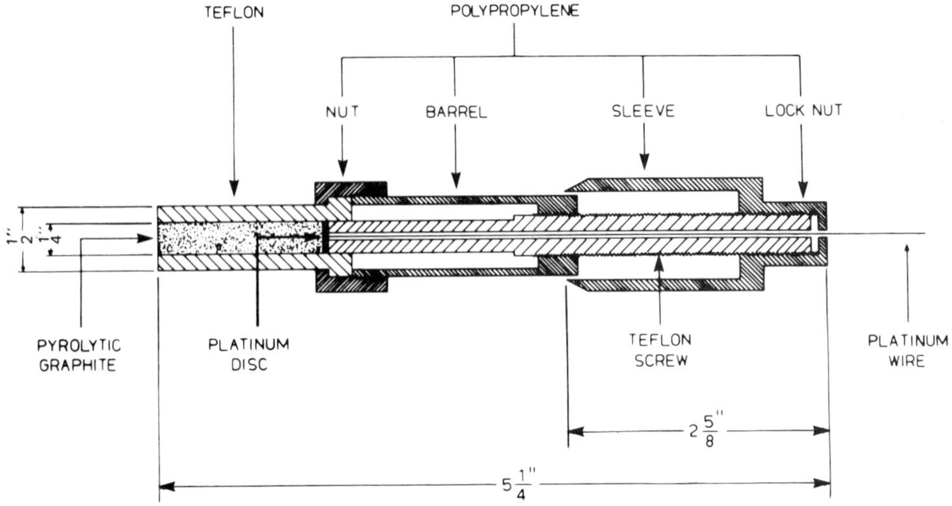


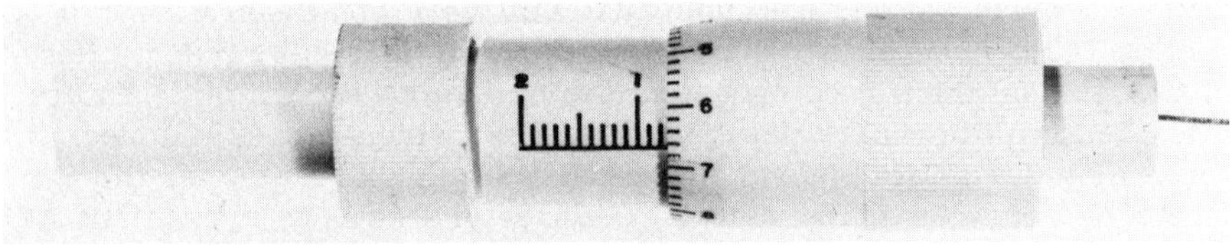

Figure 19.11 Details of the pyrolytic graphite electrode assembly. [Reprinted with permission from H. B. Mark, Jr. and F. J. Berlandi, *Anal. Chem.* *36*:2062 (1964). Copyright 1964 American Chemical Society.]

These spray coatings protect the deposited film from being mechanically rubbed off the graphite disk during the handling that occurs on irradiation and counting. Such coatings (pure polymeric hydrocarbons) yield no appreciable activity on short irradiation. The coated electrode with the deposited film is cleaved from the pyrolytic graphite cylinder as described above. The irradiation procedure and counting follow standard practice.

As mentioned above, good flux monitors are critical in order to get quantitative analysis in neutron activation. As shown below, the unique properties of the pyrolytic graphite can be employed to make ideal flux monitors and exceptional accuracy for trace elements can be obtained using this NAET method.

*Preparation of Neutron Flux Monitors*

Thin metal films of known amounts on pyrolytic graphite disks have been prepared by electrodeposition from standard metal ion solution by exactly the same potentiostatic method as discussed above for the preconcentration step [84].

In Table 19.5 typical sample results that illustrate the reproducibility of depositing a constant amount of material on each of a series of electrode disks are shown. The counting and irradiation times are also listed in Table 19.5, as well as the statistical standard error due to the randomness of decay and the experimental error for the series of disks is listed with the data.

This method of preparing flux monitors is also applicable for a variety of irradiation times, which is necessary to be of general usefulness. The amount of material deposited on the electrode must be conveniently variable electrochemically.

Two major nonchemical factors also influence the ultimate degree of precision: counting statistics and microgeometric flux inhomogeneity. The microgeometric flux variations are due to self-shielding and actual variations in the flux profile of the core.

The same disks can be stored and reused at a later date without any trouble in handling. Counting geometry is very reproducible, and only with the $^{64}$Cu positron emitter must special care be taken in mounting the samples for counting with the metal film facing the detector.

The technique also allows for calculated deposition of metals from microgram to milligram amounts. A large series of disks can easily be prepared with identical amounts of metal film to be used as a comparator in NAA. This eliminates the need for weight correction factors in calculations when intercomparing samples. If perhaps the disks are used as thermal neutron flux comparators, the self-absorption by the metal film can be kept to a minimum since the deposited film is thinner than can be obtained with any commercially available foil or wire. The large surface area over which the metal is deposited allows for a more accurate picture of the flux traveling through a relatively large portion of the sample. It should be noted that the fact that the metal films of the unknown and monitors can be placed face to face in the rabbit, the flux density in both films will be identical during the irradiation and no error will be introduced by the variation of $\Phi$ in the reactor.

### *Potential In Situ Application to Seawater Analysis*

An interesting and especially well suited analytical problem for which this combination technique can be easily applied is the analysis of trace metals in seawater (or almost any naturally occurring aqueous fluid). The direct activation of seawater, which contains a large amount of $Na^+$ and $Cl^-$, results in such a large activity corresponding to $^{24}$Na and $^{38}$Cl that the determination of the trace metal ion constituents is virtually impossible without separation of Na and Cl prior to counting. However, as neither Na nor Cl can be electrodeposited at

**Table 19.5** Flux Monitors

*Copper*

Irradiated 10 min [$4 \times 10^{12}$ neutrons/($cm^2 \cdot s$)]; counted for 10 min.

| Disk Number | Peak channels (counts) | |
|---|---|---|
| 1 | 12,803 | Average 12,840 ± 277 |
| 2 | 12,720 | Relative error ± 2.2% |
| 3 | 12,750 | Random counting |
| 4 | 13,370 | standard error = |
| 5 | 12,560 | 0.88% |
| 1 | 29,640 | Average 28,930 ± 541 |
| 2 | 28,120 | Relative error ± 1.9% |
| 3 | 28,320 | Random counting |
| 4 | 29,690 | standard error = |
| | | 0.59% |

*Gold*

Irradiated 1 min [$4 \times 10^{12}$ neutrons/($cm^2 \cdot s$)]; counted for 1 min.

| Disk Number | Counts/ min | |
|---|---|---|
| 364 | 145,814 | Average 146,295 ± 2095 |
| 365 | 146,554 | Relative error ± 1.4% |
| 366 | 144,352 | Random counting |
| 367 | 144,455 | standard error = |
| 368 | 146,505 | 0.26% |
| 369 | 150,090 | |

*Silver*

Irradiated 1 min [$4 \times 10^{12}$ neutrons/($cm^2 \cdot s$)]; counted for 15 min.

| Disk Number | Peak channels (counts) | |
|---|---|---|
| 242 | 19,124 | Average 19,369 ± 320 |
| 243 | 19,008 | Relative error ± 1.65% |
| 244 | 19,564 | Random counting |
| 245 | 19,261 | standard error = |
| 247 | 19,889 | 0.72% |
| 236 | 80,981 | Average 79,367 ± 1310 |
| 237 | 78,250 | Relative error ± 1.65% |
| 238 | 80,443 | Random counting |
| 239 | 78,023 | standard error = 0.35% |
| 240 | 79,142 | |

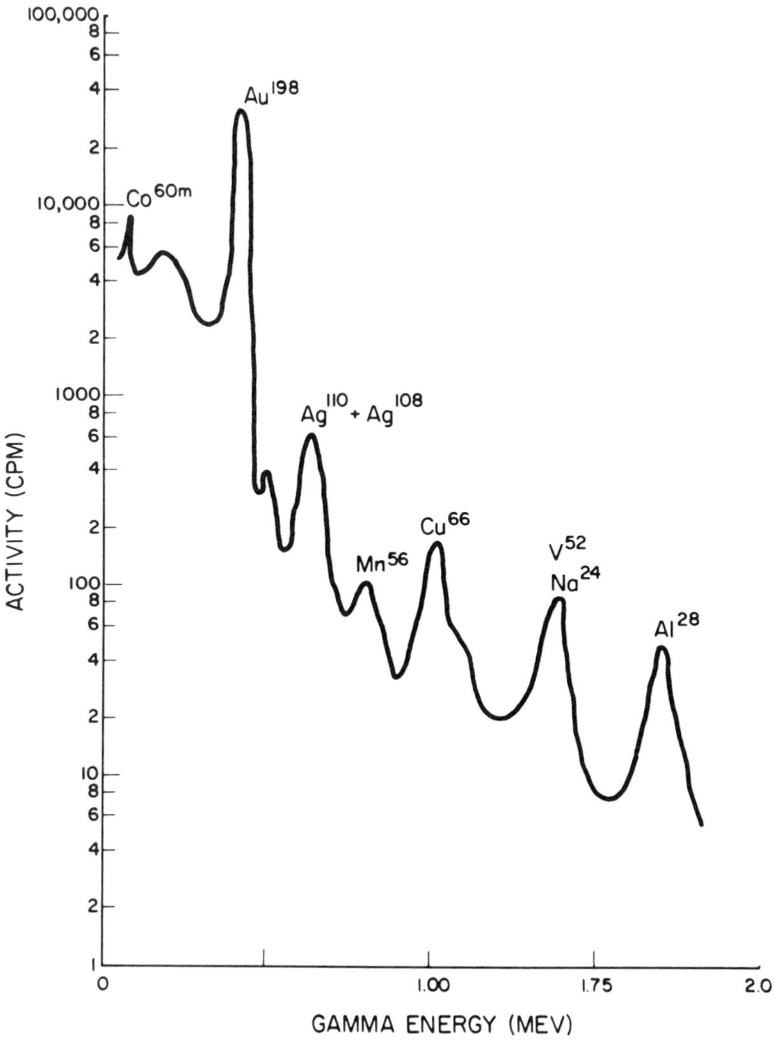


Figure 19.12 Gamma spectrum of artificial seawater.

a graphite electrode from aqueous solution, combination electrodeposition-spectroscopic techniques completely circumvent this problem. Figure 19.12 shows the gamma spectrum of artificial seawater containing 0.1 M NaCl, $\sim 5 \times 10^{-7}$ M $Cu^{2+}$, $\sim 5 \times 10^{-7}$ $Mn^{2+}$, $\sim 10^{-8}$ M $Ag^{+}$, and $\sim 10^{-9}$ M $Au^{3+}$ (this solution is not seawater, but simply a standard solution prepared in the laboratory to illustrate the point) which was electrolyzed 20 min at $-1.0$ V vs. SCE (at this potential, all these species deposit simultaneously). The resulting disk plus film was irradiated and counted after irradiation. As for this gamma spectrum, it is reproducible, and can be employed for the quantitative determination of these constituents. Note that the gamma spectrum for V and Al also appear, as this disk of pyrolytic graphite was from the same batch in which V and Al were present as impurities. The small Na peak of the gamma spectrum probably results from a small amount of $Na^+$ that diffused into the graphite lattice through defects or crevices in the ab line exposed to the surface. This technique can also be applied to the determination of trace metal ions in biological fluids such as blood, which also contains a large concentration of ions, such as $Na^+$, $K^+$, and $Cl^-$, and so on, which would interfere with conventional activation analysis techniques but which do not deposit during electrolysis.

It has been shown in trace analysis studies of seawater that plastic or glass sample containers both adsorb trace metal ions from the sample and/or contribute by surface dissolution other metal ions to the solutions. Thus the sample taken from the natural environment (unless extreme precautions are taken) is difficult to analyze because of the time-dependent effect on concentration, related simply to the nature of the container used to hold the sample.

One other problem also arises; the nature of the sample with respect to metal ion valence and/or degree of complexation can be expected to vary with pressure, light, and temperature. At first glance the only solution to this sampling problem, which arises from the simple need to maintain the sample in its initial environment or at least under the same conditions as it was when taken, is that the analysis in some way be made in situ and directly and the system not be removed or disturbed from its exact environment. However, complete in situ analysis presents many technical problems. However, for in situ preconcentration by electrodeposition, a very simple watertight device has been built which could be used for feasibility studies in relatively shallow water (surface to about 300 ft). With this limitation, it is relatively easy to construct a simple Lucite watertight box similar to the types used to house underwater cameras to house such a device,

A major problem of this type of deposition unit is that the necessity to achieve rapid film depositions of such trace metals requires mass transfer by stirring. It is also necessary to use a closed (fixed) volume type of electrolysis cell, and exhaustive electrolysis must be employed. In this manner, the preconcentration method thus becomes independent of experimental and environmental parameters, in particular, temperature and stirring and hence the diffusion coefficient that may fluctuate uncontrolled throughout the course of nonexhaustive preconcentration methods.

As such a unit must operate under pressure, a magnetically coupled arrangement that avoids the need of fitting a watertight rotor shaft through the Lucite box is used to drive a rotating electrode to provide rapid stirring. A prototype design actually built uses a magnet in the auxiliary electrode assembly [84]. Electrical contact to the electrode is made through epoxy-impregnated spectroscopic graphite rods which also act as bearings for the rotating electrode assembly.

The prototype is designed so that the graphite working electrode can be easily removed for analysis and replaced, and therefore it is held stationary during a deposition run. The cell convection diffusion (stirred) process is controlled by rotating the ridged (for more efficient convection) auxiliary electrode. Three auxiliary electrode materials, wax-impregnated graphite, platinum, and tantalum have been investigated using the artificial seawater solution. It was found for a graphite auxiliary electrode that the percentage of Cu(II) removed from solution becomes essentially independent of both the time and stirring parameters of the depositions. Only 92-95% of the Cu(II) is removed. It was found that the maximum percentage of Cu(II) that can be removed from solution is controlled by some steady-state phenomenon of the overall electrodeposition process itself related to the generation of soluble chlorine at the auxiliary electrode by the oxidation of chlorine ions, a phenomenon similar to that Anderson and Tallman observed for epoxy-impregnated graphite electrodes in chloride media [86]. The chlorine, in return, reoxidizes some of the deposited Cu(0) on the surface of the working electrode.

It was shown, then that it is necessary to shift this *chlorine overpotential* by using a different anode electrode substrate. Platinum and tantalum were then investigated as a possible auxiliary material capable of overcoming the chlorine problem.

The performance of the platinum and tantalum metal auxiliary electrodes was then evaluated in a 1 ppm Cu(II) solution and the results are seen in Figure 19.13A and B. The platinum electrode resulted in the removal of approximately 85% of the Cu(II) from solution for a 30-min deposition time (Fig. 19.13A), while the tantalum electrode gave essentially 100% deposition of the Cu(II), as seen in Figure 19.13B. A series of depositions for a 10.0-ppm solution gave the same results with essentially 100% of Cu(II) deposited for a 30-min deposition time. Binary systems containing Pb(II) and Cu(II) gave the same results.

This type of cell shows potential for in situ preconcentration sampling techniques in seawater, but also shows the complexity of both engineering and chemical complications required by in situ operations, such as a simple instrument used as an electrodeposition unit. However, the requirements of valid sampling techniques for environmental analysis necessitates the development of in situ techniques.

## C. Other Spectroscopic Methods

Holcombe and Sacks have used potentiostatic electrodeposition of such metals as Hg, Cd, Pb, and Ni on fine silver wire [87] as a preconcen-

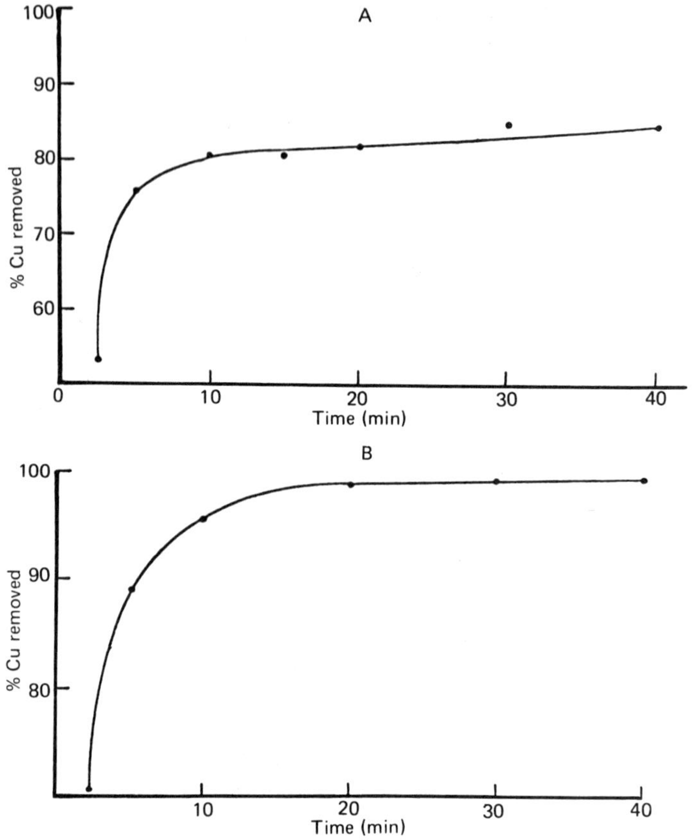


Figuer 19.13 The percentage of Cu removed from a 1-ppm solution with time: (A) for a platinum auxiliary electrode; (B) for a tantalum auxiliary electrode.

tration step and measured the deposit by recording the time-resolved emission spectra resulting from an exploding silver wire technique [88]. Absolute detection limits were in the ppb range [87]. Vassos et al. employed a constant-current (galvanostatic) electrodeposition of Co, Cr, Co, Hg, Ni, and Zn in the ppm for preconcentration on graphite electrodes which were then analyzed by x-ray fluorescence. They found that galvanostatic deposition gave better results for freshwater samples, which have high solution resistance [89]. Fairless and Bard [90], Lund and co-workers [91], and Batley and Matousek [92] have used controlled-potential electrolysis for preconcentration on various electrodes for the various metals, principally in seawater, with subsequent analysis of the films by atomic absorption spectroscopy in most cases, using the electrode itself as the heating element. The detection limits were reported to be in the ppb range.

Shieffer and Blaedel have devised a tubular electrode system which operates directly in a flow system for preconcentration by controlled-potential electrodeposition [93]. Berlandi and Mark [94] have described an adjustable thin-layer cell mounted for electrodeposition from highly radioactive small-volume samples [94]. Subsequent analyses of the radioactive films were carried out by gamma-ray and/or Mössbauer spectroscopy and autoradiography. Cowen et al. also used an electrodeposition step for subsequent $\alpha$ radiation counting [95]; others have used it in conjunction with x-ray fluorescence measurement [95,96] and microprobe analysis [97].

## REFERENCES


1. E. Barendrecht, in *Electroanalytical Chemistry*, Vol. 2 (A. J. Bard, ed.), Marcel Dekker, New York, 1967.
2. T. R. Copeland and R. K. Skogerboe, *Anal. Chem. 46*:1257A (1974).
3. F. Vydra, K. Stulik, and E. Julakova, *Electrochemical Stripping Analysis*, Halsted Press, New York, 1976.
4. Kh. A. Brainina, *Talanta 18*:513 (1971).
5. B. L. Dennis, J. L. Moyers, and G. S. Wilson, *Anal. Chem. 48*:1611 (1976).
6. H. A. Laitinen and N. H. Watkins, *Anal. Chem. 47*:1352 (1975).
7. G. E. Batley and T. M. Florence, *J. Electroanal. Chem. 55*:23 (1974).
8. G. W. Schieffer and W. J. Blaedel, *Anal. Chem. 49*:49 (1977).
9. G. W. Schieffer and W. J. Blaedel, *Anal. Chem. 50*:99 (1978).
10. L. Anderson, D. Jagner, and M. Josefson, *Anal. Chem. 54*:1371 (1982).
11. D. Yaniv and M. Ariel, *J. Electroanal. Chem. 129*:301 (1981).
12. J. A. Wise, W. R. Heineman, and P. T. Kissinger, unpublished results, 1983.
13. E. O. Martins and G. Johansson, *Anal. Chim. Acta 140*:29 (1982).
14. K. E. Haapakka and J. J. Kankare, *Anal. Chim. Acta 117*:367 (1980).
15. H. B. Hanekamp, W. H. Voogt, P. Bos, and R. W. Frei, *Anal. Chim. Acta 118*:81 (1980).
16. W. T. DeVries and E. Van Dalen, *J. Electroanal. Chem. 8*:366 (1964).
17. W. T. DeVries, *J. Electroanal. Chem. 9*:448 (1965).
18. W. T. DeVries and E. Van Dalen, *J. Electroanal. Chem. 14*:315 (1967).
19. W. H. Reinmuth, *Anal. Chem. 33*:185 (1961).

20. Z. Stojek, B. Stepnik, and Z. Kublik, *J. Electroanal. Chem. 74*:277 (1976).
21. Z. Stojek and Z. Kublik, *J. Electroanal. Chem. 77*:205 (1977).
22. P. Ostapczuk and Z. Kublik, *J. Electroanal. Chem. 93*:195 (1978).
23. R. A. Osteryoung and J. H. Christie, *Anal. Chem. 46*:351 (1974).
24. G. Gillain, G. Duyckaerts, and A. Disteche, *Anal. Chim. Acta 106*:23 (1979).
25. S. P. Perone and T. R. Mueller, *Anal. Chem. 37*:2 (1965).
26. A. Zirino and M. L. Healy, *Environ, Sci. Technol. 6*:243 (1972).
27. N. Velghe and A. Claeys, *J. Electroanal. Chem. 35*:229 (1972).
28. U. Eisner, J. A. Turner, and R. A. Osteryoung, *Anal. Chem. 48*:1608 (1976).
29. S. Bruckenstein and J. W. Bixler, *Anal. Chem. 37*:786 (1965).
30. D. Jagner and A. Graneli, *Anal. Chim. Acta 83*:19 (1976).
31. D. Jagner, *Anal. Chem. 50*:1924 (1978).
32. M. Kozlovsky and A. Zebreva, in *Progress in Polarography*, Vol. 3, (P. Zuman and L. Meites, ed.), Wiley-Interscience, New York, 1972, p. 166.
33. A. F. Trotman-Dickenson (ed.), *Comprehensive Inorganic Chemistry*, Vol. 3, Pergamon Press, London, 1973, pp. 283-285.
34. W. R. Matson, D. K. Roe, and D. E. Carritt, *Anal. Chem. 37*:1594 (1965).
35. T. R. Copeland, J. H. Christie, R. A. Osteryoung, and R. K. Skogerboe, *Anal. Chem. 45*:2171 (1973).
36. H. Siegerman and G. O'Dom, *Am. Lab. 4*(6):59 (1972).
37. T. M. Florence, *J. Electroanal. Chem. 27*:273 (1970).
38. R. G. Clem, G. Litton, and L. D. Ornelas, *Anal. Chem. 45*:1306 (1973).
39. M. Stulikova, *J. Electroanal. Chem. 48*:33 (1973).
40. A. M. Hartley, A. G. Hiebert, and J. A. Cox, *J. Electroanal. Chem. 17*:81 (1968).
41. T. L. Marple and L. B. Rogers, *Anal. Chem. 25*:1351 (1953).
42. L. Ramaley, R. L. Brubaker, and C. G. Enke, *Anal. Chem. 35*:1088 (1963).
43. S. A. Moros, *Anal. Chem. 34*:1584 (1962).
44. R. G. Clem, *Anal. Chem. 47*:1778 (1975).
45. R. G. Clem and A. F. Sciamanna, *Anal. Chem. 47*:276 (1975).
46. J. E. Anderson and D. E. Tallman, *Anal. Chem. 48*:209 (1976).
47. Z. Galus, *CRC Crit. Rev. Anal. Chem. 4*:359 (1975).
48. M. S. Shuman and G. P. Woodward, Jr., *Anal. Chem. 48*:1979 (1976).
49. T. R. Copeland, R. A. Osteryoung, and R. K. Skogerboe, *Anal. Chem. 46*:2093 (1974).

50. S. T. Crosmun, J. A. Dean, and J. R. Stokely, *Anal. Chim. Acta 75*:421 (1975).
51. T. M. Florence, *J. Electroanal. Chem. 35*:237 (1972).
52. T. M. Florence, *Anal. Chim. Acta 141*:73 (1982).
53. G. E. Batley, *Anal. Chim. Acta 124*:121 (1981).
54. W. Lund and M. Salberg, *Anal. Chim. Acta 76*:131 (1975).
55. P. Figura and B. McDuffie, *Anal. Chem. 51*:120 (1979).
56. T. E. Edmonds, P. Guogang, and T. S. West, *Anal. Chim. Acta 120*:41 (1980).
57. G. A. Bhat and J. H. Weber, *Anal. Chim. Acta 141*:95 (1982).
58. J. T. Kinard, *Anal. Lett. 10(14)*:1147 (1977).
59. M. Oehme, W. Lund, and J. Jonsen, *Anal. Chim. Acta 100*:389 (1978).
60. W. Lund and R. Eriksen, *Anal. Chim. Acta 107*:37 (1979).
61. G. Chittleborough and B. J. Steel, *Anal. Chim. Acta 119*:235 (1980).
62. N. K. Konanur and G. W. vanLoon, *Talanta 24*:184 (1977).
63. Z. J. Karpinski, A. Polosak, and Z. Kublik, *Anal. Chim. Acta 120*:55 (1980).
64. D. C. Johnson and R. E. Allen, *Talanta 20*:305 (1973).
65. P. Kiekens, R. M. H. Verbeeck, H. Donche, and E. Temmerman, *Anal. Chim. Acta 127*:251 (1981).
66. T. Monama and G. Duyckaerts, *Anal. Lett. 12*:219 (1979).
67. S. Forbes, G. P. Bounds, T. S. West, *Talanta 26*:473 (1979).
68. K. Shimizu and R. A. Osteryoung, *Anal. Chem. 53*:584 (1981).
69. R. S. Sadana, *Anal. Chem. 55*:304 (1983).
70. D. Jagner, M. Josefson, S. Westerlund, and K. Aren, *Anal. Chem. 53*:1406 (1981).
71. J. K. Christensen, L. Kryger, and N. Pind, *Anal. Chim. Acta 141*:131 (1982).
72. J. F. Coetzee, A. Hussam, and T. R. Petrick, *Anal. Chem. 55*:120 (1983).
73. A. Hu, R. E. Dessy, and A. Graneli, *Anal. Chem. 55*:320 (1983).
74. H. B. Mark, Jr., *J. Pharm. Belg. 25*:367 (1970); H. B. Mark, Jr., R. J. Boczkowski, and K. E. Paulsen, *In Situ Sampling Techniques for Trace Analysis in Water,* Second Joint Conference on Sensing of Environmental Pollutants, Instrument Society of America, Pittsburgh, Pa., 1973, p. 295.
75. H. J. Bowen and D. Gibbons, *Radioactive Analysis,* Clarendon Press, Oxford, 1963.
76. H. B. Mark, Jr., G. A. Rechnitz, and R. A. Grienke, *Kinetics in Analytical Chemistry,* Wiley-Interscience, New York, 1968.
77. F. Lux, *Radiochem. Acta 1*:20 (1962).
78. A. R. Nisbet and A. J. Bard, *J. Electroanal. Chem. 6*:332 (1963).

79. *General Electric Pyrolytic Graphite Engineering Handbook*, Special Allows Section, Metallurgical Product Dept., General Electric Corp., Detroit, Mich., 1963.
80. F. J. Miller and H. F. Zittel, *Anal. Chem. 35*:1866 (1963).
81. H. A. Laitinen and P. R. Rhodes, Analytical Division, 145th Meeting, American Chemical Society, New York, 1963, Abstracts, p. 8B.
82. A. L. Beilby, W. Brooks, Jr., and G. L. Lawrence, *Anal. Chem. 36*:22 (1964).
83. P. J. Elving, I. Fried, and W. R. Turner, 3rd International Congress on Polarography, July, 1964.
84. E. E. Brooks, Ph.D. dissertation, University of Cincinnati, 1976.
85. B. H. Vassos, R. F. Hirsch, and H. Letterman, *Anal. Chem. 45*:792 (1973).
86. J. E. Anderson and D. E. Tallman, *Anal. Chem. 48*:209 (1976).
87. J. A. Holcombe and R. D. Sacks, *Spectrochimica Acta 28B*:451 (1973).
88. R. D. Sacks and J. A. Holcombe, *Appl. Spectrosc. 28*:518 (1974).
89. B. H. Vassos, R. F. Hirsch, and H. Letterman, *Anal. Chem. 45*:792 (1973).
90. C. Fairless and A. J. Bard, *Anal. Lett. 5*:433 (1972); *Anal. Chem. 45*:2289 (1973).
91. W. Lund and B. V. Larsen, *Anal. Chim. Acta 70*:299 (1974); *72*:57 (1974); W. Lund and M. Salberg, *ibid. 76*:131 (1975); Y. Thomassen, B. V. Larsen, F. J. Langmyhr, and W. Lunde, *ibid. 83*:103 (1976); W. Lund, B. V. Larsen, and N. Gundersen, *ibid. 81*:319 (1976); and W. Lund and R. Bye, *ibid. 110*:279 (1979).
92. B. E. Batley and J. P. Matousek, *Anal. Chem. 49*:2031 (1977).
93. G. W. Shieffer and W. J. Blaedel, *Anal. Chem. 49*:49 (1977).
94. F. J. Berlandi and H. B. Mark, Jr., *Nucl. Appl. 6*:409 (1969).
95. J. P. Cowen, V. F. Hodge, and T. R. Folsoni, *Anal. Chem. 49*:494 (1977).
96. K. Wundt, H. Duschnor, and K. Starke, *Anal. Chem. 51*:1485 (1979).
97. R. Bock, E. Zimmer, and G. Weichbrodt, *Z. Anal. Chim. 100*:641 (1979).

# 20

# Constant-Current Coulometry

DAVID J. CURRAN / Department of Chemistry, University of Massachusetts, Amherst, Massachusetts



## I. INTRODUCTION

The idea of in situ generation of known amounts of chemical reagents is an intriguing one which has occupied the attention of many chemists. Electrochemists, in particular, have been notably successful in this regard. The basis for this is, of course, Faraday's law of electrolysis, which states that the quantity of electricity passed (coulombs) is directly proportional to the amount of chemical reaction (equivalents) which takes place at an electrode. As we have seen in Chapter 3, a mathematical expression for this concept is readily apparent.

$$Q = \int_0^t i\,dt = (F)(\text{equivalents}) \tag{20.1}$$

where Q is the number of coulombs, i the current in amperes, t the time in seconds. This equation is fundamental to both controlled-potential and constant-current coulometry. It is appealing to hold the current constant and measure the time for electrolysis in which case Equation 20.1 may be written

$$\text{equivalents} = \frac{Q}{F} = \frac{it}{F} \tag{20.2}$$

From Equation 20.2 it is seen that the experimental parameters in a constant-current coulometric experiment are current and time and it is possible to identify the following components of an appropriate apparatus: an electrolysis cell, a current source, a method of measuring elapsed time, and a switching arrangement to control current and the clock. Applications of constant current electrochemistry are widespread and include titrimetry, kinetic studies, process stream analysis, and other areas (see Chap. 4).

## II. COULOMETRIC GENERATION OF REAGENTS

### A. Constant Current

*Electrochemical Considerations*

Let the half-reaction at the generator electrode be the following:

$$\text{Precursor} \pm ne \rightleftharpoons \text{reagent}$$

To use Faraday's law for the purpose at hand, the generator electrode reaction must be limited to the one of interest to assure that the charge consumed by the desired electron transfer reaction is essentially equal to the charge applied to the cell. Difficulties arise because it is not possible independently to control both the current and the electrode potential and because only the current applied to the cell is known directly in the constant-current experiment. A knowledge of the current-voltage curves for the precursor and the supporting electrolyte will provide valuable information in the design of practical experiments.

For constant-current work, the cell requires an auxiliary electrode in addition to the generator electrode. Frequently, a separate compartment is used for the auxiliary electrode to prevent mass transfer between the two electrodes (see Chap. 12). The addition of a reference electrode will permit current-voltage curves to be obtained under conditions identical to those prevailing during the constant-current experiment. The limiting current $i_\ell$ of this hydrodynamic voltammogram will be directly proportional to the concentration of precursor in the electrolyte. If a series of precursor solutions are prepared and the current-voltage curves obtained for each of them, a family of curves will result in which the limiting current would become progressively smaller as the concentration of the precursor decreased. Such a family of curves for a reversible system can be represented by a single curve by normalizing the axes of the graph as shown in

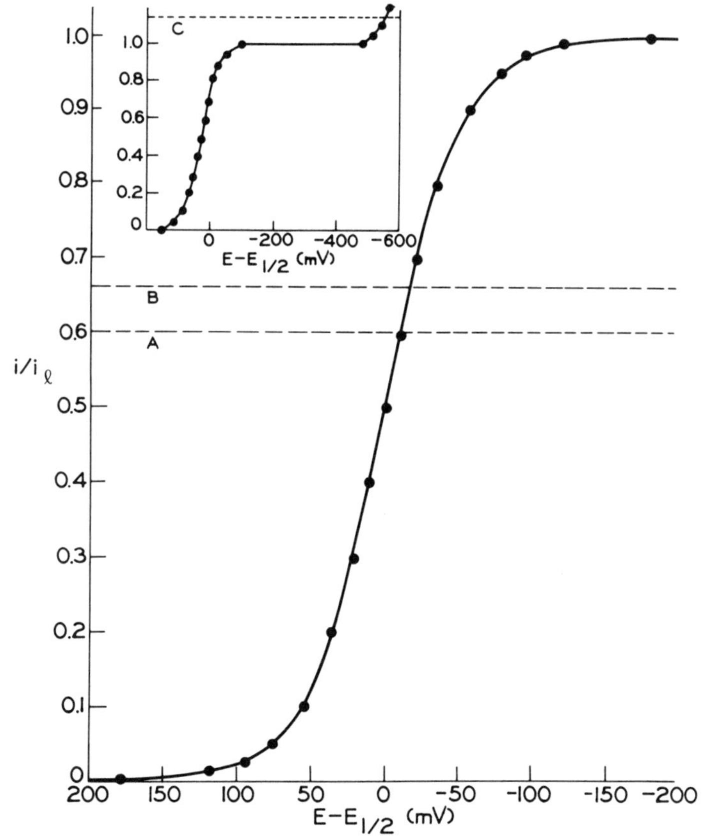


**Figure 20.1** Normalized current-voltage curve. Calculated from the equation $E = E_{1/2} + (RT/nF) \ln (i_\ell - i)/(i)$, where $i_\ell$ is the limiting current.

Figure 20.1, where the vertical axis is expressed as the ratio of the current at any point on the current-voltage curve to the limiting current, $i/i_\ell$, and the horizontal axis is plotted as $E - E_{1/2}$. Only faradaic current is considered; it is assumed that there is no background current.

If a constant current i equal to 60% of the limiting current found for the most concentrated solution of precursor (i.e., $i = 0.6\, i_\ell$) is applied to the cell, then the *ratio* $i/i_\ell$ is 0.6 and is represented by line A of Figure 20.1. The potential of the electrode at the time the constant current is applied would be the potential at the intersection of line A with the curve. Now, consider a second precursor solution which is only 90% as concentrated as the first one. The new limiting current, $i_\ell'$, would be nine-tenths of the original limiting current, or $i_\ell' = 0.9\, i_\ell$. However, in the constant-current experiment, the applied current does not change, so the new ratio, $i/i_\ell'$, is larger than the former ratio and can be calculated as

$$\frac{i}{i_\ell'} = \frac{0.6 i_\ell}{0.9 i_\ell} = 0.667 \tag{20.4}$$

This ratio is shown by line B in the figure and the electrode potential has a slightly more negative value at the new intersection. As the solution of precursor becomes more dilute, the potential of the generator electrode continues to shift negatively. As long as the applied current is smaller than the limiting current corresponding to the prevailing concentration of precursor, the electrode potential is governed by the precursor half-reaction and the electrode process is limited to a single redox reaction. If the applied current is greater than the limiting current, the electrode potential is forced to shift until a second process occurs to help carry a portion of the current. This condition is shown in the insert of Figure 20.1. Line C represents a current ratio greater than 1 and the precursor reaction alone is unable to support the current applied to the cell. This intolerable situation is usually avoided by using a high concentration of precursor.

The current efficiency of an electrode reaction refers to the fraction of the current applied to the cell which actually contributes to the faradaic reaction of interest. The magnitude of the applied current relative to any residual (background) current $i_r$ that might exist is therefore of great importance. In any real cell, there will be background current which can arise from any or all of the following: oxidation or reduction at the working electrode of chemical impurities in solution, adsorption processes at the electrode, and charge or discharge of the electrical double layer at the electrode. A discussion of these processes as they relate to constant-current coulometric titrations is found in the work by Meites [1]. In practice, the magnitude of the residual current at any potential can be determined from a residual current-voltage curve obtained in the constant-current cell containing only the supporting electrolyte. The percent current efficiency is defined as

$$\%\text{C.E.} = \frac{100(i_{\text{applied}} - i_r)}{i_{\text{applied}}} \tag{20.5}$$

In general, the residual current is a function of the electrode potential. However, the dependence is often slight over a limited range of potential. Many applications of constant-current coulometry are performed in a manner in which the reagent generated from the precursor is consumed by a chemical reaction that regenerates the precursor. The concentration of precursor is thereby nearly constant and the potential of the generator electrode remains nearly constant during electrolysis. A measurement of the residual current at the potential of the generator electrode will serve for the calculation of the current efficiency.

The applied current must be 1000 times the residual current to achieve a current efficiency of 99.9%. In many cases, a ratio of applied current to residual current of $10^3$ is reasonable for applied currents down to about 10 μA using generator electrode areas of about 0.1 $cm^2$. Currents in excess of a few hundred milliamperes are seldom used in constant-current coulometry because the solubility limit of the precursor is reached and/or the experiment may be over too quickly to permit accurate measurement of the time. Heating effects ($i^2R$) are also a problem when high currents are used.

### *Instrumentation*

A constant-current source can be constructed from a series combination of a battery and a resistance network (see Chap. 6). Such an arrangement is shown in Figure 20.2a. The current with the load shorted is equal to $E/R_{adj}$. Viewed as an electrical network, a constant-current source should have a very high internal resistance. This may be seen from the Norton's equivalent circuit for Figure 20.2a. Let $R_{adj}$ be set to 100 KΩ and E = 100 V. The Norton circuit is shown in Figure 20.2b. Since the load $R_L$ is placed in parallel with the Norton's equivalent resistance, it is clear that the 1 mA of current will only pass through the load resistor if $R_N$ is very much larger than the load. If $R_L$ were 100 Ω, the load current would be 0.999 mA, which is acceptably close to 1.000 mA. When the load is an electrochemical cell, the situation is more complex. The circuit is shown in Figure 20.2c. Let $E_{app}$ be the voltage applied to the cell. From Kirchhoff's voltage law,

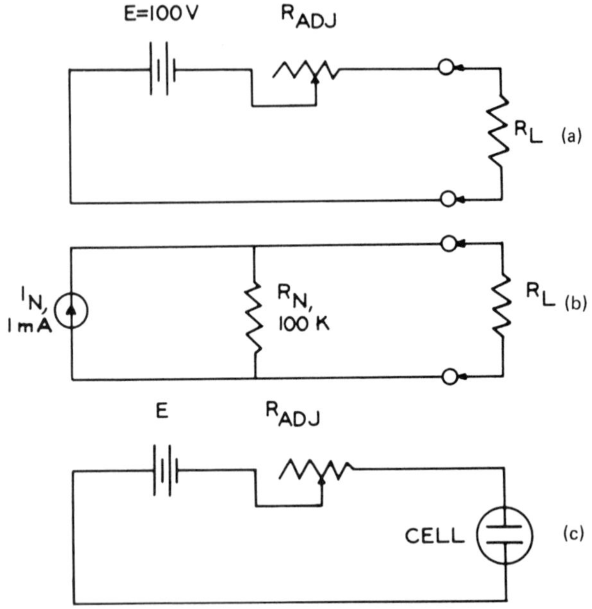


Figuer 20.2 Constant-current source using a battery and series resistor. (a) Dummy (resistor) load; (b) Norton's equivalent circuit for part a; (c) electrolysis cell as the load.

$$E = iR_{adj} + E_{app} \tag{20.6}$$

Now the applied voltage is the sum of the back emf of the cell, $E_{back}$, and the iR drop through it, or

$$E_{app} = E_{back} + iR_{cell} \tag{20.7}$$

The back emf, or galvanic voltage, is the difference between the anode and cathode potentials, which are each the sum of a reversible potential given by the Nernst equation, and an activation overpotential term $\eta$, which serves in the case of the cathode to shift the potential toward more reducing values and in the case of the anode toward more oxidizing potentials. Thus

$$\begin{aligned} E_{back} &= E_{anode} - E_{cathode} \\ &= E_a^{o\prime} + \frac{RT}{nF} \ln \frac{(O_1)}{(R_1)} + \eta_a - E_c^{o\prime} - \frac{RT}{nF} \ln \frac{(O_2)}{(R_2)} - \eta_c \end{aligned} \tag{20.8}$$

where redox couple 1 has been arbitrarily chosen as potential determining at the anode. Substituting Equations (20.7) and (20.8) in Equation (20.6) yields

$$E = iR_{adj} + iR_{cell} + (E_a^{o'} - E_c^{o'}) + (\eta_a - \eta_c) + \frac{RT}{nF} \ln \frac{(O_1)(R_2)}{(R_1)(O_2)} \tag{20.9}$$

Although E may be constant, $R_{cell}$, the overvoltage term, and the logarithmic term are subject to change during electrolysis and the current must therefore change. One recourse is to choose a supply voltage large enough that any changes can be neglected. A better solution is to monitor the current and make adjustments to correct for any changes. Operational amplifiers (OAs) are ideally suited for this purpose (see Chap. 6).

A typical OA circuit for constant-current coulometry is shown in Figure 20.3. If no significant amplifier input current is drawn, the cell current is identical to $i_{in}$. The latter is constant if the constant-voltage source and $R_{adj}$ are stable. The OA will automatically adjust the voltage applied to the cell to maintain the cell current equal to the input current. Perhaps the greatest source of instability is temperature change. Constant-voltage sources in integrated circuit packages which are quite stable to temperature changes are commercially available. An example is the Analog Devices Model AD 2702L (Analog Devices Inc., Route One Industrial Park, P.O. Box 280, Norwood, MA 02062), which delivers ±10.000 ± 0.0025 V at 0-10 mA and has a temperature coefficient of 3 ppm/°C. Metal film resistors have temperature coefficients better than ±0.1% and their use is recommended.

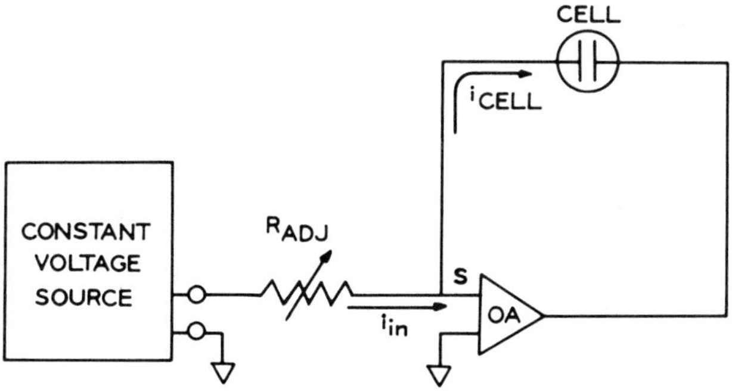


Figure 20.3 Operational amplifier constant-current source. [Adapted from J. Janata and H. B. Mark, Jr., *Electroanal. Chem.* 3:11 (1969).]

Table 20.1 Heat Generated in Resistors

| i (mA) | R (kΩ) | Power (W) | Time to generate 1 cal of heat (s) |
|---|---|---|---|
| 1 | 10 | 0.01 | 24 |
| 10 | 1 | 0.1 | 2.4 |
| 100 | 0.1 | 1.0 | 0.24 |

Assuming a 10-V output from the constant-voltage source, the values of $R_{adj}$ shown in Table 20.1 were calculated for currents covering the range 1-100 mA. Some appreciation of the heating problem can be gained from the last two columns of the table. Resistors which should have a much higher wattage rating than the power to be dissipated should be used with either a heat sink or good access to air. The accuracy of the components is not as important as their stability since the current can be calibrated, although this may not be necessary with high-quality components. In any case, it is entirely possible to generate currents that approach ±0.01% in stability, accuracy, and reproducibility. Inexpensive (<\$1) commercially available integrated-circuit operational amplifiers are of a quality to match these performance figures.

## B. Time Measurement and Switching

### *Electromechanical Timers*

Electromechanical devices for measuring elapsed time are highly developed. An example is the Model S-1-3H-60, manufactured by Standard Electric Time Corp. (80 Logan St., Springfield, MA 01109), which will totalize to 60 min, has a smallest graduation of 0.01 s, and is capable of a precision of ±0.005 s per operation if purchased with a dc clutch. Lingane [2] has discussed the use of synchronous motor timers or stopclocks for coulometric titrations and has pointed out that they are subject to two main sources of error: variations in the line frequency, and undershoot or overshoot due to motor inertia. The latter problem is minimized quite satisfactorily with solenoid brakes such as the dc clutch. If line frequency variations prove troublesome, they can be reduced substantially by operating the timer from a frequency-regulated power supply. It would appear that the state of the art is such that reliable operation of electromechanical timers cannot be achieved much below 5 ms of elapsed time. Lingane has also pointed out that while the relative error per operation of the timer can be very small, the errors are cumulative for successive operations. In many instances, the reagent is generated incrementally and it is an

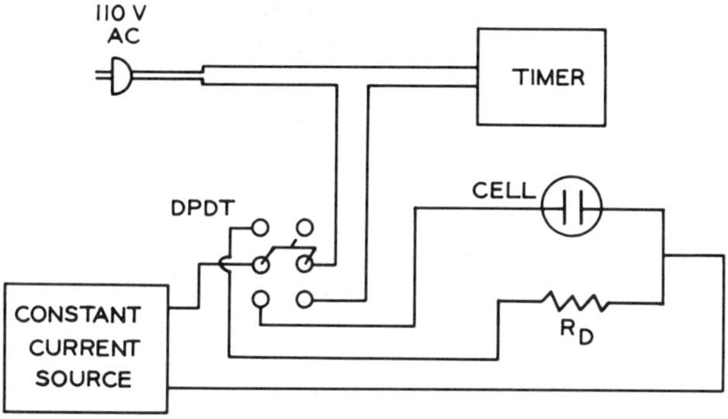


Figure 20.4 Block diagram of an apparatus for constant-current coulometry. A double-throw, double-pole switch is used to turn the timer and electrolysis current on and off. [Adapted from G. W. Ewing, in *Am. Lab.* *13*(6):16-22 (1981). Copyright 1981 International Scientific Communications, Inc.]

advantage to use a timer having a precision that is much better than the precision desired in the final result.

An arrangement of a timer in an apparatus for constant-current coulometry is shown in Figure 20.4. The double-pole, double-throw switch is perfectly adequate for manual operation. Automation is achieved by replacing the DPDT switch with an electrically actuated one. The actuating signal is produced manually or is derived from some concentration-dependent transducer associated with the cell such as a redox indicator electrode or a photodiode. The current is switched between the cell and a dummy resistor $R_D$ which has about the same resistance as the cell. In this way the constant-current source is always working into a load under nearly the same conditions at all times. For a number of years, these design features have been the basis of instrumentation reported in the literature.

*Digital Timing*

Solid-state replacements for electrochemical timers are commercially available which offer improvements in resolution. The Standard Electric Time Model STD-11-2 has four-pushbutton controls to select maximum time intervals of 9.999, 99.99, 999.9, and 9999 s. Time measurements to the nearest millisecond are possible. Details of the internal circuitry of this timer are not available but the manufacturer states that the timer is synchronized to the line frequency using phase-locked-loop techniques [3]. It would seem that the frequency of the internal oscillator is dependent on the line frequency and therefore that the timer is subject to the same type of frequency error as the electromechanical unit. To take full advantage of the capability of measuring time intervals to the nearest millisecond, it is necessary to use solid-state switch techniques. Appropriate circuit designs are provided by the manufacturer for the unit mentioned above. This timer, and others like it, can be used directly in the apparatus shown in Figure 20.4.

Another approach to timing which uses solid-state circuits involves a crystal oscillator as the basis for the time measurement. The resonant frequency of such oscillators is temperature dependent and subject to aging effects, but the state of the art is highly developed. Crystal-controlled clock oscillators having a maximum error of ±1 ppm at 25°C are commercially available from a number of manufacturers. Some units are temperature compensated and have better stability than the figure above, with aging rates of only 1 ppm per year. Oven-controlled crystal oscillators can have aging rates of $1 \times 10^{-9}$ per day and a temperature variation in the output of only $1 \times 10^{-9}$ over a 50°C temperature range. Such oscillators, used in conjunction with digital counting circuits, provide an excellent method for the time interval measurement in constant-current work. With adequate switching, timing to the nearest microsecond is routinely possible, resolution to six digits is not unusual, and very high accuracy can be achieved.

A block diagram of a counting circuit arranged for time interval measurements is shown in Figure 20.5. Individual digital circuits will not be discussed here and the reader is referred to several references [4-6]. A good review of a number of counting circuits is also available [7]. A general understanding of digital time interval measurements can be gained from the block diagram in Figure 20.5. Pulses from the crystal oscillator are sent directly to a gate and to a time base (scalar) unit which consists of decade counting units (DCUs).

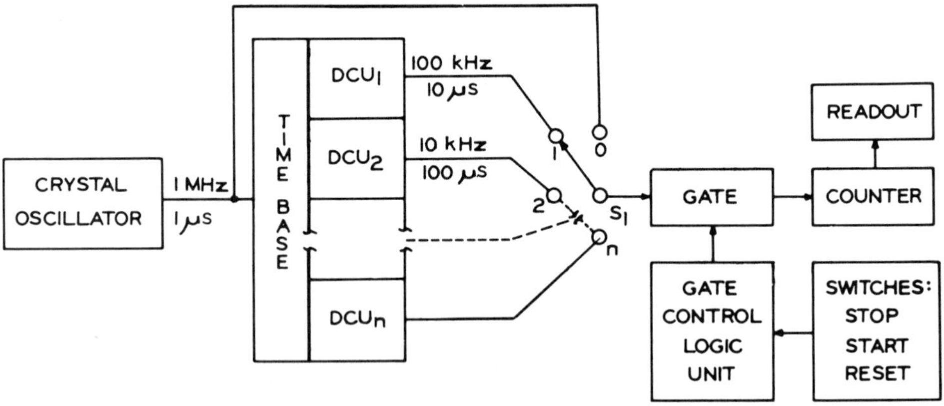


Figure 20.5 Block diagram of a solid-state timer.

Each DCU is a divide-by-10 circuit and the oscillator output is sequentially divided by 10 up to n times.

In the time domain, if the oscillator (clock) puts out a pulse every microsecond, the output of the first DCU is a 10-μs pulse train and so on through n decade counting units. When the gate is open, pulses are passed on to the counter, where they are summed. The gate is controlled by a logic signal generated in the gate control logic unit. Manual switches may initiate this signal. A numerical readout displays the count accumulated by the counter. Switch $S_1$ serves as a range switch. With a six-digit counter and readout, 999,999 counts or 0.999999 s can be accumulated in the readout with $S_1$ in position 0. Switch position 1 corresponds to a maximum readout of 9.99999 s, and so forth, to switch position n, where $0.999999 \times 10^n$ s can be accumulated. It might seem that the gate control logic unit could be eliminated and the gate operated directly by the switches. This is not feasible when it is realized that one count could correspond to 1 μs and manual switching is not possible in so short a time. Flip-flops or other circuits are triggered by the manual switches to provide the control signals to the gate. Once the time measurement has begun, it can be terminated automatically or manually.

The basic time interval counter is more useful for coulometry if a preset capability is added to it. Such counters are commercially available and are also readily constructed from digital circuits. This counter will automatically stop counting at a preselected time. In addition, a logic-level signal is available to indicate the state of the counter: for example, logic level HI for counting and logic level LO for not counting. The change in logic level occurs at the time counting stops and can be used in conjunction with a logic-operated solid-state switch to turn the cell current on and off. This provides a direct way to switch the cell current in 1 μs or less.

A block diagram of a complete apparatus for constant-current coulometry using an operational amplifier constant-current source, solid-state switch, and digital timing is shown in Figure 20.6. The solid-state switch occupies a central position in the diagram to emphasize that it is a key element in the circuit design because it provides the link between the analog and digital parts of the circuit. The actuating circuit is logic operated, or gated, while the switching circuit operates on the analog current. Solid-state switches are available from a number of manufacturers and can be obtained in a variety of configurations such as single-pole, single-throw (SPST), DPST, DPDT, and others. Make-before-break and break-before-make switching action is provided and units compatible with nearly all logic families (TTL, CMOS, etc.) are manufactured. Switching times are typically less than 1 μs and the complementary metal oxide semiconductor (CMOS) switches can be obtained with switching times of less than 10 ns. A useful introduction to analog switches can be found in Ref. 8 and the subject is discussed further in Ref. 9. The electrolysis cell, dummy

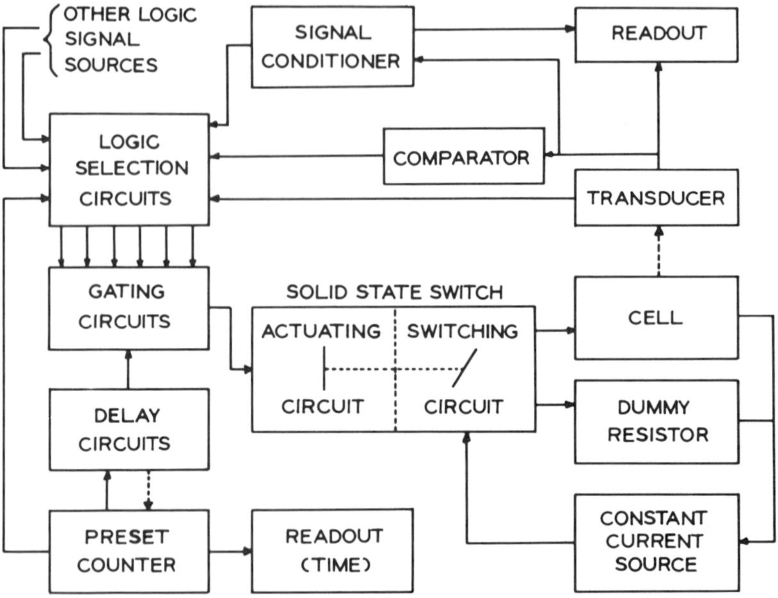


Figure 20.6 Block diagram of an all-solid-state apparatus for constant-current coulometry.

resistor, and current source appear in the lower right-hand side of Figure 20.6. Some concentration-dependent transducer may be associated with the cell, such as a potentiometric indicator electrode or a phototransistor. Its output can be presented on a readout, either directly or after signal conditioning (amplification, waveshaping, etc.). Its output can also be used to switch the cell current on and off. In principle, the transducer output could be digital, in which case the output is sent to a logic selection network, where a choice is made of which logic signal or signals will be sent to the gating circuits and on to the analog switch. It is more likely that the transducer output is an analog signal. In this case, a comparator circuit can be used for the analog-to-digital conversion or, perhaps, signal conditioning is required which will have both an analog output for the readout and a digital output to operate the switch.

The block diagram in Figure 20.6 represents a modern implementation of the classical apparatus illustrated in Figure 20.4. The solid state approach has the advantage of having the capability to use the time variable for an active role in the experiment. The output of the preset counter can be used to activate the current switching circuit through the logic selector and gates. This is a necessity for subsecond electrolysis times. Furthermore, the counter can also be used to trigger delay circuits which can override all other signals to the gates controlling the input to the actuating circuit of the solid-state switch. Once the delay circuits are activated, they also control the counter. In this way, a "waiting" period of time can automatically be introduced into the electrolysis experiment. One might, for example, want to wait for equilibrium to be established or wait for an induction period in a kinetics experiment. The system is versatile and a logic signal from any other source can be used to activate the solid-state switch. Electrolysis times from several microseconds to thousands of seconds are readily measured with this apparatus. Jaycox and Curran have described a coulometric titrator based on these design considerations [10].

## III. COULOMETRIC TITRATIONS

### A. Introduction

A cell for internal generation of titrant is easily constructed. Often it is necessary to remove dissolved oxygen and a closed vessel equipped with a gas dispersion tube and a vent is provided. Nitrogen is usually used for deoxygenation. The working electrode compartment is separated from the counter electrode compartment by a glass frit. Materials for the working and counter electrodes can be any of those commonly used in electrochemistry: mercury, platinum, gold, and carbon in its various forms as well as active electrode materials such as silver. A magnetic stirring bar is used to mix the titrant with the solution. If silver ion and chloride ion can be tolerated, the two-compartment cell can be eliminated and a silver-silver chloride counter electrode can be placed directly into the solution to be titrated. A method of end-point detection must be provided.

### B. Primary and Secondary Processes

In a constant-current coulometric titration, either the electrogenerated reagent reacts with an unknown substance in a stoichiometric or reproducible manner, or the unknown itself reacts directly at an electrode. The latter, called a primary coulometric titration, is rarely found in practice because the concentration of unknown continuously decreases during the electrolysis and it is difficult to maintain 100% current efficiency. The former approach, known as secondary coulometric titrations, is used most of the time.

### C. Titration Efficiency

If the reaction at the generator electrode is given by Equation 20.10a and the titration reaction by Equation 20.10b:

$$ox_2 + ne \rightleftharpoons red_2 \qquad (20.10a)$$

$$ox_1 + red_2 \rightleftharpoons red_1 + ox_2 \qquad (20.10b)$$

where $ox_1$ is the titrate and $red_2$ is the titrant, the electrode potential of couple 1 must be more positive than the electrode potential of couple 2 (IUPAC sign convention) for the titration reaction to be thermodynamically feasible. There are interesting consequences of this in terms of the electrochemistry of the coulometric titration, and they can be seen by referring to the insert of Figure 20.1. Initially, the solution contains two oxidants. The titrate, $ox_1$, is more easily reduced than the precursor of the titrant, $ox_2$, and therefore the first wave shown in the figure corresponds to the reduction of the titrate, while the second wave (only the foot of the wave is shown) corresponds to the reduction of the precursor. If the current applied to the cell is represented by line C, it is clear that most of the current is initially carried by the titrate. The current efficiency is less than 100% for either couple 1 or couple 2. However, the charge passed is the same whether the titrate is reduced directly at the electrode or reduced by the titrant, $red_2$. Thus 100% *titration* efficiency can be achieved although the current efficiency is less than 100%. In fact, it is possible for the current efficiency for the precursor reaction to be less than 100% due to background and have the titration efficiency maintained at an acceptable level when a large percentage of the titrate reacts directly at the electrode. Suppose that 85% of the titrate undergoes direct electroreduction with 100% current efficiency and the remaining 15% is titrated by the titrant, which is generated with 90% current efficiency. The titration efficiency is $0.85 + (0.9 \times 0.15) = 0.985$. An error of 1.5% might be entirely acceptable in some cases.

## D. Internal and External Generation

The common technique in coulometric titrations is to generate the titrant at an electrode physically placed within the coulometric electrolysis cell, or internally. This approach sometimes fails. An obvious case is when an electroactive material present in the sample undergoes a reaction at the generator electrode concurrently with the precursor. It is also possible to foul the generator electrode with surface-active materials such as proteins. These problems can be eliminated by generating the titrant in a stream of electrolyte which flows into the titration vessel. The operation becomes entirely analogous to that of a buret and some means of controlling the flow rate and stopping the flow are usually required [11]. It is physically impossible for the titrate to undergo direct electrochemical reaction and the titrant generation reaction must proceed at 100% current efficiency. Lingane has discussed a number of cell designs for external titrant generation [12], and much of the information on electrochemical detectors in Chapter 22 is useful in this regard.

## E. End-Point Detection

Some method is required to signal when the amount of titrant generated is equivalent to the amount of unknown present, and all of the end-point detection methods used in volumetric titrimetry are, in principle, applicable to coulometric titrations. A list that covers most of the published coulometric titration procedures is given in Table 20.2. It is beyond our scope here to describe any of these in detail because each of these methods is a subject for discussion in its own right. Discussions of the equations for a number of types of titration curves are found in texts by Lingane [13], Butler [14], and Laitinen and Harris [15].

Several practical features relating to end-point detection are worthy of comment. In most coulometric titration cells, the ratio of solution volume to generator electrode area is large. The solution must be stirred to mix the titrant with the solution and local excesses of titrant are common, particularly in the vicinity of the end point. The end-point detection system is easily fooled by this nonequilibrium condition and false detector signals arise. Such signals are seldom mistaken for equilibrium conditions since they do not persist, but advantage can be taken of them to introduce end-point anticipation into the titration. A false signal corresponding to the end-point condition would shut off the generator current for a short time to allow the system to reach equilibrium. To enhance this effect, the end-point detector is placed in a position where it is bathed by the solution coming directly from the generator electrode. At the same time, it is necessary with electrometric end-point detection methods to avoid placing the indicator electrodes in the current path between the generator electrodes, since interaction between the electrode pairs can occur.

Different experimental approaches are possible with the same end-point detection method. For example, the titration curve can be

Table 20.2 Selected End-Point Detection Methods for Coulometric Titrations

| | |
|---|---|
| Visual indicators | Potentiometric |
| Colorimetric | Amperometric |
| Spectrophotometric | Conductometric |
| Pressurometric | |

plotted and the end point determined graphically. First and/or second derivative curves can be plotted or the derivatives obtained electronically. Another approach is to titrate to a predetermined end-point signal. This technique is very useful with coulometric titrations and many examples, especially those involving potentiometric end-point detection, are found in the literature. The most widely applicable way of selecting the end-point potential is to acquire the data for the complete titration curve with a known sample. The time required to reach the end point is known beforehand and examination of the titration curve will reveal the potential corresponding to this time. Several advantages are realized from the use of a predetermined end-point signal with coulometric titrations: (1) the supporting electrolyte can often be pretitrated to the predetermined potential to remove reactive impurities, (2) samples can be added successively to the same supporting electrolyte, (3) coulombic error due to double-layer charging or discharging may be minimized, and (4) the ease of automation is enhanced. With respect to the latter, end-point detection becomes operationally a matter of a null comparison measurement, and the operational amplifier comparator is nicely suited to this task [16]. The third advantage listed occurs because the potential of the generator electrode is the same at the beginning and at the end of the titration, but this will be true only if the structure of the double layer remains unchanged (i.e., surfactant, for example, must be absent). The pretitration technique is widely applicable but can be subject to error. In the titration of a weak acid, the presence of unsuspected weak acids of different acid strengths from the sought-for acid would lead to an incorrect pretitration procedure. Also, it is not generally true that end-point detection methods are independent of the initial concentration of analyte, and the dilution produced by the successive addition of samples may be enough to shift the correct end point from the predetermined one. In any event, the correctness of the pretitration technique can be checked by determining known amounts of the substance to be determined.

End-point detection remains an active research area in coulometric titrations. Using a titrator based on the concepts described in Figure 20.6, Jaycox and Curran [10] were able to titrate as little as 26 ng of Mn(VII) automatically with electrogenerated Fe(II) to a precision and accuracy of 2% relative using potentiometric end-point detection to a preset end-point potential. The same apparatus, operated in a manual mode, was used to titrate V(V) solutions as dilute as 2.1 ppb, for which the total elapsed electrolysis time to reach the end point was 20.5 ms. Cadwgan and Curran [17] described a single-beam fluorescence detector for use with the same titrator. Acid was titrated with electrogenerated base in the presence of dichlorofluorescein, a fluorescent indicator. Only the conjugate base form of this indicator fluoresces. At a solution pH equal to the $pK_a$ of the indicator, the ratio of acidic to basic forms is 0.5 and slight changes in pH will produce large changes in the emission light intensity. An arbitrary null point pH was chosen where the indicator was present primarily as the conjugate base. This was accomplished by adding small volumes of dilute HCl and/or KOH. The photodiode output voltage at the null point pH was set on the reference side of the comparator. A sample of acid was added to the cell with a microburet and the solution titrated back to the null point pH. Amounts of hydrogen ion, as $KH(IO_3)_2$, from 1284 to 12.84 neq were automatically determined with errors of a few percent in most cases and precision was respectively, ±0.5 and ±8.8% relative. It was estimated that addition of the smallest samples to the cell produced a change in hydrogen ion concentration of 0.31 ppb for a solution of about pH. 5. This corresponded to a pH change of 0.013 unit, which would be difficult to determine with an expanded scale pH meter. Very small changes in pH can be measured with this fluorescence detector and the working pH region can be selected by choosing a fluorescent indicator of the appropriate $pK_a$.

With the same coulometric titrator, de Soto Perera and Curran [18] were able to titrate as little as 24.2 ppb of As(III) (or $3.22 \times 10^{-7}$ M solution) with electrogenerated bromine using thin-layer hydrodynamic biamperometric end-point detection. A sandwich-type thin-layer cell containing two identical platinum electrodes (see Chaps. 3,12 and 22) was placed in a closed loop attached to the coulometric cell. Solution was pumped through the flow loop at a rate of about 16 mL/min with a peristaltic pump and the solution in the coulometric cell was stirred magnetically. The potential difference across the thin-layer platinum electrodes was 170 mV and the titrations were performed incrementally. Plots of thin-layer cell current vs. time of electrolysis were reverse L shaped, since a response is obtained for this titration only in the presence of excess bromine. Solutions about 800 ppb in As(III) were determined with a relative error and precision of 0.1%. The method was also applied to the coulometric titration of some phenolic steroids with electrochemically generated bromine [19]. The titration curves were recorded continuously, and end points were obtained by graphical extrapolation of the linear segments.

## F. Titrants

### *Survey*

Many titrants have been electrogenerated, most of them in aqueous solution. A list is given in Table 20.3 which was compiled from biannual reviews appearing in *Analytical Chemistry* between 1968 and 1982 [20-26a] and from the text by Lingane [27]. A number of the common volumetric titrants are included as well as titrants that are unstable or difficult to prepare volumetrically. In aqueous solution the strongest oxidant is silver(II) (E°, about 2 V) and the strongest reductant is uranium(III) (E° = −0.63 V). The former is thermody-

Table 20.3 Electrogenerated Titrants

| | | |
|---|---|---|
| A. *Oxidants* | | |
| Bromine | Cerium(IV) | Octacyano-molybate ion |
| Hypobromite ion | Permanganate ion | Vanadium(IV) |
| Bromine monochloride | Chromium(VI) | Vanadium(V) |
| Iodine | Silver(II) | Lead tetraacetate |
| Chlorine | Manganese(III) | Oxygen |
| Iron(III) | Mercury(II) | Quinone |
| Ferricyanide ion | Gold(II) | Silver(I) |
| B. *Reductants* | | |
| Iron(II) | Titanium(III) | Vanadium(IV) |
| Iron(II)-EDTA | Copper(I) | Vanadium(III) |
| Ferrocyanide ion | Hexacyanomangan-ese(I) | Uranium(III) |
| Molybdenum(V) | Tin(II) | Iodide ion |
| Octacyanomolybdenum(V) | Chromium(II) | Uranium(IV) |
| C. *Precipitants and Complexing Agents* | | |
| Silver(I) | Copper(II) | Iodide ion |
| Mercury(I) | Lanthanum(III) | Fluoride ion |
| Lead(II) | Mercury(II) | Sulfide ion |
| Bismuth(III) | Ferricyanide ion | Thiocyanate ion |
| Aluminum(III) | Ferrocyanide ion | EDTA |
| D. *Bases* | | |
| Aqueous and nonaqueous media | | |
| E. *Acids* | | |
| Aqueous and nonaqueous media | | |
| F. *Miscellaneous* | | |
| Tungsten (II) in LiCl-KCl eutectic | | |
| Karl Fischer reagent | | |

*Source*: Compiled from Refs. 20 to 27.

namically capable of oxidizing water and the latter similarly capable of being oxidized by hydrogen ion. The electrogeneration of these titrants requires that the kinetics of these other reactions be slow. Bromine is a moderately strong oxidizing agent ($E° = 1.065$ V) and its electrogeneration is quite free of side reactions in an aqueous solvent. It can be easily generated with 100% current efficiency at a platinum electrode and is a titrant of major importance because of its reactions in both inorganic and organic chemistry. It has wide scope in the determination of organics because substitution reactions, addition reactions and others are possible. Specific examples of its application to the determination of pharmaceuticals are given in Chapter 21. Another important titrant is Karl Fischer reagent, which is used for the direct titration of water and indirectly for the determination of many organics which contain functional groups which will produce or consume water in a stoichiometric or quantitative manner [28]. Iodine is electrochemically produced from an iodide salt in anhydrous methanol containing sulfur dioxide and pyridine. The titration reaction written with py representing pyridine is

$$H_2O + I_2 + SO_2 + CH_3OH + 3py \longrightarrow 2pyH^+I^- + pyHOSO_2OCH_3 \tag{20.11}$$

*Uranium Electrochemistry*

When hydrogen ion is involved in a half-reaction, such as $ox + mH^+ + ne \rightleftharpoons red$, the potential is pH dependent, as the Nernst equation below shows.

$$E = E° + \frac{RT}{nF} \ln \frac{a_{ox}}{a_{red}} - \frac{m}{n}(0.0591)\text{pH} \tag{20.12}$$

At unit activities of the oxidant and reductant, the potential depends only on pH and the slope of the line for a plot of potential vs. pH is governed by the ratio m/n. Potential-pH diagrams are a concise means to display the redox properties of a system. We will take uranium as an example. The +6, +5, +4, and +3 oxidation states are known in aqueous solution. The determination of +6 uranium by coulometric titration has been investigated by many workers and the lower oxidation states have all been used as coulometric titrants. Hydrolyzed uranium species exist in a noncomplexing solution, but the chemistry is simplified considerably if the discussion is limited to solutions more acidic than about pH 4. Some of the half-reactions to be considered are listed below.

$E°$ vs. NHE:

$$+0.063 \qquad UO_2^{2+} + e \rightleftharpoons UO_2^{+} \qquad (20.13)$$

$$+0.246 \qquad UO_2^{2+} + 4H^{+} + 2e \rightleftharpoons U^{4+} + 2H_2O \qquad (20.14)$$

$$+0.429 \qquad UO_2^{+} + 4H^{+} + e \rightleftharpoons U^{4+} + 2H_2O \qquad (20.15)$$

$$-0.635 \qquad U^{4+} + e \rightleftharpoons U^{3+} \qquad (20.16)$$

The standard potentials given are those cited or calculated by Farrington and Lingane [29]. The only species of importance that is not shown in these half-reactions is $UOH^{3+}$. The predominant form of U(IV) in solutions more acidic than pH 1 is the simple aquated ion, but in the pH region 1-4, $UOH^{3+}$ predominates. Thus the following half-reactions also need to be considered:

$$UO_2^{2+} + 3H^{+} + 2e \rightleftharpoons UOH^{3+} + H_2O \qquad (20.17)$$

$$UO_2^{+} + 3H^{+} + e \rightleftharpoons UOH^{3+} + H_2O \qquad (20.18)$$

$$UOH^{3+} + e \rightleftharpoons U^{3+} + OH^{-} \qquad (20.19)$$

The potential-pH diagram shown in Figure 20.7 summarizes the information contained in these seven half-reactions. At any pH more acidic than about 1.7, the U(VI), U(V) couple is analogous to the Cu(II), Cu(I) couple in that the lower oxidation state is the stronger oxidant and therefore U(V) is expected to disproportionate to U(VI) and U(IV). The diagram also predicts that U(VI) should be directly reduced to U(IV) in solutions more acidic than pH 1.7 but to U(V) in more basic solutions. However, the diagram contains no information about any overpotentials for these reactions. A fairly large overpotential is required for the +6, +4 reaction, while the +6, +5 couple behaves reversibly. As a result, reduction directly to U(IV) occurs in solutions more acidic than about pH 1. Solutions of U(V) are readily oxidized by dissolved oxygen and air must be excluded, but deaeration is not necessary for U(IV) solutions. For many years, the strongest reductant in analytical use has been Cr(II) ($E° = -0.41$ V). The +4, +3 uranium couple has a potential at least 0.2 V more negative than the +3, +2 chromium couple and U(III) is a strong reductant indeed.

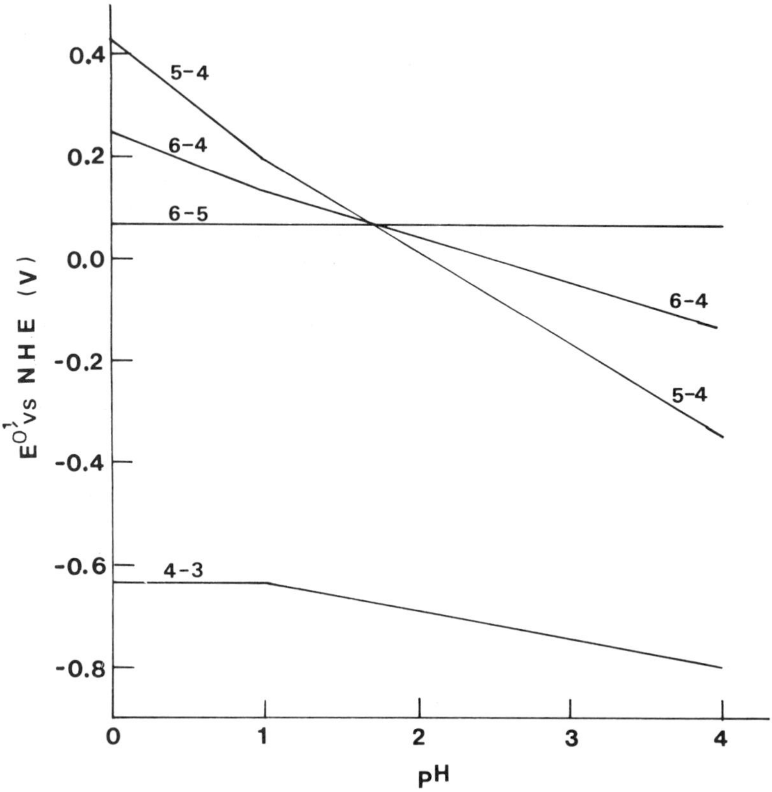


Figure 20.7 Potential-pH diagram for the uranium system in the pH region 0-4. (Adapted from Ref. 2.)

*Coulometric Titration of Uranium(VI)*

Uranium(VI) is a rather weak oxidant and its direct determination requires a strong reductant. Lingane and co-workers developed methods using electrogenerated titanous ion, first at a mercury cathode [30], and later at platinum electrodes [31]. In the latter case, a supporting electrolyte 8 M in sulfuric acid and 0.6 N in titanyl sulfate was recommended. At this acidity, U(VI) is reduced to U(IV). To overcome the slow rate of reaction, a small amount of Fe(II) was added to the solution. The Fe(III) produced by the reaction of Fe(II) with U(V) is reduced by Ti(III). Biamperometric end-point detection using platinum electrodes was used. Quantities of uranium from 8 to 225 mg were determined with an average error of less than +0.05%. Taylor and Marinenko later modified the procedure, using a 9 M $H_2SO_4$ supporting electrolyte, 1 M in titanyl ion (see Ref. 32 and the reference cited therein). They adopted the procedure of titrating beyond the end point, waiting for equilibrium, and then generating excess Ti(III) in small increments. Using amperometric end-point detection, the end point was found from the slope of the titration curve beyond the end point. One-gram samples of uranium metal were determined with a standard deviation of 0.08%. All ions reducible by Ti(III) interfere. Lingane and Kennedy applied their procedure to the titration of mixtures of uranium and vanadium. The standard potential of the $VO_2^+$, $VO^{2+}$ couple is +1.000 V and that of the $VO^{2+}$, $V^{3+}$ couple is +0.361 V. The former is sufficiently positive of the U(VI), U(V) potential that V(V) is titrated first. The V(IV) so formed is titrated simultaneously with the U(VI). The first end point yields the amount of vanadium present and the amount of uranium is obtained by difference, using the second end point. Solutions containing 10-100 ppm of vanadium and 180-1000 ppm of uranium were determined with an average error close to ±0.2% for both elements.

Other constant-current coulometric titrations of U(VI) have been indirect. Carson reduced U(VI) to U(IV) by passing the solution of the +6 ion containing 2 M HBr through a lead-amalgam reductor column directly into an air-free titration vessel [33]. The column was washed with a 3 M HBr solution containing ferric bromide. The latter is reduced to Fe(II). Titrant was then generated at a platinum anode to oxidize the U(IV) to U(VI). The titration was carried out at 95°C to increase the rate of the reaction. It is not clear if electrogenerated bromine or Fe(III), or some combination of both, is the actual titrant. Lingane argues in favor of Fe(III) on the basis of formal potentials and relative concentrations [34]. Carson recommended constant-current potentiometric end-point detection, which involves passage of a small constant current through a pair of identical indicator electrodes, the potential difference between them being measured. This technique is often appropriate for titration reactions involving irreversible couples. Seven-milligram amounts of uranium were titrated with a precision and accuracy of ±0.3% and 30-μg amounts were determined to ±6%. Several approaches have been devised to avoid the end-point detection difficulties due to the irreversibility of the U(VI), U(IV) couple, but like Carson's work, they all involve an initial reduction step to convert U(VI) to U(IV). Furman and co-workers [35] reported that a Jones reductor (zinc-amalgam) was unsatisfactory because a variable amount of U(III) was produced. A 3% cadmium amalgam was inconvenient because it is a liquid. A solid amalgam containing about 26% cadmium was finally used. Malinowski [36] developed a coulometric finish to a method described by Pszonicki [37]. Uranium(VI) is reduced primarily to U(III) by metallic aluminum in the presence of cadmium ion. The U(III) is oxidized to U(IV) by the addition of phosphoric and hydrochloric acids. This oxidation by protons should also be applicable with the Jones reductor. Goode and co-workers used the Ti(III) reaction to reduce U(VI) [38]. The excess Ti(III) was removed by reaction with a solution of sulfamic and nitric acids. In the work by both the Furman and Goode groups, the U(IV) was converted back to U(VI) by reaction with ferric ion and the equivalent amount of Fe(II) so produced was determined by coulometric generation of Ce(IV). The overall scheme is presented in the following equations:

$$\mathrm{U(VI)} + 2e \xrightarrow{\text{chemical reductant}} \mathrm{U(IV)} \qquad (20.20)$$

$$\mathrm{U(IV)} + 2\mathrm{Fe(III)} \rightleftharpoons \mathrm{U(VI)} + 2\mathrm{Fe(II)} \qquad (20.21)$$

$$\mathrm{Ce(III)} \rightleftharpoons \mathrm{Ce(IV)} + e \text{ (platinum or gold anode)} \qquad (20.22)$$

$$\mathrm{Ce(IV)} + \mathrm{Fe(II)} \rightleftharpoons \mathrm{Ce(III)} + \mathrm{Fe(III)} \qquad (20.23)$$

The procedure of Goode et al. was developed to provide a highly precise determination of uranium and it succeeded admirably since a precision of ±0.004% was achieved. The accuracy was reportedly better than 0.01%, but the method does require an accurately known solution of Ti(III). The Malinowski method offers another alternative. The U(IV) is oxidized by addition of an amount of solid primary standard $K_2Cr_2O_7$ which is in excess by a few milligrams. Ferric chloride is added to the solution and the excess dichromate is determined by reaction with electrogenerated Fe(II). Excellent results were found for the analysis of $U_3O_8$ obtained from NBS. A precision of ±3 parts in $10^5$ was reported and an error of 7 parts in $10^4$.

The initial reduction of U(VI) shown in equation [20] would be simplified operationally if the metallic phase were eliminated. Davies and Gray [39] demonstrated that this step could be carried out homogeneously with excess Fe(II) in a strong phosphoric acid solution which also contained nitric and sulfamic acids. The excess Fe(II) was oxidized to Fe(III) by reaction with the $HNO_3$ after addition of molybdate catalyst, and the resulting $HNO_2$ was removed by reaction with the sulfamic acid, the reaction producing $N_2$. This feature was necessary to prevent nitrite catalysis of the nitric acid oxidation of Fe(II) and U(IV). Goldbeck and Lerner [40] later modified the method to incorporate a coulometric step. A solution of vanadium(IV) was added to the uranium solution following the conversion of excess Fe(II) to Fe(III) and vanadium(V) was electrogenerated at gold or platinum electrodes. The end point was detected potentiometrically. Because of the possibility of V(IV) reacting with U(IV) to yield V(III) and U(VI), it was not clear whether the electrogenerated V(V) was being used to titrate U(IV) or V(III), but this was of no consequence since the same number of equivalents of V(V) would be required for the titration in either case. Solutions placed in the electrolysis cell typically contained about 100 mg of uranium and the determinations were precise and accurate to better than 0.1%. The method has been automated (a serious consideration in view of the radioactivity involved)

[41] and scaled down to 5- to 40-mg samples [42]. Chitnis et al. [43] have developed a method which combines features of the Malinowski and Goldbeck-Lerner procedures. Instead of adding V(IV) to the reduced U(VI) solution, these authors add a known excess of $K_2Cr_2O_7$ to oxidize the U(IV) and titrate the dichromate left with Fe(II) electrogenerated from the Fe(III) present in solution.

*Titrations with Electrogenerated Uranium Species*

Although the first publication appeared in 1955, few papers have appeared subsequently concerning the constant-current electrogeneration of uranium species. Shults, et al. used electrogenerated U(IV) to titrate dichromate ion and ceric ion [44]. Platinum electrodes were used as the generator cathode, isolated anode, and as the indicator electrode for a potentiometric end point. The sulfuric acid supporting electrolyte (0.1-0.5 M) was 0.1-1M in uranyl sulfate. Pretitration was used and solution volumes were about 7 mL. Milligram quantities of Ce(IV) and 100-μg quantities of chromium as dichromate were titrated with average errors of a few ppt, using current densities in the range 1-5 $mA/cm^2$. The Fe(III), U(IV) reaction was investigated but was too slow for a satisfactory titration. In later work, Edwards and Kern [45] found that by adjusting the solution pH to 1.5-2.5, the Fe(III), U(IV) reaction was rapid owing to the fact that the reductant is U(V). Results were accurate to ±2 ppt.

The most interesting electrogenerated uranium oxidation state is U(III) because it is the strongest reductant thus far successfully produced in aqueous solution for titration purposes. In 1972, Farrington and Lingane [29] described a procedure carried out at a mercury cathode. Titrations were performed under a nitrogen atmosphere. A potentiometric end point (mercury indicator electrode) was used. Uranium(IV) was titrated, so we have come full circle in this discussion of the constant-current coulometry of uranium. Convincing evidence was presented that the potential determining couple in the presence of excess U(VI) (i.e., prior to the end point) is the $UO_2^{2+}$, $UO_2^{+}$ couple. Oxygen had to be rigorously excluded from the titration cell. Because some hydrogen ion is reduced along with U(IV), the hydrochloric acid concentration should not exceed about 0.1 M, to maintain 99.9% current efficiency in a solution 44 mM in U(IV). Even better current efficiency can be attained at higher concentrations of U(IV). The oxidation of U(III) by hydrogen ion is another source of error in titrations with this reductant. At reasonable acidities, the reaction is pseudo-first order and the results of kinetic studies by Peretrukhin et al. [46] convinced Farrington and Lingane that the electrogeneration of U(III) was feasible. At 25°C, the rate constant is $1.43 \times 10^{-7}\ s^{-1}$ in 0.2 M HCl. The rate increases dramatically in the presence of high chloride concentrations or at more basic pH, but these conditions are easily avoided in the coulometric titration. It was found that the oxidation is catalyzed at a mercury surface. Nevertheless, milligram amounts of U(VI) could be titrated with an error of about +3% and several hundred milligrams were determined with less than +0.1% error. All of the titrations required about 600 s, and it would appear that milligram amounts could be titrated with better accuracy if shorter electrolysis times were used.

## G. Use of Computers

Advantages of coulometric titrations over volumetric titrations include (1) elimination of the need for standard solutions, (2) the use of unstable reagents, (3) higher accuracy and precision, (4) ease of adding small increments of titrant, and (5) ease of automation. All of these statements were true prior to the appearance of computers, but the development of computerized instrumentation holds the promise that the latter three advantages will be exploited more fully. A few examples from the literature will be presented to illustrate this. In some cases the objective has been automation. Minicomputers, programmable desktop calculators, and microcomputers have all been used, usually in conjunction with commercially available constant-current sources, pH meters, and so on. Keller and co-workers [47] used a Hewlett-Packard Model 2114B computer to control a Metrohm Model E 211A Coulometer. A DVM was interfaced to the computer for measurements using potentiometric acid-base end-point detection. A time-base generator was available for waiting loops, and switches were provided so more than one electrolysis cell could be used. Ebel et al. [48] used a Hewlett-Packard programmable desk calculator (HP 9830), together with a Metrohm Model E 524 Coulostat and a digital pH meter, to construct an automatic coulometric titrator. Wu and Malmstadt [49] incorporated a coulometric option in a versatile titrator which could also perform volumetric titrations and which employed spectrophotometric, potentiometric, or amperometric end-point detection. Their system was based on a microcomputer using the Intel 8080A microprocessor (Intel Corp., Santa Clara, California). A very interesting example is the automatic carbon and hydrogen determination reported by Houde et al. [50]. The sample is flash combusted in a stream of oxygen and the water produced is frozen in a trap, while the $CO_2$ is absorbed by a solution of 0.2 M tetrabutylammonium bromide in a mixture of 90 mL of DMSO, 10 mL of $H_2O$, and 3 mL of ethanolamine, contained in a coulometric titration cell. Oxides of nitrogen are removed from the gas train by a $MnO_2$ packing. It was suggested that the $CO_2$ reacts with the primary amine to yield the corresponding carbamic acid, which is then titrated with electrogenerated base. The cold trap containing the water is heated and the water vapor passed over solid carbonyldiimidazole which results in a quantitative yield of $CO_2$ which is then determined as above. A combination of a dual water trapping line, two titration cells, and a multiport valve allowed the carbon from one sample and the hydrogen from the previous one to be simultaneously determined.

The entire process was placed under computer control using a Hewlett-Packard 9825A desktop computer with 15K of memory. The only operator time required was for weighing the samples, which are then placed in an automatic sample tray. The computer was interfaced to the Mettler balance, so the sample weight and identification were stored in memory. The computer controlled the oxygen pressure, the temperature of the combustion furnace, heating and cooling of the traps, operation of the valve, and the coulometric cells, the latter through a commercial coulometer and potentiometric titration to a preset neutralization point. Data were received by the computer and the results after calculation were printed out as percent carbon and hydrogen. The sample tray held 23 samples and 4 h was required to complete the analyses. Two sets of 23 samples were run during a working day and a third set was started at the end of the day and left to run overnight. Results were precise to 0.1% for carbon and 1% for hydrogen. This work shows clearly the control capabilities of the computer. It is worth noting that automation is often cost effective by virtue of the manual labor replaced, but severe demands are placed on equipment in terms of reliability over long periods of continual use.

A second major area of computer use is interactive control of the experiment. Harrar and Pomernacki [51] describe a programmable constant-current source based on a Kepco Model 40-0-5B Power Supply (Kepco, Flushing, New York). Digital control signals are converted to analog with a 14-bit digital-to-analog converter (DAC) and the analog signal determines the magnitude of the current. The current range available was 0-500 mA and the digital resolution was 30 $\mu$A. The current can be set either by a computer or by front-panel switches. Further details of the circuits can be obtained from the original article. The stability of this current source is impressive. After initial calibration and zeroing, the measured current changed from 500.01 mA to 500.16 mA over a time span of $1\frac{1}{2}$ years. The percent change for the same time period for a current of about 12 mA was +0.06. No recalibration or zero adjustments were made. An electrode potential interface based on a differential measurement is described in the same article. The computer used was a Digital Electronics Corporation PDP-8/e.

The significance of instrumentation like the above lies in the problem of overshooting the end point. This presents a difficult task in the design of automatic titrators for the following reasons: (1) the titrant may not be mixed rapidly with the solution, (2) the rate of the titration reaction may be slow, (3) the end-point detection system may be slow to respond, and (4) time delays may occur in electronic circuits. The net result is that the detector response does not keep up with the addition of titrant and too much titrant is added. One approach to solving this problem is to slow down the rate of addition of titrant as the end point is approached. In the case of coulometric titrators, this is done by using smaller and smaller currents as the titration proceeds. A computer can be used to initiate the current adjustments based on the data it receives from the end-point detection system. This is particularly convenient when the titration is performed to a preset end-point signal. An example of this is found in the work of Earle and Fletcher [52]. Their titrator was based on the Intel 8008, an early 8-bit microprocessor. For acid-base titrations, the applied current was reduced linearly with the difference in pH, $\Delta$pH, between the measured value and the end-point pH. An algorithm compared $\Delta$pH with a set of rate functions specified in units of mA/pH. The magnitude of the current was then computed by multiplying $\Delta$pH by the rate function. This process was repeated every 65.536 ms. The coulombs passed were computed for each of these time intervals and summed until the end point was reached. The result was then used by the microprocessor to calculate the concentration of the unknown. The maximum initial current was 10.24 mA and the minimum was 0.1 mA or 10% of the initial current, whichever was larger. A delay time between 0.1 and 100 s could be selected as a wait period to ensure that equilibrium had been established at the end point.

The titrator was designed for process stream analyses and a unique ceramic slider valve was developed for sampling the stream. A similar valve was used to deliver supporting electrolyte containing the titrant precursor to the cell. The sliders were solenoid driven. The solution in the working compartment of the cell was stirred by a magnetically coupled impeller driven by a motor. All of these peripherals were under computer control. A digital readout was used to display data and readout from the titrator and keyboard entry of operating parameters was provided. Among the acids and bases titrated were HCl, NaOH, THAM, and KHP, all in a supporting electrolyte of 0.2 M $Na_2SO_4$. Solutions of the latter two were 0.01 M and the former two were 0.1, 0.01, and 0.001 M. The results of several hundred titrations of each of these solutions were precise and accurate to ±0.2% relative. The titrator was also capable of consecutive end-point determinations and an astonishing number of alkalinity determinations of tap water were made, the number of titrations being about 300,000. All results obtained agreed with a standard method to within ±0.5%.

Very little attention appears to have been directed thus far toward utilizing the timing capabilities of computers along the lines suggested by the scheme presented in Figure 20.6. Much of the hardware required to implement the digital timing could be replaced by software. A combination of programmed timing and programmed current might produce an effective and efficient titrator.

## H. The Future

The evidence cited in this chapter demonstrates that constant-current coulometric titrations are playing a significant role in chemical analysis. The electron is an excellent primary standard and it can be dispensed and measured with relative ease. For these reasons it is important that the method be pushed to the determination of submicrogram amounts of materials on a routine basis. Microcomputers will play an increasing role in this effort. More sensitive methods of end-point detection need to be developed. A major weakness, common to titrimetric methods, is a lack of selectivity. Perhaps this can be improved by using coulometric titrations in tandem with separation techniques. Finally, more attention should be focused on work in nonaqueous solvents.

## REFERENCES

1. L. Meites, *Polarographic Techniques*, 2nd Ed., Wiley-Interscience, New York, 1965, pp. 550-554.
2. J. J. Lingane, *Electroanalytical Chemistry*, 2nd ed., Interscience, New York, 1958, pp. 511-513.
3. *Publ. 272*, Standard Electric Time, Division of Johnson Service Co., Springfield, Mass.
4. H. V. Malmstadt and C. G. Enke, *Digital Electronics for Scientists*, W. A. Benjamin, Menlo Park, Calif. 1969.
5. H. V. Malmstadt, C. G. Enke, and S. R. Crouch, *Electronics and Instrumentation for Scientists*, Benjamin-Cummings, Menlo Park, Calif., 1981.
6. T. P. Sifferlan and V. Vartanian, *Digital Electronics with Engineering Applications*, Prentice-Hall, Englewood Cliffs, N.J., 1970.
7. R. Herzog, *TEKscope*, *4*(2):6 (1972), Tektronix, Inc., P.O. Box 500, Beaverton, Oreg.
8. D. Fullagar, *Electron. Des.* *13*(6/21):98 (1973).
9. Ref. 5, Chaps. 6 and 7.
10. L. B. Jaycox and D. J. Curran, *Anal. Chem.* *48*:1061 (1978).
11. D. D. DeFord, C. J. Johns, and J. N. Pitts, *Anal. Chem.* *27*:938, 941 (1951).
12. Ref. 2, pp. 522-528.
13. *Ibid.*, 2, Chaps. V to VII.
14. J. N. Butler, *Ionic Equilibria: A Mathematical Approach*, Addison-Wesley, Reading, Mass., 1964.
15. H. A. Laitinen and W. E. Harris, *Chemical Analysis*, 2nd ed., McGraw-Hill, New York, 1975.
16. H. V. Malmstadt, C. G. Enke, and S. R. Crouch, *Electronic Measurements for Scientists*, W. A. Benjamin, Menlo Park, Calif., 1974, pp. 74-76.

17. G. E. Cadwgan, Jr. and D. J. Curran, *Microchim. Acta,* 461 (1977).
18. M. A. de Soto Perera and D. J. Curran, *Anal. Chim. Acta* *119*:251 (1980).
19. *Ibid.*, 263.
20. A. J. Bard, *Anal. Chem.* *40*:64R (1968).
21. *Ibid.* *42*:22R (1970).
22. D. G. Davis, *Anal. Chem.* *44*:79R (1972).
23. *Ibid.* *46*:21R (1974).
24. J. T. Stock, *Anal. Chem.* *48*:1R (1976).
25. *Ibid.* *50*:1R (1978).
26. *Ibid.* *52*:1R (1980).
26a. *Ibid.* *54*:1R (1982).
27. Ref. 2, pp. 536-616.
28. J. Mitchell, Jr., and D. M. Smith, *Aquametry*, Interscience, New York, 1948.
29. G. C. Farrington and J. J. Lingane, *Anal. Chim. Acta* *60*:175 (1972).
30. J. J. Lingane and R. T. Iwamoto, *Anal. Chim. Acta* *13*:465 (1955).
31. J. H. Kennedy and J. J. Lingane, *Anal. Chim. Acta* *18*:240 (1958).
32. G. W. C. Milner and G. Phillips, in *Electroanalytical Chemistry*, Vol. 10 (H. W. Nürnberg, ed.), Wiley, New York, 1974, p. 78.
33. W. N. Carson, *Anal. Chem.* *25*:466 (1953).
34. Ref. 2, pp. 543-544.
35. N. H. Furman, C. H. Bricker, and R. V. Dilts, *Anal. Chem.* *25*:442 (1953).
36. J. Malinowski, *Talanta,* *14*:263 (1967).
37. L. Pszonicki, *Talanta,* *13*:403 (1966).
38. G. C. Goode, J. Harrington, and W. T. Jones, *Anal. Chim. Acta* *37*:445 (1967).
39. W. Davies and W. J. Gray, *Talanta* *11*:1203 (1964).
40. C. G. Goldbeck and M. W. Lerner, *Anal. Chem.* *44*:594 (1972).
41. K. Lewis, Report, 1980, *NBL-294,* 14 pp. Available from the National Technical Information Service. *Chem. Abstr.* *93*:1250484 (1980).
42. W. G. Mitchell and K. Lewis, *NBS Spec. Publ. (U.S.)* *589*:140-146 (1980); *Chem. Abstr.* *93*:125139d (1980).
43. R. T. Chitnis, S. G. Talnikar, and A. H. Paranjare, *J. Radioanal. Chem.* *59*:15 (1980).
44. W. D. Shults, P. F. Thomason, and M. T. Kelley, *Anal. Chem.* *27*:1750 (1955).
45. K. W. Edwards and D. M. Kern, *Anal. Chem.* *28*:1876 (1956).
46. V. F. Peretrukhin, N. N. Krot, and A. D. Gel'man, *Radiokhim.* *12*:96 (1970).

47. U. Keller, J. Padel, H. Gamsjager, and P. W. Schindler, *Chimia 27*:90 (1977).
48. S. Ebel, J. Hocke, and S. Kalb, *Fresenius' Z. Anal. Chem. 288*:183 (1977).
49. A. H. B. Wu and H. V. Malmstadt, *Anal. Chem. 50*:2090 (1978).
50. M. Houde, J. Champy, and R. Furminieux, *Microchem. J. 24*:300 (1979).
51. J. E. Harrar and C. L. Pomernacki, *Chem. Instrum. (N.Y.) 7*:229 (1976).
52. W. E. Earle and K. S. Fletcher III, *Chem. Instrum. (N.Y.) 7*:101 (1976).

# 21

# Application of Electrochemistry to Pharmaceutical Analysis

MARVIN A. BROOKS / Hoffmann-La Roche Inc., Nutley, New Jersey



## I. INTRODUCTION

Analytical chemistry plays a critical role in the development of a compound from its synthesis to its marketing as part of a drug formulation. The instrumental methods for quantitation which are most commonly used in a pharmaceutical laboratory fall into four basic categories: chromatographic, spectrophotometric, electrochemical, and radiometric analysis. Among these four, electrochemical analysis is the least often employed. This is historically attributable to the lack of trained personnel and dependable commercial instrumentation. The relative neglect of electroanalytical chemistry is indeed unfortunate since many problems of pharmaceutical interest can easily be solved with a high degree of accuracy and precision employing this approach.

It will be the purpose of this chapter to examine the various problems presented to the analytical pharmaceutical chemist and to explore the possible role of electroanalytical chemistry in solving these problems.

## A. Development of a Drug

Analytical methods are required to assure the purity of starting materials and can be used to monitor a synthetic process to determine if it has reached completion. After synthesis, analysis is required to determine percent yield and to separate, isolate, and quantitate by-products of the reaction.

Analytical research receives the drug after synthesis and characterizes its physicochemical parameters to ensure that all pharmacological testing is done with pure material and to establish specifications so that future production lots will yield reproducible in vivo response. These specifications will include the purity of drug, identification and quantitation of other products of synthesis and degradation, the $pK_a$ of the drug, and physical characterizations such as thermal stability, optical isomers, surface properties, particle size, and crystalline form. These studies will include the stability and solubility of the drug in aqueous media as a function of pH.

Upon completion of the analytical profile, the drug is released for preliminary pharmacological and toxicological testing in laboratory animals. At this stage of development, the drug is usually administered in a simple preliminary formulation by parenteral and oral routes, to determine its pharmacological activity, its relative safety, and dose-response characteristics. During toxicological testing, blood, urine, and tissue samples are collected and analyzed to correlate both pharmacological response and toxicity to the blood level. These data aid in the assessment of adequate absorption, dose-response behavior, metabolic effects such as accumulation, enzyme induction or attainment of steady-state conditions, intra- and interspecies variations, and the relationship of drug levels to symptomatology [1].

In addition to toxicological studies, various in vitro studies are performed to determine if any physicochemical factors such as solubility, dissolution or release rate, permeability, gastrointestinal degradation, and rapid biotransformation will affect the drug's oral absorption characteristics after administration of a given dose in a pharmaceutical dosage form [2]. Assays must be devised to assess the solubility and dissolution rate of drugs as a function of pH within the physiological pH 1-8 range at 37°C, and to evaluate the drugs' permeability characteristics across the intestinal mucosa. The utility of these in vitro studies has been reported by Kaplan [3]. Metabolic transformation is evaluated in vitro by incubating the drug in a liver microsomal enzyme fraction (9000 × g supernatant) at 37°C, which was separated from the liver homogenates [4]. Analytical methodology is needed for these studies to quantitate the rate of biotransformation, and to isolate and quantitate the metabolites formed.

The design of a suitable dosage form at a clinically effective and safe dosage level is based on the data obtained from the pharmacological and toxicological studies together with biopharmaceutical data gained from the in vitro studies. Such dosage forms are evaluated for bioavailability following oral administration to laboratory animals at a relatively high dose (typically 10 mg/kg body weight in the dog). The bioavailability, rate, and extent of absorption of this formulation are assessed by comparing its blood level, urine, and fecal elimination profile to that obtained from an intravenous and oral solution of the drug at the same dosing level [3]. The dosage form exhibiting optimal bioavailability is used for dosing animals in toxicity studies, in several animal species, in which blood, urine, and tissue levels of drug and/or metabolites are monitored. If the drug is shown to be safe, it is then ready for limited testing in humans. An Investigational New Drug application (IND) is filed with the Food and Drug Administration (FDA) which describes the drug formulation (list of components, quantitative composition), the analytical procedures employed to establish and maintain standards of identity, strength, and purity, and all the preclinical information pertaining to the administration of the drug to laboratory animals.

With approval from the FDA the drug proceeds into phase 1 clinical testing, where it is administered to humans to determine its safe dosage range, pharmacological activity, side effects, metabolism, and distribution, absorption, and elimination characteristics. From these studies, samples of blood, urine, and feces are collected and analyzed to assess the preliminary bioavailability and pharmacokinetic profile in humans. (Pharmacokinetics is defined as the study of the kinetics of absorption, distribution, metabolism, and excretion of drugs and other endogenous substances.) During phase 2 and phase 3 testing of a drug for the safety and effectiveness in the diagnosis, treatment, or prophylaxis of a given disease or condition, biological fluids are collected to obtain comprehensive pharmacokinetic information and to correlate such data with activity of the drug. At the completion of the clinical trials, all the data are accumulated about the drug from its initial synthesis to its clinical testing and submitted to the FDA as a New Drug Application (NDA) for approval of the drug as a marketable product.

Subsequent to marketing, constant quality control monitoring of the product from lot to lot must be maintained to provide the patient with a reproducible product. To guarantee this reproducibility in the dosage form, it is necessary to develop assays that will analyze reagents, bulk chemicals, and intermediates which go into making the dosage form, and also to obtain and analyze representative samples of the dosage form.

In summary, it is apparent that many analytical techniques at varying levels of sensitivity are required to solve analytical-pharmaceutical problems. Analysis of reagents, bulk chemicals, intermediates, and the final dosage form requires rapid, noncomplicated assays of relatively low sensitivity which (due to the large number of samples involved) can be readily automated. These assays use titrations, spectrophotometric techniques, or high performance liquid chromatography (HPLC). Analysis of by-products from synthesis and degradation products, the assay of drugs from in vitro testing, and the assay of drugs in biological fluids from high-dose (≥10 mg/kg body weight) animal studies require assays of intermediate detection limit (typically about 1.0 μg/mL of sample) which are fairly specific to be able to distinguish possible metabolites or breakdown products from the parent drug. Procedures that satisfy these requirements include gas chromatography employing a flame ionization detector, HPLC employing an ultraviolet detector, and thin-layer scanning densitometry. Thin-layer chromatography (TLC) followed by elution of the compounds from the plate and spectrophotometric, spectrofluorometric, or voltammetric quantitation is often used.

The most demanding of the pharmaceutical analytical assays are those which require the measurement of drugs and their metabolites in biological fluids in humans following the administration of low doses of drugs. These assays must be of high specificity and capable of measuring blood levels of drugs and their metabolites with subnanogram detection limit, and the urinary excretion of drugs at levels with a detection limit of 50-100 ng/mL urine. Gas liquid chromatography (GLC) with electron capture, nitrogen flame ionization or mass spectrometric detection, HPLC with ultraviolet or fluorescence detection, and radioimmunoassay are the most useful procedures in measuring these blood levels. Voltammetry, coupled with TLC or HPLC to ensure specificity, has recently been shown to be extremely useful in measuring blood levels and urinary excretion of drugs following low doses. This is a particularly valuable technique for the determination of polar metabolites which cannot be readily derivatized for gas chromatography.

## B. Electroanalytical Methods

The choice of the appropriate technique to solve an analytical-pharmaceutical problem is often controlled by the sample matrix, and the amount of preparation that is required before the analytical measurement can be made. If the material is an organic bulk substance or intermediate, it is usually possible to dissolve the compound in a suitable organic solvent and analyze the sample by TLC, GLC or HPLC. A dosage form may be dissolved in an aqueous medium, filtered, and analyzed directly by an electrochemical technique or extracted into an organic solvent and analyzed chromatographically or spectrophotometrically. A biological specimen cannot usually be analyzed directly (an exception will be discussed in the voltammetric techniques section) but rather requires selective extraction and cleanup, separation of the parent drug from any metabolites and/or other substances present, and a sensitive and specific detection system for accurate quantitation. A detailed discussion of each step required for the analysis of drugs in biological fluids has been reported by de Silva [5].

The final selection of the analytical technique is dependent on the intrinsic physicochemical properties of the pure compound or its easily formed derivatives. Thus compounds or their derivatives should be processed through an analytical screening program to examine their spectrophotometric, chromatographic, and electrochemical properties which might be amenable to chemical analysis. Based on the signal response per nanogram or microgram of a compound, a projected limit of sensitivity per milliliter or gram of sample is calculated. Many analytical procedures can be eliminated at this stage of method development.

The use of spectrophotometric and chromatographic techniques has been extensive through the pharmaceutical literature, with little or no emphasis on electrochemical techniques. This discussion will now turn to the potential use of electroanalytical chemistry to solve analytical-pharmaceutical problems.

In the analytical screen described above, the four electrochemical techniques—voltammetry (i.e., direct-current, alternating-current, fast-scan, and pulse polarography along with voltammetry at solid electrodes), coulometry (constant current and potential), potentiometric titrations (including the use of ion-specific electrodes), and amperometric titrations—are most commonly employed. Each of these electroanalytical techniques and their practical range of usefulness is presented in Figure 21.1. Voltammetric methods can be employed in a wide concentration range and can be used for the analysis of drugs in dosage forms at low sensitivities, and the analysis of drugs in biological fluids at highest sensitivities. Constant-current coulometry possesses intermediate sensitivity and can measure drug levels in biological fluids following only relatively high dosages of drugs. The techniques of constant-potential coulometry, amperometry, and potentiometry are restricted by their limits of sensitivity to the measurement of drugs in dosage forms and bulk materials.

In the analysis of drugs in dosage forms, electroanalytical chemistry has been shown to be an exceptional method, and quite often superior to classical wet methods or spectrophotometric methods. For the most part this is due to the great ease of sample preparation and lack of interferences from incipients in the dosage form. Dosage forms are typically pulverized (if necessary), dissolved in an aqueous solvent, filtered, and analyzed directly by one of the four common

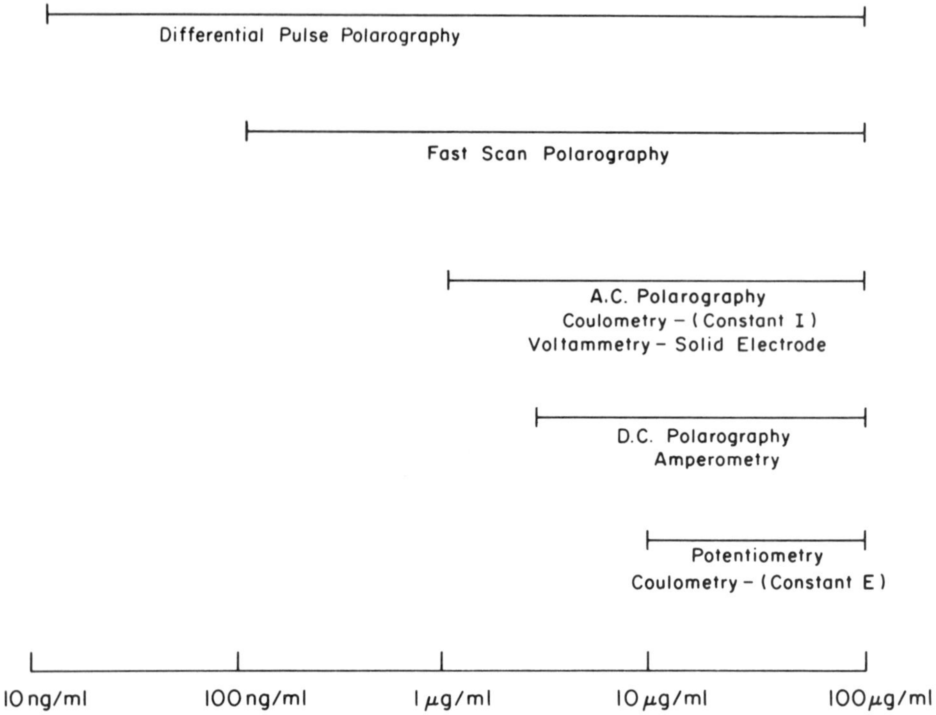


**Figure 21.1** Range of practical usefulness of electroanalytical techniques.

electroanalytical methods. Coulometric methods have been shown to be particularly useful since they are "absolute" methods requiring no chemical standards. More recently, fast-scan and differential pulse polarography have become extremely useful for the measurement of drugs in biological fluids and tissues. This can be directly attributed to the introduction of highly sensitive commercial instrumentation which yields easily interpretable data for routine quantitation at low levels. The establishment of absolute specificity is the major limitation of these two highly sensitive polarographic techniques. In an in vivo study the assay must be able to measure the compound administered without any interferences from the biological matrix or any metabolic products. To ensure specificity, TLC separation is often coupled with polarographic measurements. These TLC separation steps often introduce time-consuming and tedious spotting, developing, and elution procedures into the electrochemical assays. To avoid some of the problems associated with TLC, a liquid chromatograph can be coupled directly to an electrochemical detector. This approach has been demonstrated to have utility for a number of pharmaceutical problems (see Chap. 22).

As with any analytical procedure the reproducibility and accuracy of an electroanalytical assay must be verified for the intact compound and/or a chemical derivative used for quantitation. Methods developed should be uncomplicated for use on a routine basis by the general scientific community.

The discussion will now center on electroanalytical procedures commonly used and demonstrate their utility in solving analytically oriented pharmaceutical problems in the various stages of drug development. The examples chosen are by no means a comprehensive review for each technique. Since discussions of the pharmacological activity will be at best very limited and the structures of all drugs will not be presented, several texts are recommended [6,7]. The purpose of this chapter is to suggest the types of problems that can be solved using electrochemical techniques. However, it should be noted that although most problems in pharmaceutical analysis are not amenable to an electrochemical approach, there are a large number of examples where electrochemistry is the method of choice.

## II. COULOMETRIC METHODS

### A. Controlled-Potential Coulometry

Controlled-potential coulometry (Chap. 3, Sec. IV.F) has been used in pharmaceutical analysis for determination of pure compounds and compounds in their dosage form, and in studies to determine the nature of electrode reactions. Controlled-potential coulometry as an analysis tool has exhibited its greatest utility in the measurement of inorganic ions and nitro and halogenated compounds [8-11]. The number of applications to pharmaceutical problems are few, which attests both to unfamilarity of the technique and its lack of specificity.

The amine oxides of atropine, brucine, 3-bromopyridine, scopolamine, strychnine, and chlordiazepoxide have been assayed at a level of $8 \times 10^{-5}$ M in solvents of varying pH by this technique using a mercury pool cathode [12]. Solvents utilized were sulfuric acid-potassium sulfate systems of varying pH and an acetic acid-sodium acetate system of pH 4.63. Recoveries ranged from 98 to 100% for the six compounds investigated at the $5 \times 10^{-5}$ M level. Smaller samples could not be used because of weighing errors, and larger samples led to unacceptably long coulometric reduction periods. Pyridine-N-oxide, 2,6-lutidine-N-oxide, and quinoline-N-oxide gave meaningless recoveries since their amine oxide reduction overlapped the catalytic hydrogen wave.

An assay for the determination of vitamin $K_3$ (2-methyl-1,4-naphthoquinone) by a combination of both constant-potential and constant-current coulometry has been reported [13]. The assay requires the two-electron reduction of the compound to the corresponding hydroquinone at a mercury pool electrode (E = −0.60 V vs. SCE) in acetate buffer pH 5.9, followed by the coulometric titration of the reduction product with electrogenerated Ce(VI). The method is preferable to the standard method requiring preliminary reduction to the quinone by zinc dust in acid medium, followed by titration with standard Ce(IV) solution. It is capable of low-level determination (1-2 mg) of this vitamin in pharmaceuticals, biological fluids, and foods.

The N-substituted phenothiazines, which are employed as antipsychotic drugs, have been measured by controlled-potential coulometric oxidative analysis in sulfuric acid at the rotating platinum wire mesh electrode [14]. The compounds could be quantitatively oxidized to a cation radical or to a sulfoxide by selection of suitable acid concentrations and applied potentials. Electroreduction of the free radicals occurs at approximately +0.25 V vs. SCE on a platinum electrode. In the case of the sulfoxides, a single two-electron reduction step occurred at ca. −0.95 V vs. SCE on a mercury pool cathode. The determinations showed good reproducibility and an accuracy of ca. ± 0.1% was obtained for sample concentrations of $10^{-3}$ M or greater. The compounds examined were chlorpromazine, chlorpromazine sulfoxide, promethazine, promazine, triflupromazine, trifluoperazine, prochlorperazine, and thioridazine (see Fig. 21.2).

The oxidation of ascorbic acid at a platinum electrode in potassium acid phthalate buffer (pH 6) has been reported at +1.09 V vs. SCE [15]. Under these conditions 15-100 mg of ascorbic acid can be determined with an error of ±0.7 mg with no interference from citric, succinic, and tartaric acids.

Controlled-potential coulometry is quite often used as an ancillary technique to polarography to calculate the number of electrons involved in a polarographic reduction. The determination of an n value for some antibacterial aminoacridines using this technique demonstrated that with the exception of 5-aminoacridine, all other substituted aminoacridines undergo a one-electron reduction which corresponds to the first step of the current-voltage curve [16]. This fact substantiated the formation of an intermediate stable semiquinone radical which is of great importance in the relation to the free-radical mechanism of bacteriostasis by the aminoacridines. Analysis was performed in 0.5 M potassium sulfate at a mercury pool cathode using 1-5 mg of sample.

The coulometric method was also applied to the determination of the number of electrons involved in the two-step polarographic reduction of chlordiazepoxide [17]. The reduction of a $2.5 \times 10^{-5}$ M solution of the compound in 0.1 N $H_2SO_4$ yielded n values of 1.97 and 4.06, respectively, at a mercury pool cathode employing working electrode potentials of −0.45 V and −0.90 V vs. a Ag/AgCl wire reference electrode.

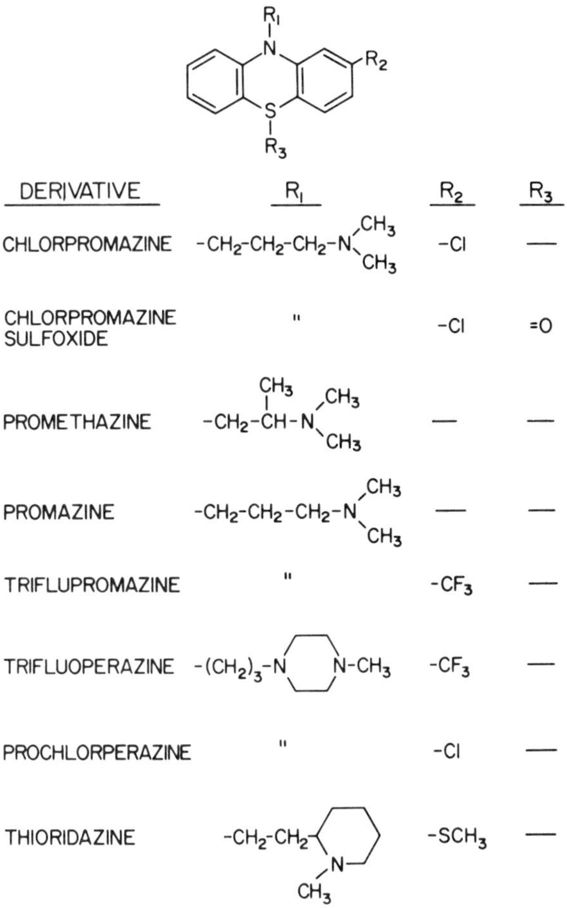


Figure 21.2 Phenothiazines assayed by controlled-potential coulometry. [From Ref. 14, reprinted with permission from F. H. Merkel and C. A. Discher, *Anal. Chem.* *36*:1640 (1964). Copyright 1964 American Chemical Society.]

These experiments clearly demonstrated that two electrons are involved in each of the two polarographic steps, which correspond to the reduction of the amine oxide and the 4,5-azomethine functional group.

The mechanism of reduction of an N-oxide, the antibiotic myxin, 6-methoxy-1-phenazinol-5,10-dioxide, has also been examined [18]. Below pH 3, the protonated species of myxin is reduced in a single four-electron step to 6-methoxy-1-phenazinol, whereas between pH 3 and 9, two two-electron steps are observed which are most likely due to the formation of an intermediate, which is in turn reduced to

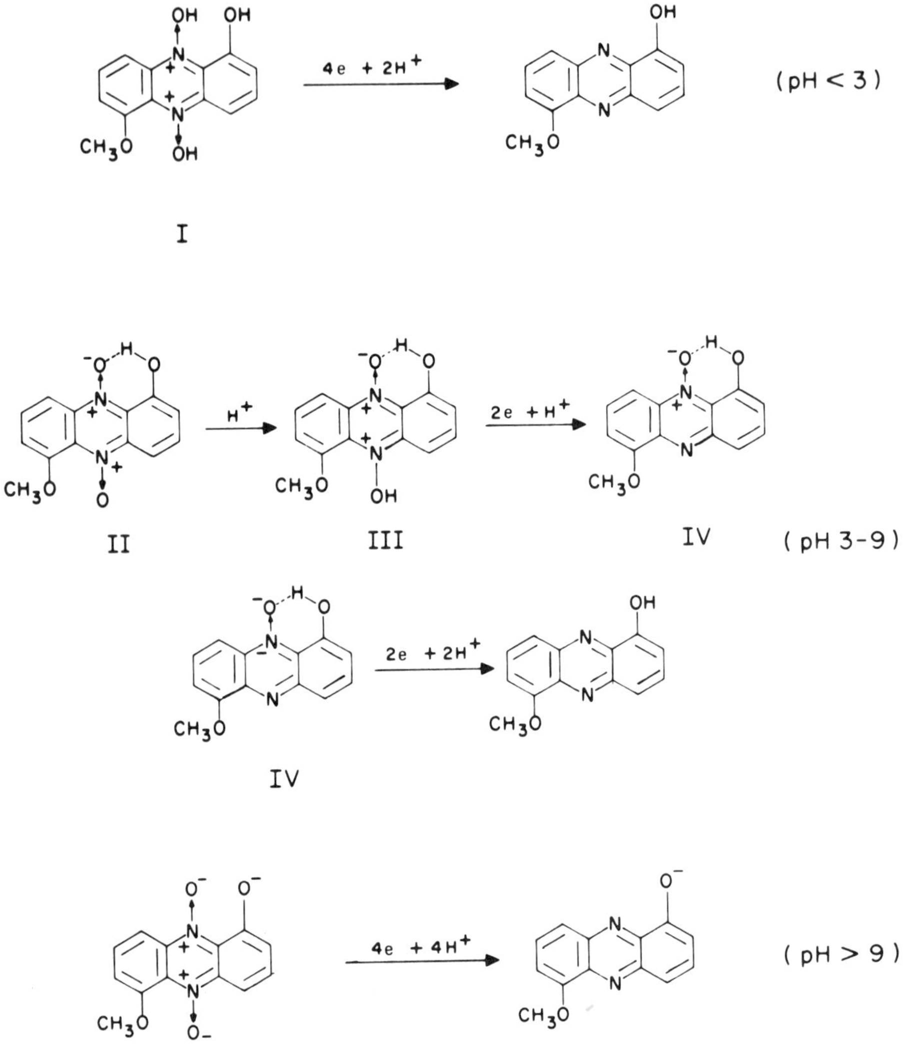


Figure 21.3 Mechanism of reduction of the antibiotic myxin. (From Ref. 18.)

6-methoxy-1-phenazinol. At pH values greater than 9, a single wave is observed, apparently due to the four-electron reduction of the anion to 6-methoxyphenazinol. The postulated mechanism is shown in Figure 21.3.

## B. Constant-Current Coulometry (Coulometric Titration)

Coulometric titrations can be conveniently carried out in the concentration range 1 μg-100 mg/mL for practically any titration which can be done volumetrically. Since current and time can be determined easily and accurately, the titrations are generally very accurate and precise (Chap. 20). In the microgram range, coulometric titrations show clear advantages over volumetric techniques, in that they eliminate problems of reagent stability and standardization. However, at concentrations greater than 100 mg/mL coulometric titrations have problems, such as high currents needed for completing the titration in a reasonable amount of time and the extremely high molar strength of generating electrolytes needed to sustain 100% current efficiency. In such cases, dilution of the sample is required.

An advantage of the technique is the use of an electrical standard to replace chemical standards and the problems associated with their preparation and stability. The coulometric titration also permits the generation of reagents such as copper(I) or bromine which are difficult to employ as standard solution, or others, such as silver(II) or chlorine, which are virtually impossible to use in any other way. A disadvantage of the coulometric titration is its lack of specificity.

Coulometric titrations have been employed for redox, acid-base, precipitation, and complexation titrations of organics and inorganics in both aqueous and nonaqueous solvents [8-11]. The basic theory [19] and applications [20] to the analysis of pharmaceuticals have been described. The present discussion will be divided into compounds which are determined by generation of a halogen and those determined by generation of other reagents. Due to the sensitivity restrictions of the technique, very few examples have been reported for the use of coulometric titrations of drugs in biological materials. Most research has been directed toward use of the titration technique to determine bulk materials and drugs in dosage forms.

### *Generation of Halogens*

The halogens are the most widely employed oxidation-reduction titrants for coulometric titrations. Bromine can be easily generated with 100% current efficiency from a solution consisting of dilute acid and 0.2 M sodium bromide. Bromine is generated at the platinum anode, while hydrogen is generated at a platinum cathode which is isolated in a fritted glass tube. Chlorine, a much stronger oxidizing agent than bromine, can be generated in solutions of acids and chloride salts; however, strict control of reagent concentration is required to guarantee a current efficiency of 100%. Iodine, the weakest oxidizing agent, can be generated from iodide solutions ranging from strongly acid to approximately pH 8 with 100% current efficiency.

The titration of pharmaceuticals employing electrogenerated halogens has been extensively reported by Kalinowska and Kalinowski and their associates. Their work has been recently reviewed by Kalinowska [21], in which the compounds analyzed were classified according to the electrogenerated halogen used. Isoniazid, cyclobarbital, hexobarbital, folic acid, and the salts of p-aminosalicylic acid were determined with electrogenerated chlorine. Reserpine was reported to consume 4 mol of electrogenerated chlorine and was determined in the range 100-300 μg in dosage forms. Other compounds determined in pure form or in dosage forms in the range 0.2-1 mg by electrogeneration of chlorine included caffeine, theobromine, theophylline, carbutamide, benzylpenicillin, phenylbutazone, and methionine. Procaine, pontocaine, thiopental, isoniazid, ethylmorphine, codeine, quercetin, p-aminobenzoic acid, oxytetracycline, chlorotetracycline, sulfonamides, sulfanilic acid, and phenylbutazone have been titrated with electrogenerated $Br_2$ from an acidic solution of a bromide salt. The electrogeneration of the excess hypobromate ion in alkaline solution of bromide salts was reported for the back-titration of methylthiouracil with $As_2O_3$ and $KMnO_4$ in pure form and tablets. Bromine titration was also used for the titration of benzylpenicillin after alkaline hydrolysis. Electrogenerated iodine was employed for the coulometric titration of thiamine in capsules and suspensions (3-9 mg), sodium noraminopyrine methanesulfonate, ascorbic acid, and benzylpenicillin after alkaline hydrolysis and phenoxymethylpenicillin. Most of the work of these authors employs four platinum electrodes: two for generation and two for a biamperometric end-point determination. Kalinowski has also reported assays for the indirect coulometric titrations of papaverine [22] and strychnine [23]. These methods [22-23] require the precipitation of the compound to be assayed with excess Reinecke salt (ammonium tetrathiocyanodiammonochromate) followed by titration of the excess salt with electrogenerated chlorine.

Procarbazine [N-isopropyl-α-(2-methylhydrazino)-p-toluamide] used in the treatment of Hodgkin's disease has been determined with electrogenerated bromine [24]. The assay is not specific in the presence of most intermediates and decomposition products since the electrogenerated bromine not only oxidizes the hydrazo group but also substitutes onto the phenyl ring. A recent report described the coulometric iodination of the same compound. This approach is more specific in that only oxidation of the hydrazo moiety to the azo group occurred with no ring substitution [25]. Iodine was generated from 0.5 M KI in a buffered sodium bicarbonate solution (pH 8.4). Determination of procarbazine hydrochloride in dosage forms was evaluated. The average for eight determinations was 51.31 mg per capsule (102.6% of claim, based on 50 mg). The standard deviation was ±0.17 mg per capsule or ±0.3%.

An automated constant-current coulometric system employing electrogenerated iodine for assay of ascorbic acid or sodium ascorbate has been reported [26]. Using this apparatus, 25 samples of 30 mg could be determined in 2.5 h with an accuracy and precision of ±0.3%. The automated system demonstrates accuracy and precision which is equivalent to or exceeds the manual USP method, is significantly more rapid, and eliminates the need for preparation, standardization, and storage of titrant.

Considerable attention has been given to the coulometric determination of sulfa drugs with electrogenerated bromine. Sixteen different sulfa drugs and two anesthetics, benzocaine and tetracaine, have been assayed directly with electrogenerated bromine [27]. Other workers have found that these compounds react too slowly with bromine and describe methods where the drug is first reacted with excess bromine, and then the excess bromine is determined coulometrically with Cu(I) [28], or employing arsenic trioxide in a coulometric back titration [29].

Seven phenothiazines have been titrated with electrogenerated bromine in the range 0.1-3 mg with an average error of less than 1% [30]. The assay utilizes the oxidation of the phenothiazine in two one-electron steps to a cation free radical and sulfoxide, respectively.

The analysis of prostaglandin $F_{2\alpha}$ content of pharmaceutical preparations with triangle programmed bromimetric titration in flowing systems has recently been described [31]. Using this technique a sample solution stream and chemical reagent stream are mixed and the reaction is monitored biamperometrically.

The use of constant current coulometry for the determination of drugs in biological fluids are few, basically due to sensitivity restriction. Monforte and Purdy [32] have reported an assay for two allylic barbituric acid derivatives, sodium seconal and sodium sandoptal, with electrogenerated bromine as the titrant and biamperometry for end-point detection. Quantitative bromination required an excess of bromine; hence back titration with standard arsenite was performed. The assay required the formation of a protein-free filtrate of serum with tungstic acid, extraction into chloroform, sample cleanup by back extraction, followed by coulometric titration with electrogenerated bromine. The protein precipitation step resulted in losses of compound due to coprecipitation. The recovery of sodium seconal and sodium sandoptal carried through the serum assay was approximately 81 and 88%. Samples in the concentration range 7.5-50 μg/mL serum were analyzed by this procedure.

### *Miscellaneous Reagents*

Other typical reagents generated for coulometric titrations are hydrogen and hydroxyl ions, redox reagents such as ceric, cuprous, ferrous, chromate, ferric, manganic, stannous, and titanous ions, precipitation reagents such as silver, mercurous, mercuric, and sulfate ions, and complex-formation reagents such as cyanide ion and EDTA [8-10].

The coulometric titration of weak bases with electrogenerated $H^+$ has been successful for nicotinamide, sodium salicylate, metronidazole, sodium veronal, isonicotinic acid, and other compounds, in 0.1 M sodium perchlorate in a mixture of acetic acid and acetic anhydride (1:6) [33]. The end point of the titration was noted photometrically using a malachite green indicator. Approximately 5 mg of each compound was determined with an average deviation of ±0.3%.

The most commonly employed oxidation reagent for the determination of organics is the ceric ion. Employing a generating reagent of 3 M sulfuric acid and 0.05 M cerous sulfate, ceric ion was generated at a platinum electrode for the determination of phenothiazines [30, 34]. The same compounds were also determined using electrogenerated manganic ion [30]. Results for the titrations of phenothiazines employing electrogenerated bromine (discussed above), ceric ion, and manganic ion demonstrated that quantities of 0.1-3 mg could be determined with an average error of less than ±1%. Vitamin $K_3$ (2-methyl-1,4-naphthoquinone) was also determined by the same authors using electrogenerated ceric ion [13]. The quinone was first coulometrically reduced at a mercury pool electrode or with zinc and then titrated with electrogenerated ceric ion.

Complex formation coulometric titrations of pharmaceuticals have been reported employing electrogenerated mercuric ion and silver ion. Thiols and disulfides, such as methimazole, cysteine, glutathione, and thiopentone sodium, were titrated with mercuric ion generated by oxidation of a mercury pool electrode [35]. A recent study reports the coulometric determination of barbiturates, including the sodium salts of phenobarbital, barbital, sandoptal, and secobarbital, with $Hg^{2+}$ employing a biamperometric end-point detection system [36]. The compounds were determined in a range of approximately 0.2-2 μeq in a supporting electrolyte of 0.5 M $NaClO_4$ or $KNO_3$ in 20% acetone. Structural analysis of the coulometrically formed product of sodium phenobarbital indicated that a polymeric compound containing continuous Hg-N linkages was formed. Precipitation determination of organic bases and alkaloids as their tetraphenylborates has been described [37]. Two to ten milligrams of these compounds, including papaverine, atropine, pilocarpine, thiamine, sparteine, yohimbine, procaine, strychnine, tetracaine, eserine, quinine, and brucine, were precipitated with an alkaline sodium tetraphenylborate reagent, and separated from the excess reagent. The precipitates were dissolved in a solution of 0.4 M sodium nitrate in 30% v/v acetic acid. The tetraphenylborate content of the precipitate was determined with $Ag^+$ ion generated from a silver anode.

## III. VOLTAMMETRIC METHODS

Whereas potentiometric and coulometric titration methods are fairly universal and can be applied to a wide variety of problems, voltammetric methods are distinctly limited to easily reducible or oxidizable compounds. On the other hand, voltammetric methods are clearly the most useful for low-level quantitation and have a wide linear dynamic range (see Fig. 21.1). Voltammetry is defined here as electrolysis at a microelectrode which is limited by the mass transfer rate at which molecules move from the body of the solution to the electrode. The current is followed as a function of the potential applied using three-electrode cells of 1-50 mL working volume. For most reductions the working electrode is a dropping mercury electrode (DME) or a hanging mercury drop electrode (HMDE). Solid electrode voltammetry which is typically employed for anodic (oxidation) processes can use a rotating platinum electrode (RPE), a wax-impregnated graphite electrode (WIGE), a carbon paste electrode (CPE), or a glassy carbon electrode (GCE).

Since the voltammetric techniques demonstrate a large linear dynamic range ($10^{-3}$-$10^{-8}$ M), any one method can readily be applied to the analysis of pharmaceuticals in bulk and in dosage forms. Solid electrode voltammetry (which excludes polarography) has an inherent detection limit of approximately $10^{-4}$-$10^{-5}$ M and is consequently limited to these applications. Small-amplitude polarographic techniques (ac and pulse) can be applied to the measurement of impurities in bulk and dosage forms, and to the assay of drugs and metabolites in biological fluids following high-dose administration where levels are at least $10^{-6}$ M (1-μg/mL sample). Only the techniques of fast-scan polarography and differential pulse polarography with limits of detection of approximately 100-200 and 10-20 ng/mL of biological fluids, respectively, are capable of measuring drugs in biological fluids following therapeutic administration on a routine basis. Levels as low as several ng/mL of some pharmaceuticals have been measured by coupling a liquid chromatograph with a low-volume electrochemical dedector (LCEC) (see Chap. 22).

The appeal of the voltammetric methods of analysis for organic compounds is attributable to their simplicity and rapidity. An extremely large number of organic compounds are either directly reducible at the DME or oxidizable at a solid electrode or can be readily derivatized chemically to yield electroactive derivatives. The functional groups which show excellent voltammetric properties include the nitro, nitroso, quinone, azo, azoxy, azomethine, activated carbonyls, and activated double bonds [38]. For the most part, the voltammetric activity of a compound can be deduced from an examination of its functional groups. However, to ensure that electrochemical activity of a compound is not missed, the analytical screen should include voltammetric scans of the compound in aqueous supporting electrolytes (it is convenient to use 0.1 M HCl, 0.1 M NaOH, and buffers with pH values of 3, 5, 7, 9, and 11) using both the DME and glassy carbon electrode. If the compound is insoluble in aqueous media, then organic cosolvents such as alcohols, dioxane, acetonitrile, dimethyl sulfoxide, and dimethylformamide must be employed. Occasionally, pure nonaqueous solvents of high dielectric constant, such as dimethylformamide, acetonitrile, dimethyl sulfoxide, pyridine, or glacial acetic acid, containing tetraalkyl salts may be required to attain highly negative potentials (greater than −1.8 V vs. SCE) without supporting electrolyte decomposition. Generally, nonaqueous polarography is not successful at the microgram level because of impurities in the supporting electrolytes or solvents which tend to distort or mask the voltammetric peak. The reproducibility of determinations in nonaqueous media is also problematic at these low levels. In the event that no electroactivity is discovered in aqueous or nonaqueous media, derivatization methods can be used to transform the compound into an electroactive moiety. Derivatizations most commonly used include nitration, nitrosation, condensation, addition, substitution, oxidation, hydrolysis, and complex formation.

The number of voltammetric assays far outnumbers all other electroanalytical assays reported for pharmaceuticals [39-45]. In this section the applications of voltammetric assays will be discussed in reference to the various stages of drug development. Particular emphasis will be placed on the determination of drugs in biological fluids, an important area which has recently come of age through the advent of commercial instrumentation capable of performing pulse polarography and LCEC.

### A. Synthesis and Characterization of Pure Compound

Voltammetry can be employed to monitor a reaction in progress or to measure the percentage yield at completion. An example of this type of study was the use of dc polarography to monitor the formation of N-desmethyldiazepam from the condensation reaction of 5-chloro-2-aminobenzophenone with ethyl glycinate [46]. The N-desmethyldiazepam was subsequently methylated to yield diazepam, the compound contained in the tranquilizer Valium (see Fig. 21.4). Polarography was performed on an aliquot of the condensation reaction in a McIlvaine buffer (pH 3.8) to determine the amount of N-desmethyldiazepam formed by simultaneous measurement of the decrease in wave height for the carbonyl in the benzophenone and the increase in wave height for the 4,5-azomethine in the N-desmethyldiazepam. The two analytical waves were separated by 200 mV. A similar procedure was described for monitoring the synthesis of chloramphenicol [47]. A dc polarographic assay was used to measure three nitro-function intermediates in the presence of each other.

Pulse polarography has been used to characterize the products of the photochemical oxidation of 3,4-dihydro-1-isoquinolineacetamide to 1-isoquinolineacetamide (see Fig. 21.5) [48]. After photochemical oxidation the polarographic activity which the compound had previously

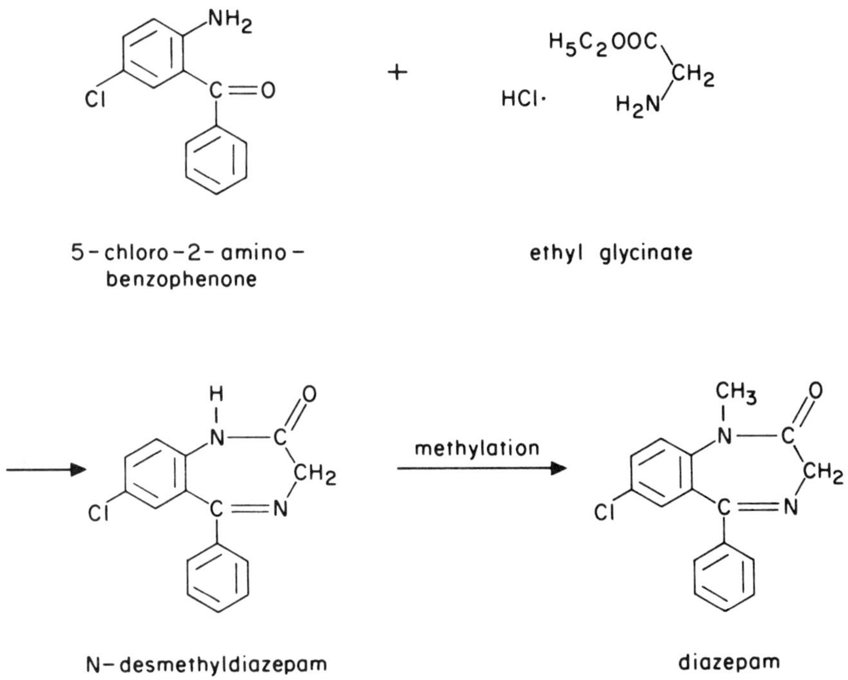


Figure 21.4 Synthesis of diazepam. (From Ref. 46.)

demonstrated in acid and neutral supporting electrolytes was destroyed. This was in agreement with the development of fluorescence activity due to the formation of the unsaturated derivative.

The determination of quinicine, the oxidative decomposition product of quinine, has been reported by polarography [49]. Quinicine was determined in the presence of a 1000-fold excess of quinine in quinine salts and injectibles. The assay was rendered specific for quinicine in the presence of quinine, by the reduction of the carbonyl group, which is present only in quinicine. Polarography ($E_{1/2} = -0.59$ V vs. SCE) was performed in 0.2 M acetate buffer (pH 4.4) on a solution of 10 μg/mL of quinicine.

## B. Analytical Research

Voltammetric research on pharmaceutical compounds is quite extensive. Therefore, this section attempts to demonstrate the wide variety of molecules which can be determined by voltammetric techniques with selected examples.

Chloramphenicol, a broad-spectrum antibiotic, has probably received more attention in the polarographic literature than any other

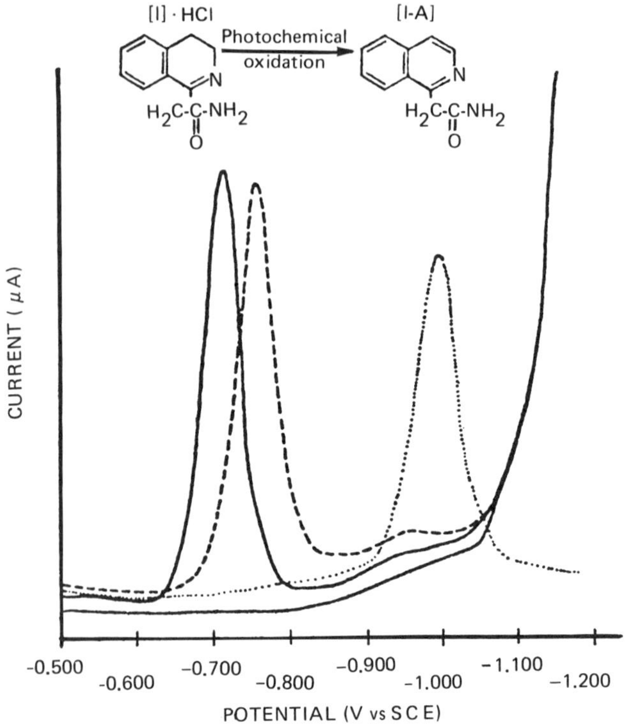


**Figure 21.5** Differential pulse polarography of 3,4-dihydro-1-isoquinolineactamide [I]·HCl and isoquinoleacetamide. —— 1.0 N HCl, --- 0.1 N HCl, ···· 1.0 M KCl/1.0 M $PO_4^{3-}$ (pH = 7.0). (From Ref. 48, reproduced with permission of the copyright owner.)

pharmaceutical. The aromatic nitro group is quite easily reduced. Studies [50] employing polarography, cyclic voltammetry, and constant potential coulometry have suggested that the drug undergoes a slow two-electron reduction followed by a fast two-electron reduction of hydroxylamine. At more negative potentials the hydroxylamine group is further reduced to the arylamine. In pH 4.7 acetate buffer with 0.003% decylamine present, the drug produces a well-defined four-electron polarographic wave which can be used for quantitation over the range from 0.3-600 μg/mL.

The 1,4-benzodiazepine class of pharmaceuticals (see Fig. 21.6), which are in widespread clinical use today, all possess intrinsic electrochemical activity which renders them ideal for voltammetric studies. Senkowski et al. [51] were the first to perform a comprehensive study of this class of compounds by polarography and demonstrated that the 1,4-benzodiazepines show a wave for the two-electron reduction of the 4,5-azomethine "electrophore" in a supporting electrolyte of 0.1 M HCl containing 20% methanol. Secondary polarographic waves were attributed to the reduction of the 1,2-azomethine and $N_4$ oxide in chlordiazepoxide, and to the reduction of the nitro group in nitrazepam. Factors that influenced the azomethine reduction such as pH or substituents were also reported. Another study [52] of the polarographic activity of oxazepam demonstrated that the compound in acidic media (pH <6) showed a four-electron step corresponding to the simultaneous reduction of the 4,5-azomethine bond and reductive cleavage of the hydroxyl group in the 3-position of the 1,4-diazepine ring. The acid-base equilibrium and polarographic reduction mechanism of six, 1,4-benzodiazepine derivatives [53] and some of their metabolic products [54] have been reported. The general range of analytical utility for the compounds was reported as $10^{-4}$-$10^{-7}$ M. Brooks et al. [55] have studied the effect of substituent groups and the ease of reduction of

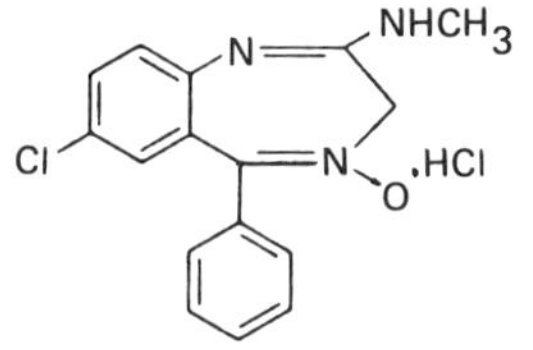


Chlordiazepoxide HCl

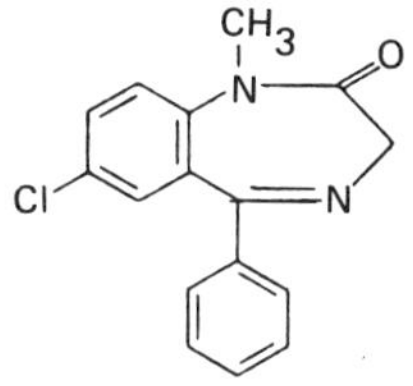


Diazepam

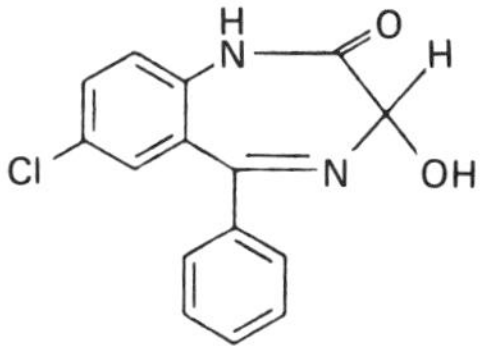


Oxazepam

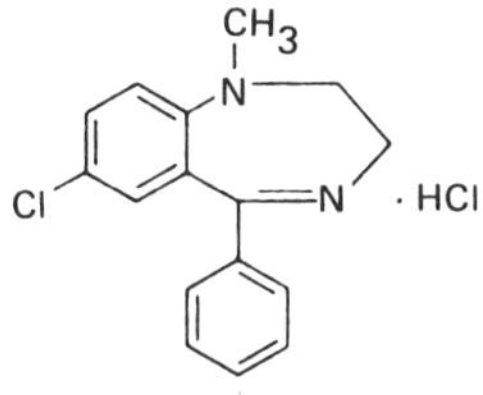


Medazepam HCl

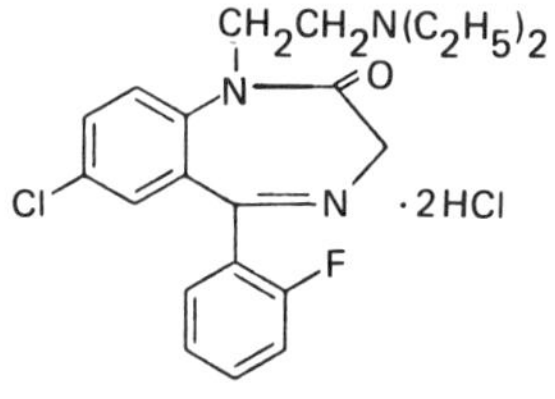


Flurazepam HCl

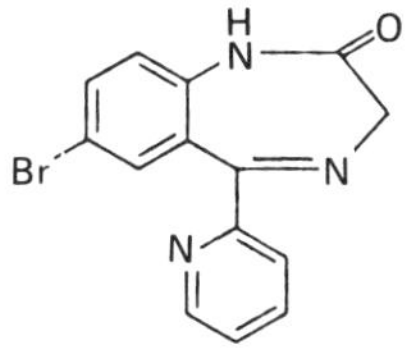


Bromazepam

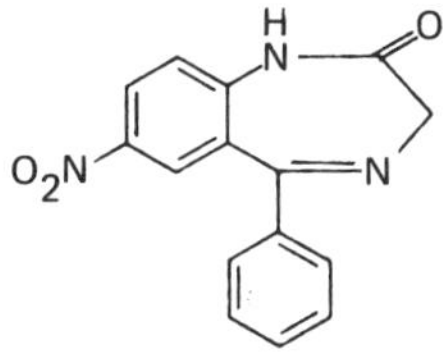


Nitrazepam

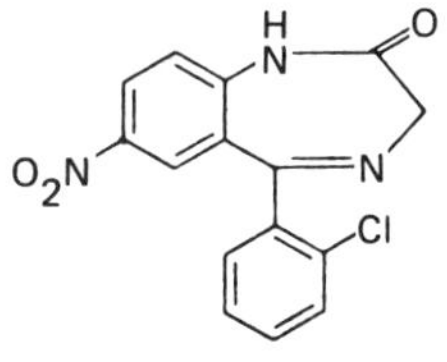


Clonazepam

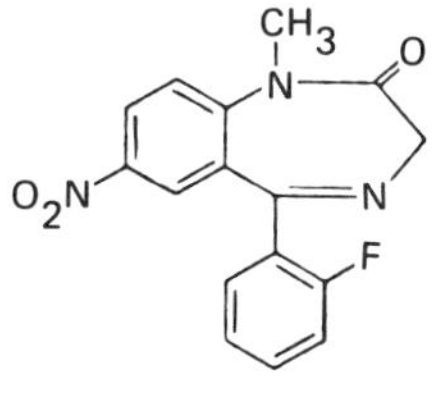


Flunitrazepam

**Figure 21.6** Structural formulas of some 1,4-benzodiazepines.

the 4,5-azomethine on 1,4-benzodiazepine derivatives related to chlordiazepoxide, medazepam, and N-desmethyldiazepam in pH 3.0 and 7.0 phosphate buffer as supporting electrolytes. A pulse polarogram of the acetyl derivative of chlordiazepoxide is shown in Figure 21.7. A good Hammett correlation was obtained for all three series of compounds. Polarographic data were also obtained for two groups of 1,4-benzodiazepines structurally related to medazepam and N-desmethyldiazepam with heterocyclic substituents in the 5-position.

An investigation of the dc and ac polarographic response of 24 pharmaceuticals in an aprotic solvent system has been described [56]. Barbiturates, salicylates, corticosteroids, and alkaloids were examined in acetonitrile-tetrabutylammonium perchlorate. The majority of the compounds yielded one-electron reversible waves in this solvent which

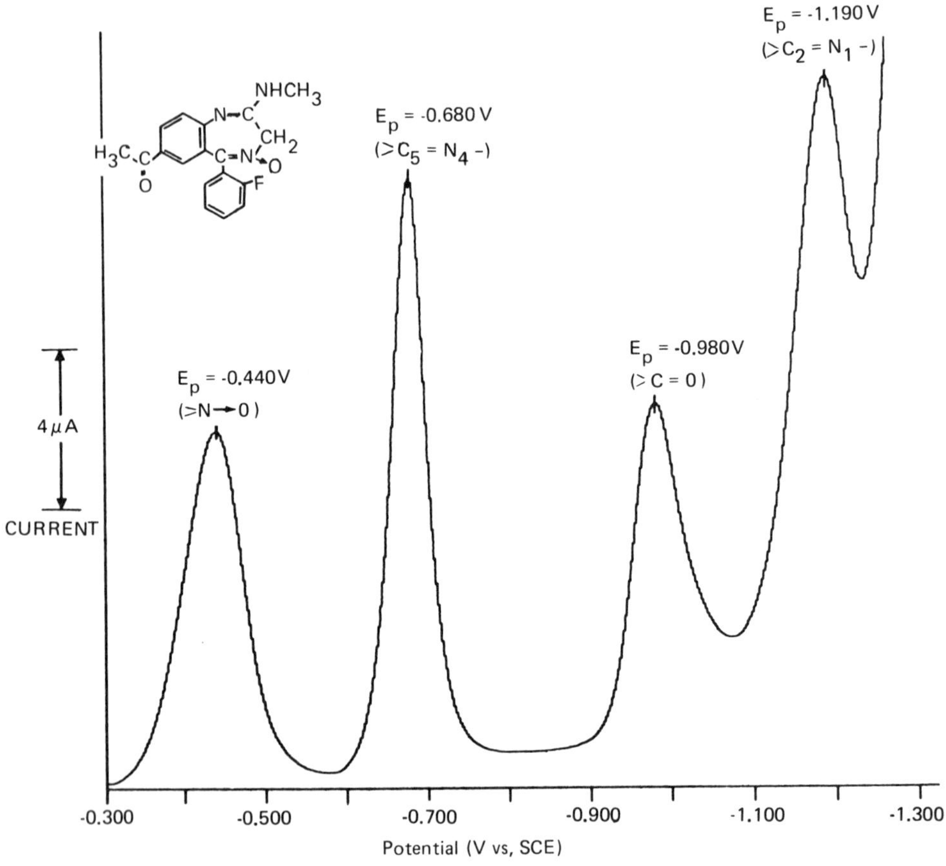


Figure 21.7 Differential pulse polarography for the 7-acetyl analog of chlordiazepoxide. (From Ref. 55.)

were suitable for analytical purposes, in contrast to the corresponding aqueous solution responses, which were irreversible or nonexistent. Data presented for each compound include the dc half-wave potential, fundamental harmonic ac peak potential, the potential of the second harmonic ac minimum, and the apparent n value for each polarographic technique. The detection limit of dc and fundamental harmonic ac polarography was typically $5 \times 10^{-5}$ M, while that for the second harmonic ac polarography was one order of magnitude lower.

The polarography of steroids has received considerable attention [39]. A comparative study of ac and dc polarography as a method of analysis for several $\Delta^4$-3-ketosteroids, including testosterone, methyltestosterone, and progesterone, in a supporting electrolyte of pH 1.3 buffer and tetrabutylammonium iodide in 50% ethanol was described [57]. Results showed that the lowest practical detection limit for both techniques was $3.3 \times 10^{-5}$ M, and that ac polarography was more precise. A combined polarographic/thin-layer chromatographic (TLC) assay has been described for the separation and determination of some corticoids in mixtures [58]. The method has the distinct advantage of employing elution of the TLC plate directly into the polarographic cell with a specially designed suction apparatus (see Fig. 21.8) and thus

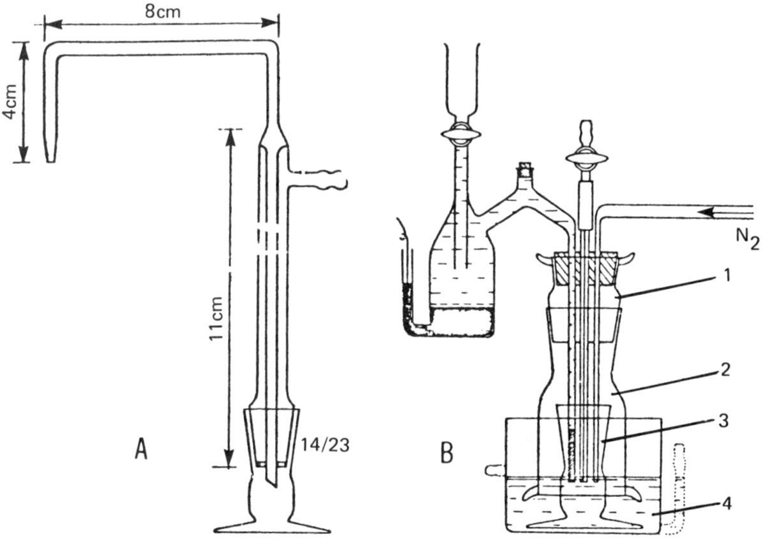


Figure 21.8 (A) Suction equipment for TLC. (B) Equipment for polarography of small volumes: 1, upper part of ground joint; 2, mantle of the ground joint; 3, electrolysis cell; 4, water container sealing the inner space. (From Ref. 58.)

avoids time-consuming filtration, separation, and evaporation procedures.

The dc polarography of a large group of pteridines in pH 2.0 phosphate buffer has been reported [59]. Three groups of pteridines, including 2,4,7-triaminopteridines, 7-substituted-4-amino-2-aryl-5-pteridine carboxamides, and 4,7-diamino-2-aryl-5-pteridine carboximides, demonstrated polarographic reduction which was found to occur in two steps, to be pH dependent, and involve two electrons per step.

Other dc and ac polarographic assays have been reported for warfarin [60], tetramethylthiuram disulfide employing the reduction of the S-S bond [61], and hexachlorophene using reductive cleavage of the halogram atom [62]. Another report dealt with dc polarographic assay of the p-fluoro-substituted butyrophenone derivatives, triperidol, fluanisone, and haloperidol [63].

The classical dc polarography of vitamins A, $B_1$, $B_2$, $B_6$, $B_{12}$, and C, nicotinamide, tocopherols, and naphthoquinones has been reviewed [39]. Other studies have examined in detail the cyclic voltammetry of vitamin $B_{12}$ employing rapid scan voltammetry at the DME [64] and the HMDE [65]. Vitamin $B_{12}$ is complexed with trivalent cobalt ion at the heterocyclic nitrogen atoms. As a result of the complexation, a catalytic hydrogen wave is formed for the compound. In addition to the catalytic wave, a wave corresponding to the reduction of the trivalent cobalt to the monovalent state is observed.

Derivatization techniques such as nitration, nitrosation, and others have commonly been employed for the analysis of pharmaceuticals. Benactyzine [66,67], antipyrine [68], 4-hydroxybenzoic acid ester [69], phenacetin [70], phenobarbital [71], diphenylhydantoin [71], meprobamate [66], glutethimide [72], and phenothiazines [73] have been determined as their nitro derivatives. Nitrosation has been employed to convert phenothiazine to an electroactive derivative, which could be determined with a sensitivity limit to 50 ppm [74]. Chlorpromazine has been assayed as its sulfoxide by cathode ray polarography after oxidation with saturated aqueous bromine [75]. Phenelzine was determined at a level of $4 \times 10^{-4}$ M as an acetone derivative using a 4% acetone solution containing 0.01% gelatin [76]. Pulse polarographic determination of sulfonamides, including sulfanilamide, sulfathiazole, sulfaguanidine, and sulfacetamide, as their diazo derivatives was described [77]. The compounds are reacted with sodium nitrate in acid solution, sulfamic acid is added, the solution is made basic, and the coupling is complete with 1-naphthol to yield the polarographically active azo dye. The lower limit of detection was about $5 \times 10^{-8}$ for sulfanilamide, sulfathiazine, and sulfacetamide, and about $1.2 \times 10^{-7}$ M for sulfaguanidine (see Fig. 21.9). These values are approximately twofold greater than those of the spectrophotometric techniques. Epicillin [78] and ampicillin [79] have been recently assayed by dc or pulse polarography as the 2-hydroxy-3-phenyl-6-methylpyrazine derivative following acid hydrolysis in the presence of formaldehyde.

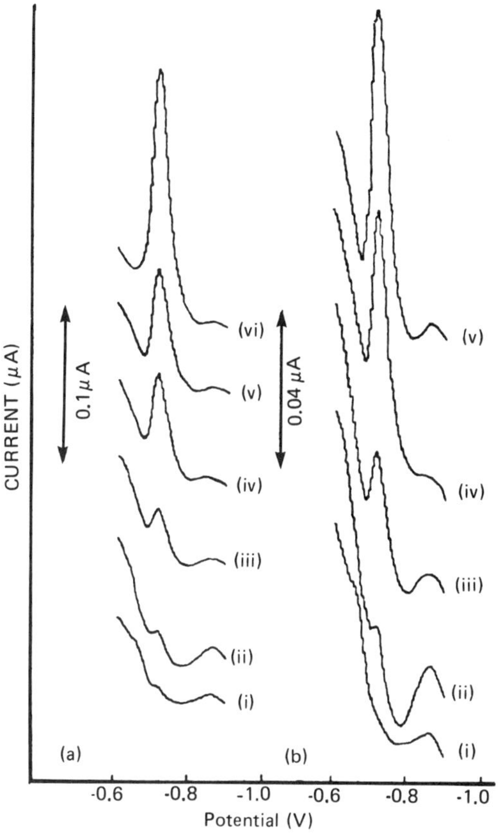


Figure 21.9 Differential pulse polarography of sulfathiozole derivatives; sensitivities: (a) 0.5 μA, (b) 0.2 μA; concentrations (x $10^{-8}$ M): i, 4; ii, 12; iii, 20; iv, 32; v, 40; and vi, 80. (From Ref. 77.)

The voltammetric examination of the oxidation of phenothiazines in strong acid using various solid electrodes [14,80-83] has been reported by several groups. Fourteen phenothiazine tranquilizers were found to yield regular, reproducible, well-defined anodic voltammetric waves [80] employing the gold electrode. The report describes the effect of pH, concentration, and temperature on the position and size of the waves as well as the effects of substituents in the 10- and 2-positions. The quantitative oxidation of this series of compounds to a free radical or to a sulfoxide at the rotating platinum electrode has been described [14,81]. The results indicated that in 9 N sulfuric acid two successive one-electron oxidations occurred, involving a stable semiquinone free-radical intermediate and a sulfoxide. The voltammetric behavior of a series of phenothiazines using a silicon-rubber-based graphite electrode in a variety of aqueous and non-aqueous supporting electrolytes has been described [82,83]. Other pharmaceuticals, such as methyldopa, phenacetin, p-aminosalicyclic acid, and amidopyrine, were also determined with the same electrode [82]. Recent studies describe the voltammetric determination of phenylbutazone at the glassy carbon electrodes [84] and morphine at platinum and graphite electrodes [85].

### C. Stability Testing

Voltammetric techniques are quite useful for the direct measurement of the stability of some pharmaceuticals in aqueous solution. Depending on the compounds involved, voltammetry can be used to either monitor the decomposition of a drug and/or the formation of a decomposition product. This type of study is typically performed by applying a potential to the working electrode which is on the voltammetric plateau for the starting material or end product to be monitored. The resulting current vs. time data provide the desired kinetic information.

The polarographic reduction of cephaloridine, cephalothin, and cephalosporin C each yield one wave which is both pH and concentration dependent [86]. The reduction of these compounds and a related derivative, 3-(5-methyl-1,3,4-thiadiazol-2-ylthiomethyl)-7-[2-(3-sydnone)acetamido]-3-cephem-4-carboxylic acid-sodium salt [I] have been described [87]. The reduction of [I] yields two waves. The first is believed to be the two-electron reductive elimination of the 3-position substituent, and the second is believed to be the six-electron reduction of the sydnone acetamido 7-position substituent, common to all cephalosporin C derivatives. The first wave can be used to follow the acid, base, and β-lactamase degradation of [I], which involves the loss of the 3-position substituent.

Polarography has also been applied to the determination of ethacrynic acid in the presence of its principal degradation product, a dimer [88]. Ethacrynic acid contains an electroactive α, β-unsaturated ketone with $E_{1/2} = -1.28$ V vs. SCE in pH 8 phosphate buffer, whereas the dimer degradation product does not have this functional group. By monitoring the decrease in wave height of the parent compound, the stability of ethracrynic acid in dosage forms was evaluated.

The hydrolysis of azomethine (imine) bonds is a fairly common mechanism for the degradation of pharmaceuticals. The acid hydrolysis of thiacetazone [89] and p-chlorobenzaldoxime (PCBO) [90] can easily be monitored employing polarographic reduction of the azomethine in each of the compounds. The latter compound is extremely interesting in that it was inactive as a muscle relaxant when administered orally but active when administered parenterally. The data obtained in this

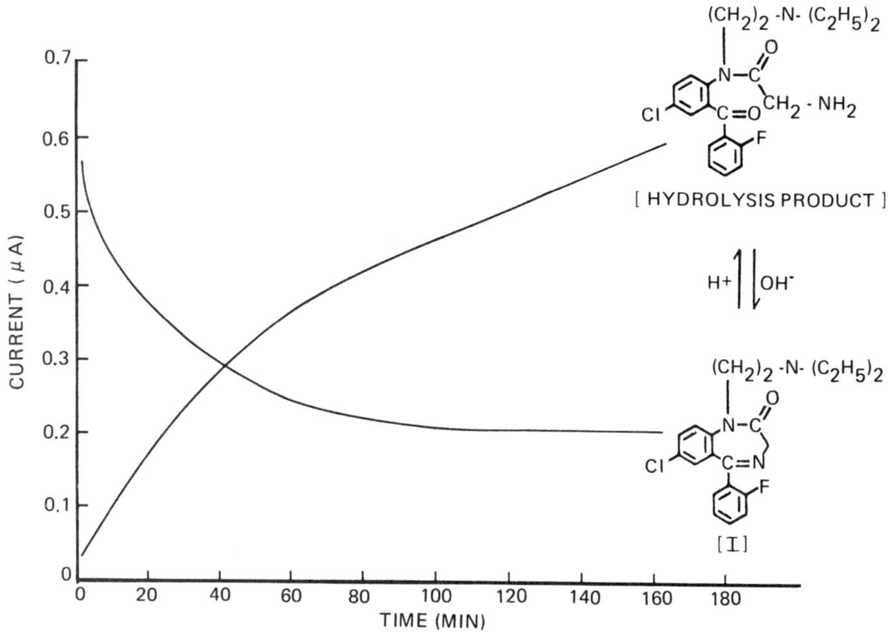


Figure 21.10 Time course plot of the hydrolysis of flurazepam in 0.1 N $H_2SO_4$ determined by differential pulse polarography. (From Ref. 91.)

study showed that PCBO was hydrolyzed to p-chlorobenzaldehyde at the pH of the stomach and thus accounted for the loss of activity following oral dosing of the drug.

Flurazepam and two of its metabolites have demonstrated reversible hydrolysis of the 4,5-azomethine [91]. In acid solution this reaction may be monitored employing pulse polarography to measure the decrease in concentration of the 4,5-azomethine of the 1,4-benzodiazepine or increase in carbonyl concentration of the "open" benzophenone (see Fig. 21.10).

## D. Assays of Dosage Forms

The voltammetric analysis of drugs in pharmaceutical products is by far the most common use of electrochemistry for analytical-pharmaceutical problems. As a rule many of the active constituents of formulations in contrast to excipients can be readily oxidized or reduced. Thus sample preparation usually consists of dissolving out the active ingredient from the particular formulation with a suitable solvent (nonaqueous, if necessary) and performing direct analysis on an aliquot of this solution. The specificity of the method is usually excellent because the compound can be identified by its voltammetric half-wave potential. Examples of polarographic determination of organic compounds in pharmaceutical preparations include many classes of drugs: tranquilizers, sedatives, hypnotics, antibiotics, steroids, antihistamines, diuretics, muscle relaxants, anticoagulants, and others.

A rapid polarographic determination of the antihistamine chlorpheniramine maleate in tablets based on the two-electron reduction of the compound at −0.7 V vs. Ag-AgCl in acid has been described [92]. The assay requires the extraction of the compound from a tablet into the 0.5 M sulfuric acid, dilution, and direct polarographic assay of an aliquot of this solution. The assay is accurate, sensitive, and since it does not require the removal of insoluble matter, is more rapid than the spectrophotometric procedures.

Disulfiram [bis(diethylthiocarbamoyl)disulfide] used in the treatment of alcoholism can be assayed directly by pulse polarography in an aliquot of a solution of a ground-tablet dissolved in ethanol-acetate buffer (pH 4.5) [93]. A mechanism for the electrode process was proposed involving the reaction of the disulfiram with the mercury drop to form an insoluble mercuric salt, which then underwent reduction at the electrode surface.

The development of a simple fast-scan polarographic method for the determination of the $\Delta^4$-3-ketosteroid flurandrenolone in pharmaceutical preparations has been reported [94]. The polarographic peak due to the reduction of the carbon-fluorine bond is measured in ointments and creams to determine concentrations as low as 0.01% w/w. Recent pulse polarographic procedures have been described for progesterones [95], $\Delta^4$-3-ketosteroids [96], hydrocortisone [97], and flucytosine [98] in pharmaceutical preparations.

The determination of vitamins in pharmaceutical preparations continues to receive considerable attention. The voltammetric oxidation of vitamin A at a carbon paste electrode in the presence of vitamin E, a potential source of error in the assay, has been described [99,100]. Other assays involve the polarographic determination of niacinamide [101-103], menadione (vitamin $K_3$) [104], riboflavin (vitamin $B_2$) [105], thiamine, riboflavin, and nicotinamide in multivitamin preparations [106], and multivitamins [107].

The use of a tubular carbon electrode (TCE) for the electrochemical oxidative determination of ascorbic acid [108], L-dopa [109], and methyldopa [110] in dosage forms has been described. The flow system, electrode assembly, and electrochemical instrumentation required for these assays are shown in Figure 21.11. The method is based on continuous analysis in flowing streams by oxidation of the compound of interest at the TCE. Comparisons to colorimetric methods show this procedure to require fewer manipulative steps and to give comparable precision and accuracy. In the determination of ascorbic acid [108] no interferences were noted by antioxidants due to their lower concentra-

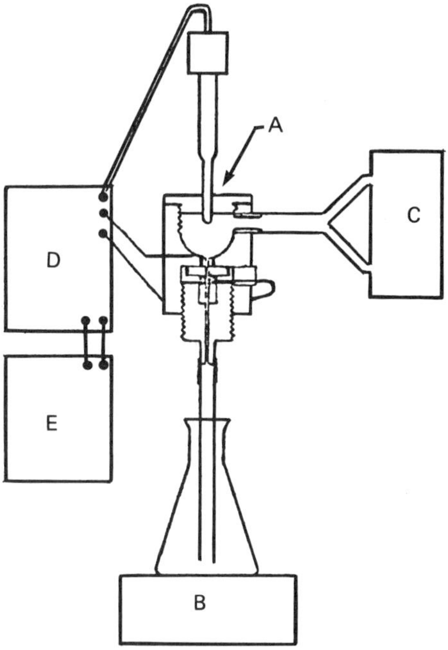


Figure 21.11 Flow system and tubular carbon electrode. A, electrode assembly; B, magnetic stirrer; C, pump; D, polarograph; E, recorder. (From Ref. 108, reproduced with permission of the copyright holder.)

tion (usually 1:10,000) and greater oxidation potential. No interferences were noted from other vitamins in multivitamin preparations. Although the process was not automated in these studies, it is suggested that the method may be readily incorporated into automated or semiautomated systems because it employs continuous analysis on a flow stream of sample.

Lorazepam, a member of the 1,4-benzodiazepine class of compounds, has been determined in dosage forms using a sensitive and versatile automated procedure capable of performing voltage-scanning polarographic analysis [111]. The system interfaces a polarograph with a Technicon continuous-flow system to permit the analysis of samples at a high rate with a broad selection of electrode configurations. Fifteen tablet or capsule formulations of lorazepam per hour can be analyzed at the 0.5-mg level with a standard deviation of ±1.4%.

Sulfenazone shows two polarographic waves on reduction at the DME due to reduction in the pyridazine nucleus and the antipyrine substituent of the molecule. An automated procedure has been described for the determination of the compound after extraction using dc or ac polarography ($E_{1/2} = -1.5$ V vs. Ag/AgCl) at a rapidly dropping DME (0.1-0.9 s) in a 140-μl Plexiglas cell [112]. The supporting electrolyte is phosphate buffer (pH 8.4) employing a Tween-80 surface-active agent. The system is capable of determining 60 samples per hour. The automatic and simultaneous determination of methylesculetol derivatives by polarography and of rutine by colorimetry in pharmaceutical preparations has been described using a modification of the equipment used for the determination of sulfenazone [113].

## E. In Vitro Studies

Voltammetric studies have been used to correlate pharmacological activity of a group of compounds with minor changes in their structure. A correlation of the polarographic behavior of 1,2-dimethyl-3-arylpyrazolinium salts and their antidepressant activity has recently been reported [114]. Results indicated that making substituent changes at the 3-aryl position would change the $E_{1/2}$ consistent with a Hammett type of relationship. A good correlation of $E_{1/2}$ values and antidepressant activity was also demonstrated. The authors conclude that apart from other factors which might affect the total drug-receptor interaction, the antidepressant activity of these salts is associated with a redox effect.

In similar studies, the reduction potentials of 1,2-benzothiazole derivatives in DMF were correlated to antimycotic activity [115], and those of 1,3,5,7-substituted-1,3-dihydro-2H-1,4-benzodiazepin-2-ones and benzodiazepine-2-thiones in DMF-water solutions were correlated to inhibition of orientation reactions and to protection from electric shock [116].

Dissolution rate testing is an important aspect of the development of a drug. It can serve as a guide to formulation development and may be correlated to the absorbability potential of some poorly soluble drugs. A review of the role of in vitro dissolution testing in the development of a drug has been published [117]. Voltammetric methods are ideal for these studies since they are selective and the electrode used to monitor the rate of dissolution may be placed directly in the dissolution apparatus obviating the need for repetitive sampling.

The silicone-rubber-based graphite electrode has been employed as an appropriate voltammetric sensor to measure the rates of dissolution of several oxidizable pharmaceutical compounds [118]. The recirculating dissolution measuring arrangement with this sensing electrode and an Ag/AgCl reference are shown in Figure 21.12. A peristaltic pump continuously samples the system in the thermostatically controlled dissolution vessel. A suitable constant potential is applied between the two electrodes and the current vs. time curve is recorded. An example of this system was demonstrated for the dissolution of aminopyrine tablets and promethazine tablets. A similar design ex-

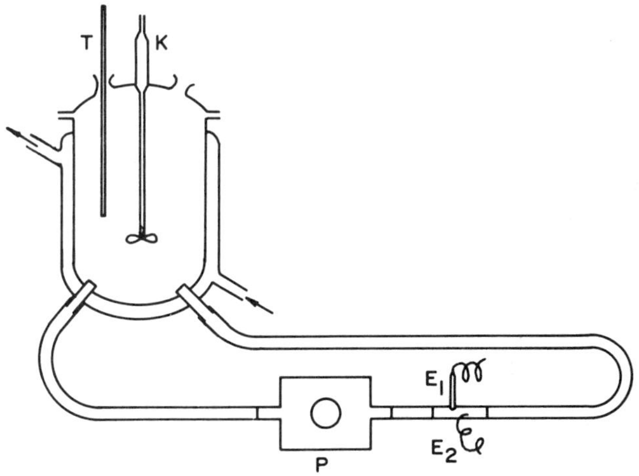


Figure 21.12 Measuring setup for investigation of the dissolution of drugs from pharmaceutical preparations. T, thermometer; K, stirrer; P, peristaltic pump; $E_1$, silicone-rubber-based electrode; $E_2$, silver-silver chloride reference electrode. (From Ref. 118.)

perimental set-up using a dropping mercury electrode in a flow cell has recently been applied to the determination of drug dissolution of chlordiazepoxide, trimethoprim, ornidazole, and isoniazid using polarography [119]. The dissolution rate of L-dopa in 0.1 M HCl has been determined employing a tubular carbon electrode [109]. A 1-liter sample was pumped through a filter unit to the TCE and returned to the beaker. After setting the electrode potential at +0.9 V vs. SCE and establishing a baseline current, a tablet was added and the limiting current was recorded as a function of time.

The in vitro everted sac technique has been used for screening the permeability parameters of drug substances associated with their in vivo absorbability [120]. The time of onset of permeability across the sac and the cumulative amount of drug transferred per unit concentration of drug in mucosal solution were shown to relate to potential permeability limited absorption characteristics of a drug substance. The system utilizes a 10-cm segment of rat ileum, which is everted, cannulated, and suspended in a solution of drug in pH 7.4 Krebs-Ringer bicarbonate buffer. The amount of drug passing across the intestinal mucosa is then determined by serial sampling at various time intervals. The solutions removed from the everted sac have been analyzed directly polarographically for the determination of the permeability characteristics of diazepam [121]. The polarographic assay has an advantage of not requiring the extraction of the compound from the buffer to remove interfering UV-absorbing substances.

The pharmacological response exerted by a chemotherapeutic agent may be related to the concentration of unbound drug in the plasma water. The extent to which a drug is bound to plasma albumin may influence its distribution, onset of activity, and rate of metabolism and excretion. A recent study has investigated the degree of binding to serum albumin for four series of 1,4-benzodiazepines homologs employing scintillation radiometry and pulse polarography for quantitation [122]. The procedure consisted of placing drug and plasma in a dialysis membrane sac and suspending this sac in an isotonic buffer solution for a fixed interval of time at 37°C. After completion of the incubation, the solutions both inside and outside the dialysis membrane were removed, made alkaline, and extracted with ether, the residue of which was analyzed by pulse polarography in 0.1 M HCl or pH 7.0 phosphate buffer. The pulse method was rapid and sensitive (≈0.5 μg/mL of solution) and allowed a large number of compounds to be determined under the same conditions, in contrast to gas chromatography, which would have required specific conditions for each compound.

### F. Determination of Drugs in Biological Fluids

The use of voltammetric methods for the determination of drugs in biological fluids has only become practical in the last several years due to the advent of commercial instrumentation capable of measuring concentration with limits of $10^{-7}$-$10^{-8}$ M [123,124]. The techniques of pulse polarography, single-sweep polarography, and liquid chromatography with amperometric detection (LCEC) are the only techniques capable of routinely determining drugs at levels of less than $10^{-6}$ M (typically 10 μg/mL of biological fluids). Solid electrode voltammetry does not have sufficient sensitivity to measure levels below 10 μg/mL of biological fluids unless micro cells (⩽10 μL) are used in conjunction with preconcentration. The use of LCEC to measure drugs in biological fluids is discussed in Chapter 22.

The earlier works using polarography for the determination of drugs in biological fluids employed direct analysis of the biological material. Determinations of 2-ethyl-4-thioureidopyridine, for example, were performed directly in plasma without any additional supporting electrolyte or buffering [125]. The dc polarographic determination of chlordiazepoxide [126] and diazepam [127] in plasma, and urine diluted with the appropriate supporting electrolyte, was suggested by Oelschlager et al.

There are two basic shortcomings of these direct assays: (1) they are usually only sufficiently sensitive to measure the drug after the ingestion of an overdosage, and (2) they are inherently nonspecific in that compounds of similar structure, metabolites, and biological materials may interfere and thus yield erroneous results. In order for polarographic assays to yield analytically meaningful results for the determination of drugs in biological fluids, selective extraction, cleanup, and separation of the parent drug from any metabolites and/or other substances present must be employed prior to polarographic assay. Such extraction procedures are routine in bioanalytical laboratories, so they do not result in serious inconvenience. After the administration of many drugs it is quite common to find high percentages of the dose excreted in urine. For example if 50% of a 50-mg dose is recoverable in the urine in a 1000-mL volume over an excretion period of 0-24 h, the resulting concentration of drug in the urine sample would be 25 μg/mL. Pulse polarography is ideally suited for measurements in this concentration range.

A pulse polarographic method to determine dantrolene sodium and its major metabolites in urine after ethyl acetate extraction has been

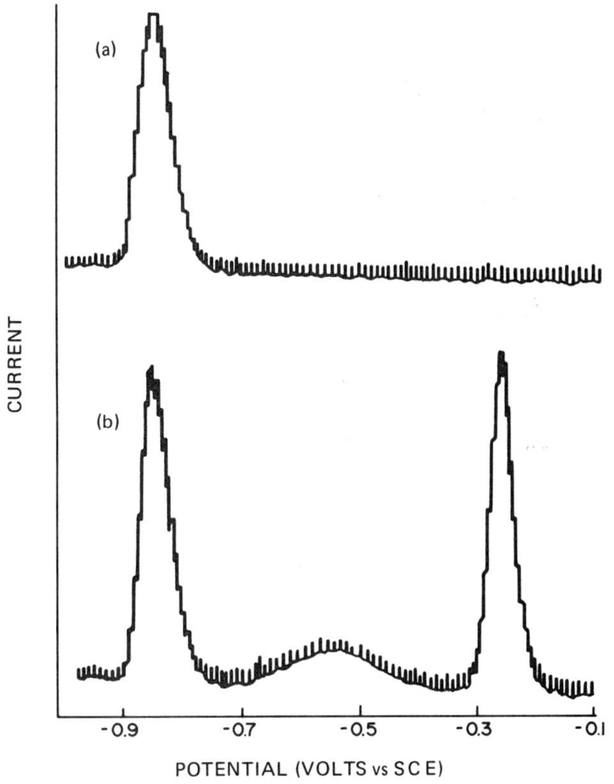


Figure 21.13 Differential pulse polarography of (a) reduced dantrolene, (b) dantrolene, and nonreduced metabolites. (From Ref. 129, reproduced with permission of the copyright holder.)

reported [128]. The ethyl acetate is brought to a residue and the dantrolene plus the total extractable metabolites are analyzed for reduction of the azomethine linkage at −0.86 V in a DMF-acetate buffer (pH 4.0). The nitro compounds are simultaneously determined in the same media as dantrolene equivalents from the reduction of the nitro group at −0.26 V (see Fig. 21.13). The difference between the two determinations represents the metabolites not containing the nitro

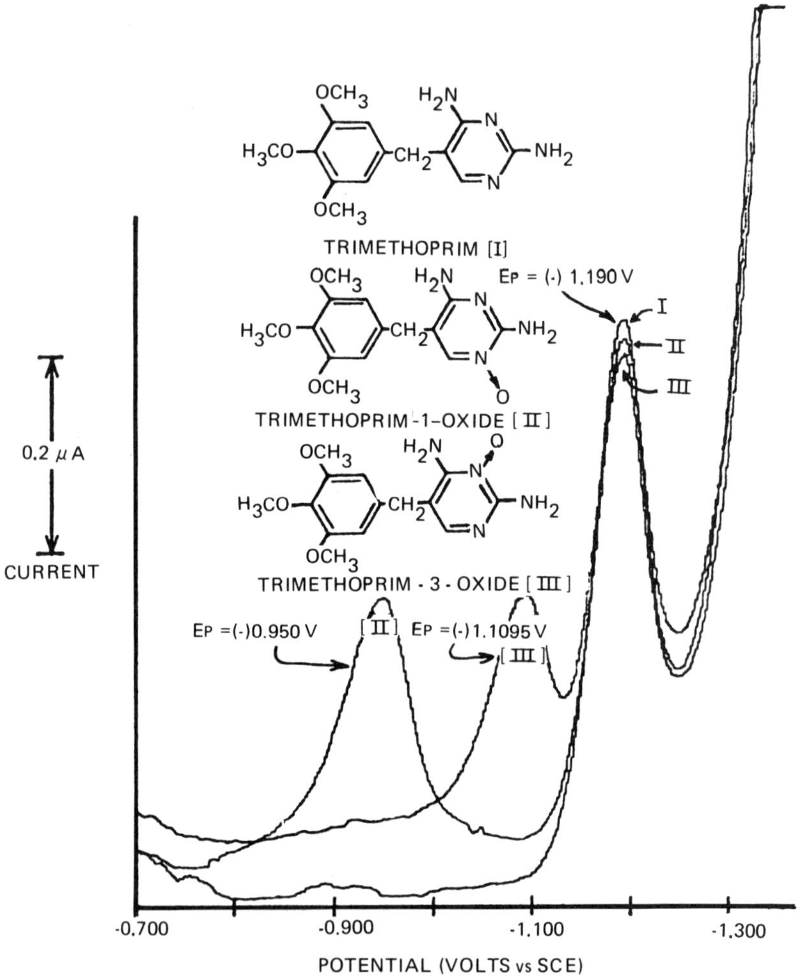


Figure 21.14 Differential pulse polarography of trimethoprim and its N-oxide metabolites in 1 M phosphate buffer (pH 3). (From Ref. 131, reproduced with permission of the copyright holder.)

group. Levels as low as 0.1 μg/mL can be determined for either functional group.

Differential pulse polarography has been employed for the determination of trimethoprim and its N-oxide metabolites in urine of humans, dog, and rat [129]. By combining a TLC separation of a urinary extract with polarography, a high degree of specificity was obtained to denote quantitatively the preferred metabolic N oxidation of each species. The pulse polarographic determination in pH 3.0 phosphate buffer is particularly specific because each compound can be distinguished by a peak due to the N-oxide and azomethine reductions (see Fig. 21.14).

Polarographic methods have been extremely useful for the determination of the urinary excretion of the 1,4-benzodiazepines. An assay that employs selective solvent extraction and acid hydrolysis of diazepam and its major metabolites, N-desmethyldiazepam and oxazepam, to their respective benzophenones has been employed to measure the urinary excretion of diazepam [130]. A pulse polarographic assay has been reported which will measure the urinary excretion of bromazepam following a single 12-mg dose [131]. The assay employs selective extraction of bromazepam and the 2-amino-5-bromobenzoylpyridine metabolite from the deconjugated metabolites, 3-hydroxybromazepam and 2-amino-3-hydroxy-5-bromobenzoylpyridine, into separate diethyl ether fractions. The residues of the respective extracts are dissolved in phosphate buffer (pH 5.4) and analyzed by pulse polarography, which yields two distinct well-resolved reduction peaks for the 4,5-azomethine functional group of the benzodiazepine-2-one and for the carbonyl function group of the benzoylpyridine in each fraction (see Fig. 21.15). Pulse polarographic assays have also been reported for the determination of the urinary excretion of 7-chloro-1,3-dihydro-5-(2'-chlorophenyl)-2H-1,4-benzodiazepine-2-one [132], clonazepam [133], and flurazepam [91]. The assay for flurazepam involves the determination of the three major metabolites of flurazepam employing a combination of selective extraction, TLC, and pulse polarography and was used to measure the urinary excretion following a 90-mg dose. A direct assay with a detection limit of 5 μg/mL which employs 0.1 mL of urine diluted with phosphate buffer (pH 7.0) is also described as a rapid means of measuring total benzodiazepines to confirm ingestion of flurazepam. The pulse assays [91,131-134] for the 1,4-benzodiazepine-2-ones typically have a detection limit of 0.1-0.2 μg/mL, which is more than sufficient to determine the urinary excretion of these compounds.

The use of polarographic assays for the measurement of blood levels is the most demanding on the detection limitations of the technique. Pulse polarography and LCEC are the only electrochemical methods presently available for routine determination of drugs at levels below 1.0 μg/mL of blood.

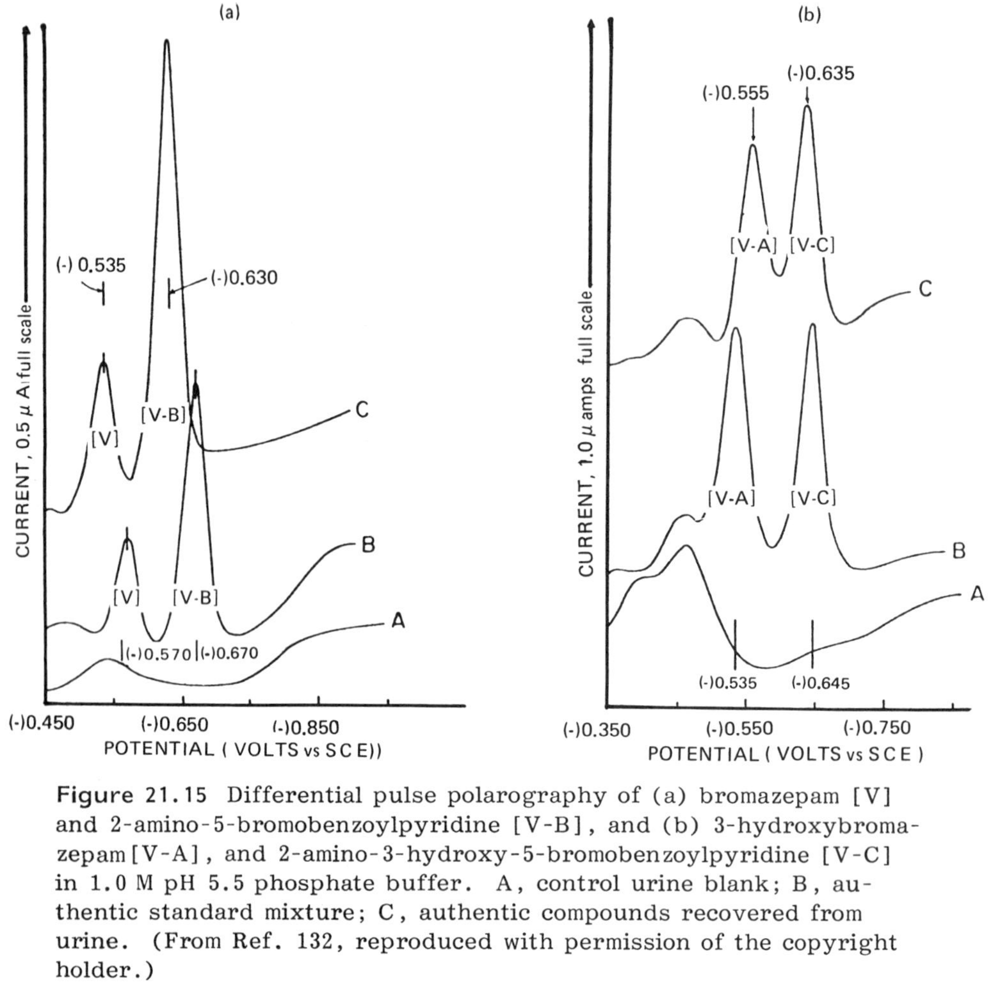


**Figure 21.15** Differential pulse polarography of (a) bromazepam [V] and 2-amino-5-bromobenzoylpyridine [V-B], and (b) 3-hydroxybromazepam [V-A], and 2-amino-3-hydroxy-5-bromobenzoylpyridine [V-C] in 1.0 M pH 5.5 phosphate buffer. A, control urine blank; B, authentic standard mixture; C, authentic compounds recovered from urine. (From Ref. 132, reproduced with permission of the copyright holder.)

The determination of diazepam in plasma with detection limits of 0.03-0.2 μg/mL by cathode ray [135] and pulse polarography [121,136] has been described. The first reported pulse assay [121] differs from the other polarographic assays [135,136] in that a more polar solvent is employed to ensure quantitative extraction, followed by TLC separation and determination of diazepam and its major blood metabolite, N-desmethyldiazepam, to ensure specificity.

The pulse polarographic determination of chlordiazepoxide and its metabolites in plasma has also been described [137]. Chlordiazepoxide and its metabolites are extracted from serum buffered to pH 9.0 followed by a TLC separation, elution, and final quantitation in

0.5 M $H_2SO_4$. The detection limit of the assay of 0.05-0.1 μg of each compound per milliliter of serum using a 2-mL sample is sufficiently useful for pharmacokinetic studies.

Pulse polarography has also been used to measure dantrolene in plasma [128] and trimethoprim [138] and nitroimidazoles in blood [139, 140]. The overall recovery of the trimethoprim assay was reported to be 81.7 ± 6.3% (SD) with a detection limit of 0.5-0.75 μg/mL of blood. There was no interference from the sulfamethoxazole which is administered simultaneously. Glibornuride [141], phenobarbital, and diphenylhydantoin [71] have all been determined as their nitro derivatives after extraction from blood. The recovery of phenobarbital and diphenylhydantoin from blood was 72.3 ± 6.5% (SD) and 76.6 ± 2.3% (SD), respectively, with a detection limit of 1-2 μg/mL. A modified assay [71] for the determination of both compounds in blood with TLC separation was also described.

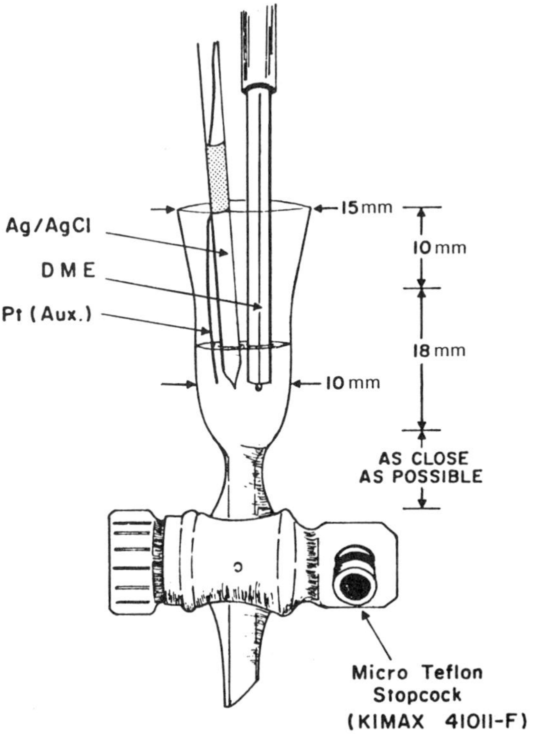


Figure 21.16 Schematic diagram of the 0.5-ml-capacity micro cell for pulse polarographic studies. [From Ref. 144, reprinted with permission from M. A. Brooks and M. R. Hackman, *Anal. Chem.* 47:2059 (1975). Copyright 1975 American Chemical Society.]

Recent work has demonstrated that pulse voltammetry at the carbon paste electrode can be applied to the assay of drugs in body fluids [142,143]. Methods based on the electrooxidation of theophylline [142] and acetaminophen [143] were described at concentrations of approximately 5-10 of μg/mL plasma.

Pulse polarographic studies have recently been described using a microcell of 0.5 ml capacity, which analyzed two 1,4-benzodiazepines with the lowest detection limit reported to date: 10-20 ng/ml of blood [144]. Detailed construction of the cell and electrode assembly were also described (shown in Fig. 21.16). Further miniaturization of this type of three electrode cell is not practical; hence further increases in sensitivity will have to rely on electrochemical detector flow cells of microliter capacity such as those used in conjunction with liquid chromatography (see Chap. 22).

## REFERENCES

1. C. B. Coutinho, J. A. Cheripko, T. Crews, B. H. Min, and A. C. Levy, Abstract, Society of Toxicology Meeting, Williamsburg, Va., March 9-13, 1975.
2. S. A. Kaplan, *Drug Metab. Rev. 1*:15 (1972).
3. S. A. Kaplan, in *Current Concepts in the Pharmaceutical Sciences: Dosage Form Design and Bioavailability* (J. Swarbrick, ed.), Lea & Febiger, Philadelphia, 1974, pp. 1-30.
4. M. A. Schwartz and E. Postma, *Biochem. Pharmacol. 17*:2443 (1968).
5. J. A. F. de Silva, in *Current Concepts in the Pharmaceutical Sciences: Biopharmaceutics* (J. Swarbrick, ed.), Lea & Febiger, Philadelphia, 1970, pp. 203-264.
6. J. R. DiPalma, *Drill's Pharmacology in Medicine*, 4th ed., McGraw-Hill, New York, 1971.
7. W. C. Cutting, *Handbook of Pharmacology*, 5th ed., Meredith Corporation, New York, 1969.
8. K. Abresch and I. Claassen, *Coulometric Analysis*, Franklin Publishing Co., Englewood, N.J., 1966.
9. G. W. C. Milner and G. Phillips, *Coulometry in Analytical Chemistry*, Pergamon Press, New York, 1967.
10. W. C. Purdy, in *Electroanalytical Methods in Biochemistry*, McGraw-Hill, New York, 1965, pp. 223-276.
11. G. Patriarche, *Contribution de l'analyse coulométrique*, Éditions Arscia S. A., Bruxelles, 1963.
12. R. W. Janssen and C. A. Discher, *J. Pharm. Sci. 60*:798 (1971).
13. G. Patriarche and J. J. Lingane, *Anal. Chim. Acta 49*:241 (1970).
14. F. H. Merkel and C. A. Discher, *Anal. Chem. 36*:1639 (1964).

15. K. S. V. Santhanam and V. R. Krishnan, *Anal. Chem. 33*:1493 (1961).
16. F. P. Wilson, C. G. Butler, P. H. B. Ingle, and H. Taylor, *J. Pharm. Pharmacol. 12*:220T (1960).
17. E. Jacobsen and T. V. Jacobsen, *Anal. Chim. Acta 55*:293 (1971).
18. J. J. Donahue and S. Oliveri-Vigh, *Anal. Chim. Acta 63*:415 (1973).
19. M. L. Girard, F. Rousselet, P. Levillain, and H. Fouye, *Ann. Pharm. Fr. 26*:535 (1968).
20. K. Stulik and J. Zyka, *Chemist-Analyst 55*:120 (1966).
21. Z. E. Kalinowska, *Pharmazie 22*:1 (1967).
22. K. Kalinowski and A. Olech, *Acta Pol. Pharm. 25*:171 (1968); *Chem. Abstr. 69*:46099 (1968).
23. K. Kalinowski, *Chem. Anal. 15*:277 (1970); *Chem. Abstr. 73*:38596U (1971).
24. H. Beral and V. Stoicescu, *Pharm. Zentralhalle 108*:469 (1969).
25. S. Oliveri-Vigh, J. J. Donahue, J. E. Heveren, and B. Z. Senkowski, *J. Pharm. Sci. 60*:1851 (1971).
26. S. A. Moros, C. M. Hamilton, J. E. Heveran, J. J. Donahue, and S. Oliveri-Vigh, *J. Pharm. Sci. 64*:1229 (1975).
27. M. L. Girard, F. Rousselet, H. Fouye, and P. Levillain, *Ann. Pharm. Fr. 27*:173 (1967).
28. S. Ebel and S. Kalb, *Arch. Pharm. (Weinheim) 307*:2 (1974).
29. R. L. Charles and A. M. Knevel, *J. Pharm. Sci. 54*:1678 (1965).
30. G. Patriarche, *Mikrochim. Acta, 1970,* 950.
31. Zs. Feher, G. Nagy, K. Toth, and E. Pungor, *Analyst (London) 104*:560 (1979).
32. J. R. Monforte and W. C. Purdy, *Anal. Chim. Acta 52*:433 (1970).
33. V. J. Vaygand and R. Mihaylovic, *Talanta 16*:1311 (1969).
34. G. J. Patriarche and J. J. Lingane, *Anal. Chim. Acta 49*:25 (1970).
35. C. A. Mairesse-Ducarmois, J. L. Vandenbalak, and G. J. Patriarche, *J. Pharm. Belg. 28*:300 (1973); *Anal. Abstr. 26*:3470 (1974).
36. J. R. Monforte and W. C. Purdy, *Anal. Chim. Acta 52*:25 (1970).
37. G. J. Patriarche and J. J. Lingane, *Anal. Chim. Acta 37*:455 (1967).
38. P. Zuman, *Organic Polarographic Analysis*, Pergamon Press, Elmsford, N.Y., 1964.
39. P. Zuman and M. Brezina, in *Progress in Polarography*, Vol. 2 (P. Zuman and I. M. Kolthoff, eds.), Wiley-Interscience, New York, 1962, pp. 687-701.

40. W. C. Purdy, in *Electroanalytical Methods in Biochemistry*, McGraw-Hill, New York, 1965, pp. 148-174.
41. R. J. Gayan, in *Oscillographic Polarography in Methods in Pharmacology*, Vol. 2: *Physical Methods* (C. F. Chignell, ed.), Appleton-Century-Crofts, New York, 1972, pp. 443-463.
42. J. E. Page, *J. Pharm. Pharmacol. 4*:1 (1952).
43. H. Hoffmann and J. Volke, in *Advances in Analytical Chemistry and Instrumentation*, Vol. 10: *Electroanalytical Chemistry* (H. W. Nurnberg, ed.), Wiley, New York, 1974, pp. 287-334.
44. M. R. Smyth and W. F. Smyth, *Analyst (London) 103*:529 (1978).
45. W. F. Smyth, *Polarography of Molecules of Biological Significance*, Academic Press, London, 1979.
46. S. Arizan, R. Simionovici, and V. Voinov, *Pharmazie 24*:746 (1969).
47. D. Dumanovic, J. Volke, and R. Jovanovic, *J. Assoc. Off. Analy. Chem. 54*:884 (1971).
48. J. A. F. de Silva, N. Strojny, and N. Munno, *J. Pharm. Sci. 62*:1066 (1973).
49. M. Girard and F. Rousselet, *Ann. Pharm. Fr. 20*:109 (1962); *Anal. Abstr. 9*:3872 (1962).
50. K. Fossdal and E. Jacobsen, *Anal. Chim. Acta 56*:105 (1971).
51. B. Z. Senkowski, M. S. Levin, J. R. Urbigikit, and E. G. Wollish, *Anal. Chem. 36*:1991 (1964).
52. H. Oelschlager, J. Volke, G. T. Lim, and U. Bremer, *Arch. Pharm. Berl. 303*:364 (1970).
53. J. M. Clifford and W. F. Smyth, *Z. Anal. Chem. 264*:149 (1973).
54. J. Barrett, W. F. Smyth, and J. P. Hart, *J. Pharm. Pharmacol. 26*:9 (1974).
55. M. A. Brooks, J. J. Bel Bruno, J. A. F. de Silva, and M. R. Hackman, *Anal. Chim. Acta 74*:367 (1975).
56. A. L. Woodson and D. E. Smith, *Anal. Chem. 42*:242 (1970).
57. J. L. Spahr, and A. M. Knevel, *J. Pharm. Sci.* 55:1020 (1966).
58. J. Hakl, *J. Electroanal. Chem. 11*:37 (1966).
59. M. Lapidus and M. E. Rosenthale, *J. Pharm. Sci. 55*:555 (1966).
60. T. Komura, S. Morishita, C. Aikawa, and Y. Ueda, *Jpn. Anal. 18*:943 (1969); *Anal. Abstr. 19*:4374 (1970).
61. M. J. D. Brand and B. Fleet, *Analyst (London) 95*:1023 (1970).
62. E. Jacobsen and T. Rojahn, *Anal. Chim. Acta 61*:320 (1972).
63. J. Volke, L. Wasilewska, and A. Ryvolova-Kejharova, *Pharmazie 26*:399 (1971).

64. S. L. Tackett and J. W. Ide, *J. Electroanal. Chem.* *30*:510 (1971).
65. P. G. Swetik and D. G. Brown, *J. Electroanal. Chem.* *51*:433 (1974).
66. V. Vorel, J. Prokes, and V. Dolezal, *Soudni Lek.* *5*:49 (1961); *Anal. Abstr.* *9*:3419 (1962).
67. J. Vachek, *Cesk. Farm.* *10*:187 (1961); *Anal. Abstr.* *8*:4340 (1961).
68. H. Oelschlager and D. Hamel, *Arch. Pharm. Berl.* *302*:847 (1969).
69. S. Tammilehto and M. Perala, *Pharm. Acta Helv.* *46*:351 (1971).
70. H. Oelschlager, *Arch. Pharm. Berl.* *296*:7 (1963).
71. M. A. Brooks, J. A. F. de Silva, and M. R. Hackman, *Anal. Chim. Acta* *64*:165 (1973).
72. A. Danek and H. Strozik, *Diss. Pharm. Warsz.* *18*:519 (1966); *Anal. Abstr.* *15*:429 (1968).
73. A. G. Dumortier and G. J. Patriarche, *Z. Anal. Chem.* *264*:153 (1973).
74. Y. I. Tur'yan, T. V. Merkyvkova, and O. V. Bogdanova, *Zh. Anal. Khim.* *25*:384 (1970); *Anal. Abstr.* *21*:1455 (1971).
75. G. S. Porter, *J. Pharm. Pharmacol., Suppl.* *16*:24T (1964).
76. L. Schlitt, M. Rink, and M. vonStackelberg, *J. Electroanal. Chem.* *13*:10 (1967).
77. A. G. Fogg and Y. Z. Ahmed, *Anal. Chim. Acta* *70*:241 (1974).
78. J. A. Squella and L. J. Nunez-Vergara, *Acta Pharm. Suec.* *16*:339 (1979).
79. A. G. Fogg and N. M. Fayad, *Anal. Chim. Acta* *113*:91 (1980).
80. P. Kabasakalian and J. McGlotten, *Anal. Chem.* *31*:431 (1959).
81. F. H. Merkle and C. A. Discher, *J. Pharm. Sci.* *53*:620 (1964).
82. E. Pungor, Z. Feher, and G. Nagy, *Magy. Kem. Foly.* *77*:298 (1971); *Anal. Abstr.* *22*:350 (1972).
83. E. Pungor, Z. Feher, and G. Nagy, *Acta Chim. Hung.* *70*:207 (1971); *Anal. Abstr.* *23*:747 (1972).
84. H. K. Chan and A. G. Fogg, *Anal. Chim. Acta* *109*:341 (1979).
85. B. Proksa and L. Molnar, *Anal. Chim. Acta* *97*:149 (1978).
86. I. F. Jones, J. E. Page, and C. T. Rhodes, *J. Pharm. Pharmacol., Suppl.* *20*:45S (1968).
87. D. A. Hall, *J. Pharm. Sci.* *62*:980 (1973).
88. E. M. Cohen, *J. Pharm. Sci.* *60*:1702 (1971).
89. Y. Asah, *Chem. Pharm. Bull. Jpn.* *11*:930 (1963); *Anal. Abstr.* *11*:4514 (1964).
90. E. R. Garrett, *J. Pharm. Sci.* *51*:410 (1962).
91. J. A. F. de Silva, C. V. Puglisi, M. A. Brooks, and M. R. Hackman, *J. Chromatogr.* *99*:461 (1974).

92. E. Jacobsen and K. Hogbert, *Anal. Chim. Acta 71*:157 (1974).
93. D. G. Prue, C. R. Warner, and B. T. Kho, *J. Pharm. Sci. 61*:249 (1972).
94. M. J. Heasman and A. J. Wood, *J. Pharm. Pharmacol., Suppl. 23*:176S (1971).
95. L. G. Chatten, R. N. Yadav, S. Binnington, and R. E. Moskalyk, *Analyst (London) 102*:323 (1977).
96. R. N. Yadav and F. W. Teare, *J. Pharm. Sci. 67*:436 (1978).
97. E. Jacobsen and B. Korvald, *Anal. Chim. Acta 99*:255 (1978).
98. F. W. Teare, R. N. Yadav, and M. Spino, *J. Pharm. Sci. 67*:1642 (1978).
99. S. S. Atuma and J. Linquist, *Analyst (London) 98*:886 (1973).
100. S. S. Atuma, J. Linquist, and K. Lundstrom, *Analyst (London) 99*:683 (1974).
101. A. Y. Taira, *J. Assoc. Off. Anal. Chem. 57*:910 (1974).
102. E. Jacobsen and K. B. Thorgersen, *Anal. Chim. Acta 71*:175 (1974).
103. J. M. Moore, *J. Pharm. Sci. 58*:1117 (1969).
104. A. M. Mayo and O. Nudelman, *J. Assoc. Off. Anal. Chem. 56*:1460 (1973).
105. B. Breyer and T. Biegler, *J. Electroanal. Chem. 1*:453 (1959/60).
106. M. E. Schertel and A. J. Sheppard, *J. Pharm. Sci. 60*:1070 (1971).
107. J. Ballantine and A. D. Woolfson, *J. Pharm. Pharmacol. 32*:353 (1980).
108. W. D. Mason, T. D. Gardner, and J. T. Stewart, *J. Pharm. Sci. 61*:1301 (1972).
109. W. D. Mason, *J. Pharm. Sci. 62*:999 (1973).
110. J. T. Stewart, H. C. Loo, and W. D. Mason, *J. Pharm. Sci. 63*:954 (1974).
111. L. F. Cullen, M. P. Brindle, and G. J. Paperiello, *J. Pharm. Sci. 62*:1708 (1973).
112. A. Cinci and S. Silvestri, *Farmaco, Ed. Prat. 27*:28 (1972).
113. P. B. Brunai, A. Cinci, and S. Silvestri, *Farmaco, Ed. Prat. 27*:89 (1972).
114. N. M. Omar and N. A. El-Rabbat, *Can. J. Pharm. Sci. 9*:57 (1974).
115. V. Riganti and G. Spini, *Farmaco Ed. Sci. 28*:243 (1973).
116. A. V. Bogatskii, S. A. Andronati, V. P. Gul'tyai, Y. I. Vikhlyaev, A. F. Galatin, Z. I. Zhilina, and T. A. Klygul, *J. Gen. Chem. USSR 41*:1364 (1971).
117. S. A. Kaplan, "The Role of In Vitro Dissolution Testing in the Development of a New Drug," Arden Housing Conference on Dissolution Technology, February 11, 1973.
118. Zs. Feher, G. Nagy, K. Toth, and E. Pungor, *Analyst 99*:699 (1974).

119. M. R. Hackman and M. A. Brooks, *J. Pharm. Sci. 67*:842 (1978).
120. S. A. Kaplan and S. Cotler, *J. Pharm. Sci. 61*:1361 (1972).
121. M. A. Brooks, J. A. F. de Silva, and M. R. Hackman, *Am. Lab. 5*(12):23 (1973).
122. R. W. Lucek and C. B. Coutinho, *Mol. Pharmacol. 12*:621 (1976).
123. J. A. F. de Silva and M. A. Brooks, in *Drug Fate and Metabolism*, Vol. 2 (E. Garrett and J. Hirtz, eds.), Marcel Dekker, New York, 1978, pp. 1-48.
124. M. A. Brooks, in *Analytical Chemistry Symposia Series* (W. F. Smyth, ed.), Vol. 2: *Electroanalysis in Hygiene, Environmental, Clinical and Pharmaceutical Chemistry* (Proceedings of Conference), Elsevier Scientific, Amsterdam, 1980, pp. 287-298.
125. P. O. Kane, *Nature 183*:1674 (1959).
126. H. Oelschlager, *Arch. Pharm. Berl. 296*:396 (1963).
127. H. Oelschlager, J. Volke, and E. Kurek, *Arch. Pharm. Ber. 297*:431 (1964).
128. P. L. Cox, J. P. Heotis, D. Polin, and G. M. Rose, *J. Pharm. Sci. 58*:987 (1969).
129. M. A. Brooks, J. A. F. de Silva, and L. D'Arconte, *J. Pharm. Sci. 62*:1395 (1973).
130. R. Dugal, G. Caille, and S. F. Cooper, *Union Med. Can. 102*:2491 (1973).
131. J. A. F. de Silva, I. Bekersky, M. A. Brooks, R. E. Weinfeld, W. Glover, and C. V. Puglisi, *J. Pharm. Sci. 63*:1440 (1974).
132. J. A. F. de Silva, I. Bekersky, and M. A. Brooks, *J. Pharm. Sci. 63*:1943 (1974).
133. J. A. F. de Silva, C. V. Puglisi, and N. Munno, *J. Pharm. Sci. 63*:520 (1974).
134. M. A. Brooks and J. A. F. de Silva, *Talanta 22*:849 (1975).
135. D. J. Berry, *Clin. Chim. Acta 32*:235 (1971).
136. E. Jacobsen, T. V. Jacobsen, and T. Rojahn, *Anal. Chim. Acta 64*:473 (1973).
137. M. R. Hackman, M. A. Brooks, J. A. F. de Silva, and T. S. Ma, *Anal. Chem. 46*:1075 (1974).
138. M. A. Brooks, J. A. F. de Silva, and L. M. D'Arconte, *Anal. Chem. 45*:263 (1973).
139. J. A. F. de Silva, N. Munno, and N. Strojny, *J. Pharm. Sci. 59*:201 (1970).
140. M. A. Brooks, L. D'Arconte, and J. A. F. de Silva, *J. Pharm. Sci. 65*:1 (1976).
141. J. A. F. de Silva and M. R. Hackman, *Anal. Chem. 44*:1145 (1972).
142. J. W. Munson and H. Abdine, *Talanta 25*:221 (1978).
143. J. W. Munson and H. Abdine, *J. Pharm. Sci. 67*:1775 (1978).
144. M. A. Brooks and M. R. Hackman, *Anal. Chem. 47*:2059 (1975).

# 22

# Electrochemical Detection in Liquid Chromatography and Flow Injection Analysis

PETER T. KISSINGER / Department of Chemistry, Purdue University, and Bioanalytical Systems, Inc., West Lafayette, Indiana



## I. INTRODUCTION

### A. The State of the Art

Liquid chromatography with electrochemical detection (LCEC) is coming into widespread use for the trace determination of easily oxidizable and reducible organic compounds. Detection limits at the 0.1-pmol level have been achieved for a number of oxidizable compounds. Due to problems with dissolved oxygen and electrode stability, the practical limit of detection for easily reducible substances is currently about 10-fold less favorable. As with all detectors, such statements of the minimum detectable quantity must only be considered with the proverbial grain of salt. Detector performance varies widely with the analyte and the chromatographic conditions.

The modern interest in electrochemical detectors for liquid chromatography was stimulated by the recognition that this technique was ideal for the study of aromatic metabolism in the mammalian central nervous system. Most of the papers published during the past eight years have focused on the applications of the LCEC technique to neurochemical problems. Since the first commercial detectors became available in 1974, a number of other areas of application have been explored. A running bibliography of LCEC applications is frequently updated and provides a useful overview of current applications [1]. The basic concepts of LCEC have been recently reviewed in several places [2-5].

Liquid chromatography (LC) and hydrodynamic electrochemistry are, for the most part, very compatible technologies which in combination yield important advantages for a number of trace determinations. In order of decreasing importance, the three major advantages are selectivity, low detection limits, and low cost. The use of modern LC for residue determinations requires a selective detector with a rapid response time, wide dynamic range, and low active dead volume ($<20$ $\mu$L). Because electrochemistry is a surface technique, small-volume transducers ($<1$ $\mu$L) can easily be constructed which meet these criteria. Figure 22.1 illustrates the basic concept. A redox reaction supplies (or removes) electrons to/from an electronic conductor (the electrode). The resulting current is proportional to the concentration of reactant passing through the cell.

Direct-current amperometry (the measurement of electrochemical current in response to a fixed electrode potential) continues to be the most widely used finite current electrochemical technique. Popular applications include end-point detection in volumetric and coulometric titrations, and the measurement of oxygen activity in gases and liquids. Recently, a number of publications have pointed to the ways in which amperometry can be used to measure the activity of redox enzymes or the concentration of substrates for redox enzymes [6]. In the latter case, the enzyme itself is used as a reagent in the solution, trapped behind or in a membrane, or even covalently bound to the electrode surface. It is likely that these developments will ultimately be incorporated in LC detectors.

It is now clear that electrochemical measurements can often have significant advantages over the classical spectroscopic approaches. Amperometry can be more specific and therefore lower detection limits are often feasible. Because electrochemical detectors do not require an optical carrier, they can be much less expensive than UV absorption or fluorescence detectors. This is especially true when one con-

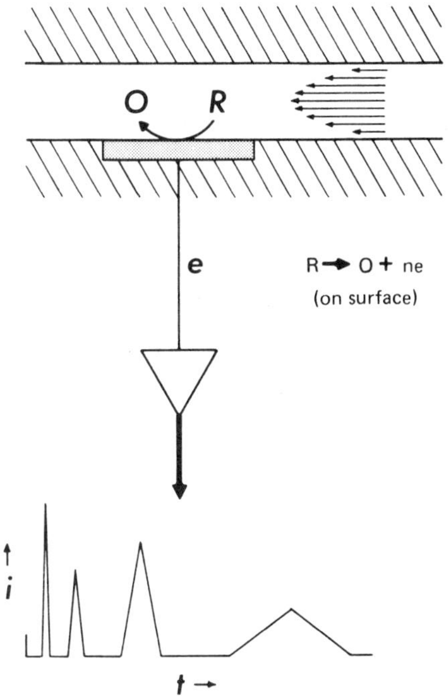


Figure 22.1 Schematic view of thin-layer amperometric detection.

siders that electrochemical detectors are inherently tunable without need for such things as monochrometers or filters. On the other hand, there can be significant problems with reliability, and more often than not, there is a lack of acceptance by chemists weaned on Beer's law. Amperometric methods in biochemistry are just beginning to be commercialized and it is now almost certain that they will come into widespread use.

Another recent development is the advent of pulse amperometry in which the potential is repeatedly pulsed between two values. The current at each potential or the difference between these two currents ("differential pulse amperometry") can be used to advantage for a number of applications. Similar advantages can result from the simultaneous monitoring of two (or more) electrodes poised at different potentials. In the remainder of this chapter it will be shown how the basic concepts of amperometry can be applied to various liquid chromatography detectors. There is not one universal electrochemical detector for liquid chromatography, but rather a family of different devices which have advantages for particular applications. In some cases the selectivity of hydrodynamic amperometry is such that it can be used without a prior chromatographic separation, as in flow injection analysis (FIA). Although such applications are rare, they utilize the same transducer technology developed for LCEC.

## B. Fundamental Principles

As described in Chapter 3, amperometry is normally carried out in stirred (or flowing) solutions or at a rotated electrode. Current is measured as the compound undergoes an oxidation or reduction at the electrode held at a fixed operating potential (strictly speaking, the *potential difference* between the electrode and the solution is held constant). Hydrodynamic voltammetry, a steady-state technique from which amperometry is derived, is used to select the operating potential. Thermodynamic data (e.g., tables of E° values), although quite valuable for some purposes, are often insufficient for determining the operating potential for most amperometric methods. Electrode kinetics, which depend on a number of factors (electrode material, choice of electrolyte, etc.), also plays a significant role in practice. Often a greater potential is required for amperometry than would be predicted on the basis of equilibrium thermodynamics.

LCEC is a special case of steady-state hydrodynamic chronoamperometry. In LCEC, the concentration changes as the chromatographic zones flow past the detector. The electrode is operated in the limiting current region for the eluted compounds even though the concentration varies as the zones enter and leave the detector compartment. It is important to note that the volume of solution in the active region of the typical electrochemical detector (ca. 1 μL) is very small compared to the volume occupied by the typical chromatographic zone (ca. 1 mL).

The resolution of the column provides much of the selectivity in LCEC and therefore the practical limitations of amperometry are circumvented to a large extent. Nevertheless, amperometry is more often than not used to improve the selectivity of an LC method. Compounds that oxidize or reduce at low potentials can be detected with great selectivity.

Adjustment of the electrode potential and the use of complex excitation functions (pulse amperometry) can improve the selectivity of amperometric detection. However, amperometry in standard solutions is one thing and amperometry in complex samples is quite another. In the real world, direct amperometric detection without prior cleanup is useless in most applications. Proteins, lipids, and industrial surfactants can contribute to electrode fouling and coating, often rendering the electrode inoperative. This problem has been overcome to a large degree by (1) the addition of membranes over the electrode surface which can exclude molecules on the basis of permeability (2) the use of various extraction techniques to "clean up" the sample, and (3) the introduction of a liquid chromatography column prior to the amperometric transducer to separate the desired component from possible surface contaminants or other reactive compounds that might interfere. With these points in mind, amperometry can be a very reliable detection approach for even the most complex mixtures.

## II. TRANSDUCER DESIGN

### A. Electrode Materials

The choice of electrode material is more critical in LCEC than for the usual electroanalytical experiment, primarily due to the mechanical ruggedness and long-term stability required. Carbon paste (an admixture of graphite powder and a dielectric material) remains a popular choice as an electrode material for LCEC. Unfortunately, it will not stand up to most nonaqueous solvents, nor will it work well under the high-volume flow rates characteristic of recent chromatographic separations. While "carbon paste" can be used in nonaqueous solvents if formulated using high-molecular-weight waxes or various polymers (e.g., Kel-F) as the dielectric binder, vitreous ("glassy") carbon has been generally found to be more satisfactory. Vitreous carbon has excellent mechanical and electrical properties and is relatively free of impurities. It can be brought to a high surface polish using standard metallographic techniques. Besides its solvent resistance, glassy carbon has a more favorable background current at negative potentials than the usual carbon paste formulations, making it possible to carry out more difficult reductions.

It is not yet possible to make detailed comparisons between different electrode materials because organic electron transfer reactions at surfaces are not well understood at a molecular level. Reactions favorable (fast, reversible) at one surface may be unfavorable at another surface with (apparently) similar properties to the first. Perhaps the most important characteristic of an electrode for hydrodynamic amperometry is "the background current and noise at a potential where the compound of interest reacts at a mass transfer limited rate." The fact that one electrode gives a lower background current than another at a given potential is not by itself indicative of a better electrode. The electron-transfer rates for the analyte(s) of interest might also be lower and the resulting signal-to-noise ratio could be much worse. At the present time, it would appear that both glassy carbon and carbon paste have advantages in given situations. Nevertheless, the superior solvent resistance of glassy carbon will in many cases decide the issue.

A number of other carbon materials have been used for electrochemical detection recently [6]; however, at this point none of these appear to have a clear advantage over the electrodes described above. Nevertheless, these alternative materials certainly do "work" and there is little doubt that we will continue to see additional entries as the search for the ideal electrode continues.

Mercury is the electrode material of choice for many electrochemical reductions and some unique oxidations (see Chap. 9). We have explored the use of both small mercury pools and amalgamated gold disks in thin-layer amperometry. Other workers have used pools in a capillary tube [7] and amalgamated platinum wire [8]. In 1979 Princeton Applied Research introduced a unique approach based on their Model 303 Static Mercury Drop Electrode (see Sec. II.F). Our laboratory and MacCrehan et al. [9] have focused on the use of amalgamated gold disks. This approach results in an inexpensive, easily prepared, and mechanically rigid electrode which can be used in conventional thin-layer cells (Sec. II.C) of the type manufactured by Bioanalytical Systems.

Metal electrodes other than mercury have been used for a few inorganic LCEC applications, but appear to be unsuitable for most organic applications. Metal oxide electrodes (including thin-film semiconductors) show some promise, but nothing of substance has yet been published with regard to LCEC.

It would appear certain that the most important need in LCEC is the development of improved electrode materials. It may be possible in the near future to design an electrode that will give superior performance for certain classes of compounds. Modifying electrode surfaces by covalent attachment of various ligands or electron-transfer catalysts (perhaps even enzymes) may well provide the key to better amperometric devices for all sorts of analytical purposes. Research in the area of chemically modified electrodes (CMEs) has been reviewed [6,10]. Those interested in improving the performance of electrochemical detectors would do well to study these developments in detail.

## B. Mobile-Phase Limitations

All detectors place limitations on mobile-phase composition to some degree; however, in electrochemical detection one must be conscious of the fact that a complex surface reaction is involved which depends on both the physical and chemical properties of the medium. Considerable effort may be required to optimize an LCEC determination. Both column and detector must be considered together. In fact, the entire chromatograph may be considered part of the electrochemical cell. Fortunately, it is possible to make some generalizations regarding chromatographic mobile phases compatible with the electrochemical detector. For all practical purposes nonpolar organic solvents cannot be used because of their inability to support a significant ionic strength, necessary for conductivity. This rules out classical "normal-phase" separations on silica or alumina columns.

The majority of LC separations performed today are reverse phase, including ion exchange and ion-pair methods. The mobile phases are generally aqueous solutions, with or without organic modifiers (e.g., methanol, acetonitrile, tetrahydrofuran). Retention is altered by adjusting the modifier concentration, the pH, or the ionic strength, or by adding ion-pairing agents. In some cases temperature control can also be very useful. Mobile phases of this type are generally ideal for electrochemistry, which requires that the mobile phase be able to carry an ionic current, be chemically and electrochemically inert in the desired potential range, have a convenient liquid temperature range, and dissolve the analyte of interest.

When buffered solutions are used they are usually in the moderate to low ionic strength range (0.010-0.10 M) to provide adequate conductivity while minimizing the contribution to the background current of impurities found in the buffer reagents. Trace impurities found in the mobile phase are especially a problem when reductive electrochemical detection techniques are employed. Very minute amounts of electrochemically reducible metal ions, which apparently cannot be totally eliminated during the manufacturing process, become a very major source of baseline problems.

Electrolysis of the mobile phase prior to the LC pump can be an effective way to minimize such problems and may be an essential procedure when very low detection limits are required. Good reagents and pure solvents are clearly the best solution to this problem.

An even greater problem in reductive LCEC determinations is the removal of dissolved oxygen from the mobile phase. As in dc polarography, dissolved oxygen can cause serious problems. If it is not removed from the mobile phase and sample prior to injection, and precautions are not made to prevent oxygen from reentering the LC system after deoxygenating, the utility of this detection scheme is greatly limited. Even at relatively low working electrode potentials, very large background currents are produced by the reduction of oxygen; however, the severity of the problem depends on the electrode material (mercury or carbon).

Totally nonaqueous mobile phases (such as acetonitrile, DMF, and DMSO) can be used with electrochemical detection but require salts such as tetrabutylammonium hexafluorophosphate or tetrafluoroborate. Using these totally nonaqueous solvents can give a distinct advantage when attempting to detect very difficult to oxidize or reduce compounds. Many times the potential limits in a particular electrochemical system are determined by solvent breakdown (e.g., $H^+$ reduction, $H_2O$ oxidation). In aqueous solutions, the limits are from about +1.2 to −1.2 V, depending on the electrode. In dry acetonitrile, with either of the salts mentioned above, the limits can be extended from +3 to −3 V. Many reactions inaccessible in water become facile at high potentials; examples include oxidation of alcohols, aromatic hydrocarbons, and reduction of activated alkenes. The major problem in using these reactions for trace determinations is the difficulty in eliminating water from the mobile phase. Water must be removed to the ppm level to avoid major background current problems. In addition, the higher impedance of nonaqueous electrolytes makes noise pickup more of a problem than for water. These problems are not easily solved, but the overall utility of electrochemical detection could be greatly expanded by the advent of nonaqueous LCEC.

## C. Thin-Layer Cells

The most popular electrochemical detectors to date have been based on the amperometric conversion of analyte in a thin-layer cell. The basic functioning of this mode of detection is depicted schematically in Figure 22.2. As illustrated, an electrochemically active substance passes over an electrode held at potential sufficiently great (positive or negative) for an electron transfer (either an oxidation or reduction) to occur. An amperometric current is produced which is proportional to the concentration of analyte entering the thin-layer cell. As the concentration changes as a function of time due to elution of the analyte from the chromatographic column, so does the current; thus the usual chromatographic trace is recorded when current is displayed as a func-

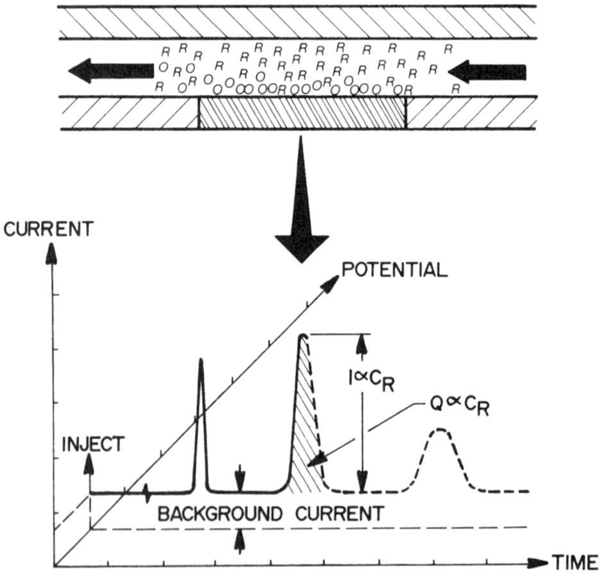


Figure 22.2 At any instant in time, the current in an LCEC experiment reflects the rate of conversion of reactant (R) to product (O).

tion of time. Thin-layer cells have excellent flow characteristics and a high ratio of electrode surface area to solution volume. In contrast to tubular electrodes (see below), the thin-layer geometry is suitable for essentially all electrode materials and provides for convenient surface polishing. Thin-layer cells can be used as either amperometric or coulometric detectors, depending on the mass transport efficiency of the cell (related to electrode area, channel thickness, and flow rate). All faradaic electrochemistry ultimately depends on Faraday's law:

$$Q = nFN$$

where the number of coulombs (Q) passed is directly proportional to the number of moles (N) converted to product in the chromatographic zone, number of electrons (n) involved in the reaction, and the Faraday constant ($F = 9.65 \times 10^4$ C/eq). The instantaneous current is given by

$$i_t = \frac{dQ}{dt} = nF\frac{dN}{dt} = 9.65 \times 10^4 \times \text{(equivalents converted per second)}$$

A current of $10^{-11}$ A (easily measured in practice) therefore corresponds to a conversion rate of $10^{-16}$ eq/s. The detection limit actually achieved will; of course, depend on the chromotographic efficiency, the detector conversion efficiency, and the baseline noise. Unfortunately, these three are not independent.

Amperometric detectors are those operated in the limit where the concentration of the analyte entering and leaving the detector cell is the same (i.e., conversion efficiency is 0%). Coulometric detectors are those where the concentration of the entering analyte is reduced to zero as it exits (i.e., conversion efficiency is 100%). For practical purposes, amperometric detection may involve a conversion efficiency up to ca. 5% and for coulometric detection a conversion efficiency above 95%. In between, the technique may be termed quasi-amperometric. These designations are, of course, arbitrary and depend on the specific situation. From a mathematical point of view, an amperometric determination is one where the current output is dependent on the cube root of the linear velocity across the electrode surface as described by Levich's hydrodynamic equations for laminar flow. Conversely, the current response for a coulometric detector is directly proportional to the velocity of the flowing solution. Everything in between is in the quasi-amperometric region. While the mathematics describing these things is quite interesting, it is beyond the scope of this chapter.

The question arises as to which mode, amperometric or coulometric, is better suited for electrochemical detection for LC. At present, the amperometric detectors have the advantage of being easier to construct and have lower detection limits. To obtain a coulometric response, the electrode must have a large surface area. This can be achieved in practice only at the expense of a larger dead volume and residence time. Such cells are difficult to construct. The electrodes are usually beds made from conducting particles or fibers to increase the surface area-to-volume ratio (see Sec. II.E). These are not often practical for routine use. While coulometric detection does give an absolute measurement for those compounds completely resolved chromatographically, this is not often a significant advantage for LC. As the electrode surface area is increased to improve efficiency, each added increment contributes proportionately less to the total amount converted, but approximately equal to the background current from solvent, electrolyte, and electrode breakdown. Detector noise increases with the increasing background current developed by these large electrode areas and the signal-to-noise ratio decreases. Although less material is converted in the amperometric mode, lower detection limits are attainable. It should, however, be obvious that extremely low conversion efficiencies are also not desirable because the current becomes extremely small and noise from the electronics will take over to defeat us. There does not appear to be an optimum cell geometry which will hold up for all applications, but performance will generally be adequate in the quasi-amperometric mode.

The iR drop commonly associated with thin-layer electrochemical cells has not been a significant problem for most LCEC applications because aqueous mobile phases of relatively high ionic strength have been used. For the commercial version of this cell, a linear range from 10 pg to 200 ng injected is easily achieved for the typical separation of small molecules (typically MW $\simeq$ 200) with capacity factors between about 2 and 5. Obviously, many factors can influence the detector response in a given case, including volume flow rate, channel thickness, chromatographic efficiency, ionic strength, applied potential, electron transfer rate constant, temperature, and background current (chromatographic "baseline"). When thin-layer detectors are used in nonaqueous solvents the iR-drop problem can be very severe due to the poor conductivity. In addition, severe iR problems can exist when operating at extremes of potential (positive and negative) where the background current is very great (often far greater than the chromatographic peak heights). Although the background current is always "bucked out" by the electronics and is therefore not "seen" on the chart recorder, it can be the major contributor to the iR drop.

Although electronic compensation for this problem is possible by using positive feedback, the problem cannot be completely solved by this means due to the fact that a significant portion of the iR drop exists along the electrode surface rather than perpendicular to the electrode as in the usual electrochemical experiment. The current between the working and auxiliary electrodes passes along the thin-layer channel, including that portion adjacent to the electrode (Fig. 22.3A). Therefore, the potential between the working electrode and the solution is not uniform across the face of the electrode. The potential difference at the downstream edge of the electrode will be closest to the value controlled by the potentiostat, whereas the upstream potential may be insufficient to oxidize or reduce the compounds of interest! The best solution to this problem is to position the auxiliary electrode opposite the working electrode so that the charge passes perpendicular to the working-electrode surface (Fig. 22.3B). This geometry ensures both that the uncompensated iR drop will be extremely small (due to the thinness of the layer) and that the potential will be uniform at all points on the working-electrode surface. A more detailed treatment of this problem requires rather complex mathematics; however, the influence on the design of practical detector cells is important and can be easily understood from a qualitative point of view.

Figure 22.4 illustrates five cell geometries which we have investigated for a number of years and are currently using for various applications. Figure 22.4A illustrates the conventional geometry with both the reference (R) and auxiliary (A) electrodes mounted downstream. This design is relatively inexpensive, but can suffer from iR drop problems as described above. It is important to recognize that for this design the three-electrode system does not compensate for *any* significant solution resistance, but does have the advantage of not drawing current through the reference electrode. At low currents the auxiliary electrode is not needed and the cell can be operated with the reference electrode serving as a classical counter electrode (as in

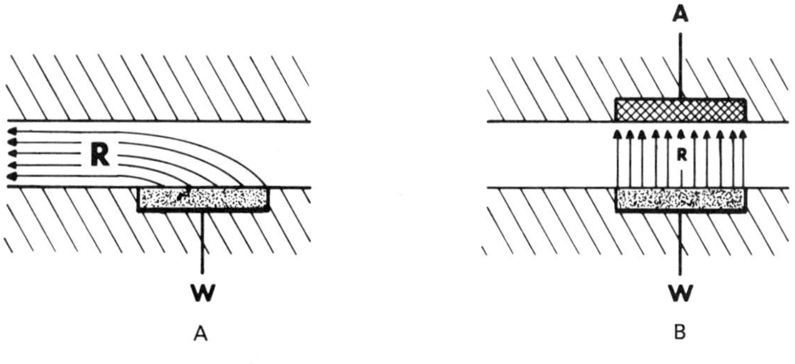


Figure 22.3 Thin-layer experiments are subject to large iR drops under certain conditions (A). Placement of the auxiliary electrode across the channel from the working electrode (B) minimizes this problem.

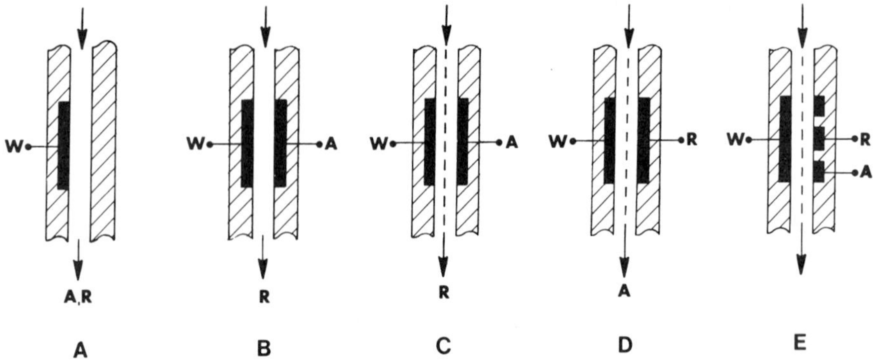


Figure 22.4 Five geometries for thin-layer electrochemical cells. Relative placement of the working (W), reference (R), and auxiliary (A) electrodes is shown.

two-electrode polarography or Clark amperometric oxygen electrodes). This cell has the disadvantage that it is not suitable for collection of fractions (especially radiolabeled metabolites from metabolism studies).

Figure 22.4B illustrates a simple modification in which a planar auxiliary electrode is placed opposite the working electrode. This is an excellent design for many applications and commercial cells of this type have been available for several years (Bioanalytical Systems). The thin-layer channel serves as a salt bridge to the reference electrode, but since no current passes along the thin layer, there is no resulting iR drop in the solution. In most instances the transit time for molecules along the electrode face is short compared to the diffusion time *across* the thin layer of solution. Therefore, no interference occurs between chemical events taking place at the working and auxiliary electrodes. Some attention should be paid to this problem with large electrode areas, slow flow rates, and very thin gaskets, in which case the geometry shown in Figure 22.4C can be advantageous. The problem here is that cell construction and maintenance becomes awkward when a porous diffusion barrier must be incorporated in the cell to isolate the auxiliary electrode.

Placing the reference electrode probe opposite the working electrode in the wall of the thin channel can best be accomplished by use of an ionic junction constructed from a porous material which permits ions to migrate but blocks the flow of electrolyte solution (Fig. 22.4D). Porous Vycor glass ("thirsty" glass, Corning 7930 glass) or ion exchange membranes (DuPont Nafion 811) are useful for ths purpose. With the potential-sensing element across the stream from the working electrode the uncompensated solution resistance is small (but not uniform) and the auxiliary electrode can be placed downstream with relatively little ill effect. The compensated resistance is, however, quite large and the current must pass along the layer of solution adjacent to the electrode.

Configurations 22.4B and C are superior on fundamental grounds because the current density across the electrode will not be significantly influenced by iR drop and the potential of the auxiliary electrode will remain low due to the low impedance between A and W.

Configuration 22.4E is perhaps most satisfactory from a fundamental point of view. This design also provides for convenient collection of fractions. Nevertheless, construction difficulties presently make this cell less desirable for routine work. In summary, configurations 22.4A and B are the most practical approach to rugged cells capable of routine use for many months. With the geometry illustrated by B the linear range of the detector extends to over six orders of magnitude even in mobile phases with poor conductivity. In general, configuration 22.4A can be recommended for most trace applications and 22.4B for those cases where either a high background current or

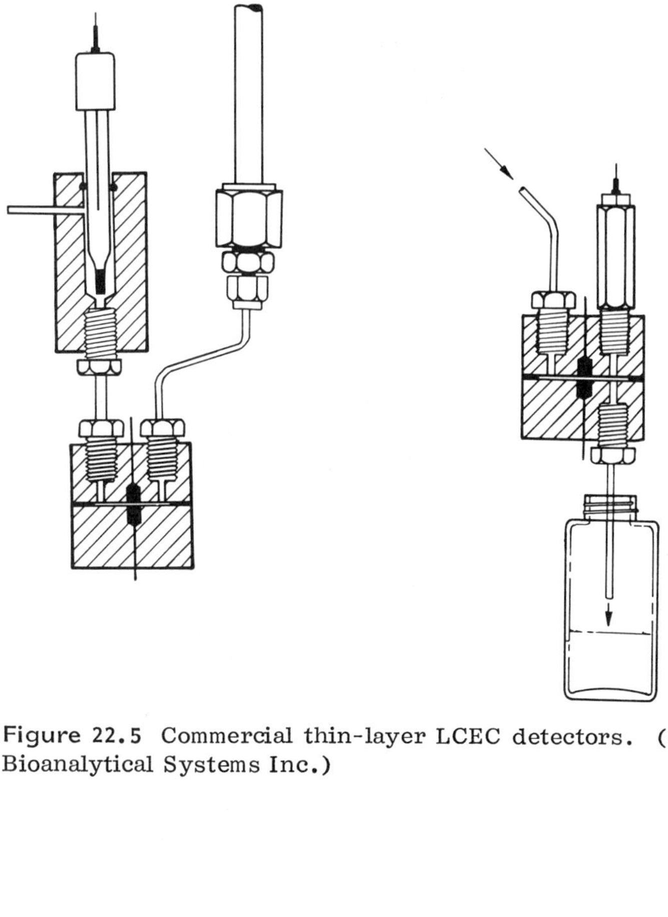

Figure 22.5 Commercial thin-layer LCEC detectors. (Courtesy of Bioanalytical Systems Inc.)

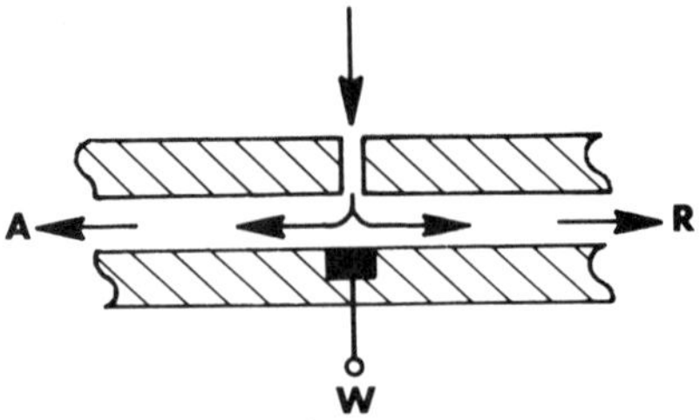


Figure 22.6 Wall-jet electrochemical cell. (From Ref. 11.)

high resistance causes significant iR problems. Figure 22.5 illustrates two commercially available detectors of this general type.

While thin-layer cells have most commonly been used with flow parallel to the electrode surface as described above, several commercial detectors have employed the so-called wall-jet principle and operate with flow perpendicular to the electrode surface as illustrated in Figure 22.6. This apparently is intended to provide more effective mass transfer to the electrode surface; however, in contests to establish superior performance, conventional thin-layer cells have consistently come out the winner. Whether this is a result of engineering or theory (or both) has not been determined.

## D. Tubular Electrodes

Flow through electrochemical detectors based on a cylindrical geometry have been used for many years for a variety of purposes. For 20 years Walter Blaedel and his co-workers at the University of Wisconsin have had primary responsibility for stimulating interest in so-called tubular electrodes, although several other groups have done pioneering work as well [6]. Figure 22.7 illustrates three tubular electrode configurations. All three geometries have been used for LCEC applications and all three are now commercially available or will be in the near future. Electrochemists like open tubular electrodes (Fig. 22.7A) because the mass-transfer mathematics for the cylindrical geometry is relatively easily solved. The problem with this approach to LCEC detectors is that the surface of the electrode is relatively difficult to polish (compared to planar thin-layer cells), some electrode materials cannot be easily used (e.g., mercury or carbon paste), and it is not possible to use a large surface area-to-volume ratio without encountering a severe iR drop (since the auxiliary and reference electrodes cannot be easily placed in an optimum position). In many cases these problems are not particularly significant and satisfactory results have been obtained.

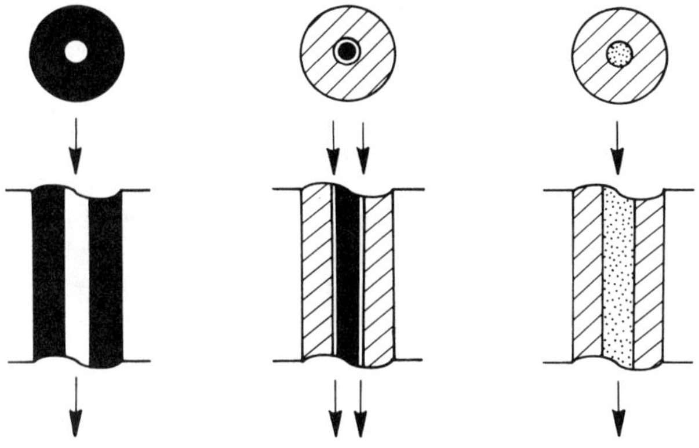


Figure 22.7 Tubular working electrodes for hydrodynamic electrochemistry: (A) open tubular electrode; (B) wire electrode in capillary; (C) packed-bed tubular electrode.

The wire in a capillary geometry (Fig. 22.7B) has been used in two forms by electrochemists. In one case the capillary is an impervious diffusion barrier and in the other case a capillary is used which is capable of transporting ions in a radial direction. In the latter situation, the auxiliary electrode can be positioned in a solution outside the capillary in order to minimize uncompensated iR drop. Porous Vycor class (Corning 7930 glass) and ion exchange membranes (DuPont Nafion) have been used for the capillary, but in both cases construction difficulties prohibit the production of reliable devices capable of withstanding a variety of mobile phases. A wire placed in ordinary glass or in a hole drilled in a chemically resistant plastic is not quite so elegant, but will work very well for most low-current LCEC applications. Johnson and his group at Iowa State have reported a very simple cell which functions in the quasi-amperometric mode (a typical conversion efficiency of 35%) using a platinum wire electrode [12]. While centering the wire in the capillary would help the theoreticians, this is not necessary for a practical device and greatly simplifies construction of the detector cell.

## E. Packed-Bed Electrodes

Tubular packed-bed electrodes are extremely useful for electrosynthetic purposes and adapting them for coulometric LCEC detection is a worthwhile endeavor as well (Fig. 22.7C). Johnson and his students have also pioneered in this area [6]. As for the wire in a capillary, one can distinguish between electrodes packed in an impervious tube (usually a metal or graphite) and packed in a porous ionic conductor (usually porous Vycor). If the particles in the bed are small (typically metal chips or graphite powder), coulometric behavior can be achieved in a relatively short bed (a few millimeters) at typical LC flow rates as long as the current is low (always true for trace analysis). Under these conditions an impervious bed works fine and the auxiliary and reference electrodes can be conveniently placed downstream (as shown in Fig. 22.4 for a thin-layer cell). If a high material throughput is necessary (higher concentration at a higher linear velocity), it is essential to operate the cell with current perpendicular rather than parallel to the axis to minimize iR drop. In such cases a porous tube is necessary. An example of a coulometric flow cell of this type is shown in Figure 22.8. This cell has been used for synthetic purposes and to remove electroactive contaminants from LC mobile phases.

Packed-bed electrodes need not be cylindrical. Takata and Muto reported on a rectangular design some years ago in which a bed of carbon fibers was used for the electrode material [13]. While a number of

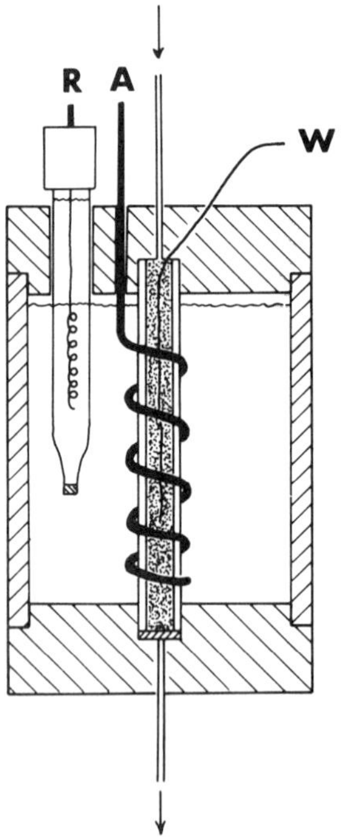


Figure 22.8 Flow-through packed-bed coulometric reactor.

innovative applications were reported, for the reasons described earlier, coulometric cells do not provide detection limits competitive with what can be achieved using amperometric detectors.

## F. Dropping Mercury Electrodes

While various dropping mercury electrodes (DMEs) have long been proposed as LC detectors, it is fair to say that they have never been put to significant practical use for this purpose. Kemula innovated in this area in 1952 [14]. As the efficiency of LC columns improved it became essential to reduce the volume of DME-based detector cells and significant early work was published by Koen et al. [15], Stillman and Ma [16], and Fleet and Little [11], among others. Very recently, several papers have appeared which advocate the use of a rapidly dropping DME using a nonvertical (or even horizontal) capillary. Michel and Zatka were apparently the first to try this [17] and subsequent reports have appeared from Frei's group in Amsterdam [18,19] and from Kemula's group in Warsaw [20]. The active volumes of these devices is adequately small for modern liquid chromatography. Unfortunately, the very concept of a dropping mercury electrode is fraught with problems of both a fundamental (high double-layer charging due to changing surface area) and a practical nature (clogged capillaries, mercury toxicity). At this date, the minimum detectable quantities achieved with DMEs are not competitive with stationary electrodes. Nevertheless, the renewable surface and the high hydrogen overpotential of pure mercury are decided advantages which may give the DME a competitive edge for some applications. I personally doubt that the DME will last as a competitive analytical device (for any purpose) beyond the turn of the century. Even now, there usually is a better way to carry out analytical electrochemical measurements.

## G. Multiple-Electrode Detectors

While most practical LCEC experiments will continue to be carried out with transducers incorporating a single working electrode, it is a relatively simple matter to monitor the current at several working electrodes simultaneously. The electronics needed to accomplish this are very straightforward (Chap. 6) and in principle any number of electrodes could be used. In practice, the use of more than two electrodes is rather awkward and expensive. Figure 22.9A schematically depicts a thin-layer cell with two working electrodes in parallel, with a corresponding auxiliary electrode across the channel and a reference electrode downstream. Using this arrangement, one can monitor the amperometric current at two different applied potentials. This is perfectly analogous to the "dual-wavelength UV absorption detector" and the output is typically plotted using a dual-pen recorder. This

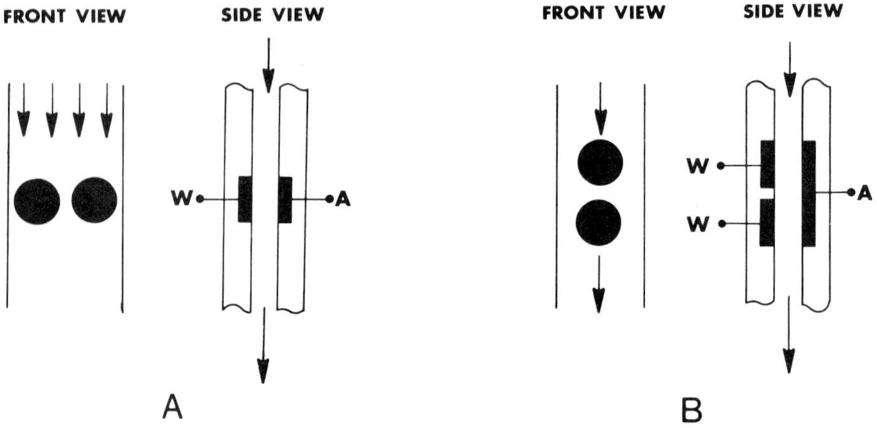


Figure 22.9 Dual parallel (A) and dual series (B) thin-layer amperometric transducers.

arrangement is particularly useful when one wishes to quantitate a very easily oxidized (or reduced) substance in the presence of others which react at higher energies. The selectivity for the easily reacted substance can thus be excellent and the other compounds detected as well. The parallel dual-electrode arrangement can also provide good qualitative information in that the ratio of the response at each electrode will be characteristic of a particular hydrodynamic voltammogram (just as the absorbance ratio at two wavelengths is characteristic of a given UV spectrum).

Figure 22.9B illustrates another arrangement, the series dual-electrode transducer. This device is used in the same manner as the classical ring-disk electrode (see Chap. 3). The products of the upstream electrode are monitored at the downstream electrode. This can be useful to enhance selectivity when the electrolysis products can be detected in a more favorable potential region than was necessary to carry out the original reaction. In addition, some electrochemical reactions are "chemically irreversible" and some are "chemically reversible," and therefore the dual-electrode arrangement can eliminate the response to irreversible processes while enhancing the response to reversible reactions. Both the parallel and series arrangements provide most of the advantages of differential pulse operation without the disadvantage of enhancing double-layer charging current. Instrumentation for LCEC based on these ideas is available from Bioanalytical Systems and a unique cell consisting of coulometric reactors in series is available from Environmental Sciences Associates.

Figure 22.10 illustrates a third dual-electrode arrangement which permits enhancement of the response by reversible redox cycling.

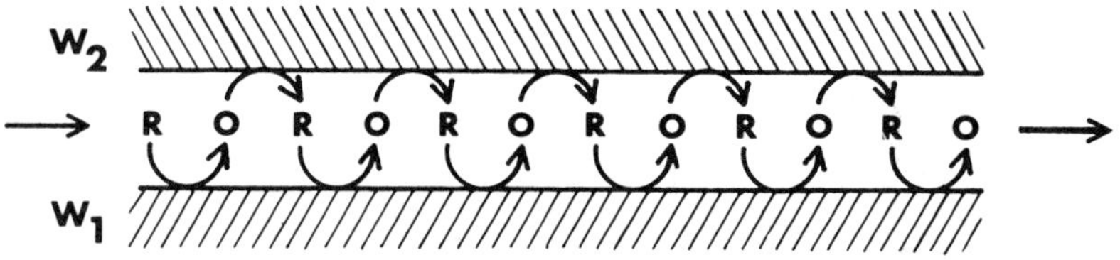


Figure 22.10 Parallel opposed thin-layer amperometric transducer.

Many more electrons are therefore transferred than would be the case with a single electrode and the current is amplified dramatically. This concept does not work well with conventional LC columns because the volume flow rate is too large to permit a significant number of redox cycles. Nevertheless, the concept is certainly interesting and as reverse-phase capillary columns are developed it may well have some practical value. A detailed treatment of multiple-electrode LCEC has been published [5].

## H. Performance Criteria

The terms "sensitivity" (or "response factor") and "detection limit" (or "minimum detectable quantity") are frequently confused and used in a manner that is impossible to interpret. It is now generally accepted that sensitivity should refer to output/input. In the case of a chemical determination this is generally a measured signal (current, voltage, absorbance, etc.) divided by an amount (weight in grams, or moles of substance). For electrochemical detection, sensitivity for a given substance under given chromatographic conditions is described in terms of "peak current per injected equivalents" (nA/neq or nA/nmol for a one-electron transfer). Other expressions are convenient for routine use, such as nA/ng, but make it more difficult to compare results under different conditions. For thorough work, the sensitivity in terms of "peak area (coulombs) per injected equivalents" is a useful measure because it permits direct calculation of the conversion efficiency of the cell.

Sensitivity by itself does not tell us anything about the minimum detectable quantity. It is also necessary to evaluate the baseline noise. When comparing two sets of conditions, the one giving the highest sensitivity does not necessarily result in the lowest detection limit.

It is useful to evaluate the peak-to-peak baseline noise, $\Delta i_b$, over a time period about 10 times the width of the chromatographic peak. The frequency content of the noise relative to that of the chromatographic peak is very important since frequencies much higher or lower can usually be removed. The high frequencies are easily filtered out and the very low frequencies are, in effect, baseline drift, which is eliminated when the peak height is quantitated. The amount

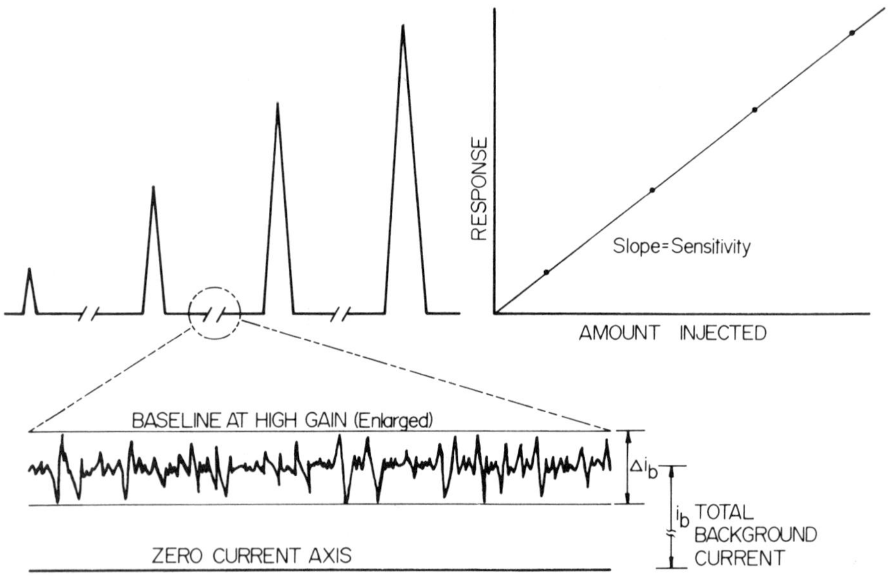


Figure 22.11 Establishing detection limits in LCEC or FIA requires measurement of both the response slope (sensitivity) and noise. Sensitivity alone does not tell the story.

injected which would give a peak height equal to $\Delta i_b$ is clearly below the limit of usefulness. An amount that would give a peak height of five times $\Delta i_b$ is useful under many circumstances. I recommend that LCEC users report sensitivity for each peak of interest and $\Delta i_b$. With this information it is possible for the reader to predict a reasonable detection limit for each substance being determined. Figure 22.11 illustrates the procedure required to obtain the necessary data.

## III. APPLICATIONS REVIEW

### A. Introduction

LCEC systems are now used for a wide variety of applications, many of which have been published [1]. At present (early 1983), several papers reporting analytical results using LCEC are appearing every week. Space does not permit a thorough review; however, it is possible to generalize by considering the classes of compounds that have most frequently been studied: phenols, aromatic amines, thiols, quinones, and nitro compounds. By analogy with LCUV, it is frequently desirable to obtain an electrochemical "spectrum" of a compound to assess its suitability for LCEC. Cyclic voltammetry (CV) is the electrochemical equivalent of spectroscopy. It is useful to carry out CV experiments in several possible mobile phases since electrochemical reactions are very media dependent.

## B. Oxidative Applications

### *Phenols*

Phenolic substances are for the most part readily oxidized at a graphite electrode. The oxidation potentials for phenols vary widely with structure and some (hydroquinones and catechols) are far more readily oxidized than others (cresols). Many compounds of biological interest (catecholamines, [21], pharmaceuticals, plant phenolics) and industrial interest (antioxidants, antimicrobials, agricultural chemicals [22]) are phenolic and trace determination based on LCEC is now quite popular.

### *Aromatic Amines*

Like phenols, aromatic amines are oxidized at a graphite electrode over a wide range of oxidation potentials. Some compounds (phenylenediamines, benzidines, and aminophenols) are ideal candidates due to their very low oxidation potentials and numerous applications have been developed (some published, many not) [1,22].

### *Thiols*

Thiols ("sulfhydryls" or "mercaptans") are very easily oxidized to disulfides in solution, but this thermodynamically very favorable redox reaction occurs only very slowly at most electrode surfaces (e.g., glassy carbon). LCEC methods for thiols therefore usually depend on the unique behavior of these compounds at a mercury electrode surface at about +0.10 V (a very low potential). The reaction involves formation of a stable complex between the thiol and mercury. In a formal sense, the mercury and not the thiol is oxidized. This approach has been used to determine the amino acid cysteine, the tripeptide glutathione, and the pharmaceuticals penicillamine and captopril. Recently Allison and Shoup have developed a unique approach whereby both thiols and their corresponding disulfides can be determined in a single chromatographic experiment [23]. A series dual-electrode detector was used. Besides thiols, many other sulfur-containing compounds are good candidates for LCEC.

### *Miscellaneous Oxidizable Compounds*

A number of unique substances have been studied by oxidative LCEC. Being an excellent reducing agent, ascorbic acid is easily detected with excellent selectivity in very complex samples (food products, biological fluids, multivitamin products). Similarly, uric acid is readily detected in biological materials. The important enzyme cofactor NADH is readily oxidized at carbon electrodes and LCEC methods for this biochemical are beginning to appear now. Some heterocycles of pharmaceutical interest (phenothiazines, imipramine) are also uniquely applicable.

## C. Reductive Applications

### *Quinones*

Quinones are among the best behaved organic compounds to undergo redox reactions in aqueous solutions. There are a reasonably large number of synthetic and natural products containing the quinone moiety and many of these are excellent candidates for selective determination by LCEC. Unfortunately, some of the most important of these compounds are extremely hydrophobic due to the presence of long hydrocarbon side chains and are therefore quite difficult to study by reverse-phase LC.

### *Nitro Compounds*

Nitrobenzene was one of the first organic compounds studied by classical polarography in the 1920s and it is not surprising that nitro (and nitroso) compounds have been among the most extensively investigated by both organic and analytical electrochemists. Aromatic nitro and nitroso compounds are very readily reduced at both carbon and mercury electrodes, but other compounds such as nitrate esters, nitramines, and nitrosamines are often good candidates as well. Many pharmaceuticals, explosives [25], agricultural chemicals [22], and important industrial intermediates fall into these classes and a number of LCEC methods have been developed. Often, the selectivity is extremely good in biological and environmental samples because the nitro group is rare in nature and few other organic compounds are so easily reduced. Reagents containing the aromatic nitro group have frequently been used to derivatize amines, aldehydes, ketones, and carboxylic acids (etc.) to improve their characteristics for determination by absorption spectroscopy. The same or closely related reagents are now being used to provide an electrochemically reactive handle for many of these same compounds [5,26]. Although it is true that aldehydes and ketones can be electrochemically reduced and alkyl amines and carboxylic acids can be electrochemically oxidized, the energy required to carry out these well-known reactions appears to be too great to permit development of a successful LCEC trace determination procedure without derivatization.

### *Miscellaneous Reducible Compounds*

A number of organometallic compounds show promise for LCEC study and a few have already been examined in detail (especially mercury alkyls) [9]. More highly conjugated organic compounds such as $\alpha,\beta$-unsaturated ketones and imines are occasionally good candidates, but at this time UV detectors frequently outperform electrochemical detectors for such systems. At this writing there have been only a few reported LCEC studies of metal ions or metal complexes. Perhaps the major reason for this is that very little modern LC has been carried out on such things using any detector. It is tough to compete with atomic spectroscopy techniques for determination of most elements, but as "speciation" becomes more important it seems likely that more LCEC methods will be developed for metal complexes.

### D. Postcolumn Reaction Detectors

The limitations of electrochemical detection can be overcome in some cases by the use of postcolumn reactions. A reagent (e.g., an enzyme) may be mixed with the column effluent to convert the sample components of interest into electroactive compounds. To date, it has been more common to measure a change in the reagent itself. The most thoroughly investigated system involves the use of a two-stage electrochemical reaction detector in which the reagent itself is generated at constant current in one thin-layer cell and the consumption of reagent is monitored downstream using a conventional thin-layer amperometric transducer. Probably the most generally applicable system of this type involves generation of bromine from bromide and detection of bromine consumption after addition to alkenes or substitution on aromatics [27].

It is also possible to use postcolumn mixing to permit the chromatography to be carried out with purely organic mobile phases. A miscible solvent containing a supporting electrolyte is added just prior to the detector cell. The additional complexity of postcolumn reactions always detracts somewhat from the convenience of direct measurement of mobile-phase constituents. On the other hand, the added selectivity and/or sensitivity can provide important advantages for certain analytical problems. This subject has recently been reviewed [28].

## IV. CONCLUSION

Electrochemical detectors for liquid chromatography have reached a level of maturity in that thousands of these devices are in routine daily use for a variety of mundane purposes. Nevertheless, the technology is advancing rapidly in several respects. Multiple electrode detectors are being developed for more specialized applications. Small-volume transducers are being explored for capillary and micropacked columns. Finally, new electrode materials with unique properties are likely to afford improved sensitivity and selectivity for important applications.

It is important for the reader to recognize that there is no single optimal electrochemical detector, but rather a whole family of devices that can be adapted to solve a variety of problems. To make the best choice, there is no substitute for a little knowledge of the principles of electrochemical instrumentation. Even more important is an understanding of the *chemistry* involved when a redox active molecule ventures into the high-field region near an electrode-solution interface.

## REFERENCES


1. *Recent Reports on Liquid Chromatography/Electrochemistry* (R. E. Shoup, ed.), BAS Press, West Lafayette, Ind., 1982.
2. P. T. Kissinger, *Anal. Chem. 49*:447A (1977).
3. R. J. Rucki, *Talanta 27*:147 (1980).
4. P. T. Kissinger, *J. Chem. Educ., 60*:308 (1983).
5. *Principles and Applications of Liquid Chromatography/Electrochemistry* (P. T. Kissinger, ed.), BAS Press, West Lafayette, Ind., 1984.
6. W. R. Heineman and P. T. Kissinger, *Anal. Chem. 50*:166R (1978); *52*:138R (1980).
7. D. L. Rabenstein and R. Saetre, *Anal. Chem. 49*:1036 (1977).
8. R. C. Buchta and L. J. Papa, *J. Chromatogr. Sci. 14*:213 (1976).
9. W. A. MacCrehan, R. A. Durst, and J. M. Bellama, *Anal. Lett. 10*:1175 (1977).
10. R. W. Murray, *Acc. Chem. Res. 13*:135 (1980).
11. B. Fleet and C. J. Little, *J. Chromatogr. Sci. 12*:747 (1974).
12. R. C. Koile and D. C. Johnson, *Anal. Chem. 51*:741 (1980).
13. Y. Takata and G. Muto, *Anal. Chem. 45*:1864 (1973).
14. W. Kemula, *Roca. Chem. 26*:281 (1952).
15. J. G. Koen, J. F. K. Huber, H. Poppe, and G. den Boef, *J. Chromatogr. Sci. 8*:192 (1970).
16. R. Stillman and T. S. Ma, *Microchim. Acta,* 491 (1973).
17. L. Michel and A. Zatka, *Anal. Chim. Acta 105*:109 (1979).
18. H. B. Hanekamp, P. Bos, and R. W. Frei, *J. Chromatogr. 186*:489 (1979).
19. H. B. Hanekamp, W. H. Voogt, P. Bos, and R. W. Frei, *J. Liq. Chromatogr. 3*:1205 (1980).
20. W. Kutner, J. Debowski, and W. Kemula, *J. Chromatogr. 191*:47 (1980).
21. P. T. Kissinger, C. S. Bruntlett, and R. E. Shoup, *Life Sci. 28*:455 (1981).
22. P. T. Kissinger, K. Bratin, W. P. King, and J. R. Rice, *ACS Symp. Ser. 136*:57 (1980).
23. L. A. Allison and R. E. Shoup, *Anal. Chem. 55*:8 (1983).
24. M. H. Joseph and P. Davies, *Curr. Sep. 4*(4):62 (1982).

25. K. Bratin and R. C. Briner, *Curr. Sep.* *2*(3):1 (1980).
26. P. T. Kissinger, K. Bratin, G. C. Davis, L. A. Pachla, *J. Chromatogr. Sci.* *17*:137 (1979).
27. W. P. King and P. T. Kissinger, *Clin. Chem.* *26*:1484 (1980).
28. *Curr. Sep.* *4*(4):58-78 (1982).

# 23

# Photoelectrochemistry and Electrochemiluminescence

HIROYASU TACHIKAWA* and LARRY R. FAULKNER / Department of Chemistry, University of Illinois at Champaign-Urbana, Urbana, Illinois



## I. INTRODUCTION

In this chapter we examine phenomena in which photons are active participants in electrochemical processes. We therefore distinguish our present concern from that of spectroelectrochemistry, which involves the use of photons as nonperturbing probes aiding observations

---

*Present address: Department of Chemistry, Jackson State University, Jackson, Mississippi

of electrode reactions. As a participant the photon may serve as a product or "reactant." On one hand, it may emerge as emission from homogeneous chemical reactions involving electrolytically produced substances. Phenomena of this first sort we call *electrochemiluminescence* (ECL). Alternatively, the photon may stimulate the electrode process by photolyzing solutes to yield electroactive species, by ejecting electrons from the electrode into the solution phase, by altering the distribution of charge carriers within a semiconductor electrode, or by creating electroactive excited states of photoreceptors adsorbed onto electrode surfaces. It is proper to label this domain *photoelectrochemistry*. Both areas have received much attention over the past 20 years, because they bear on some key fundamental questions and because they may give rise to some valuable applications. Our goal here is to describe basic phenomena and to outline briefly the experimental approaches one can use to study them.

## II. PHOTOELECTROCHEMISTRY AT SEMICONDUCTOR ELECTRODES

### A. Semiconductor Electrodes

There is now much interest in light-stimulated electrode processes at semiconductors in contact with electrolytes, because they may offer fairly efficient means for converting light to electrical or chemical energy. These opportunities arise because one can change the free energy of electrons at the semiconductor surface to a great extent by illumination, and one can stabilize the high-energy state resulting in this way. According to the band theory of solids [1,2], semiconductors exhibit a forbidden electron energy region (called the *band gap*) between the *valence band,* which is a set of states that is nearly filled with electrons, and a *conduction band,* a second set that is nearly empty of them. The magnitudes of the band gaps for several common semiconductors are shown in Table 23.1. In order for an *intrinsic semiconductor* (e.g., pure Ge or Si) to show conductivity, electrons must jump across the band gap from the valence band to the conduction band. Either heat or light can supply the necessary energy. When an electron is promoted in this way, it is mobile in the conduction band and acts as a charge carrier. Additionally, it leaves behind a hole in the valence band. Continual electron rearrangements in that band cause the hole to move from site to site, hence it behaves like a positively charged, mobile particle. In an intrinsic semiconductor, equal numbers of the two carriers always exist.

Besides the intrinsic semiconductors, there are also *n-type* and *p-type* impurity-doped (*extrinsic*) semiconcuctors. To obtain the former, one adds to the intrinsic material an electron donor (e.g., P in Si) which has electrons that are promoted easily by thermal processes into the conduction band. Conductivity is consequently dom-

Table 23.1 Band Gaps of Common Semiconductors

| Semiconductor | Band gap[a] (eV) | Wavelength[b] (nm) |
|---|---|---|
| Ge | 0.67 | 1900 |
| Si | 1.14 | 1090 |
| GaAs | 1.43 | 870 |
| GaP | 2.26 | 550 |
| CdS | 2.42 | 510 |
| $TiO_2$ | 3.03[c] | 410 |
| ZnO | 3.2 | 385 |

[a]T = 300 K unless otherwise noted.
[b]Absorption edge corresponding to band-gap transition.
[c]$T \to 0$ K.
*Source*: Ref. 1, p. 364.

inated by these electrons, and they are termed the *majority carriers*. In a p-type semiconductor, the dopant is an electron acceptor (such as Al in Si) upon which valence electrons are easily trapped by thermal equilibrium. Thus holes are created in the valence band and conductivity increases. Obviously, the holes are the majority carriers in this case. The density of charge carriers in semiconductors is $10^{13}$-$10^{17}$ $cm^{-3}$. For comparison, the metallic value is about $10^{22}$ $cm^{-3}$, and a 0.01 M electrolyte solution has about $10^{19}$ $cm^{-3}$.

When a semiconductor electrode is brought into contact with an electrolyte [3-9], a double layer generally develops. However, the charge on the electrode itself does not reside strictly on the surface, as for a metal; instead, it extends from the surface into the bulk for an appreciable distance (100-10,000 Å). There is an exponential potential distribution inside this *space-charge region*. The very existence of the space charge implies that one of the charge carriers exists more favorably in the surface region than in the bulk semiconductor. For example, when the electrode carries a positive charge, it is clear tht electrons find the surface region a high-energy location. (Remember their energetically unfavorable interaction with the like charges on the solution side.) This situation is often depicted as an upward bending of the valence and conduction bands, as shown in Figure 23.1a. Obviously, a negative charge reverses the direction of bending. At the potential of zero charge, there is no bending; hence this potential is called the *flat-band potential* (Fig. 23.1b).

A subtle, but important feature of semiconductor electrochemistry is that the energies of the band edges (bottom of the conduction band and top of the valence band) at the electrode-solution interface are often virtually independent of the electrode's potential vs. an external reference [2-10]. Modulating the potential of the electrode causes the

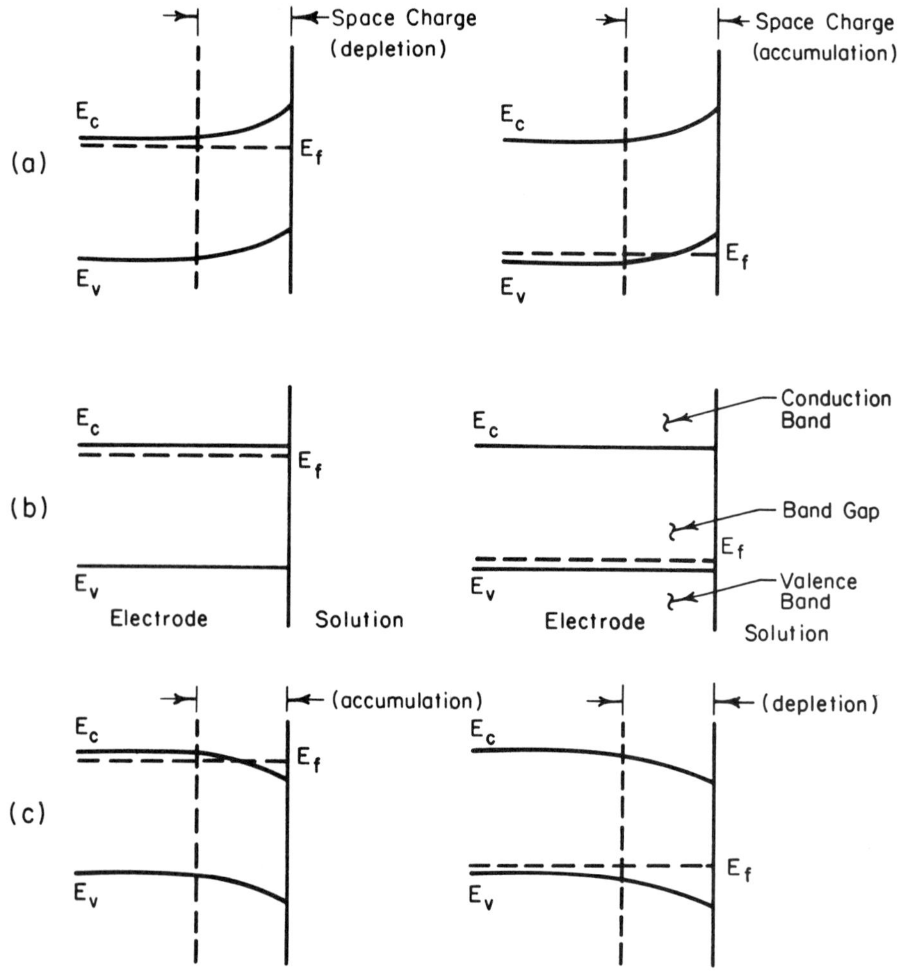


Figure 23.1 Energy bands of semiconductors at various potentials in contact with electrolyte. Left column is for n type and right column for p type (a) Top pair for $E > E_{fb}$; (b) middle pair for $E = E_{fb}$; (c) bottom pair for $E < E_{fb}$. The vertical axis is electron energy, and the horizontal axis represents distance into the semiconductor from the electrode-electrolyte interface. $E_c$, bottom of the conduction band; $E_v$, top of the valence band; $E_f$, Fermi level.

band edges to shift in the interior of the electrode, but not at the surface. As a consequence, changes in potential are wholly manifested as changes in the voltage drop across the space-charge layer. Thus the chief effect of changes in potential is to regulate the *populations* (not the *energies*) of holes and electrons at the surface, where they are eligible for electrode reactions.

This basic idea permeates most of the literature on electrochemistry at semiconductor electrodes. Fundamentally, it rests on the condition that the carrier density within the semiconductor is much smaller than the concentration of ions in the adjacent solution, so that the formation of a double layer requires that one draw charges from a much thicker region in the semiconductor than in solution. The voltage drops within the space-charge layer and within the solution side of the double layer are the agents establishing the interfacial structure. Any change in potential requires a corresponding change in the charge on the double layer. Since it is much harder to gather that charge within the semiconductor than within the solution, most of the potential change must appear across the space-charge layer. The interfacial potential difference between the semiconductor and the bulk of the solution hardly changes at all; thus the energy of an electron or a hole at the surface is not altered by the change in potential.

Quite recently, there has been growing realization that this behavior does not always apply [10-14]. The concern is over the effects of *surface states* [2,15], which are electronic energy levels within the band gap caused by the peculiarities of the surface. Since the lattice terminates at the surface, conditions there must be different from those in the bulk semiconductor, and one could easily see that dangling bonds, adsorbed species, oxides, and so forth, could produce local electronic energy levels not found in the bulk. Usually, the chemical nature of these states is not known, but they can have enormous impact on electrochemical behavior, because they lie directly on the route of charge flow through the interface. They can trap electrons and holes, and they act as *recombination centers* where electron-hole annihilation is mediated [2]. Many oddities in electrochemistry at semiconductors are ascribed to these states [3-14].

Of concern to us here is the effect of surface states on the potential distribution through the interfacial region. Because surface states offer lower energies to electrons and holes than would normally be available within the conduction and valence bands, respectively, the carriers can become trapped there. They therefore represent reservoirs into which charge can be deposited and from which it can be withdrawn. If they exist in sufficient numbers, a change in potential need not be devoted to gathering charge from the bulk of the semiconductor, because the necessary changes in charge can be effected from the surface states. Then changes in potential appear at the electrode solution interface, and the *energies* of carriers at the interface move with potential. For reasons that we shall not explore, this kind of behavior is called *Fermi-level* pinning [10-14].

We must, nevertheless, consider the concept of the *Fermi level* $E_f$, which is very important in this field [2]. The simplest definition is that the Fermi level is the electron energy where the probability of occupation of a single state at statistical-mechanical equilibrium is 50%. Levels below $E_f$ have higher occupation probabilities than this figure, and higher levels are less likely to be occupied. Since the transition from nearly unit probability of occupation to almost zero probability takes place over a range of energies only a few kT wide (kT = 25.7 meV at 25°C), $E_f$ can be regarded as the dividing line between filled and empty states.

In the bulk of an n-type semiconductor, $E_f$ lies just below the conduction band. This situation arises from the donor states (from the dopant), which lie near the conduction band edge. Note that at low temperatures these states would be entirely filled and the conduction band would be empty, so $E_f$ would be situated between the donor level and the band edge. At normal temperatures, electrons are ionized from the donor levels into the conduction band. The value of $E_f$ then shifts slightly, but it remains near the conduction band edge. Similarly, for a p-type semiconductor, $E_f$ lies just above the valence band edge.

The Fermi level can be shown to be the partial molar free energy (i.e., the electrochemical potential) of electrons within a phase, thus it must be constant throughout a phase at equilibrium [2], as shown in Figure 23.1. There is a linear relationship between the potential of an electrode and its Fermi level. Varying the potential negatively and positively amounts to tuning the Fermi level up and down, respectively, on the electron energy axis. That, in turn, affects the occupancy of states near the electrochemical interface and the availability of carriers there. The bulk of the semiconductor is unaffected by the events and conditions at the interface; hence the relationship between $E_f$ and the band edges in the bulk is not a function of potential. When a change in potential causes a change in $E_f$, the band edges in the interior must shift correspondingly. Frames (a) and (c) show the effects resulting from a shift in potential from the flat-band condition (Fig. 23.1b) to more positive and negative values, respectively. If the edges at the interface are fixed in energy, the bands must bend as shown, and the space-charge field is established.

It is very important to understand the qualitative difference in the condition of a semiconductor electrode on opposite sides of the flat-band potential [2-10]. Figure 23.1 provides a summary of the possibilities for both p- and n-type electrodes. In both cases the space charge is positive for $E > E_{fb}$ and negative for $E < E_{fb}$. For an n-type material in the negative region, that space charge is created by the excess of majority carriers. They are said to form an *accumulation layer*, in which the electrical conductivity is higher than in the bulk. On the other hand, taking the n-type electrode to the positive side of $E_{fb}$ drains the interfacial zone of majority carriers. The positive space charge arises primarily from the positively ionized donor sites left behind by the n carriers. They are not mobile, so they do not contribute to conductivity. Holes can be expected to accumulate in the space-charge zone, but there are not many holes in the material, as a consequence of the doping. In this case, the space-charge region is a *depletion layer*. It has few carriers and low conductivity. By exactly analogous reasoning, one finds that a p-type electrode has an accumulation layer in the positive region and a depletion layer in the negative zone. Thus the behavior of p-type and n-type materials is complementary. The difference between depletion and accumulation of majority carriers in the interfacial region has enormous impact on electrochemical behavior, as we shall now discover.

## B. Electrolytic Processes Stimulated by Optical Excitation of the Electrode

When a semiconductor electrode operates with an accumulation layer, the electrochemical behavior usually resembles that of a metal, and no effect ordinarily arises from photoirradiation. These generalizations apply because the electrochemistry is dominated by the majority carriers, which are abundant in the interfacial region and are not made much more so by irradiation [3-10]. On the other hand, the behavior with a depletion layer at the interface is quite different. In the dark, electrochemistry proceeds only at very slow rates, often invisibly, because the depletion layer offers very few carriers that can engage in electrode processes. Irradiation increases the carrier population and promotes electrochemistry in quite a special way [3-10]. Consider the parts of Figure 23.1 showing depletion layers, and imagine the creation of an electron-hole pair by absorption of a photon within either of those layers. Note that the band bending (i.e., the electric field) in each case splits the n-p pair so as to send the majority carrier to the bulk and the *minority carrier to the interface*. The minority carrier can be viewed as a high-energy species that is capable of electrochemistry that cannot be done by the majority carriers. In general, then, n-type electrodes at potentials more positive than $E_{fb}$ act as photoanodes. The light-generated holes are collected at the interface and extract electrons from species in solution. In complement, p-type electrodes operating at more negative potentials than $E_{fb}$ act as photocathodes.

These principles are illustrated by some experiments first performed by Honda's group [16-18]. Single crystals of n-$TiO_2$ (rutile) were used as electrodes and a 500-W high-pressure mercury or xenon lamp was used as a light source. Figure 23.2 is a set of current-potential curves at $TiO_2$ immersed in 0.5 M KCl solution adjusted to pH 4.7 via an acetate buffer. When there is not irradiation (curve 3), there is almost no anodic current (ca. 1 $\mu A/cm^2$), even when the

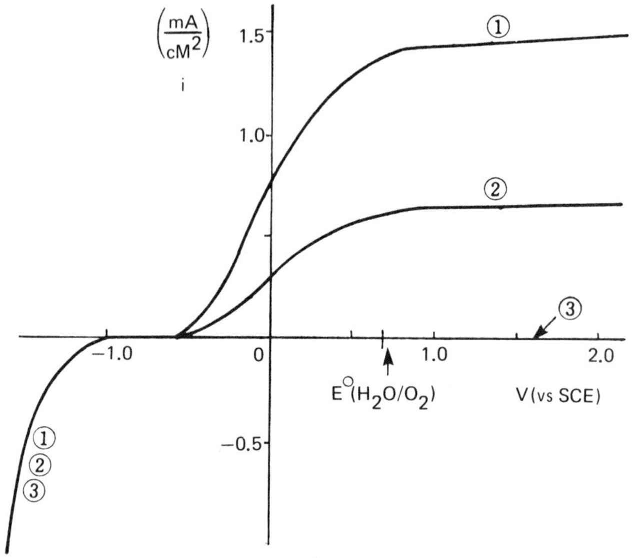


Figure 23.2 Current-voltage curves at a $TiO_2$ electrode in a pH 4.7 buffer. 1, Relative intensity of irradiation, 100%; 2, relative intensity, 50%; 3, without irradiation. (From Ref. 18.)

electrode potential exceeds +4.0 V vs. SCE. In the negative direction, hydrogen evolution is observed, although there is a substantial overvoltage. When the light is shined on the $TiO_2$ surface (curves 1 and 2), a large anodic current due to oxygen generation,

$$4p^+ + 2H_2O \longrightarrow O_2 + 4H^+ \tag{23.1}$$

starts to flow near −0.5 V, which is far more negative than the formal potential for the process (0.7 V vs. SCE). In other words, the light drives this process at a *thermodynamic underpotential.* The energy of each photon is converted to a hole, which is a potent oxidant. Therefore, less electrical energy is needed to complete the oxidation of water. Saturation of the current is observed at more positive potentials, and the relation between the saturation current and the light intensity is almost linear. As Figure 23.3 shows, the anodic current flows only when light of wavelength shorter than 415 nm is used for exposure. Since the band gap of $TiO_2$ is 3.0 eV (the energy of 415-nm light), these observations indicate that the participating species in the oxygen evolution are the holes created by photoexcitation of electrons from the valence band. The space charge

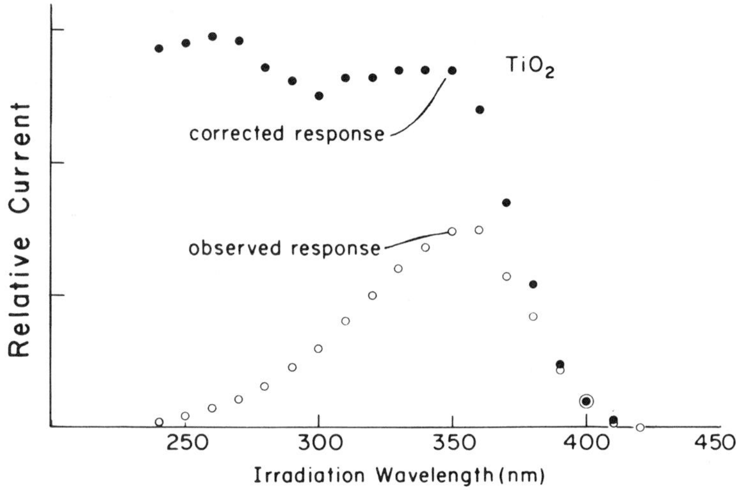


Figure 23.3 Dependence of anodic current at $TiO_2$ on the wavelength of irradiation. The observed response falls off toward shorter wavelengths because the light intensity decreases in that region. Correction for this effect gives the curve defined by the filled circles. (Adapted from Ref. 19.)

is positive in the positive potential region, hence the bands are bent upward, as shown in Figure 23.1a. Thus the n carriers move toward the external circuit, and the p carriers tend to concentrate at the interface, where they can undergo reaction with water by Equation 23.1. Note that without the upward band bending, the photogenerated n-p pair would face nearly certain annihilation through recombination; hence the hole lifetime would be too short to allow electrochemical activity.

Honda's work triggered a very large research effort [5-10] to find ways of using photoelectrochemical cells to convert photonic energy into electrical or chemical forms. The obvious driving force is technological interest in solar energy. The fundamental principles are illustrated by the cell shown in Figure 23.4, involving an n-$TiO_2$ single crystal anode [17-19]. It was coupled through an external load to a Pt black cathode. When a 1-$cm^2$ surface of the semiconductor was illuminated by a xenon lamp, oxygen evolved at $TiO_2$ and reduction products, possibly including hydrogen, were generated at Pt. Wrightons' group [20] showed that the generation of hydrogen at the cathode can be made stoichiometric if n-$SiTiO_3$ is substituted for the n-$TiO_2$. The basic processes occurring in these cells can be summarized as follows:

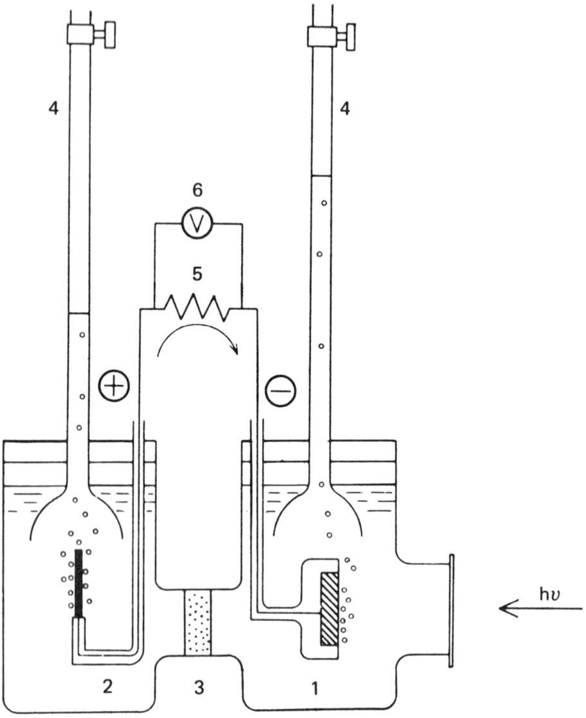


Figure 23.4 Schematic diagram of electrochemical photocell: 1, $TiO_2$ electrode; 2, platinum electrode; 3, diaphragm (agar salt bridge); 4, gas buret; 5, load resistance; 6, voltmeter. (From Ref. 18.)

1. Formation of holes in the valence band of $TiO_2$ by light
2. Oxidation of water by the holes
3. Movement of the n carriers through the load to the Pt electrode, where they can reduce protons to hydrogen gas

In effect, light can supply the energy required to split water:

$$H_2O + 2h\nu \longrightarrow (1/2)O_2 + H_2 \qquad (23.2)$$

The illuminated cells operate galvanically if the light-induced underpotential is large enough to shift the potential of the anode to a value more negative than the operating potential of the cathode. Then such a cell can convert light to both electrical energy (dissipated in the load) and chemical energy (stored as $H_2$ and $O_2$). In the dark, one can create the products only by supplying the energy electrically.

Photoanodes, featuring n-type semiconductors, are quite generally susceptible to undesirable photocorrosion porcesses [3-10,21, 22]. The photogenerated holes collected at a semiconductor-solution interface represent, in chemical terms, broken bonds among lattice constituents. If they are not filled rapidly by oxidation of some species in solution, they will cause oxidation of the lattice. For example, n-type CdS degrades according to

$$CdS + 2p^+ \longrightarrow Cd^{2+} + S \tag{23.3}$$

where the cadmium ion is lost to the solution and elemental sulfur remains behind [23]. Similarly, n-GaAs corrodes in basic solution to $Ga(OH)_4^-$ and $AsO_2^-$ [24]. The oxide semiconductors n-$TiO_2$ and n-$SiTiO_3$ are resistant to this sort of degradation but are not immune [16,19,20,25]. Unsurprisingly, p- and n-Si build up layers of $SiO_2$ upon exposure to water, especially in the presence of light. These reactions almost always lead to passivation of the surface or degradation of performance by creating electrical barriers and recombination centers; hence a good deal of attention has been directed toward *stabilization* of semiconductor electrodes.

The universal strategy is to add a reagent that is more easily oxidized than the electrode itself [21]. For example, Wrighton's group has shown that n-CdS can be stabilized by an electrolyte formally containing 1 M NaOH, 1 M $Na_2S$, and 1 M S. The photoanodic activity is then confined to oxidation of solution-phase sulfide species. The electrodes themselves suffer no detectable weight losses even after many hours of high-intensity illumination [26]. Similarly, selenide-polyselenide electrolytes can stabilize n-CdS, n-CdSe, and n-GaAs [24].

Novel kinds of photoelectrochemical cells have exploited that scheme [19,24]. At Bell Laboratories, Heller and co-workers have extensively developed a system featuring an n-GaAs photoanode, a selenide-polyselenide aqueous phase, and a carbon cathode [24]. In operation, light is absorbed within the space-charge region of the GaAs. The resulting electron-hole pairs are split apart, with the holes going to the interface, where they oxidize selenide, according to, for example,

$$2Se^{2-} + 2p^+ \longrightarrow Se_2^{2-} \tag{23.4}$$

The electrons move into the bulk GaAs, then into the external circuit, through a load, to the carbon cathode, where they reduce polyselenide, (e.g., by reversal of Equation 23.4). Since the anodic and cathodic processes exactly compensate, no net chemical change occurs in the cell. It is, however, able to supply power to the load. Devices such as this are receiving scrutiny as alternatives to silicon photovoltaics for the generation of electric power. The story of their development is a particular interesting example of the interplay between technology and science.

One can also stabilize an electrode by attaching redox centers directly to the electrode surface. For example, Wrighton's group has treated n-Si by first oxidizing the surface slightly to produce a thin (ca. 20 Å) layer of hydrated $SiO_2$, which is then exposed to a silanizing reagent having ferrocene as redox center [27]. This procedure provides surface-bound ferrocene, which can intercept holes by oxidation to ferricinium,

$$Fe(Cp)_2 + p^+ \longrightarrow Fe(Cp)_2^+ \qquad (23.5)$$

Once the hole is trapped in this way, it will not degrade the lattice of the semiconductor (essentially because some of its oxidizing power has been sacrificed), but it still can be used to oxidize species in solution, which undergo electron transfer to ferricenium. Surface modification is also interesting for attachment of electrocatalysts to semiconductors. For example, Wrightons' group [28] has bound alkylviologen centers within a polymeric phase attached to p-Si, which is then impregnated with Pt by electroreduction of $PtCl_6^{4-}$. The viologen intercepts photogenerated electrons from the p-Si and apparently passes them to Pt centers, which catalytically evolve hydrogen from aqueous solutions.

If one is interested in production of chemicals or fuels by photoelectrochemistry, then semiconducting powders may be appealing. Here one relies on the idea that a single particle can carry both the photoelectrode and its counterelectrode, so that the mechanical superstructure of the cell can be eliminated. Bard's group, for example, examined platinized n-$TiO_2$ powder [6,29-31]. The interpretive scheme is shown in Figure 23.5. Upon illumination, anodic reactions take place at exposed $TiO_2$ by collection of holes in the manner discussed above. The electrons are transferred to the Pt phase, where they serve to reduce components of the solution. Suspensions of such powders have been used to oxidize cyanide [6,30] and to carry out Kolbe electrolysis of carboxylic acids to alkanes [6,31]. Very recently, Grätzel's group [32] have reported that $TiO_2$ particles coated with both Pt and $RuO_2$ can split water to $H_2$ and $O_2$. They propose that light absorption in the $TiO_2$ phase generates carriers, which are then separated so that the electrons go to Pt, where they evolve $H_2$, and the holes go to $RuO_2$, where they oxidize water to $O_2$. This proposal is interesting, and possibly quite significant, but the details of the process are quite unclear at present.

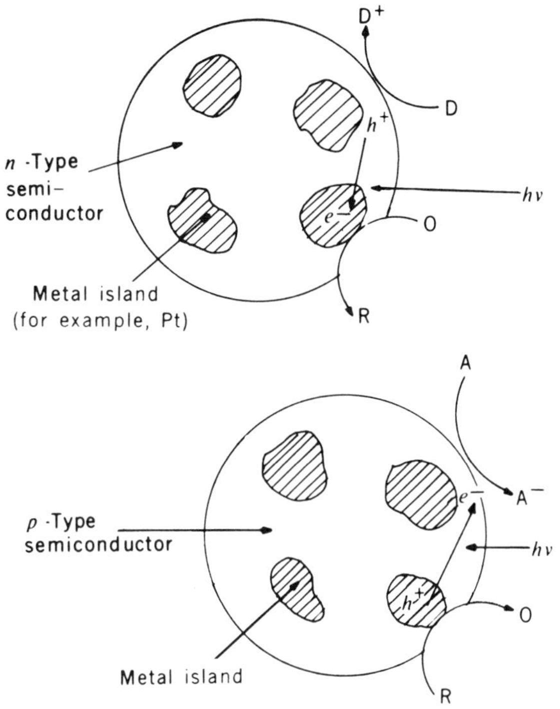


Figure 23.5 Schematic diagram of photoelectrocatalysis on metallized semiconductor particles. (From Ref. 8, copyright 1980 American Association for the Advancement of Science.)

## C. Electrolytic Behavior of Adsorbed, Excited Photoreceptors

If a semiconductor electrode such as an n-ZnO crystal is illuminated in the presence of a small amount of dye (e.g., $5 \times 10^{-5}$ M rhodamine B), one can observe an enhanced photocurrent in the anodic region. If illumination is carried out by a tunable, monochromatic source, it is possible to obtain a photocurrent spectrum like that shown in Figure 23.6. An apparatus like the one depicted in Figure 23.7 may be used to make these measurements. After removing dissolved $O_2$ by nitrogen purging, one sets the potential of the semiconductor at a value in the region of limiting photocurrent (ca. +0.5 V vs. SCE) in the absence of dye. This current is due simply to the photoevolution of oxygen as described in Section II.B. If a sensitizing molecule (rhodamine B in this case) is added to the cell, there is an additional photocurrent whenever illumination occurs at a wavelength absorbed by the sensi-

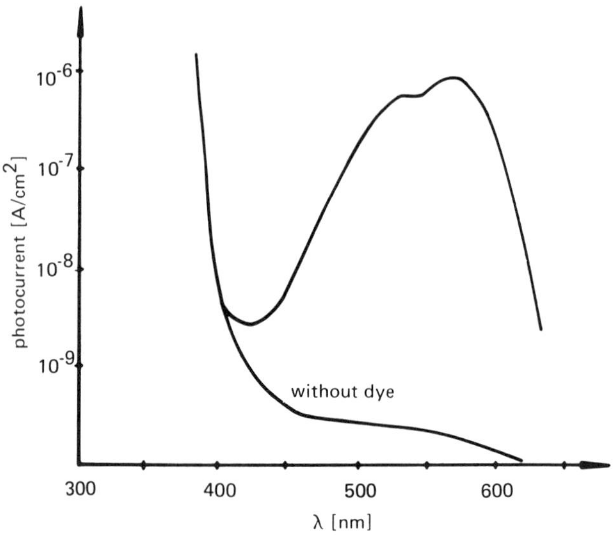


Figure 23.6 Sensitized and unsensitized photocurrent spectrum at a ZnO electrode (doped with Cu). Electrolyte: 1 M KCl, pH 2; dye: $10^{-4}$ M rhodamine B. [From Ref. 46, reprinted with permission from H. Gerischer, *Photochem. Photobiol. 16*:243 (1972). Copyright 1972 Pergamon Press Ltd.]

tizer. In the long-wavelength region the background photocurrent is often quite small; hence the photocurrent spectrum is essentially identical to the absorption spectrum of the sensitizer.

This phenomenon seems to arise from charge-transfer reactions between the semiconductor and excited dye molecules adsorbed on its surface. In this connection it is important to realize that electron transfer reactions across *any* interface, metal or semiconductor, are thought to occur readily only between electrode states and solute states at approximately the same energy level. This idea results directly from the radiationless occurrence of charge transfer on a Franck-Condon time scale. Returning now to the excited dye adsorbed onto the semiconductor, one realizes that if the band gap is large compared to the dye's excitation energy, electron transfer between the dye and the electrode may involve the highest normally filled level or the excited level of the dye, but usually not both. One of the dye levels will not be electroactive because it will match some energy in the electrode's forbidden region.

For our specific case, rhodamine B on n-Zn, at +0.5 V vs. SCE, there is no dye electroactivity without illumination, yet an anodic current oxidizes the dye when the light shines. The absence of the dark reaction implies that the highest occupied orbital of the dye is at an energy in the band gap or the valence band (which is nearly devoid of carriers). The photocurrent then arises because the absorption raises a dye electron to an excited level that apparently overlaps the conduction band. Thus the net effect is injection of electrons into the upper band. Hole neutralization cannot explain the observations, because light of smaller energy than the band-gap energy will create the effect. Moreover, hole neutralization is generally fully accounted for by the saturation level of $O_2$ generation. The case is therefore convincing that the system can be depicted according to Figure 23.8.

It is interesting to compare this behavior with that expected at a metal electrode. If the Fermi level lies above the highest normally filled dye level, but below the lowest excited level, no dark redox process will occur. If the dye is then excited, oxidation can occur by removal of an electron from the excited level; but simultaneously, electron transfer from the filled metal levels will reduce the vacancy in the highest normally filled level. The net result would be quenching of the excited state by the electrode, but no external current could be measured. These consequences follow from the absence of a band gap in the metal.

It has been found that the photosensitized electrochemical process is also influenced by other substances dissolved in solution. For ex-

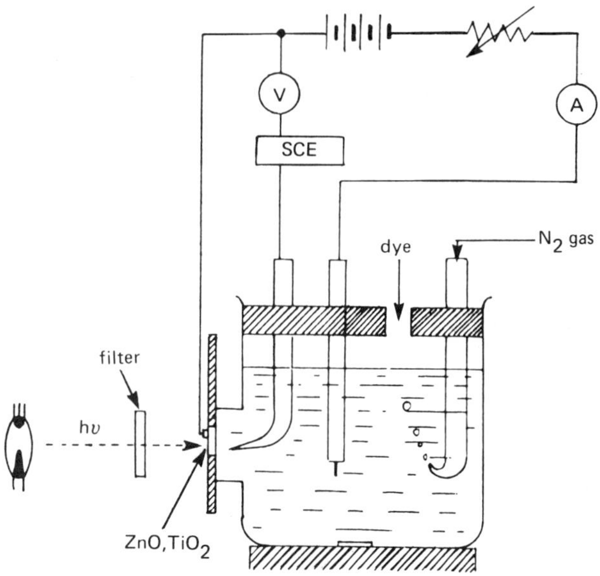


Figure 23.7 Measurement apparatus for electrochemical spectral sensitization. (From Ref. 18.)

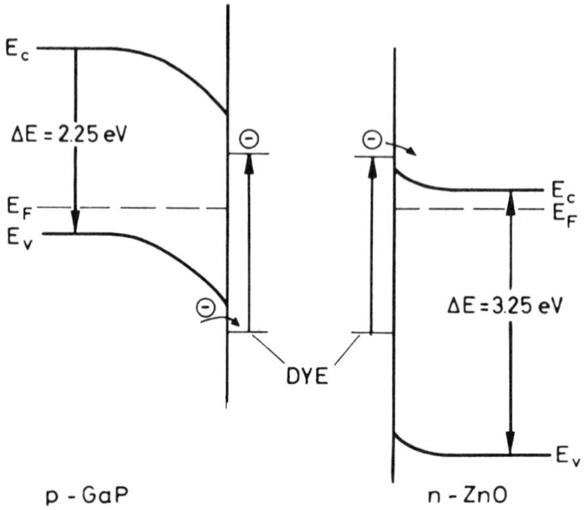


**Figure 23.8** Comparison between GaP and ZnO surfaces. [From Ref. 45, reprinted with permission from R. Memming and H. Tributsch, *J. Phys. Chem.* 75:562 (1971). Copyright 1971 American Chemical Society.]

ample, the photocurrent from the rhodamine B/n-ZnO system increases when reducing agents such as hydroquinone are added to the system [33]. Apparently, the hydroquinone serves to restore the adsorbed rhodamine B to its active reduced, ground-state form so that it can undergo repeated absorption acts leading to charge transfer. Several competing mechanisms are now being investigated for this *supersensitization* phenomenon. In contrast, oxygen often reduces (or *desensitizes*) the photocurrent. It seems to interfere by quenching the excited dye species in competition with electron transfer.

Photosensitized electrochemical reactions using ZnO single crystals with other dyes and solutes have been studied extensively by Gerischer and Tributsch [33,34], Hauffe and co-workers [35] and other groups [36-38]. Photosensitization effects with n-CdS [39,40] and n-$TiO_2$ [41] are reported to be similar to those with n-ZnO. Recently, this type of work has been extended to include covalently bound sensitizers on n-$SnO_2$ and n-$TiO_2$ [42-44].

For p-type semiconductors, such as GaAs, GaP, and $Cu_2O$, sensitized photoelectrochemical behavior is complementary to that seen with n-type material. In other words, electrons are transferred from the valence band to the dye, hence holes are injected into the electrode, and a cathodic current flows. Arguments like those presented above lead to a picture like that presented in Figure 23.8, which emphasizes the electrochemical inactivity of the excited level. Results are displayed in Figure 23.9 for p-GaP with N,N'-diethylpseudocyanine as sensitizer. In this system, $O_2$ acts as a supersensitizer. Major studies of p-type electrodes have been carried out by Memming and Gerischer and their co-workers [39,45,46].

Quantum efficiencies for the sensitized photoelectrochemical reactions have been estimated at about 10%, although the electrode surface, the dye, and the redox agents in solution have influence on the values [47].

Arden and Fromherz [48-50] have recently reported elegant work involving monolayers of surfactants transferred to electrodes by the Langmuir-Blodgett method. Some results are shown in Figure 23.10.

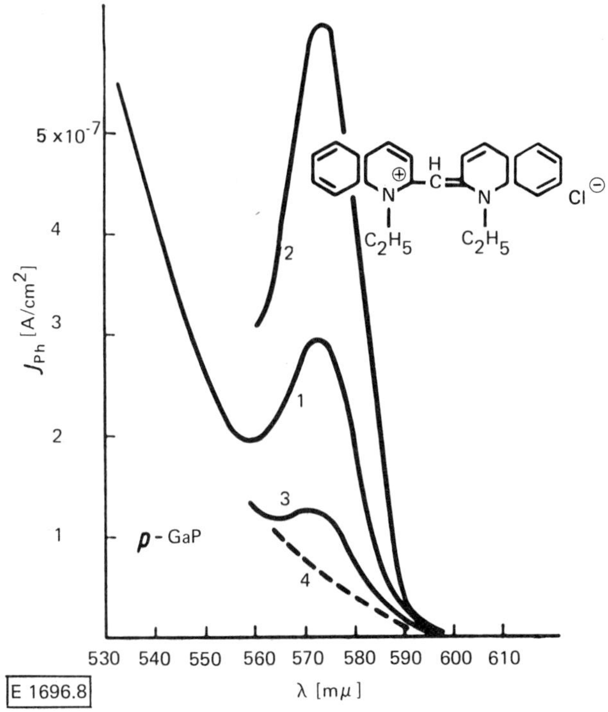


Figure 23.9 Spectral dependence of the photocurrents at a p-type GaP electrode with sensitization by N,N'-diethylpseudoisocyanine [Cathodic photocurrent, electrolyte: 1 M KCl + N,N'-diethylpseudoisocyanine chloride (1 ml of $10^{-2}$ M ethanol to 10 ml of water)] 1, $N_2$ flushing; 2, $O_2$ flushing; 3, addition of piperidine ($10^{-2}$ M); 4, trace of the background photocurrent at p-type GaP without dye. (From Ref. 39.)

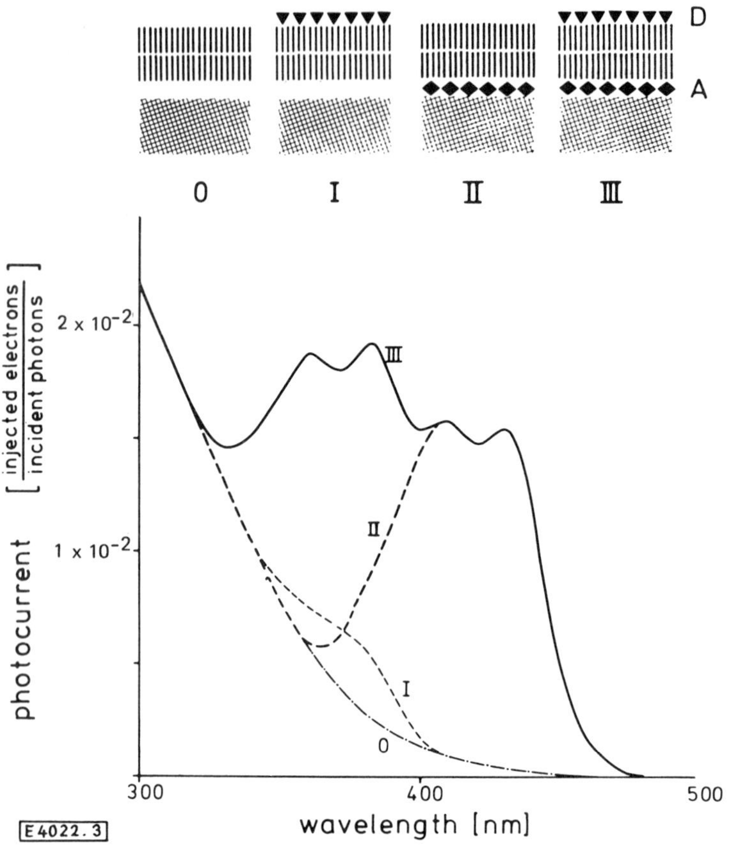


E4022.3

Figure 23.10 Photocurrent per incident photon for various monolayer assemblies on $In_2O_3$. Configuration 0 is for a bilayer of surfactant molecules without dye deposited on the electrode (represented as the lower shaded area). Configuration I involves the donor dye in the outer layer and configuration II features the acceptor dye in the inner layer. In each of these cases, the remaining layer is made up of a surfactant without dye. Configuration III has both dyes together in the arrangement shown above. The electrolyte was 5 mM borate buffer of pH 10 with 0.5 M allylthiourea as supersensitizer. (From Ref. 50.)

Two separate cationic dyes, thiacyanine (A) and oxacyanine (D), were synthesized with $C_{18}$ alkyl chains, so that they would form oriented monolayers on water-air interfaces. By standard techniques, these were transferred alternatively or in sequence to n-$In_2O_3$ electrodes as shown at the top of Figure 23.10. The particular orientations shown there allow the polar head groups of A and D to interact, respectively, with the polar oxide surface and the aqueous solution, while hydrophobic paraffinic segments interact with each other in the interior of the bilayer. Much optical evidence can be cited for the existence and stability of this type of ordered arrangement [51].

The results in Figure 23.10 show that chromophore A sensitizes photocurrent (curve II), whereas chromophore D has little effect by itself (curve I). The apparent reason for the difference is the remoteness of chromophore A. In the tandem arrangement (III), however, one finds that light absorbed by D (at wavelengths below 380 nm) gives rise to a significant photocurrent. Arden and Fromherz interpret these results to imply that energy absorbed by D is transferred (by the well-known Förster process) to A, which then undergoes charge transfer with the electrode. Without the A underlayer, D cannot sensitize photocurrent; and without the D overlayer, A cannot produce significant effects below 380 nm. Luminescence from the D layer can also be observed, and by exploiting it, together with the electrochemical data, Fromherz and Arden have produced some very detailed pictures of interfacial charge-transfer dynamics.

In this short space, we have not been able to give the extremely active field of semiconductor photoelectrochemistry coverage even approaching completeness. We have edited and simplified in hopes of conveying the gist of research in the area. For readers desiring more complete discussion, the available reviews [3-10,52] must be consulted.

## III. PHOTOEMISSION FROM METAL ELECTRODES

For almost 20 years it has been known that cathodic currents are stimulated by the illumination of metal electrodes [53-57]. These currents are often strongly enhanced by the presence in the solution of known electron scavengers such as $N_2O$ or $H_3O^+$. Research carried out by Barker [57-59], Pleskov [70-62], Delahay [63], and their coworkers indicates rather strongly that the impinging photons eject electrons from the electrode surface. They appear to travel some distance (ca. 50 Å) before becoming solvated [53,57,60,63]. If a scavenger is present, the solvated (usually aquated) electrons may react with it irreversibly. For example,

$$e_{aq} + N_2O + H^+ \longrightarrow OH\cdot + N_2 \qquad (23.4)$$

The photoelectrons are then lost permanently from the electrode, and in this case the cathodic current is actually enhanced further by faradaic conversion of $OH\cdot$ to $OH^-$ [53,57,58]. Without the scavenger, the photocurrent is small, because the photoelectrons return to the electrode, and the net charge transfer is almost zero.

Naturally, one must supply a minimum energy to remove an electron from the electrode; thus there is a *red-limit* wavelength above which photoejection is very improbable [32,53]. Since the Fermi level depends on electrode potential, this limit shifts to shorter wavelengths as the potential becomes more positive. It is clear that for monochromatic radiation, there is a threshold potential for photoemission and that the emission becomes more probable at more negative potentials [57,60].

A variety of experimental approaches have been developed for studies of these phenomena [53,57-59,61,63]. Early techniques advanced by Barker's group featured irradiation of a DME by chopped light from low-, medium-, and high-pressure mercury lamps [57]. Pulse polarographic monitoring circuitry synchronized to the light chopper was used for measurement of the photocurrent. Note, however, that the time scale of a photoemission experiment is inherently very short. The diffusion time for return of the ejected electrons is on the order of 100 ns; hence there is a great potential for studies of very fast electrode processes. Because one cannot realize this potential with low-frequency chopping techniques, the emphasis has shifted toward flash illumination methods.

Barker later described some work that involved apparatus like that shown in Figure 23.11. Light was supplied to a continuously renewed mercury pool electrode by a Q-switched, frequency-doubled ruby laser with a pulse width of ~15 ns. The electrode was set initially at any desired potential by a simple polarizing circuit, whose response was slow enough that the electrode's reaction to the flash could be monitored as a coulostatic transient, $\Delta E$ (measured with respect to the initial potential) vs. time. The difference in charge with respect to the initial condition is straightforwardly related to $\Delta E$,

$$\Delta q = C_d \, \Delta E \tag{23.5}$$

where $C_d$ is the differential capacitance of the interface. Since $\Delta E$ is only a few millivolts, $C_d$ is effectively constant for a given experiment. Useful information could be obtained for times as short as a few tens of nanoseconds after the flash.

The solution chemistry that can be studied by photoemission techniques is largely that instigated by solvated electrons. It therefore has much in common with the domain of pulse radiolysis, although the photoemission method advantageously avoids extensive radiolytic damage to the solution components. For electrochemistry, there is a very useful ability to generate, near the electrode, radicals, and other intermediates of electrochemical importance. Purely electrochemical experiments often cannot provide the insight into electrode processes that one can obtain in this way.

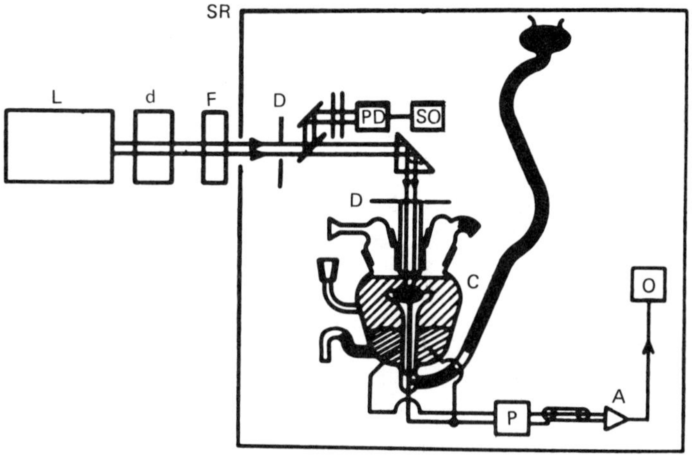


Figure 23.11 Apparatus for photoemission measurements. L, Q-switched ruby laser; d, frequency doubler; F, $CuSO_4$ solution filter; SR, screened room; D, diaphragm; PD, photodiode; SO, storage oscilloscope; C, cell; P, polarizing circuit; A, wideband amplifier; O, oscilloscope. The mercury pool working electrode is renewed continuously from the reservoir at the upper right. (From Ref. 59.)

An important example relates to hydrogen discharge on Hg [53, 56,57,59,62]. It has been advocated that this process occurs in two steps, with H· as an intermediate:

$$H_3O^+ + e \xrightarrow{\text{slow}} H\cdot + H_2O \tag{23.6}$$

$$H\cdot + e + H_3O^+ \longrightarrow H_2 + H_2O \quad \text{(low pH)} \tag{23.7}$$

or

$$H\cdot + e + H_2O \longrightarrow H_2 + OH^- \quad \text{(high pH)} \tag{23.8}$$

However, electrochemical studies alone have not been able to verify the participation of these reactions, because one cannot unambiguously isolate the electrochemistry of H· within the whole discharge process. The Pleskov and Barker groups have very cleverly untangled the web by relying on photoelectrons to create H· homogeneously [58,59,62],

$$H_3O^+ + e_{aq} \longrightarrow H\cdot + H_2O \tag{23.9}$$

Thus they have been able to study hydrogen atom electrochemistry cleanly and in impressive detail. The atoms seem to be adsorbed strongly on the electrode, and from the adsorbed state they undergo very rapid anodic or cathodic electrochemistry, depending on potential.

## IV. ELECTROCHEMICAL MONITORING OF PHOTOLYTIC INTERMEDIATES

Photochemistry often involves species that are electrochemically interesting. Excited states, radicals, and metals with light-influenced oxidation states are very common intermediates or products; thus it is not surprising that numerous attempts have been made to study photochemical processes with electrochemical tools [64,65].

Experimental techniques are conveniently divided according to the way in which the irradiation is carried out. The simplest approach is to shine a fairly intense source continuously on the solution, so that the overall photochemical transformation occurs approximately at steady state. Almost any electrometric method might serve to monitor the intermediates or products. During the first half of this century, potentiometric measurements were often made in this way [64], but these *photopotential* studies have been almost wholly supplanted by faradaic approaches, which have superior sensitivity and selectivity. Berg and his co-workers first introduced the faradaic methods, with emphasis on polarography and amperometry at the DME, in the early 1960s [66,67]. Kuwana's review gives good coverage to their work through about 1966 and to the earlier work on photopotentials [64].

There are several drawbacks to experiments done with continuous irradiation. High light fluxes cannot be introduced without serious heating; thus the concentrations of intermediates are low. Moreover, all intermediates are present simultaneously, since the reaction proceeds steadily. This situation makes it difficult to discern the order in which sequential intermediates appear, and it may create interference problems from overlapping electrochemical responses.

The desire for temporal resolution of photolysis led to the development of flash methods. In these experiments [68] the solution is exposed to a short (ca. 10 $\mu$s width) burst of light at high intensity (several hundred joules dissipated in the flash lamp). Absorption by the photoactive solute creates a high initial concentration of the primary intermediate. Its decay with time often leads to the rise and fall of other transient species that appear later in the reaction scheme. Because these time dependences tell much about the photolysis mechanism, flash methods are immensely valuable to photochemistry and have become very common. Usually, the intermediates are followed by UV or visible absorption spectroscopy. Berg and Schweiss were first to implement electrochemical monitoring [69], but Perone and his co-workers have been particularly active since the middle 1960s in the development and application of the technique [65,70-74].

In the most straightforward approach, one observes the current transient arising from the electrolysis of a flash-generated species at an electrode held at constant potential. However, there are two serious complications:

1. Electrolysis occurring before the flash may deplete the electrode region of the photoactive substance or generate obtrusive products.
2. The current decay is not linked simply to a time-dependent bulk concentration of the electroactive substance. Superimposed on that time function is the normal Cottrell decay from the diffusion process.

Perone's group surmounted these difficulties through a technique called time-delayed potentiostatic analysis [70,71]. Figure 23.12 depicts its operation. Before the flash and after it for a delay $\tau$, the working electrode is disconnected from ground by an open FET switch. At time $\tau$, the switch closes, electrolysis begins, and for a brief period thereafter, one sees a current decay that is essentially a Cottrell relation undistorted by reactant disappearance. An observation proportional to the bulk concentration of the transient species can be obtained by sampling the current at a fixed interval following $\tau$. Repeating the experiment with different potentials or different $\tau$ values enables one to derive current-potential curves for a sequence of delay

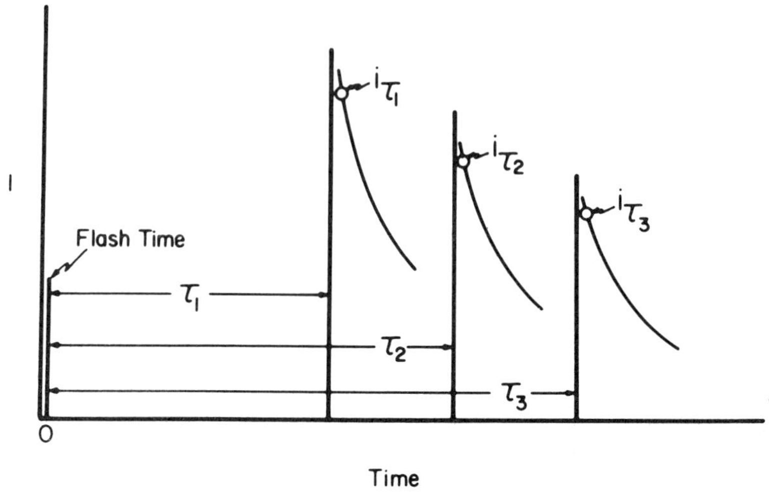


Figure 23.12 Current-time curves observed with time delayed potentiostatic analysis. $\tau_1$, $\tau_2$, and $\tau_3$ refer to three different delay times used for three different experiments. (From Ref. 65.)

times; hence there is both qualitative and quantitative information about the transients. Observations as soon as 50 μs after the flash can be made in this way. The value of this qualitative information was demonstrated by a study of ferrioxalate photolysis [72,73]. Using simultaneous electrochemical and optical monitoring of flash transients [74], Perone and co-workers were able to make some strong inferences about the chemical nature of a key intermediate, and they found evidence for some additional mechanistic steps that could not be discerned solely from optical data.

The techniques described here for monitoring flash-photolyzed solutions are clearly applicable to other situations in which the solution's composition is modified suddenly. Henglein and co-workers have used them in some interesting electrochemical studies of free radicals generated by pulse radiolysis [75,76]. They have been able to resolve events on a $10^{-5}$-s time scale.

Interest in continuous irradiation has been revived by the promising rotating electrode systems developed by two groups. Johnson and Resnick have used an RRDE design in which the disk was replaced by a quartz window, through which light was directed downward [77]. Photolytic products and intermediates are therefore swept past a Pt ring electrode, where they can be detected [77,78]. Albery et al., used a RDE design featuring a transparent thin-film Pt disk, through which the photolytic flux was directed [79]. These workers have used their apparatus to carry out a detailed study of intermediates in the oxidation-reduction processes involving anthraquinones and hydroanthraquinones. The use of these rotated electrodes offers some advantages. The measurements are simpler than with flash methods, and for strongly absorbing solutions they feature a localized photolysis in a region near the electrode. Moreover, they hold out an intriguing promise for studying photolytic processes in substances that are produced electrochemically.

## V. ELECTROCHEMILUMINESCENCE

### A. Chemical Background

Research has uncovered a general tendency for very fast, energetic electron transfer reactions to yield electronically excited products that may reveal themselves via luminescence [9,64,80-87]. A prototype of this chemistry is the reaction between the anion and cation radicals of 9,10-diphenylanthracene (DPA) in acetonitrile:

$$DPA^{\overset{+}{\cdot}} + DPA^{\overset{-}{\cdot}} \longrightarrow {}^{1}DPA^{*} + DPA \qquad (23.10)$$

The blue emission accompanying this reaction is easily identified with DPA's first excited singlet state $^1DPA^*$, which clearly must be an ultimate reaction product.

Theories of electron transfer have rationalized redox excitation as a manifestation of the Franck-Condon principle [88-90]. The actual redox act in the DPA(+)/DPA(−) reaction occurs very quickly; hence the production of ground-state DPA molecules would require nearly instantaneous mechanical excitation of the molecular frames by the full reaction energy of 3.2 eV. Because the system has a limited ability to accept mechanical energy on a short time scale, very fast, energetic reactions may follow paths that limit the mechanical excitation of the products. Raising DPA to $^1DPA^*$ requires about 3.0 eV; hence Reaction (23.10) is nearly thermoneutral. Very little energy must be taken up in mechanical modes; thus we can see why that path competes significantly with the more exothermic production of ground-state molecules.

Various kinds of experimental evidence indicate that the DPA(+)/DPA(−) reaction does indeed yield its light by direct population of the emitting state. Other cases are not so straightforward. Consider the oxidation of $DPA^{\overline{\cdot}}$ by the cation radical of N,N,N',N'-tetramethyl-p-phenylenediamine (TMPD). The free energy of that redox process, which also produces light from $^1DPA^*$, is only 2.0 eV per transferred electron. Thus the energy emitted in a single photon greatly exceeds the energy available from one redox event. The emitting state must arise by a more complicated mechanism, and much evidence points to a path involving redox-excited triplets, $^3DPA^*$:

$$DPA^{\overline{\cdot}} + TMPD^{\dot{+}} \longrightarrow {}^3DPA^* + TMPD \tag{23.11}$$

$$^3DPA^* + {}^3DPA^* \longrightarrow {}^1DPA^* + DPA \tag{23.12}$$

Reaction (23.12), called *triplet-triplet annihilation*, effectively localizes the energy from two electron transfers onto one DPA molecule.

One should not gain the impression from the foregoing that redox excitation is restricted to reactions between aromatic radical ions in aprotic solvents. Studies of such reactions have indeed dominated research because the optical and electrochemical properties of many aromatics are well known, but there are numerous cases of redox excitation outside this chemical domain. For example, singlet oxygen seems to arise from oxidations of superoxide ion in acetonitrile [91]. Similarly, luminescent tris(2,2'-bipyridyl)ruthenium(II) can arise in at least three ways: (1) from a kind of ion annihilation in $CH_3CN$ [92] or DMF,

$$Ru(bpy)_3^{+3} + Ru(bpy)_3^{+1} \longrightarrow [Ru(bpy)_3^{+2}]^* + Ru(bpy)_3^{+2} \quad (23.13)$$

(2) from reduction of $Ru(bpy)_3^{+3}$ with radiolytically produced aquated electrons [93], and (3) from reduction of the same species in aqueous base containing hydrazine [94].

Experimental work has amply demonstrated the generality of redox excitation, and it has uncovered many details of the mechanisms by which chemiluminescence arises. Thus attention has been partly turned toward the quantitative aspects: the probabilities of photonic emission per redox event and the excitation yields for various energetically accessible product states. Understanding the factors governing the distribution of excitation yields is an intriguing fundamental issue, because it bears on the details of the intermolecular interaction giving rise to the redox process. Opportunities to focus sharply on the details of solution phase reactions are very rare; hence it is not surprising that redox excitation distributions are receiving serious theoretical and experimental study.

## B. Basic Experimental Considerations

Experimental work is often hampered by limited reactant stability. Any redox process that is sufficiently energetic to produce electronic excitation necessarily involves very potent redox agents. Only in a few cases is it feasible to isolate the reactants in bulk and react them by a flow method. More often they are electrolytically generated as needed from parent substances such as neutral aromatics. Electrochemical generation offers several advantages:

1. Electrode processes at controlled potential can be carried out selectively and with high current efficiency.
2. One has flexibility in the time domain, geometry, and sequencing of reactant generation.
3. It is possible to obtain accurate theoretical time profiles of the homogeneous redox rate. These are important for quantitative measurements and some mechanistic studies.

In the interests of improved electrochemical background limits and reactant stability, it is important to employ solvents that are as free as possible of nucleophiles and proton sources. Special attention always goes to the removal of water. The most important media are carefully purified acetonitrile, dimethylformamide, benzonitrile, and tetrahydrofuran. Popular supporting electrolytes are tetra-n-butylammonium perchlorate (TBAP) and fluoroborate ($TBABF_4$). Solutions

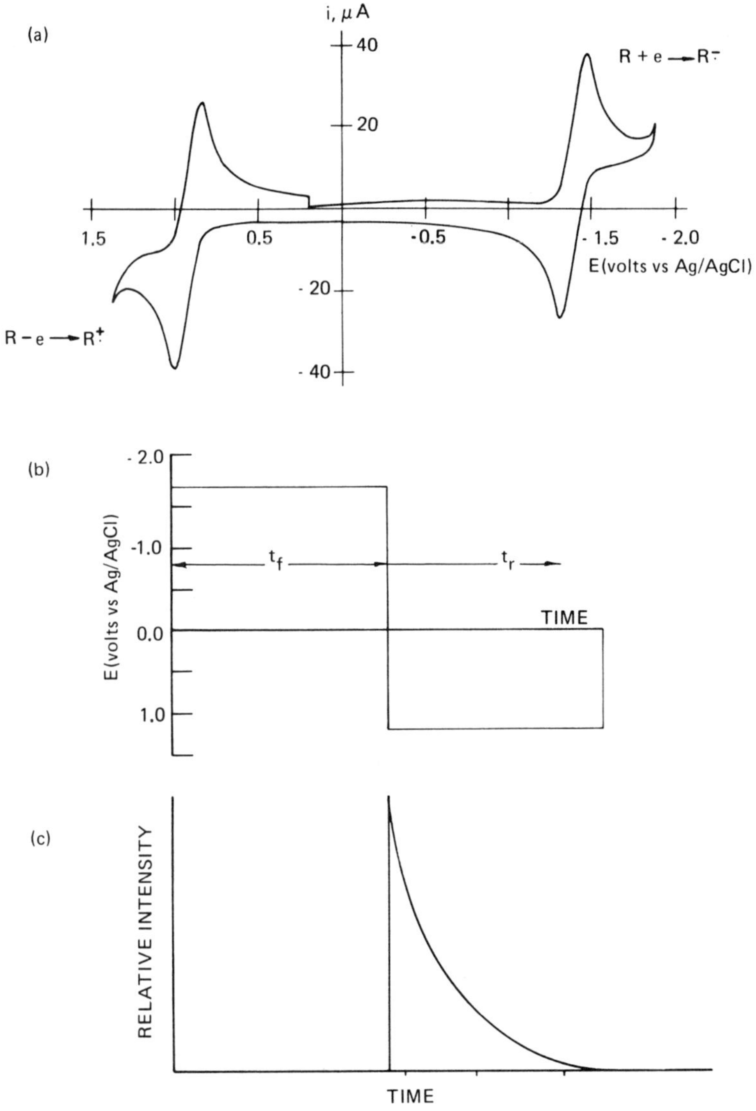


**Figure 23.13** ECL generation by the step technique. The Ag/AgCl, KCl (satd.) reference is −0.045 V vs. SCE. (a) Cyclic voltammetric curve (0.593 mM rubrene in benzonitrile with 0.1 M TBAP); (b) working electrode potential program; (c) emission intensity vs. time.

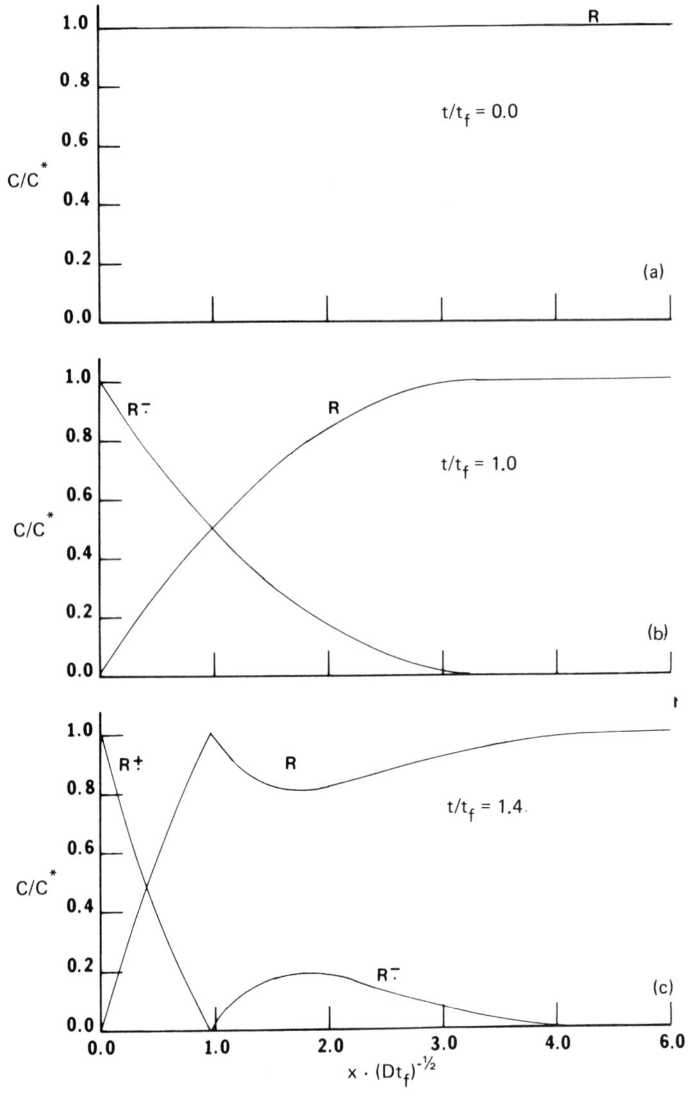


Figure 23.14 Concentration profiles in the generation of ECL by the step method. The initial bulk concentrations of R is represented as C*. (From Ref. 84.)

are usually prepared by vacuum line methods (Chap. 14) or in a dry box (Chap. 15) to exclude oxygen from the systems and to avoid contamination by water.

The most widely used technique for producing pairs of reactants involves the sequential generation of the oxidant and reductant at a single electrode [83,84,87]. Figure 23.13 displays the essentials of one such experiment in which the rubrene anion-cation annihilation is to be studied. The working electrode is a small Pt disk held initially at the rest potential in an unstirred solution of rubrene in benzonitrile. It is then carried through the potential program shown in Figure 23.13b. First the anions, then the cations, are produced. Concentration profiles for the electroactive substances are depicted in Figure 23.14. During the second step, there is a zone at which anions and cations diffuse together and react (see Fig. 23.14c). Thus emission arises during the second step, but it decays as the initial reactant is consumed. Obviously, this experiment could be performed so as to produce a single transient as described, or one could continue indefinitely the alternate generation of reactants to yield a train of pulses.

An alternative to sequential generation is to use two separate electrodes to generate the reactants, then to move them together for reaction. An elegant implementation involves the rotating ring-disk electrode [84,95]. Consider, for example, an investigation of the rubrene reaction discussed above. One could produce the cation at the disk and the anion at the ring. The convective-diffusion pattern would then bring them together for reaction near the face of the ring. A steady reaction rate can be achieved by this method.

The steady-state method has advantages in its freedom from double-layer charging and in the simplicity of light and current measurements. The transient method is mechanically simpler and does not require a dual potentiostat (Chap. 6). Moreover, the rate of electrolysis is smaller, hence solutions may suffer more slowly from the buildup of contaminants arising from side reactions. Either method can be treated theoretically via digital simulation to obtain time dependences of redox reaction rates, which are needed in some diagnostic and efficiency studies.

## C. Types of Experiments

Among the most important qualitative studies of any system are investigations of its emission spectra. One might wish to identify the emitting species or examine the spectral effects of certain additives. Spectral recording requires a light level that appears steady to the photometric instrument, which normally is a conventional fluorescence spectrometer. Steady-state generation of reactants straightforwardly satisfies this requirement. The continuous sequential method will also suffice if the period between light pulses is short compared to the photometric time constant.

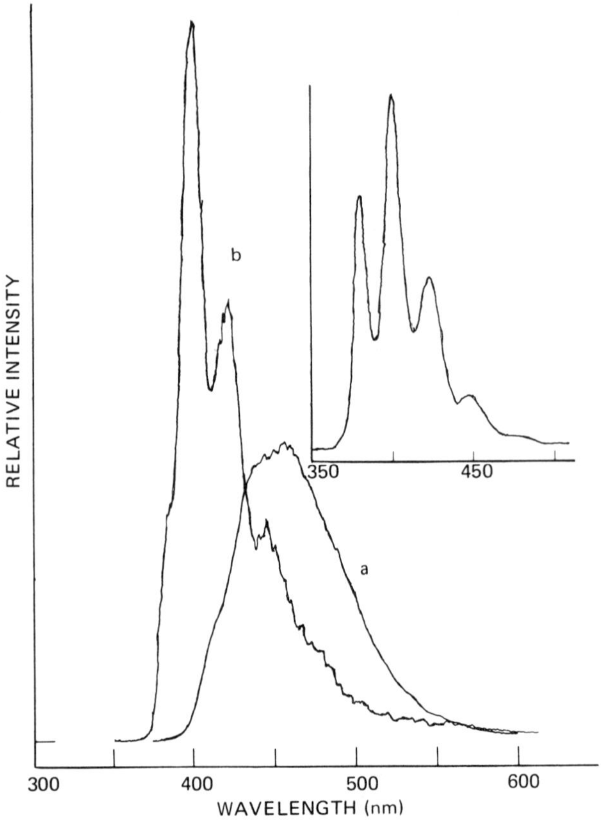


Figure 23.15 a, Chemiluminescence spectrum obtained from electrolysis of a DMF solution containing 1 mM fluoranthene and 1 mM 10-MP. Alternating steps at −1.75 V and +0.88 V vs. SCE were used. b, Chemiluminescence spectrum under the same conditions, but with 1 mM anthracene added. Inset shows anthracene fluorescence spectrum for a $10^{-5}$ M DMF solution. Reabsorption reduces the 0,0 intensity in (b). [From Ref. 96, adapted with permission from D. J. Freed and L. R. Faulkner, *J. Am. Chem. Soc. 93*:2097 (1971). Copyright 1971 American Chemical Society.]

One might use spectral recording in experiments in which one adds a triplet energy acceptor to the reacting system. Results from a study of the fluoranthene (−)/10-methylphenothiazine(+) reaction are shown in Figure 23.15 [83,96]. This process is an energetically deficient system that is thought to yield its blue fluoranthene-like emission via redox-produced fluoranthene triplets:

$$FA^{\overline{\cdot}} + 10\text{-}MP^{\overset{+}{\cdot}} \longrightarrow {}^{3}FA^{*} + 10\text{-}MP \tag{23.14}$$

$$^{3}FA^{*} + {}^{3}FA^{*} \longrightarrow {}^{1}FA^{*} + FA \tag{23.15}$$

$$^{1}FA^{*} \longrightarrow FA + h\nu \tag{23.16}$$

Figure 23.15, curve a, displays the emission spectrum. Anthracene (An) can be added to this system without electrochemical interference; yet it alters the spectrum to the anthracene-like distribution of Figure 23.15, curve b. Apparently, the exothermic energy transfer

$$An + {}^{3}FA^{*} \longrightarrow {}^{3}An^{*} + FA \tag{23.17}$$

quenches $^{3}FA^{*}$ completely, and emission arises from $^{1}An^{*}$ produced by annihilation of anthracene triplets.

A quantitative variant of this interception approach is useful for evaluating the yields of triplet states from certain redox processes [83,84]. Consider, for example, the FA(−)/10-MP(+) reaction in the presence of trans-stilbene. No emission is seen because the following quenching process occurs:

$$^{3}Fl^{*} + \text{trans-stilbene} \longrightarrow Fl + {}^{3}(\text{stilbene})^{*} \tag{23.18}$$

$$^{3}(\text{stilbene})^{*} \longrightarrow \begin{cases} \text{cis-stilbene} \\ \text{trans-stilbene} \end{cases} \tag{23.19}$$

Coulometric generation of the reactants in bulk provides a known number of redox events plus a quantity of cis-stilbene that can be determined chromatographically. Since the cis-trans isomerization yields are known from photochemical studies, one can calculate the efficiency with which $^{3}Fl^{*}$ is formed. Triplet yields ranging from 2 to 80% have been recorded for various systems in this way [83,84].

Another interesting quantitative question is the probability of emission per redox event, $\phi_{ECL}$. Evaluating this parameter is very difficult because one requires *absolute* luminescence measurements. That is, one must know the total number of photons emitted at every wavelength in every direction. Absolute studies are complicated by geometric factors, variations of photometric response with wavelength, effects due to reabsorption of light before it emerges from solution, and uncertain reflectivity of the electrode. Many approaches to the problem have appeared in the literature [82-84, 97-100]. In every such study, $\phi_{ECL}$ is obtained by comparing the photonic output with the faradaic charge used to generate the reactants. Steady-state and transient methods have both played important roles and $\phi_{ECL}$ values ranging from 0.001% to about 20% have been reported.

The single-pulse transient experiment described above has also been studied extensively as a tool for mechanistic investigations. Feldberg was first to see the diagnostic value in the shape of the light decay curve [101,102], and he showed that it could be used to discriminate between systems producing singlets directly (S route) and those proceeding through triplet intermediates (T route). More recent theoretical and experimental work has explored the extraction of quantitative excitation yield information from the shapes of absolute decay curves [86,87,103-107]. Feldberg also noted that the curve shape for S-route emission could be used to evaluate the homogeneous electron transfer rate constant if the time domain of the experiment could be made sufficiently short [101]. These rate parameters are very difficult to obtain because the reactions are so fast, yet they are of much current theoretical interest. Van Duyne's group followed Feldberg's lead by carrying out some difficult time-correlated single-photon-counting experiments involving 10-μs step widths [90]. They reported some rate data for the DPA(+)/DPA(−) reaction and compared the results with predictions from several theories.

Mechanistic questions have also been addressed via studies of magnetic field effects on electrogenerated chemiluminescence intensity [82-87,108]. Figure 23.16 contrasts the magnetic behavior for two systems. There is no significant field effect on DPA(+)/DPA(−) emis-

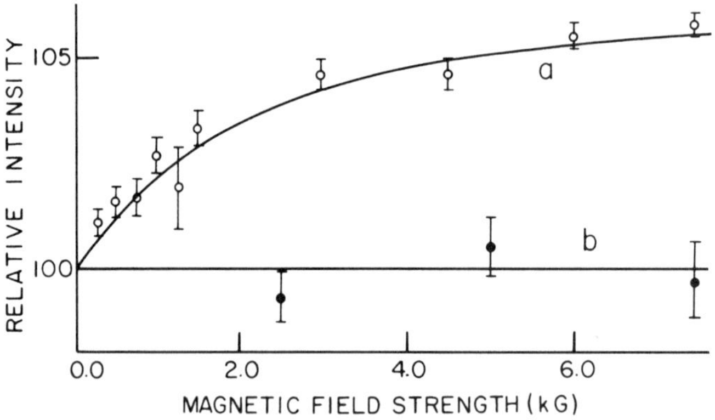


**Figure 23.16** Magnetic field effects on ECL intensity. a, From DPA(−)/TMPD(+). b, From DPA(+)/DPA(−). Light was generated at Pt by a train of alternating steps. [Adapted with permission from L. R. Faulkner and A. J. Bard, *J. Am. Chem. Soc. 91*:209 (1969). Copyright 1969 American Chemical Society.]

sion, but luminescence from DPA(−)/TMPD(+) is enhanced by the field. This enhancement apparently arises from field-dependent rate constants for triplet-triplet annihilation and for triplet quenching by radical ions. Both of these processes involve changes in spin angular momentum and, according to a theory of Merrifield [109], ought to have field-dependent encounter reaction probabilities. Extensive experimental work generally supports the view that a magnetic enhancement will generally appear whenever emission arises via the T route and no effect will be seen when the S route applies [82-87].

In this area, as in semiconductor photoelectrochemistry, we have been able only to set forth major concepts and experimental tools. The chemical issues, which are often fascinating, are explored much more fully in several up-to-date reviews [83-87].

## APPENDIX: STRUCTURES OF CITED MOLECULES

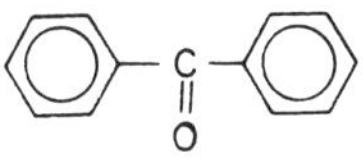


BP

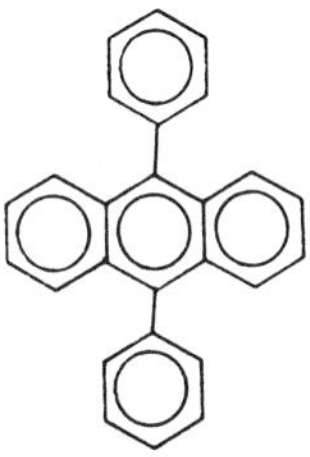

DPA

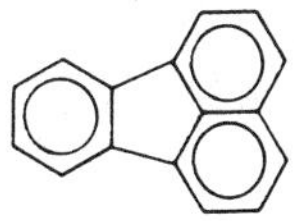

Fluoranthene

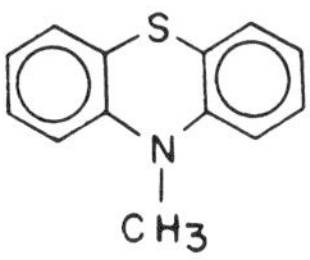


10-MP

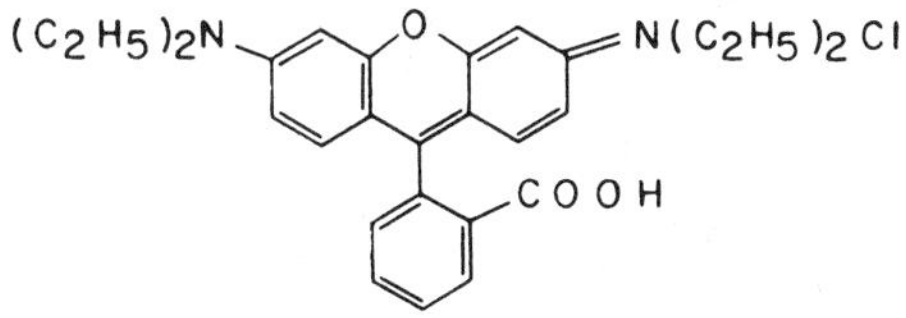


Rhodamine B

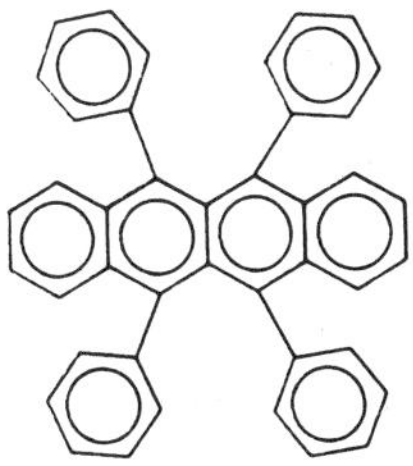

Rubrene

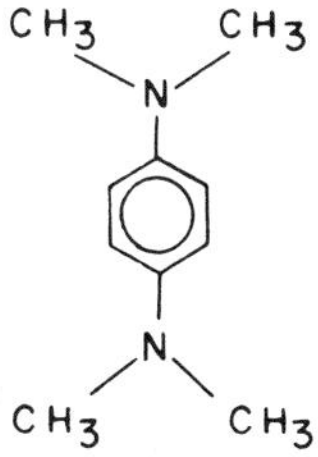


TMPD

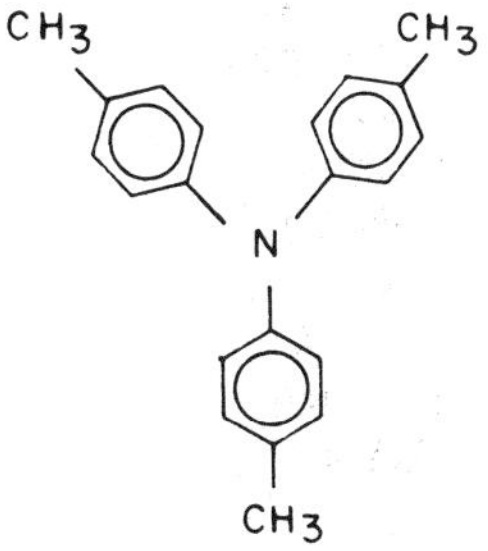


TPTA

## ACKNOWLEDGMENTS

We are grateful to A. J. Bard, G. C. Barker, H. Gerischer, K. Honda, R. Memming, S. P. Perone, and M. S. Wrighton for offering comments on this chapter and for granting us permission to reproduce material from their publications. This work was supported generously by Grants MPS-75-05361 and CHE-78-00584 from the National Science Foundation.


## REFERENCES


1. C. Kittel, *Introduction to Solid-State Physics*, 4th ed., Wiley, New York, 1971.
2. S. M. Sze, *Physics of Semiconductor Devices*, Wiley, New York, 1969.
3. V. A. Myamlin and Yu. V. Pleskov, *Electrochemistry of Semiconductors*, Plenum Press, New York, 1967.
4. H. Gerischer, in *Physical Chemistry*, Vol. IXA (H. Eyring, D. Henderson, and W. Jost, eds.), Academic Press, New York, 1970, Chap. 5.
5. A. J. Nozik, *Annu. Rev. Phys. Chem. 29*:189 (1978).
6. A. J. Bard, *Science 207*:139 (1980).
7. S. R. Morrison, *Electrochemistry at Semiconductor and Oxidized Metal Electrodes*, Plenum Press, New York, 1980.
8. A. Heller, *Acc. Chem. Res. 14*:154 (1981).
9. A. J. Bard and L. R. Faulkner, *Electrochemical Methods*, Wiley, New York, 1980, Chap. 14.
10. A. J. Nozik, *Faraday Discuss. Chem. Soc. 70*:7 (1980).
11. A. J. Bard, A. B. Bocarsly, F.-R. Fan, E. G. Walton, and M. S. Wrighton, *J. Am. Chem. Soc. 102*:3671 (1980).
12. A. B. Bocarsly, D. C. Bookbinder, R. N. Dominey, N. S. Lewis, and M. S. Wrighton, *J. Am. Chem. Soc. 102*:3683 (1980).
13. F.-R. Fan and A. J. Bard, *J. Am. Chem. Soc. 102*:3677 (1980).
14. A. J. Bard, F.-R. Fan, A. S. Gioda, G. Nagasubramanian, and H. S. White, *Faraday Discuss. Chem. Soc. 70*:19 (1980).
15. J. Bardeen, *Phys. Rev. 71*:717 (1947).
16. A. Fujishima, K. Honda, and S. Kikuchi, *J. Chem. Soc. Jpn. (Ind. Chem. Soc.) 72*:108 (1969).
17. A. Fujishima and K. Honda, *Nature 238*:37 (1972).
18. A. Fujishima and K. Honda, *J. Photochem. Soc. Jpn. (Nippon Shashin Gakkaishi) 37*:303 (1974).
19. M. S. Wrighton, D. S. Ginley, P. T. Wolczanski, A. B. Ellis, D. L. Morse, and A. Linz, *Proc. Natl. Acad. Sci. USA 72*:1518 (1975).

20. M. S. Wrighton, A. B. Ellis, P. T. Wolczanski, D. L. Morse, H. B. Abrahamson, and D. S. Ginley, *J. Am. Chem. Soc.* *98*:2774 (1976).
21. A. J. Bard and M. S. Wrighton, *J. Electrochem. Soc.* *124*:1706 (1977).
22. H. Gerischer, *Faraday Discuss. Chem. Soc.* *70*:137 (1980).
23. R. Williams, *J. Chem. Phys.* *32*:1505 (1960).
24. A. Heller and B. Miller, *Adv. Chem. Ser.* *184*:215 (1980).
25. L. A. Harris and R. H. Wilson, *J. Electrochem. Soc.* *123*:1010 (1976).
26. A. B. Ellis, S. W. Kaiser, and M. S. Wrighton, *J. Am. Chem. Soc.* *98*:1635 (1976).
27. M. S. Wrighton, A. B. Bocarsly, J. M. Bolts, M. G. Bradley, A. B. Fischer, N. S. Lewis, M. C. Palazzotto, and E. G. Walton, *Adv. Chem. Ser.* *184*:269 (1980).
28. D. C. Bookbinder, J. A. Bruce, R. N. Dominey, N. S. Lewis, and M. S. Wrighton, *Proc. Natl. Acad. Sci. USA* *77*:6280 (1980).
29. T. Freund and W. P. Gomes, *Catal. Rev.* *3*:1 (1969).
30. S. N. Frank and A. J. Bard, *J. Phys. Chem.* *81*:1484 (1977).
31. B. Kraeutler and A. J. Bard, *J. Am. Chem. Soc.* *100*:2239, 5985 (1978).
32. M. Grätzel, *Faraday Discuss. Chem. Soc.* *70*:311 (1980); *Acc. Chem. Res.* *14*:376 (1981).
33. H. Gerischer and H. Tributsch, *Ber. Bunsenges. Phys. Chem.* *72*:437 (1968).
34. H. Tributsch and H. Gerischer, *Ber. Bunsenges. Phys. Chem.* *73*:251 (1969); H. Tributsch, *Ibid.* *73*:582 (1969); H. Tributsch and H. Gerischer, *Ibid.* *73*:850 (1969); H. Gerischer, M. E. Michel-Beyerle, F. Rebentrost, and H. Tributsch, *Electrochem. Acta.* *13*:1509; H. Gerischer, *Photochem. Photobiol.* *16*:243 (1972).
35. K. Hauffe, H. J. Danzmann, H. Pusch, J. Rauge, and H. Volz, *J. Electrochem. Soc.* *117*:993 (1970).
36. H. Tributsch and M. Calvin, *Photochem. Photobiol.* *14*:95 (1971).
37. A. Terenin and I. Akimov, *J. Phys. Chem.* *69*:730 (1965).
38. W. P. Gomes and F. Cardon, *Ber. Bunsenges. Phys. Chem.* *75*:914 (1971).
39. H. Tributsch and H. Gerischer, *Ber. Bunsenges. Phys. Chem.* *73*:850 (1969).
40. A. Fujishima, T. Watanabe, O. Tatsuoki, and K. Honda, *Chem. Lett.*, 13 (1975).
41. A. Fujishima, E. Hayashitani, and K. Honda, *J. Inst. Ind. Sci.* Univ. Tokyo *(Seisan Kenkyo)*, *23*:363 (1971).
42. T. Osa and M. Fujihira, *Nature* *264*:349 (1976).
43. M. Fujihira, N. Ohishi, and T. Osa, *Nature* *268*:226 (1977).

44. M. Fujihira, T. Osa, D. Hursh, and T. Kuwana, *J. Electroanal. Chem. 88*:285 (1978).
45. R. Memming and H. Tributsch, *J. Phys. Chem. 75*:562 (1971).
46. H. Gerischer, *Photochem. Photobiol. 16*:243 (1972).
47. H. Tributsch, *Photochem. Phobobiol. 16*:261 (1972).
48. W. Arden and P. Fromherz, *Ber. Bunsenges. Phys. Chem. 82*:868 (1978).
49. W. Arden and P. Fromherz, *J. Electrochem. Soc. 127*:370 (1980).
50. P. Fromherz and W. Arden, *Ber. Bensenges. Phys. Chem. 84*:1045 (1980).
51. H. Bücher, K. H. Drexhage, M. Fleck, H. Kuhn, D. Möbius, F. P. Schaefer, J. Sondermann, W. Sperling, P. Tillmann, and J. Wiegand, *Mol. Cryst. 2*:199 (1967).
52. R. Memming, *Electroanal. Chem. 11*:1 (1979).
53. Yu. Ya. Gurevich, Yu. V. Pleskov, and Z. A. Rotenberg, *Photoelectrochemistry*, Plenum Press, New York, 1978.
54. Yu. V. Pleskov and Z. A. Rotenberg, *Adv. Electrochem. Eng. 11*:1 (1978).
55. M. Heyrovsky and R. G. W. Norrish, *Nature 200*:880 (1963); M. Heyrovsky, *Nature 206*:1356 (1965).
56. H. Berg and H. Schweiss, *Electrochim. Acta 9*:425 (1964).
57. G. C. Barker, A. W. Gardner, and D. C. Sammon, *J. Electrochem. Soc. 113*:1182 (1966).
58. G. C. Barker, *Ber. Bunsenges. Phys. Chem. 75*:728 (1971).
59. G. C. Barker, D. McKeown, M. J. Williams, G. Bottura, and V. Concialini, *Faraday Discuss. Chem. Soc. 56*:41 (1974).
60. Yu. V. Pleskov and Z. A. Rotenberg, *J. Electroanal. Chem. 20*:1 (1969).
61. A. Brodsky and Yu. V. Pleskov, in *Progress in Surface Sciences*, Vol. 2, Part 1 (S. G. Davison, ed.), Pergamon Press, Oxford, 1972.
62. Yu. V. Pleskov, Z. A. Rotenberg, V. V. Eletsky, and V. I. Lakomov, *Faraday Discuss. Chem. Soc. 56*:52 (1974).
63. P. Delahay and V. S. Srinivasan, *J. Phys. Chem. 70*:420 (1966).
64. T. Kuwana, in *Electroanalytical Chemistry*, Vol. 1 (A. J. Bard, ed.), Marcel Dekker, New York, 1966, Chap. 3, and references contained therein.
65. S. P. Perone and H. D. Drew, in *Analytical Photochemistry and Photochemical Analysis: Solids, Solutions, and Polymers* (J. Fitzgerald, ed.), Marcel Dekker, New York, 1971, Chap. 7, and references contained therein.
66. H. Berg and H. Schweiss, *Naturwissenschaften 47*:513 (1960).
67. H. Berg, H. Schweiss, E. Stutter, and K. Weller, *J. Electroanal. Chem. 15*:415 (1967), and references contained therein.

68. G. Porter, in *Techniques of Organic Chemistry*, Vol. 3, Part II (A. Weissberger, ed.), Wiley-Interscience, New York, 1963, Chap. XIX.
69. H. Berg and H. Schweiss, *Nature 191*:1270 (1961).
70. S. P. Perone and J. R. Birk, *Anal. Chem. 38*:1589 (1966).
71. G. L. Kirschner and S. P. Perone, *Anal. Chem. 44*:443 (1972).
72. R. A. Jamieson and S. P. Perone, *J. Phys. Chem. 76*:830 (1972).
73. J. I. H. Patterson and S. P. Perone, *J. Phys. Chem. 77*:2437 (1973).
74. J. I. H. Patterson and S. P. Perone, *Anal. Chem. 44*:1978 (1972).
75. M. Grätzel and A. Henglein, *Ber. Bunsenges. Phys. Chem. 77*:2 (1973); *77*:6 (1973); *77*:11 (1973); *77*:17 (1973).
76. A. Henglein, *Electroanal. Chem. 9*:163 (1976).
77. D. C. Johnson and E. W. Resnick, *Anal. Chem. 44*:637 (1972).
78. J. R. Lubbers, E. W. Resnick, P. R. Gaines, and D. C. Johnson, *Anal. Chem. 46*:865 (1974).
79. W. J. Albery, M. D. Archer, N. J. Field, and A. D. Turner, *Faraday Discuss. Chem. Soc. 56*:28 (1974).
80. E. A. Chandross, *Trans. N.Y. Acad. Sci., Ser. 2*:32, 571 (1969), and references contained therein.
81. D. M. Hercules, in *Physical Methods of Organic Chemistry*, Part II (A. Weissberger and B. Rossiter, eds.), Academic Press, New York, 1971, and references contained therein.
82. A. J. Bard, C. P. Keszthelyi, H. Tachikawa, and N. E. Tokel, in *Chemiluminescence and Bioluminescence* (D. M. Hercules, J. Lee, and M. Cormier, eds.), Plenum Press, New York, 1973, and references contained therein.
83. L. R. Faulkner, *Int. Rev. Sci., Phys. Chem., Ser. 2 9*:213 (1975).
84. L. R. Faulkner and A. J. Bard, *Electroanal. Chem. 10*:1 (1977).
85. F. Pragst, *Z. Chem. 18*:41 (1978).
86. L. R. Faulkner, *Methods Enzymol. 57*:494 (1978).
87. R. S. Glass and L. R. Faulkner, in *Chemical and Biological Generation of Excited States* (W. Adam and G. Cilento, eds.), Academic Press, New York, 1982, Chap. 6.
88. R. A. Marcus, *Annu. Rev. Phys. Chem. 15*:155 (1964).
89. R. A. Marcus, *J. Chem. Phys. 43*:2654 (1965); *52*:2083 (1970).
90. R. P. Van Duyne and S. F. Fischer, *Chem. Phys.* 5:183 (1974).
91. E. A. Mayeda and A. J. Bard, *J. Am. Chem. Soc. 95*:6223 (1973).

92. N. E. Tokel-Takvoryan, R. E. Hemingway, and A. J. Bard, *J. Am. Chem. Soc. 95*:6582 (1973).
93. J. E. Martin, E. J. Hart, A. W. Adamson, H. Gafney, and J. Halpern, *J. Am. Chem. Soc. 94*:9238 (1972).
94. D. M. Hercules and F. E. Lytle, *J. Am. Chem. Soc. 88*:4745 (1966).
95. J. T. Maloy, K. B. Prater, and A. J. Bard, *J. Am. Chem. Soc. 93*:5959 (1971).
96. D. J. Freed and L. R. Faulkner, *J. Am. Chem. Soc. 93*:2097 (1971).
97. C. P. Keszthelyi, N. E. Tokel-Takvoryan, and A. J. Bard, *Anal. Chem. 47*:249 (1975).
98. P. R. Michael and L. R. Faulkner, *Anal. Chem. 48*:1188 (1976).
99. P. R. Michael and L. R. Faulkner, *J. Am. Chem. Soc. 99*:7754 (1977).
100. W. L. Wallace and A. J. Bard, *J. Phys. Chem. 83*:1350 (1979).
101. S. W. Feldberg, *J. Am. Chem. Soc. 88*:390 (1966).
102. S. W. Feldberg, *J. Phys. Chem. 70*:3928 (1966).
103. L. R. Faulkner, *J. Electrochem. Soc. 124*:1724 (1977).
104. J. L. Morris, Jr. and L. R. Faulkner, *J. Electrochem. Soc. 125*:1079 (1978).
105. J. D. Luttmer and A. J. Bard, *J. Phys. Chem. 85*:1155 (1981).
106. R. S. Glass and L. R. Faulkner, *J. Phys. Chem. 85*:1160 (1981).
107. R. S. Glass and L. R. Faulkner, *J. Phys. Chem. 86*:1652 (1982).
108. L. R. Faulkner, H. Tachikawa, and A. J. Bard, *J. Am. Chem. Soc. 94*:691 (1972).
109. R. E. Merrifield, *J. Chem. Phys. 48*:4318 (1968).

# 24

# Principles and Techniques of Electrochemical-Electron Spin Resonance Experiments

IRA B. GOLDBERG / Rockwell International Science Center, Thousand Oaks, California

TED M. McKINNEY / Riverside, California

# I. MAGNETIC RESONANCE IN ELECTROCHEMICAL STUDIES

Techniques that provide definitive identification of intermediate or product species can be a valuable adjunct in the study of complicated electrochemical reaction sequences. Almost every imaginable analytical method has been used, and spectroscopic techniques have proven to be particularly valuable; each particular method contributes a unique set of data for the experimentalist to interpret. Conversely, it should be recognized that electrochemistry has also aided spectroscopists by enabling them to prepare and study species that might otherwise be inaccessible.

This chapter provides a review of the principles of electron spin resonance (ESR) as they apply to electrochemical studies. ESR is a spectroscopic method that is sensitive to materials with unpaired electrons, such as organic and inorganic free radicals, radical anions and cations, and the complexes of various transition metal ions; many substances of biological significance fall into these categories.

As originally postulated by Michaelis, most electrode processes involve the transfer of *one* electron to or from an "electroactive" material. In such experiments, either the parent compound or the electrode product contains at least one unpaired electron and it is said to be "paramagnetic." The electrochemical reduction or oxidation of neutral organic molecules represents a typical system:

$$R^\circ + e \longrightarrow R^{\overline{\cdot}} \quad \text{(anion radical)} \qquad (24.1a)$$

$$R^\circ - e \longrightarrow R^{\overset{+}{\cdot}} \quad \text{(cation radical)} \qquad (24.1b)$$

Often the initial electrode product will undergo subsequent chemical or electron-transfer reactions, but the stability of the species depends greatly on the reactivity of the medium in which it is produced.

Nuclear magnetic resonance (NMR) is an allied technique which has many formal similarities to ESR. It is widely used for product analysis, as in studies leading to the development of electrosynthetic methods. The intrinsic sensitivity of NMR is several orders of magnitude lower than that of ESR, making simultaneous NMR-electrochemical experiments unattractive.

The remainder of this chapter presents an elementary discussion of the theory, instrumentation, and practice of ESR-electrochemical studies. We recite the usual disclaimers about limitations of space to explain that the subject cannot be covered comprehensively here. The selected bibliography at the end of the chapter is broken down into broad categories to guide the interested reader to specific topics. The student who wishes a more thorough discussion of the general subject at an elementary level may find McKinney's review helpful.

## II. MOLECULAR ORBITALS IN ELECTROCHEMISTRY AND ESR

Molecular orbital (MO) theory is widely used in the current literature to describe both organic [1] and inorganic chemical systems. The MOs of a given molecule can be calculated by a variety of methods at different levels of sophistication. Increased accuracy is gained at the expense of longer computation time, but the results all have the following properties in common:

1. Each MO is characterized by a particular energy level (an "eigenvalue" of the MO calculation) measured on a convenient energy scale. The zero of energy is fixed at some arbitrary point, analogous to the zero point of potential (reference electrode) in an electrochemical experiment.
2. Each MO represents a specific spatial distribution of probability for the electrons in that orbital. This distribution is known as the "eigenfunction" for that orbital. This information can be used to evaluate the probability of finding a given electron in some particular volume of space (e.g., in the neighborhood of a certain atomic nucleus).

The electrochemical behavior of a molecule is directly related to the MO energy levels, while experimental ESR parameters correlate with the electron probability distributions.

We restrict our attention in this chapter to the simple but widely used Hückel MO (HMO) method for calculating orbitals for $\pi$-electron and aromatic molecules [1]. The HMO scheme assumes that a conjugated $\pi$-electron molecule consists of a network of $sp^2$-hybridized carbon atoms lying in a plane and each participating atom i has a 2p electron in an atomic orbital, $\phi_i$, perpendicular to this plane. Linear combinations of these atomic orbitals (LCAO) results in the molecular $\pi$ wavefunctions, $\psi_j$, each of which has a discrete energy, $E_j$. In terms of the parameters used in HMO computations,

$$E_j = \alpha + \lambda_j \beta^\circ_{CC} \tag{24.2}$$

where $\alpha$ is a reference point and can be ignored; then $\lambda_j$ becomes a measure of the energy of the jth MO in units of $\beta^\circ_{CC}$, which can be estimated from experimental data. For example, $\beta^\circ_{CC}$ has been assigned values between $-2.04$ and $-3.13$ eV for electrochemical oxidations and between $-2.37$ and $-2.72$ eV for electrochemical reductions [2]. This range of values reflects the crudeness of the HMO model itself since electron correlations, electron repulsions, and solvation changes are neglected.

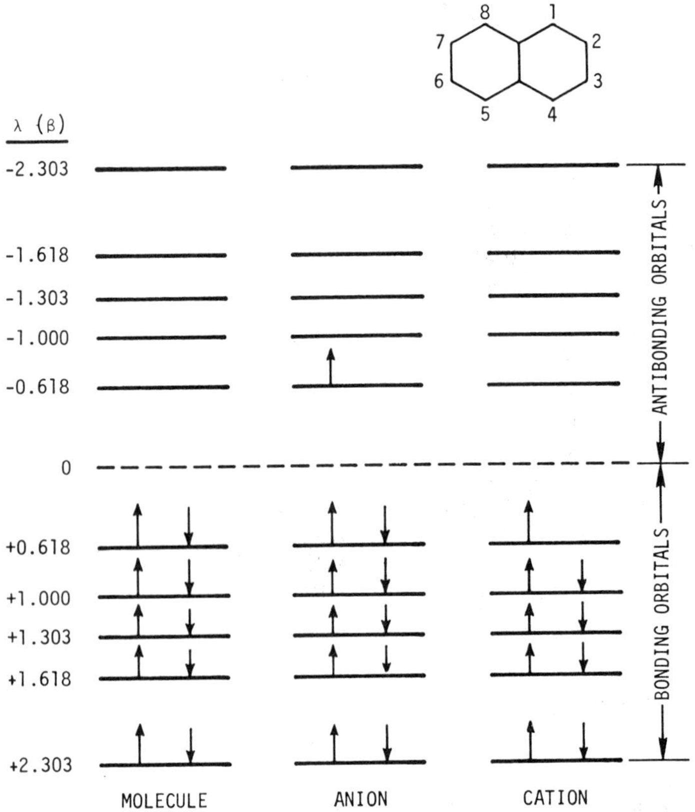


Figure 24.1 Hückel molecular orbital (HMO) energy levels and electron occupation for naphthalene, its anion radical, and its cation radical. HMO calculations employ $\beta = -2.5$ eV as the unit of energy so that $+\beta$ values correspond to negative or binding energy.

The energy levels correspond to bonding ($E_j < \alpha$), antibonding ($E_j > \alpha$), and in some molecules nonbonding ($E_j = \alpha$) orbitals. For example, the naphthalene molecule has only bonding and antibonding orbitals as shown in Figure 24.1. The net binding energy of the $\pi$-electron system can be evaluated by summing the energy of each electron in the occupied orbitals. The reader can demonstrate that energy must be supplied to the naphthalene molecule to form either the anion or cation radical; that is, the change in free energy is positive for either reaction. The half-wave potential is closely related to the value $\lambda_j$ of the molecular orbital of the anion or cation. The relationship

$$E_{1/2} = A + B\lambda_j \tag{24.3}$$

has been well established [2,3]. A includes the difference between the zero points of the MO calculations and the reference electrode potential, and B is the empirical value of $\beta^{o}_{CC}$, expressed in electron volts.

The spatial distribution of electrons in the jth MO is related to the numerical coefficients $c_i$ of the atomic orbitals $\phi_i$ whose linear combination yields the $\pi$ orbital, $\psi_j$:

$$\psi_j = c_{j1}\phi_1 + c_{j2}\phi_2 + \cdots + c_{jn}\phi_n = \sum_{i=1}^{n} c_{ji}\phi_i \tag{24.4}$$

When $\psi_i$ is a normalized wavefunction the atomic orbital coefficients fulfill the condition

$$\sum_{i=1} c_{ji}^2 = 1 \tag{24.5}$$

and when $\psi_k$ denotes the particular MO containing an unpaired electron,

$$\rho_i = c_{ki}^2 \tag{24.6}$$

where $\rho_i$ is the probability of the unpaired electron being at atom i. The quantity $\rho$ is known as the unpaired electron density or spin density (advanced treatments [4] make quantitative distinction between these two terms) and its importance in the interpretation of ESR data will become apparent below.

## III. PRINCIPLES OF ESR

### A. Spin Resonance

Electrons and some nuclei have intrinsic angular momentum, which can be visualized as rotation of the particle, hence the trivial name "spin." This spin causes the charged particle to behave as a small magnet which is characterized by a magnetic moment, $\underset{\sim}{\mu}$. For electrons the unit of measure is the Bohr magneton,

$$\mu_B = \frac{|e|h}{2m_e c} = 9.274078 \times 10^{-21} \text{ erg/G*} \quad (24.7a)$$

(e is the electron charge, h is Planck's reduced constant, $m_e$ is the electron mass, and c the speed of light); for nuclei the unit is the nuclear magneton

$$\mu_N = \frac{|e|h}{2M_p c} = 5.050824 \times 10^{-24} \text{ erg/G*} \quad (24.7b)$$

The magnetic moment of a free electron is related to its spin angular momentum $\underset{\sim}{S}$ as

$$\underset{\sim}{\mu}_e = -g_e \mu_B \underset{\sim}{S} \quad (24.8a)$$

where the negative sign indicates opposite vector orientations of $\underset{\sim}{\mu}_e$ and $\underset{\sim}{S}$, and $g_e = 2.00232$ is the free-electron g factor. Electrons on isolated atoms may also possess orbital angular momentum which augments the spin magnetic moment. This orbital contribution tends to be *quenched* by strong electrostatic fields such as those present in the covalent bonds of organic molecules or metal-ligand complexes. However, the quenching is seldom complete and the net magnetic moment has some contribution from the residual orbital angular momentum. This can be accommodated by rewriting Equation 24.8a as

$$\mu_S = g\mu_B S \quad (24.8b)$$

and viewing the g factor as a parameter that is characteristic of the particular species.

All of the reported g values for organic radicals fall between 2.0007 (nitrosobenzene cation radical) and 2.0120 (tetraiodo-p-benzosemiquinone). Meticulous experimental technique is required to measure g values to high precision so that organic species are usually characterized in terms of other ESR parameters (see below). A theoretical treatment has been developed by Stone [5] to explain the g values of organic radicals in terms of the results of HMO calculations,

---

*At the time this article is being revised, a change from cgs to SI units has been recommended. Most data in the literature are given in terms of cgs units as in this review. A conversion table with recommended units is given in Appendix II.

and simplified versions of this treatment can be found in the general references at the end of this chapter. Ions of the higher transition series, particularly the rare earth ions, exhibit g values that differ greatly from $g_e$. This fact can be very useful in identifying such species. Although these g values are well understood, the calculations are quite lengthy and the reader is again referred to the general texts and the review papers on metal ions at the end of the chapter.

Many metal ions have more than one unpaired electron; similarly, certain organic species (diradicals, triplet, and higher states) may have several unpaired spins. But the presence of several electron spin dipoles in close proximity on a molecule in solution broadens the spectral lines so drastically that its ESR spectrum cannot usually be observed in solution. Consequently, such species have little relevance to ESR-electrochemical studies and we will limit out attention to doublet (one unpaired electron) states. Exceptions to this are $Mn^{2+}$ and sometimes $Cr^{3+}$ with stong ligand coordination. However, in solution, these may be treated as if they had only one unpaired electron, although the ESR absorption is much greater than expected for a one-electron species.

The measurable component of electron spin has the magnitude of $\hbar/2$, which produces two allowed quantum spin states, $m_s = +1/2$ (designated as $|\alpha\rangle$ spin, by convention, and $m_s = -1/2$ ($|\beta\rangle$ spin). When a free electron is placed in a strong magnetic field $\underset{\sim}{H}$* its magnetic moment $\mu_s$ behaves like a tiny bar magnet. It will line up either parallel or antiparallel to $\underset{\sim}{H}$.* The energy of this interaction is

$$E = \underset{\sim}{\mu}_s \cdot \underset{\sim}{H} = \pm \frac{1}{2} g\mu_B H \qquad (24.9)$$

where the plus sign corresponds to the energy of $|\alpha\rangle$ spin and the minus to $|\beta\rangle$ spin. As shown in Figure 24.2a the energy difference between the two levels depends linearly on the strength of the applied field,

$$\Delta E = g\mu_B H \qquad (24.10)$$

Transitions between the two electron-spin states can be induced by application of electromagnetic radiation of the correct frequency (i.e., $\nu = \Delta E/h$). In practice, it is simpler to hold the frequency constant while scanning the magnetic field to produce the condition for resonant

*H rather than B is often used to designate magnetic field. B is, in fact, correct (see Appendix II).

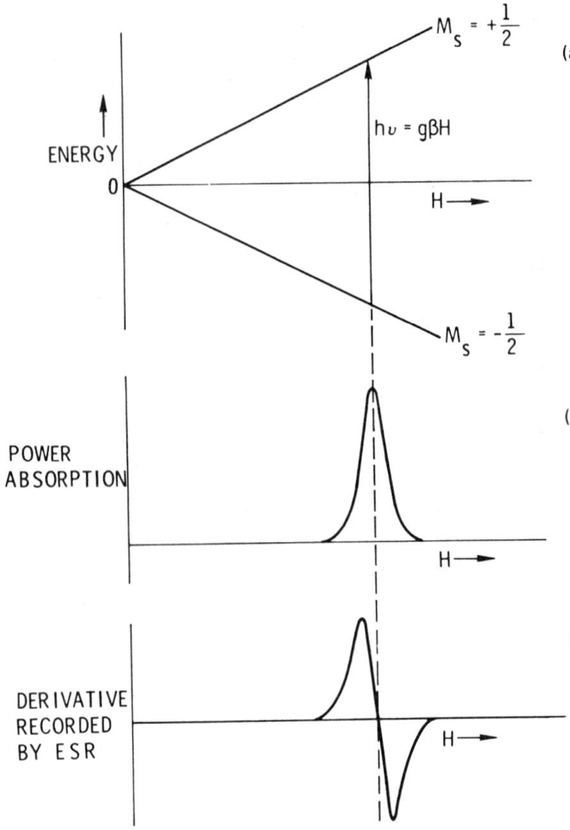


Figure 24.2 (a) Energy level of diagram of an unpaired electron in a magnetic field. (b) Absorption at constant frequency and swept field. (c) Derivative signal recorded by ESR.

absorption of energy; see Figure 24.2b and c. The majority of ESR experiments are carried out with radiation at a frequency of about 9.5 GHz. For g values close to 2, this corresponds to a field strength of about 3400 G and the electron energy levels are separated by approximately $6.3 \times 10^{-17}$ erg.

Typical samples for ESR spectroscopy contain about $10^{17}$ spins. The relative populations of the two spin states are given by the Boltzmann distribution,

$$\frac{N_\alpha}{N_\beta} = \exp\left(\frac{-\Delta E}{kT}\right) \tag{24.11}$$

where k is the Boltzmann constant and T the absolute temperature. At room temperature there are about 998 spins in the higher-energy level for every 1000 in the lower level. When the frequency of the electromagnetic radiation and the applied field satisfy the resonance condition

$$h\nu_0 = g\mu_B H_0 \tag{24.12}$$

the radiation excites both absorption and emission transitions with equal probabilities. The observation of a net absorption of energy requires that the Boltzmann distribution remain essentially undisturbed. Fortunately, there are pathways other than stimulated emission which permit spins in the upper state to return to the lower state. This phenomenon, known as spin relaxation, has received much attention but cannot be described in any detail here (see the references at the end of the chapter). For doublet-state species, the relaxation mechanisms are often relatively inefficient. When such a sample is subjected to excessive levels of microwave power, the two spin states tend to become equally populated and the net absorption of energy approaches zero. This phenomenon is known as power saturation.

Similar considerations are often mentioned in discussions of NMR spectroscopy because the two techniques are formally quite similar. Indeed, the reader might well inquire what the nuclei are doing while the unpaired electron spins are undergoing transitions. A simple calculation (with $\mu_N$ from Eq. 24.7b and $g_p = 5.585$ for the proton g value) reveals that the proton resonance frequency at 3400 G is about 14.5 MHz [i.e., in the much longer wavelength, radio-frequency (RF) region of the electromagnetic spectrum]. So although the external field does produce polarization of the proton spins, the radiation frequency used in ESR is completely ineffectual in causing nuclear transitions.

## B. Hyperfine Interactions

Nuclear spin does, in fact, play a significant role in ESR. It can interact with the unpaired electron spin to give rise to the hyperfine splittings (HFSs) which greatly increase the amount of information that can be extracted from an ESR spectrum.

When a paramagnetic sample contains magnetic nuclei, the alignment of these nuclear magnetic moments relative to the external field produces small local fields which are capable of splitting the electron-spin energy levels into electron-nuclear hyperfine levels. The simplest example of this appears in the ESR spectrum of hydrogen *atoms*, H· (ls configuration). Protons have a nuclear spin of $I = 1/2$ (in units of $\hbar$) and a positive magnetic moment. This generates a magnetic field,

$H_{local}$, which adds to, or subtracts from, the applied field at the electron. Since it is the total field strength that determines the spacing of the electron spin energy levels, the resonance condition, Equation 24.12, is achieved when $H_{total} = H_0$, or

$$\frac{h\nu_0}{g\mu_B} = H_0 = H_{external} \pm H_{local} \tag{24.13}$$

These arguments are summarized in Figure 24.3. The lower portion of the figure shows that the ESR spectrum of H· atoms consists of two hyperfine lines separated by 506 G. This separation is a measure of the strength of the coupling between the electron and nuclear magnetic moments in this species (i.e., the value of $H_{local}$ experienced by the unpaired electrons). It is called the hyperfine splitting constant (HFSC), usually symbolized by $a_i$, where i refers to a set of nuclei.

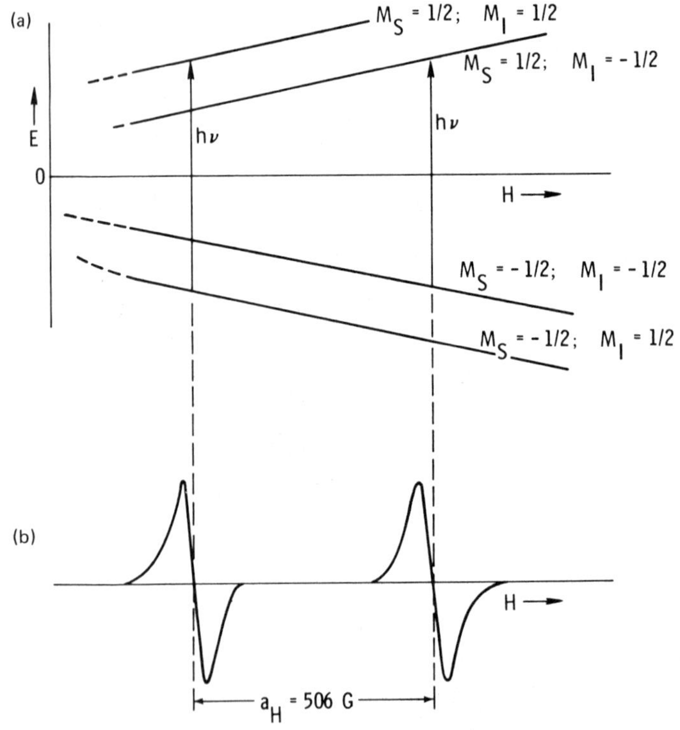


Figure 24.3 (a) Energy level of hydrogen atom in a magnetic field. (b) ESR spectrum of hydrogen atom.

Actually, the word "constant" is slightly misleading: $a_H$ = 506 G is indeed a constant for H· atoms in the 1s configuration; but H· atoms in the 2s state display $a_H$ = 63 G. So the constants have different values for different species or for different configurations of the same species.

In general, a nucleus of spin I will produce 2I + 1 lines in the ESR spectrum. Thus a single $^{14}N$ nucleus (I = 1) yields a three-line spectrum; $^{23}Na$ (I = 3/2) yield four lines, and so on. Many common nuclei (e.g., $^{12}C$ and $^{16}O$) have zero nuclear spin and do not interact with the unpaired electron.

Usually, an organic radical will have more than one magnetic nucleus. In such instances, the ESR spectrum can display rich hyperfine structure which is useful both for identifying the species and for extracting information about the electronic structure of the radical. The origin of such hyperfine patterns is straightforward, but it is most easily understood in terms of concrete examples.

Consider the methyl radical, $H_3C\cdot$, which has three geometrically equivalent protons. Their spins combine to produce four distinct values for the local field:

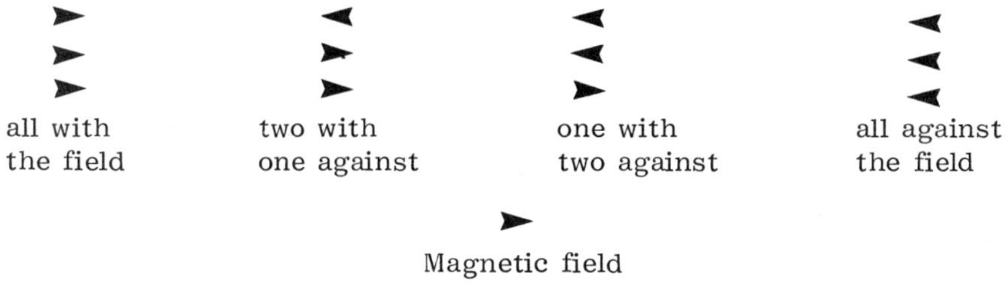


But the statistical weights for these combinations are in the ratio 1:3:3:1:

So the experimental spectrum will contain four lines with relative intensities in this ratio (see Fig. 24.4). The separation between adjacent lines is constant (to first order) and is $a_H$ = 23 G.

In general, n equivalent nuclei of spin I will produce an ESR multiplet consisting of (2nI + 1) equally spaced lines. For spins of I = 1/2 the intensity ratios are given by the coefficients of the binomial

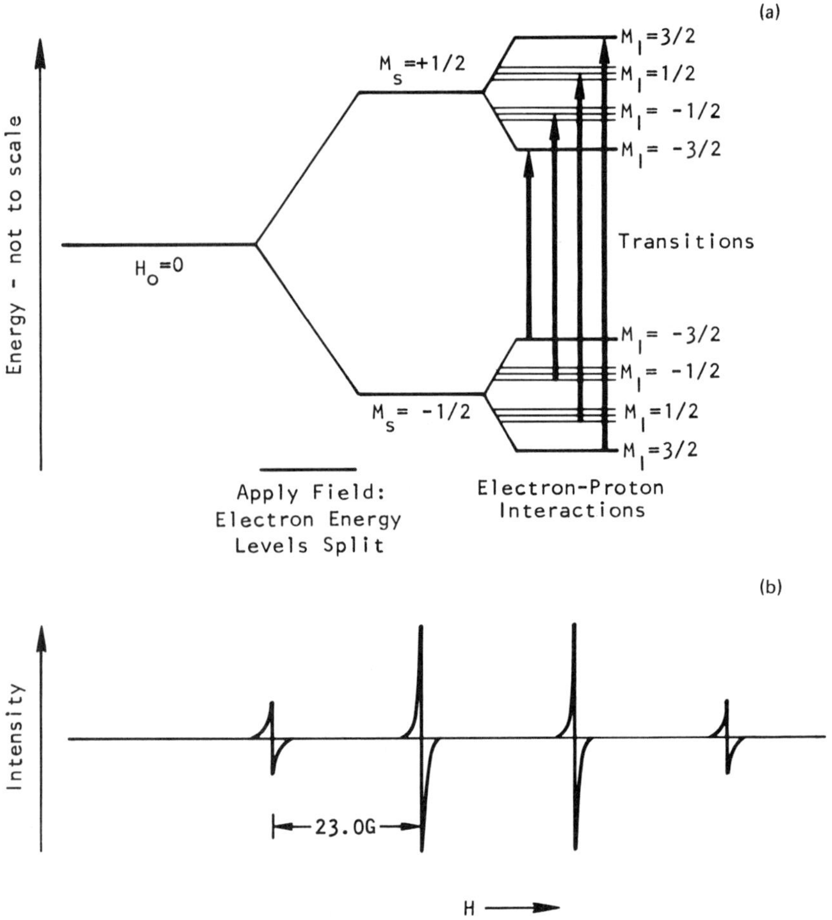


Figure 24.4 (a) Energy-level diagram of the methyl radical. (b) ESR spectrum of the methyl radical.

expansion $(a + b)^n$; but for $I > 1/2$ the formula for intensities is much more complicated. For modest values of n it is easy to find the intensity ratios by sketching a branching diagram; thus for two equivalent $^{14}N$ nuclei the first contributes a 1:1:1 triplet and each of these lines is subsequently split by the other nucleus into secondary 1:1:1 triplets of the same spacing. So the overall pattern is a quintet of 1:2:3:2:1 intensity.

Branching diagrams are often crucial to obtaining an assignment of a spectrum when a radical contains more than one set of equivalent nuclei. The spectrum of the naphthalene anion is shown in Figure 24.5

together with a diagram assigning each line position and intensity. This radical contains two sets of protons of four spins each, with $a_H$ = 4.95 G and $a_H$ = 1.82 G. The first set produces a quintet of intensity 1:4:6:4:1 and each line of this pattern is subsequently split into another quintet of intensity 1:4:6:4:1. Note that the intensity of each line of the secondary quintet is determined by the intensity of the line from which it branches.

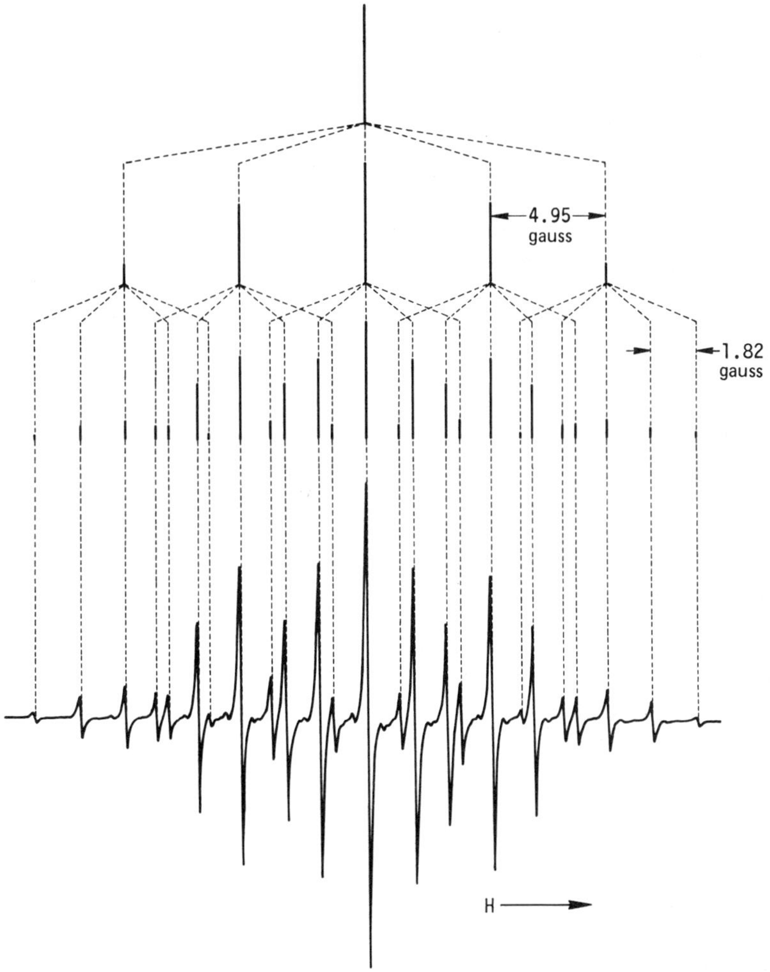


**Figure 24.5** Spectrum of the naphthalene radical anion and a reconstruction of the spectrum.

When a radical contains several different sets of nuclei, the task of assigning the splitting constants can become arduous. The procedure is discussed in many textbooks on ESR.

The local field produced by a nuclear magnetic moment can interact with the electron spin by two different mechanisms. The first and most obvious is the familiar interaction of two magnetic dipoles separated by a distance vector $\underset{\sim}{r}$; the strength of interaction depends on the magnitude of $|\underset{\sim}{r}|$. Accordingly, the dipolar interaction is *anisotropic* and its magnitude changes for different orientations of the radical relative to the direction of the external field. For fluid samples in a low-viscosity solvent the radicals tumble freely so that the dipolar interaction averages to zero and need not be considered in electrochemical systems.

The other mechanism is called the Fermi contact interaction and it produces the *isotropic* splittings observed in solution-phase ESR spectra. Electrons in spherically symmetric atomic orbitals (s orbitals) have finite probability in the nucleus. (Mössbauer spectroscopy is another technique that depends on this fact.) Of course, the strength of interaction will depend on the particular s orbital involved. Orbitals of lower-than-spherical symmetry, such as p or d orbitals, have zero probability at the nucleus. But an unpaired electron in such an orbital can acquire a fractional quantity of s character through hybridization or by polarization of adjacent orbitals (configuration interaction). Some simple cases will be described below.

### *Proton Splittings*

*Spin polarization.* The methyl radical $H_3C\cdot$ is a planar species with $C_{3v}$ symmetry. The unpaired electron resides in a carbon 2p orbital and the C-H bonds are $sp^2$ hybrids. The unpaired electron has zero probability in the plane of the hydrogen nuclei; yet the protons display an isotropic splitting of 23 G, so we must rationalize this experimental observation.

The same phenomenon that leads to Hund's rule of maximum multiplicity in atoms (i.e., quantum mechanical exchange stabilization) produces polarization of the electron spins in the C-H $\sigma$ bond. In a valence-bond treatment, the bond is comprised of one electron from a carbon $sp^2$ orbital and another from a hydrogen 1s orbital. Exchange forces act to polarize the $sp^2$ electron so that its spin is parallel to the unpaired spin in the carbon 2p orbital; this leaves the hydrogen 1s electron oriented in the opposite direction. Of course, the separation into atomic orbitals is far from complete, but this mechanism does place net spin (of opposite sign to the unpaired spin) in a spherical orbital centered on the adjacent atom (see Fig. 24.6).

This spin polarization mechanism was first advanced by McConnell and Chesnut [6], who proposed the equation

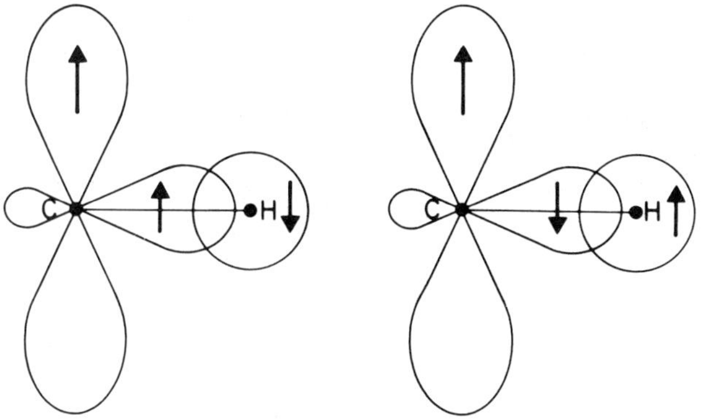


Figure 24.6 Mechanism of hyperfine interactions by spin polarization in the CH fragment.

$$a_H = -Q\rho_C^{\pi} \quad (24.14)$$

to explain the hydrogen HFSC for aromatic hydrocarbons in general. The negative sign indicates that the spin density at the proton is oriented opposite to the unpaired spin in the carbon p orbital. (However, there is no simple experimental means for distinguishing positive and negative coupling constants.) Q is a semiempirical parameter on the order of 24 G and $\rho_C^{\pi}$ is the *spin density* in an orbital of $\pi$ symmetry on the carbon adjacent to the hydrogen atom. This concept is clarified by examination of the data in Table 24.1, where the spin densities can be evaluated by inspection. For radicals of lower symmetry, MO calculations are required.

Table 24.1 Proton Hyperfine Splittings in High-Symmetry Radicals

| Radical | Splitting $a_H$ | Q(G) | $\rho_C^{\pi}$ | $a_H$ (Calc. Eq. 15) |
|---|---|---|---|---|
| $CH_3\cdot$ | −23.04 | −23.04 | 1 | −27.0 |
| $C_5H_5\cdot$ | −5.98 | −29.9 | 1/5 | −5.40 |
| $C_6H_6^{\bar{\cdot}}$ | −3.75 | −22.5 | 1/6 | −4.14 |
| $C_7H_7\cdot$ | −3.91 | −27.4 | 1/7 | −3.85 |
| $C_8H_8^{\bar{\cdot}}$ | −3.21 | −25.7 | 1/8 | −3.17 |

It will be noted that Q does vary among the radicals in Table 24.1. Even more interesting is the fact that those molecules which form anion radicals upon reduction and cation radicals upon oxidation usually display larger proton splittings in the cationic species. Colpa and Bolton [7] proposed a more sophisticated equation based on the fact that negative charge in the p-type orbital will cause the orbital to expand and thus decrease the exchange interaction. Their relationship is of the form

$$(a_H)_i = (Q - K\varepsilon_i)(\rho_C^\pi)_i \qquad (24.15)$$

where Q and K have the empirical values of −27.0 and −12.9 G, respectively, and $\varepsilon_i$ is the net excess charge in the p orbital on the ith carbon atom (to which $H_i$ is bonded) (e.g., for benzene anion radical $\varepsilon_i = -1/6$).

*Hyperconjugation.* Spin polarization is rapidly attenuated by interposition of tetrahedral carbon atoms between the atom carrying the unpaired electron and nuclear spin. Yet protons of a methyl group often display large splittings. For example, the ethyl radical $CH_3CH_2\cdot$ exhibits splittings from two protons, $a_H$ = 22.4 G and three methyl protons of $a_H$ = 26.9 G. A hyperconjugative mechanism, familiar from organic chemistry, is invoked to explain such observations:

$$H_3C - \dot{C}H_2 \rightleftharpoons \overset{H\cdot}{H_2C} = CH_2 \qquad (24.16)$$

A more formal MO treatment of this process suggests that hyperconjugation can be viewed as a direct transfer of electron spin from the carbon p orbital to an s orbital centered on hydrogen (Fig. 24.7); the efficiency of transfer depends on the position of the so-called β proton relative to the p orbital of the adjacent trigonal carbon. An equation analogous to Equation 24.14 describes the coupling of β protons.

$$a_H = (B_0 + B_2 \cos^2 \theta)\rho_C^\pi \qquad (24.17)$$

where $B_0$ (≃ −1 to +5 G) is the contribution from residual spin polarization and $B_2$ (≃ +40 to +53 G) is the direct-transfer term; the angle θ is defined in Figure 24.7b. The presence of the $\cos^2 \theta$ term sometimes allows extraction of steric information about rigid radicals or the preferred orientation of the β protons of bulky substituents. For a

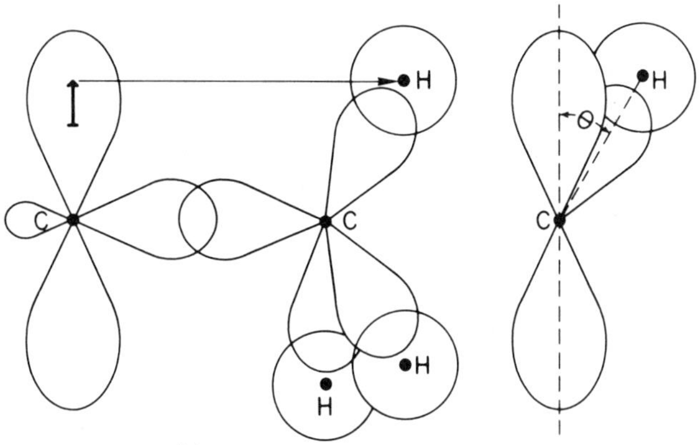


Figure 24.7 Mechanism of hyperfine interaction by hyperconjugation.

freely rotating group, such as methyl, the time average value of $\cos^2 \theta$ is 0.5.

*Other Nuclei*

Magnetic nuclei associated with orbitals in the π-electron system can also exhibit HFSs. The interactions that describe the HFSC of such nuclei involve electron exchange between adjacent atoms as well as between the unpaired electron and the core electrons of the particular atom. These expressions are complicated, and are derived on theoretical grounds in Refs. 8 and 9, and empirical parameters for many heteroatoms are tabulated in Refs. 4, 9, and 10.

## IV. ESR INSTRUMENTATION

### A. The Basic ESR Experiment

Figure 24.8 shows a comparison between the components of a rudimentary ESR spectrometer and the corresponding elements of a more familiar apparatus for visible spectrometry. In ESR the source of excitation radiation is a microwave device called a klystron. The microwaves would disperse in free space and must therefore be conducted to the sample by waveguide or coaxial cable. The sample, contained in the sample tube, is held in a microwave cavity between the poles of a magnet. The detector is usually a diode that produces a dc output proportional to the level of RF power incident on it. The dc from the detector is displayed on a recorder or oscilloscope. In principle the ESR spectrometer could be operated in a mode analogous

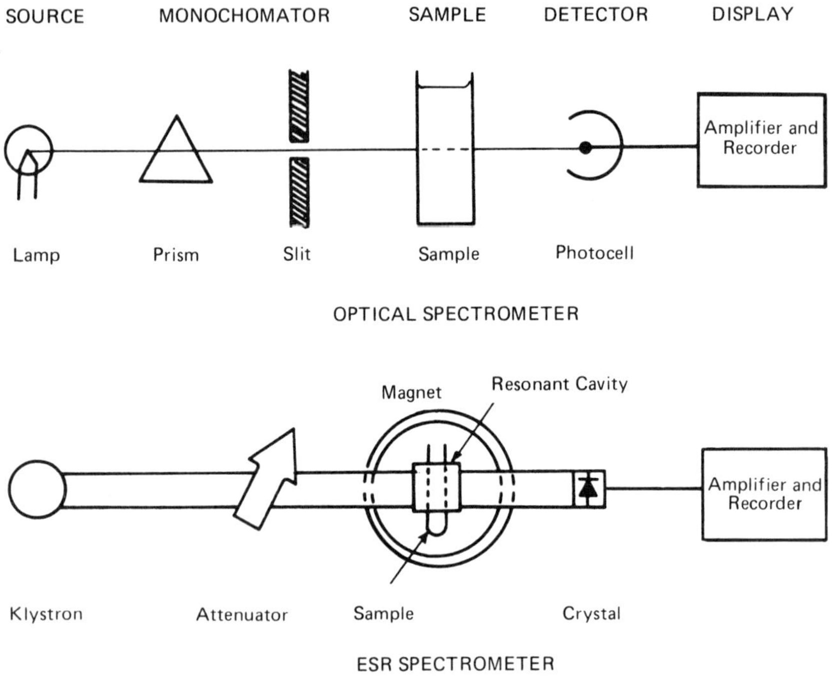


Figure 24.8 Comparison of an optical spectrometer with a simple ESR spectrometer.

to visible spectrometers (i.e., by varying the frequency of the klystron while holding the magnetic field fixed). In practice the klystron frequency is fixed because the microwave cavity acts as a high-Q tuned circuit and it is far easier to vary the magnetic field to attain the resonance condition. Although some instruments can operate in the mode outlined in Figure 24.8, it is not used widely because of its poor sensitivity. Various modifications involving modulation of the magnetic field and lock-in (phase-sensitive) or superheterodyne detection give much better results and are used in most research instruments.

## B. Components of the Basic Instrumentation

An understanding of ESR instrumentation requires a general knowledge of the operation of microwave components and magnet systems. A brief description of these will be given below. Detailed discussions of the design of ESR spectrometers can be found in the references, particularly those books by Alger, Poole, and Wilmshurst listed in Appendix I.

*Klystrons.* The most commonly used radiation source is a klystron; these tubes are available at discrete frequencies between 2.5 and 220 GHz. Many klystrons can be tuned over a range up to ±3% of the nominal frequency by a control which varies the physical dimensions of a resonant cavity inside the tube. Finer adjustment of the frequency is achieved by varying the voltage applied to the resonator and reflector electrodes. Thermal stability is obtained by immersion of the entire tube in an oil bath, or by water or air cooling. A feedback circuit provides automatic frequency control (AFC) to continuously correct the output frequency to the resonance frequency of the cavity. The power output of the klystrons used in ESR spectrometers is generally about 300-700 mW. The most widely used frequency for ESR spectrometers is 9.5 GHz.

*Isolators.* Reflection of microwave power back into the klystron is prevented by an isolator. This consists of a strip of ferrite material which passes microwaves in one direction only and is necessary to stabilize the klystron frequency.

*Attenuators.* Klystrons operate at constant power which can be attenuated by insertion of a piece of "lossy" material into the waveguide. The position determines the degree of attenuation. Another type of device relies on polarization of the microwave radiation.

*Waveguide.* The microwave radiation is usually conveyed to the sample and detector by waveguide, which is a hollow, rectangular tube made of aluminum, copper, or brass plated with silver or iridium.

*Cavities.* The sample is contained in a resonant cavity which is a structure capable of supporting a standing-wave pattern. The cavity is analogous to a tuned circuit (e.g., a parallel RLC combination) used at lower frequencies and must be matched to the klystron frequency range. A measure of the quality of the cavity is its "Q" or "Q factor," which is defined as

$$Q = 2\pi \frac{\text{energy stored in cavity}}{\text{energy lost per cycle}} \tag{24.18}$$

and the sensitivity of the spectrometer is directly proportional to the value of Q. The standing wave is composed of both magnetic and electric fields at right angles to each other. Two commonly used cavities are the rectangular $TE_{102}$ cavity and the cylindrical $TE_{011}$ cavity (Fig. 24.9). It is the magnetic field component which stimulates ESR transitions so that the sample should be placed where the intensity of the RF magnetic field is maximum. The electric field, however, also interacts with the sample and if it is "lossy" (i.e., absorbs microwave energy) the Q of the cavity may be drastically decreased. There-

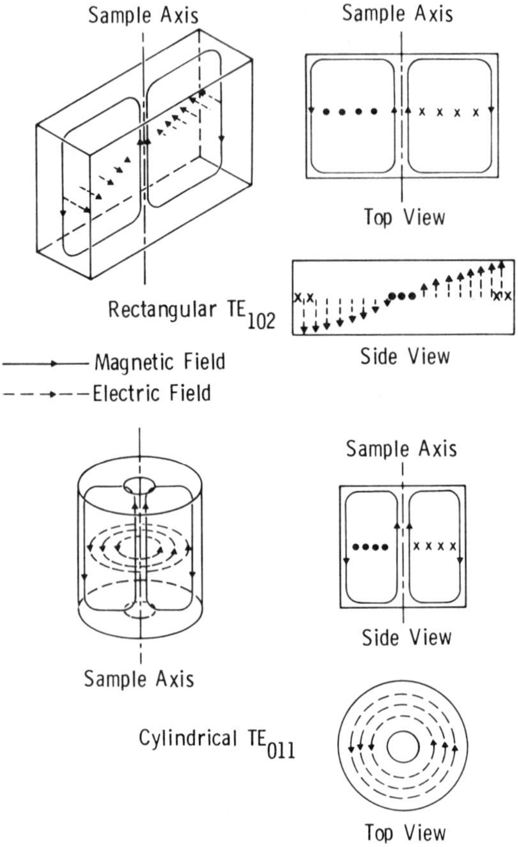


Figure 24.9 Magnetic and electric field distribution inside rectangular and cylindrical cavities.

fore, both the geometry and placement of the sample are critical. For 9.5-GHz spectrometers, rectangular flat cells with a thickness of about 0.25 mm (sample volume of 0.05 ml) are often used for aqueous samples in rectangular cavities. Tubing of 2-4 mm i.d. (sample volume about 0.15-0.5 ml) can be used for low-loss samples in a rectangular cavity or with moderately lossy samples in a cylindrical cavity. The solvents used in electrochemical studies encompass a wide range of dielectric constants and the experimental geometry should be optimized accordingly.

There may be special advantages in performing electrochemical-ESR experiments at frequencies lower than 9-9.5 GHz. Because cavities are larger, cells can occupy a greater volume. Many solvents exhibit smaller dielectric loss at 1-2 GHz than at 9 GHz. Dielectric loss is the absorption of microwave power by the liquid in the ESR cavity. This causes a reduction of the power which induces the electron spin transitions and hence reduces spectrometer sensitivity. Also, the detector sensitivity increases at lower frequency, and since a greater sample size is possible, a greater signal-to-noise ratio may be obtained. However, very little work has been done in this frequency region.

*Couplers and matching screws.* Coupling microwaves between the components can be achieved by a variety of methods. Frequently, an iris or slot of variable size is used. Matching of waveguide elements (analogous to impedance matching in conventional circuits) is accomplished by using screws or stubs which can be inserted into the waveguide or across the coupling iris or moved in lateral slots of the waveguide.

*Microwave bridges.* In principle ESR absorption could be observed as a small decrease in a large background signal. But a preferable arrangement utilizes a microwave bridge which functions like the familiar Wheatstone bridge. This permits much more efficient amplification of the ESR signal.

*Magnets.* An electromagnet capable of producing fields of at least 5000 G is required for ESR at 9.5 GHz, which is the most common frequency for ESR systems. For organic radicals in solution, the homogeneity of the field should be better than about 50 mG over the sample region. The homogeneity requirements are not so severe for transition metal ions due to their wider spectral lines.

The ESR spectrum is recorded by slowly varying the magnetic field through the resonance condition. In most instruments the magnetic field is regulated by using a feedback circuit to sense deviations of the field strength from the anticipated value and to correct for these discrepancies. Accordingly, the field sweep involves the use of a slow ramp voltage which is compared with the output of a field-strength sensor (e.g., a Hall effect probe or a rotating coil) which generates a signal proportional to the actual field strength.

*Modulation coils.* It is easier to discriminate against noise in an ac, rather than dc, signal so most ESR spectrometers employ magnetic field modulation to render the absorption as an ac waveform. This modulation is produced by supplying an adjustable alternating current to coils oriented parallel to the magnetic field coils. For low-frequency modulation (400 Hz or less) the coils can be mounted outside the cavity or perhaps on the magnet pole pieces. At higher frequencies (1 kHz or more) the modulation field cannot penetrate metal effectively, so either the coils must be mounted inside the cavity or the cavity must be constructed of a nonmetallic material (e.g., quartz or plastic with a thin plating of silver to support the standing microwave pattern).

*Detectors.* Most ESR spectrometers use a diode to rectify microwave energy to a signal that can be manipulated by conventional amplifier circuits. The magnitude of the diode output depends on the incident power, so the fluctuations resulting from field modulated absorption by the sample exit from the diode as a pulsating current. Microwave rectification is inefficient at low power levels and noisy at high levels, so the diode performance is optimized by application of a constant-bias power level. The silicon crystal detectors (point contact diodes) which were used originally produce a fairly high noise level which is inversely proportional to the frequency of the detected signal (1/f noise); to minimize this effect the signal was generally detected with 100-kHz field modulation or by superheterodyne methods. However, 100-kHz modulation introduces absorption sidebands at ±36 mG and these distort the main absorption signal if the line width is less than about 60 mG (typical of organic radicals in solution).

More recently, Schottky-barrier diodes and backward diodes have been used as detectors. These do not require as much power to bias the diode to its optimum output and thus permit observation of ESR at lower incident power levels. They also have a much lower 1/f noise characteristic so that modulation frequencies between 6 and 25 kHz (equivalent to 2- to 9-mG sidebands) can yield the same sensitivity that 100 kHz provides with silicon diodes.

## C. Phase-Sensitive Detection

Although some spectrometers operate on the superheterodyne principle, the majority employ field modulation and phase-sensitive detection. Figure 24.10 shows how this method converts the absorption line to the first derivative display which is characteristic of ESR. The modulation field (represented by the sine waves enclosed in the vertical bands) is superimposed on the strong field being swept by the electromagnet; the net field thus produces a degree of microwave absorption which varies sinusoidally. The amplitude of this variation is proportional to the *slope* of the ESR absorption line at the nominal value of the sweep field (represented by the sine waves in the horizontal bands). The phase of the pulsating current produced by the detector is compared with the phase of the field modulation current; as the field sweeps through the absorption maximum a phase shift of 180° occurs. The output of the phase-sensitive detector is further amplified to produce the final presentation shown at the bottom of the figure.

One of the problems associated with this technique is that when the amplitude of the modulation field is too large, the detected line shape becomes distorted and decreases the resolution as well as the signal amplitude in extreme cases.

The components described above are assembled in a typical spectrometer as shown in Figure 24.11.

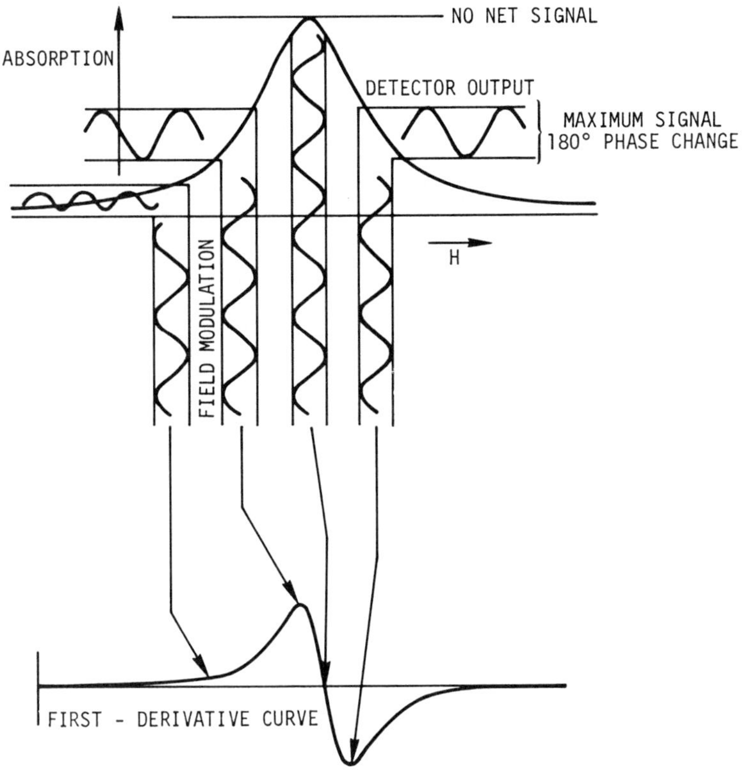


Figure 24.10 Result of field modulation and phase-sensitive detection in ESR spectrometers.

## D. Spectrometer Sensitivity

The sensitivity of a spectrometer is proportional to the magnetic susceptibility of the particular sample and to the values of various instrumental parameters. Neglecting the susceptibility term, the minimum detectable number of spins, $N_{min}$, is proportional to

$$N_{min} \propto \frac{\Delta \nu}{Q_0 \eta \nu_0} \left( \frac{\Delta f \; kT}{2P_0} \right)^{1/2} \qquad (24.19)$$

where $\Delta \nu$ is the line width of the absorption, $\nu_0$ the frequency of the absorption, $Q_0$ the Q of the cavity off resonance, $\Delta f$ the bandwidth of the spectrometer response, k is Boltzmann's constant, T is the absolute

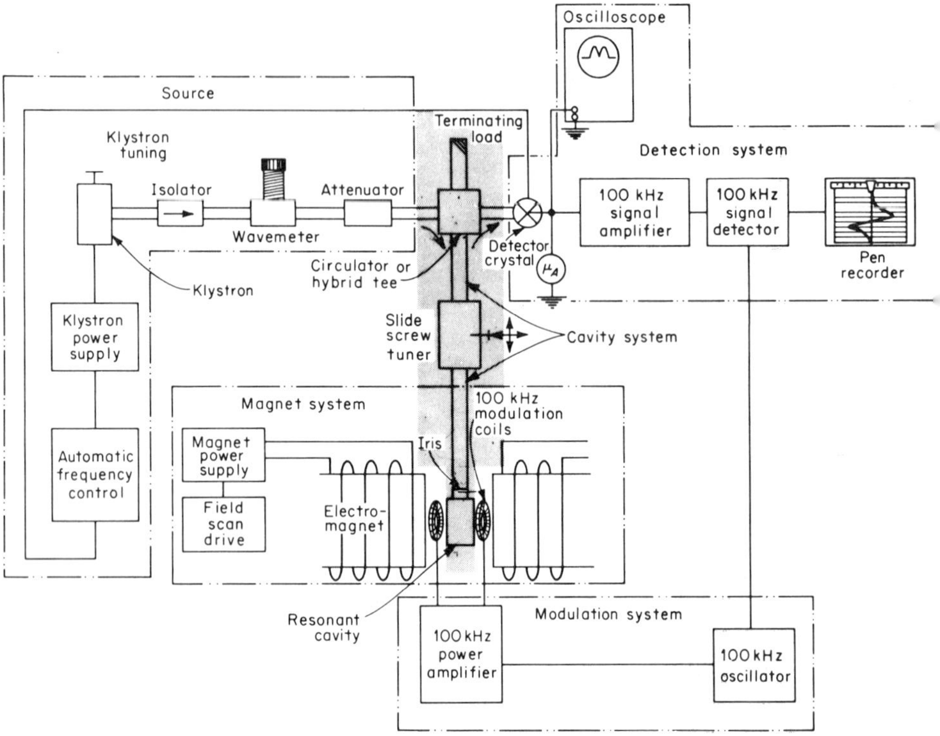


Figure 24.11 Block diagram of an ESR spectrometer. (Courtesy of J. R. Bolton, University of Ontario, London, Ontario.)

temperature of the detector, $P_0$ the incident power, and $\eta$ the filling factor, which is the power density integrated over the sample divided by the power density integrated over the cavity volume.

Several parameters in Equation 24.19 can be adjusted to optimize the sensitivity of the spectrometer. For example, the incident power can be increased provided that the absorption is not saturated; and the extent of RC filtering can be increased (i.e., the $\Delta f$ decreased). For small samples it is sometimes feasible to use a higher-frequency (higher-field) spectrometer and thus increase the population difference between the two energy levels (cf. Equation 24.11). But it can be shown that if the same geometry is maintained, the minimum detectable *concentration* is theoretically proportional to $\nu_0^{-1/2}$, so for electrochemical experiments the sensitivity actually decreases at higher frequencies, all other factors being equal.

Finally, the filling factor and the cavity Q must be considered. For a given sample, the best signal-to-noise ratio is obtained when the presence of the sample reduces $Q_0$ to about two-thirds that of the Q of the empty cavity. The dielectric behavior of the sample puts some constraint on the filling factor and one chooses the cavity geometry which provides the maximum $\eta$ with the minimum reduction in Q. As a general rule, rectangular geometry provides the best performance for lossy samples such as aqueous solutions.

## V. EXPERIMENTAL METHODS OF ESR-ELECTROCHEMICAL STUDIES

### A. General Description

It is convenient to subdivide ESR-electrochemical experiments into two broad categories. The first, called "external generation," involves electrolysis outside the ESR cavity, transfer of the radicals to the cavity, and observation of the ESR spectrum. This category includes both experiments that require a discrete transfer step and those in which the electrolyzed solution is pumped through the cavity for either continuous or stopped-flow measurements. The second classification is called "internal," "in situ," or "intra muros" generation. It involves placement of a working electrode directly inside the ESR cavity and permits recording the ESR signal with minimal time lapse. This method requires certain design compromises that detract from both the electrochemical and ESR performance of the cell.

Some experiments are designed primarily to obtain the spectra of radical species; electrochemistry provides a convenient means for producing the radicals. Other investigations are designed to elucidate the electrochemical behavior of a system and ESR provides a simple method for detecting the presence, estimating the lifetimes, or determining the kinetics of the free-radical intermediates. The following sections describe many different kinds of electrochemical-ESR experiments. The details of a given experiment are often determined by the particular goals of the study.

*Oxygen removal.* ESR and electrochemistry share a common concern about removal of oxygen from the experimental solutions but for different reasons. In addition to the possibility of chemical reactions between oxygen and the free radicals, there is also the problem of magnetic interactions between the paramagnetic (ground-state triplet) $O_2$ molecules and the unpaired spins of interest. The resultant line broadening can cause loss of resolution in the ESR spectrum. Methods for deoxygenating ESR samples run the gamut from simple nitrogen sparging to sophisticated vacuum techniques.

#### *External Generation of Radicals*

The use of external generation requires long-term stability of the radicals because of the significant time lapse between generation and

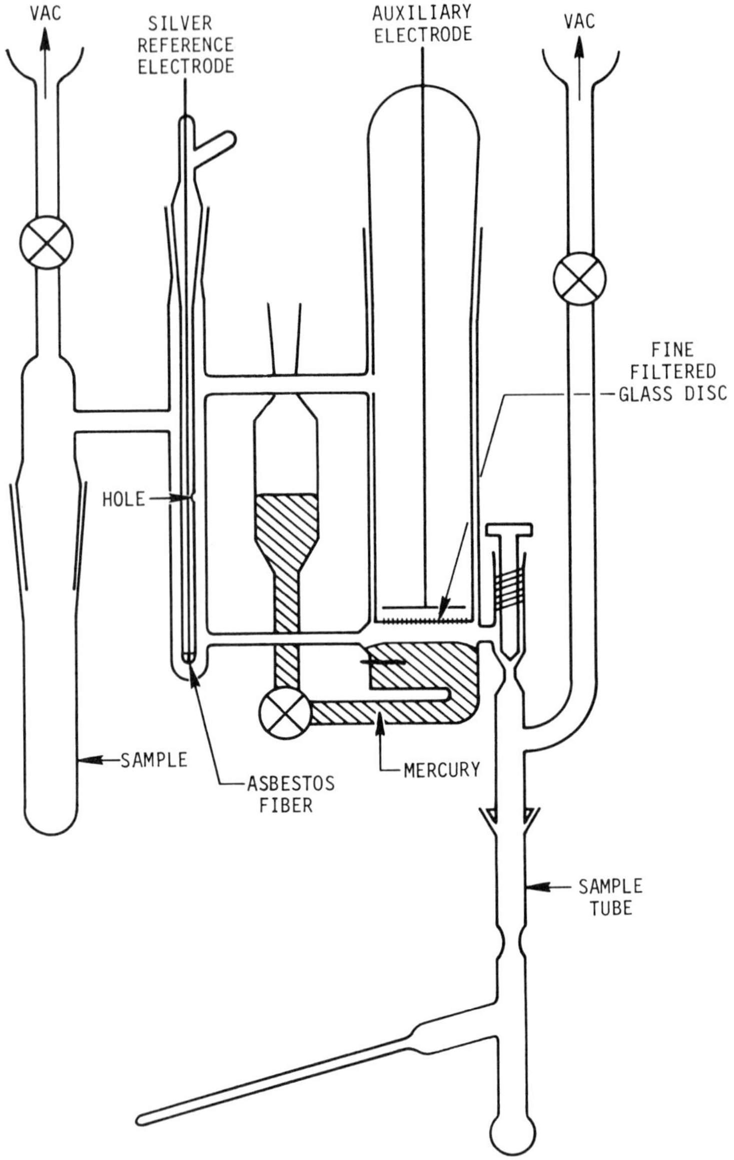


Figure 24.12 Cell used for generation of radicals outside the spectrometer for ESR detection.

observation. Nonetheless, this method has distinct advantages in certain situations. For example, the solution can be exhaustively electrolyzed. This produces the maximum signal intensity from the sample and diminishes complications due to electron exchange reactions (see Sec. VI.F). Moreover, since the ESR tube is essentially independent of the electrochemical glassware, optimum design of both cells can be achieved. This means that the electrode materials and geometry can be chosen freely, with a view toward efficient electrolysis and negligible uncompensated cell resistance. Similarly, the only constraint on the ESR sample tube is that it provide the optimum balance between cavity Q and filling factor for the particular solvent system employed.

Accordingly, external generation is suitable for studies of stable free radicals when it is desirable to optimize signal and resolution, such as in relaxation studies and in the correlation of ESR parameters with MO calculations. Fraenkel and co-workers [11] developed the cell shown in Figure 24.12 used in studies of spin-density distribution in nitriles, semiquinones, and nitroaromatic anion radicals. Oxygen is pumped away from the sample and supporting electrolyte and all subsequent operations are performed under high-vacuum conditions. The rather complicated manipulations are described in the original paper, but the important points for this discussion are that the parallel orientation of the working and auxiliary electrodes results in uniform current density, the potential can be controlled accurately, and the fine frit prevents mixing of auxiliary electrode products with the solution of free radicals. Upon completion of the generation step, the sample is transferred to the evacuated ESR tube, the solution is frozen, and the ESR cell is torch-sealed to prevent aerial contamination. Distillation of solvent between the rerervoir bulb and the ESR tube permits investigation of concentration effects, and also permits an optimum signal to be obtained.

### *Internal Generation*

Electrochemical generation directly inside the microwave cavity permits observation of radicals that might otherwise be too unstable for observation. This method was originated by Maki and Geske [12], using a cell similar to that shown in Figure 24.13 to generate nitrobenzene anion in acetonitrile. These early experiments were designed primarily to detect the presence of radicals and to obtain ESR spectra. For similar systems in which electrochemical behavior is relatively uncomplicated there need be little concern about the precise value of the working electrode potential as long as secondary reactions do not occur. Application of a voltage (corresponding to a point on the diffusion current plateau of the polarographic wave) between the reference and working electrodes produces a faradaic current which decays rapidly as the semi-infinite diffusion layer is established. Within minutes the current drops to a few microamperes so that $E_{effective}$ (= $E_{applied} - iR$)

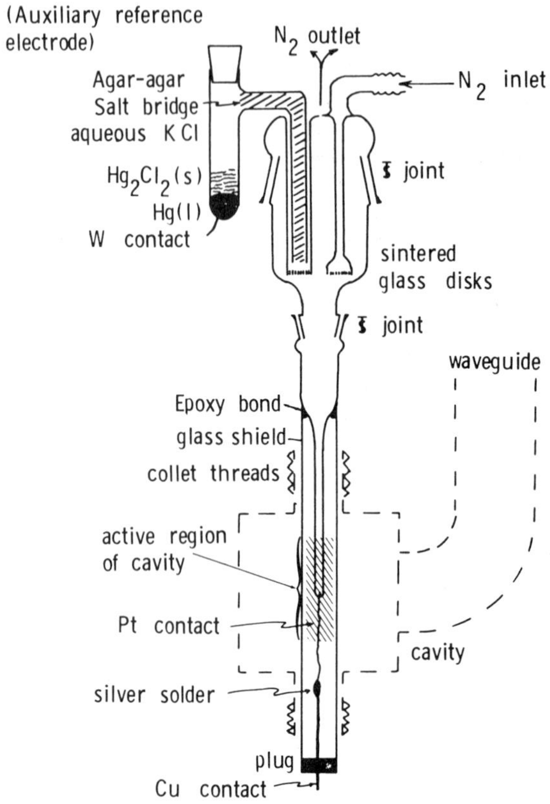


Figure 24.13 Cell used for generation of radicals with spectrometer.

approaches the desired value. (The cell of Figure 24.13 has a resistance of about 2 kΩ when filled with a 0.1 M solution of tetraalkylammonium perchlorate in acetonitrile.) The usual concentration of substrate is millimolar or less, so that the total quantity of charge transferred is only of the order of 1 C. Hence the massive, low-resistance reference electrode is not significantly polarized.

Adams and co-workers [13] and others [14] developed a flat cell suitable for aqueous solution studies. The Varian electrolysis apparatus shown in Figure 24.14 is modeled after their design and has been used widely because most ESR spectrometers come equipped with a $TE_{102}$ cavity. Even when using aqueous solutions, the cell resistance is high due to the thin cross section of electrolyte near the working electrode, and this problem is compounded with most nonaqueous solvents. With a mercury working electrode filling the lower half of the flat cuvette, the current density is nearly uniform and the current-

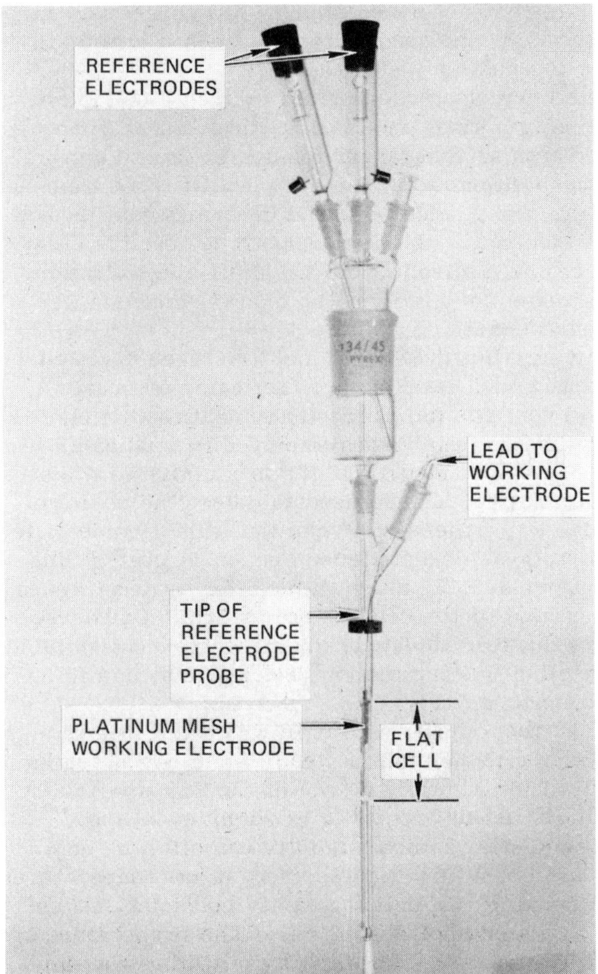


Figure 24.14 Commercial flat cell for internal generation of radical ions. (Courtesy of Varian Associates, Palo Alto, California.)

time response is similar to that described above. (In three-electrode controlled-potential operation, the secondary reference electrode in Fig. 24.14 functions as the auxiliary electrode.)

The area of the mercury electrode surface in such a cell is only about 0.45 $mm^2$. Thus only a small amount of radical can be generated. To increase the surface area so that larger quantities can be generated, a foil or wire mesh electrode is used in place of the mercury. However, in this configuration, large potential gradients can develop along the length of the electrode. Thus the electrode reaction does not occur uniformly over the electrode, and resulting concentration gradients can promote convection, making the overall electrode behavior impossible to characterize.

Nonetheless, many significant ESR studies have been accomplished with flat cells. Most commercial versions are fabricated of quartz since Pyrex is lossy and contains paramagnetic impurities that can be detected at the high modulation amplitudes employed in studies of transition metal ions. Since the majority of the loss is due to the solvent, and low modulation amplitudes are used to detect the narrower lines of organic radicals, less expensive Pyrex flat cells are adequate. An even cheaper version, made of a flattened section of heat-shrinkable Teflon, has been reported [15] to be suitable for systems where the dimensions and placement of the cell are not critical. Of course, when irradiation studies are contemplated, quartz is the most suitable material. A cell for simultaneous irradiation, electrolysis, and ESR observation has been described [16].

There are many electrochemical systems in which the free radical is not sufficiently stable to maintain a steady-state concentration over the interval required for a complete scan of the ESR spectrum. A scan of at least 10 min is usually required to obtain even a moderately well resolved spectrum. Radical stability can often be enhanced by generating and observing the spectrum at low temperatures. A dewar jacketed tube running through the cavity holds the sample tube, which is bathed in a stream of chilled gas. The temperature of the sample can be held within close tolerances by a heating element that warms the gas to the desired temperature. (The same apparatus can be used for high-temperature studies when radical stability is not a problem.) Since the cavity dewar is cylindrical, a sample tube similar to that in Figure 24.13 (modified by removal of the supporting shield tube) has been developed for these studies. For example, we have used a 4-mm-o.d. Pyrex tube connected by a ground-glass joint to a reservoir similar to that shown in Figure 24.13. A platinum foil was sealed in the tube so that it is in the sensitive region of the cavity. The lead that penetrates the Pyrex seal is brought out of the cavity through the top of the Dewar insert [17]. A more sophisticated system incorporating a helical electrode was used in order to obtain a large surface area with minimum loss due to the electrode and solution [18]. A cell designed specifically for electrolytic studies in liquid ammonia has also been described [19].

### B. Some Problems of Quantitative Measurements

Precise analytical measurements with ESR are difficult even though the technique is sensitive to low concentrations of paramagnetic species. The heterogeneous nature of electrochemical charge transfer makes the problem even more complicated, and only a few detailed investigations have been published. Nearly all of these have employed a flat cell in a $TE_{102}$ cavity and the following discussions will be based on this configuration.

*Sample Position*

Figure 24.15 shows the ESR signal amplitude as a function of sample position when a small sample of paramagnetic material is moved along

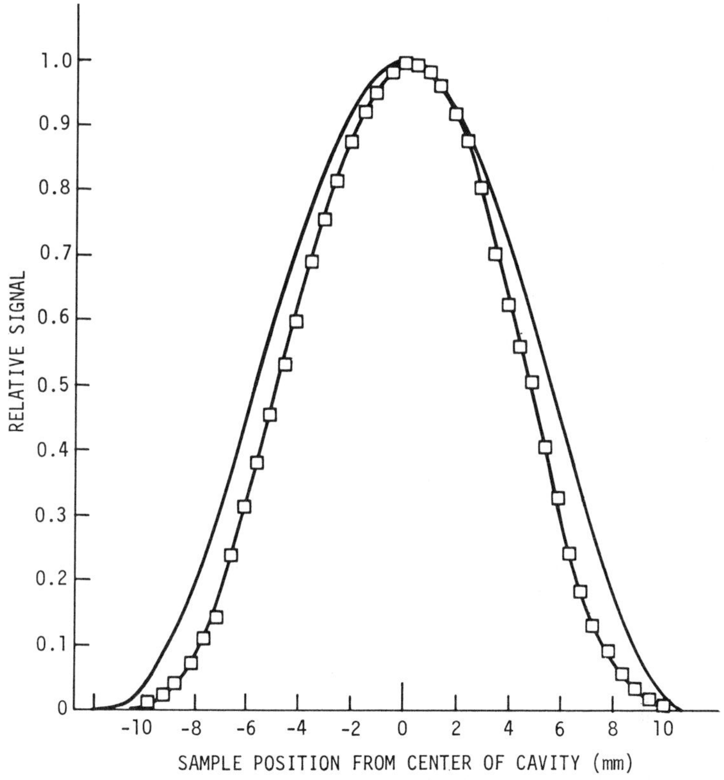


**Figure 24.15** Signal intensity as a function of the placement of a small paramagnetic sample along the axis of a $TE_{102}$ mode cavity. Squares are experimental points; solid line represents $\cos^2$ dependence of the RF field strength.

the vertical axis of the cavity. Clearly, the sensitivity is greatest when the sample is at the geometric center. The change in response when the sample is moved from side to side in the plane of maximum $H_{rf}$ is much less than the up-and-down variation. The implication of Figure 24.15 is obvious: even for a homogeneous solution of radicals in a flat cell, about 80% of the signal comes from those spins within 0.5 cm of the center of the cavity, and the determination of the absolute number of spins present is difficult. Moreover, when electrolytic generation is employed, the greatest concentration of radicals occurs at the edge of the electrode which is nearest to the auxiliary electrode, and it is important that this region of high concentration be near the center of the cavity. However, reproducible placement and generation of radicals in electrolytic cells is difficult.

### *Line Widths*

The ultimate sensitivity of an ESR spectrometer is often quoted as

$$N_{min} = N_0(\Delta H) \tag{24.20}$$

(cf. Eq. 24.19), where the smallness of $N_0$ is a measure of the design quality and performance of the system and $\Delta H$ is the peak-to-peak width of a one-line spectrum.

The important functionality of Equation 24.20 is that the sensitivity is less for broad lines. We shall describe in a later section how electron exchange reactions cause lines to broaden when the electrochemical reaction is incomplete (see Sec. VI.F).

### *Hyperfine Structure*

Equation 24.20 applies only to a one-line spectrum and the sensitivity decreases when the radical sample exhibits HFS. As a simple case, reconsider the $H_3C\cdot$ spectrum, with its four lines; one of the outside lines represents only one-eighth of the total number of radicals in the sample because the 1:3:3:1 intensity ratios reflect the statistical weights of the various nuclear spin configurations. Thus the detection sensitivity falls off rapidly with progressive HFS.

## C. Dynamics of In Situ Generation

### *Controlled-Potential Operation*

*Quiescent solution.* When the applied potential of the working electrode is set onto the plateau of the voltammetric curve, the concentration of electrochemical reactant is depleted at the electrode surface with concommitant increase of the product concentration. A diffusion profile (concentration gradient) is established which extends progressively farther out into solution as the electrolysis continues and the current decays as $t^{-1/2}$. For cells in which there is high electrolyte resistance, the largest proportion of the electrolysis current flows through that portion of the working electrode which is closest to the auxiliary electrode so that the thickness of the diffusion layer is not uniform [20]. Local thermal and density gradients evolve so that diffusion is augmented by convective transfer which produces a steady-state current on the order of tens of microamperes. The ESR signal level becomes constant after several minutes if the radicals are sufficiently stable. Although the concentration profile is essentially indeterminate, the electrolysis is efficient because the working electrode potential is controlled. The current-carrying edge of the electrode should be positioned such that it is about three-fourths of the way into the cavity.

*Flowing solutions.* For unstable radicals the rate of transfer of parent material to the electrode is not sufficient to maintain a steady-state radical concentration in quiescent solutions. But the ESR signal can be maintained by forcing a controlled flow of fresh solution through the electrolysis cuvette, with the rate of flow determining the thickness of the diffusion layer. The nonuniform current distribution over the electrode surface makes quantitative measurements difficult. If the direction of solution flow is downward, the current-carrying edge of the electrode should be placed slightly above the cavity center so that the radicals will be swept into the sensitive region.

### Controlled-Current Operation

*Quiescent solutions.* Coulometry at constant current provides a simple method for measuring the quantity of electrogenerated species as long as the reaction proceeds with 100% current efficiency. However, this condition breaks down with depletion of the electroactive material in the diffusion layer (cf. chronopotentiometric transitions). For low values of the applied current, the thermal and density gradients supplement diffusion sufficiently to sustain electrolysis without a potential shift to a different reaction. This mode of radical generation has been employed successfully in the study of stable species.

At higher values of the applied current, the effective potential across the double layer ultimately increases enough to produce solvent decomposition. But the ohmic potential drop through the solution along the electrode in a cell such as Figure 24.14 prevents this from occurring uniformly over the electrode surface. This cell was modeled in computer simulations [21]. One set of results is shown schematically in Figure 24.16 for the constant-current reduction of anthraquinone, which exhibits the following couples:

$$R + e \longrightarrow R^{\overline{\cdot}} \qquad E_{1/2} = -0.83 \text{ V (vs. SCE)} \tag{24.21a}$$

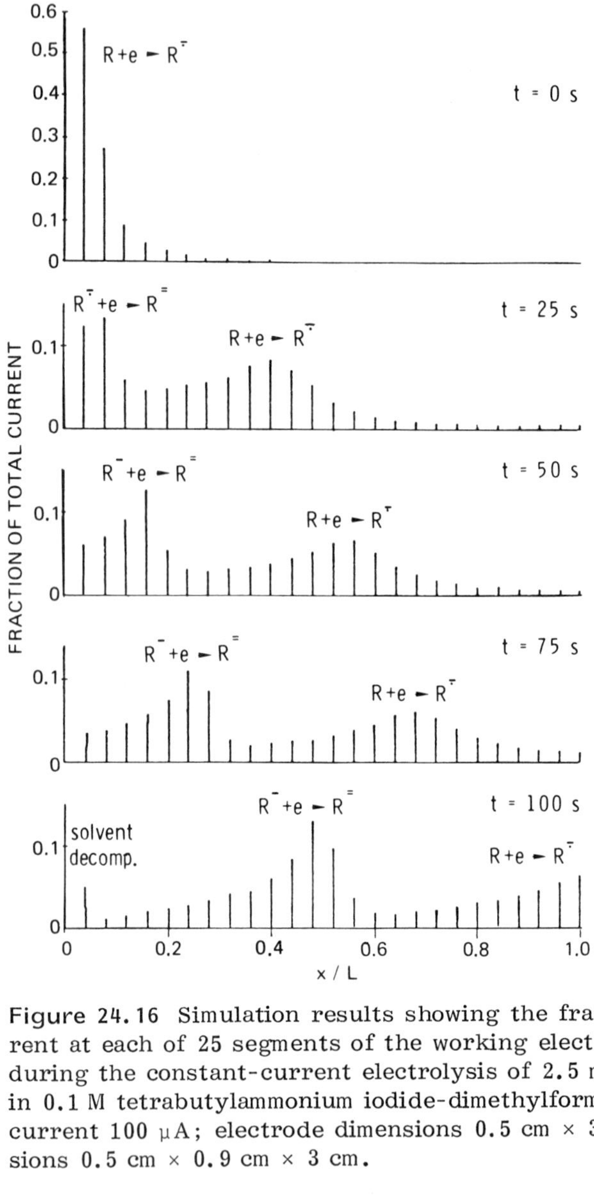


Figure 24.16 Simulation results showing the fraction of the total current at each of 25 segments of the working electrode at different times during the constant-current electrolysis of 2.5 mM anthraquinone and in 0.1 M tetrabutylammonium iodide-dimethylformamide solution. Total current 100 μA; electrode dimensions 0.5 cm × 3 cm; flat cell dimensions 0.5 cm × 0.9 cm × 3 cm.

$$R^{\overset{-}{\cdot}} + e \longrightarrow R^{2-} \qquad E_{1/2} = -1.4 \text{ V (vs. SCE)} \tag{24.21b}$$

$$\text{Solvent} + ne \longrightarrow \text{products} \qquad E \text{ more negative than } -2.7 \text{ V (vs. SCE)} \tag{24.21c}$$

When the current pulse is initiated (t = 0), all of the current goes to produce radical anion; after 25 s the concentration of parent material is depleted at the end of the electrode closest to the auxiliary electrode (X/L = 0) and production of the dianion commences there while anion radical continues to be produced at the middle of the electrode. Further depletion causes the shifts in current profile shown at 50 and 75 s. At 100 s, the current in the portion of the working electrode closest to the counter electrode again increases with the onset of solvent decomposition. Figure 24.17 compares the resulting time-dependent ESR signal intensity with that of an ideal system for which the current efficiency of radical production is 100%. The real system shows a rapid rise in ESR signal when the current is interrupted. This is caused both by homogeneous solution reaction, $R + R^{2-} \rightarrow 2R^{\overset{-}{\cdot}}$, as well as the heterogeneous reactions shown in the figure. The latter reactions produce an open-circuit current within the electrode in response to galvanic potentials arising from local concentration gradients along the electrode surface.

*Flowing solutions.* When higher current densities are used, the mass transfer can be supplemented by flowing solution through the

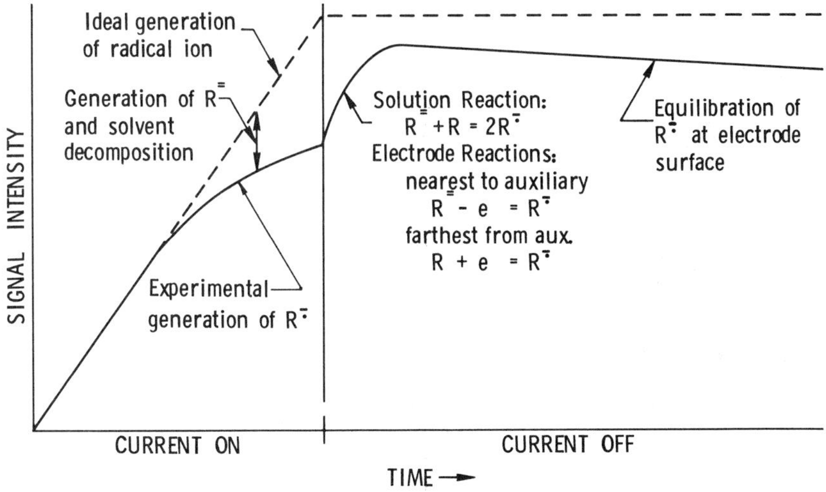


Figure 24.17 Time dependence of the ESR signal during and following generation by constant current.

cuvette. Dohrmann and Vetter [22] explored this arrangement to determine the minimum lifetime, reported as $\tau/\Delta H$ ($k = 1/\tau$ is the rate for first-order radical decay and $\Delta H$ is the peak-to-peak ESR line width), of electrogenerated radicals that could be detected. Their approximate calculations yielded values from $10^{-3}$ to $10^{-5}$ s/G, depending on considerations of flow rate, current density, electrode area, electrolyte conductance, and substrate concentration, for a spectrum of one line.

### *SEESR Techniques (Simultaneous Electrochemistry-Electron Spin Resonance)*

Because of the difficulties described above, electroanalytical studies are usually performed separately from radical generation studies. But a flat cell has been designed [23] (Fig. 24.18) to permit simultaneous monitoring of the electrochemical and ESR response of a free-radical system (SEESR). The auxiliary electrode extends along the edges of the working electrode, which diminishes the problems of iR drops and provides better uniformity of current density than is possible with conventional electrode placement. This cell is used primarily for short-term (on the order of seconds) electrochemical experiments, such as cyclic voltammetry or current-pulse electrolysis, while recording the ESR signal intensity. When a current pulse is interrupted before the chronopotentiometric transition time, the decay of the ESR signal provides a measure of the radical instability, and the shape of the decay response can be related to kinetic parameters. Working curves can be calculated [24], using different assumptions about the mechanism

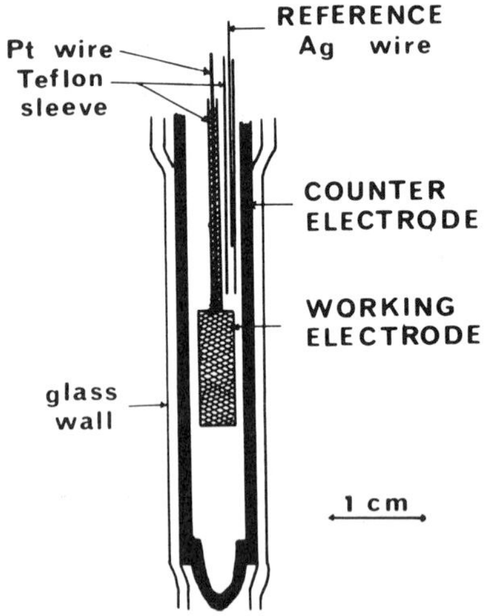


Figure 24.18 Flat cell used to minimize iR losses.

and rate constant, and comparison with experimental observations permits extraction of kinetic data. This has been demonstrated for some dimerization reactions following the electron transfer step [25].

Coulometry is generally difficult with in situ ESR experiments because of convection and iR drop in the long, thin layer of solution. In order to determine the current efficiency of carbazole anion formation, an in situ measurement was carried out [63,64]. Carbazoles can both form soluble products or polymerize forming a conductive layer on the electrode. Results of the measurement showed that there was between 0.2 and 2% current efficiency for radical formation depending on the carbazole used. The system described, however, can only be used for coulometry over short periods of time because of the problems discussed above.

A cell design for a $TE_{011}$ mode cavity which minimizes cell resistance and uses a helical working electrode with the counter electrode in the center was also designed [26]. If the helix is wound so that the spacings between the loops are of the same order as the thickness of the wire, the microwave field does not penetrate to the inside of the helix. As a result, the material generated at the counter electrode is not detected by the ESR spectrometer. This principle was also employed in the design of the low-temperature cell mentioned previously [18].

An electrode configuration similar to that in Figure 24.18 has been designed for cylindrical samples [27] in a $TE_{012}$ cavity. It is available commercially but the smaller working electrode area requires use of flowing solution and signal averaging techniques (Sec. V.E). A more ambitious approach to the iR drop-current density problem has been advanced by designing a cell containing a set of discrete electrodes, with each being controlled by a separate potential or current source [28]. All of these cells have yielded encouraging results for model systems, but the field of SEESR measurements remains to be exploited.

### D. Electrochemical ESR Flow Cells: Design and Methodology

The need for flowing solutions with in situ generation was mentioned above. A simple cell fulfilling these requirements is shown in Figure 24.19a. Since the concentration of free radicals within the cavity can be evaluated by a potentiostatic reversal method [29], this cell is also useful for kinetic studies [30].

The cell shown schematically in Figure 24.19b permits external generation, followed by ESR detection. The solution can either be recirculated to the electrolysis cell or discarded after observation. Umemoto [31] used a similar apparatus to generate moderately stable radicals coulometrically, followed by stopped-flow measurements of the

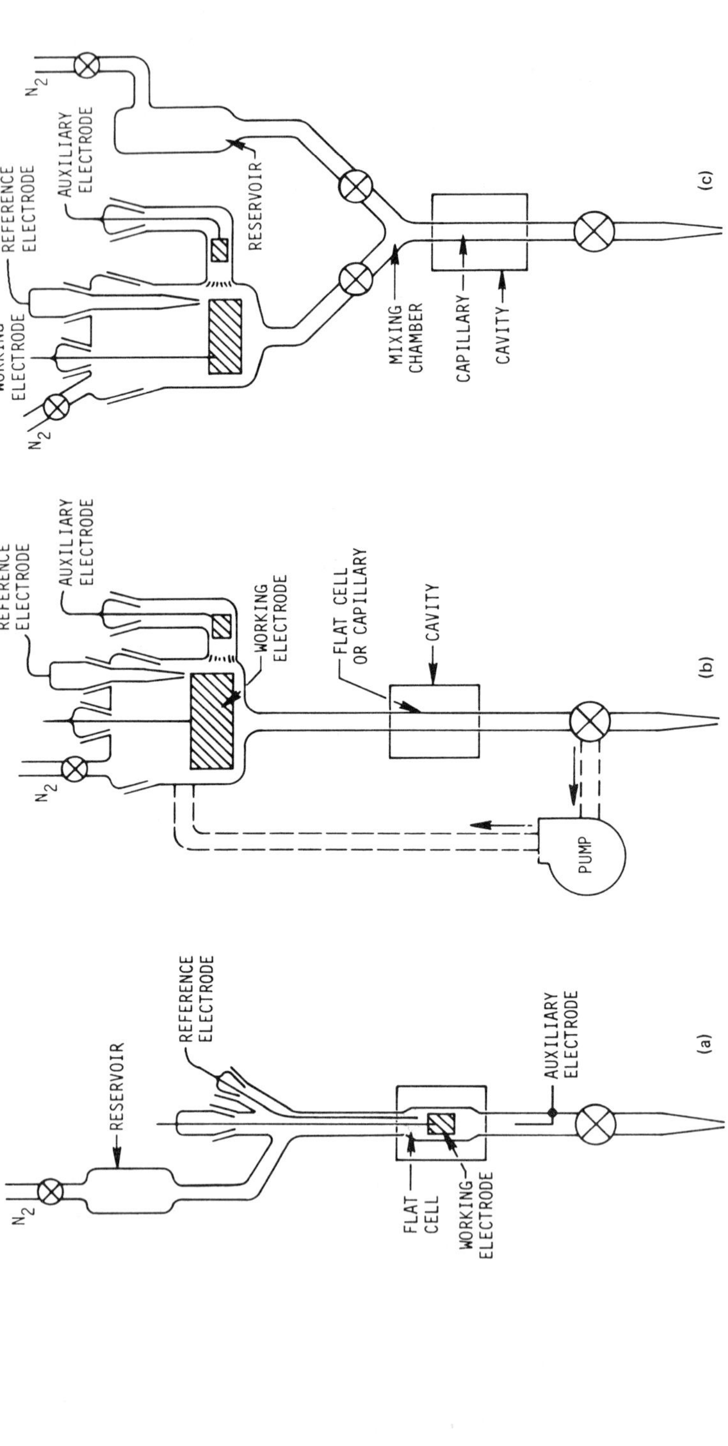


Figure 24.19 ESR-electrochemical cells for flowing electrolytes: (a) gravity-flow system with internal generation; (b) recirculating electrolyte with external generation; (c) flow-mixing cell to study kinetics of reactions with electrochemically generated radicals.

decay kinetics. Forno [32] used a more elaborate recirculating system with two electrochemical cells in series. The unstable product of the first electrolysis was pumped to the second electrolysis cell, where it was converted to a free radical and thence to the cavity for observation (Sec. VI.A).

The cell diagrammed in Figure 24.19c has been used extensively by Kastening to study the reaction mechanisms and kinetics when a stable electrochemically generated species is mixed with other reactants [33]. The calculations that were used to reduce the experimental data to kinetic parameters included the effects of flow rate, reaction mechanism, rate constant, and the dependence of the ESR spectrometer nonuniform sensitivity to the position of the radicals in the cavity. Unfortunately, the correction for the latter effect was introduced as the idealized cosine-squared function (solid line in Fig. 24.15) rather than the actual sensitivity variation that occurs with magnetic field modulation. It is important to be aware of this difficulty when performing experiments to determine accurate kinetic parameters.

External tubular electrodes have been used with flowing solutions in some recent electrochemical-ESR studies by Albery and co-workers [34-37]. The electrode was placed above the ESR cavity and the solution is allowed to flow past the electrode into the cavity. This electrode was fabricated by drilling a 1-mm-diameter hole through a 2-mm section of gold rod, and mounting between two lengths of 1-mm-i.d. capillary tubing. The experimental performance of this electrode was compared to the mathematical model for stable radicals [35], and for radicals that decay by first [36] and second [37] order processes. Many of these mathematical analyses require numerical solutions. Recent interest in photoelectrochemistry has spurred the development of a photoelectrochemical ESR cell [59]. This cell is constructed in an analogous way to the flat cell. The semiconductor or metallic working electrode is mounted on one wall and the solution flows through a channel ($0.4 \times 6.0 \times 30$ mm) between the electrode and the outer cell wall. Electrical contact to the working electrode is made by a fine wire placed through a hole in the wall behind the electrode. The outside optically transparent wall is cemented to the channel containing the working electrode. The counter electrode was a Pt mesh placed below the cell, and the reference electrode was placed above the cell. A theoretical model was developed to predict the ESR signal as a function of solution flow rate through the cell in the absence of homogeneous reactions. In the first case, a constant concentration of radical at the electrode was assumed, and in the second case, a uniform current density at the electrode was assumed. Performance of the cell was consistent with the theoretical predictions, using the photoelectrochemical reduction of fluorescein in aqueous media, which has been well established [60].

### E. Data Acquisition and Signal Averaging

Data processing techniques are extremely useful in both pure ESR and electrochemical-ESR studies. Details of the ESR computer interface are unique to each system and to the goals of each experiment. Since the theory and methodology of these digital operations are similar to those described elsewhere in this book, the discussion will not be reiterated here. There are numerous examples of signal averaging for kinetic measurements and for spectral accumulation using rapid scans. Short-lived species may be studied by these techniques.

### F. Alternating-Current Generation

One of the earliest ESR-electrochemical attempts to utilize alternating current to generate radicals [38] was of questionable success, probably due to the low sensitivity of spectrometers of that period. Recently, under extreme conditions Denat and Gosse [39] observed that a signal could be obtained at potential modulation frequencies of 0.5 to 1000 Hz if high voltages (up to 1.3 kV) were used in high-resistance media. This method, when applied to practical systems, may allow enhancement of the signal-to-noise ratio, or permit determination of kinetics by modulation spectroscopy.

## VI. SELECTED APPLICATIONS OF ESR AND ELECTROCHEMICAL MEASUREMENTS

Attempts to reduce anthracene with an alkali metal in acetonitrile causes solvent decomposition, whereas controlled-potential electrolysis produces stable anion radicals. Thus the working electrode of a coulometric cell can be considered as a continuously adjustable "reagent," capable of producing a wide variety of radical species in diverse solvent systems. The versatility of electrochemical ESR methods is best illustrated by citing a few specific examples from the extensive literature. More complete compilations appear in the reviews listed in Appendix I, but the studies mentioned below provide some appreciation for the techniques.

### A. Identification of Radicals

The literature (see reviews) describes abundant examples in which the observed radical differs from the parent compound by only one electron. It should not be assumed that these studies represent a trivial application of electrochemical-ESR. Rather, the results have greatly enriched our understanding of electronic structure and molecular dynamics.

In more complicated cases, it may not be known whether the detected radicals are generated at the electrode or if the initial species reacts quickly to form other paramagnetic products. In some cases the electrolysis initiates an intramolecular rearrangement so that the structure of the radical differs from that of the parent compound. Electrochemistry alone does not provide direct information of the structure on the product.

One of the earliest applications of combined ESR and electrochemical measurements was the study of nitroalkane reductions [40]. Cyclic voltammetry revealed irreversible reduction waves, but some anodic peaks were observed on the reverse scan and were attributed to reaction intermediates. In situ generation produced an initial spectrum attributable to the nitroalkane anion radical, but after some time the dialkylnitroxide spectrum was detected. This information, combined with analysis of the products formed during bulk electrolysis, suggested the following reaction sequence:

$$RNO_2 + e \longrightarrow RNO_2^{\overline{\cdot}}$$ (spectrum observed)

$$RNO_2^{\overline{\cdot}} \xrightarrow{\text{slow}} R\cdot + NO_2^-$$ (no spectrum, nitrite ion isolated)

$$R\cdot + RNO_2^{\overline{\cdot}} \longrightarrow R_2NO_2^{\overline{\cdot}}$$ (product isolated)

$$2R\cdot + RNO_2 \longrightarrow R_2NOR$$ (product isolated)

$$R_2NO_2^- \xrightarrow{\text{HS}} R_2NO\cdot + OH^-$$ (spectrum observed)

Similar studies [41,42] on the cathodic reduction of halogenated nitrobenzenes revealed the following mechanism:

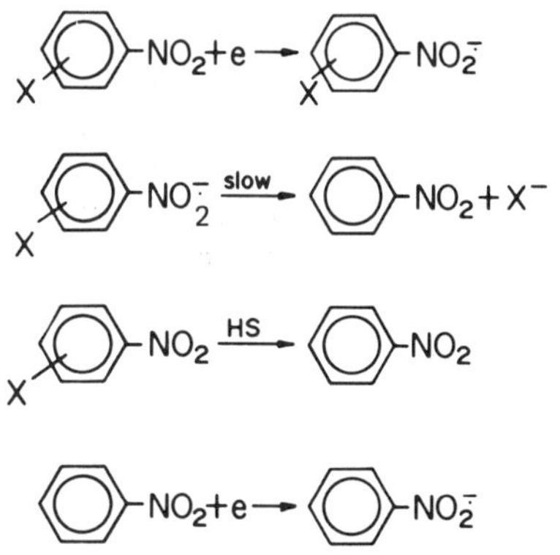

(spectra observed for X = I and X = m-Br)

(no spectra)

(diamagnetic product)

(spectrum observed)

The rates of elimination followed the order: I < m-Br < o-Br < o-Cl < p-Br < p-Cl.

The tetraphenylallene system represents a particularly elegant confluence of chemistry, electrochemistry, and ESR. The carbanion of 1,1,3,3,-tetraphenylpropene is formed spontaneously in alkaline medium. Or it can be formed by the two-electron reduction of 2-ethoxy-1,1,3,3-tetraphenyl-prop-1-ene (at 2.2 V) or of tetraphenylallene (at −2.11 V). The tetraphenylallyl radical is then produced by anodic oxidation of the carbanion at −0.95 V. Its ESR spectrum was obtained by using the two-stage flow cell described earlier [32].

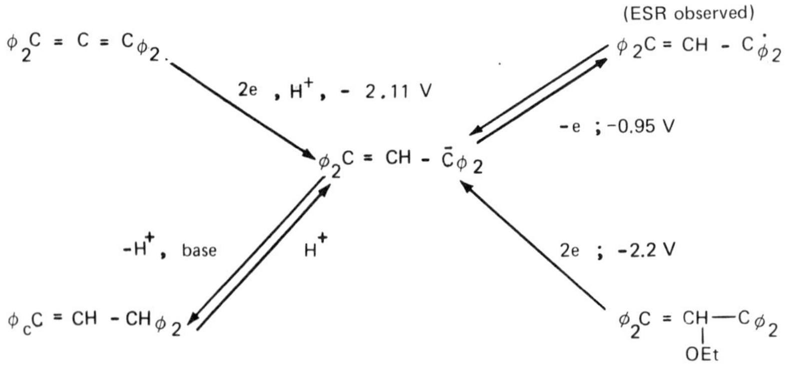


## B. Kinetic Measurements

Kinetic measurements by ESR are difficult, primarily because of the instrumental considerations mentioned in Sections V.B and V. C. Nonetheless, the specificity of ESR toward paramagnetic species has prompted workers to measure reaction rates of radicals generated by both in situ and external electrolysis, using both static and flowing solutions.

Umemoto [31] used external generation, followed by stopped-flow detection, to study the protonation of anthracene, anthraquinone, and benzophenone anions in dimethylformamide-water solutions using an apparatus similar to Figure 24.19c. The measured half-lives were about 1½ min or more. In the case of anthracene, the decomposition rates agreed with those obtained from polarographic measurements and the following scheme was proposed.

$$R + e \rightleftharpoons R\overset{-}{\cdot}$$

$$R\overset{-}{\cdot} + H_2O \longrightarrow RH^{\cdot} + OH^-$$

$$RH^{\cdot} + e \longrightarrow RH^{-} \qquad \text{(at electrode)}$$

$$RH^{-} + R^{\overline{\cdot}} \longrightarrow RH^{-} + R \qquad \text{(in solution)}$$

$$RH^{-} + H_2O \longrightarrow RH_2 + OH^{-}$$

However, the anions of anthraquinone and benzophenone appeared to undergo additional steps corresponding to disproportionation of the radicals followed by protonation of the dianion:

$$2R^{\overline{\cdot}} \rightleftharpoons R^{2-} + R$$

$$R^{2-} + 2H_2O \longrightarrow RH_2 + 2OH^{-}$$

and the ESR data did not agree with polarographic measurements. In the same vein, studies of the protonation rates of nitrobenzene anions, using a variety of flow and in situ techniques, were not consistent among themselves and will not be discussed here.

The cell diagrammed in Figure 24.18 has been used to study the dimerization reactions of olefin anions [24]. The radicals were generated by constant-current electrolysis using a pulse of short duration (with respect to the radical half-life). The decay of the ESR signal was compared with digital simulations assuming various mechanisms and rate constants. In order to distinguish the correct mechanism it was necessary to perform the individual experiments at different values of the applied current. Fresh solution was transferred to the electrochemical-ESR cuvette for each such measurement. Rate constants obtained in this fashion agreed with values obtained from cyclic voltammetry, double potential step, and rotating ring-disk electrode experiments. The dominant mechanism appeared to involve dimerization of two anion radicals followed by rapid protonation.

The kinetics of the reaction of nitrobenzene anion radicals at pH 9.3 was studied using a flow cell and varying the velocity and concentration of the solution [37]. The analysis showed that unsubstituted nitrobenzene anions may undergo a second-order reaction of the form

$$H_2O + 2\phi NO_2^{\overline{\cdot}} \xrightarrow[\text{slow}]{} \phi NO_2 + \phi NO + 2OH^{-}$$

$$2H_2O + \phi NO + 2\phi NO_2^{\overline{\cdot}} \xrightarrow[\text{fast}]{} 2\phi NO_2 + \phi NNOH + 2OH^{-}$$

where the rate of the slow reaction was 45 $M^{-1}\ s^{-1}$, which is comparable to rates found by other ESR-electrochemical measurements [33, 43].

Cells such as those described in Refs. 20, 21, and 26 are particularly suited to study of short-lived intermediates requiring in situ generation at accurately controlled potentials. When a conventional electrochemical cell was used to study the Kolbe synthesis oxidation of triphenylacetic acid [44] it was concluded that the initially formed radical was triphenylacetoxyl ($\phi_3$-CCOO·), based on the assignment of two para- and 4 ortho-proton splittings. A more careful study [45] using the cell described in Ref. 20 showed that it is in fact the triphenylmethyl radical which is formed initially; the identity of the other species was not established, although it is clearly not the acetoxyl radical.

## C. Spin Trapping for Identification of Intermediates

Spin trapping is a method by which unstable radicals are allowed to react with a diamagnetic "trap" so that they form a secondary, more stable radical called the spin adduct [46,47]. Nitrones, such as phenyl-t-butylnitrone (Equation 24.22), and nitroso compounds, (Equation 24.23) are generally used as traps. Nitrones offer greater stability than nitroso compounds; however, in the nitrone, the radical group is incorporated at a position farther from the nitroxyl group than the nitroso trap, so that distinctive spectral information is lost. Some ambiguity may result in the use of these traps. For example, the phenyl-t-butylnitrone was employed to study the electrolysis of water [48]. Benzyl-t-butyl nitroxide was formed during the reduction of water; however, this may be the result either of direct hydrogenation or the abstraction of a proton after formation of the radical anion [49]. Furthermore, during oxidation of water the nitrone might cyclize [49], giving anomalously large proton splittings rather than a hydroxy adduct as originally supposed [48].

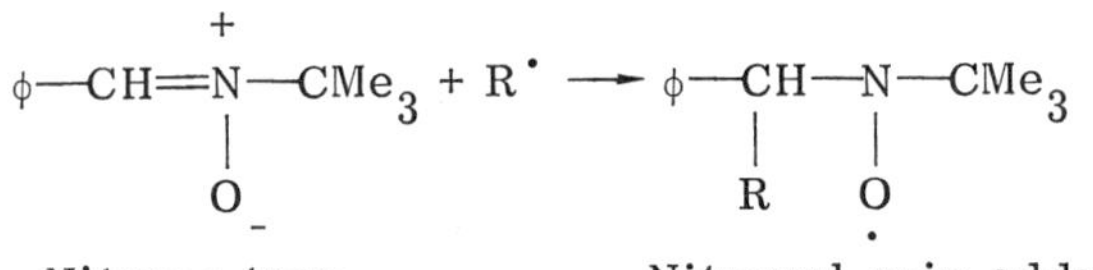
(24.22)

Nitrone trap Nitroxyl spin adduct

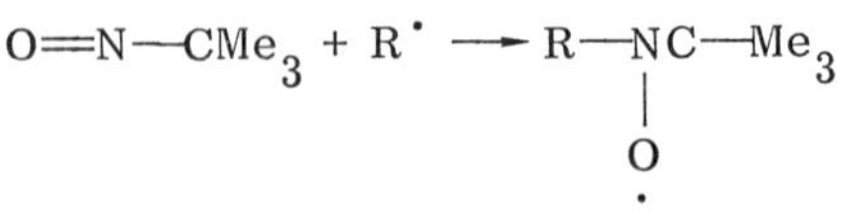
(24.23)

Nitroso trap Nitroxyl spin adduct

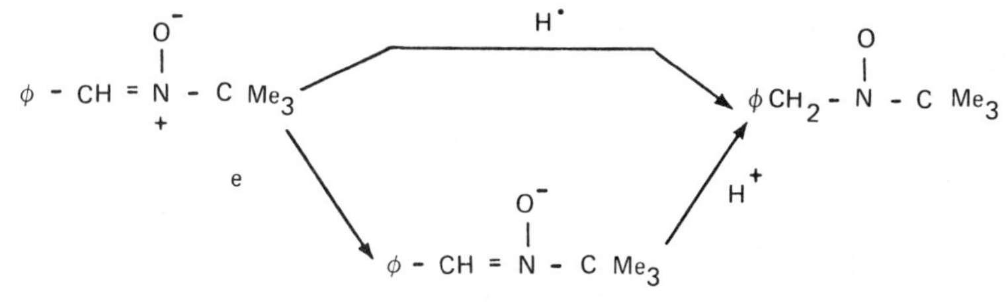

Electrochemical studies in nonaqueous media [50] indicate that phenyl-t-butyl nitrone is not electroactive between +1.5 and −2.4 V vs. SCE. Studies of the reduction of phenyldiazonium salts in the presence of the nitrone gave ESR spectra of the phenyl adduct. However, even here, a two-electron reduction may give the phenyl anion which can add to the nitrone to give phenyl-t-butyl hydroxylamine, which can be reoxidized to the nitroxide. Chemical tests were used to show that the phenyl radical, rather than the anion, was produced.

Studies of the reactions of supporting electrolytes solely by electrochemical means is very difficult because of the extreme irreversibility of their oxidation and reduction. In one study [61], tetrabutylammonium tetrabutylboride oxidation ($E_{pa}$ + 0.35 V) was studied using a phenyl-t-butylnitrone (Eq. 24.22) spin trap. In deoxygenated solutions, the n-butyl radical adduct was confirmed. When the solutions were not deoxygenated, mixtures of the n-butyl- and n-butoxy-radical adducts were found. These radicals can be distinguished by their different nitrogen and β-proton hyperfine splittings, which are 14.88 and 3.06 G, and 13.80 and 2.27 G, respectively, for the butyl and butoxy spin adducts. The results show that the alkyl radicals form on oxidation of the tetralkylboride anion, and that alkyl radical subsequently reacts with $O_2$.

While the spin-trap method has a great deal to offer in terms of mechanism diagnostics, the experimenter must be cautious to confirm the nature and origin of the spin adduct, and the electroinactivity of the spin trap or spin adduct in the potential range used in the experiments. In this regard, the electrochemical characteristics of a variety of nitrones and nitroso-t-butane were measured [62] in aqueous and acetonitrile solutions. The potential ranges of phenyl-t-butylnitrone and nitrosobutane were, respectively, +1.47 to −2.40 V vs. SCE and +1.82 to −1.77 V vs SCE. However, nitrosobutane exhibited dimer formation in the anodic wave at +1.45 V in acetonitrile. In aqueous media, no oxidation waves were observed for these materials, but both exhibited reductions at about −1.8 V.

## D. The Use of ESR in Structural Studies

We mentioned earlier that the half-wave potential of an electrode reaction can be related to the energy of the HMO containing the unpaired electron, Equation 24.3. In addition, we also noted that the ESR spectrum is dependent on the coefficients of this molecular orbital. $\beta^{\circ}_{CC}$ (Eq. 24.2) is usually called the resonance integral and it represents the energy of a $\pi$ bond in the aromatic system. If we consider 9,10-diphenylanthracene as an example, the phenyl groups are not coplanar with the anthracene nucleus because of steric interactions between protons on the adjacent rings. This twist of angle $\theta$ diminishes the resonance energy of the bond between the phenyl and anthryl carbon atoms. In the HMO approximation the resonance integral, $\beta_{rs}$, between those atoms becomes

$$\beta_{rs} = \beta^{\circ}_{CC} \, |\cos \theta| \qquad (24.24)$$

The value of $\theta$ depends on the steric bulk of both groups. For example, phenyl is less bulky than 2,4,6-tri-t-butylphenyl.

Accordingly, HMO theory predicts that the half-wave potential and the HFSC of the anthryl protons should approach those of the unsubstituted anthracene ions when substitution causes $\theta$ to approach 90°. But for smaller values of $\theta$, oxidation or reduction is facilitated and the unpaired electron can delocalize onto the substituent and produce additional hyperfine splittings.

The value of $\theta$ in a particular radical can be estimated by comparison of the experimental half-wave potential and HFSC with the results obtained from a series of HMO calculations using different assumed values of $\theta$. A large number of phenyl-substituted aromatic compounds [51] and ethylenes [52] have been treated in this fashion. Similar evidence for the twisting of the nitro group in nitroaromatic anion radicals is summarized in Ref. 10. Restricted rotation of alkyl substituents is also discussed in Ref. 10, but this torsion does not significantly affect the electrochemical behavior.

## E. Inorganic Systems

Inorganic radicals and transition metal ions typically exhibit broad lines, and hence diminished sensitivity in the EPR method. Consequently, when dealing with small quantities of paramagnetic material, it is often more difficult to detect inorganic species. Several important studies have been reported, however. Kastening's study [53] of the reduction of $SO_2$ in dimethylformamide showed that an equilibrium was established between the $SO_2^{\overline{\cdot}}$ radical ion and the dimer $S_2O_4^{\overline{\cdot}}$,

$$SO_2 + SO_2^{\overline{\cdot}} \rightleftharpoons S_2O_4^{\overline{\cdot}}$$

in which $K \sim 320\ M^{-1}$. The data show that $(SO_2)_2SO_2^{\bar{\cdot}}$ and higher adducts are not likely.

Gardiner and Casey [54] observed the radicals, $O_3^{\bar{\cdot}}$, $O_2^{\bar{\cdot}}$, and $MO_2^{\bar{\cdot}}$, which are formed on various electrode surfaces during oxidation processes on different materials in KOH solutions. Studies of copper dissolution in pyrophosphate [55] resulted in polymeric copper pyrophosphate complexes.

### F. Homogeneous Electron-Transfer Reactions

The occurrence of homogeneous electron-transfer reactions between the radical ion and parent molecule,

$$R\overset{-}{\cdot} + R \rightleftharpoons R + R\overset{-}{\cdot} \qquad (24.25a)$$

or

$$R\overset{+}{\cdot} + R \rightleftharpoons R + R\overset{+}{\cdot} \qquad (24.25b)$$

has some important implications for electrochemical-ESR experiments. These reactions are usually characterized by large rate constants [$k_2 \sim 10^7$-$10^9$ L/(mol·s)]. When the electron is transferred, the new radical that is formed will generally have a different configuration of nuclear spins, so that the ESR resonance condition shifts to a different hyperfine line. If the absolute rate is large, these shifts are manifested as abnormally broad ESR lines with concommitant loss of spectral resolution.

Consider a series of ESR spectra of a radical species in the presence of increasing amounts of the parent material. A dilute solution of the radical in the absence of parent shows a well-resolved spectrum. As parent compound is added, the individual hyperfine lines begin to broaden with onset of the electron transfer reaction. At higher concentrations of parent, the electron transfer rate increases. When the rate becomes very rapid, the average hyperfine field, $H_{local}$, begins to average toward zero and the individual hyperfine lines eventually merge into a single, broad ESR signal. Continued addition of parent causes the breadth of this line to decrease. In this fast-exchange limit the second-order rate constant can be evaluated [56,57]. The reader is directed to Refs. 57 or 58 for details of this procedure since it is beyond the scope of this chapter.

The method of determination of the exchange rate is particularly suited to in situ electrogeneration methods, since one can add different, high concentrations of parent material to the cell, coulometrically generate a small amount of radical ion, and determine the exchange-narrowed line width. The occurrence of exchange reactions (Eq. 24.25) can also be a problem in routine electrochemical ESR studies, for unless the radical generation step is essentially complete, the spectral resolution will be less than ideal.

It is fitting to close this chapter by noting that Equations 24.25a and b do not represent chemical reactions in the usual sense. Indeed, there would be little reason for even contemplating them were it not for the ESR evidence. But closer scrutiny reveals an intriguing similarity with the transport mechanism (mobility) of hydrogen ion in water and it might not be idle to speculate about the role of Reaction 24.25 in biological processes. We can also speculate on those oxidation and reduction processes of importance in energy storage or in biological systems, which may some day be elucidated by the use of electrochemical-ESR methodology.

## APPENDIX I: SELECTED BIBLIOGRAPHY

### Books


R. S. Alger, *Electron Paramagnetic Resonance: Techniques and Applications*, Wiley-Interscience, New York, 1967.

N. M. Atherton, *Electron Spin Resonance*, Hilger and Watts, London, 1975.

A. Carrington and A. D. McLachlan, *Introduction to Magnetic Resonance*, Harper & Row, New York, 1967.

F. Gerson, *High Resolution Electron Spin Resonance Spectroscopy*, Wiley, London, 1970.

R. J. Myers, *Molecular Magnetism and Magnetic Resonance Spectroscopy*, Prentice Hall, Englewood Cliffs, N.J., 1973.

C. P. Poole, *Electron Spin Resonance*, Wiley-Interscience, New York, 1967.

J. E. Wertz and J. R. Bolton, *Electron Spin Resonance; Elementary Theory and Practical Applications*, McGraw-Hill, New York, 1972.

T. H. Wilmshurst, *Electron Spin Resonance Spectrometers*, Adam Hilger, London, 1967 (Instrumentation).


### Reviews

#### *Electrochemistry and ESR*


R. N. Adams, *J. Electroanal. Chem.* *8*:252 (1964). "Application of Electron Paramagnetic Resonance Techniques in Electrochemistry."

I. B. Goldberg and A. J. Bard, in *Magnetic Resonance in Chemistry and Biology* (J. Herak, ed.), Marcel Dekker, New York, 1975. "The Application of Electron Spin Resonance Spectroscopy to Electrochemistry."

B. Kastening, *Chem. Ing. Tech.* *42*:290 (1970). "Application of Electron Spin Resonance in Organic Electrochemistry."

B. Kastening, *Adv. Anal. Chem. Instrum./Electroanalytical Chemistry* (R. N. Reilley and H. W. Nurnberg, eds.), Wiley, New York, 1974, Vol. 20, p. 421.

T. M. McKinney, *Electroanal. Chem.* *10*:97 (1977). "Electron Spin Resonance and Electrochemistry."

*Relaxation Processes in Solutions*

P. W. Atkins, *Adv. Mol. Relaxation Processes* *2*:121 (1972). "Dynamics of Electron Spin Relaxation in Solution."

A. Hudson and G. R. Luckhurst, *Chem. Rev.* *69*:191 (1969). "The Electron Resonance Lineshapes of Radicals in Solution."

W. B. Lewis and L. O. Morgan, *Transition Met. Chem.* *33* (1968). "Paramagnetic Relaxation in Solution."

P. D. Sullivan and J. R. Bolton, *Adv. Magn. Reson.* *4*:39 (1970). "The Alternating Line-Width Effect."

*Organic Radicals in Solution*

A. Carrington, *Q. Rev.* *17*:67 (1963). "Electron Spin Resonance Spectra of Aromatic Radicals and Radical-Ions."

D. H. Geske, *Prog. Phys. Org. Chem.* *4*:125 (1967). "Conformation and Structure as Studied by Electron Spin Resonance Spectroscopy."

A. Hudson and K. D. J. Root, *Adv. Magn. Reson.* *5*:1 (1971). "Halogen Hyperfine Interactions."

E. T. Kaiser and L. Levan (eds.), *Radical Ions*, Wiley, New York, 1968. Chap. 1 to 8, especially.

S. P. Solodovnikov and A. I. Prokof'ev, *Russ. Chem. Rev.* *39*:591 (1970). "The Electronic Structure of Radical Anions."

*Inorganic Ions*

B. A. Goodman and J. B. Raynor, *Adv. Inorg. Chem. Radiochem.* *13*:235 (1970).

G. F. Kokoszka and G. Gordon, *Tech. Inorg. Chem.* *7*:151 (1968). "Electron Paramagnetic Resonance."

E. Konig, in *Physical Methods in Advanced Inorganic Chemistry* (H. A. O. Hill and P. Day, eds.), Wiley, New York, 1968, Chap. 7. "Electron Paramagnetic Resonance."

H. A. Kuska and M. T. Rogers, in *Radical Ions* (E. T. Kaiser and L. Kevan, eds.), Wiley, New York, 1968. "Electron Spin Resonance of First Row Transition Metal Complex Ions."

B. R. McGarvey, *Transition Met. Chem.* *3*:89 (1966). "Electron Spin Resonance of Transition-Metal Complexes."

Teh Fu Yen (ed.), *Electron Spin Resonance of Metal Complexes*, Plenum Press, New York, 1969.

*Tabulations of ESR Data*

B. H. J. Bielski and J. M. Gebecki, *Atlas of Electron Spin Resonance Spectra*, Academic Press, New York, 1967. (Has many errors; check each reference.)

K. W. Bowers, *Adv. Magn. Reson. 1*:317 (1965). "Electron Spin Resonance of Radical Ions."

*Landolt-Bornstein Numerical Data and Functional Relationships in Science and Technology*, Springer-Verlag, New York, 1965. Group II: Atomic and Molecular Physics. Vol. 1: *Magnetic Properties of Free Radicals*. Vol. 2: *Magnetic Properties of Coordination and Organometallic Transition Metal Ions*, 2nd ed. Vols. 1-6, 1980.

*General*

*Specialist Periodical Reports, Electron Spin Resonance*, Vols. 1-6, Chemical Society, London.

## APPENDIX II: UNITS

### Magnetic Field Units

The current units of magnetic field are given by the SI system. Previously, the cgs system has been used for most ESR work. To understand the units, the magnetic flux, often called the magnetic field B, is related to the magnetic field strength H by the equation

$$B = \mu H$$

where B and H are both vectors and $\mu$ is the permeability. In fact, the value of B is important in magnetic resonance since that is the

**Table A24.1** Magnetic Units

| Quantity | Symbol | Name: SI | Name: cgs |
|---|---|---|---|
| Magnetic flux | B | Tesla (T) | Oersted (Oe) |
| Magnetic field strength | H | Amperes/meter | Gauss (G) |
| Permeability in vacuo | $\mu_0$ | $1.25664 \times 10^{-6}$ henries/meter | 1 |

parameter which influences the sample. A comparison of values is given in Table A24.1. The confusion in units results from the fact that in vacuo and approximately in air, B and H are numerically equal in the cgs system. However, the tesla is related to the oersted. The conversion factor is that $1\ T = 10^4\ Oe$.

Furthermore, in most ESR literature, hyperfine structure is given in gauss. In principle, since this parameter is determined from a spectrometer, the correct units should be oersted. Frequency units are recommended for fine and hyperfine structure. Some conversion factors are given below.

$$B\ (T) = B\ (Oe)/10^4$$

$$A\ (MHz) = 2.80247\ (g/g_e)\ a\ (Oe)$$

$$g_e = 2.00231928$$

$$A\ (MHz) = 2.997924 \times 10^4\ A/c\ (cm^{-1})$$

$$A\ (MHz) = 2.80247 \times 10^4\ (g/g_e)\ a\ (T)$$

where $a$ represents the hyperfine splitting with the magnetic units in parentheses. The conversion factors of $a$ to $A$ also apply to other parameters derived from ESR.

Other useful constants are

$$c = 2.997925 \times 10^8\ m/s$$

$$\mu_B = 9.27408 \times 10^{-24}\ J/T$$

$$\mu_p = 1.410617 \times 10^{-26}\ J/T$$

## REFERENCES


1. A. Streitweiser, Jr., *Molecular Orbital Theory for Organic Chemists*, Wiley, New York, 1961.
2. See *ibid.*, pp. 173-188, and R. Zahrodnik and C. Parkanyi, *Talanta 12*:1289 (1965).
3. K. Mobius, *Z. Naturforsch. 20a*:1093, 1102 (1965).
4. J. R. Bolton, in *Radical Ions* (E. T. Kaiser and L. Levan, eds.), Wiley-Interscience, New York, 1968, pp. 1-32.
5. A. J. Stone, *Proc. R. Soc. A271*:424 (1963); *Mol. Phys. 6*:311, 509 (1963).

6. H. M. McConnell and D. B. Chesnut, *J. Chem. Phys.* *28*:107 (1958).
7. J. R. Bolton, *J. Chem. Phys.* *43*:309 (1965); J. P. Colpa and J. R. Bolton, *Mol. Phys.* *6*:273 (1963).
8. M. Karplus and G. K. Fraenkel, *J. Chem. Phys.* *35*:1312 (1961).
9. K. D. Sales, *Adv. Free Radical Chem.* *3*:139 (1969).
10. T. M. McKinney, *Electroanal. Chem.* *10*:97 (1977).
11. P. H. Reiger, I. Bernal, W. H. Reinmuth, and G. K. Fraenkel, *J. Am. Chem. Soc.* *85*:683 (1963); P. H. Reiger and G. K. Fraenkel, *J. Chem. Phys.* *37*:2795 (1962); *37*:2811 (1963); *39*:609 (1963); J. R. Bolton and G. K. Fraenkel, *ibid.* *40*:3307 (1964).
12. A. H. Maki and D. H. Geske, *J. Chem. Phys.* *30*:1356 (1960); D. H. Geske and A. H. Maki, *J. Am. Chem. Soc.* *82*:2671 (1960); A. H. Maki and D. H. Geske, *ibid.* *83*:1852 (1961); A. H. Maki and D. H. Geske, *J. Chem. Phys.* *33*:825 (1960).
13. L. H. Piette, P. Ludwig, and R. N. Adams, *J. Am. Chem. Soc.* *83*:2671 (1961); L. H. Piette, P. Ludwig and R. N. Adams, *ibid.* *84*:4212 (1962).
14. T. Kubota, H. Miyazaki, and K. Nishikida, *Annu. Rep. Shionogi Res. Lab.* *16*:1 (1966).
15. J. F. Ambrose, D. Dillard, A. K. Carpenter, and R. F. Nelson, *Anal. Chem.* *42*:814 (1970).
16. C. M. Lang, J. Harbour, and A. V. Guzzo, *J. Phys. Chem.* *75*:2861 (1971).
17. I. B. Goldberg and A. J. Bard, in J. N. Herak and K. J. Adamic, *Magnetic Resonance in Chemistry and Biology*, Marcel Dekker, New York, 1975, Chap. 10.
18. H. Ohya-Nishiguchi, *Bull. Chem. Soc. Jpn.* *52*:1064 (1979).
19. D. H. Levy and R. J. Myers, *J. Chem. Phys.* *41*:1062 (1962).
20. I. B. Goldberg and A. J. Bard, *J. Electroanal. Chem.* *38*:313 (1971).
21. I. B. Goldberg, A. J. Bard, and S. W. Feldberg, *J. Phys. Chem.* *76*:2550 (1972).
22. J. K. Dohrmann and K. J. Vetter, *J. Electroanal. Chem.* *20*:23 (1969).
23. I. B. Goldberg and A. J. Bard, *J. Phys. Chem.* *75*:3281 (1971).
24. I. B. Goldberg and A. J. Bard, *J. Phys. Chem.* *78*:290 (1974).
25. I. B. Goldberg, D. Boyd, R. Hirasawa, and A. J. Bard, *J. Phys. Chem.* *78*:295 (1974).
26. R. D. Allendoerfer, G. A. Martinchek, and S. Bruckenstein, *Anal. Chem.* *47*:890 (1975).
27. R. Hirasawa, T. Mukaibo, H. Hasegawa, Y. Kanda, and T. Maruyama, *Rev. Sci. Instrum.* *39*:935 (1968); R. Hirasawa, T. Mukaibo, H. Hasegawa, N. Odan, and T. Maruyama, *J. Phys. Chem.* *72*:2541 (1968).

28. B. Kastening, B. Gostisa-Mihelcic, and J. Divisek, *Disc. Faraday Soc.* *56*:341 (1973); B. Kastening, J. Divisek, B. Gostisa-Michelcic, and H. G. Muller, *Z. Phys. Chem.* *87*:125 (1973).
29. R. Koopman and H. Gerischer, *Ber. Bunsenges, Phys. Chem.* *70*:118 (1966).
30. R. Koopman, *Ber. Bunsenges. Phys. Chem.* *70*:121 (1966); R. Koopman and H. Gerischer, *ibid.* *70*:127 (1966).
31. K. Umemoto, *Bull. Chem. Soc. Jpn.* *40*:1058 (1967).
32. A. E. J. Forno, *Chem. Ind.* *1728* (1968).
33. B. Kastening, *Coll. Czech. Chem. Commun.* *30*:4033 (1965); B. Kastening, *Z. Anal. Chem.* *224*:296 (1967); B. Kastening and S. Vavricka, *Ber. Bunsenges. Phys. Chem.* *72*:27 (1968).
34. W. J. Albery, B. A. Coles, A. M. Couper, and K. M. Garnett, *J. Chem. Soc., Chem. Commun.* 198 (1974).
35. W. J. Albery, B. A. Coles, and A. M. Couper, *J. Electroanal. Chem.* *65*:901 (1975).
36. W. J. Albery, R. G. Compton, A. T. Chadwick, B. A. Coles, and J. A. Kenkait, *J. Chem. Soc. Faraday Soc. 1,* *76*:1391 (1980).
37. W. J. Albery, A. T. Chadwick, B. A. Coles, and N. A. Hampson, *J. Electroanal. Chem.* *75*:229 (1977).
38. R. J. Collier and M. E. Peover, *Nature* *220*:155 (1968).
39. A. Denat and B. Gosse, *C.R. Acad. Sci., Sec. C,* *290*:227 (1980).
40. A. K. Hoffman, W. G. Hodgson, D. L. Maricle, and W. H. Jura, *J. Am. Chem. Soc.* *86*:631 (1964).
41. T. Kitagawa, T. P. Layloff, and R. N. Adams, *Anal. Chem.* *35*:822 (1963).
42. T. Fujinaga, Y. Deguchi, and K. Umemoto, *Bull. Chem. Soc. Jpn.* *37*:822 (1964).
43. D. Kolb, W. Wirths, and H. Gerischer, *Ber. Bunsenges. Phys. Chem.* *73*:148 (1969).
44. N. B. Kondrikov, V. V. Orlov, V. I. Ermakov, and M. Ya. Fioshin, *Sov. Electrochem.* *8*:920 (1972).
45. R. D. Goodin, J. C. Gilbert, and A. J. Bard, *J. Electroanal. Chem.* *59*:163 (1975).
46. M. J. Perkins, *Chem. Soc. Spec. Publ.* 24, 1970, p. 97.
47. E. G. Janzen, *Acc. Chem. Res.* *4*:31 (1971).
48. P. H. Kasai and D. McLeod, *J. Phys. Chem.* *82*:619 (1978).
49. E. G. Janzen, R. L. Dudley, E. G. Davis, and C. A. Evans, *J. Phys. Chem.* *82*:2445 (1978).
50. A. J. Bard, J. C. Gilbert, and R. D. Goodin, *J. Am. Chem. Soc.* *96*:620 (1974).
51. L. O. Wheeler, K. S. V. Santhanam, and A. J. Bard, *J. Phys. Chem.* *71*:2223 (1967); L. S. Marcoux, A. Lomax, and A. J. Bard, *J. Am. Chem. Soc.* *92*:243 (1970) (see this paper

for a complete list of phenyl-substituted hydrocarbons that have been studied); L. O. Wheeler and A. J. Bard, *J. Phys. Chem.* *71*:4513 (1967); A. J. Bard, K. S. V. Santhanam, J. T. Maloy, J. Phelps, and L. O. Wheeler, *Discuss. Faraday Soc.* *45*:167 (1968).

52. J. A. Valenzuela and A. J. Bard, *J. Phys. Chem.* *72*:286 (1968); D. H. Geske and M. V. Merritt, *J. Am. Chem. Soc.* *91*:6921 (1969).
53. B. Kastening and B. Gostisa-Mihelcic, *J. Electroanal. Chem.* *100*:801 (1979).
54. C. L. Gardiner and E. J. Casey, *Can. J. Chem.* *52*:930 (1974).
55. M. S. Shapnik, K. A. Zinkicheva, A. V. Ilyasov, and G. S. Vozdvizhenskii, *Sov. Electrochem* *8*:289 (1974).
56. C. S. Johnson, Jr., in *Advances in Magnetic Resonance*, Vol. 1 (J. S. Waugh, ed.), Academic Press, New York, 1965, p. 33.
57. R. Chang and C. S. Johnson, Jr., *J. Amer. Chem. Soc.* *88*: 2338 (1966); C. S. Johnson, Jr., and J. B. Holz, *J. Chem. Phys.* *50*:4420 (1969).
58. B. A. Kowert, L. Marcoux and A. J. Bard, *J. Am. Chem. Soc.* *94*:5538 (1972).
59. B. A. Coles and R. G. Compton, *J. Electroanal. Chem.*, *144*:87 (1983).
60. S. K. Yig and N. R. Bannerjee, *Electrochim. Acta.* *16*:157 (1971).
61. E. E. Bancroft, H. N. Blount, and E. G. Janzen, *J. Am. Chem. Soc.* *101*:3692 (1979).
62. G. L. McIntire, H. N. Blount, H. J. Stronks, R. V. Shetty, and E. G. Janzen, *J. Phys. Chem.* *84*:916 (1980).
63. P. Kubacek, *Coll. Czech. Chem. Commun.* *46*:40 (1981).
64. S. Kolar and P. Kubacek, *Chem. Listy* *72*:1294 (1978).

# Index

## D

## E

## F

K



L